The Calderón Problem
An Introduction

GRADUATE STUDIES IN MATHEMATICS 253

The Calderón Problem

An Introduction

Joel Feldman

Mikko Salo

Gunther Uhlmann

AMERICAN MATHEMATICAL SOCIETY

Providence, Rhode Island

2020 *Mathematics Subject Classification.* Primary 35R30, 35-01, 35D30, 35J15, 35P25, 35Q60, 35R25, 30C62, 78A46, 81U40.

For additional information and updates on this book, visit
www.ams.org/bookpages/gsm-253

Library of Congress Cataloging-in-Publication Data

Names: Feldman, Joel S., 1949- author | Salo, Mikko author | Uhlmann, Gunther, 1952- author
Title: The Calderón problem : An introduction / Joel S. Feldman, Mikko Salo, Gunther A. Uhlmann, authors.
Description: Providence, Rhode Island : American Mathematical Society, [2025] | Series: Graduate studies in mathematics, 1065-7339 ; volume 253 | Includes bibliographical references.
Identifiers: LCCN 2025027819 | ISBN 9780821898840 hardback | ISBN 9781470481971 paperback | ISBN 9781470481988 ebook
Subjects: LCSH: Inverse problems (Differential equations) | Inverse scattering transform | Differential equations, Elliptic | AMS: Partial differential equations – Miscellaneous topics – Inverse problems | Partial differential equations – Instructional exposition (textbooks, tutorial papers, etc.) | Partial differential equations – Generalized solutions – Weak solutions | Partial differential equations – Elliptic equations and systems – Second-order elliptic equations | Partial differential equations – Spectral theory and eigenvalue problems – Scattering theory | Partial differential equations – Equations of mathematical physics and other areas of application – PDEs in connection with optics and electromagnetic theory | Partial differential equations – Miscellaneous topics – Improperly posed problems | Functions of a complex variable – Geometric function theory – Quasiconformal mappings in the plane | Optics, electromagnetic theory – General – Inverse scattering problems | Quantum theory – Scattering theory – Inverse scattering problems
Classification: LCC QA378.5 .F45 2024
LC record available at https://lccn.loc.gov/2025027819

Graduate Studies in Mathematics ISSN: 1065-7339 (print); 2376-9203 (online)

DOI: https://doi.org/10.1090/gsm/253

∞ The paper used in this book is acid-free and falls within the guidelines established to ensure permanence and durability.
Visit the AMS home page at `https://www.ams.org/`

10 9 8 7 6 5 4 3 2 1 30 29 28 27 26 25

To our families, and more generally,
to the many people who have supported us

Contents

Preface

Inverse problems are those where from "external" observations of a hidden, "black box" system (a patient's body, nontransparent industrial object, interior of the Earth, etc.) one needs to recover the unknown parameters of the system. Such problems lie at the heart of contemporary scientific inquiry and technological development. Applications include a vast variety of medical as well as other (geophysical, industrial, radar, sonar) imaging techniques, which are used for early detection of cancer and pulmonary edema, location of oil and mineral deposits in the Earth's interior, creation of astrophysical images from telescope data, finding cracks and interfaces within materials, shape optimization, model identification in growth processes and modeling in the life sciences among others.

Many inverse problems arise from a physical situation modeled by a partial differential equation. The inverse problem is to determine information on coefficients of the equation given some information about the solutions. Analysis of such problems brings together diverse areas of mathematics such as complex analysis, differential geometry, harmonic analysis, integral geometry, microlocal analysis, numerical analysis, optimization, partial differential equations, probability etc. Inverse problems is a fertile area for interaction between pure and applied mathematics.

A prototypical example of an inverse boundary value problem (BVP) for an elliptic equation is the by now classical Calderón problem, forming the basis of Electrical Impedance Tomography (EIT). Calderón proposed the problem in the mathematical literature in 1980. In EIT, one attempts to determine the electrical conductivity of a medium by making voltage and current measurements at the boundary of the medium. The information is encoded in the Dirichlet-to-Neumann (DN) map associated to the conductivity equation. EIT arises in several applications including geophysical prospection (the original motivation of Calderón) and in medical imaging. Since the original work of Calderón there has been remarkable progress on this problem.

This book includes a thorough account of many important developments in the mathematical theory of the Calderón problem. It is intended for graduate students that are familiar with elements of real, complex and functional analysis roughly at the level of [**Rud87**, §1–10]. We will also need basic properties of weak solutions of partial differential equations, Sobolev spaces and Fourier transform. These will be developed as we go along, and more detailed treatments are given in the appendices of this book.

Organization. We briefly summarize the contents of this book. In Chapter 1 we give a motivation to Calderón's inverse problem as well as an introduction to other related inverse problems like optical tomography and inverse scattering. A list of selected open questions is also included.

Chapter 2 gives a precise formulation of Calderón's problem. We give the minimal amount of facts on Sobolev spaces and weak solutions needed to define the DN map and state the problem rigorously. Also in this chapter one can find the reduction of the Calderón problem to a study of the DN map associated to the Schrödinger equation, together with basic integral identities.

In Chapter 3 we show that one can determine, in a stable and constructive fashion, the conductivity and its normal derivative at the boundary from the DN map. We also determine the full Taylor series of a smooth conductivity at boundary points, and consider boundary determination for the anisotropic Calderón problem involving matrix conductivities.

In Chapter 4 we begin by analyzing the linearized inverse problem at a constant conductivity, which is the case originally studied by Calderón. Then we construct complex geometrical optics (CGO) solutions for the conductivity equation which have been the basis of many developments in the theory of inverse boundary problems for elliptic equations. We use these solutions to prove, in dimensions $n \geq 3$, uniqueness for C^2 conductivities from the DN map, to establish stability estimates for the inverse problem, and to develop a procedure for reconstructing the conductivity from the DN map.

The two-dimensional case is considered in detail in Chapter 5. We prove a uniqueness result for bounded measurable conductivities via a reduction to Beltrami equations, thus connecting the problem to the theory of quasiconformal mappings. The proof involves CGO solutions and an analysis of the related $\partial_{\bar{k}}$-equation and scattering transform. Then we temporarily add the assumption that the conductivity is C^2 and prove uniqueness via suitable decay estimates. Returning to bounded measurable conductivities, we discuss relevant topological methods and conclude the proof via subexponential growth estimates for CGO solutions. Some arguments are only sketched.

The results described in Chapters 4 and 5 concern the case when the DN map is measured on the whole boundary. Chapter 6 describes several results for the case when the measurements are made on part of the boundary. A basic tool is to construct CGO solutions vanishing on an open subset of the boundary. This is done in dimension three or greater for special geometries using a reflection method. Some other cases are treated using Carleman estimates. A discussion of the unique continuation principle is also included.

Finally in Chapter 7 we establish basic facts on the stationary approach to scattering theory for potentials. Scattering measurements are defined in terms of the scattering amplitude. Then we consider an application of the methods developed in the previous chapters to inverse scattering problems with full data or at a fixed frequency.

Auxiliary material that may be useful for reading this book has been collected in four appendices. In Appendix A we assemble certain standard facts on functional analysis including compact operators. Appendix B discusses basic properties of the Fourier transform and tempered distributions. We give a more detailed treatment of the Sobolev spaces $H^s(\Omega)$ and $H^s(\partial\Omega)$ in Appendix C. Finally, Appendix D includes a discussion of second order elliptic equations and regularity results.

The literature on the Calderón problem is extensive, and it has not been possible to cover all recent results in this introductory textbook. However, in each chapter we have included a Notes section that discusses important related developments and gives further references. The bibliography is not by any means comprehensive. We have only attempted to give a number of selected references that are most closely related to the topics in this book, and we apologize for any omissions. We hope that the Notes sections will be helpful to those readers who wish to study further.

There are a few other textbooks that discuss aspects of the Calderón problem. The comprehensive monograph [**Isa17**] considers several inverse problems including the Calderón problem, but many proofs are only outlined or omitted. Uniqueness and stability results are treated in [**Cho09**] (in French). The two-dimensional case is discussed in detail in [**AIM09**], computational aspects are considered in [**MS12**] and the fixed frequency inverse scattering problem in [**CK19, Iso20, Kir11**]. We also mention the introductory lecture notes [**Sal08**]. We hope that the present textbook, with a detailed development of many fundamental results and their prerequisites, will be a useful addition to the literature.

As core material for a short graduate course we suggest covering §1.1, §2.1–§2.8, §3.1–§3.2 as well as §4.1–§4.4. For a full-semester course one could also cover Chapters 5 and 6 and other topics if time permits. Exercises may be found interspersed with the presentation.

Acknowledgements. The authors are grateful to the many students who have studied or attended courses based on earlier drafts of this textbook. We are also grateful to several colleagues who have taken time to comment on the book. Their questions and suggestions have considerably improved the presentation. Particular thanks go to Haim Grebnev, Katya Krupchyk, Ru-Yu Lai, Li Li, Leo Tzou, Jenn-Nan Wang, Mike Wendels, Ruirui Wu, Jingni Xiao, Yang Yang, and Ting Zhou. We would like to thank the Department of Mathematics at the University of Jyväskylä and at the University of Washington for the opportunity to teach courses on these topics several times and for their support. We are grateful to Samuli Siltanen for providing references to §1.5, and to Kari Astala for helpful suggestions in §5.6 and in particular for the argument in Remark 5.6.4. We also extend our gratitude to Tommi Brander, Andoni Garcia, Joonas Ilmavirta, David Johansson, Shubham Jaiswal, Jere Lehtonen, Henri Lipponen, and Janne Nurminen for their comments and help with various LaTeX formatting issues during the writing of this book. Sergei Gelfand and the AMS staff have provided constant support, and we thank them for their efforts.

J. F. was partially supported by the Natural Sciences and Engineering Research Council of Canada. M. S. was partly supported by the Research Council of Finland (Centre of Excellence in Inverse Modelling and Imaging and FAME Flagship, grant numbers 312121 and 359208) during the writing of this book. G. U. was partly supported by NSF, the Robert R. and Elaine F. Phelps Endowed Professorship at UW, Walker Endowed Professorship at UW, a Si-Yuan Professorship at IAS, HKUST, a FIDIPRO Professorship from the Academy of Finland, the Chair d'Excellence Fondation Mathématiques de Paris, and the Excellence in Teaching Endowed Professorship at UC Irvine.

Chapter 1

Introduction

In this introduction we discuss a number of imaging methods for which the Calderón problem is relevant. In each situation we have a medium whose internal properties are unknown, and the objective is to determine internal properties of the medium from various measurements (electrical, optical, or acoustic) on its boundary or far away. Along the way we will state some of the main theorems to be proved in this book.

1.1. Electrical impedance tomography

The one-dimensional case. Consider a simple electric circuit consisting of two components: a resistor given by a metal wire occupying the interval $0 \leq x \leq \ell$ on the real line, and a voltage source attached to the resistor's terminals at $x = 0$ and $x = \ell$. We denote by $u(x)$ the voltage at x. By Ohm's law, the voltage difference between the points at x and $x + h$ is the current, I, flowing through the wire times the resistance between x and $x + h$. If the resistance density (or resistivity) $\rho(x)$ at each point x on the wire is continuous, then Ohm's law takes the form

$$u(x + h) - u(x) = -I\rho(x')h$$

for some x' between x and $x + h$. Dividing across by h and taking the limit $h \to 0$, we get the differential form of Ohm's law,

$$u'(x) = -I\rho(x).$$

There are no sources or sinks of charge inside the wire, so the current I is a constant. We may eliminate it from the equation just by dividing $\rho(x)$ across and differentiating. In terms of the conductivity $\gamma(x) = \frac{1}{\rho(x)}$, we have

$$\gamma(x)u'(x) = -I \implies \big(\gamma(x)u'(x)\big)' = 0. \tag{1.1}$$

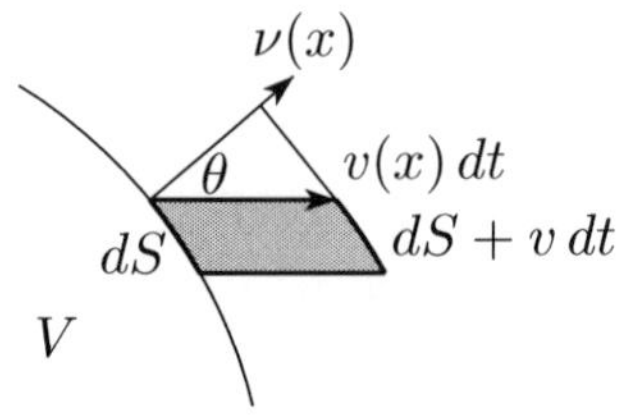

Figure 1.1. Derivation of Kirchhoff's law.

Now suppose that the conductivity $\gamma(x)$ of the wire is unknown to us, and we may only measure the voltages and currents at the ends of the wire. That is, we may only measure $u(0)$, $u(\ell)$, $\gamma(0)u'(0)$ and $\gamma(\ell)u'(\ell)$. By (1.1), $\gamma(x)u'(x)$ is a constant and so takes the value $\gamma(0)u'(0)$ everywhere. Thus

$$u'(x) = \gamma(0)u'(0)\frac{1}{\gamma(x)} \implies u(\ell) - u(0) = \gamma(0)u'(0)\int_0^\ell \frac{1}{\gamma(x)}\,dx.$$

Consequently, the only property of the wire that one can determine by measurements at the ends of the wire is the total resistance $\int_0^\ell \frac{1}{\gamma(x)}\,dx$.

Derivation of the conductivity equation. Replacing the wire by a two- or higher-dimensional body changes the picture completely. In $\mathbb{R}^n$, $n \geq 2$, the current $J(x)$ is a vector and Ohm's law takes the form

$$J(x) = -\gamma(x)\nabla u(x). \tag{1.2}$$

Assuming that there are still no sources or sinks of charge within the body, the net rate of charge flow across the boundary ∂V of any region V must vanish, so we have Kirchhoff's law

$$\int_{\partial V} J(x)\cdot\nu(x)\,dS = 0, \tag{1.3}$$

where $\nu(x)$ is the outward unit normal to ∂V at x. To derive this condition, concentrate on the charge that, at time t, is on an infinitesimal piece, dS, of the surface of V. If this charge has velocity $v(x)$, then at the end of an infinitesimal time interval dt it has moved to a surface element that is the translate by $v(x)\,dt$ of dS. In Figure 1.1, this surface element is denoted $dS + v\,dt$. The charge that has left V through dS during this time interval now fills a tube whose ends are dS and $dS + v\,dt$. In the above (two-dimensional) picture, the tube has height $dS\cos\theta$ and width $|v(x)|\,dt$. Hence the tube has volume $|v(x)|\cos\theta\,dS\,dt = v(x)\cdot\nu(x)\,dS\,dt$. If the charge density at x is $\kappa(x)$, the tube contains charge $\kappa(x)v(x)\cdot\nu(x)\,dS\,dt = J(x)\cdot\nu(x)\,dS\,dt$. The total charge that leaves V during the time interval dt is $dt\int_{\partial V} J(x)\cdot\nu(x)\,dS$.

Using the zero flux condition (1.3) and the Gauss divergence theorem, we have

$$\int_V \nabla \cdot J(x)\, dx = 0.$$

This is true for all regions V. So, assuming that $\nabla \cdot J(x)$ is continuous, Kirchhoff's law can be rewritten as

$$\nabla \cdot J(x) = 0.$$

Combining this with Ohm's law (1.2) yields the equation

$$\nabla \cdot \big(\gamma(x)\nabla u(x)\big) = 0.$$

This equation will be called the *conductivity equation.*

The Calderón problem. Suppose now that we have a conductor filling a region Ω and that we apply a voltage f on the boundary $\partial\Omega$ and measure the current that then flows out of the region. By measuring the rate at which charge is leaving various parts of $\partial\Omega$, we are measuring the current flux through $\partial\Omega$, which determines the quantity $\gamma(x)\nabla u(x) \cdot \nu(x) = \gamma(x)\partial_\nu u(x)$ on $\partial\Omega$, where $\partial_\nu u(x)$ is the *normal derivative* of u at $x \in \partial\Omega$,

$$\partial_\nu u(x) = \nabla u(x) \cdot \nu(x).$$

For a given γ and f, we will see in Chapter 2 that the boundary value problem

$$\begin{cases} \nabla \cdot (\gamma(x)\nabla u(x)) = 0 & \text{in } \Omega, \\ u = f & \text{on } \partial\Omega, \end{cases}$$

has a unique solution u in Ω. Let $\Lambda_\gamma(f)$ be the function $\gamma\partial_\nu u|_{\partial\Omega}$ on the boundary that results from a given γ and f. The map $f \mapsto \Lambda_\gamma(f)$, which clearly depends linearly on f, is called the *Dirichlet-to-Neumann map* (DN map). This map encodes the electrical boundary measurements for all possible functions f on the boundary. More precisely, we will see in Chapter 2 that the DN map Λ_γ is a bounded linear map between two Sobolev spaces on $\partial\Omega$,

$$\Lambda_\gamma : H^{1/2}(\partial\Omega) \to H^{-1/2}(\partial\Omega).$$

The inverse problem of Calderón, also-called the *inverse conductivity problem* or the *inverse problem of Electrical Impedance Tomography* (EIT), is to determine the conductivity function γ from the knowledge of the map Λ_γ (that is, from the knowledge of $\Lambda_\gamma(f)$ for all $f \in H^{1/2}(\partial\Omega)$).

Formal dimension count. When dealing with inverse problems it is often informative to do a formal variable count. The Calderón problem for a domain in $\mathbb{R}^n$ asks to determine γ, which is a function depending on n variables in general, from the DN map Λ_γ. Pretend for a minute that the boundary $\partial\Omega$ contains only a finite number, m, of points and call the value of f at the kth boundary point f_k and the value of $\Lambda_\gamma(f)$ at the jth boundary point $\Lambda_\gamma(f)_j$. Then the map $f \mapsto \Lambda_\gamma(f)$ would correspond to a linear map taking $f \in \mathbb{R}^m$ to $\Lambda_\gamma(f) \in \mathbb{R}^m$, having of the form

$$\Lambda_\gamma(f)_j = \sum_{k=1}^{m} \lambda_{j,k} f_k,$$

where $\lambda_{j,k}$ is the current that results at j^{th} boundary point when a unit voltage is applied at the kth boundary point. The analogous formula for the true, continuous, boundary $\partial\Omega$ is

$$\Lambda_\gamma(f) = \int_{\partial\Omega} \lambda_\gamma(x,y) f(y)\, dS(y),$$

where dS is the surface measure on $\partial\Omega$ and $\lambda_\gamma(x,y)$ the current density that results at x when a unit voltage is applied at y. Knowing the DN map $\Lambda_\gamma(f)$ for all applied surface voltages f is equivalent to knowing the integral kernel $\lambda_\gamma(x,y)$ for all $x, y \in \partial\Omega$. This is a function of $2n-2$ variables, or more precisely a distribution defined on a manifold of dimension $2n-2$.

Thus, in the inverse problem of Calderón we are hoping to determine a function of n variables (the conductivity γ) from a function of $2n-2$ variables (the integral kernel λ_γ of the DN map). For $n = 1$, this problem is formally underdetermined since γ is a function of more variables than λ_γ, and we have already seen that for $n = 1$ the DN map cannot possibly determine the conductivity. For $n = 2$ the inverse problem is formally well-determined (γ and λ_γ are both functions depending on two variables), while for $n \geq 3$ the problem is formally overdetermined since the data has more degrees of freedom than the quantity that we wish to recover. This heuristic variable count suggests that in large dimensions there may be redundancy in the data, and this sometimes (but certainly not always) means that the inverse problem may be easier to solve in high dimensions.

Different aspects of the Calderón problem. In connection with any inverse problem, there are a number of different questions that are of interest. In the following we give a list of theorems proved in this book addressing these questions. The first result states that knowledge of the DN map determines the unknown coefficient γ uniquely on the boundary. This is often a first step in determining γ in the interior.

Theorem 1.1.1 (Boundary uniqueness). *Let $\Omega \subset \mathbb{R}^n$, $n \geq 2$, be a bounded C^1 domain, and let $\gamma_1, \gamma_2 \in C(\overline{\Omega})$ be positive functions. If $\Lambda_{\gamma_1} = \Lambda_{\gamma_2}$, then $\gamma_1|_{\partial\Omega} = \gamma_2|_{\partial\Omega}$.*

The next result states that, under certain conditions, the DN map uniquely determines the conductivity in the interior. The cases $n \geq 3$ and $n = 2$ will require different proofs, as suggested by the variable count.

Theorem 1.1.2 (Interior uniqueness). *Let $\Omega \subset \mathbb{R}^n$ be a bounded C^2 domain, and let γ_1, γ_2 be positive functions in $C^2(\overline{\Omega})$ if $n \geq 3$ and in $L^\infty(\Omega)$ if $n = 2$. If $\Lambda_{\gamma_1} = \Lambda_{\gamma_2}$, then $\gamma_1 = \gamma_2$ in Ω.*

The first proof that we will obtain for the interior uniqueness result is not constructive, that is, it does not yield an algorithm for computing the values of γ in Ω from the knowledge of Λ_γ. However, with extra work we can also give a constructive procedure.

Theorem 1.1.3 (Reconstruction). *Let $\Omega \subset \mathbb{R}^n$ be a bounded C^2 domain, $n \geq 3$, and let $\gamma \in C^2(\overline{\Omega})$ be a positive function. There is a convergent algorithm for determining the function γ from knowledge of Λ_γ.*

In practice one would like to have an efficient numerical implementation of the algorithm. This is out of the scope of this book, and is a challenging topic for several reasons. First, the imaging method is very diffuse and it is difficult to obtain high-resolution images, and secondly the problem is *ill-posed* in the sense that small errors in the measurements may lead to a large error in the reconstructed image.

The next stability result quantifies the degree of ill-posedness in the Calderón problem and states that, under *a priori* bounds on the coefficients, the inverse map taking Λ_γ to γ has a logarithmic modulus of continuity. It has also been proved that the logarithmic modulus is optimal, and it cannot be improved to a Hölder or Lipschitz type continuity in general.

Theorem 1.1.4 (Stability). *Let $\Omega \subset \mathbb{R}^n$ be a bounded C^∞ domain, $n \geq 3$, and let γ_j, $j = 1, 2$, be two positive functions in the Sobolev space $H^{s+2}(\Omega)$ with $s > n/2$, satisfying*

$$M^{-1} \leq \gamma_j \leq M, \qquad \left\|\gamma_j\right\|_{H^{s+2}(\Omega)} \leq M.$$

There are constants $C = C(\Omega, n, M, s) > 0$ and $\sigma = \sigma(n, s) \in (0, 1)$ such that

$$\|\gamma_1 - \gamma_2\|_{L^\infty(\Omega)} \leq \omega(\left\|\Lambda_{\gamma_1} - \Lambda_{\gamma_2}\right\|_{H^{1/2}(\partial\Omega) \to H^{-1/2}(\partial\Omega)}),$$

where ω is a modulus of continuity satisfying

$$\omega(t) \leq C\,|\log t|^{-\sigma}, \quad 0 < t < 1/e.$$

The previous results considered the case of full data, where one has access to measurements on the whole boundary $\partial\Omega$. In practice this is often not possible, and for instance in geophysical imaging one can only cover a tiny part of the Earth's surface with measurement devices. It is therefore of interest to consider partial data problems. For the Calderón problem, this could mean that one has knowledge of $\Lambda_\gamma(f)$ on some subset of the boundary for functions f supported in some subset of the boundary. The next theorem gives a result of this type for the special case where the part of the boundary that is inaccessible to measurements is part of a sphere.

Theorem 1.1.5 (Partial data). *Let $\Omega \subset \mathbb{R}^n$ be a bounded C^∞ domain where $n \geq 3$, and let γ_1 and γ_2 be two positive functions in $C^2(\overline{\Omega})$. Assume that $\Omega \subset B$ for some open ball B in $\mathbb{R}^n$, let $\Gamma_0 = \partial\Omega \cap \partial B$, and let $\Gamma = \partial\Omega \setminus \Gamma_0$. Assume also that $\partial B \setminus \partial\Omega \neq \emptyset$. If*

$$\Lambda_{\gamma_1} f|_\Gamma = \Lambda_{\gamma_2} f|_\Gamma \quad \textit{for all } f \in H^{1/2}(\partial\Omega) \textit{ with } \operatorname{supp}(f) \subset \Gamma,$$

then $\gamma_1 = \gamma_2$ in Ω.

There are certainly many other important aspects of inverse problems besides the ones listed above. A standard one is the problem of range characterization, which asks to find a set of necessary and sufficient conditions for a map acting on functions on $\partial\Omega$ to be the DN map of some conductivity. Not much is known about range characterization for the Calderón problem at the time of writing this. Also, we have only considered the idealized mathematical problem where one can make infinitely many measurements (for all possible boundary voltages f) at infinitely many points on the boundary. On the practical side of things, of course only finitely many measurements at finitely many points are available and discrete versions of the problem need to be studied. One also needs a careful modeling of how the measurements are implemented with electrodes. There will always be some noise in the measured DN map, and the stability result above may not apply since the noisy version of the true DN map may not be a DN map corresponding to some conductivity. Therefore, an improved stability analysis possibly including a regularization strategy for the numerical algorithm would be of interest. These topics are outside the scope of this book.

The anisotropic Calderón problem. For certain materials, which are called *isotropic*, if you apply a voltage $u(x)$, then the current at x is in the direction opposite to the voltage gradient, $\nabla u(x)$, and has magnitude proportional to the magnitude of the voltage gradient, with the constant of proportionality called the conductivity and denoted $\gamma(x)$. So, for isotropic materials, $J(x) = -\gamma(x)\nabla u(x)$. The results mentioned previously in this section are concerned with isotropic materials. However, there are more complicated *anisotropic* materials where the current at x need not be parallel to $\nabla u(x)$ and the magnitude of the current depends on the direction as well as the magnitude of $\nabla u(x)$. For these materials, $J(x) = -\gamma(x)\nabla u(x)$, but with $\gamma(x)$ being an $n \times n$ matrix, rather than just a number. In general, $\gamma(x)$ is a positive definite, symmetric, $n \times n$ matrix. The conductivity equation still has the form

$$\nabla \cdot (\gamma(x)\nabla u(x)) = 0 \text{ in } \Omega.$$

Theorem 1.1.2 showed that, for $n \geq 2$, the map Λ_γ does indeed determine an isotropic conductivity satisfying suitable assumptions. However, it cannot possibly determine anisotropic conductivities for the following obvious reason. Let $\Psi : \bar{\Omega} \to \bar{\Omega}$ be any diffeomorphism that is the identity map on $\partial\Omega$ and set

$$\tilde{\gamma} = \Psi_* \gamma = \left[\frac{1}{|\det(D\Psi)|}(D\Psi)\gamma(D\Psi)^t\right] \circ \Psi^{-1}, \qquad \tilde{u} = u \circ \Psi^{-1},$$

where $D\Psi = [\partial_k \Psi_j]_{j,k=1}^n$ is the Jacobian matrix (the matrix of first partial derivatives) of Ψ. In fact, a change of variables, see Lemma 3.3.6, shows that

$$\text{(1.4)} \qquad \begin{cases} \nabla \cdot (\gamma \nabla u) = 0 & \text{in } \Omega, \\ u = f & \text{on } \partial\Omega \end{cases} \iff \begin{cases} \nabla \cdot (\tilde{\gamma} \nabla \tilde{u}) = 0 & \text{in } \Omega, \\ \tilde{u} = f & \text{on } \partial\Omega. \end{cases}$$

Let u_f and $\tilde{u}_f$ denote the solutions of the left- and right-hand boundary value problems of (1.4), respectively. By definition, $\Lambda_\gamma f = \gamma \nabla u_f \cdot \nu|_{\partial\Omega}$ and

$$\Lambda_{\tilde{\gamma}} f = \tilde{\gamma} \nabla \tilde{u}_f \cdot \nu|_{\partial\Omega}.$$

If Ψ is the identity map in some neighborhood of $\partial\Omega$, then $D\Psi(x) = \text{Id}$, $\tilde{\gamma}(x) = \gamma(x)$ and $\tilde{u}_f(x) = u_f(\Psi^{-1}(x)) = u_f(x)$ for all x in that neighborhood of $\partial\Omega$. Thus it would follow that

$$\text{(1.5)} \qquad \Lambda_{\Psi_*\gamma} = \Lambda_\gamma.$$

The equality (1.5) remains true if the diffeomorphism Ψ only satisfies $\Psi|_{\partial\Omega} = \text{Id}$, see Lemma 3.3.6. It is known that, for $n = 2$, Λ_γ determines an anisotropic conductivity up to a diffeomorphism that fixes all boundary points. It is conjectured that this is also true for $n \geq 3$.

Exercises and examples.

Example 1.1.6. Here is a simplified example in which an isotropic conductivity is computed from a DN map. The region is the square $\Omega = \{ (x, y) \in \mathbb{R}^2 \mid 0 < x, y < 1 \}$. To reduce the number of variables that we are dealing with, we assume that the conductivity is a function of x only. As $\gamma(x)$ is a function only of a single variable, we hope to be able to determine it by measuring just one function of a single variable. We choose to measure the current $k(x) = \gamma(x)\frac{\partial u}{\partial y}|_{y=0}$ at the base of the square that results from applying the boundary voltage function specified in Figure 1.2.

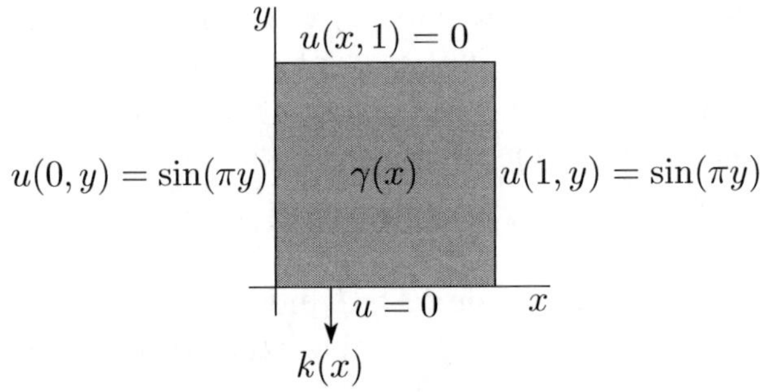

Figure 1.2. Setting of Example 1.1.6.

So our boundary value problem is

$$(1.6)\qquad \begin{array}{lll} \text{(a)} & \nabla\cdot\left[\gamma(x)\nabla u(x,y)\right]=0 & \text{in } \Omega, \\ \text{(b)} & u(0,y)=u(1,y)=\sin(\pi y) & \text{for all } 0\le y\le 1, \\ \text{(c)} & u(x,0)=u(x,1)=0 & \text{for all } 0\le x\le 1. \end{array}$$

The standard technique for solving the boundary value problem (1.6) is to Fourier expand $u(x,y)=\sum_{n=1}^{\infty}a_n(x)\sin(n\pi y)$. From the boundary condition (b), we would expect to only need the $n=1$ term. So we look for a solution of the form $u(x,y)=a(x)\sin(\pi y)$. Boundary condition (c) is satisfied for all functions $a(x)$. Boundary condition (b) is satisfied if and only if $a(0)=a(1)=1$. The differential equation (a) is satisfied if and only if

$$\begin{aligned} 0 &= \nabla\cdot\big(\gamma(x)a'(x)\sin(\pi y)\,,\,\pi\gamma(x)a(x)\cos(\pi y)\big) \\ &= \sin(\pi y)\big[\big(\gamma(x)a'(x)\big)'-\pi^2\gamma(x)a(x)\big], \end{aligned}$$

which is the case if and only if

$$(1.7)\qquad \big(\gamma(x)a'(x)\big)'-\pi^2\gamma(x)a(x)=0 \qquad \text{for all } 0<x<1.$$

We imagine that we have measured

$$k(x)=\gamma(x)\frac{\partial u}{\partial y}\Big|_{y=0}=\gamma(x)\pi a(x)\cos(\pi y)\,\Big|_{y=0}=\pi\gamma(x)a(x)$$

and that we wish to determine $\gamma(x)$. We can do so by inserting $\gamma(x) = \frac{k(x)}{\pi a(x)}$ into (1.7) and solving for a, which is assumed to be a positive function:

$$\begin{aligned}\Big(k(x)\frac{a'(x)}{a(x)}\Big)' = \pi^2 k(x) &\implies \frac{d}{dx}\Big[k(x)\frac{d}{dx}\ln a(x)\Big] = \pi^2 k(x)\\ &\implies k(x)\frac{d}{dx}\ln a(x) = \pi^2\int_0^x k(t)\,dt - \pi^2 C\\ &\implies \ln a(x) = \pi^2\int_0^x \frac{1}{k(s)}\Big[\int_0^s k(t)\,dt - C\Big]\,ds + D\end{aligned}$$

for some constants C and D. To satisfy the boundary condition $a(0) = 1$, we need $D = 0$ and to satisfy $a(1) = 1$, we need

$$C = \Big[\int_0^1 \frac{1}{k(s)}\,ds\Big]^{-1}\Big[\int_0^1 \frac{1}{k(s)}\int_0^s k(t)\,dt\,ds\Big].$$

This determines $a(x)$ and hence $\gamma(x) = \frac{k(x)}{\pi a(x)}$.[1]

Exercise 1.1.7. Prove (1.4) by integrating the first equation against a test function $\varphi \in C_c^\infty(\Omega)$, using the divergence theorem and making a suitable change of variables.

Exercise 1.1.8. Determine the DN map Λ_γ when $\Omega = \{\, x \in \mathbb{R}^2 \mid |x| < 1 \,\}$ and the conductivity is constant, i.e., $\gamma(x) \equiv 1$.

Exercise 1.1.9. Let $\Omega = (-\infty, 0) \times S^1$. Functions on Ω can be identified with those functions $u(x,\theta)$ that are defined for $x < 0$ and all $\theta \in \mathbb{R}$ and that are periodic of period 2π in θ. The gradient operator for Ω is $\nabla = \big(\frac{\partial}{\partial x}, \frac{\partial}{\partial \theta}\big)$. Find the DN map Λ_γ when $\gamma(x,\theta) \equiv 1$. Assume that the potentials $u(x,\theta)$ must remain bounded in the limit $x \to -\infty$.

Exercise 1.1.10. Let Ω be a bounded open subset of $\mathbb{R}^n$. Assume that the divergence theorem is applicable in Ω. Let $\gamma(x)$ be a C^∞ function all of whose derivatives are bounded on $\overline{\Omega}$. Suppose that the complex numbers λ, μ and the, not identically zero, functions $\varphi, \psi \in C^2(\overline{\Omega})$ obey

$$\begin{aligned}\nabla\cdot(\gamma(x)\nabla\varphi(x)) &= \lambda\varphi(x) && \text{in } \Omega,\\ \varphi(x) &= 0 && \text{on } \partial\Omega,\\ \nabla\cdot(\gamma(x)\nabla\psi(x)) &= \mu\psi(x) && \text{in } \Omega,\\ \psi(x) &= 0 && \text{on } \partial\Omega.\end{aligned}$$

[1] If you are worried about dividing by k in the integrals, you shouldn't be. We know that $0 \le u \le 1$ on $\partial\Omega$. By the maximum principle, this implies that $0 < u < 1$ in the interior of Ω. This in turn forces $\frac{\partial u}{\partial y} \ge 0$ when $y = 0$. In fact, by the strong maximum principle [**Eva10**, §6.4.2], $\frac{\partial u}{\partial y} > 0$ for $y = 0$, which ensures that $k(x) > 0$ for all $0 \le x \le 1$.

We say that φ and ψ are eigenfunctions for the differential operator $u \mapsto \nabla \cdot [\gamma\nabla u]$ with Dirichlet boundary conditions on $\partial\Omega$. The numbers λ and μ are the corresponding eigenvalues.

(a) Prove that $\lambda, \mu \in \mathbb{R}$.

(b) Prove that if $\lambda \neq \mu$ then φ and ψ are orthogonal in $L^2(\Omega)$. In other words, prove that $\int_\Omega \varphi(x)\overline{\psi(x)}\,dx = 0$.

(c) Suppose that $\gamma(x) > 0$ for all $x \in \overline{\Omega}$. Prove that $\lambda, \mu < 0$.

(d) Let $\mathcal{H}$ be the closure of the subspace of $L^2(\Omega)$ spanned by the eigenfunctions for the differential operator $u \mapsto \nabla \cdot [\gamma\nabla u]$ with Dirichlet boundary conditions on $\partial\Omega$. Prove that there is an orthonormal basis for $\mathcal{H}$ consisting of real–valued eigenfunctions.

Exercise 1.1.11. Let Ω and γ be as in Exercise 1.1.10. Assume that we already know

- an orthonormal basis for $L^2(\Omega)$ consisting of C^2 eigenfunctions for the differential operator $u \mapsto \nabla \cdot [\gamma\nabla u]$ with Dirichlet boundary conditions on $\partial\Omega$. Call the eigenfunctions and corresponding eigenvalues $\{\varphi_\ell\}_{\ell\in\mathbb{N}}$ and $\{\lambda_\ell\}_{\ell\in\mathbb{N}}$, and
- a linear map $E : C^\infty(\partial\Omega) \to C^\infty(\overline{\Omega})$ such that $(Ef)(x) = f(x)$ for all $x \in \partial\Omega$.

Find the DN map Λ_γ for the conductivity γ.

Exercise 1.1.12. Apply the method of Exercise 1.1.11 to find the DN map for $\{\, x \in \mathbb{R}^2 \mid |x| < 1 \,\}$ with conductivity $\gamma \equiv 1$. You may assume that a suitable orthonormal basis exists.

Exercise 1.1.13. Let Ω be a bounded open subset of $\mathbb{R}^n$. Let $\gamma \in C^2(\overline{\Omega})$ be bounded away from zero. Find $q \in C(\overline{\Omega})$ and $\beta \in C^2(\overline{\Omega})$ such that

$$\nabla \cdot [\gamma\nabla u] = 0 \iff (-\Delta + q)v = 0 \text{ for } v = \beta u.$$

Exercise 1.1.14. Let Ω be a bounded open subset of $\mathbb{R}^n$. Let Λ_γ denote the DN map for the conductivity $\gamma(x)$. Let $\beta(x)$ be a C^∞ function on $\overline{\Omega}$ all of whose derivatives are bounded. Compute $\frac{d}{dt}\Lambda_{\gamma+t\beta}(f)\big|_{t=0}$. Assume that we already know

- a complete orthonormal basis for $L^2(\Omega)$ consisting of C^2 eigenfunctions for the differential operator $u \mapsto \nabla\cdot[\gamma\nabla u]$ with Dirichlet boundary conditions on $\partial\Omega$. Call the eigenfunctions and corresponding eigenvalues $\{\varphi_\ell\}_{\ell\in\mathbb{N}}$ and $\{\lambda_\ell\}_{\ell\in\mathbb{N}}$, and
- the solution to the boundary value problem $\nabla\cdot[\gamma\nabla u] = 0$ in Ω, $u = f$ on $\partial\Omega$. Call the solution $u_0(x)$.

1.2. Optical tomography

Optical tomography is concerned with the determination of spatially varying optical absorption and scattering properties of a medium by measuring the response of the medium to transmitted near-infrared light. This has been proposed as a diagnostic tool in medicine. The standard model for propagation of photons is the *radiative transfer equation*, also-called the linear Boltzmann equation. The main point in this section is to indicate that in the *diffusion approximation*, which is often used to model highly scattering media, the optical tomography problem essentially reduces to the inverse problem of Calderón.

Consider photons propagating in a medium $\Omega \subset \mathbb{R}^n$ where absorption and scattering may occur, and assume that there are no sources of photons in Ω. Consider the unit sphere

$$S^{n-1} = \{\, x \in \mathbb{R}^n \mid |x| = 1 \,\}.$$

The function $\phi(x, v, t)$, modeling the density of photons at point $x \in \Omega$ moving in direction $v \in S^{n-1}$ at time t, solves the radiative transfer equation

$$\left(\frac{1}{c}\frac{\partial}{\partial t} + v \cdot \nabla_x + \mu_{tr}(x)\right)\phi(x, v, t) = \mu_s(x)\int_{S^{n-1}} p(v \cdot v')\phi(x, v', t)\, dv'.$$

Here c is the speed of light, $\mu_{tr} = \mu_a + \mu_s$, where $\mu_a, \mu_s \in C(\overline{\Omega})$ are the absorption and scattering coefficients, and $p(v, v') = p(v \cdot v')$ with $p \in C([-1, 1])$ is the scattering kernel representing the probability of photons traveling in direction v to scatter in direction v'. We will also make use of the photon density

$$\Phi(x, t) = \int_{S^{n-1}} \phi(x, v, t)\, dv$$

and the current $J = (J_1, \ldots, J_n)$, where

$$J_k(x, t) = \int_{S^{n-1}} v_k \phi(x, v, t)\, dv.$$

In the diffusion approximation, the following two assumptions that are valid in predominantly scattering media are made:

(1) The density $\phi(x, v, t)$ is only weakly dependent on the direction v.

(2) The current $J(x, t)$ is constant with respect to time.

Assume for simplicity that $\Omega \subset \mathbb{R}^2$, so unit vectors can be written as $v_\theta = (\cos\theta, \sin\theta)$ for $\theta \in [0, 2\pi]$. We formally expand $\phi(x, v_\theta, t)$ in complex Fourier series with respect to θ,

$$\phi(x, v_\theta, t) = \sum_{k=-\infty}^{\infty} \phi_k(x, t) e^{ik\theta}.$$

A simple computation shows that

$$v_\theta \cdot \nabla_x \phi = e^{i\theta} \partial \phi + e^{-i\theta} \bar{\partial} \phi,$$

where ∂ and $\bar{\partial}$ are the complex derivatives

$$\partial = \frac{1}{2}\left(\frac{\partial}{\partial x_1} - i\frac{\partial}{\partial x_2}\right), \quad \bar{\partial} = \frac{1}{2}\left(\frac{\partial}{\partial x_1} + i\frac{\partial}{\partial x_2}\right).$$

Exercise 1.2.1. Prove the above representation for $v_\theta \cdot \nabla_x \phi$.

Thus, the radiative transfer equation may formally be written as

$$(1.8) \quad \left(\frac{1}{c}\frac{\partial}{\partial t} + e^{i\theta}\partial + e^{-i\theta}\bar{\partial} + \mu_{tr}\right) \sum_{k=-\infty}^{\infty} \phi_k(x,t) e^{ik\theta}$$
$$= \mu_s \int_{S^1} p(v \cdot v') \phi(x, v', t)\, dv'.$$

To express also the right-hand side as a series in $e^{i\theta}$, we will introduce the Chebyshev polynomials T_k. We use the following properties [**MH03**]: one has

$$(1.9) \qquad T_k(\cos\theta) = \cos(k\theta),$$

and the set $\{T_0/\sqrt{\pi}\} \cup \{\sqrt{2/\pi}T_k\}_{k=1}^{\infty}$ is an orthonormal basis of the L^2 space on $(-1,1)$ with weight $1/\sqrt{1-t^2}$. We write the scattering kernel as the Chebyshev polynomial expansion

$$p(t) = \sum_{k=0}^{\infty} p_k T_k(t), \quad t \in (-1,1).$$

We have $p_0 = 1/2\pi$ by the following exercise.

Exercise 1.2.2. Assume that p is a probability density function in the sense that

$$\int_0^{2\pi} p(v_\theta \cdot v_{\theta'})\, d\theta' = 1 \text{ for all } \theta \in [0, 2\pi].$$

Show that $p_0 = 1/2\pi$.

Then formally, by using (1.9),

$$\begin{aligned}
&\int_0^{2\pi} p(v_\theta \cdot v_{\theta'})\phi(x, v_{\theta'}, t)\, d\theta' \\
&= \sum_{j=0}^{\infty} \sum_{l=-\infty}^{\infty} \int_0^{2\pi} p_j T_j(\cos(\theta - \theta'))\phi_l(x,t)e^{il\theta}\, d\theta' \\
&= \sum_{j=0}^{\infty} \sum_{l=-\infty}^{\infty} \int_0^{2\pi} p_j \frac{e^{ij(\theta-\theta')} + e^{-ij(\theta-\theta')}}{2}\phi_l(x,t)e^{il\theta'}\, d\theta' \\
&= 2\pi p_0\phi_0(x,t) + \pi \sum_{j=1}^{\infty} p_j\left[e^{ij\theta}\phi_j + e^{-ij\theta}\phi_{-j}\right].
\end{aligned}$$

We insert the last expression in (1.8) and consider the three equations obtained by collecting the $e^{ik\theta}$ terms for $k = 0, 1, -1$. In these equations, motivated by the assumption that ϕ is only weakly dependent on v, we also make the approximation that

$$\phi_k = 0 \quad \text{for } |k| \geq 2.$$

The resulting equations are

$$\left(\frac{1}{c}\frac{\partial}{\partial t} + \mu_{tr}(x)\right)\phi_0 + \bar{\partial}\phi_1 + \partial\phi_{-1} = 2\pi\mu_s p_0\phi_0, \tag{1.10}$$

$$\left(\frac{1}{c}\frac{\partial}{\partial t} + \mu_{tr}(x)\right)\phi_1 + \partial\phi_0 = \pi\mu_s p_1\phi_1, \tag{1.11}$$

$$\left(\frac{1}{c}\frac{\partial}{\partial t} + \mu_{tr}(x)\right)\phi_{-1} + \bar{\partial}\phi_0 = \pi\mu_s p_1\phi_{-1}. \tag{1.12}$$

Now, note that the photon density and current are given by

$$\Phi(x,t) = 2\pi\phi_0(x,t)$$

and

$$\begin{aligned}
J(x,t) &= \left(\int_0^{2\pi} \frac{e^{i\theta} + e^{-i\theta}}{2}\phi(x, v_\theta, t)\, d\theta, \int_0^{2\pi} \frac{e^{i\theta} - e^{-i\theta}}{2i}\phi(x, v_\theta, t)\, d\theta\right) \\
&= 2\pi\left(\frac{\phi_1 + \phi_{-1}}{2}, \frac{i(\phi_1 - \phi_{-1})}{2}\right).
\end{aligned}$$

Consequently,

$$2\pi\phi_{-1} = J_1 + iJ_2, \quad 2\pi\phi_1 = J_1 - iJ_2.$$

Using these expressions in (1.10)–(1.12) and taking suitable combinations, and using that $p_0 = 1/2\pi$, we get the two equations

(1.13)
$$\left(\frac{1}{c}\frac{\partial}{\partial t} + \mu_a(x)\right)\Phi + \nabla_x \cdot J = 0,$$

(1.14)
$$\left(\frac{1}{c}\frac{\partial}{\partial t} + \mu_a(x) + (1 - \pi p_1)\mu_s(x)\right)J + \frac{1}{2}\nabla_x\Phi = 0.$$

Exercise 1.2.3. Verify (1.13)–(1.14).

We now make the approximation that J is constant in time,

$$\frac{\partial J}{\partial t} = 0.$$

The equation (1.14) becomes

$$J = -D(x)\nabla_x\Phi,$$

where $D = 1/(2(\mu_a + (1 - \pi p_1)\mu_s))$ is the diffusion coefficient. Note that this equation is similar to Ohm's law (1.2). Inserting this expression for J into (1.13), we obtain

$$-\nabla_x \cdot D(x)\nabla_x\Phi + \mu_a(x)\Phi + \frac{1}{c}\frac{\partial\Phi}{\partial t} = 0.$$

Finally, suppose that our transmitters emit light at a given frequency $\omega > 0$, so that

$$\Phi(x, t) = u(x)e^{i\omega t}.$$

Then u will solve the equation

(1.15)
$$-\nabla \cdot D\nabla u + \left(\mu_a + \frac{i\omega}{c}\right)u = 0 \text{ in } \Omega.$$

A similar equation may be derived in dimensions $n \geq 3$, if the Fourier series in the previous argument are replaced by expansions in spherical harmonics.

The inverse problem in diffuse optical tomography (with infinitely many measurements) is then to determine properties of D and μ_a from the DN map associated to the equation (1.15). This problem is analogous to the inverse conductivity problem.

1.3. Inverse scattering

Suppose that we are interested in a system in which sound waves, for example, scatter off of some obstacle. Let $p(x, t)$ be the pressure at position x and time t. In (a somewhat idealized) free space, p obeys the wave equation $\frac{\partial^2 p}{\partial t^2} = c^2\Delta p$, where c is the speed of sound. We shall assume that in most of the world, c takes a constant value c_0. But we introduce an obstacle by allowing c to depend on position in some compact region. We further allow for some absorption in that region. Then p obeys the damped wave equation

$$\frac{\partial^2 p}{\partial t^2} + \gamma(x)\frac{\partial p}{\partial t} = c(x)^2 \Delta p,$$

where $\gamma(x)$ is the damping coefficient of the medium at x. If the solution p has fixed (temporal) frequency, then $p(x,t) = \operatorname{Re}\left[u(x)e^{-i\omega t}\right]$ where u satisfies

$$\Delta u + \frac{\omega^2}{c(x)^2}\left[1 + i\frac{\gamma(x)}{\omega}\right] u = 0.$$

Outside of some compact region, the coefficient in brackets is constant:

$$\frac{\omega^2}{c(x)^2}\left[1 + i\frac{\gamma(x)}{\omega}\right] = \frac{\omega^2}{c_0^2} = k^2, \qquad \text{where} \qquad k = \frac{\omega}{c_0} > 0.$$

If we define the *index of refraction* $n(x)$ by

$$n(x) = \frac{c_0^2}{c(x)^2}\left[1 + i\frac{\gamma(x)}{\omega}\right],$$

then

$$\Delta u + k^2 n(x) u = 0 \tag{1.16}$$

with $n(x) = 1$ outside of some compact region. For concreteness, we restrict to the three-dimensional space $\mathbb{R}^3$ for the rest of this section. We first consider two special cases.

Example 1.3.1 (Free space). In the absence of any obstacle, we have $\Delta u + k^2 u = 0$ on all of $\mathbb{R}^3$. Any function $u = e^{ik\theta\cdot x}$, where θ is a unit vector, is a solution that represents a plane wave moving in direction θ. It is also true that any solution of this equation (say, in the class of tempered distributions) can be obtained as a superposition of these plane waves when interpreted in the right way.

Example 1.3.2 (Point source). If we have free space everywhere except at the origin and we have a unit point source at the origin, then

$$\Delta u + k^2 u = \delta(x) \quad \text{in } \mathbb{R}^3,$$

where the Dirac delta function $\delta(x)$ is a distribution (generalized function) that is determined formally by the properties that $\delta(x) = 0$ for all $x \neq 0$ and $\int_{\mathbb{R}^3} \delta(x)\,dx = 1$. A rigorous version of "$\Delta u + k^2 u = \delta(x)$" is provided in Exercise 1.3.3. Except at the origin, where there is a singularity, we still have $\Delta u + k^2 u = 0$. The point source generates expanding spherical waves. So u should be a function of $r = |x|$ only and obey

$$u''(r) + \frac{2}{r}u'(r) + k^2 u(r) = 0 \quad \text{in } \{r > 0\}.$$

This equation is easily solved by changing variables to $v(r) = ru(r)$, which obeys

$$v''(r) + k^2 v(r) = 0 \quad \text{in } \{r > 0\}.$$

So $v(r) = \alpha \sin(kr) + \beta \cos(kr)$ and $u(r) = \alpha \frac{\sin(kr)}{r} + \beta \frac{\cos(kr)}{r}$. To be an outgoing (rather than incoming) wave, we require that $u(r) = \alpha' \frac{e^{ikr}}{r}$. (Note that $e^{ikr}e^{-i\omega t}$ is constant on $r = \frac{\omega}{k}t$, which is a sphere that is expanding with speed c_0.) To give the Dirac delta function on the right-hand side of $\Delta u + k^2 u = \delta(x)$, we need $u(x) = -\frac{e^{ik|x|}}{4\pi|x|}$.

Exercise 1.3.3. Set $\Phi(x) = -\frac{e^{ik|x|}}{4\pi|x|}$.

(a) Prove that $\Delta\Phi(x) + k^2\Phi(x) = 0$ for all $x \neq 0$.

(b) Let B_ε be the ball of radius ε centered on the origin and let dS be the surface measure on ∂B_ε. Prove that, for any continuous function $\psi(x)$,

$$\lim_{\varepsilon \to 0+} \int_{\partial B_\varepsilon} \frac{\psi(x)}{|x|^p}\, dS = \begin{cases} 4\pi\psi(0) & \text{if } p = 2 \\ 0 & \text{if } p < 2. \end{cases}$$

(c) Let $\psi(x) \in C_c^\infty(\mathbb{R}^3)$. Prove that

$$\int_{\mathbb{R}^3} \Phi(x)[\Delta\psi(x) + k^2\psi(x)]\, dx = \psi(0).$$

Now let us return to the general case. We want to think of a physical situation in which we send a plane wave $u^i(x) = e^{ik\theta\cdot x}$ in from infinity. This plane wave shakes up the obstacle which then emits a bunch of expanding spherical waves $\frac{e^{ik|x-y|}}{|x-y|}$ emanating from various points y in the obstacle. So the full solution is of the form

$$u(x) = u^i(x) + u^s(x),$$

where the scattered wave, u^s, obeys the "radiation condition"

$$\frac{\partial}{\partial r} u^s(x) - iku^s(x) = O\left(\frac{1}{|x|^2}\right) \quad \text{as} \quad |x| \to \infty. \tag{1.17}$$

This condition is chosen to allow outgoing waves $\frac{e^{ik|x-y|}}{|x-y|}$ but not incoming waves $\frac{e^{-ik|x-y|}}{|x-y|}$.

Let, as in Exercise 1.3.3,

$$\Phi(x) = -\frac{e^{ik|x|}}{4\pi|x|}.$$

Since $\delta(x-y)$ is the kernel of the identity operator, the equality

$$(\Delta_x + k^2)\Phi(x-y) = \delta(x-y)$$

says, roughly, that $u(x) \mapsto \int \Phi(x-y)u(y)\,dy$ is the inverse of the map $u(x) \mapsto (\Delta + k^2)u(x)$ for functions that obey the radiation condition. We can exploit this to convert (1.16), (1.17) into an equivalent integral equation

$$\Delta u + k^2 n(x)u = 0 \iff \Delta u + k^2 u = k^2\big(1-n(x)\big)u$$
$$\iff \Delta u^s + k^2 u^s = k^2\big(1-n(x)\big)u$$

since $\Delta u^i + k^2 u^i = 0$. As u^s obeys the radiation condition, we have

$$u^s(x) = k^2 \int_{\mathbb{R}^3} \Phi(x-y)\big(1-n(y)\big)u(y)\,dy,$$

so that

$$u(x) = u^i(x) + k^2 \int_{\mathbb{R}^3} \big(1-n(y)\big)\Phi(x-y)u(y)\,dy. \tag{1.18}$$

This is called the *Lippmann–Schwinger equation.* Observe that it is of the form $u = u^i - Ku$ or $(\mathrm{Id}+K)u = u^i$ where K is the linear operator $u(x) \mapsto k^2 \int \Phi(x-y)\big(n(y)-1\big)u(y)\,dy$. This operator is compact (if you impose the appropriate norms) and so behaves much like a finite-dimensional matrix. If K has operator norm smaller than one, which is the case if $k^2(n-1)$ is small enough, then $\mathrm{Id}+K$ is trivially invertible and the equation $(\mathrm{Id}+K)u = u^i$ has a unique solution. Even if K has operator norm larger than or equal to one, $(\mathrm{Id}+K)u = u^i$ fails to have a unique solution only if K has eigenvalue minus one. One can show that this is impossible in the present setting (see Theorem 7.2.1). Thus, one can prove the following result.

Theorem 1.3.4 (Outgoing solutions). *If* $n \in C^2(\mathbb{R}^3)$, $n(x) - 1$ *is compactly supported, and* $\operatorname{Im} n(x) \geq 0$, *then* (1.16), (1.17) *has a unique solution.*

For $|y|$ bounded and $|x|$ large, $\Phi(x-y)$ has the asymptotic behaviour

$$\Phi(x-y) = -\frac{e^{ik|x|}}{4\pi|x|}e^{-ik\hat{x}\cdot y} + O\left(\frac{1}{|x|^2}\right). \tag{1.19}$$

Thus when the incident plane wave is $u^i(x;\theta) = e^{ik\theta\cdot x}$, the corresponding solution $u(x) = u(x;\theta)$ given by (1.18) satisfies

$$u(x;\theta) = u^i(x;\theta) + \frac{e^{ik|x|}}{4\pi|x|}u_\infty(\hat{x};\theta) + O\left(\frac{1}{|x|^2}\right), \tag{1.20}$$

where

$$u_\infty(\hat{x};\theta) = -k^2 \int_{\mathbb{R}^3} e^{-ik\hat{x}\cdot y}\big(1-n(y)\big)u(y;\theta)\,dy.$$

If we are observing the scattered wave from vantage points far from the obstacle, i.e., when $|x|$ is very large, we will only be able to measure $u_\infty(\hat{x};\theta)$ for $\hat{x} \in S^2$. The quantity $u_\infty(\hat{x};\theta)$ is called the *far field pattern*, or *scattering amplitude*, of n corresponding to the incident wave $u^i(x) = e^{ik\theta\cdot x}$. Assuming that we can send incident waves at a fixed frequency $k > 0$ from all possible directions θ, and that we can measure the corresponding far field patterns for all $\hat{x}$, the inverse problem may be formulated in the following way.

Fixed frequency inverse scattering problem. Given $u_\infty(\hat{x};\theta)$ for all $\hat{x},\theta \in S^2$ and for fixed $k > 0$, can we determine $n(x)$?

The answer is yes, as shown by the following theorem.

Theorem 1.3.5. *Fix $k > 0$. If $n_1, n_2 \in C^2(\mathbb{R}^3)$ with $n_1 - 1, n_2 - 1$ of compact support and $u_{1,\infty}(\hat{x};\theta) = u_{2,\infty}(\hat{x};\theta)$, for all $\hat{x},\theta \in S^2$, then $n_1 = n_2$.*

We can get a rough idea why this theorem is true by looking at the Born approximation. In this approximation u^s is ignored in the computation of u_∞ so that

$$\begin{aligned} u_\infty(\hat{x};\theta) &\approx -k^2 \int e^{-ik\hat{x}\cdot y}(1-n(y))u^i(y;\theta)\,dy \\ &= -k^2 \int e^{-ik(\hat{x}-\theta)\cdot y}(1-n(y))\,dy. \end{aligned}$$

If we measure $u_\infty(\hat{x};\theta)$, then, in this approximation, we know the Fourier transform of $1-n(y)$ on the set $\{\, k(\hat{x}-\theta) \mid \hat{x},\theta \in S^2 \,\}$, which is exactly the closed ball of radius $2k$ centered on the origin in $\mathbb{R}^3$. Since $1-n(y)$ is of compact support, its Fourier transform is analytic. So knowledge of the Fourier transform on any open ball uniquely determines it.

It turns out that the scattering amplitude for n at a fixed frequency is an analog of the DN map, except that the measurements are made far away (on the sphere at infinity) instead of on the boundary of a domain. We shall discuss a quantum mechanical analog of the above classical inverse scattering problem in Chapter 7. The methods for dealing with that problem will be very similar to those applied to the Calderón problem.

Exercise 1.3.6. Prove (1.19).

Exercise 1.3.7. Let $f \in C_c^\infty(\mathbb{R}^3)$. Prove that

$$F(x) = \int_{\mathbb{R}^3} \Phi(x-y)f(y)\,dy$$

obeys $\Delta F + k^2 F = f$ and the radiation condition.

1.4. Open questions

Here we give a selection of questions related to the Calderón problem that are open, to the authors' best knowledge, at the time of writing this. Known results and related references will be given in the Notes sections of the subsequent chapters. In each question below, Ω will be a bounded domain in $\mathbb{R}^n$ with smooth boundary.

We begin with the anisotropic Calderón problem concerning matrix conductivities as discussed in §1.1. It is customary to consider an equivalent formulation in terms of Riemannian metrics (see §3.6).

Question 1.4.1 (Anisotropic problem, $n \geq 3$). Let g_1, g_2 be two smooth Riemannian metrics on $\overline{\Omega}$. If $\Lambda_{g_1} = \Lambda_{g_2}$, is it true that $g_2 = F^* g_1$ for some diffeomorphism $F : \overline{\Omega} \to \overline{\Omega}$ with $F|_{\partial\Omega} = \mathrm{Id}$?

Another open question is the partial data problem.

Question 1.4.2 (Partial data, $n \geq 3$). Let $\gamma_1, \gamma_2 \in C^\infty(\overline{\Omega})$ be positive, and let $\Gamma \subset \partial\Omega$ be a nonempty open set. If $\Lambda_{\gamma_1} f|_\Gamma = \Lambda_{\gamma_2} f|_\Gamma$ for any $f \in C^\infty(\partial\Omega)$ with $\mathrm{supp}(f) \subset \Gamma$, is it true that $\gamma_1 = \gamma_2$?

In the full data problem, uniqueness is known for Lipschitz continuous conductivities but the optimal regularity is not fully understood.

Question 1.4.3 (Low regularity, $n \geq 3$). Does Λ_γ determine a positive Hölder continuous conductivity γ uniquely, or can one find two positive Hölder continuous conductivities $\gamma_1 \not\equiv \gamma_2$ such that $\Lambda_{\gamma_1} = \Lambda_{\gamma_2}$?

The previous questions have been answered when $n = 2$. However, range characterization in the Calderón problem is open in any dimension.

Question 1.4.4 (Range characterization). Can one find necessary and sufficient conditions for a map $T : C^\infty(\partial\Omega) \to C^\infty(\partial\Omega)$ to satisfy $T = \Lambda_\gamma$ for some positive $\gamma \in C^\infty(\overline{\Omega})$?

One can also study inverse problems for systems. An example is given by the stationary elasticity system, for linear isotropic bodies Ω whose elastic properties are characterized by two Lamé coefficients $\lambda, \mu \in C^\infty(\overline{\Omega})$. This system describes infinitesimal displacements $u = (u_1(x), \ldots, u_n(x))$ of Ω resulting from boundary displacements f. It is given by

$$\begin{cases} Lu = 0 & \text{in } \Omega, \\ u = f & \text{on } \partial\Omega, \end{cases}$$

where the elastic operator is

$$Lu = \mathrm{div}(C(Du + (Du)^t)).$$

Here $\mathrm{div}(\sigma)_j = \sum_{k=1}^n \partial_{x_k} \sigma_{j,k}$, $Du = [\partial_{x_k} u_j]_{j,k=1}^n$ is the Jacobi matrix of u, and C is the *stiffness tensor* acting on a matrix $\varepsilon = [\varepsilon_{j,k}]_{j,k=1}^n$ by

$$(C\varepsilon)_{j,k} = \lambda \ \mathrm{tr}(\varepsilon)\delta_{j,k} + 2\mu\varepsilon_{j,k},$$

where $\operatorname{tr}(\varepsilon) = \sum_{j=1}^n \varepsilon_{j,j}$ is the trace. The Legendre–Hadamard ellipticity of L amounts to the conditions

$$\lambda + 2\mu > 0, \qquad \mu > 0.$$

Under the above conditions the bilinear form associated to L is positive definite. Thus the above Dirichlet problem is well-posed and one obtains a boundary deformation map

$$\Lambda_{\lambda,\mu} : C^\infty(\partial\Omega)^n \to C^\infty(\partial\Omega)^n, \quad \Lambda f = C(Du + (Du)^t) \cdot \nu|_{\partial\Omega}.$$

Uniqueness in the inverse problem of determining the pair (λ, μ) from $\Lambda_{\lambda,\mu}$ is known if one of the Lamé coefficients is known or close to a constant.

Question 1.4.5 (Elasticity, $n \geq 3$). Let $\lambda_j, \mu_j \in C^\infty(\overline{\Omega})$ satisfy $\lambda_j + 2\mu_j > 0$ and $\mu_j > 0$. If $\Lambda_{\lambda_1,\mu_1} = \Lambda_{\lambda_2,\mu_2}$, is it true that $\lambda_1 = \lambda_2$ and $\mu_1 = \mu_2$?

Another variation is to consider nonlinear equations. We state an open question related to the p-conductivity equation

$$\operatorname{div}(\gamma|\nabla u|^{p-2}\nabla u) = 0 \text{ in } \Omega.$$

Here $1 < p < \infty$. For $p = 2$ this is the linear conductivity equation, and for $\gamma \equiv 1$ this is the so-called p-Laplace equation. If $\gamma \geq c > 0$, the Dirichlet problem for this equation is well-posed with weak solutions in the Sobolev space $W^{1,p}(\Omega)$. In a suitable weak sense, one can define a nonlinear DN map

$$\Lambda_\gamma^p : u|_{\partial\Omega} \mapsto \gamma|\nabla u|^{p-2}\nabla u \cdot \nu|_{\partial\Omega}.$$

Question 1.4.6 (p-conductivity equation, $n \geq 3$). Let $\gamma_1, \gamma_2 \in C^\infty(\overline{\Omega})$ be positive and $1 < p < \infty$ with $p \neq 2$. If $\Lambda_{\gamma_1}^p = \Lambda_{\gamma_2}^p$, is it true that $\gamma_1 = \gamma_2$?

We conclude with a question related to inverse scattering. In Chapter 7 it is proved that the scattering amplitude $a_q(\lambda_0, \cdot, \cdot)$ at a fixed frequency λ_0 uniquely determines a compactly supported potential q. When q is not compactly supported the problem becomes more delicate.

Question 1.4.7 (Scattering for a frequency interval). Let $q_1, q_2 \in \mathcal{S}(\mathbb{R}^n)$ be real valued Schwartz functions. If $a_{q_1}(\lambda, \cdot, \cdot) = a_{q_2}(\lambda, \cdot, \cdot)$ for λ in some interval $[\lambda_1, \lambda_2]$ with $\lambda_1 < \lambda_2$, is it true that $q_1 = q_2$?

1.5. Notes

The inverse problem of Calderón has a rich history. The electrical imaging method discussed in §1.1 goes back to the 1910s within the context of geophysics. In that setting the method is known as *Electrical Resistivity Tomography* (ERT), and it is used for studying structures in the interior of the Earth by electrical measurements on the surface. The oil field services company Schlumberger was founded in 1926 for this purpose [**Sch82, OD09**]. Early mathematical studies may be found in [**Lan33, Sli33**]. In the 1940s this problem was studied by Tikhonov [**Tih49**], also known for his pioneering work on topology including Tychonoff's theorem on products of compact spaces and Tychonoff's fixed point theorem. The fundamental Tikhonov regularization method for solving ill-posed problems is also named after him.

Calderón was interested in the electrical imaging problem in the 1940s as well, motivated by the possibility of locating oil deposits while he was working for the Argentinian state-owned oil company YPF (Yacimientos Petrolíferos Fiscales). He did not publish his results at that time, and only made them available much later in the short article [**Cal80**] that became a starting point for much of the modern theory of this problem. Further historical information may be found in [**Cal08**].

In [**Cal80**] Calderón gave an elegant mathematical formulation of the problem, solved its linearized version and gave an approximate reconstruction method. He also introduced the idea of using exponential solutions, which have become widely used in related inverse problems. Exponential solutions had appeared as well in the work of Faddeev [**Fad16**] on inverse scattering. A related early result involving exponential solutions is [**Had53**, §2.18] where it was proved that the Cauchy problem for the Laplace equation is ill-posed. Ill-posedness is a fundamental feature of many inverse problems including the Calderón problem.

Electrical impedance tomography (EIT) has various potential applications in medical imaging. These include pulmonary imaging and ventilation monitoring [**CGLA09, Fre00, RSSN$^+$24**], improved detection of breast cancer [**KBK$^+$08**], or classification and monitoring of stroke [**AÇL$^+$20, TPD$^+$24**]. Commercial devices are available, such as the Dräger PulmoVista 500 device with 16 electrodes, launched in 2011 for ventilation and respiratory monitoring. In industrial imaging EIT has been employed in process tomography [**Yor01**] and nondestructive testing of concrete structures [**KSL$^+$10, HSPG14**]. Further information and references on these applications may be found in [**MS12, AH21**].

There is a large body of literature on computational methods related to the Calderón problem and EIT. We refer the readers to [**MS12**, §12–16 and in particular Table 14.1] for a discussion and references up to 2012. More recent references may be found in [**AH21**, §6–7] or [**TNH$^+$25**].

Textbooks considering different aspects of the Calderón problem were listed in the introduction. For survey articles on the Calderón problem see [**Uhl92, Syl96, Uhl99, Bor02, Uhl09, Uhl14a, Uhl14b**]. Further survey articles are given in the Notes sections in the subsequent chapters.

The theorems stated in §1.1 will be proved and their background will be discussed in the later chapters. As we mentioned, range characterization for the Calderón problem is an open question except for the case of the geometric Calderón problem in the unit disk [**Sha11**]. Example 1.1.6 is discussed in [**CDJ63, KV84b**].

Our derivation of the diffusion approximation in optical tomography in §1.2 follows the survey [**Arr99**]. An alternative derivation may be found in the survey [**AS09**].

The acoustic inverse scattering problem in $\mathbb{R}^3$ discussed in §1.3 is studied in more detail in [**CK19**], where also Theorems 1.3.4 and 1.3.5 are proved. We will prove analogous results for quantum inverse scattering in Chapter 7.

Chapter 2

Formulation of the Calderón Problem

In this chapter we formulate rigorously the inverse boundary value problem for the conductivity equation

$$\operatorname{div}(\gamma \nabla u) = 0$$

and for the related Schrödinger equation

$$(-\Delta + q)u = 0.$$

We will begin by collecting basic facts from multivariable calculus. Then we continue with a discussion of weak solutions of the corresponding Dirichlet problems, the definition of boundary measurements in terms of the Dirichlet-to-Neumann map (DN map for short), and basic properties of the DN map. We will also derive useful integral identities that allow us to relate boundary measurements of solutions to integrals involving solutions in the interior. The section will end with a reduction of the inverse boundary value problem for conductivity equation to an inverse problem for Schrödinger equation, and with a brief discussion of the Neumann-to-Dirichlet map.

Convention. All functions will be complex valued unless stated otherwise. When proving inequalities, we will denote by C a generic positive constant that may change from line to line. The constant C will be independent of relevant quantities, such as the function u when proving an inequality like $\|Tu\|_{L^2} \leq C\,\|u\|_{L^2}$ for a linear operator $T : L^2 \to L^2$.

2.1. Cutoff functions

In the course of the book we will frequently need to work locally in small sets and then patch up these local constructions into global ones. Therefore, in this preliminary section we will discuss smooth cutoff functions and partitions of unity that are the standard tool in such local arguments.

Definition 2.1.1 (Multi-index notation, C^k and L^p spaces, $O(\,\cdot\,)$ and $o(\,\cdot\,)$). Let Ω be an open set in $\mathbb{R}^n$ (not necessarily bounded). A *multi-index* is an n-tuple $\alpha = (\alpha_1, \ldots, \alpha_n) \in \mathbb{N}_0^n$, where $\mathbb{N}_0 = \{0, 1, 2, \ldots\}$. We write $|\alpha| = \alpha_1 + \ldots + \alpha_n$. If $f : \Omega \to \mathbb{C}$ is k times continuously differentiable and $\alpha \in \mathbb{N}_0^n$ with $|\alpha| \leq k$, we write

$$\partial^\alpha f = \partial_{x_1}^{\alpha_1} \cdots \partial_{x_n}^{\alpha_n} f, \qquad \partial_{x_j} = \frac{\partial}{\partial x_j}.$$

The set of continuous functions in Ω is denoted by $C(\Omega) = C^0(\Omega)$. If $k \in \mathbb{N}_0$, we denote by $C^k(\Omega)$ the set of functions $f : \Omega \to \mathbb{C}$ whose partial derivatives up to order k are continuous in Ω, i.e., $\partial^\alpha f \in C(\Omega)$ whenever $|\alpha| \leq k$. By $C_c^k(\Omega)$ we denote the set of functions in $C^k(\Omega)$ whose support is a compact subset of Ω. The set $C^k(\overline{\Omega})$ consists of the functions in $C^k(\Omega)$ whose partial derivatives up to order k extend continuously to $\overline{\Omega}$. By $C^\infty(\Omega)$, $C_c^\infty(\Omega)$, and $C^\infty(\overline{\Omega})$ we denote the corresponding sets of infinitely many times continuously differentiable functions.

If $A \subset \mathbb{R}^n$ is measurable and $1 \leq p < \infty$, the set $L^p(A)$ is the Lebesgue space consisting of measurable functions $f : A \to \mathbb{C}$ (with functions that agree almost everywhere identified) such that

$$\|f\|_{L^p(A)} = \left(\int_A |f(x)|^p \, dx \right)^{1/p} < \infty.$$

The space $L^2(A)$ is equipped with the sesquilinear inner product

$$(f, g)_{L^2(A)} = \int_A f(x)\overline{g(x)} \, dx.$$

The space $L^\infty(A)$ consists of measurable functions with

$$\|f\|_{L^\infty(A)} = \operatorname{ess\,sup}_{x \in A} |f(x)| < \infty.$$

If $\Omega \subset \mathbb{R}^n$ is an open set, then $L_c^p(\Omega)$ consists of those functions in $L^p(\Omega)$ whose support is compact in Ω, and $L_{\mathrm{loc}}^p(\Omega)$ is the set of measurable functions $f : \Omega \to \mathbb{C}$ such that $f|_K \in L^p(K)$ for any compact subset K of Ω.

We will also use the "big O" and "small o" notations

$f = O(h)$ as $|h| \to 0$, to mean that $|f| \leq C|h|$ when $|h|$ is small;

$g = o(h)$ as $|h| \to 0$, to mean that $|g|/|h| \to 0$ as $|h| \to 0$.

As examples of the above definition, we have

$$C^\infty(\Omega) = \{\, f : \Omega \to \mathbb{C} \mid \partial^\alpha f \in C(\Omega) \text{ whenever } \alpha \in \mathbb{N}_0^n \,\}$$

and

$$C_c^\infty(\Omega) = \{\, f \in C^\infty(\Omega) \mid \operatorname{supp}(f) \text{ is a compact subset of } \Omega \,\},$$

where the support is defined by

$$\operatorname{supp}(f) = \Omega \setminus \{\, x \in \Omega \mid f = 0 \text{ a.e. in some neighborhood of } x \,\}.$$

The next definition gives examples of functions in $C_c^\infty(\mathbb{R}^n)$, in particular showing that this set is not empty.

Definition 2.1.2 (Mollifiers). Define the function

$$\eta(x) = \begin{cases} C\exp\left(\frac{1}{|x|^2-1}\right), & |x| < 1, \\ 0, & |x| \geq 1, \end{cases}$$

where the constant C is chosen so that $\int_{\mathbb{R}^n} \eta(x)\,dx = 1$. If $\varepsilon > 0$ define the mollifier

$$\eta_\varepsilon(x) = \varepsilon^{-n}\eta(x/\varepsilon).$$

Exercise 2.1.3. Prove that η and η_ε are in $C_c^\infty(\mathbb{R}^n)$.

Note that $\eta_\varepsilon \in C_c^\infty(\mathbb{R}^n)$ satisfies

$$\int_{\mathbb{R}^n} \eta_\varepsilon(x)\,dx = 1, \qquad \operatorname{supp}(\eta_\varepsilon) = \overline{B(0,\varepsilon)}.$$

If ε is small, the function η_ε looks like a sharp peak at the origin with area under the peak equal to one. It is thus a smooth approximation of the Dirac delta function at the origin, and it can be used to approximate a locally integrable function by smooth functions.

The *convolution* of two functions $f, g : \mathbb{R}^n \to \mathbb{C}$ is the function $f * g$ defined by

$$(f * g)(x) = \int_{\mathbb{R}^n} f(y)g(x-y)\,dy = \int_{\mathbb{R}^n} f(x-y)g(y)\,dy.$$

The convolution $f * g$ is well defined when $f, g \in L^1(\mathbb{R}^n)$, and in this case $f * g \in L^1(\mathbb{R}^n)$ essentially by Fubini's theorem. If $f \in L^1_{\mathrm{loc}}(\mathbb{R}^n)$ and $g \in L^1_c(\mathbb{R}^n)$, then similarly $f * g \in L^1_{\mathrm{loc}}(\mathbb{R}^n)$. Additionally, if g is smooth, then also $f * g$ is smooth as proved in the next theorem.

Theorem 2.1.4 (Mollifications). *If $f \in L^1_{\mathrm{loc}}(\mathbb{R}^n)$, define for $\varepsilon > 0$ the mollifications of f by*

$$f_\varepsilon = f * \eta_\varepsilon.$$

(a) *$f_\varepsilon \in C^\infty(\mathbb{R}^n)$ for any $\varepsilon > 0$, and $\partial^\alpha f_\varepsilon = f * \partial^\alpha \eta_\varepsilon$.*

(b) $\operatorname{supp}(f_\varepsilon) \subset \{x \in \mathbb{R}^n\,;\, \operatorname{dist}(x, \operatorname{supp}(f)) \leq \varepsilon\}$.

(c) *If* $f \in C^0(\mathbb{R}^n)$*, then* $f_\varepsilon \to f$ *uniformly on compact sets in* $\mathbb{R}^n$ *as* $\varepsilon \to 0$.

(d) *If* $f \in L^p(\mathbb{R}^n)$ *where* $1 \leq p < \infty$*, then* $f_\varepsilon \to f$ *in* L^p *as* $\varepsilon \to 0$.

Proof.

(a) Since $\eta_\varepsilon \in C_c^0(\mathbb{R}^n)$ and $|f|$ is integrable on compact sets, the function

$$f_\varepsilon(x) = \int_{\mathbb{R}^n} f(y)\eta_\varepsilon(x-y)\,dy$$

is well defined for all $x \in \mathbb{R}^n$. Fix $x \in \mathbb{R}^n$ and note that for any $h \in B(0,1)$,

$$f_\varepsilon(x+h) - f_\varepsilon(x) = \int_{\mathbb{R}^n} f(y)\,[\eta_\varepsilon(x+h-y) - \eta_\varepsilon(x-y)]\,dy.$$

The integral is over the set of all y for which $|x-y| \leq \varepsilon$ or $|x+h-y| \leq \varepsilon$. This is a compact set (depending on x and ε) that we denote by K. Also, since η_ε is C^∞, we have the Taylor expansion

$$\eta_\varepsilon(x-y+h) = \eta_\varepsilon(x-y) + \nabla\eta_\varepsilon(x-y)\cdot h + \frac{1}{2}\sum_{j,k=1}^n \frac{\partial^2\eta_\varepsilon}{\partial x_j \partial x_k}(x-y+h')h_j h_k,$$

where $h' \in \mathbb{R}^n$ is some point on the line segment between 0 and h. It follows that

$$f_\varepsilon(x+h) - f_\varepsilon(x) = \int_K f(y)\left[\nabla\eta_\varepsilon(x-y)\cdot h + \sum_{j,k=1}^n b_{jk}(y;x,h)h_j h_k\right]dy,$$

where $|b_{jk}(y;x,h)| \leq C$ uniformly over $y \in K$ and $h \in B(0,1)$. By dominated convergence we obtain, as $h \to 0$,

$$f_\varepsilon(x+h) - f_\varepsilon(x) = \sum_{j=1}^n \left[\int_{\mathbb{R}^n} f(y)\partial_j\eta_\varepsilon(x-y)\,dy\right]h_j + O(h^2).$$

This shows that f_ε is differentiable. Repeating the argument for higher-order derivatives gives that $f_\varepsilon \in C^\infty$ and $\partial^\alpha f_\varepsilon = f * \partial^\alpha \eta_\varepsilon$.

(b) Clear since $\mathrm{supp}(\eta_\varepsilon) \subset \overline{B(0,\varepsilon)}$.

(c) Let $x \in K$, where $K \subset \mathbb{R}^n$ is compact, and fix $\varepsilon_0 > 0$. Since $\int \eta_\varepsilon\,dy = 1$, we have

$$\begin{aligned} |f_\varepsilon(x) - f(x)| &= \left|\int_{\mathbb{R}^n} f(x-y)\eta_\varepsilon(y)\,dy - \int_{\mathbb{R}^n} f(x)\eta_\varepsilon(y)\,dy\right| \\ &\leq \int_{\mathbb{R}^n} |f(x-y) - f(x)|\,\eta_\varepsilon(y)\,dy. \end{aligned}$$

Note that the integral is over $\overline{B(0,\varepsilon)}$. Since f is continuous, it is uniformly continuous on compact sets, which implies that there is $\delta_0 > 0$ such that

$$|f(x-y) - f(x)| < \varepsilon_0 \quad \text{whenever } x \in K \text{ and } |y| < \delta_0.$$

This shows that $\|f_\varepsilon - f\|_{L^\infty(K)} < \varepsilon_0$ whenever $\varepsilon < \delta_0$.

(d) To prove L^p convergence, we use the fact that $C_c(\mathbb{R}^n)$ is dense in $L^p(\mathbb{R}^n)$ if $1 \le p < \infty$ (see [**Rud87**, Theorem 3.14]). Fix $\varepsilon_0 > 0$ and choose some $g \in C_c(\mathbb{R}^n)$ with

$$\|f - g\|_{L^p(\mathbb{R}^n)} < \varepsilon_0/3.$$

By the triangle inequality

$$\|f_\varepsilon - f\|_{L^p} \le \|f_\varepsilon - g_\varepsilon\|_{L^p} + \|g_\varepsilon - g\|_{L^p} + \|g - f\|_{L^p}.$$

The supports of g and g_ε are contained in some compact set K, and

$$\|g_\varepsilon - g\|_{L^p(\mathbb{R}^n)} = \|g_\varepsilon - g\|_{L^p(K)} \le C_K \|g_\varepsilon - g\|_{L^\infty(K)}.$$

By part (c), there is $\delta_0 > 0$ such that $\|g_\varepsilon - g\|_{L^p} < \varepsilon_0/3$ whenever $\varepsilon < \delta_0$. Thus, for $\varepsilon < \delta_0$,

$$\|f_\varepsilon - f\|_{L^p} \le \|(f - g)_\varepsilon\|_{L^p} + 2\varepsilon_0/3.$$

The result will now follow if we can show that for all $h \in L^p(\mathbb{R}^n)$,

$$\|h_\varepsilon\|_{L^p} \le \|h\|_{L^p}.$$

This is a direct consequence of the Minkowski inequality in integral form (Exercise 2.1.5). The usual Minkowski inequality reads

$$\Big\|\sum_{j=1}^N f_j\Big\|_{L^p} \le \sum_{j=1}^N \|f_j\|_{L^p},$$

and the integral form is the same inequality but where the sums are replaced by integrals. Thus we have

$$\begin{aligned}
\|h_\varepsilon\|_{L^p} &= \Big\|\int \eta_\varepsilon(y) h(\cdot - y)\,dy\Big\|_{L^p} \\
&\le \int \eta_\varepsilon(y)\,\|h(\cdot - y)\|_{L^p}\,dy \\
&= \|h\|_{L^p} \int \eta_\varepsilon(y)\,dy \\
&= \|h\|_{L^p}.
\end{aligned}$$

□

Exercise 2.1.5 (Minkowski inequality in integral form). If (X,μ) and (Y,ν) are σ-finite measure spaces, $F : X \times Y \to \mathbb{C}$ is measurable, and $1 \le p < \infty$, prove that

$$\left(\int_X \left(\int_Y |F(x,y)|\,d\nu(y)\right)^p d\mu(x)\right)^{\frac{1}{p}} \le \int_Y \left(\int_X |F(x,y)|^p\,d\mu(x)\right)^{\frac{1}{p}} d\nu(y).$$

Theorem 2.1.4(d) implies that $C^\infty(\mathbb{R}^n) \cap L^p(\mathbb{R}^n)$ is dense in $L^p(\mathbb{R}^n)$ when $p < \infty$. This can be improved as follows.

Lemma 2.1.6. *If $U \subset \mathbb{R}^n$ is an open set, then $C_c^\infty(U)$ is dense in $L^p(U)$ when $1 \le p < \infty$.*

Proof. As in the proof of Theorem 2.1.4, we note that $C_c(U)$ is dense in $L^p(U)$ if $1 \le p < \infty$ [**Rud87**, Theorem 3.14]. Given any $f \in C_c(U)$, the functions f_ε for $\varepsilon > 0$ sufficiently small are in $C_c^\infty(U)$ and approximate f in $L^p(U)$ arbitrarily well. This proves the lemma. □

Mollifications yield an immediate proof of a variant of the du Bois–Reymond lemma, which will be fundamental for the definition of weak derivatives.

Theorem 2.1.7 (du Bois–Reymond lemma). *Let $U \subset \mathbb{R}^n$ be an open set and $f \in L^1_{\mathrm{loc}}(U)$. If*

$$\int_U f\varphi \, dx = 0 \qquad \text{for all } \varphi \in C_c^\infty(U),$$

then $f = 0$ almost everywhere in U.

Proof. Let $x \in U$ and choose $B = B(x,r)$ with $\overline{B} \subset U$. Define $\tilde{f} = \chi_B f \in L^1(\mathbb{R}^n)$, where χ_B is the characteristic function of the set B (that is, $\chi_B = 1$ on B and $\chi_B = 0$ elsewhere). If $y \in B$, then $\tilde{f} * \eta_\varepsilon(y) = 0$ for ε small by the assumption. Moreover, $\tilde{f} * \eta_\varepsilon \to \chi_B f$ in L^1 as $\varepsilon \to 0$ by Theorem 2.1.4. Thus $f = 0$ almost everywhere near x. □

The next result shows how mollification can be used for creating smooth cutoff functions with specified behavior.

Theorem 2.1.8 (Smooth cutoff function). *Let $K \subset U \subset \mathbb{R}^n$, where K is compact and U is open. There exists a function $\zeta \in C_c^\infty(U)$ with $0 \le \zeta \le 1$ in U and $\zeta = 1$ on K.*

Proof. Choose a compact set L with $K \subset \mathrm{int}(L) \subset L \subset U$. The characteristic function χ_L is in $L^1(\mathbb{R}^n)$. Since L is compact and $\mathbb{R}^n \setminus U$ is a closed set disjoint from L, there exists $\varepsilon > 0$ so that the set $\{\, x \in \mathbb{R}^n \mid \mathrm{dist}(x,L) \le \varepsilon \,\}$ is strictly contained in U. By Theorem 2.1.4 we have $\chi_L * \eta_\varepsilon \in C_c^\infty(U)$. By further decreasing ε we have $\chi_L * \eta_\varepsilon = 1$ on K, and $\zeta = \chi_L * \eta_\varepsilon$ satisfies the required properties. □

The following theorem shows that one can create a finite set of smooth cutoff functions adapted to a cover $\{V_1, \ldots, V_N\}$ of a compact set K. It will be used frequently in this book.

Theorem 2.1.9 (Partition of unity). *Let $K \subset \mathbb{R}^n$ be compact and let $K \subset \bigcup_{j=1}^N V_j$, where V_j are open sets. There exist functions $\zeta_j \in C_c^\infty(V_j)$ such that $0 \le \zeta_j \le 1$*

and

$$\sum_{j=1}^{N} \zeta_j \le 1 \text{ in } \mathbb{R}^n, \qquad \sum_{j=1}^{N} \zeta_j = 1 \text{ on } K.$$

Definition 2.1.10. In the setting of Theorem 2.1.9, we say that $\{\zeta_j\}_{j=1}^N$ is a *partition of unity* on K subordinate to the cover $\{V_j\}_{j=1}^N$.

Exercise 2.1.11. Prove Theorem 2.1.9.

2.2. C^k domains and integration by parts

In this section we give a brief discussion on one of the most fundamental and useful methods in mathematical analysis, namely *integration by parts*. In the most classical case, this amounts to the fundamental theorem of calculus: if f is a continuously differentiable real valued function on an interval $[a, b]$, then

$$\int_a^b f'(t)\,dt = f(b) - f(a).$$

Observe that this formula relates information in the interior of the domain (the integral of f' over the interval) to information on the boundary (the "boundary integral", or the sum of values of f at the endpoints taken with opposite signs).

The integration by parts formulas in this section are multidimensional generalizations of the fundamental theorem of calculus. They underlie the notion of weak solutions for partial differential equations, and in fact the theory of weak solutions essentially amounts to taking the integration by parts formula as an *axiom* rather than a theorem. Integration by parts is also particularly useful in inverse problems, since it allows one to relate measurements at the boundary to information in the interior just as in the one-dimensional case above.

It is natural to first discuss the multidimensional domains over which we integrate. To have a reasonable boundary integral, we need to assume some regularity of the boundary of the domain. By definition, a *domain* in $\mathbb{R}^n$ is an open connected subset of $\mathbb{R}^n$. Much of the time connectedness will not be required and we will work with bounded open sets.

Our definition of sets with C^k boundary is given in terms of mappings Φ that flatten the boundary locally. Such a mapping is required to be a C^k *diffeomorphism*, which means that Φ is bijective and the component functions of both Φ and Φ^{-1} are C^k.

Definition 2.2.1 (C^k boundary)**.** Let $\Omega \subset \mathbb{R}^n$ be a bounded open set and let $k \in \mathbb{Z}_+$.

(a) We say that Ω has C^k boundary (or that $\partial\Omega$ is C^k or that $\partial\Omega \in C^k$) if, for each $p \in \partial\Omega$, there is an open neighborhood $U = U(p)$ of p and a C^k

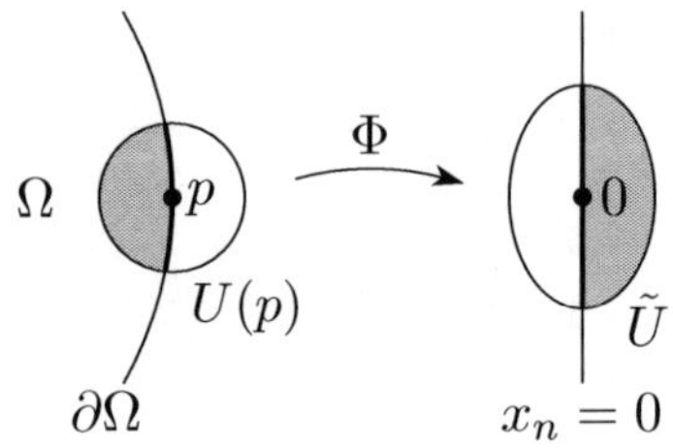

Figure 2.1. Boundary flattering map Φ.

diffeomorphism $\Phi = \Phi_p : U \to \tilde{U}$ onto some open set $\tilde{U} \subset \mathbb{R}^n$ such that $\Phi(p) = 0$ and, as in Figure 2.1,

$$\Phi\big(U(p) \cap \Omega\big) = \{\, x \in \tilde{U} \mid x_n > 0 \,\},$$
$$\Phi\big(U(p) \cap \partial\Omega\big) = \{\, x \in \tilde{U} \mid x_n = 0 \,\}.$$

(b) We say that Ω has smooth boundary (or that $\partial\Omega$ is smooth or that $\partial\Omega \in C^\infty$) if each Φ_p, $p \in \partial\Omega$, of (a) is a C^∞ diffeomorphism.

(c) We call the system $\big(U(p), \Phi_p\big)_{p\in\partial\Omega}$ a coordinate system for $\partial\Omega$.

If $\partial\Omega$ has C^1 boundary, there is a well-defined tangent space $T_p(\partial\Omega)$ at each point p of $\partial\Omega$: if $(U_q, \Phi_q)_{q\in\partial\Omega}$ is a coordinate system, a basis for this space is given by $\{\dot{\gamma}_1(0), \dots, \dot{\gamma}_{n-1}(0)\}$ where $\gamma_\alpha : (-\varepsilon, \varepsilon) \to \mathbb{R}^n$ are the C^1 curves (here $\varepsilon > 0$ is sufficiently small, $\alpha = 1, \dots, n-1$, and e_α is the αth coordinate vector)

$$\gamma_\alpha(t) = \Phi_p^{-1}(te_\alpha). \tag{2.1}$$

There are two other equivalent ways of looking at sets with C^k boundary. We will mention them here since they will be useful below. The first way expresses $\partial\Omega$ locally as the graph of a C^k function. Below, by an orthonormal coordinate system $x = (x_1, \dots, x_n)$ at p we mean that we identify points $p + \sum_{j=1}^n x_j v_j$ with x, where $\{v_1, \dots, v_n\}$ is an orthonormal basis of $\mathbb{R}^n$.

Theorem 2.2.2 (Local graph representation). *Let Ω be a bounded open set in $\mathbb{R}^n$. Then Ω has C^k boundary if and only if for any point $p \in \partial\Omega$, there exist $r > 0$, an orthonormal coordinate system $x = (x', x_n)$ with origin at p, and a C^k function $h : \mathbb{R}^{n-1} \to \mathbb{R}$ such that*

$$\Omega \cap B(p,r) = \{\, x \in B(p,r) \mid x_n > h(x') \,\},$$
$$\partial\Omega \cap B(p,r) = \{\, x \in B(p,r) \mid x_n = h(x') \,\}.$$

Proof. Suppose $(U(q), \Phi_q)$ is a coordinate system for $\partial\Omega$, fix $p \in \Omega$, and let $\Phi = \Phi_p$. We may translate and rotate the coordinate systems so that $p = 0$,

$\Phi(0) = 0$, and the tangent space to $\partial\Omega$ at 0 is spanned by $\{e_1, \dots, e_{n-1}\}$. Since the nth component of Φ vanishes on $\partial\Omega$, $\Phi^n(q) = 0$ for $q \in \partial\Omega$, we have

$$\Phi^n(\gamma_\alpha(t)) = 0,$$

where γ_α are the curves (2.1). Differentiating in t and evaluating at $t = 0$ implies that

$$\partial_\alpha \Phi^n(0) = 0, \quad \alpha = 1, \dots, n-1.$$

Since Φ is a C^k diffeomorphism ($k \geq 1$), the Jacobian matrix $D\Phi(0)$ is invertible. This implies that $\partial_n \Phi^n(0) \neq 0$. Changing x_n to $-x_n$ if necessary, we may assume that $\partial_n \Phi^n(0) > 0$.

By the implicit function theorem, there is a C^k function h defined near the origin in $\mathbb{R}^{n-1}$ such that near 0 we have

$$\Phi^n(x', x_n) = 0 \quad \Longleftrightarrow \quad x_n = h(x').$$

We also have the Taylor expansion of $\Phi^n(x', \cdot\,)$ at $x_n = h(x')$,

$$\Phi^n(x', x_n) = \Phi^n(x', h(x')) + \partial_n \Phi^n(x', h(x'))(x_n - h(x')) + o(x_n - h(x'))$$

as $x_n \to h(x')$. Here $\Phi^n(x', h(x')) = 0$ and $\partial_n \Phi^n(x', h(x')) > 0$ for x' sufficiently close to 0 by continuity. Thus, if x is close to 0, the conditions $x_n > h(x')$ and $\Phi^n(x', x_n) > 0$ are equivalent. This shows that locally Ω is given by $\{\, (x', x_n) \mid x_n > h(x') \,\}$, and from this it follows that $\partial\Omega$ is locally given by $\{\, (x', x_n) \mid x_n = h(x') \,\}$.

For the converse, it is sufficient to observe that the maps $\Phi(x', x_n) = (x', x_n - h(x'))$ are boundary flattening C^k diffeomorphisms in suitable neighborhoods of the related boundary points. □

The second equivalent way of defining a C^k boundary is based on boundary defining functions.

Theorem 2.2.3 (Boundary defining function). *Let Ω be a bounded open set in $\mathbb{R}^n$. Then Ω has C^k boundary if and only if there is a C^k function $\rho : \mathbb{R}^n \to \mathbb{R}$ such that*

$$\Omega = \{\, x \in \mathbb{R}^n \mid \rho(x) > 0 \,\},$$
$$\partial\Omega = \{\, x \in \mathbb{R}^n \mid \rho(x) = 0 \,\},$$

and $\nabla\rho \neq 0$ on $\partial\Omega$.

Sketch of Proof. If $p \in \partial\Omega$ and if $\Omega \cap B(p, r) = \{\, x \in B(p, r) \mid x_n > h(x') \,\}$, the function ρ in $B(p, r)$ can be taken to be $\rho(x', x_n) = x_n - h(x')$. We can construct the global boundary defining function ρ from the local expressions by using a suitable partition of unity as in Theorem 2.1.9. □

Exercise 2.2.4. Prove Theorem 2.2.3 in detail.

We can use any of the above definitions to define $C^l(\partial\Omega)$ functions if Ω has C^k boundary and $l \leq k$.

Definition 2.2.5. If Ω is a bounded open set with C^k boundary and if $l \leq k$, we say that $f : \partial\Omega \to \mathbb{C}$ is C^l and write $f \in C^l(\partial\Omega)$ if there is a coordinate system $\big(U(p), \Phi_p\big)_{p\in\partial\Omega}$ such that for any $p \in \partial\Omega$, $f \circ \Phi_p^{-1}$ is a C^l function in the open subset $\Phi_p(U(p) \cap \partial\Omega)$ of $\mathbb{R}^{n-1}$.

There are two quantities associated with C^k domains that will be used frequently. The first is the outer unit normal vector of $\partial\Omega$.

Definition 2.2.6 (Outer unit normal and normal derivative)**.** Let Ω have C^1 boundary and let ρ be a boundary defining function for $\partial\Omega$. The *outer unit normal vector* of $\partial\Omega$ at a point $p \in \partial\Omega$ is

$$\nu(p) = -\frac{\nabla\rho(p)}{|\nabla\rho(p)|}.$$

The *normal derivative* of $u \in C^1(\overline{\Omega})$ at $p \in \partial\Omega$ is

$$\partial_\nu u(p) = \nabla u(p) \cdot \nu(p).$$

The definition of ν is independent of the choice of boundary defining function. We write the vector field $\nu : \partial\Omega \to \mathbb{R}^n$ in terms of its components as $\nu = (\nu_1, \ldots, \nu_n)$, where each ν_j is a function in $C^{k-1}(\partial\Omega)$ if the boundary is C^k. If Ω is locally given as the set $\{x_n > h(x')\}$, the outer unit normal has the expression

$$\nu(x', h(x')) = \frac{(\nabla_{x'}h(x'), -1)}{(1 + |\nabla_{x'}h(x')|^2)^{1/2}}.$$

One has $\nu(p) \cdot v = 0$ for any $v \in T_p(\partial\Omega)$.

Exercise 2.2.7. Verify the claims about ν in the preceding paragraph.

The second useful quantity on a C^k boundary $\partial\Omega$ is its (Euclidean) surface measure dS, induced by the usual Lebesgue measure dx in $\mathbb{R}^n$. The surface measure is a constant multiple of the $(n-1)$-dimensional Hausdorff measure restricted to $\partial\Omega$. Another way to obtain this measure is as follows.

Theorem 2.2.8 (Surface measure)**.** *Let Ω be a bounded open set with C^1 boundary. There is a unique positive Borel measure on $\partial\Omega$, acting on functions $f \in C^0(\partial\Omega)$ by $f \mapsto \int_{\partial\Omega} f\,dS$, such that*

$$\int_{\partial\Omega} f\,dS = \lim_{\delta\to 0} \frac{1}{2\delta} \int_{V_\delta} \tilde{f}(x)\,dx,$$

where $V_\delta = \{\, x \in \mathbb{R}^n \mid \mathrm{dist}(x, \partial\Omega) < \delta \,\}$ and $\tilde{f}$ is any function in $C^0(\mathbb{R}^n)$ with $\tilde{f}|_{\partial\Omega} = f$.

If $\partial\Omega$ is expressed as the graph $x' \mapsto (x', h(x'))$ near some $p \in \partial\Omega$ and if $f \in C^0(\partial\Omega)$ is supported sufficiently close to p, then

$$\int_{\partial\Omega} f\, dS = \int_{\mathbb{R}^{n-1}} f(x', h(x'))(1 + |\nabla h(x')|^2)^{1/2}\, dx'.$$

Here $\mathrm{dist}(x, \partial\Omega) = \inf_{p\in\partial\Omega} |x - p|$. A proof of Theorem 2.2.8 is outlined in the following exercise.

Exercise 2.2.9. Let Ω be a bounded open set with C^1 boundary.

(a) Show that if $\mathrm{dist}(x, \partial\Omega) = |x - p|$ where $p \in \partial\Omega$, then $x - p$ is parallel to $\nu(p)$. (You can use the curves (2.1).)

(b) Use the boundary normal coordinates in Theorem 3.3.1 to show that $|V_\delta| \leq C\delta$ and that the limit in Theorem 2.2.8 does not depend on the choice of $\tilde{f}$.

(c) Use a partition of unity to show that there is a linear operator

$$E : C^0(\partial\Omega) \to C^0(\mathbb{R}^n)$$

such that $Ef|_{\partial\Omega} = f$ for any $f \in C^0(\partial\Omega)$ and $Ef \geq 0$ whenever $f \geq 0$.

(d) Use the previous two items and the Riesz representation theorem for measures [**Rud87**, Theorem 2.14] to show that dS is a well-defined positive Borel measure.

(e) Use boundary normal coordinates and a change of variables in the integral over V_δ to prove the last formula in Theorem 2.2.8.

We are now ready to state the integration by parts formulas that will be used in this book. Most of them are equivalent, and all are consequences of the next result.

Theorem 2.2.10 (Gauss–Green formula). *If $\Omega \subset \mathbb{R}^n$ is a bounded open set with C^1 boundary and if $u \in C^1(\overline{\Omega})$, then*

$$\int_\Omega \partial_j u\, dx = \int_{\partial\Omega} u\nu_j\, dS$$

for $j = 1, \ldots, n$.

Proof. For each $p \in \partial\Omega$, we choose an orthonormal coordinate system (x', x_n), a ball $B(p, r)$ and a C^1 function $h : \mathbb{R}^{n-1} \to \mathbb{R}$ such that $\Omega \cap B(p, r) = \{ x \in B(p, r) \,|\, x_n > h(x') \}$. These balls cover the compact set $\partial\Omega$, and there is a finite subcover $\{B_k\}_{k=1}^N$. Choose some open set $B_0 \subset \Omega$ such that the sets $\{B_k\}_{k=0}^N$

cover $\overline{\Omega}$. By Theorem 2.1.9, we may find a partition of unity $\{\zeta_k\}_{k=0}^N$ on $\overline{\Omega}$ subordinate to the cover $\{B_k\}$. Then

$$\int_\Omega \partial_j u \, dx = \sum_{k=0}^N \int_\Omega \partial_j(\zeta_k u) \, dx.$$

Since $\zeta_0 u \in C_c^1(\Omega)$, we have

$$\int_\Omega \partial_j(\zeta_0 u) \, dx = \int_{\mathbb{R}^n} \partial_j(\zeta_0 u) \, dx = \int_{\mathbb{R}^{n-1}} \int_{-\infty}^{\infty} \partial_j(\zeta_0 u) \, dx_j \, dy,$$

where y contains all the variables in x except for x_j. The fundamental theorem of calculus implies that

$$\int_{-\infty}^{\infty} \partial_j(\zeta_0 u)(x_1, \ldots, x_n) \, dx_j = \lim_{R\to\infty} \int_{-R}^{R} \partial_j(\zeta_0 u) \, dx_j = 0.$$

Thus $\int_\Omega \partial_j(\zeta_0 u) \, dx = 0$.

Let now $v = \zeta_k u$, where $1 \leq k \leq N$, and write $B_k = B(p, r)$ and $\Omega \cap B_k = \{\, x \in B_k \mid x_n > h(x') \,\}$. Since $\zeta_k \in C_c^\infty(B_k)$, we have

$$\int_\Omega \partial_j v \, dx = \int_{\mathbb{R}^{n-1}} \int_{h(x')}^{\infty} \partial_j v(x) \, dx_n \, dx'.$$

Choose a function $\psi \in C^\infty(\mathbb{R})$ with $\psi(t) = 0$ for $t < 0$ and $\psi(t) = 1$ for $t > 1$. (Such a function can be obtained by mollifying the function which is zero for $t < 1/2$ and equals one for $t > 1/2$.) Define also $\rho(x) = x_n - h(x')$ in B_k. It follows from the properties of ψ and the Leibniz rule that

$$\begin{aligned}\int_\Omega \partial_j v \, dx &= \lim_{\varepsilon\to 0} \int_{\mathbb{R}^n} \partial_j v(x) \psi\left(\frac{\rho(x)}{\varepsilon}\right) dx \\ &= \lim_{\varepsilon\to 0} \int_{\mathbb{R}^n} \left[\partial_j\left(v\psi\left(\frac{\rho}{\varepsilon}\right)\right) - v\psi'\left(\frac{\rho}{\varepsilon}\right)\frac{\partial_j \rho}{\varepsilon}\right] dx.\end{aligned}$$

Since $v\psi(\rho/\varepsilon) \in C_c^1(\mathbb{R}^n)$, the integral $\int_{\mathbb{R}^n} \partial_j(v\psi(\rho/\varepsilon)) \, dx$ vanishes by the fundamental theorem of calculus. Consequently,

$$\int_\Omega \partial_j v \, dx = -\lim_{\varepsilon\to 0} \int_{\mathbb{R}^n} v\psi'\left(\frac{\rho}{\varepsilon}\right)\frac{\partial_j \rho}{\varepsilon} \, dx.$$

Using that $\psi'(t) = 0$ for $t < 0$ and $t > 1$ and writing $x_n = h(x') + \varepsilon t$, we may write

$$\begin{aligned}\int_\Omega \partial_j v \, dx &= -\lim_{\varepsilon\to 0} \int_{\mathbb{R}^{n-1}} \int_{h(x')}^{h(x')+\varepsilon} v(x)\psi'\left(\frac{x_n - h(x')}{\varepsilon}\right)\frac{\partial_j \rho(x)}{\varepsilon} \, dx_n \, dx' \\ &= -\lim_{\varepsilon\to 0} \int_{\mathbb{R}^{n-1}} \int_0^1 v(x', h(x') + \varepsilon t)\psi'(t)\partial_j \rho(x', h(x') + \varepsilon t) \, dt \, dx'.\end{aligned}$$

We may take the limit by dominated convergence. Then the fact that

$$\int_0^1 \psi'(t)\,dt = \psi(1) - \psi(0) = 1$$

yields

$$\int_\Omega \partial_j v\,dx = -\int_{\mathbb{R}^{n-1}} v(x', h(x'))\partial_j\rho(x', h(x'))\,dx'.$$

From the definitions of the outer unit normal and surface measure, we see that

$$\nu_j(x', h(x')) = -\partial_j\rho(x', h(x'))/\,|\nabla\rho(x', h(x'))| \quad \text{and}$$
$$dS(x') = |\nabla\rho(x', h(x'))|\;dx'.$$

This shows that

$$\int_\Omega \partial_j(\zeta_k u)\,dx = \int_{\partial\Omega} \zeta_k u\nu_j\,dS, \quad k = 1, \dots, N.$$

The result follows by summing over k from 0 to N and using the fact that $\{\zeta_k\}$ is a partition of unity. □

A number of fundamental integration by parts identities can be easily derived from the Gauss–Green formula.

Theorem 2.2.11. *Let Ω have C^1 boundary.*

(a) (*Integration by parts*). *If $u, v \in C^1(\overline{\Omega})$, then*

$$\int_\Omega u\partial_j v\,dx = -\int_\Omega (\partial_j u)v\,dx + \int_{\partial\Omega} uv\nu_j\,dS.$$

(b) (*Divergence theorem*). *If $F : \overline{\Omega} \to \mathbb{R}^n$ is a C^1 vector field, then*

$$\int_\Omega \operatorname{div}(F)\,dx = \int_{\partial\Omega} F\cdot\nu\,dS.$$

(c) (*Green's identity*). *If $u \in C^1(\overline{\Omega})$, $v \in C^2(\overline{\Omega})$, then*

$$\int_{\partial\Omega} u\partial_\nu v\,dS = \int_\Omega \nabla u\cdot\nabla v\,dx + \int_\Omega u\Delta v\,dx.$$

(d) (*Green's identity*). *If $u, v \in C^2(\overline{\Omega})$, then*

$$\int_{\partial\Omega} (u\partial_\nu v - v\partial_\nu u)\;dS = \int_\Omega (u\Delta v - v\Delta u)\;dx.$$

Exercise 2.2.12. Prove Theorem 2.2.11.

2.3. Sobolev spaces

Let Ω be a bounded open subset of $\mathbb{R}^n$, and let $\gamma \in L^\infty(\Omega)$ be a positive function representing the electrical conductivity of the medium Ω. We wish to consider a suitable function space for solutions of the conductivity equation

$$\operatorname{div}(\gamma \nabla u) = 0 \qquad \text{in } \Omega.$$

The function spaces that will be used for this purpose are called *Sobolev spaces*, and they turn out to be appropriate for describing weak solutions of a large class of partial differential equations. In this section we develop a minimal amount of Sobolev space theory that is sufficient for formulating the Calderón problem precisely. A much more detailed treatment of Sobolev spaces is given in Appendix C.

If $\gamma \in C^1(\overline{\Omega})$ and $u \in C^2(\overline{\Omega})$, then one can interpret the equation $\operatorname{div}(\gamma \nabla u) = 0$ in the classical pointwise sense since the derivatives exist pointwise. Solutions in $C^2(\overline{\Omega})$ are often called *classical solutions*. However, when the conductivity γ is only in $L^\infty(\Omega)$ the expression $\operatorname{div}(\gamma \nabla u)$ requires some interpretation. Also, our aim is to use energy methods and Hilbert space theory to produce solutions, and for this purpose it is more natural to use spaces based on $L^2(\Omega)$ rather than the C^k spaces.

To define weak solutions of the equation $\operatorname{div}(\gamma \nabla u) = 0$, one first needs the concept of weak derivatives of functions that may not be differentiable in the classical sense. Weak derivatives will be defined via a suitable test function space.

Definition 2.3.1 (Test functions). The elements of $C_c^\infty(\Omega)$, that is, infinitely differentiable functions $\varphi : \Omega \to \mathbb{C}$ with compact support in Ω, are called *test functions*.

Motivation 2.3.2. To motivate the definition of weak derivatives, let $\Omega \subset \mathbb{R}^n$ be a bounded open set with C^1 boundary and let $u \in C^k(\overline{\Omega})$. Consider the classical αth derivative of u, $\partial^\alpha u$, where $\alpha \in \mathbb{N}_0^n$ is a multi-index with $|\alpha| \leq k$. If $\varphi \in C_c^\infty(\Omega)$ is a test function, integrating by parts repeatedly using Theorem 2.2.11(a) implies that

$$\int_\Omega u \partial^\alpha \varphi \, dx = (-1)^{|\alpha|} \int_\Omega (\partial^\alpha u) \varphi \, dx.$$

We used the fact that φ and its derivatives vanish near $\partial\Omega$, so there are no boundary terms. Now the left-hand side is well defined if $u \in L^1_{\mathrm{loc}}(\Omega)$. In this case we say that u has αth weak partial derivative in $L^1_{\mathrm{loc}}(\Omega)$ if the above identity remains true for all test functions φ, when $\partial^\alpha u$ on the right-hand side is replaced by some locally integrable function v.

Definition 2.3.3 (Weak derivatives). Let $\Omega \subset \mathbb{R}^n$ be open, let $u, v \in L^1_{\mathrm{loc}}(\Omega)$, and let $\alpha \in \mathbb{N}_0^n$ be a multi-index. We say that v is the αth *weak partial derivative* of u, written

$$v = \partial^\alpha u,$$

if

$$\int_\Omega u \partial^\alpha \varphi \, dx = (-1)^{|\alpha|} \int_\Omega v \varphi \, dx \qquad \text{for all} \quad \varphi \in C_c^\infty(\Omega).$$

By Theorem 2.1.7, the αth weak partial derivative (whenever it exists) is uniquely defined as an L^1_{loc} function. However, the weak derivative can be understood as an object that is integrated against test functions, as opposed to the classical pointwise notion of derivatives:

- Pointwise derivative: $\partial_j u(x) = \lim_{h \to 0} \frac{u(x+he_j)-u(x)}{h}$ for all $x \in \Omega$.
- Weak derivative: $\int_\Omega (\partial_j u) \varphi \, dx = -\int_\Omega u \partial_j \varphi \, dx$ for all $\varphi \in C_c^\infty(\Omega)$.

Having given the definition of weak derivatives, we proceed to discuss the spaces of functions relevant for weak solutions of the equation $\mathrm{div}(\gamma \nabla u) = 0$ in Ω.

Definition 2.3.4 (Sobolev space $H^k(\Omega)$). If $\Omega \subset \mathbb{R}^n$ is open and $k \in \mathbb{N}_0$, the *Sobolev space* $H^k(\Omega)$ consists of all functions $u \in L^2(\Omega)$ for which the weak partial derivative $\partial^\alpha u$ is in $L^2(\Omega)$ whenever $\alpha \in \mathbb{N}_0^n$ and $|\alpha| \leq k$. We equip this space with the inner product

$$(u, v)_{H^k(\Omega)} = \sum_{|\alpha| \leq k} (\partial^\alpha u, \partial^\alpha v)_{L^2(\Omega)}$$

and with the norm

$$\|u\|_{H^k(\Omega)} = (u, u)_{H^k(\Omega)}^{1/2}.$$

In connection with the Calderón problem in two dimensions in Chapter 5, we will also need L^p based Sobolev spaces $W^{1,p}(\Omega)$ where $1 \leq p \leq \infty$.

Definition 2.3.5 (Sobolev space $W^{k,p}(\Omega)$). If $\Omega \subset \mathbb{R}^n$ is open, $k \in \mathbb{N}_0$ and $1 \leq p \leq \infty$, the *Sobolev space* $W^{k,p}(\Omega)$ consists of all functions $u \in L^p(\Omega)$ for which the weak partial derivative $\partial^\alpha u$ is in $L^p(\Omega)$ whenever $\alpha \in \mathbb{N}_0^n$ and $|\alpha| \leq k$. We equip this space with the norm

$$\|u\|_{W^{k,p}(\Omega)} = \left[\sum_{|\alpha| \leq k} \|\partial^\alpha u\|_{L^p(\Omega)}^p \right]^{\frac{1}{p}}.$$

Note that $W^{k,2}(\Omega) = H^k(\Omega)$. The $W^{k,p}$ spaces will mainly be used in Chapter 5, and for now we will restrict ourselves to the spaces $H^k(\Omega)$.

Examples of Sobolev functions are given next. Exercise 2.3.7 shows that some rather pathological functions can still belong to $H^1(\Omega)$.

Example 2.3.6.

(a) If $\Omega \subset \mathbb{R}^n$ is a bounded open set, the space $C^k(\overline{\Omega})$ is contained in $H^k(\Omega)$.

(b) If $\Omega \subset \mathbb{R}^n$ is open, $C_c^\infty(\Omega)$ is contained in $H^k(\Omega)$ for all $k \geq 0$.

(c) If $\Omega = B(0,1) \subset \mathbb{R}^n$, then the function

$$u(x) = |x|^{-\alpha}, \qquad |x| < 1$$

is in $H^1(\Omega)$ if and only if $\alpha < n/2 - 1$. Indeed, this function has gradient

$$\nabla u(x) = -\alpha\,|x|^{-\alpha-2}\,x, \qquad x \in \Omega.$$

A computation in polar coordinates shows that $x \mapsto |x|^{-\beta}$ is integrable near 0 if and only if $\beta < n$. Using these facts, it is not hard to see that the weak gradient of u is equal to the pointwise gradient of u and that $u \in H^1(\Omega)$ for any $\alpha < n/2 - 1$.

Exercise 2.3.7. If $\Omega = B(0,1) \subset \mathbb{R}^n$, $n \geq 3$, give an example of a function $u \in H^1(\Omega)$ that is not essentially bounded in any open subset of Ω. Give a similar example for $n = 2$. Can you find an example of this type for $n = 1$?

The next result shows that $H^k(\Omega)$ is a Hilbert space, as the notation already suggests.

Theorem 2.3.8. *$H^k(\Omega)$ is a Hilbert space for each $k \in \mathbb{N}_0$.*

Exercise 2.3.9. Prove Theorem 2.3.8.

Exercise 2.3.10 (Pointwise multipliers). If $\Omega \subset \mathbb{R}^n$ is a bounded open set and $a \in W^{k,\infty}(\Omega)$, $u \in H^k(\Omega)$, show that $au \in H^k(\Omega)$. Show also that

$$\partial^\alpha(au) = \sum_{\beta \leq \alpha} \binom{\alpha}{\beta} \partial^\beta a \partial^{\alpha-\beta} u, \quad \alpha \in \mathbb{N}_0^n,\ |\alpha| \leq k,$$

where $\beta \leq \alpha$ means that $\beta_j \leq \alpha_j$ for $j = 1, \ldots, n$, $\binom{\alpha}{\beta} = \frac{\alpha!}{\beta!(\alpha-\beta)!}$, and $\alpha! = \alpha_1! \cdots \alpha_n!$. (Hint: use induction on $|\alpha|$. It is enough to do the case $k = 1$ if the general case is difficult.)

To deal with boundary value problems for the equation $\mathrm{div}(\gamma\nabla u) = 0$ in Ω, we need to consider the space $H_0^1(\Omega)$ consisting of those functions in $H^1(\Omega)$ which vanish on the boundary $\partial\Omega$ in some sense. Functions in $C_c^\infty(\Omega)$ always vanish near the boundary, which motivates the following definition.

Definition 2.3.11 (The spaces $H_0^1(\Omega)$ and $H^{-1}(\Omega)$). We denote by $H_0^1(\Omega)$ the closure of the set $C_c^\infty(\Omega)$ in $H^1(\Omega)$. Its dual space is denoted by

$$H^{-1}(\Omega) = (H_0^1(\Omega))^* = \{F : H_0^1(\Omega) \to \mathbb{C} \text{ bounded linear functional}\}.$$

Intuitively, one can think of these spaces as follows: $H_0^1(\Omega)$ is the set of all $u \in H^1(\Omega)$ such that $u|_{\partial\Omega} = 0$ (in a certain sense), and $H^{-1}(\Omega)$ is the space of generalized functions having -1 derivatives in $L^2(\Omega)$ (thus $L^2(\Omega)$ is a strict subspace of $H^{-1}(\Omega)$).

Remark 2.3.12. There is one potentially confusing point in the previous definition: Hilbert space theory tells that any Hilbert space is isomorphic to its dual, so one might wonder why the dual space $H^{-1}(\Omega)$ is needed. In fact, the Riesz representation theorem (Theorem A.2.15) shows that any $F \in H^{-1}(\Omega)$ can be uniquely represented in the form

$$F(v) = (v, w)_{H_0^1(\Omega)}, \quad v \in H_0^1(\Omega),$$

for some $w \in H_0^1(\Omega)$. The point is that this representation involves the $H_0^1(\Omega)$ inner product, whereas the definition of weak derivatives is given in terms of the L^2 inner product. The above representation can be written in the weak sense as

$$F(v) = (v, w)_{L^2(\Omega)} + (\nabla v, \nabla w)_{L^2(\Omega)} = (v, w - \Delta w)_{L^2(\Omega)}.$$

Thus $H^{-1}(\Omega)$ can be identified with the set $\{w - \Delta w \,;\, w \in H_0^1(\Omega)\}$, if these functions are understood to act on $H_0^1(\Omega)$ functions with respect to the L^2 inner product. In this interpretation $H^{-1}(\Omega)$ contains all functions in $L^2(\Omega)$, since any function $g \in L^2(\Omega)$ gives rise to a bounded linear functional on $H_0^1(\Omega)$ by

$$g(v) = \int_\Omega gv \, dx, \quad v \in H_0^1(\Omega).$$

Theorem 2.3.13. *$H_0^1(\Omega)$ and $H^{-1}(\Omega)$ are Hilbert spaces, and $H_0^1(\Omega)$ is a closed subspace of $H^1(\Omega)$.*

Proof. This follows from basic facts in Hilbert space theory (see Appendix A). □

We proceed to describe Sobolev spaces on the boundary $\partial\Omega$ that will serve as appropriate function spaces for boundary values of weak solutions. As mentioned above, we think of $H_0^1(\Omega)$ as the set of those functions in $H^1(\Omega)$ whose boundary value, also-called *trace*, on $\partial\Omega$ vanishes. In the same spirit, we consider two functions $u, v \in H^1(\Omega)$ to have the same boundary value on $\partial\Omega$ if $u - v \in H_0^1(\Omega)$. This motivates the following definition of an abstract trace space of $H^1(\Omega)$.

Definition 2.3.14 (Abstract trace space $H^{1/2}(\partial\Omega)$). Define $H^{1/2}(\partial\Omega)$ as the quotient space

$$H^{1/2}(\partial\Omega) = H^1(\Omega)/H_0^1(\Omega).$$

The elements of $H^{1/2}(\partial\Omega)$ are the equivalence classes

$$[u] = \{u + \varphi \,;\, \varphi \in H_0^1(\Omega)\},$$

where u runs through all elements of $H^1(\Omega)$. Also define the (abstract) *trace operator*

$$R : H^1(\Omega) \to H^{1/2}(\partial\Omega), \quad Ru = [u].$$

We also write $u|_{\partial\Omega} = Ru$.

It is proved below that if Ω is a bounded open set with C^1 boundary, the abstract space $H^{1/2}(\partial\Omega)$ can be identified with a subspace of $L^2(\partial\Omega)$. Here $L^2(\partial\Omega)$ is the space of square integrable functions on $\partial\Omega$ with respect to surface measure dS. A precise description of $H^{1/2}(\partial\Omega)$ is obtained via the Fourier transform, and this will be discussed in detail in Appendix C.

At this point, we only motivate the notation $H^{1/2}(\partial\Omega)$ with an example of a function $u \in H^1(\Omega)$ whose boundary value $u|_{\partial\Omega}$ is in $L^2(\partial\Omega)$ but not in $H^1(\partial\Omega)$. In this example, one can heuristically think that $u|_{\partial\Omega}$ has half a derivative in $L^2(\partial\Omega)$.

Example 2.3.15. Let $\Omega \subset \mathbb{R}^n$ be a bounded open set with C^∞ boundary such that $\Omega \subset \{x_n > 0\}$ and $B(0,1) \cap \{x_n > 0\} \subset \Omega$, and consider the function in Example 2.3.6(c),

$$u(x) = |x|^{-\alpha}, \qquad x \in \Omega.$$

We fix $\alpha < n/2 - 1$, so that $u \in H^1(\Omega)$.

Writing $x = (x', x_n)$, the pointwise restriction of u to $\partial\Omega \cap \{x_n = 0\}$ is

$$u(x', 0) = |x'|^{-\alpha}.$$

The pointwise gradient of this function is

$$\nabla_{x'} u(x', 0) = -\alpha\, |x'|^{-\alpha-2}\, x'.$$

Thus, the function $x' \mapsto u(x', 0)$ is in $L^2(\partial\Omega \cap \{x_n = 0\})$ (since $\alpha < \frac{n-1}{2}$), but its pointwise gradient is L^2 integrable only if $\alpha < \frac{n-1}{2} - 1$. Heuristically, interpolating the expressions for $u(x', 0)$ and $\nabla_{x'} u(x', 0)$ suggests that the absolute value of the "fractional gradient" $|\nabla_{x'}^{\theta} u(x', 0)|$ would behave like $|x'|^{-\alpha-\theta}$. This is always L^2 integrable if $0 \le \theta \le 1/2$, suggesting that $u|_{\partial\Omega}$ has half a derivative in $L^2(\partial\Omega)$ but not a full derivative in general.

One benefit of the abstract definition of $H^{1/2}(\partial\Omega)$ is that this definition is valid without any regularity assumptions on the boundary $\partial\Omega$. For many results considered in this book, this abstract setup is actually sufficient to formulate the corresponding inverse problems and their solutions. We now describe some further properties of the abstract space $H^{1/2}(\partial\Omega)$. Note first the orthogonal decomposition

$$H^1(\Omega) = H_0^1(\Omega) \oplus H_0^1(\Omega)^{\perp}.$$

This is valid since $H_0^1(\Omega)$ is a closed subspace of $H^1(\Omega)$. The next result gives the standard Hilbert space structure on $H^{1/2}(\partial\Omega)$.

Theorem 2.3.16. *The orthogonal projection $P : H^1(\Omega) \to H_0^1(\Omega)^\perp$ induces a bijective linear map*

$$S : H^{1/2}(\partial\Omega) \to H_0^1(\Omega)^\perp, \quad S([u]) = P(u).$$

The space $H^{1/2}(\partial\Omega)$ becomes a Hilbert space when equipped with the inner product

$$([u],[v])_{H^{1/2}(\partial\Omega)} = (S([u]), S([v]))_{H^1(\Omega)}, \qquad u, v \in H^1(\Omega)$$

and with the norm

$$\|[u]\|_{H^{1/2}(\partial\Omega)} = \|S([u])\|_{H^1(\Omega)} = \inf_{v \in H_0^1(\Omega)} \|u+v\|_{H^1(\Omega)}, \qquad u \in H^1(\Omega).$$

Proof. The map S is well defined since $P(u+v) = P(u)$ for $v \in H_0^1(\Omega)$. If $S([u]) = 0$, then $P(u) = 0$ so $u \in H_0^1(\Omega)$ and $[u] = 0$, and given $w \in H_0^1(\Omega)^\perp$ we have $w = P(w) = S([w])$. Thus S is linear and bijective.

The inner product on $H^{1/2}(\partial\Omega)$ is clearly sesquilinear, conjugate symmetric, and positive definite. If $([u_j])$ is a Cauchy sequence, then $(S[u_j]) = (P(u_j))$ is Cauchy in $H^1(\Omega)$ and converges in $H^1(\Omega)$. Since $H_0^1(\Omega)^\perp$ is closed it follows that $P(u_j) \to P(u)$ in $H^1(\Omega)$ for some $u \in H^1(\Omega)$, so that $[u_j] \to [u]$ in $H^{1/2}(\partial\Omega)$. □

Theorem 2.3.17 (Right inverse of abstract trace operator). *There is a bounded linear map*

$$E_{\partial\Omega} : H^{1/2}(\partial\Omega) \to H^1(\Omega)$$

that satisfies

$$RE_{\partial\Omega} f = f, \qquad f \in H^{1/2}(\partial\Omega).$$

In particular, for any $f \in H^{1/2}(\partial\Omega)$ there is $v_f \in H^1(\Omega)$ with

$$\left\|v_f\right\|_{H^1(\Omega)} \leq C\|f\|_{H^{1/2}(\partial\Omega)}, \quad v_f|_{\partial\Omega} = f.$$

Proof. It is enough to take $E_{\partial\Omega}([u]) = P(u)$ for $u \in H^1(\Omega)$. Then

$$RE_{\partial\Omega}([u]) = [P(u)] = [u]$$

and $\|E_{\partial\Omega}([u])\|_{H^1(\Omega)} = \|P(u)\|_{H^1(\Omega)} = \|[u]\|_{H^{1/2}(\partial\Omega)}$. □

Let us finally define the negative order Sobolev space $H^{-1/2}(\partial\Omega)$ as a dual space.

Definition 2.3.18. Define $H^{-1/2}(\partial\Omega)$ as the Hilbert dual

$$H^{-1/2}(\partial\Omega) = (H^{1/2}(\partial\Omega))^* = \{g : H^{1/2}(\partial\Omega) \to \mathbb{C} \text{ bounded linear functional}\}.$$

To conclude this section we will indicate that if Ω is a bounded domain with C^1 boundary, then the abstract space $H^{1/2}(\partial\Omega)$ can be identified with a concrete subspace of $L^2(\partial\Omega)$ (a precise description is given in Appendix C). This boils down to proving a trace theorem for the space $H^1(\Omega)$, which we will do next.

Theorem 2.3.19 (Trace theorem). *Let $\Omega \subset \mathbb{R}^n$ be a bounded open set with C^1 boundary. There is a unique bounded linear map*

$$T : H^1(\Omega) \to L^2(\partial\Omega)$$

such that $T(u) = u|_{\partial\Omega}$ for all $u \in C^1(\overline{\Omega})$.

Proof. We first construct a C^1 vector field $F : \overline{\Omega} \to \mathbb{R}^n$ with $F \cdot \nu|_{\partial\Omega} \geq 1/2$. By Theorem 2.2.3, Ω has a C^1 boundary defining function $\rho : \mathbb{R}^n \to \mathbb{R}$ such that $\nabla\rho|_{\partial\Omega} \neq 0$ and $\nu = -\nabla\rho/|\nabla\rho||_{\partial\Omega}$. Let $\chi \in C_c^\infty(\mathbb{R}^n)$ be a cutoff function such that $\chi = 1$ near $\partial\Omega$ and $\nabla\rho \neq 0$ near $\mathrm{supp}(\chi)$ (the existence of χ follows from Theorem 2.1.8). Define the vector field

$$G : \mathbb{R}^n \to \mathbb{R}^n, \quad G(x) = -\chi(x)\frac{\nabla\rho(x)}{|\nabla\rho(x)|}.$$

Then $G|_{\partial\Omega} = \nu$, but G is in general only C^0 and not C^1. To rectify this, we let η_ε be the mollifier appearing in Theorem 2.1.4, and define

$$F_\varepsilon = G * \eta_\varepsilon.$$

Here the mollification is componentwise, i.e., $(F_\varepsilon)_j = G_j * \eta_\varepsilon$ for $1 \leq j \leq n$. Now each F_ε is C^∞, and by Theorem 2.1.4 one has $F_\varepsilon|_{\partial\Omega} \to \nu$ uniformly on $\partial\Omega$ as $\varepsilon \to 0$. We may thus take $F = F_\varepsilon$ where ε is chosen so small that $F \cdot \nu|_{\partial\Omega} \geq 1/2$, and F is C^1 (even C^∞).

For any $u \in C^1(\overline{\Omega})$, which may be taken real valued for simplicity, the divergence theorem (Theorem 2.2.11) implies that

$$\begin{aligned}\frac{1}{2}\int_{\partial\Omega} u^2\,dS &\leq \int_{\partial\Omega} (F\cdot\nu)u^2\,dS \\ &= \int_\Omega \mathrm{div}(Fu^2)\,dx \\ &= \int_\Omega \mathrm{div}(F)u^2\,dx + 2\int_\Omega F\cdot u\nabla u\,dx.\end{aligned}$$

Taking absolute values and using Young's inequality, $2ab \leq a^2 + b^2$, we obtain

$$\begin{aligned}\frac{1}{2}\int_{\partial\Omega} u^2\,dS &\leq \|\mathrm{div}(F)\|_{L^\infty}\int_\Omega u^2\,dx + 2\|F\|_{L^\infty}\int_\Omega |u|\,|\nabla u|\,dx \\ &\leq C\int_\Omega (u^2 + |\nabla u|^2)\,dx.\end{aligned}$$

This inequality is valid for all $u \in C^1(\overline{\Omega})$. Since $C^1(\overline{\Omega})$ is dense in $H^1(\Omega)$ by Lemma C.2.20, the result follows using the B.L.T. theorem, (Theorem A.2.13). □

If Ω has C^1 boundary, it will be proved in Theorem C.2.1 that

$$H_0^1(\Omega) = \{ u \in H^1(\Omega) \mid Tu = 0 \}.$$

It follows that

$$H^{1/2}(\partial\Omega) = T(H^1(\Omega)) \tag{2.2}$$

and thus $H^{1/2}(\partial\Omega)$ is a subspace of $L^2(\partial\Omega)$. Moreover, any function $h \in L^2(\partial\Omega)$ can be identified with the element $T_h \in H^{-1/2}(\partial\Omega)$ defined by

$$T_h : H^{1/2}(\partial\Omega) \to \mathbb{R}, \quad T_h(f) = \int_{\partial\Omega} hf\, dS.$$

Thus $L^2(\partial\Omega)$ is a subspace of $H^{-1/2}(\partial\Omega)$.

2.4. Compact Sobolev embedding

We will need only one nontrivial fact about Sobolev spaces to obtain weak solutions to the partial differential equations that we are interested in. This is the fact that the inclusion map $i : H_0^1(\Omega) \to L^2(\Omega)$ is compact, or in other words, every bounded sequence in $H_0^1(\Omega)$ has a subsequence that converges in $L^2(\Omega)$. After showing this we will have access to powerful tools in the theory of compact operators, such as the Fredholm alternative and the spectral theorem, in the analysis of weak solutions. Compact operators are discussed in more detail in §A.3.

The fundamental principle that allows us to extract a convergent subsequence is the Arzelà–Ascoli theorem. This result is an extension of the fact that any bounded sequence of complex numbers has a convergent subsequence.

Theorem 2.4.1 (Arzelà–Ascoli). *Let (X, d) be a compact metric space, and let (f_j) be a sequence of functions $X \to \mathbb{C}$. Assume that (f_j) is pointwise bounded and equicontinuous, that is, $\sup_{j\in\mathbb{Z}_+} |f_j(x)| < \infty$ for each $x \in X$ and for any $\varepsilon > 0$ there is $\delta > 0$ such that*

$$|f_j(x) - f_j(y)| < \varepsilon \quad \text{whenever } d(x,y) < \delta \text{ and } j = 1, 2, \ldots.$$

Then (f_j) has a subsequence that converges uniformly on X.

Exercise 2.4.2. Prove Theorem 2.4.1 in the case where additionally (X, d) is separable. (Hint: choose a countable dense subset $\{x_l\}_{l=1}^{\infty}$ of X, and first find a subsequence that converges at each x_l. Proofs may be found in [**RS80**, Theorem 1.28] or [**Con90**, Theorem 3.8].)

Theorem 2.4.3 (Compact Sobolev embedding). *If $\Omega \subset \mathbb{R}^n$ is a bounded open set, then the inclusion map $i : H_0^1(\Omega) \to L^2(\Omega)$ is compact.*

Proof. Suppose that $(u_j) \subset H_0^1(\Omega)$ is a bounded sequence, so

$$\left\|u_j\right\|_{H^1(\Omega)} \leq C, \qquad j = 1, 2, \ldots.$$

Since $C_c^\infty(\Omega)$ is dense in $H_0^1(\Omega)$, for any j there is $\varphi_j \in C_c^\infty(\Omega)$ with

$$\left\|u_j - \varphi_j\right\|_{H^1(\Omega)} < 1/j.$$

Then also (φ_j) is a bounded sequence in $H_0^1(\Omega)$, and if we can find a subsequence (φ_{j_k}) that converges in $L^2(\Omega)$ then also (u_{j_k}) will converge in $L^2(\Omega)$ to the same limit. Consequently, we may assume that $(u_j) \subset C_c^\infty(\mathbb{R}^n)$, $\mathrm{supp}(u_j) \subset \Omega$ for each j, and

$$\left\|u_j\right\|_{H^1(\mathbb{R}^n)} \leq C, \qquad j = 1, 2, \ldots. \tag{2.3}$$

Assume for the moment that instead of (2.3) we have uniform bounds for $p = \infty$,

$$\left\|u_j\right\|_{L^\infty(\mathbb{R}^n)} + \left\|\nabla u_j\right\|_{L^\infty(\mathbb{R}^n)} \leq C, \qquad j = 1, 2, \ldots.$$

Then the sequence $(u_j|_{\overline{\Omega}})$ would be pointwise bounded. It would also be equicontinuous since, for $x, y \in \overline{\Omega}$, the mean value theorem would give

$$|u_j(x) - u_j(y)| \leq \left(\sup_{t \in [0,1]} |\nabla u_j(x + t(y - x))| \right) |x - y| \leq C\,|x - y|\,.$$

By the Arzelà–Ascoli theorem we could find a uniformly convergent subsequence on $\overline{\Omega}$.

The device that will be used to pass from the uniform bound (2.3) to the case $p = \infty$ described above is mollification. Define

$$u_j^\varepsilon = u_j * \eta_\varepsilon.$$

We compute

$$\begin{aligned} u_j^\varepsilon(x) - u_j(x) &= \int_{\mathbb{R}^n} \eta_\varepsilon(y)\left[u_j(x-y) - u_j(x)\right] dy \\ &= \int_{\mathbb{R}^n} \eta_\varepsilon(y)\left[\int_0^1 \frac{d}{dt}u_j(x-ty)\,dt\right] dy \\ &= \int_{\mathbb{R}^n}\int_0^1 \eta_\varepsilon(y)\nabla u_j(x-ty)\cdot(-y)\,dt\,dy \\ &= -\varepsilon\int_{B(0,1)}\int_0^1 \eta(y)\nabla u_j(x-t\varepsilon y)\cdot y\,dt\,dy. \end{aligned}$$

The Minkowski inequality in integral form (Exercise 2.1.5) implies that

$$\left\|u_j^\varepsilon - u_j\right\|_{L^2(\mathbb{R}^n)} \le \varepsilon\int_{B(0,1)}\int_0^1 \eta(y)\left\|\nabla u_j(\,\cdot\, - t\varepsilon y)\right\|_{L^2(\mathbb{R}^n)} |y|\;dt\,dy.$$

Since $\left\|\nabla u_j(\,\cdot\, - t\varepsilon y)\right\|_{L^2} = \left\|\nabla u_j\right\|_{L^2}$, the uniform bound (2.3) shows that

$$\left\|u_j^\varepsilon - u_j\right\|_{L^2(\mathbb{R}^n)} \le C'\varepsilon, \qquad j = 1, 2, \ldots$$

for some $C' > 0$. The point is that these bounds are uniform with respect to j.

We will now prove the theorem by showing that (u_j) has a subsequence that is Cauchy in $L^2(\Omega)$. Fix $\varepsilon_0 > 0$, and choose ε so small that

$$\left\|u_j - u_j^\varepsilon\right\|_{L^2(\mathbb{R}^n)} \le \varepsilon_0/3, \qquad j = 1, 2, \ldots.$$

For this ε, the sequence (u_j^ε) is uniformly bounded and equicontinuous. In fact, by (2.3) and Cauchy–Schwarz we have

$$\left|u_j^\varepsilon(x)\right| = \left|\int \eta_\varepsilon(x-y)u_j(y)\,dy\right| \le \|\eta_\varepsilon(x - \,\cdot\,)\|_{L^2}\left\|u_j\right\|_{L^2} \le C_\varepsilon$$

and similarly

$$\left|\nabla u_j^\varepsilon(x)\right| = \left|\int \nabla\eta_\varepsilon(x-y)u_j(y)\,dy\right| \le \|\nabla\eta_\varepsilon(x - \,\cdot\,)\|_{L^2}\left\|u_j\right\|_{L^2} \le C_\varepsilon,$$

where the constants are uniform over $x \in \mathbb{R}^n$ and $j = 1, 2, \ldots$. The Arzelà–Ascoli theorem shows that there is a subsequence $(u_{j_k}^\varepsilon)_{k=1}^\infty$ that converges uniformly on $\overline{\Omega}$. It follows that

$$\begin{aligned} \left\|u_{j_k} - u_{j_l}\right\|_{L^2(\Omega)} &\le \left\|u_{j_k} - u_{j_k}^\varepsilon\right\|_{L^2(\Omega)} + \left\|u_{j_k}^\varepsilon - u_{j_l}^\varepsilon\right\|_{L^2(\Omega)} + \left\|u_{j_l}^\varepsilon - u_{j_l}\right\|_{L^2(\Omega)} \\ &\le 2\varepsilon_0/3 + C_\Omega\left\|u_{j_k}^\varepsilon - u_{j_l}^\varepsilon\right\|_{L^\infty(\Omega)}. \end{aligned}$$

Here we used that Ω is bounded.

It follows that for any $\varepsilon_0 > 0$ there is a subsequence (u_{j_k}) such that

$$\limsup_{k,l\to\infty} \left\| u_{j_k} - u_{j_l} \right\|_{L^2(\Omega)} \leq \varepsilon_0.$$

We apply this argument with $\varepsilon_0 = 1$ to obtain a subsequence $(u_j^{(1)})$ of (u_j) with

$$\limsup_{k,l\to\infty} \left\| u_k^{(1)} - u_l^{(1)} \right\|_{L^2(\Omega)} \leq 1.$$

Now, repeat the argument for $\varepsilon_0 = 1/2$, but with (u_j) replaced by the sequence $(u_j^{(1)})$, to obtain a further subsequence $(u_j^{(2)})$ with

$$\limsup_{k,l\to\infty} \left\| u_k^{(2)} - u_l^{(2)} \right\|_{L^2(\Omega)} \leq \frac{1}{2}.$$

We continue this for $\varepsilon_0 = \frac{1}{3}, \frac{1}{4}, \ldots$ and use the diagonal procedure to obtain a sequence (v_m), where $v_m = u_m^{(m)}$, that is a subsequence of the original (u_j) and satisfies

$$\limsup_{k,l\to\infty} \|v_k - v_l\|_{L^2(\Omega)} = 0.$$

By the Cauchy criterion, we have found a subsequence that converges in $L^2(\Omega)$. $\square$

As the first application of compact Sobolev embedding, we prove a Poincaré inequality that will be crucial in showing existence of weak solutions. The proof is quite general and adapts to many situations, but it does not give any bounds on the constant C. A direct proof is given in Proposition 2.4.7.

Theorem 2.4.4 (Poincaré inequality). *Let $\Omega \subset \mathbb{R}^n$ be a bounded open set. There is a constant $C > 0$ such that*

$$\|u\|_{L^2(\Omega)} \leq C \|\nabla u\|_{L^2(\Omega)}, \qquad u \in H_0^1(\Omega).$$

Proof. By density it is enough to prove this for all $u \in C_c^\infty(\Omega)$. We argue by contradiction and assume that for any $k \in \mathbb{Z}_+$ there is $u_k \in C_c^\infty(\Omega)$ with

$$\|u_k\|_{L^2(\Omega)} > k \|\nabla u_k\|_{L^2(\Omega)}.$$

By dividing each u_k by $\|u_k\|_{L^2(\Omega)}$, we may assume that

$$\|u_k\|_{L^2(\Omega)} = 1, \quad \|\nabla u_k\|_{L^2(\Omega)} < \frac{1}{k}. \tag{2.4}$$

Then (u_k) is a bounded sequence in $H_0^1(\Omega)$, and by compact Sobolev embedding there is a subsequence, also denoted by (u_k), converging to some u in $L^2(\Omega)$. By (2.4), we also have $\nabla u_k \to 0$ in $L^2(\Omega)$.

We claim that $u \in H_0^1(\Omega)$ and $\nabla u = 0$ in the weak sense. In fact, if $\varphi \in C_c^\infty(\Omega)$ then

$$\int_\Omega u \partial_j \varphi \, dx = \lim_{k\to\infty} \int_\Omega u_k \partial_j \varphi \, dx = -\lim_{k\to\infty} \int_\Omega (\partial_j u_k)\varphi \, dx = 0$$

and so $\nabla u = 0$. (There are no boundary terms in the integration by parts since $u_k, \varphi \in C_c^\infty(\Omega)$.) Thus $u_k \to u$ in $H^1(\Omega)$, and since $H_0^1(\Omega)$ is a closed subspace we have $u \in H_0^1(\Omega)$.

It is proved in Exercise 2.4.5 that any $u \in H_0^1(\Omega)$ with vanishing gradient must be identically zero. But then

$$0 = \|u\|_{L^2(\Omega)} = \lim_{k\to\infty} \|u_k\|_{L^2(\Omega)} = 1,$$

and we have arrived at a contradiction. □

Exercise 2.4.5 (Functions with $\nabla u = 0$). Show that any $u \in H^1(\Omega)$ whose weak gradient vanishes is constant on each connected component of Ω. If additionally $u \in H_0^1(\Omega)$, show that $u = 0$.

Remark 2.4.6. The Poincaré inequality implies that $\|\nabla \cdot\|_{L^2(\Omega)}$ is an equivalent norm on $H_0^1(\Omega)$:

$$C^{-1}\|u\|_{H^1(\Omega)} \leq \|\nabla u\|_{L^2(\Omega)} \leq C\|u\|_{H^1(\Omega)}, \quad u \in H_0^1(\Omega).$$

(The first inequality follows from Poincaré, and the second one is trivial.) This will be useful for the existence of weak solutions.

The following result gives a more precise Poincaré inequality with an elementary proof.

Proposition 2.4.7 (Poincaré inequality in a strip). *Let α be a unit vector in $\mathbb{R}^n$, and let Ω be an open set contained in the infinite strip $\{a < \alpha \cdot x < b\}$, where $a < b$. Then*

$$\|u\|_{L^2(\Omega)} \leq \frac{b-a}{\sqrt{2}} \|\alpha \cdot \nabla u\|_{L^2(\Omega)}, \qquad u \in H_0^1(\Omega).$$

Proof. We may take $\alpha = e_n$, so $\Omega \subset \{a < x_n < b\}$. If $u \in C_c^\infty(\Omega)$ and $a < x_n < b$, we have

$$\begin{aligned} |u(x', x_n)| &= \left| \int_a^{x_n} \partial_n u(x', t)\, dt \right| \\ &\leq \int_a^{x_n} |\partial_n u(x', t)| \cdot 1 \, dt \\ &\leq \left(\int_a^b |\partial_n u(x', t)|^2 \, dt \right)^{1/2} (x_n - a)^{1/2}. \end{aligned}$$

Squaring this inequality and integrating gives

$$\int_{\mathbb{R}^{n-1}} \int_a^b |u(x', x_n)|^2 \, dx_n \, dx' \le \|\partial_n u\|^2_{L^2(\{a<x_n<b\})} \int_a^b (x_n - a) \, dx_n$$
$$\le \frac{(b-a)^2}{2} \|\partial_n u\|^2_{L^2(\{a<x_n<b\})} .$$

The result follows since $C_c^\infty(\Omega)$ is dense in $H_0^1(\Omega)$. □

For later purposes it will be helpful to have a version of compact Sobolev embedding for functions in $H^1(\Omega)$, not just $H_0^1(\Omega)$.

Theorem 2.4.8 (Compact Sobolev embedding). *If $\Omega \subset \mathbb{R}^n$ is a bounded open set with C^1 boundary, then the inclusion map $i : H^1(\Omega) \to L^2(\Omega)$ is compact.*

Exercise 2.4.9. Prove Theorem 2.4.8. You may use the extension theorem, Lemma C.2.19, and multiplication by a suitable cutoff function to reduce matters to Theorem 2.4.3.

The following two variants of the Poincaré inequality are stated for H^1 functions. They can be shown by using Theorem 2.4.8 and by adapting the proof of Theorem 2.4.4.

Exercise 2.4.10 (Poincaré–Wirtinger inequality). Let $\Omega \subset \mathbb{R}^n$ be a bounded connected open set with C^1 boundary. Show that there is a constant $C > 0$ such that

$$\|u - (u)_\Omega\|_{L^2(\Omega)} \le C \|\nabla u\|_{L^2(\Omega)} , \qquad u \in H^1(\Omega).$$

Here $(u)_\Omega = \frac{1}{|\Omega|} \int_\Omega u(x) \, dx$ is the average of u over Ω.

Exercise 2.4.11 (Poincaré inequality). Let $\Omega \subset \mathbb{R}^n$ be a bounded connected open set with C^1 boundary, and let Γ be a nonempty open subset of $\partial\Omega$. Show that there is a constant $C > 0$ such that

$$\|u\|_{L^2(\Omega)} \le C(\|\nabla u\|_{L^2(\Omega)} + \|Ru\|_{L^2(\Gamma)}), \qquad u \in H^1(\Omega).$$

Here $R : H^1(\Omega) \to L^2(\partial\Omega)$ is the trace operator in Theorem C.2.1.

We conclude with a compact Sobolev embedding theorem for $W^{1,p}(\Omega)$, which will only be needed in Chapter 5.

Theorem 2.4.12 (Compact Sobolev embedding for $W^{1,p}(\Omega)$). *If $\Omega \subset \mathbb{R}^n$ is a bounded open set with C^1 boundary and $1 < p < \infty$, then the inclusion map $i : W^{1,p}(\Omega) \to L^p(\Omega)$ is compact.*

Exercise 2.4.13. Prove Theorem 2.4.12 by adapting the proof of Theorem 2.4.3. You may assume that there is a bounded extension operator

$$W^{1,p}(\Omega) \to W^{1,p}(\mathbb{R}^n).$$

2.5. Weak solutions

In this section, Ω will be a bounded open subset of $\mathbb{R}^n$ (no regularity of the boundary $\partial\Omega$ is required). Consider a second order differential operator L, acting on functions u on Ω, given by

$$Lu = -\sum_{j,k=1}^{n} \frac{\partial}{\partial x_j}\left(a^{jk}\frac{\partial u}{\partial x_k}\right) + qu. \tag{2.5}$$

Here the coefficients are assumed to satisfy the following conditions:

$$\begin{cases} a^{jk}, q \in L^\infty(\Omega) \text{ are real valued} \\ a^{jk} = a^{kj} \text{ for all } j,k = 1,\ldots,n, \\ \sum_{j,k=1}^n a^{jk}(x)\xi_j\xi_k \geq c\,|\xi|^2 \text{ for a.e. } x \in \Omega \text{ and all } \xi \in \mathbb{R}^n, \end{cases} \tag{2.6}$$

where $c > 0$. The last condition is an *ellipticity condition* for the operator L, and it ensures that the operator L will have similar properties as the Laplace operator $-\Delta = -\sum_{j=1}^n \partial_{x_j}^2$.

Later in this book, we will mostly consider L to be one of the following special cases:

(1) The conductivity operator

$$Lu = -\operatorname{div}(\gamma\nabla u),$$

where $\gamma \in L^\infty(\Omega)$ has a positive lower bound.

(2) The anisotropic conductivity operator

$$Lu = -\operatorname{div}(\gamma\nabla u),$$

where $\gamma = (\gamma^{jk})_{j,k=1}^n$ satisfies $\gamma^{jk} = \gamma^{kj} \in L^\infty(\Omega)$ and

$$\sum_{j,k=1}^{n} \gamma^{jk}(x)\xi_j\xi_k \geq c\,|\xi|^2.$$

(3) The Schrödinger operator

$$Lu = (-\Delta + q)u,$$

where $q \in L^\infty(\Omega)$.

Remark 2.5.1. The ellipticity condition implies a similar condition for complex vectors: if $A = (a^{jk}(x))_{j,k=1}^n$ and $\zeta \in \mathbb{C}^n$, then writing $\zeta = \xi + i\eta$, where $\xi, \eta \in \mathbb{R}^n$ and using the symmetry of A gives

$$A\zeta \cdot \bar{\zeta} = A(\xi + i\eta)\cdot(\xi - i\eta) = A\xi\cdot\xi + A\eta\cdot\eta.$$

The ellipticity condition thus implies

$$\sum_{j,k=1}^{n} a^{jk}(x)\zeta_j\bar{\zeta}_k \geq c\,|\zeta|^2 \quad \text{for a.e. } x \in \Omega \text{ and all } \zeta \in \mathbb{C}^n. \tag{2.7}$$

We will use this stronger condition below.

Remark 2.5.2. Notice that if $a^{jk} \in C^1(\overline{\Omega})$, one can write Lu in the form

$$Lu = -\sum_{j,k=1}^{n} a^{jk}\partial_{jk}u - \sum_{j,k=1}^{n} (\partial_j a^{jk})\partial_k u + qu.$$

This operator is said to be in *nondivergence form*, while the operator (2.5) is in *divergence form*. We will consider divergence form operators in this section since they are better suited to the energy method related to weak solutions.

Motivation 2.5.3 (Weak solutions). Suppose Ω has C^1 boundary, L has C^1 coefficients, and $u \in C^2(\overline{\Omega})$ is a classical solution of $Lu = F$ in Ω which satisfies the boundary condition $u|_{\partial\Omega} = f$, where $F \in L^2(\Omega)$ and $f \in C^0(\partial\Omega)$. Multiplying the equation $Lu = F$ by $\bar{v}$, where $v \in C^1(\overline{\Omega})$ satisfies $v|_{\partial\Omega} = 0$, an integration by parts implies that

$$\int_\Omega \left(\sum_{j,k=1}^{n} a^{jk}\partial_j u \overline{\partial_k v} + qu\bar{v} \right) dx = \int_\Omega F\bar{v}\,dx.$$

Now, the left-hand side makes sense for any $u, v \in H^1(\Omega)$ (note however that the integration by parts above made use of the vanishing of v on $\partial\Omega$). We can use this identity to define weak solutions of the equation $Lu = F$ in Ω with $u|_{\partial\Omega} = f$. More generally, we can consider any right-hand side F that is a continuous linear functional on $H_0^1(\Omega)$.

Definition 2.5.4. Let L be the differential operator (2.5). The sesquilinear form related to L is given by

$$B[u,v] = \int_\Omega \left(\sum_{j,k=1}^{n} a^{jk}\partial_j u \overline{\partial_k v} + qu\bar{v} \right) dx, \quad u, v \in H^1(\Omega). \tag{2.8}$$

If $F \in H^{-1}(\Omega)$ and $f \in H^{1/2}(\partial\Omega)$, we say that a function $u \in H^1(\Omega)$ is a weak solution of the Dirichlet problem

$$\begin{cases} Lu = F & \text{in } \Omega, \\ u = f & \text{on } \partial\Omega \end{cases}$$

if one has

$$B[u,v] = F(\bar{v}) \quad \text{for all } v \in H_0^1(\Omega)$$

and if $Ru = f$, where R is the trace operator in Definition 2.3.14.

Remark 2.5.5. The condition $u = f$ on $\partial\Omega$ is called the *Dirichlet boundary condition*. It is understood in an abstract sense, having the following equivalent interpretations:

(a) $Ru = f$, where R is the trace operator,
(b) $u - v \in H_0^1(\Omega)$ for some $v \in H^1(\Omega)$ with $Rv = f$,
(c) $u - v \in H_0^1(\Omega)$ for any $v \in H^1(\Omega)$ with $Rv = f$.

We are ready to give the first solvability result for boundary value problems.

Theorem 2.5.6 (Weak solutions for $q \geq 0$). *Let Ω be a bounded open set in $\mathbb{R}^n$, and let L be as in (2.5), (2.6). Assume in addition that*

$$q \geq 0 \quad \text{a.e. in } \Omega.$$

For any $F \in H^{-1}(\Omega)$ and $f \in H^{1/2}(\partial\Omega)$, there is a unique $u \in H^1(\Omega)$ that solves the Dirichlet problem

$$\begin{cases} Lu = F & \text{in } \Omega, \\ u = f & \text{on } \partial\Omega. \end{cases}$$

There is a constant C, depending only on n, Ω, the ellipticity constant and the L^∞ norms in (2.6), such that

$$\|u\|_{H^1(\Omega)} \leq C\big(\|F\|_{H^{-1}(\Omega)} + \|f\|_{H^{1/2}(\partial\Omega)} \big).$$

This result follows readily from the next lemma.

Lemma 2.5.7. *If $s \in \mathbb{R}$ is a constant such that $q+s \geq 0$ almost everywhere, then the sesquilinear form*

$$B_s[u, v] = B[u, v] + s(u, v)_{L^2(\Omega)}$$

is an inner product on the space $H_0^1(\Omega)$ that induces a norm equivalent to the original one:

$$C^{-1} \|u\|_{H^1(\Omega)}^2 \leq B_s[u, u] \leq C \|u\|_{H^1(\Omega)}^2, \quad u \in H_0^1(\Omega).$$

Proof. It is clear that the map $(u, v) \mapsto B_s[u, v]$ is sesquilinear, and

$$B_s[u, v] = \overline{B_s[v, u]}$$

since $a^{jk} = a^{kj}$ and a^{jk}, q, s are real. The ellipticity condition (2.7) and the assumption that $q + s \geq 0$ imply that

$$B_s[u, u] = \int_\Omega \left(\sum_{j,k=1}^n a^{jk} \partial_j u \overline{\partial_k u} + (q+s)\,|u|^2 \right) dx \geq c \int_\Omega |\nabla u|^2\, dx.$$

Thus $B_s[u, u] \geq 0$, and if $B_s[u, u] = 0$ then $\nabla u = 0$ a.e. and thus $u = 0$ by Exercise 2.4.5. We have proved that $B_s[\,\cdot\,, \cdot\,]$ is an inner product on $H_0^1(\Omega)$.

The triangle inequality and the fact that $a^{jk}, q \in L^\infty(\Omega)$ imply that

$$B_s[u,u] \leq C_{a^{jk},q,s} \int_\Omega \left(|\nabla u|^2 + |u|^2\right) \leq C \|u\|_{H^1(\Omega)}^2, \quad u \in H_0^1(\Omega).$$

Moreover, the previous argument and the Poincaré inequality show that

$$B_s[u,u] \geq c \|\nabla u\|_{L^2(\Omega)}^2 \geq c \|u\|_{H^1(\Omega)}^2, \quad u \in H_0^1(\Omega).$$

Thus $B_s[\,\cdot\,,\,\cdot\,]$ gives an equivalent norm on $H_0^1(\Omega)$. □

Proof of Theorem 2.5.6. Consider first the case of zero boundary values, where we want to solve

$$\begin{cases} Lu = F & \text{in } \Omega, \\ u = 0 & \text{on } \partial\Omega. \end{cases}$$

The assumption that $q \geq 0$ together with Lemma 2.5.7 show that $B[\,\cdot\,,\,\cdot\,] = B_0[\,\cdot\,,\,\cdot\,]$ is an inner product on $H_0^1(\Omega)$ giving a norm equivalent to the original one. It follows that the inner product space $(H_0^1(\Omega), B[\,\cdot\,,\,\cdot\,])$ has the same Cauchy sequences as the original $H_0^1(\Omega)$ and therefore is a Hilbert space. Also, F is a bounded linear functional on this space, satisfying

$$|F(v)| \leq \|F\|_{H^{-1}(\Omega)} \|v\|_{H^1(\Omega)} \leq C \|F\|_{H^{-1}(\Omega)} B[v,v]^{1/2}, \quad v \in H_0^1(\Omega).$$

The Riesz representation theorem (Theorem A.2.15) implies that there exists a unique function $u \in H_0^1(\Omega)$, whose norm is equal to the norm of F as a bounded linear functional on $(H_0^1(\Omega), B[\,\cdot\,,\,\cdot\,])$, such that

$$B[u,v] = F(\bar{v}), \quad v \in H_0^1(\Omega).$$

Since this function satisfies $Ru = 0$, we have found the unique solution of our boundary value problem. The solution also satisfies

$$\|u\|_{H^1(\Omega)} \leq C \|F\|_{H^{-1}(\Omega)}.$$

We move to the case of nonzero boundary values, and want to find $u \in H^1(\Omega)$ with

$$B[u,v] = F(\bar{v}) \quad \text{for } v \in H_0^1(\Omega), \quad Ru = f.$$

Choose $e_f \in H^1(\Omega)$ with $e_f|_{\partial\Omega} = f$ and

$$\left\|e_f\right\|_{H^1(\Omega)} \leq C \|f\|_{H^{1/2}(\partial\Omega)}$$

(this is possible by Theorem 2.3.17). Writing $u = e_f + \tilde{u}$, the boundary value problem is equivalent with finding $\tilde{u} \in H_0^1(\Omega)$ that satisfies

$$B[\tilde{u},v] = F(\bar{v}) - B[e_f,v] \quad \text{for } v \in H_0^1(\Omega), \quad R\tilde{u} = 0.$$

The map $\tilde{F} : w \mapsto F(w) - B[e_f,w]$ is a bounded linear functional on $H_0^1(\Omega)$, since by the triangle inequality and Cauchy–Schwarz

$$|B[e_f,w]| \leq C \int_\Omega \left(|\nabla e_f|\,|\nabla w| + |e_f|\,|w|\right) dx \leq C \left\|e_f\right\|_{H^1(\Omega)} \|w\|_{H^1(\Omega)}.$$

It follows that

$$\|\tilde{F}\|_{H^{-1}(\Omega)} \leq C(\|F\|_{H^{-1}(\Omega)} + \|f\|_{H^{1/2}(\partial\Omega)}).$$

The result for zero boundary values proved above shows that there is a unique solution $\tilde{u} \in H_0^1(\Omega)$ satisfying

$$\|\tilde{u}\|_{H^1(\Omega)} \leq C(\|F\|_{H^{-1}(\Omega)} + \|f\|_{H^{1/2}(\partial\Omega)}).$$

Thus, the original boundary value problem for u also has a unique solution with

$$\|u\|_{H^1(\Omega)} \leq C(\|F\|_{H^{-1}(\Omega)} + \|f\|_{H^{1/2}(\partial\Omega)}).$$

One can check that the constant C only depends on the stated quantities. □

Since the conductivity operator $-\operatorname{div}(\gamma\nabla u)$ is of the form (2.5) with $q = 0$, the previous result implies the basic solvability result for the conductivity equation. Next we wish to deal with the case where q may be negative somewhere. This is different from the previous case since the Dirichlet problem is not always uniquely solvable, and there may be nontrivial solutions of the equation $(-\Delta + q)u = 0$ with $u|_{\partial\Omega} = 0$. This is due to the existence of eigenfunctions, as illustrated by the following example.

Example 2.5.8. Let $\Omega = (0, \pi) \subset \mathbb{R}$, and consider the Schrödinger operator $-\Delta + q$ in Ω in the special case where the function q is a constant, $q = -\lambda$ where $\lambda > 0$. A function u solves $(-\Delta + q)u = 0$ with vanishing boundary values if

$$u''(x) + \lambda u(x) = 0 \text{ for } 0 < x < \pi, \qquad u(0) = u(\pi) = 0.$$

The general solution to the ordinary differential equation $u''(x) + \lambda u(x) = 0$ is

$$u(x) = A\sin\left(\sqrt{\lambda}\,x\right) + B\cos\left(\sqrt{\lambda}\,x\right).$$

The boundary condition $u(0) = 0$ is satisfied if and only if $B = 0$. With $B = 0$, the boundary condition $u(\pi) = 0$ is satisfied if and only if either $A = 0$ or $\sin\left(\sqrt{\lambda}\,\pi\right) = 0$. The latter condition is equivalent to

$$\sqrt{\lambda} \in \mathbb{Z} \iff \lambda = k^2, \quad k \in \mathbb{Z}_+.$$

Thus there is a nontrivial solution $u \in H^1(\Omega)$ with vanishing boundary value whenever $\lambda = k^2$ for some positive integer k.

The next theorem shows that there is only a countable set of eigenvalues where unique solvability of the Dirichlet problem for L may fail. Outside of these eigenvalues, we recover the same solvability result as before. The proof uses compact Sobolev embedding and the spectral theorem for compact operators.

Theorem 2.5.9 (Weak solutions). *Let Ω be a bounded open set in $\mathbb{R}^n$, and let L be as in* (2.5), (2.6).

(a) *There is a set of real numbers*

$$\mathrm{Spec}(L) = \{\lambda_j\}_{j=1}^{\infty}$$

with $\lambda_1 \leq \lambda_2 \leq \ldots \to \infty$, such that whenever $\lambda \notin \mathrm{Spec}(L)$ the boundary value problem

$$\begin{cases} Lu = \lambda u + F & \text{in } \Omega, \\ u = f & \text{on } \partial\Omega \end{cases}$$

has a unique solution $u \in H^1(\Omega)$ for any $F \in H^{-1}(\Omega)$ and $f \in H^{1/2}(\partial\Omega)$.

(b) *If $\lambda \notin \mathrm{Spec}(L)$, then the map*

$$H^1(\Omega) \to H^{-1}(\Omega) \oplus H^{\frac{1}{2}}(\partial\Omega),$$
$$u \mapsto \big(Lu - \lambda u\,,\ Ru\big)$$

is an isomorphism (1–1, onto, bounded with bounded inverse). There is a constant C independent of F and f (but depending on λ) such that

$$\|u\|_{H^1(\Omega)} \leq C\big(\,\|F\|_{H^{-1}(\Omega)} + \|f\|_{H^{1/2}(\partial\Omega)}\,\big).$$

(c) *If $\lambda \in \mathrm{Spec}(L)$, there is a nontrivial solution $u \in H_0^1(\Omega)$ to the Dirichlet problem*

$$\begin{cases} Lu = \lambda u & \text{in } \Omega, \\ u = 0 & \text{on } \partial\Omega. \end{cases}$$

The space of such solutions is finite-dimensional.

(d) *If $a \in \mathbb{R}$ is a constant such that*

$$q(x) \geq a \quad \text{a.e. in } \Omega,$$

then $\mathrm{Spec}(L) \subset (a, \infty)$.

Definition 2.5.10. The set $\mathrm{Spec}(L)$ is called the *spectrum* of the operator L with Dirichlet boundary condition. The elements of $\mathrm{Spec}(L)$ are called *Dirichlet eigenvalues.*

Proof.

(a) Consider first the case of zero boundary values. As $q \in L^\infty(\Omega)$, we may choose $s \in \mathbb{R}$ such that $q+s \geq 0$ a.e. in Ω. Let L_s be the operator obtained from L by replacing q by $q + s$, so that $L_s = L + s$, and let B_s be the corresponding sesquilinear form. Lemma 2.5.7 shows that $B_s[\,\cdot\,,\,\cdot\,]$ is an inner product on $H_0^1(\Omega)$ giving a norm equivalent to the original one.

Theorem 2.5.6 implies that the Dirichlet problem for L_s with zero boundary values has unique solutions, and the map $L_s : H_0^1(\Omega) \to H^{-1}(\Omega)$ is invertible with bounded inverse L_s^{-1} taking $F \in H^{-1}(\Omega)$ to the unique solution

$u \in H_0^1(\Omega)$ of $Lu = F$ in Ω with $Ru = 0$. We now have, for $u \in H_0^1(\Omega)$,

$$\begin{aligned} Lu = \lambda u + F \quad &\Longleftrightarrow \quad L_s u = (\lambda + s)u + F \\ &\Longleftrightarrow \quad u - (\lambda + s)L_s^{-1}u = L_s^{-1}F. \end{aligned}$$

If $\lambda \neq -s$, the last part may be equivalently written as

$$(\mu I - K)(i(u)) = \tilde{F}$$

with $\mu = \frac{1}{\lambda+s}$, $K = i \circ L_s^{-1} \circ j$, and $\tilde{F} = \mu i(L_s^{-1}F)$, where $i : H_0^1(\Omega) \to L^2(\Omega)$ and $j : L^2(\Omega) \to H^{-1}(\Omega)$ are the inclusion maps.

We claim that

$K : L^2(\Omega) \to L^2(\Omega)$ is a compact, self-adjoint, positive definite operator.

By Theorem 2.4.3 the map i is compact, and consequently K is compact (see Proposition A.3.8). It is also self-adjoint, since for $F, G \in L^2(\Omega)$, writing $v = L_s^{-1}F$ and $w = L_s^{-1}G$ and using the definition of weak solutions gives that

$$(KF, G)_{L^2} = (L_s^{-1}(F), G)_{L^2} = \overline{(G, v)_{L^2}} = \overline{B_s[w, v]}.$$

Similarly

$$(KG, F)_{L^2} = \overline{B_s[v, w]}.$$

Since B_s is conjugate symmetric, $\overline{B_s[v, w]} = B_s[w, v]$, we have $(KF, G)_{L^2} = (F, KG)_{L^2}$. Finally, K is positive definite since by the above computation

$$(KF, F)_{L^2} = B_s[v, v],$$

where B_s is a positive definite inner product.

By the spectral theorem for self-adjoint compact operators (Proposition A.3.18), $\mathrm{Spec}(K)$ is an at most countable subset of $\mathbb{R}$ that may only accumulate at 0, and each element of $\mathrm{Spec}(K)$ (except possibly 0) is an eigenvalue with finite-dimensional eigenspace. Each eigenvalue is positive since K is positive definite. Since $L^2(\Omega)$ is not finite-dimensional, $\mathrm{Spec}(K)$ is in fact countably infinite and contains 0, and we may write $\mathrm{Spec}(K) = \{\mu_j\}_{j=1}^{\infty} \cup \{0\}$ where $\mu_1 \geq \mu_2 \geq \ldots$ and $\mu_j \to 0$ as $j \to \infty$. Now, $\mu I - K$ is invertible on $L^2(\Omega)$ for all $\mu \in \mathbb{R} \setminus \mathrm{Spec}(K)$.

We now return to solvability of the Dirichlet problem. We wrote earlier that $\mu = \frac{1}{\lambda+s}$, so we define

$$\lambda_j = \frac{1}{\mu_j} - s, \qquad j = 1, 2, \ldots.$$

Then $\lambda_1 \leq \lambda_2 \leq \ldots$ and $\lambda_j \to \infty$. If $F \in H^{-1}(\Omega)$ and $\lambda + s \neq 0$, we saw above that for $u \in H_0^1(\Omega)$ one has

$$Lu = \lambda u + F \quad \Longleftrightarrow \quad (\mu I - K)(i(u)) = \mu i(L_s^{-1}(F)),$$

where $\mu = \frac{1}{\lambda+s} \notin \mathrm{Spec}(K)$. If we assume that $\lambda \in \mathbb{R}\setminus\{\lambda_j\}_{j=1}^\infty$, then $\mu \notin \mathrm{Spec}(K)$ and for any $F \in H^{-1}(\Omega)$ there is a unique solution $\tilde{u} \in L^2(\Omega)$ of

$$(\mu I - K)(\tilde{u}) = \mu i(L_s^{-1}(F)).$$

This function satisfies $\tilde{u} = \mu^{-1}L_s^{-1}\tilde{u} + L_s^{-1}F$, so $\tilde{u} = i(u)$ for some $u \in H_0^1(\Omega)$ with $\|u\|_{H^1(\Omega)} \leq C(\|\tilde{u}\|_{L^2(\Omega)} + \|F\|_{H^{-1}(\Omega)}) \leq C\,\|F\|_{H^{-1}(\Omega)}$, using that $\mu I - K$ and L_s are invertible. This shows the existence of a unique solution if $\lambda + s \neq 0$. In the remaining case where $\lambda + s = 0$, the equation is $L_s u = F$ which has unique solutions in $H_0^1(\Omega)$ by Theorem 2.5.6.

The case of nonzero boundary values is handled in the same way as in Theorem 2.5.6. This proves part (a) in the theorem.

(b) The fact that $u \mapsto (Lu - \lambda u, Ru)$ is an isomorphism follows easily. Only boundedness remains to be proved, but the inequality

$$\|Lu - \lambda u\|_{H^{-1}(\Omega)} + \|Ru\|_{H^{1/2}(\partial\Omega)} \leq C\,\|u\|_{H^1(\Omega)}$$

follows from the arguments above.

(c) If $\lambda = \lambda_j \in \mathrm{Spec}(L)$, then the proof in part (a) shows that the equation $Lu = \lambda u$ for $u \in H_0^1(\Omega)$ is equivalent with

$$(\mu I - K)(i(u)) = 0$$

for $\mu = \mu_j = \frac{1}{\lambda+s}$. Since $\mu_j \in \mathrm{Spec}(K)$ and K is compact, there is a nontrivial finite-dimensional space consisting of those $\tilde{u} \in L^2(\Omega)$ with

$$(\mu I - K)\tilde{u} = 0.$$

This space is contained in $H_0^1(\Omega)$ because any such $\tilde{u}$ satisfies $\tilde{u} = \mu^{-1}L_s^{-1}\tilde{u}$, and gives rise to a finite-dimensional space of solutions to $Lu = \lambda u$ in Ω.

(d) If $q \geq a$ a.e., then we may choose $s = -a$ above and each eigenvalue satisfies $\lambda_j > a$ by definition. □

Exercise 2.5.11. Prove Theorem 2.5.6 in the simpler case of the Dirichlet problem for the Laplacian,

$$\begin{cases} -\Delta u = F & \text{in } \Omega, \\ u = f & \text{on } \partial\Omega, \end{cases}$$

by using Remark 2.4.6 instead of Lemma 2.5.7.

Exercise 2.5.12. Show that Theorem 2.5.9, except for part (d), remains true if L is the following operator containing first-order terms,

$$Lu = -\sum_{j,k=1}^n \frac{\partial}{\partial x_j}\left(a^{jk}\frac{\partial u}{\partial x_k}\right) + i\sum_{j=1}^n \left(b^j\frac{\partial u}{\partial x_j} + \frac{\partial}{\partial x_j}(b^j u)\right) + qu,$$

where the coefficients satisfy (2.6) and additionally $b^j \in L^\infty(\Omega)$ are real valued. The sesquilinear form corresponding to L is given by

$$B[u,v] = \int_\Omega \left(\sum_{j,k=1}^n a^{jk}\partial_j u \overline{\partial_k v} + i\sum_{j=1}^n (b^j(\partial_j u)\bar{v} - b^j u \partial_j \bar{v}) + qu\bar{v} \right) dx.$$

2.6. Higher regularity

We will next discuss higher-order regularity of solutions. The philosophy is that a solution of the second order elliptic equation $Lu = F$ should always be two derivatives smoother than the right-hand side F, unless this gain in regularity is prevented by lack of smoothness in the coefficients of L, the boundary values of u, or in the boundary $\partial\Omega$.

In order to discuss higher regularity of boundary values, we need to use the Sobolev spaces $H^s(\partial\Omega)$ consisting of functions on $\partial\Omega$ that have s derivatives in $L^2(\partial\Omega)$. These spaces are well-defined for $0 \leq s \leq k$ whenever Ω has C^k boundary, and they are studied in detail in Appendix C. The following theorem will be proved in §D.3. (Recall from Definition 2.3.5 that $W^{k,\infty}(\Omega)$ is the space of functions $u \in L^\infty(\Omega)$ with $\partial^\alpha u \in L^\infty(\Omega)$ whenever $|\alpha| \leq k$).

Theorem 2.6.1 (Elliptic regularity). *Let Ω be a bounded open set in $\mathbb{R}^n$, and let L be as in* (2.5), (2.6). *Assume that $u \in H^1(\Omega)$ solves the Dirichlet problem*

$$\begin{cases} Lu = F & \text{in } \Omega, \\ u = f & \text{on } \partial\Omega. \end{cases}$$

(a) *Assume that Ω has C^2 boundary, $a^{jk} \in W^{1,\infty}(\Omega)$, and $q \in L^\infty(\Omega)$. If $F \in L^2(\Omega)$ and $f \in H^{3/2}(\partial\Omega)$, then $u \in H^2(\Omega)$ and*

$$\|u\|_{H^2(\Omega)} \leq C(\|F\|_{L^2(\Omega)} + \|f\|_{H^{3/2}(\partial\Omega)}).$$

(b) *Let $\ell \geq 1$ and assume that Ω has $C^{\ell+2}$ boundary, $a^{jk} \in W^{\ell+1,\infty}(\Omega)$, and $q \in W^{\ell,\infty}(\Omega)$. If $F \in H^\ell(\Omega)$ and $f \in H^{\ell+3/2}(\partial\Omega)$, then $u \in H^{\ell+2}(\Omega)$ and*

$$\|u\|_{H^{\ell+2}(\Omega)} \leq C(\|F\|_{H^\ell(\Omega)} + \|f\|_{H^{\ell+3/2}(\partial\Omega)}).$$

Theorem 2.6.1 has two useful consequences. The first concerns interior regularity of solutions (see Proposition D.3.5).

Theorem 2.6.2 (Interior regularity). *Let Ω' and Ω be bounded open subsets of $\mathbb{R}^n$ with $\overline{\Omega'} \subset \Omega$, let $\ell \geq 0$, and let L be as in Theorem 2.6.1(b). There is a constant C, depending only on ℓ, Ω', Ω and L such that, for all $u \in H^1(\Omega)$ and $F \in H^\ell(\Omega)$ obeying*

$$Lu = F \qquad \text{in } \Omega$$

we have $u|_{\Omega'} \in H^{\ell+2}(\Omega')$ and

$$\|u\|_{H^{\ell+2}(\Omega')} \leq C\Big(\|F\|_{H^\ell(\Omega)} + \|u\|_{H^1(\Omega)} \Big).$$

The second consequence of Theorem 2.6.1 shows that if all quantities are C^∞, then also the weak solution u is C^∞ up to the boundary.

Theorem 2.6.3 (Smoothness up to the boundary). *Let Ω be a bounded open set in $\mathbb{R}^n$ with smooth boundary, and let L be the operator in (2.5), (2.6). Assume that $a^{jk} \in C^\infty(\overline{\Omega})$ and $q \in C^\infty(\overline{\Omega})$. If $u \in H^1(\Omega)$ solves the Dirichlet problem*

$$\begin{cases} Lu = F & \text{in } \Omega, \\ u = f & \text{on } \partial\Omega, \end{cases}$$

where $F \in C^\infty(\overline{\Omega})$ and $f \in C^\infty(\partial\Omega)$, then $u \in C^\infty(\overline{\Omega})$.

2.7. The DN map and inverse problems

Let Ω be a bounded open set in $\mathbb{R}^n$, and let $\gamma \in L^\infty(\Omega)$ be a positive function. We wish to give a precise definition of the Dirichlet-to-Neumann map (DN map) mapping a boundary voltage f to the current flux at the boundary,

$$\Lambda_\gamma : f \mapsto \gamma \partial_\nu u_f|_{\partial\Omega},$$

where u_f is the solution of the conductivity equation $\mathrm{div}(\gamma \nabla u_f) = 0$ in Ω with $u_f|_{\partial\Omega} = f$.

In fact, we can consider the more general operators (2.5),

$$Lu = -\sum_{j,k=1}^{n} \frac{\partial}{\partial x_j}\left(a^{jk} \frac{\partial u}{\partial x_k}\right) + qu,$$

where $a^{jk}, q \in L^\infty(\Omega)$ satisfy the conditions (2.6). The corresponding DN map for L would be formally given by

$$\Lambda_L : f \mapsto \sum_{j,k=1}^{n} a^{jk}(\partial_j u)\nu_k\Big|_{\partial\Omega}. \tag{2.9}$$

Motivation 2.7.1 (Definition of DN map). Assume that Ω has smooth boundary and $a^{jk}, q \in C^\infty(\overline{\Omega})$, and suppose that u_f solves the Dirichlet problem $Lu = 0$ in Ω, $u_f|_{\partial\Omega} = f$, for some $f \in C^\infty(\partial\Omega)$. Then $u_f \in C^\infty(\overline{\Omega})$ by Theorem 2.6.3, and we may define $\Lambda_L f$ by the right-hand side of (2.9) as a function in $C^\infty(\partial\Omega)$. Let $g \in C^\infty(\partial\Omega)$ and let $e_g \in C^\infty(\overline{\Omega})$ be any function such that $e_g|_{\partial\Omega} = g$. An integration by parts, using that all quantities are smooth, shows that

$$\begin{aligned}\int_{\partial\Omega} (\Lambda_L f) g \, dS &= \sum_{j,k=1}^{n} \int_{\partial\Omega} a^{jk}(\partial_j u_f) e_g \nu_k \, dS \\ &= \sum_{j,k=1}^{n} \int_{\Omega} \partial_k \left(a^{jk}(\partial_j u_f) e_g \right) dx \\ &= \int_{\Omega} \Big[\sum_{j,k=1}^{n} a^{jk} \partial_j u_f \partial_k e_g + q u_f e_g \Big] \, dx.\end{aligned}$$

In the last step we used that $Lu = 0$.

Notice that the expression on the last line is just $B[u_f, \bar{e}_g]$ where B is the sesquilinear form corresponding to L. This expression is well defined even when $a^{jk}, q \in L^\infty(\Omega)$ and $u_f, e_g \in H^1(\Omega)$. We use this observation to define the DN map in a weak sense even when the quantity $a^{jk}(\partial_j u)\nu_k$ may not be defined pointwise.

Recall that $H^{-1/2}(\partial\Omega)$ is the dual space of $H^{1/2}(\partial\Omega) = H^1(\Omega)/H^1_0(\Omega)$. If $f \in H^{-1/2}(\partial\Omega)$, we will express the duality by the notation

$$\langle f, g \rangle_{\partial\Omega} = f(g), \qquad g \in H^{1/2}(\partial\Omega).$$

If Ω has C^1 boundary and $f \in L^2(\partial\Omega)$, this reduces to the usual integral

$$\langle f, g \rangle_{\partial\Omega} = \int_{\partial\Omega} f g \, dS.$$

Theorem 2.7.2 (DN map for L). *Let Ω be a bounded open set in $\mathbb{R}^n$ and let L satisfy (2.5), (2.6). Assume that 0 is not a Dirichlet eigenvalue of L in Ω. There is a unique bounded linear map*

$$\Lambda_L : H^{1/2}(\partial\Omega) \to H^{-1/2}(\partial\Omega)$$

that satisfies

$$\langle \Lambda_L f, g \rangle_{\partial\Omega} = B[u_f, \bar{e}_g] = \int_{\Omega} \Big[\sum_{j,k=1}^{n} a^{jk} \partial_j u_f \partial_k e_g + q u_f e_g \Big] \, dx, \tag{2.10}$$

where $u_f \in H^1(\Omega)$ is the unique solution of $Lu = 0$ in Ω with $u|_{\partial\Omega} = f$, and e_g is any function in $H^1(\Omega)$ with $e_g|_{\partial\Omega} = g$.

If Ω has C^∞ boundary and $a^{jk}, q \in C^\infty(\overline{\Omega})$, then Λ_L restricts to a linear map

$$\Lambda_L : C^\infty(\partial\Omega) \to C^\infty(\partial\Omega)$$

which satisfies (2.9) for all $f \in C^\infty(\partial\Omega)$.

Proof.

(1) The first step is to show that the right-hand side of (2.10) does not depend on the particular choice of the extension e_g of g. That is, if $e_g, \tilde{e}_g$ are two functions in $H^1(\Omega)$ with $e_g|_{\partial\Omega} = \tilde{e}_g|_{\partial\Omega} = g$, then

$$B[u_f, \bar{e}_g] = B[u_f, \overline{\tilde{e}_g}].$$

But this follows from the fact that $\tilde{e}_g = e_g + \varphi$ for some $\varphi \in H_0^1(\Omega)$, since the definition of weak solutions implies $B[u_f, \bar{\varphi}] = 0$.

(2) Fix $f \in H^{1/2}(\partial\Omega)$, and define a linear functional $T_f : H^{1/2}(\partial\Omega) \to \mathbb{C}$ by

$$T_f(g) = B[u_f, \bar{e}_g], \qquad g \in H^{1/2}(\partial\Omega),$$

where $e_g \in H^1(\Omega)$ is the extension of g provided by Theorem 2.3.17 satisfying $e_g|_{\partial\Omega} = g$ and $\left\| e_g \right\|_{H^1(\Omega)} \le C \left\| g \right\|_{H^{1/2}(\partial\Omega)}$. Then by Cauchy–Schwarz

$$\begin{aligned} |T_f(g)| &\le \int_\Omega \left(\sum_{j,k=1}^n |a^{jk}| \, |\nabla u_f| \, |\nabla e_g| + |q| \, |u_f| \, |e_g| \right) dx \\ &\le C \left\| u_f \right\|_{H^1(\Omega)} \left\| e_g \right\|_{H^1(\Omega)}. \end{aligned}$$

By Theorem 2.5.9 we have

$$\left\| u_f \right\|_{H^1(\Omega)} \le C \left\| f \right\|_{H^{1/2}(\partial\Omega)}.$$

Consequently,

$$|T_f(g)| \le C \left\| f \right\|_{H^{1/2}(\partial\Omega)} \left\| g \right\|_{H^{1/2}(\partial\Omega)}.$$

Thus T_f is a bounded linear functional on $H^{1/2}(\partial\Omega)$, or in other words $T_f \in H^{-1/2}(\partial\Omega)$, and T_f has norm less than or equal to $C \left\| f \right\|_{H^{1/2}(\partial\Omega)}$. We define

$$\Lambda_\gamma : H^{1/2}(\partial\Omega) \to H^{-1/2}(\partial\Omega), \quad f \mapsto T_f.$$

This map satisfies (2.10) and is the unique map with this property.

(3) If Ω has C^∞ boundary and $\gamma \in C^\infty(\overline{\Omega})$, and if $f, g \in C^\infty(\partial\Omega)$, then $u_f \in C^\infty(\overline{\Omega})$ by Theorem 2.6.3. The computation done in Motivation 2.7.1 implies that

$$\langle \Lambda_L f, g \rangle_{\partial\Omega} = \sum_{j,k=1}^n \int_{\partial\Omega} a^{jk} (\partial_j u_f) e_g \nu_k \, dS.$$

Thus $\langle \Lambda_L f - \sum_{j,k=1}^n a^{jk} (\partial_j u_f) \nu_k|_{\partial\Omega}, g \rangle_{\partial\Omega} = 0$ for all $g \in C^\infty(\partial\Omega)$. Since $C^\infty(\partial\Omega)$ is dense in $H^{1/2}(\partial\Omega)$ (this follows from (2.2) and the fact that $C^\infty(\overline{\Omega})$

is dense in $H^1(\Omega)$, see Lemma C.2.20), this shows that

$$\Lambda_L f = \sum_{j,k=1}^{n} a^{jk}(\partial_j u_f)\nu_k|_{\partial\Omega}$$

as elements of $H^{-1/2}(\partial\Omega)$. Consequently $\Lambda_\gamma f$ can be identified with the C^∞ function $\sum_{j,k=1}^n a^{jk}(\partial_j u_f)\nu_k|_{\partial\Omega}$. □

We now obtain the DN maps for the conductivity equation, anisotropic conductivity equation, and Schrödinger equation as special cases of the previous result.

Theorem 2.7.3 (DN map for conductivity and Schrödinger equations). *Let Ω be a bounded open set in $\mathbb{R}^n$, and let $c > 0$. Assume that*

(1) *$\gamma \in L^\infty(\Omega)$ and $\gamma(x) \geq c$ for a.e. $x \in \Omega$,*
(2) *$G = (\gamma^{jk})_{j,k=1}^n$ is a symmetric matrix of $L^\infty(\Omega)$ functions and*

$$\sum_{j,k=1}^{n} \gamma^{jk}(x)\xi_j\xi_k \geq c\,|\xi|^2 \quad \text{for a.e. } x \in \Omega \text{ and for all } \xi \in \mathbb{R}^n,$$

(3) *$q \in L^\infty(\Omega)$ is real valued and 0 is not a Dirichlet eigenvalue of $-\Delta + q$ in Ω.*

There are bounded linear maps

$$\Lambda_\gamma, \Lambda_G, \Lambda_q : H^{1/2}(\partial\Omega) \to H^{-1/2}(\partial\Omega)$$

that satisfy for all $f, g \in H^{1/2}(\partial\Omega)$

$$\langle \Lambda_\gamma f, g\rangle_{\partial\Omega} = \int_\Omega \gamma \nabla u_f \cdot \nabla e_g \, dx,$$
$$\langle \Lambda_G f, g\rangle_{\partial\Omega} = \int_\Omega G \nabla v_f \cdot \nabla e_g \, dx,$$
$$\langle \Lambda_q f, g\rangle_{\partial\Omega} = \int_\Omega \left(\nabla w_f \cdot \nabla e_g + q w_f e_g\right) dx,$$

where $u_f, v_f, w_f \in H^1(\Omega)$ are the unique solutions of

$$\operatorname{div}(\gamma\nabla u) = 0, \qquad \operatorname{div}(G\nabla v) = 0, \quad \text{and} \quad (-\Delta + q)w = 0$$

in Ω with boundary value f, and e_g is any function in $H^1(\Omega)$ with $e_g|_{\partial\Omega} = g$.

If Ω has C^∞ boundary and $\gamma, \gamma^{jk}, q \in C^\infty(\overline{\Omega})$, then these maps restrict to linear maps

$$\Lambda_\gamma, \Lambda_G, \Lambda_q : C^\infty(\partial\Omega) \to C^\infty(\partial\Omega)$$

which satisfy for $f \in C^\infty(\partial\Omega)$

$$\Lambda_\gamma f = \gamma \partial_\nu u_f|_{\partial\Omega}, \quad \Lambda_G f = G\nabla u_f \cdot \nu|_{\partial\Omega}, \quad \Lambda_q f = \partial_\nu u_f|_{\partial\Omega}.$$

Proof. By our assumptions, the operators $-\operatorname{div}(\gamma\nabla\,\cdot\,)$, $-\operatorname{div}(G\nabla\,\cdot\,)$, and $-\Delta+q$ are of the form (2.5), (2.6). In the first two cases there is no zero order term and we see from Theorem 2.5.9 that 0 is not a Dirichlet eigenvalue, and for the third case this is explicitly assumed. The result follows from Theorem 2.7.2. □

Exercise 2.7.4. Assume the conditions of Theorem 2.7.2, and show that knowledge of the DN map Λ_L is equivalent to knowing the quadratic form

$$Q_L : H^{1/2}(\partial\Omega) \to \mathbb{R},$$

$$Q_L(f) = B[u_f, u_f] = \int_\Omega \Big[\sum_{j,k=1}^n a^{jk}\partial_j u_f \overline{\partial_k u_f} + q\,|u_f|^2 \Big]\,dx.$$

Physically, for the conductivity equation, the quadratic form

$$Q_\gamma(f) = \int_\Omega \gamma\,|\nabla u_f|^2\,dx$$

expresses the power needed to maintain the voltage f at the boundary.

Exercise 2.7.5. If Ω has C^2 boundary and $\gamma \in C^1(\overline{\Omega})$ is positive, show that $u_f \in H^2(\Omega)$ and that the DN map has the pointwise expression

$$\Lambda_\gamma f = \gamma\partial_\nu u_f|_{\partial\Omega}.$$

The next result shows that the DN map is a symmetric operator.

Theorem 2.7.6. *Let $\Omega \subset \mathbb{R}^n$ be a bounded open set, and let L satisfy* (2.5), (2.6). *Assume that* 0 *is not a Dirichlet eigenvalue of L in Ω. Then*

$$\langle \Lambda_L f, g\rangle_{\partial\Omega} = \langle f, \Lambda_L g\rangle_{\partial\Omega}, \quad f, g \in H^{1/2}(\partial\Omega).$$

In particular, the maps Λ_γ, Λ_G, Λ_q in Theorem 2.7.3 *also have this property.*

Proof. By definition

$$\langle \Lambda_L f, g\rangle_{\partial\Omega} = B[u_f, \bar{e}_g],$$

where $u_f \in H^1(\Omega)$ is the unique solution of $Lu = 0$ in Ω with $u_f|_{\partial\Omega} = f$, and e_g is any function $H^1(\Omega)$ with $e_g|_{\partial\Omega} = g$. We choose $e_g = u_g$, the solution of $Lu = 0$ with boundary value g. Then, since $B[\,\cdot\,,\,\cdot\,]$ is conjugate symmetric and a^{jk}, q are real valued,

$$\langle \Lambda_L f, g\rangle_{\partial\Omega} = B[u_f, \bar{u}_g] = \overline{B[\bar{u}_g, u_f]} = B[u_g, \bar{u}_f] = \langle \Lambda_L g, f\rangle_{\partial\Omega}. \qquad \square$$

We are now in a position to give mathematically precise formulations for the inverse problems considered in this book.

2.7.1. Calderón problem. Let $\Omega \subset \mathbb{R}^n$ be a bounded open set. We think of Ω as an electrical conductor, and assume that the conductivity at each point of Ω is given by a function $\gamma \in L^\infty(\Omega)$ satisfying $\gamma(x) \geq c > 0$ a.e. in Ω. By Theorem 2.7.3 there is a bounded linear map

$$\Lambda_\gamma : H^{1/2}(\partial\Omega) \to H^{-1/2}(\partial\Omega)$$

which formally associates to a function $f \in H^{1/2}(\partial\Omega)$ an element $\Lambda_\gamma f \in H^{-1/2}(\partial\Omega)$ that may be thought of as the electrical current $\gamma\partial_\nu u_f|_{\partial\Omega}$ corresponding to boundary voltage f. (If Ω has C^∞ boundary and $\gamma \in C^\infty(\overline{\Omega})$, we saw in by Theorem 2.7.3 that $\Lambda_\gamma f = \gamma\partial_\nu u_f|_{\partial\Omega}$ for any $f \in C^\infty(\partial\Omega)$ in the classical sense.)

We think that for each boundary voltage $f \in H^{1/2}(\partial\Omega)$, we can measure the corresponding current $\Lambda_\gamma f \in H^{-1/2}(\partial\Omega)$. This leads to the following inverse problem.

Calderón problem. Let $\Omega \subset \mathbb{R}^n$ be a bounded open set, and let $\gamma \in L^\infty(\Omega)$ satisfy $\gamma \geq c > 0$ a.e. in Ω. From the knowledge of the map Λ_γ, determine the function γ in Ω.

2.7.2. Inverse BVP for Schrödinger equation. Let $\Omega \subset \mathbb{R}^n$ be a bounded open set, and let $q \in L^\infty(\Omega)$. We consider an inverse problem for the equation $(-\Delta + q)u = 0$ analogous to the Calderón problem. However, in order to have a well-defined DN map we need to assume that 0 is not a Dirichlet eigenvalue of $-\Delta + q$ in Ω. If this is the case, then by Theorem 2.7.3 there is a bounded linear map

$$\Lambda_q : H^{1/2}(\partial\Omega) \to H^{-1/2}(\partial\Omega)$$

that associates to any function $f \in H^{1/2}(\partial\Omega)$ an element $\Lambda_q f \in H^{-1/2}(\partial\Omega)$ that corresponds (in the case where everything is smooth) to the normal derivative $\partial_\nu u_f|_{\partial\Omega}$, where $u_f \in H^1(\Omega)$ is the unique solution of $(-\Delta + q)u = 0$ in Ω with boundary value f. The inverse problem is as follows.

Inverse BVP for Schrödinger equation. Let $\Omega \subset \mathbb{R}^n$ be a bounded open set, let $q \in L^\infty(\Omega)$, and assume that 0 is not a Dirichlet eigenvalue of $-\Delta + q$ in Ω. From the knowledge of the map Λ_q, determine the function q in Ω.

It is also possible to formulate this problem without the assumption that 0 is not a Dirichlet eigenvalue, by using the notion of Cauchy data sets. To do this, we observe that even though the normal derivative on $\partial\Omega$ does not make sense for a general function $u \in H^1(\Omega)$, we can still define the normal derivative weakly if we assume that u is additionally a solution. The proof is similar to that of Theorem 2.7.2.

Exercise 2.7.7. If $u \in H^1(\Omega)$ is a solution of $(-\Delta + q)u = 0$ in Ω, show that the following identity defines $\partial_\nu u|_{\partial\Omega}$ as an element of $H^{-1/2}(\partial\Omega)$:

$$\langle \partial_\nu u|_{\partial\Omega}, g\rangle_{\partial\Omega} = \int_\Omega (\nabla u \cdot \nabla e_g + que_g)\,dx, \quad g \in H^{1/2}(\partial\Omega),$$

where e_g is any function in $H^1(\Omega)$ with $e_g|_{\partial\Omega} = g$. Show that if $u \in C^\infty(\overline{\Omega})$, then this definition of $\partial_\nu u|_{\partial\Omega}$ coincides with the usual one.

Definition 2.7.8. Let $\Omega \subset \mathbb{R}^n$ be a bounded open set and let $q \in L^\infty(\Omega)$. The *Cauchy data set* associated with the operator $-\Delta + q$ in Ω is the set

$$C_q = \{\, (u|_{\partial\Omega}, \partial_\nu u|_{\partial\Omega}) \mid u \in H^1(\Omega),\ (-\Delta + q)u = 0 \text{ in } \Omega \,\}.$$

By Exercise 2.7.7, C_q is a subset of $H^{1/2}(\partial\Omega)\times H^{-1/2}(\partial\Omega)$. If 0 is not a Dirichlet eigenvalue of $-\Delta+q$ in Ω, the next exercise shows that knowing the Cauchy data set C_q is equivalent to knowing the DN map Λ_q.

Exercise 2.7.9. Let $\Omega \subset \mathbb{R}^n$ be a bounded open set, let $q \in L^\infty(\Omega)$, and assume that 0 is not a Dirichlet eigenvalue of $-\Delta + q$ in Ω. Show that the Cauchy data set is the graph of the DN map acting on $H^{1/2}(\partial\Omega)$:

$$C_q = \{\, (f, \Lambda_q f) \mid f \in H^{1/2}(\partial\Omega) \,\}.$$

The next question generalizes the inverse BVP for Schrödinger equation given above.

Inverse BVP for Schrödinger equation, Cauchy data set version. Let $\Omega \subset \mathbb{R}^n$ be a bounded open set, and let $q \in L^\infty(\Omega)$. From the knowledge of the set C_q, determine the function q in Ω.

2.7.3. Anisotropic Calderón problem. Again, $\Omega \subset \mathbb{R}^n$ is a bounded open set that is thought of as an electrical conductor, but this time the conductivity at each point of Ω is given by a symmetric matrix function $G = (\gamma^{jk})_{j,k=1}^n$. For this problem we assume that Ω has C^∞ boundary and that each γ^{jk} is in $C^\infty(\overline{\Omega})$. We also assume the ellipticity condition for some $c > 0$,

$$\sum_{j,k=1}^n \gamma^{jk}(x)\xi_j\xi_k \geq c\,|\xi|^2 \quad \text{for all } x \in \Omega \text{ and for all } \xi \in \mathbb{R}^n.$$

By Theorem 2.7.3 there is a linear map

$$\Lambda_G : C^\infty(\partial\Omega) \to C^\infty(\partial\Omega)$$

which associates to a function $f \in C^\infty(\partial\Omega)$ the boundary current $\Lambda_\gamma f = G\nabla u_f \cdot \nu|_{\partial\Omega} \in C^\infty(\partial\Omega)$.

We noted in (1.5) that in the anisotropic case, one can only expect to recover the conductivity matrix G up to a diffeomorphism of $\overline{\Omega}$ that fixes the boundary. Thus the inverse problem is formulated as follows.

Anisotropic Calderón problem. From the knowledge of the map Λ_G, determine a symmetric matrix function $\hat{G}$ with elements in $C^\infty(\overline{\Omega})$ such that $\hat{G} = \Psi_* G$ for some diffeomorphism $\Psi : \overline{\Omega} \to \overline{\Omega}$ with $\Psi|_{\partial\Omega} = \mathrm{Id}$.

2.8. Integral identities and reductions

The purpose of this section is to introduce certain integral identities, often called Alessandrini identities, for differences of DN maps. These identities relate the boundary measurements to interior information about the coefficients. They will allow us to reduce uniqueness questions in inverse problems to questions about the completeness of products of (gradients of) solutions. We also give two simple reductions that will be useful later.

Proposition 2.8.1 (Integral identity for $\Lambda_{L_1} - \Lambda_{L_2}$). *Let $\Omega \subset \mathbb{R}^n$ be a bounded open set, and let L_1, L_2 be two operators of the form*

$$L_m u = -\sum_{j,k=1}^{n} \frac{\partial}{\partial x_j}\left(a_m^{jk} \frac{\partial u}{\partial x_k}\right) + q_m u,$$

where a_m^{jk}, q_m are as in (2.6) for $m = 1, 2$. Assume that 0 is not a Dirichlet eigenvalue for L_1 or L_2 in Ω. Then for any $f_1, f_2 \in H^{1/2}(\partial\Omega)$,

$$\langle(\Lambda_{L_1} - \Lambda_{L_2})f_1, f_2\rangle_{\partial\Omega} = \int_\Omega \left(\sum_{j,k=1}^{n} (a_1^{jk} - a_2^{jk})\partial_j u_1 \partial_k u_2 + (q_1 - q_2)u_1 u_2\right) dx,$$

where $u_m \in H^1(\Omega)$ is the unique solution of $L_m u_m = 0$ in Ω with boundary condition $u_m|_{\partial\Omega} = f_m$.

Proof. Let u_1 and u_2 be as described. By Theorem 2.7.3 we have

$$\langle\Lambda_{L_1} f_1, f_2\rangle_{\partial\Omega} = \int_\Omega \left(\sum_{j,k=1}^{n} a_1^{jk} \partial_j u_1 \partial_k v_2 + q_1 u_1 v_2\right) dx.$$

where v_2 is any function in $H^1(\Omega)$ with $v_2|_{\partial\Omega} = f_2$. Similarly, also using Theorem 2.7.6, we have

$$\langle\Lambda_{L_2} f_1, f_2\rangle_{\partial\Omega} = \langle\Lambda_{L_2} f_2, f_1\rangle_{\partial\Omega} = \int_\Omega \left(\sum_{j,k=1}^{n} a_2^{jk} \partial_j u_2 \partial_k v_1 + q_2 u_2 v_1\right) dx.$$

where v_1 is any function in $H^1(\Omega)$ with $v_1|_{\partial\Omega} = f_1$. Now, we may choose $v_1 = u_1$ and $v_2 = u_2$. Subtracting the two identities above and using the symmetry of a_2^{jk}, we obtain the theorem. □

As an immediate consequence of the above result, we obtain integral identities for the differences of DN maps in the case of the conductivity and Schrödinger equation.

Proposition 2.8.2 (Integral identity for $\Lambda_{\gamma_1} - \Lambda_{\gamma_2}$). *Let $\Omega \subset \mathbb{R}^n$ be a bounded open set, and let $\gamma_1, \gamma_2 \in L^\infty(\Omega)$ satisfy $\gamma_1, \gamma_2 \geq c > 0$ a.e. in Ω.*

(a) *For any $f_1, f_2 \in H^{1/2}(\partial\Omega)$ one has the integral identity*

$$\langle(\Lambda_{\gamma_1} - \Lambda_{\gamma_2})f_1, f_2\rangle_{\partial\Omega} = \int_\Omega (\gamma_1 - \gamma_2)\nabla u_1 \cdot \nabla u_2 \, dx,$$

where $u_j \in H^1(\Omega)$ is the unique solution of $\mathrm{div}(\gamma_j \nabla u_j) = 0$ in Ω with $u_j|_{\partial\Omega} = f_j$.

(b) *If $\Lambda_{\gamma_1} = \Lambda_{\gamma_2}$, then*

$$\int_\Omega (\gamma_1 - \gamma_2)\nabla w_1 \cdot \nabla w_2 \, dx = 0$$

for all $w_j \in H^1(\Omega)$ with $\mathrm{div}(\gamma_j \nabla w_j) = 0$ in Ω.

Proposition 2.8.3 (Integral identity for $\Lambda_{q_1} - \Lambda_{q_2}$). *Let $\Omega \subset \mathbb{R}^n$ be a bounded open set, let $q_1, q_2 \in L^\infty(\Omega)$, and assume that 0 is not a Dirichlet eigenvalue of $-\Delta + q_j$ in Ω.*

(a) *One has the integral identity*

$$\langle(\Lambda_{q_1} - \Lambda_{q_2})f_1, f_2\rangle_{\partial\Omega} = \int_\Omega (q_1 - q_2)u_1 u_2 \, dx, \quad f_1, f_2 \in H^{1/2}(\partial\Omega),$$

where $u_j \in H^1(\Omega)$ is the unique solution of $(-\Delta + q_j)u_j = 0$ in Ω with $u_j|_{\partial\Omega} = f_j$.

(b) *If $\Lambda_{q_1} = \Lambda_{q_2}$, then*

$$\int_\Omega (q_1 - q_2)w_1 w_2 \, dx = 0$$

for all $w_j \in H^1(\Omega)$ with $(-\Delta + q_j)w_j = 0$ in Ω.

Exercise 2.8.4. Show that Proposition 2.8.3(b) remains true in the case of Cauchy data sets: if $\Omega \subset \mathbb{R}^n$ is a bounded open set and $q_1, q_2 \in L^\infty(\Omega)$, then

$$\int_\Omega (q_1 - q_2)w_1 w_2 \, dx = \langle \partial_\nu w_1, w_2\rangle_{\partial\Omega} - \langle w_1, \partial_\nu w_2\rangle_{\partial\Omega}$$

for all $w_j \in H^1(\Omega)$ with $(-\Delta + q_j)w_j = 0$ in Ω, where the normal derivatives are interpreted as in Exercise 2.7.7. If additionally $C_{q_1} = C_{q_2}$, then

$$\int_\Omega (q_1 - q_2)w_1 w_2 \, dx = 0$$

for all such w_j.

Suppose now that γ_1, γ_2 are two conductivities such that $\Lambda_{\gamma_1} = \Lambda_{\gamma_2}$. We would like to conclude that $\gamma_1 = \gamma_2$. By Proposition 2.8.2(b), the condition $\Lambda_{\gamma_1} = \Lambda_{\gamma_2}$ implies that

$$\int_\Omega (\gamma_1 - \gamma_2)\nabla w_1 \cdot \nabla w_2 \, dx = 0$$

for any $w_j \in H^1(\Omega)$ satisfying $\mathrm{div}(\gamma_j \nabla w_j) = 0$ in Ω. By taking linear combinations, it follows that

$$\int_\Omega (\gamma_1 - \gamma_2)\left[\sum_{k=1}^N c_k \nabla w_1^{(k)} \cdot \nabla w_2^{(k)}\right] dx = 0$$

whenever $c_k \in \mathbb{C}$ and $\mathrm{div}(\gamma_j \nabla w_j^{(k)}) = 0$ in Ω. Thus the conclusion $\gamma_1 = \gamma_2$ would follow if one could show that the set

$$\{\, \nabla w_1 \cdot \nabla w_2 \mid w_j \in H^1(\Omega),\ \mathrm{div}(\gamma_j \nabla w_j) = 0 \text{ in } \Omega \,\}$$

is *complete* (meaning that its linear span, consisting of finite linear combinations, is dense) in $L^1(\Omega)$. In the same way, if $q_1, q_2 \in L^\infty(\Omega)$ and $\Lambda_{q_1} = \Lambda_{q_2}$, the conclusion $q_1 = q_2$ would follow from Proposition 2.8.3(b) if one could show that the set

$$\{\, w_1 w_2 \mid w_j \in H^1(\Omega),\ (-\Delta + q_j)w_j = 0 \text{ in } \Omega \,\}$$

is complete in $L^1(\Omega)$. In other words, uniqueness in the Calderón problem would follow from the completeness of products of gradients of solutions to conductivity equations. Similarly, uniqueness in the inverse BVP for the Schrödinger equation would follow from the completeness of products of solutions to Schrödinger equations. Most of the interior uniqueness results in this book will be proved by following this route.

Next we give a result showing that the Calderón problem for the conductivity equation can be reduced to the inverse BVP for the Schrödinger equation, provided that the conductivity has two derivatives. This is based on a simple argument, sometimes called a *Liouville transformation*, where the substitution $u = \gamma^{-1/2} w$ reduces the conductivity equation $\mathrm{div}(\gamma \nabla u) = 0$ to the Schrödinger equation $(-\Delta + q_\gamma)w = 0$, where the potential q_γ is given by

$$q_\gamma = \frac{\Delta(\gamma^{1/2})}{\gamma^{1/2}}.$$

Proposition 2.8.5 (Reduction to Schrödinger equation). *Let $\Omega \subset \mathbb{R}^n$ be a bounded open set, let $\gamma \in C^2(\overline{\Omega})$ be strictly positive, and set $q = \frac{\Delta(\gamma^{1/2})}{\gamma^{1/2}}$.*

(a) *If $f \in H^{1/2}(\partial\Omega)$, then $u \in H^1(\Omega)$ is a solution of*

$$\nabla \cdot \gamma \nabla u = 0 \text{ in } \Omega, \qquad u = f \text{ on } \partial\Omega$$

if and only if $w = \gamma^{1/2}u$ *is a solution of*

$$-\Delta w + qw = 0 \text{ in } \Omega, \qquad w = \gamma^{1/2} f \text{ on } \partial\Omega.$$

In particular, 0 is not a Dirichlet eigenvalue for $-\Delta + q$ *in* Ω *if* q *arises from a* C^2 *conductivity.*

(b) *If* Ω *has* C^∞ *boundary and* $\gamma \in C^\infty(\overline{\Omega})$*, then the DN maps* Λ_γ *and* Λ_q *for* q *as defined above are related by*

$$\Lambda_q f = \gamma^{-1/2}\Lambda_\gamma(\gamma^{-1/2} f) + \tfrac{1}{2}\gamma^{-1}(\partial_\nu \gamma) f|_{\partial\Omega}, \quad f \in C^\infty(\partial\Omega).$$

Proof.

(a) Assume first that $w \in C^2(\overline{\Omega})$. To reduce the Schrödinger equation to a conductivity equation, we attempt to find $a \in C^2(\overline{\Omega})$ such that $\nabla \cdot (\gamma \nabla (aw)) = \gamma a(\Delta w + rw)$ for some function r. By the product rule

$$\begin{aligned}\nabla \cdot (\gamma \nabla (aw)) &= \nabla \cdot (\gamma (\nabla a) w + \gamma a \nabla w) \\ &= \gamma a \Delta w + (\nabla(\gamma a) + \gamma \nabla a) \cdot \nabla w + (\nabla \cdot (\gamma \nabla a)) w.\end{aligned}$$

The first-order term will disappear if we can choose a so that

$$\nabla(\gamma a) + \gamma \nabla a = 0.$$

This is equivalent with $2\gamma \nabla a + (\nabla \gamma) a = 0$, and dividing by γa we obtain the equation

$$\nabla(2 \log a + \log \gamma) = 0.$$

Thus, by the properties of logarithms $\log(a^2\gamma)$ should be constant, so that $a = C\gamma^{-1/2}$ for some constant C.

We choose $a = \gamma^{-1/2}$, and the previous computation implies

$$\nabla \cdot (\gamma \nabla (aw)) = \gamma a \Delta w + (\nabla \cdot (\gamma \nabla a)) w.$$

Here $\gamma \nabla a = -\nabla(\gamma a)$, so we obtain

$$\nabla \cdot (\gamma \nabla (\gamma^{-1/2} w)) = \gamma^{1/2}(\Delta w - \frac{\Delta(\gamma^{1/2})}{\gamma^{1/2}} w).$$

This shows part (a) in the case where $u, w \in C^2(\overline{\Omega})$. The case where $u, w \in H^1(\Omega)$ follows either from the definition of weak solutions or by approximation, and is left as an exercise.

(b) Set $g = \gamma^{-1/2} f$ and $u_g = \gamma^{-1/2} w_f$, where w_f is the unique solution of $(-\Delta + q)w = 0$ in Ω with boundary value f. Then, by part (a), u_g is the solution of $\nabla \cdot \gamma \nabla u_g = 0$ in Ω with $u_g|_{\partial\Omega} = g$, and

$$\begin{aligned}
\Lambda_q(f) &= \sum_{j=1}^{n} (\partial_j w_f)\nu_j \Big|_{\partial\Omega} \\
&= \sum_{j=1}^{n} \partial_j(\gamma^{1/2} u_g)\nu_j \Big|_{\partial\Omega} \\
&= \sum_{j=1}^{n} \gamma^{1/2}(\partial_j u_g)\nu_j + \sum_{j=1}^{n} \tfrac{1}{2} u_g \gamma^{-1/2}(\partial_j \gamma)\nu_j \Big|_{\partial\Omega} \\
&= \gamma^{-1/2}\Lambda_\gamma(g) + \sum_{j=1}^{n} \tfrac{1}{2} g \gamma^{-1/2}(\partial_j \gamma)\nu_j \Big|_{\partial\Omega} \\
&= \gamma^{-1/2}\Lambda_\gamma(\gamma^{-1/2} f) + \tfrac{1}{2}\gamma^{-1}(\partial_\nu \gamma) f \Big|_{\partial\Omega}. \qquad \square
\end{aligned}$$

Exercise 2.8.6. Verify Proposition 2.8.5(a) in detail for $u, w \in H^1(\Omega)$.

The final reduction states that if the DN maps for two operators in some domain Ω are equal, then the DN maps corresponding to extensions of the operators to a larger domain $\tilde{\Omega}$ are also equal provided that the coefficients agree in $\tilde{\Omega} \setminus \Omega$.

Proposition 2.8.7 (Reduction to a larger domain). *Let Ω and $\tilde{\Omega}$ be bounded open sets in $\mathbb{R}^n$ with $\Omega \subset \tilde{\Omega}$. Suppose that L_1 and L_2 are two operators in Ω of the form*

$$L_m u = -\sum_{j,k=1}^{n} \frac{\partial}{\partial x_j}\left(a_m^{jk} \frac{\partial u}{\partial x_k}\right) + q_m u.$$

Let $\tilde{a}_m^{jk}, \tilde{q}_m$ be extensions of a_m^{jk}, q_m to $\tilde{\Omega}$, and denote by $\tilde{L}_1$ and $\tilde{L}_2$ the corresponding operators in $\tilde{\Omega}$,

$$\tilde{L}_m u = -\sum_{j,k=1}^{n} \frac{\partial}{\partial x_j}\left(\tilde{a}_m^{jk} \frac{\partial u}{\partial x_k}\right) + \tilde{q}_m u.$$

Assume that $a_m^{jk}, q_m, \tilde{a}_m^{jk}, \tilde{q}_m$ satisfy (2.6) for $m = 1, 2$, and assume that 0 is not a Dirichlet eigenvalue for L_1 or L_2 in Ω or for $\tilde{L}_1$ or $\tilde{L}_2$ in Ω.

If

$$\tilde{a}_1^{jk} = \tilde{a}_2^{jk} \text{ in } \tilde{\Omega} \setminus \Omega, \qquad \tilde{q}_1 = \tilde{q}_2 \text{ in } \tilde{\Omega} \setminus \Omega,$$

then for $\tilde{f}_1, \tilde{f}_2 \in H^{1/2}(\partial\tilde{\Omega})$ one has

$$\langle (\Lambda_{\tilde{L}_1} - \Lambda_{\tilde{L}_2})\tilde{f}_1, \tilde{f}_2 \rangle_{\partial\tilde{\Omega}} = \langle (\Lambda_{L_1} - \Lambda_{L_2})(\tilde{u}_1|_{\partial\Omega}), \tilde{u}_2|_{\partial\Omega} \rangle_{\partial\Omega},$$

where $\tilde{u}_m$ is the unique solution in $H^1(\tilde{\Omega})$ of $\tilde{L}_m \tilde{u}_m = 0$ in $\tilde{\Omega}$ with boundary condition $\tilde{u}_m|_{\partial\tilde{\Omega}} = \tilde{f}_m$.

In particular,

$$\Lambda_{L_1} = \Lambda_{L_2} \implies \Lambda_{\tilde{L}_1} = \Lambda_{\tilde{L}_2}.$$

Proof. By Proposition 2.8.1, and using the fact that $\tilde{a}_1^{jk} = \tilde{a}_2^{jk}$ and $\tilde{q}_1 = \tilde{q}_2$ outside of Ω, we have

$$\begin{aligned}\langle(\Lambda_{\tilde{L}_1} - \Lambda_{\tilde{L}_2})\tilde{f}_1, \tilde{f}_2\rangle_{\partial\tilde{\Omega}} &= \int_{\tilde{\Omega}} \left(\sum_{j,k=1}^n (\tilde{a}_1^{jk} - \tilde{a}_2^{jk})\partial_j\tilde{u}_1\partial_k\tilde{u}_2 + (\tilde{q}_1 - \tilde{q}_2)\tilde{u}_1\tilde{u}_2 \right) dx \\ &= \int_{\Omega} \left(\sum_{j,k=1}^n (a_1^{jk} - a_2^{jk})\partial_j\tilde{u}_1\partial_k\tilde{u}_2 + (q_1 - q_2)\tilde{u}_1\tilde{u}_2 \right) dx \\ &= \langle(\Lambda_{L_1} - \Lambda_{L_2})(\tilde{u}_1|_{\partial\Omega}), \tilde{u}_2|_{\partial\Omega}\rangle_{\partial\Omega}\end{aligned}$$

since $\tilde{u}_m|_\Omega$ solves $L_m\tilde{u}_m = 0$ in Ω. □

2.9. The ND map

It is also possible to consider other boundary measurement setups. For example, one could prescribe Neumann data and measure Dirichlet data, which gives rise to the *Neumann-to-Dirichlet map* (ND map). Another possibility is to consider Robin boundary conditions. In this section we briefly discuss the ND map, and we will show that in fact knowing the DN map is equivalent to knowing the ND map.

We consider the spaces $H_\diamond^{\pm 1/2}(\partial\Omega) = \{\, g \in H^{\pm 1/2}(\partial\Omega) \mid \langle g, 1\rangle_{\partial\Omega} = 0 \,\}$ of Sobolev functions with zero average. For $g \in H_\diamond^{-1/2}(\partial\Omega)$ consider the Neumann problem

$$\nabla \cdot (\gamma\nabla w) = 0 \text{ in } \Omega, \qquad \gamma\partial_\nu w = g \text{ on } \partial\Omega. \tag{2.11}$$

The compatibility condition $\langle g, 1\rangle_{\partial\Omega} = 0$ is natural since for smooth γ and w, integration by parts gives

$$\langle \gamma\partial_\nu w, 1\rangle_{\partial\Omega} = \int_{\partial\Omega} \gamma\partial_\nu w \, dS = \int_\Omega \nabla \cdot (\gamma\nabla w) \, dx = 0.$$

We say that $w \in H^1(\Omega)$ is a weak solution of the Neumann problem (2.11) if

$$B_\gamma[w, \bar{v}] = \langle g, v\rangle_{\partial\Omega}, \qquad v \in H^1(\Omega).$$

Here, recall that $B_\gamma[w, \bar{v}] = \int_\Omega \gamma\nabla w \cdot \nabla v \, dx$. If w is a weak solution, we see that $w + c$ is also a weak solution for any constant c (this is also suggested by the formulation (2.11)). A solution whose restriction to $\partial\Omega$ has zero average is unique, which gives rise to a well-defined ND map.

Theorem 2.9.1 (ND map). *Let $\Omega \subset \mathbb{R}^n$ be a bounded connected open set with C^1 boundary, and let $\gamma \in L^\infty(\Omega)$ satisfy $\gamma \geq c > 0$ a.e. in Ω. Given any $g \in H_\diamond^{-1/2}(\partial\Omega)$ there is a weak solution $w \in H^1(\Omega)$ of (2.11) that is unique up to* *adding a constant. Letting w_g be the unique solution with $\langle w_g|_{\partial\Omega}, 1\rangle_{\partial\Omega} = 0$, we can define the ND map as*

$$N_\gamma : H_\diamond^{-1/2}(\partial\Omega) \to H_\diamond^{1/2}(\partial\Omega), \quad g \mapsto w_g|_{\partial\Omega}.$$

Proof. The proof of solvability is analogous to the proof of Theorem 2.5.6. Consider the space $H_\diamond^1(\Omega) = \{ u \in H^1(\Omega) \mid \int_\Omega u\,dx = 0 \}$ of functions in $H^1(\Omega)$ with zero average, and consider the bilinear form

$$B_\gamma[u, v], \qquad u, v \in H_\diamond^1(\Omega).$$

As in Lemma 2.5.7 it follows that this bilinear form is an inner product on $H_\diamond^1(\Omega)$: the main point is that one needs the version of the Poincaré inequality given in Exercise 2.4.10. The map

$$F : H_\diamond^1(\Omega) \to \mathbb{C}, \quad F(v) = \langle g, v\rangle_{\partial\Omega},$$

is a bounded linear functional on $H_\diamond^1(\Omega)$. The Riesz representation theorem ensures that there is a unique $w \in H_\diamond^1(\Omega)$ satisfying

$$B_\gamma[w, \bar{v}] = \langle g, v\rangle_{\partial\Omega}, \qquad v \in H_\diamond^1(\Omega).$$

This identity remains true for $v \in H^1(\Omega)$. To see this it is enough to decompose $v = (v - (v)_\Omega) + (v)_\Omega$ where the average $(v)_\Omega$ is a constant, use the compatibility condition $\langle g, 1\rangle_{\partial\Omega} = 0$, and use the definition of B_γ. This proves the existence of a weak solution. Moreover, any function $w + c$ where c is a constant is also a solution.

If $w_1, w_2 \in H^1(\Omega)$ are weak solutions, then $B_\gamma(w_1 - w_2, \bar{v}) = 0$ for any $v \in H^1(\Omega)$, and choosing $v = \overline{w_1 - w_2}$ gives that $\nabla(w_1 - w_2) = 0$ a.e. in Ω. Since Ω was assumed connected, we see from Exercise 2.4.5 that $w_1 - w_2 = c$ for some constant c. This proves the uniqueness of weak solutions up to additive constants. Thus the solution w_g whose boundary restriction has zero average is unique, and the ND map $N_\gamma : H_\diamond^{-1/2}(\partial\Omega) \to H_\diamond^{1/2}(\partial\Omega)$ is well defined by the trace theorem. □

Physically, measuring the ND map means that we prescribe different currents at the boundary and measure the resulting voltages at the boundary. It is possible to formulate the inverse problems stated in §2.7 in terms of the ND map instead of the DN map. For example, one could ask if knowledge of the ND map N_γ allows one to determine the conductivity γ. It turns out that mathematically there is no difference, since the ND map and DN map are equivalent information.

Lemma 2.9.2 (Equivalence of ND and DN maps). *Let $\Omega \subset \mathbb{R}^n$ be a bounded connected open set with C^1 boundary, and let $\gamma \in L^\infty(\Omega)$ satisfy $\gamma \geq c > 0$ a.e. in Ω. Then the ND map $N_\gamma : H_\diamond^{-1/2}(\partial\Omega) \to H_\diamond^{1/2}(\partial\Omega)$ is the inverse of the map $\Lambda_\gamma : H_\diamond^{1/2}(\partial\Omega) \to H_\diamond^{-1/2}(\partial\Omega)$ and vice versa.*

Proof. We first observe that for any $f \in H^{1/2}(\partial\Omega)$, the definition of the DN map (see Theorem 2.7.3) yields

$$\langle \Lambda_\gamma f, 1\rangle_{\partial\Omega} = \int_\Omega \gamma \nabla u_f \cdot \nabla(1)\, dx = 0.$$

Thus in particular Λ_γ maps $H^{1/2}_\diamond(\partial\Omega)$ to $H^{-1/2}_\diamond(\partial\Omega)$. Now if $g \in H^{-1/2}_\diamond(\partial\Omega)$ and $h \in H^{1/2}_\diamond(\partial\Omega)$, we have

$$\langle \Lambda_\gamma(N_\gamma g), h\rangle_{\partial\Omega} = \langle \Lambda_\gamma(w_g|_{\partial\Omega}), h\rangle_{\partial\Omega} = \int_\Omega \gamma \nabla w_g \cdot \nabla e_h\, dx = \langle g, h\rangle_{\partial\Omega}$$

using the definition of DN map and of the weak solution of Neumann problem. Thus $\Lambda_\gamma(N_\gamma g) = g$ for $g \in H^{-1/2}_\diamond(\partial\Omega)$.

To prove that $N_\gamma(\Lambda_\gamma h) = h$, we first observe that for any $\tilde{h} \in H^{-1/2}_\diamond(\partial\Omega)$ one has

$$\langle N_\gamma g, \tilde{h}\rangle_{\partial\Omega} = \langle w_g|_{\partial\Omega}, \Lambda_\gamma(N_\gamma \tilde{h})\rangle_{\partial\Omega} = \int_\Omega \gamma \nabla w_g \cdot \nabla w_{\tilde{h}}\, dx.$$

This implies that N_γ is symmetric, i.e.,

$$\langle N_\gamma g, \tilde{h}\rangle_{\partial\Omega} = \langle g, N_\gamma \tilde{h}\rangle_{\partial\Omega}.$$

Using this together with the symmetry of Λ_γ from Theorem 2.7.4, we have

$$\langle N_\gamma(\Lambda_\gamma h), g\rangle_{\partial\Omega} = \langle h, \Lambda_\gamma(N_\gamma g)\rangle_{\partial\Omega} = \langle h, g\rangle_{\partial\Omega}.$$

This proves that $N_\gamma(\Lambda_\gamma h) = h$ for all $h \in H^{1/2}_\diamond((\partial\Omega)$, so indeed N_γ is the inverse map of $\Lambda_\gamma : H^{1/2}_\diamond(\partial\Omega) \to H^{-1/2}_\diamond(\partial\Omega)$. □

Due to the equivalence of the ND and DN maps given in Lemma 2.9.2, in this book we will consider inverse problems formulated in terms of the DN map. However, we remark that in situations such as partial data problems (see Chapter 6) this equivalence may fail and one may need to consider the DN and ND maps separately.

2.10. Notes

§2.1 includes only the most basic properties of smooth cutoff functions and convolution approximation. For a more complete treatment we refer the reader to [**Hör03a**, Chapter I].

There is a very large literature on Sobolev spaces and related function spaces. Our presentation in §2.3–§2.4 gives the minimal amount of facts needed for the formulation of the Calderón problem, and many notions like the trace space $H^{1/2}(\partial\Omega)$ are only defined in the abstract sense. Appendix C gives a detailed treatment of the $H^k(\Omega)$ and $H^{k-1/2}(\partial\Omega)$ spaces and their trace properties based on the Fourier transform. Other useful references include [**AF03, Bre11, Eva10, Leo17, Leo23, Tri83**].

Basic references on weak solutions for second order elliptic equations and higher regularity include [**Bre11**], [**Eva10**, Chapter 6] and [**GT01**].

The pioneering works [**Cal80, KV84a**] made use of the quadratic form Q_γ discussed in Exercise 2.7.4 instead of using the DN map Λ_γ. The basic integral identity for Λ_γ given in Proposition 2.8.2 is contained in [**SU86, SU87**] in a slightly different form. The analogous identity for the Schrödinger equation, Proposition 2.8.3, is given in [**Ale88**]. Identities of this type are often called Alessandrini identities.

When the boundary and coefficients are smooth, the DN map is an elliptic pseudodifferential operator of order 1. This can be proved via layer potentials [**Tay23**, §7.11], Calderón projectors [**SU88**] or factorization arguments [**LU89**]. The eigenvalues of the DN map coincide with the so-called Steklov spectrum [**GP17**].

Chapter 3

Boundary Determination

The goal of this chapter is to show that if two conductivities γ_1 and γ_2 are in $C^\infty(\overline{\Omega})$ and give rise to the same boundary measurements (i.e., $\Lambda_{\gamma_1} = \Lambda_{\gamma_2}$), then the conductivities and their normal derivatives of all orders agree on $\partial\Omega$. This was the first identifiability theorem proved for the conductivity equation and it seems to remain a necessary ingredient in many proofs of identifiability in the interior.

The critical observation is that by choosing the Dirichlet data f to be highly oscillatory and supported near a point $p \in \partial\Omega$, we can arrange that the solution to $\operatorname{div}(\gamma\nabla u) = 0$ in Ω, $u|_{\partial\Omega} = f$, is concentrated near p. Solutions of this type can be used to extract the Taylor series of the conductivity at p from the knowledge of Λ_γ.

Theorem 3.0.1. *Let Ω be a bounded domain in $\mathbb{R}^n$ with smooth boundary and let Γ be an open subset of $\partial\Omega$. Suppose that $\gamma \in C^\infty(\overline{\Omega})$ is a positive function, and that one has knowledge of the measurements*

$$\Lambda_\gamma f|_\Gamma \quad \textit{for all } f \in C^\infty(\partial\Omega) \textit{ supported in } \Gamma.$$

From this information it is possible to determine $\partial_\nu^\ell \gamma$ on Γ for any integer $\ell \geq 0$.

The precise definition of the ℓth order normal derivative $\partial_\nu^\ell \gamma$ is given below in §3.3. Note that Theorem 3.0.1 is a constructive and local result: from the knowledge of the Dirichlet-to-Neumann map on a small subset Γ of the boundary, one can constructively determine the conductivity and its normal derivatives on Γ. In particular, the following uniqueness result is an immediate corollary.

Theorem 3.0.2. *Let Ω be a bounded domain in $\mathbb{R}^n$ with smooth boundary and suppose that γ_1 and γ_2 are two positive functions in $C^\infty(\overline{\Omega})$. If*

$$\Lambda_{\gamma_1} = \Lambda_{\gamma_2}$$

then for any integer $\ell \geq 0$

$$\partial_\nu^\ell \gamma_1 = \partial_\nu^\ell \gamma_2 \quad \text{on } \partial\Omega.$$

If γ_1 and γ_2 are additionally assumed to be real-analytic, then $\gamma_1 = \gamma_2$ everywhere in $\overline{\Omega}$.

We wish to give one heuristic explanation as to why oscillating boundary data are useful in boundary determination. This explanation is based on symbol calculus for (pseudo)differential operators, which will not be used anywhere else in the book.

Recall that we are trying to determine the boundary values of the conductivity from the map

$$\Lambda_\gamma : H^{1/2}(\partial\Omega) \to H^{-1/2}(\partial\Omega).$$

As the notation suggests, this map acts like a first-order differential operator on $\partial\Omega$ in the sense that it takes away one derivative from any function that it is applied to. Pretend for the moment that $\partial\Omega = \mathbb{R}^{n-1}$. First-order differential operators on $\mathbb{R}^{n-1}$ have the form

$$A(x', D') = \sum_{j=1}^{n-1} a_j(x') D_j, \qquad x' \in \mathbb{R}^{n-1},$$

where we write $D' = (D_1, \ldots, D_{n-1})$ and $D_j = \frac{1}{i}\frac{\partial}{\partial x_j}$. This operator is characterized by its symbol, which is the function

$$a(x', \xi') = \sum_{j=1}^{n-1} a_j(x') \xi_j, \qquad x', \xi' \in \mathbb{R}^{n-1}.$$

The symbol can be obtained from the operator by testing against oscillatory functions:

$$a(x', \xi') = N^{-1} e^{-iNx'\cdot\xi'} A(x', D') e^{iNx'\cdot\xi'}, \quad N \text{ large}.$$

Now, the operator Λ_γ is not a differential operator of order one, but belongs to the more general class of *classical pseudodifferential operators* of order one. This means that the main behavior of Λ_γ is also governed by a symbol, but this time the symbol is given by an infinite asymptotic sum

$$a(x', \xi') \sim a_1(x', \xi') + a_0(x', \xi') + a_{-1}(x', \xi') + \ldots,$$

where each a_j is a smooth function that is positively homogeneous of degree j in ξ' for ξ' away from the origin. As for differential operators, the functions a_j can be obtained from the operator Λ_γ by testing against (localized) highly oscillatory functions. It turns out that one can recover $\gamma|_{\partial\Omega}$ from the principal symbol a_1, $\partial_\nu\gamma|_{\partial\Omega}$ from the next order symbol a_0, and so on. The (distributional) integral kernel $\lambda(x,y)$ of Λ_γ has an expansion corresponding to the symbol expansion,

$$\lambda(x,y) \sim \lambda_1(x,y) + \lambda_0(x,y) + \lambda_{-1}(x,y) + \ldots,$$

where $\lambda_1(x,y)$ corresponds to the strongest singularities of the kernel $\lambda(x,y)$, $\lambda_0(x,y)$ corresponds to the next strongest singularities, and so on. It turns out that from the singularities of the kernel of Λ_γ one can only recover the Taylor series of the conductivity at boundary points, but not at interior points. The values of γ at interior points are hidden in the C^∞ part of the kernel, which makes the interior uniqueness problem rather subtle.

In this chapter, instead of using the theory of pseudodifferential operators to recover the boundary values, we will employ elementary direct methods. In §3.1 and §3.2 we will show that the boundary value $\gamma|_{\partial\Omega}$ and the normal derivative $\partial_\nu\gamma|_{\partial\Omega}$ are determined by the DN map in a local and stable way. These arguments are valid also when the conductivity and the boundary have limited regularity.

The proof of Theorem 3.0.1 follows similar ideas, but is longer and requires a higher-order asymptotic construction. The proof is divided in three parts. The first step in §3.3 is to flatten the boundary near a fixed boundary point p by a suitable change of coordinates (*boundary normal coordinates*). Next, in §3.4 we construct the solutions that concentrate near the boundary point and oscillate rapidly on the flat boundary piece, the speed of oscillations depending on a large parameter $s > 0$. The third step in §3.5 is to use the boundary values ϕ_s of these solutions in the expression $\langle \Lambda_\gamma \phi_s, \bar{\phi}_s \rangle_{\partial\Omega}$, where Λ_γ is our given data and ϕ_s will be explicit oscillating functions. The Taylor series of γ can now be read off from the large s asymptotics of this expression.

We conclude this chapter by considering the anisotropic Calderón problem for matrix conductivities. In §3.6 we show that the same argument based on concentrating solutions allows us to construct the Taylor series of a smooth matrix conductivity at the boundary from the DN map, when the conductivity is written in boundary normal coordinates.

3.1. Recovering boundary values

The main result in this section states that the boundary values $\gamma|_{\partial\Omega}$ can be determined from the knowledge of the DN map Λ_γ.

Theorem 3.1.1 (Recovering γ on $\partial\Omega$). *Let Ω be a bounded open set with C^1 boundary, and let $\gamma \in C^0(\overline{\Omega})$ be positive. Given a point $x_0 \in \partial\Omega$, there exists a* *sequence of functions* $(f_M) \subset C^1(\partial\Omega)$ *for which*

$$\lim_{M\to\infty} \langle \Lambda_\gamma f_M, \bar{f}_M \rangle_{\partial\Omega} = \gamma(x_0).$$

The functions f_M *do not depend on* γ *and are supported in* $B(x_0, 1/M) \cap \partial\Omega$.

In fact, since f_M are independent of γ and are supported in small balls, the result gives a constructive method for finding $\gamma(x_0)$ from the local DN map evaluated in a small neighborhood of x_0 on $\partial\Omega$. The method also allows us to show the following stability result at the boundary.

Theorem 3.1.2 (Stability of γ on $\partial\Omega$). *Let* Ω *be a bounded open set with* C^1 *boundary, and let* $\gamma_1, \gamma_2 \in C^0(\overline{\Omega})$ *be positive. There is* $C = C(\Omega) > 0$ *such that*

$$\|\gamma_1 - \gamma_2\|_{L^\infty(\partial\Omega)} \leq C \left\| \Lambda_{\gamma_1} - \Lambda_{\gamma_2} \right\|_{H^{1/2}(\partial\Omega)\to H^{-1/2}(\partial\Omega)}.$$

Here we have used the operator norm (see also (A.1))

$$\|T\|_{X\to Y} = \sup_{\|f\|_X = 1} \|Tf\|_Y.$$

Both of these results follow immediately from the existence of special solutions to the conductivity equation that concentrate at the boundary point x_0.

Theorem 3.1.3 (Solutions concentrating at a boundary point). *Let* Ω *be a bounded open set with* C^1 *boundary, and let* $\gamma \in C^0(\overline{\Omega})$ *be positive. Given a point* $x_0 \in \partial\Omega$, *there is a sequence of solutions* $(u_M) \subset H^1(\Omega)$ *of the conductivity equation*

$$\mathrm{div}(\gamma \nabla u) = 0 \quad \textit{in } \Omega$$

such that

$$\lim_{M\to\infty} \int_\Omega \gamma \left|\nabla u_M\right|^2 dx = \gamma(x_0).$$

Further, $f_M = u_M|_{\partial\Omega}$ *are functions in* $C^1(\partial\Omega)$ *supported in* $B(x_0, 1/M) \cap \partial\Omega$ *that do not depend on* γ, *and there is* $C = C(\Omega)$ *such that*

$$\|f_M\|_{H^{1/2}(\partial\Omega)} \leq C \quad \textit{uniformly over } x_0 \in \partial\Omega \textit{ and } M \textit{ large.}$$

Proof of Theorem 3.1.1. Let u_M be as in Theorem 3.1.3. By the definition of the DN map (Theorem 2.7.3), we have

$$\langle \Lambda_\gamma f_M, \bar{f}_M \rangle_{\partial\Omega} = \int_\Omega \gamma \nabla u_M \cdot \nabla \bar{u}_M \, dx$$

and consequently

$$\lim_{M\to\infty} \langle \Lambda_\gamma f_M, \bar{f}_M \rangle_{\partial\Omega} = \lim_{M\to\infty} \int_\Omega \gamma \left|\nabla u_M\right|^2 dx = \gamma(x_0).$$ □

Proof of Theorem 3.1.2. Let u_M and v_M be solutions provided by Theorem 3.1.3 of the equations

$$\operatorname{div}(\gamma_1 \nabla u_M) = 0, \qquad \operatorname{div}(\gamma_2 \nabla v_M) = 0.$$

Since the boundary values of u_M and v_M only depend on M and $\partial\Omega$, we have $u_M|_{\partial\Omega} = v_M|_{\partial\Omega} = f_M$. Then by the definition of the DN maps,

$$\langle (\Lambda_{\gamma_1} - \Lambda_{\gamma_2}) f_M, \bar{f}_M \rangle_{\partial\Omega} = \int_\Omega \gamma_1 \left|\nabla u_M\right|^2 \, dx - \int_\Omega \gamma_2 \left|\nabla v_M\right|^2 \, dx.$$

Taking the limit as $M \to \infty$ and taking absolute values, we have

$$\begin{aligned} |\gamma_1(x_0) - \gamma_2(x_0)| &= \lim_{M\to\infty} |\langle (\Lambda_{\gamma_1} - \Lambda_{\gamma_2}) f_M, \bar{f}_M \rangle_{\partial\Omega}| \\ &\leq \lim_{M\to\infty} \left\| \Lambda_{\gamma_1} - \Lambda_{\gamma_2} \right\|_{H^{1/2}(\partial\Omega)\to H^{1/2}(\partial\Omega)} \|f_M\|^2_{H^{1/2}(\partial\Omega)}. \end{aligned}$$

Since $\|f_M\|_{H^{1/2}(\partial\Omega)}$ is bounded uniformly with respect to M and x_0, it follows that

$$\|\gamma_1 - \gamma_2\|_{L^\infty(\partial\Omega)} \leq C \left\| \Lambda_{\gamma_1} - \Lambda_{\gamma_2} \right\|_{H^{1/2}(\partial\Omega)\to H^{1/2}(\partial\Omega)}. \qquad \square$$

To prove Theorem 3.1.3, we use Theorem 2.2.2 and choose coordinates so that $x_0 = 0$ and $h : \mathbb{R}^{n-1} \to \mathbb{R}$ is a C^1 function such that for some $r > 0$,

$$\Omega = \{ x \in B(0,r) \mid x_n > h(x') \}, \qquad \partial\Omega = \{ x \in B(0,r) \mid x_n = h(x') \},$$

and moreover $h(0) = 0$ and $\nabla_{x'} h(0) = 0$. Consider the local boundary defining function

$$\rho : B(0,r) \to \mathbb{R}, \quad \rho(x) = x_n - h(x').$$

Then $\rho(0) = 0$ and $\nabla\rho(0) = e_n$. Also choose some unit tangent vector α to $\partial\Omega$ at 0, that is, $\alpha \in \mathbb{R}^n$ is a unit vector with $\alpha \cdot e_n = 0$. We wish to use oscillating boundary data $e^{iN\alpha\cdot x}$ for $x \in \partial\Omega$, where $N > 0$ is a large number, to determine the conductivity on the boundary. However, in order to focus on the value of γ at the origin, we need to multiply by a cutoff function.

We will eventually choose the boundary data to be

$$f_M = c_{M,N} \eta(Mx) e^{iN\alpha\cdot x}, \quad x \in \partial\Omega,$$

where η is a cutoff function supported in the unit ball, M and N are large numbers, and $c_{M,N}$ is a scaling constant. The boundary value f_M oscillates with period $2\pi/N$, and we should choose N so that there are many oscillations in the ball of radius $1/M$. For this reason, we will choose $N = N(M)$ such that

$$M/N = o(1) \quad \text{as } M \to \infty. \tag{3.1}$$

For example, $N(M) = M^\beta$ for $\beta > 1$ satisfies this, and we will actually fix the choice $N(M) = M^3$ in the end of the proof.

The next lemma will be useful in estimating the size of the corresponding solutions.

Lemma 3.1.4. *Let η be continuous and supported in $B(0,1)$. Then*

$$\lim_{M\to\infty} M^{n-1}N\int_\Omega \eta(Mx)e^{-2N\rho(x)}\,dx = \frac{1}{2}\int_{\mathbb{R}^{n-1}} \eta(x',0)\,dx',$$

and for M large, for some constant C only depending on η,

$$\left|\int_\Omega \eta(Mx)e^{2N\rho(x)}\,dx\right| \leq CM^{1-n}N^{-1}.$$

Proof. If M is large, changing variables $x_n = t + h(x')$ and scaling gives

$$\begin{aligned}\int_\Omega \eta(Mx)e^{-2N\rho(x)}\,dx &= \int_{\mathbb{R}^{n-1}}\int_{h(x')}^\infty \eta(Mx', Mx_n)e^{-2N(x_n-h(x'))}\,dx_n\,dx' \\ &= \int_{\mathbb{R}^{n-1}}\int_0^\infty \eta(Mx', M(t+h(x'))e^{-2Nt}\,dt\,dx' \\ &= M^{1-n}N^{-1}\int_{\mathbb{R}^{n-1}}\int_0^\infty \eta(x', \frac{M}{N}t + Mh(\frac{x'}{M}))e^{-2t}\,dt\,dx'.\end{aligned}$$

Note that

$$\lim_{M\to\infty} Mh(\frac{x'}{M}) = \lim_{s\to 0}\frac{h(sx')-h(0)}{s} = \nabla_{x'}h(0)\cdot x' = 0.$$

Since the integral over $\mathbb{R}^{n-1}$ is actually over the unit ball, dominated convergence and (3.1) imply, as $M\to\infty$, that

$$\begin{aligned}&M^{n-1}N\int_\Omega \eta(Mx)e^{-2N\rho(x)}\,dx \\ &\qquad \to \int_{\mathbb{R}^{n-1}}\int_0^\infty \eta(x',0)e^{-2t}\,dt\,dx' = \tfrac{1}{2}\int_{\mathbb{R}^{n-1}}\eta(x',0)\,dx'.\end{aligned}$$

This shows the first claim, and the second one is an immediate consequence. □

Proof of Theorem 3.1.3. The proof is elementary but somewhat long, and we will divide it in five steps.

(1) With M and N as in (3.1), we define

$$v_0(x) = \eta_M(x)h_N(x), \quad x\in\mathbb{R}^n,$$

where h_N is the complex exponential

$$h_N(x) = e^{N(i\alpha\cdot x-\rho(x))}$$

and η_M is a cutoff function

$$\eta_M(x) = \eta(Mx),$$

where $\eta\in C_c^\infty(\mathbb{R}^n)$, $0\leq\eta\leq 1$, $\eta=1$ for $|x|\leq 1/2$, and $\eta=0$ for $|x|\geq 1$. Note that $v_0\in C_c^1(\mathbb{R}^n)$ is supported in a small ball $B(0,1/M)$.

The function v_0 calls for some explanation. First of all, we will think of $\partial\Omega$ as being almost flat in small neighborhoods near 0 (this is justified since the boundary is C^1). In the case where $\partial\Omega$ is exactly flat near 0, we have $v_0(x) = \eta_M(x)e^{N(i\alpha-e_n)\cdot x}$ for M large. The exponential $e^{N(i\alpha-e_n)\cdot x}$ is harmonic in Ω:

$$\Delta(e^{N(i\alpha-e_n)\cdot x}) = N^2(i\alpha - e_n)\cdot(i\alpha - e_n)e^{N(i\alpha-e_n)\cdot x} = 0$$

since $|\alpha|^2 = 1$ and $\alpha\cdot e_n = 0$. Consequently, this exponential solves the conductivity equation $\mathrm{div}(\gamma(0)\nabla v) = 0$ with coefficient frozen at 0. Multiplying by the cutoff η_M concentrates the exponential in a small neighborhood of 0, and in this neighborhood the conductivity equation $\mathrm{div}(\gamma\nabla v) = 0$ can be approximated by the same equation with γ replaced by $\gamma(0)$. Thus, v_0 is an *approximate solution* of the conductivity equation that concentrates near the boundary point 0 and eventually allows to determine $\gamma(0)$.

(2) We establish two basic properties of v_0: as $M \to \infty$,

$$\int_\Omega |\nabla v_0|^2\,dx = O(M^{1-n}N),\text{ and} \tag{3.2}$$

$$\lim_{M\to\infty} M^{n-1}N^{-1}\int_\Omega \gamma\,|\nabla v_0|^2\,dx = c_\eta\gamma(0), \tag{3.3}$$

where $c_\eta = \int_{\mathbb{R}^{n-1}}\eta(x',0)^2\,dx'$. For the proof we first write

$$\nabla v_0 = \underbrace{N(i\alpha - \nabla\rho)\eta_M h_N}_{F_1} + \underbrace{M(\nabla\eta)(M\,\cdot)h_N}_{F_2}. \tag{3.4}$$

Lemma 3.1.4 implies that, as $M \to \infty$,

$$\begin{cases} \|F_1\|^2_{L^2(\Omega)} = O(M^{1-n}N), \\ \|F_2\|^2_{L^2(\Omega)} = O(M^{1-n}N(M/N)^2) = o(M^{1-n}N) \end{cases} \tag{3.5}$$

using that $M/N = o(1)$. This shows (3.2). The second claim follows by writing

$$\begin{aligned}\int_\Omega \gamma\,|\nabla v_0|^2\,dx &= \gamma(0)\int_\Omega |\nabla v_0|^2\,dx + \int_\Omega (\gamma-\gamma(0))\,|\nabla v_0|^2\,dx \\ &= \gamma(0)\int_\Omega |F_1|^2\,dx + \gamma(0)\int_\Omega (F_1\cdot\bar{F}_2 + \bar{F}_1\cdot F_2 + |F_2|^2)\,dx \\ &\quad + \int_\Omega (\gamma-\gamma(0))\,|\nabla v_0|^2\,dx.\end{aligned}$$

By continuity of γ we have

$$\sup_{x\in B(0,1/M)} |\gamma(x)-\gamma(0)| = o(1)\text{ as } M \to \infty. \tag{3.6}$$

Since $\mathrm{supp}(v_0) \subset B(0,1/M)$, all the terms on the right except for the first one are $o(M^{1-n}N)$ by (3.5) and Cauchy–Schwarz. Since $\nabla\rho(0) = e_n$, for the first term we have

$$\begin{aligned} M^{n-1}N^{-1}\int_\Omega |F_1|^2\,dx &= M^{1-n}N\int_\Omega \eta(Mx)^2 e^{-2N\rho(x)}(1+|\nabla\rho(x)|^2)\,dx \\ &= 2M^{1-n}N\int_\Omega \eta(Mx)^2 e^{-2N\rho}\,dx \\ &\quad + M^{1-n}N\int_\Omega \eta(Mx)^2 e^{-2N\rho}(|\nabla\rho(x)|^2-|\nabla\rho(0)|^2)\,dx. \end{aligned}$$

In the last expression, the first term has limit c_η as $M \to \infty$ by Lemma 3.1.4, and the second term is $o(1)$ since $x \mapsto |\nabla\rho(x)|^2$ is continuous near 0. We have proved (3.3).

(3) In addition to the approximate solution v_0, we will make use of the exact solution v of the conductivity equation obtained by solving the Dirichlet problem

$$\begin{cases} \operatorname{div}(\gamma\nabla v) = 0 & \text{in } \Omega, \\ v = f_0 & \text{on } \partial\Omega, \end{cases}$$

where

$$f_0 = v_0|_{\partial\Omega}.$$

Since $v_0 \in H^1(\Omega)$, we have $f_0 \in H^{1/2}(\partial\Omega)$ and the Dirichlet problem above has a unique solution $v \in H^1(\Omega)$ by Theorem 2.5.6. We also write

$$v = v_0 + v_1,$$

where $v_1 = v - v_0$ is (by Theorem 2.5.6) the unique $H^1_0(\Omega)$ solution of

$$\begin{cases} \operatorname{div}(\gamma\nabla v_1) = -\operatorname{div}(\gamma\nabla v_0) & \text{in } \Omega, \\ v_1 = 0 & \text{on } \partial\Omega. \end{cases}$$

The right-hand side $-\operatorname{div}(\gamma\nabla v_0)$ is in $H^{-1}(\Omega)$, and it acts on functions $\varphi \in H^1_0(\Omega)$ by

$$\langle -\operatorname{div}(\gamma\nabla v_0), \varphi\rangle = \int_\Omega \gamma\nabla v_0\cdot\nabla\varphi\,dx.$$

(4) Now we consider the function v_1, and claim that as $M \to \infty$

$$\int_\Omega |\nabla v_1|^2\,dx = o(M^{1-n}N). \tag{3.7}$$

This estimate justifies calling v_0 an approximate solution, since it says that the difference v_1 between the exact solution v and v_0 is asymptotically smaller than v_0. By Theorem 2.5.6, to prove (3.7) it is enough to show that

$$\|-\operatorname{div}(\gamma\nabla v_0)\|^2_{H^{-1}(\Omega)} = o(M^{1-n}N),$$

or equivalently,

$$\left|\int_\Omega \gamma\nabla v_0\cdot\nabla\varphi\,dx\right| \le o(M^{(1-n)/2}N^{1/2})\,\|\varphi\|_{H^1(\Omega)},\quad \varphi\in C_c^\infty(\Omega). \tag{3.8}$$

Let $\varphi\in C_c^\infty(\Omega)$. We begin by writing

$$\int_\Omega \gamma\nabla v_0\cdot\nabla\varphi\,dx = \gamma(0)\int_\Omega \nabla v_0\cdot\nabla\varphi\,dx + \int_\Omega(\gamma-\gamma(0))\nabla v_0\cdot\nabla\varphi\,dx.$$

Using the continuity of γ and (3.2), the second term is

$$o(M^{(1-n)/2}N^{1/2})\,\|\varphi\|_{H^1(\Omega)}.$$

For the first term, write ∇v_0 as in (3.4) and use (3.5) to obtain

$$\int_\Omega \gamma\nabla v_0\cdot\nabla\varphi\,dx = \gamma(0)\int_\Omega N(i\alpha-\nabla\rho)\eta_M h_N\cdot\nabla\varphi\,dx \\ + o(M^{(1-n)/2}N^{1/2})\,\|\varphi\|_{H^1(\Omega)}.$$

In the first term on the right write $\nabla\rho = \nabla\rho(0)+(\nabla\rho-\nabla\rho(0))$. Since $\nabla\rho$ is continuous and the integral is over $B(0,1/M)$, we get (recall that $\nabla\rho(0)=e_n$)

$$\int_\Omega \gamma\nabla v_0\cdot\nabla\varphi\,dx = \gamma(0)\int_\Omega N(i\alpha-e_n)\eta_M h_N\cdot\nabla\varphi\,dx \\ + o(M^{(1-n)/2}N^{1/2})\,\|\varphi\|_{H^1(\Omega)}.$$

Integrating by parts in the first term on the right, which is possible since φ has compact support, yields

$$\int_\Omega \gamma\nabla v_0\cdot\nabla\varphi\,dx = -\gamma(0)\int_\Omega N^2(i\alpha-e_n)\cdot(i\alpha-\nabla\rho)\eta_M h_N\varphi\,dx \\ -\gamma(0)\int_\Omega MN(i\alpha-e_n)\cdot(\nabla\eta)(M\,\cdot)h_N\varphi\,dx \\ + o(M^{(1-n)/2}N^{1/2})\,\|\varphi\|_{H^1(\Omega)}.$$

Here comes a key point in the proof: we can now use the fact that $e^{N(i\alpha-e_n)\cdot x}$ is a harmonic function, or equivalently that $(i\alpha-e_n)\cdot(i\alpha-e_n)=0$, to write

$$\int_\Omega \gamma\nabla v_0\cdot\nabla\varphi\,dx = \gamma(0)\int_\Omega N^2(i\alpha-e_n)\cdot(\nabla\rho-\nabla\rho(0))\eta_M h_N\varphi\,dx \\ -\gamma(0)\int_\Omega MN(i\alpha-e_n)\cdot(\nabla\eta)(M\,\cdot)h_N\varphi\,dx \\ + o(M^{(1-n)/2}N^{1/2})\,\|\varphi\|_{H^1(\Omega)}. \tag{3.9}$$

Recall that we want $\|\varphi\|_{H^1(\Omega)}$ on the right. For this purpose, we write

$$h_N = -\frac{1}{N}\partial_n h_N. \tag{3.10}$$

This is valid since $\partial_n \rho = 1$. Using (3.10) and integrating by parts, the second integral on the right-hand side of (3.9) becomes

$$\int_\Omega M(i\alpha - e_n) \cdot (\nabla\eta)(M\,\cdot)(\partial_n h_N)\varphi\,dx$$
$$= -\int_\Omega M(i\alpha - e_n) \cdot (\nabla\eta)(M\,\cdot)h_N \partial_n \varphi\,dx$$
$$- \int_\Omega M^2(i\alpha - e_n) \cdot (\nabla\partial_n\eta)(M\,\cdot)h_N\varphi\,dx.$$

The first integral on the right is of the form $\int_\Omega F_2 \cdot (i\alpha - e_n)\partial_n\varphi\,dx$ and is bounded by $o(M^{(1-n)/2}N^{1/2})\,\|\varphi\|_{H^1}$ by Cauchy–Schwarz and (3.5). Similarly, the second integral is bounded by $O(M^{(1-n)/2}N^{1/2}(M^2/N))\,\|\varphi\|_{L^2}$. If we make the choice

$$N(M) = M^3,$$

then this is $o(M^{(1-n)/2}N^{1/2})\,\|\varphi\|_{H^1}$.

We have proved that if $N(M) = M^3$, then

$$\int_\Omega \gamma\nabla v_0 \cdot \nabla\varphi\,dx = \int_\Omega N^2(i\alpha - e_n) \cdot (\nabla\rho - \nabla\rho(0))\eta_M h_N \varphi\,dx$$
$$+ o(M^{(1-n)/2}N^{1/2})\,\|\varphi\|_{H^1(\Omega)}.$$

Inserting (3.10) in the integral on the right and integrating by parts, this integral becomes

$$\int_\Omega N(i\alpha - e_n) \cdot (\nabla\rho - \nabla\rho(0))\eta_M h_N \partial_n\varphi\,dx$$
$$+ \int_\Omega N(i\alpha - e_n) \cdot (\nabla\rho - \nabla\rho(0))M\partial_n\eta(M\,\cdot)h_N\varphi\,dx.$$

Here we used that $\nabla\rho(x) = (-\nabla h(x'), 1)$ is independent of x_n, which justifies that one can integrate by parts with respect to x_n even though $\nabla\rho$ is only continuous. The second integral is essentially of the same form as the second integral on the right-hand side of (3.9), and the argument above shows that it is $o(M^{(1-n)/2}N^{1/2})\,\|\varphi\|_{H^1}$ (use again that $\nabla\rho$ is independent of x_n). Also the first integral is $o(M^{(1-n)/2}N^{1/2})\,\|\varphi\|_{H^1}$ by Cauchy–Schwarz, Lemma 3.1.4, and the continuity of $\nabla\rho$. This shows (3.7).

(5) We can now finish the proof of the theorem. Define

$$u_M = c_{M,N}v, \quad f_M = c_{M,N}f_0, \quad c_{M,N} = \sqrt{\frac{M^{n-1}N^{-1}}{c_\eta}}.$$

Since $f_0 = v_0|_{\partial\Omega}$, clearly f_M is in $C^1(\partial\Omega)$ and supported in $B(0, 1/M) \cap \partial\Omega$ and

$$\|f_M\|_{H^{1/2}(\partial\Omega)} \le Cc_{M,N}\,\|v_0\|_{H^1(\Omega)}.$$

We saw in (3.2) that $c_{M,N} \|\nabla v_0\|_{L^2(\Omega)} \leq C$ uniformly over M, and since the constant only depends on the choice of η and the C^1 norm of ρ it can be chosen uniform over $x_0 \in \partial\Omega$. Similarly $c_{M,N} \|v_0\|_{L^2(\Omega)} \leq C$ uniformly over M and x_0. Furthermore,

$$\int_\Omega \gamma \nabla u_M \cdot \nabla \bar{u}_M \, dx$$
$$= \frac{M^{n-1}N^{-1}}{c_\eta} \int_\Omega \gamma(|\nabla v_0|^2 + \nabla v_0 \cdot \nabla \bar{v}_1 + \nabla \bar{v}_0 \cdot \nabla v_1 + |\nabla v_1|^2)\, dx.$$

The first term satisfies by (3.3)

$$\lim_{M\to\infty} \frac{M^{n-1}N^{-1}}{c_\eta} \int_\Omega \gamma |\nabla v_0|^2 \, dx = \gamma(0).$$

The other terms may be estimated by Cauchy–Schwarz and (3.2), (3.7), so that

$$\left| \int_\Omega \gamma(\nabla v_0 \cdot \nabla \bar{v}_1 + \nabla v_0 \cdot \nabla v_1 + |\nabla v_1|^2)\, dx \right|$$
$$\leq C(\|\nabla v_0\|_{L^2(\Omega)} + \|\nabla v_1\|_{L^2(\Omega)}) \|\nabla v_1\|_{L^2(\Omega)}$$
$$= o(M^{1-n}N).$$

This proves the result. □

Remark 3.1.5. For later purposes, we make the following remarks about the proof. If Ω has C^k boundary, it is clear that the approximate solution is in $C^k(\overline{\Omega})$ and consequently $f_M \in C^k(\partial\Omega)$ and $u_M \in H^k(\Omega)$. Inspecting the proof, we have actually shown that the sequence (u_M) satisfies

$$\lim_{M\to\infty} \int_\Omega g |\nabla u_M|^2 \, dx = g(x_0)$$

for any function $g \in C^0(\overline{\Omega})$ (not just $g = \gamma$).

Exercise 3.1.6. Verify the details in Remark 3.1.5.

The end part of the proof of Theorem 3.1.3 may be simplified by using the Hardy inequality. Writing

$$\delta(x) = \operatorname{dist}(x, \partial\Omega) = \inf_{z\in\partial\Omega} |x - z|,$$

this inequality is as follows.

Theorem 3.1.7 (Hardy inequality). *Let Ω be a bounded open set with C^1 boundary. There is a constant $C > 0$ such that*

$$\|\varphi/\delta\|_{L^2(\Omega)} \leq C \|\nabla\varphi\|_{L^2(\Omega)}, \quad \varphi \in H^1_0(\Omega).$$

The following problems contain a proof of the Hardy inequality and discuss how it is used in boundary determination.

Exercise 3.1.8 (Hardy inequality on the half line). If $f \in C_c^\infty((0,\infty))$, define

$$Tf(x) = \frac{1}{x}\int_0^x f(t)\,dt, \quad x > 0.$$

For $1 < p < \infty$, prove that

$$\|Tf\|_{L^p((0,\infty))} \leq C_p \|f\|_{L^p((0,\infty))}, \quad f \in C_c^\infty((0,\infty)).$$

Exercise 3.1.9 (Hardy inequality in half space). Let

$$\mathbb{R}^n_+ = \{\, x \in \mathbb{R}^n \mid x_n > 0 \,\}.$$

If $1 < p < \infty$, prove that

$$\|u/x_n\|_{L^p(\mathbb{R}^n_+)} \leq C_p \|\partial_n u\|_{L^p(\mathbb{R}^n_+)}, \quad u \in C_c^\infty(\mathbb{R}^n_+).$$

Exercise 3.1.10. Let Ω be a bounded open set with C^1 boundary. Show that for any $x_0 \in \partial\Omega$, there is $r > 0$, a C^1 function $h : \mathbb{R}^{n-1} \to \mathbb{R}$, and a constant $c > 0$ such that $\Omega \cap B(x_0, r) = \{\, x \in B(x_0, r) \mid x_n > h(x') \,\}$ and

$$c(x_n - h(x')) \leq \delta(x) \leq x_n - h(x'), \quad x \in B(x_0, r) \cap \Omega.$$

Exercise 3.1.11. Prove Theorem 3.1.7.

Exercise 3.1.12. Consider the situation before the proof of Theorem 3.1.3, and show that if η is continuous and supported in $B(0,1)$, then

$$\left| \int_\Omega \delta(x)^k \eta(Mx) e^{-2N\rho(x)}\,dx \right| \leq CM^{1-n}N^{-k-1}.$$

Exercise 3.1.13. Give an alternative proof of Theorem 3.7, by using the Hardy inequality and Exercise 3.1.12 in the part following (3.9).

Exercise 3.1.14. In this problem we construct harmonic functions on $\mathbb{R}^n_+$ that are concentrated near the origin. We denote $x' = (x_1, \cdots, x_{n-1}) \in \mathbb{R}^{n-1}$. Let

$$P_{x_n}(x') = \frac{\Gamma(\frac{n}{2})}{\pi^{n/2}} \frac{x_n}{(x_n^2 + |x'|^2)^{n/2}}$$

be the Poisson kernel.

(a) Prove that $\Delta P_{x_n}(x') = 0$ for all $x_n > 0$. Here Δ is the Laplacian on $\mathbb{R}^n$.

(b) Prove that $\int_{\mathbb{R}^{n-1}} P_{x_n}(x')\,dx' = 1$ for all $x_n > 0$.

(c) Let $\varphi(x')$ be a bounded continuous function on $\mathbb{R}^{n-1}$. Prove that

$$\Phi(x) = \int_{\mathbb{R}^{n-1}} P_{x_n}(x' - y')\varphi(y')\,dy'$$

obeys $\Delta\Phi(x) = 0$ in $\mathbb{R}^n_+$ and

$$\lim_{\substack{x \to (z',0) \\ x_n > 0}} \Phi(x) = \varphi(z').$$

(d) Now suppose that $\varphi(x') = \partial^\alpha \psi(x')$ with $\psi \in C^{|\alpha|}(\mathbb{R}^{n-1})$ supported in the ball of radius 1 centered on the origin. Prove that there is a constant, which depends only on $|\alpha|$ and n, such that

$$|\Phi(x)| \leq \frac{C}{1+|x|^{|\alpha|+n-1}}.$$

3.2. Recovering normal derivatives

The next results imply uniqueness, reconstruction, and Hölder type stability for determining the normal derivative of γ on the boundary from the DN map.

Theorem 3.2.1 (Recovering $\partial_\nu \gamma$ on $\partial\Omega$). *Let Ω be a bounded open set with C^2 boundary, and let $\gamma \in C^1(\overline{\Omega})$ be positive. The normal derivative $\partial_\nu \gamma|_{\partial\Omega}$ can be reconstructed from the knowledge of $\Lambda_\gamma f$ for all $f \in C^2(\partial\Omega)$.*

Moreover, given a point $x_0 \in \partial\Omega$ and an open set $\Gamma \subset \partial\Omega$ containing x_0, the quantity $\partial_\nu \gamma(x_0)$ can be determined from the knowledge of $\Lambda_\gamma f|_\Gamma$ for all $f \in C^2(\partial\Omega)$ with $\mathrm{supp}(f) \subset \Gamma$.

The stability result for the normal derivative involves *a priori* bounds for the coefficients γ_j stated in terms of a constant E. Then the stability inequality will have a constant depending on E.

Theorem 3.2.2 (Stability of $\partial_\nu \gamma$ on $\partial\Omega$). *Let Ω be a bounded open set with C^2 boundary, and let $\gamma_j \in C^2(\overline{\Omega})$ for $j = 1, 2$. Let $E > 0$ be a constant so that*

$$1/E \leq \gamma_j \leq E \text{ in } \Omega,$$
$$\left\|\gamma_j\right\|_{C^2(\overline{\Omega})} \leq E.$$

Then

$$\|\partial_\nu \gamma_1 - \partial_\nu \gamma_2\|_{L^\infty(\partial\Omega)} \leq C(E) \left\|\Lambda_{\gamma_1} - \Lambda_{\gamma_2}\right\|_{H^{1/2}(\partial\Omega)\to H^{-1/2}(\partial\Omega)}^{1/3}.$$

In the previous section, we recovered the value of γ at a boundary point x_0 by using that

$$\gamma(x_0) = \lim_{M\to\infty} \int_\Omega \gamma \left|\nabla u_M\right|^2 dx = \lim_{M\to\infty} \langle \Lambda_\gamma f_M, \bar{f}_M \rangle_{\partial\Omega}.$$

Here $u_M \in H^1(\Omega)$ are special solutions to the conductivity equation which concentrate near x_0, and f_M is the boundary value of u_M. These solutions can also be used to recover the normal derivative of γ at x_0. By Remark 3.1.5, if $\gamma \in C^1(\overline{\Omega})$ and if $\alpha \in \mathbb{R}^n$ is a constant vector, then one has

$$\alpha \cdot \nabla\gamma(x_0) = \lim_{M\to\infty} \int_\Omega (\alpha \cdot \nabla\gamma) \left|\nabla u_M\right|^2 dx. \tag{3.11}$$

By choosing $\alpha = \nu(x_0)$, Theorem 3.2.1 will follow if we can somehow determine the right-hand side of the above identity from boundary measurements. This will be done by the following Rellich type identity.

Lemma 3.2.3. *Let $\Omega \subset \mathbb{R}^n$ be a bounded open set with C^2 boundary, and let $\gamma \in C^1(\overline{\Omega})$. If $u \in H^2(\Omega)$ satisfies $\operatorname{div}(\gamma \nabla u) = 0$ in Ω, then for any $\alpha \in \mathbb{R}^n$ one has*

$$\int_\Omega (\alpha \cdot \nabla \gamma) |\nabla u|^2 \, dx = \int_{\partial\Omega} (\alpha \cdot \nu)\gamma |\nabla u|^2 \, dS - 2 \int_{\partial\Omega} \mathrm{Re}((\alpha \cdot \nabla u)\gamma \overline{\partial_\nu u}) \, dS.$$

Proof. Let first $w \in C^\infty(\overline{\Omega})$. Integrating by parts, rearranging terms, and integrating by parts again gives that

$$\begin{aligned}
&\int_\Omega (\alpha \cdot \nabla \gamma) |\nabla w|^2 \, dx \\
&= \int_{\partial\Omega} (\alpha \cdot \nu)\gamma |\nabla w|^2 \, dS - \sum_{j=1}^n \int_\Omega \gamma \alpha_j \partial_j (\nabla w \cdot \overline{\nabla w}) \, dx \\
&= \int_{\partial\Omega} (\alpha \cdot \nu)\gamma |\nabla w|^2 \, dS - 2 \sum_{j=1}^n \int_\Omega \alpha_j \, \mathrm{Re}(\nabla \partial_j w \cdot \gamma \overline{\nabla w}) \, dx \\
&= \int_{\partial\Omega} (\alpha \cdot \nu)\gamma |\nabla w|^2 \, dS - 2 \sum_{j=1}^n \int_{\partial\Omega} \alpha_j \, \mathrm{Re}(\nu \partial_j w \cdot \gamma \overline{\nabla w}) \, dS \\
&\quad + 2 \sum_{j=1}^n \int_\Omega \alpha_j \, \mathrm{Re}((\partial_j w) \operatorname{div}(\gamma \overline{\nabla w})) \, dx \\
&= \int_{\partial\Omega} (\alpha \cdot \nu)\gamma |\nabla w|^2 \, dS - 2 \int_{\partial\Omega} \mathrm{Re}((\alpha \cdot \nabla w)\gamma \partial_\nu \bar{w}) \, dS \\
&\quad + 2 \int_\Omega \mathrm{Re}((\alpha \cdot \nabla w) \operatorname{div}(\gamma \nabla \bar{w})) \, dx.
\end{aligned}$$

Let now $u \in H^2(\Omega)$ solve $\operatorname{div}(\gamma \nabla u) = 0$ in Ω. Since Ω has C^2 boundary, by Lemma C.2.20 we may choose $(w_j) \subset C^\infty(\overline{\Omega})$ such that $w_j \to u$ in $H^2(\Omega)$. Multiplication by $\gamma \in C^1(\overline{\Omega})$ is a bounded operator on $H^1(\Omega)$ by Exercise 2.3.10. This shows that $\gamma \nabla w_j \to \gamma \nabla u$ in $H^1(\Omega)$ and

$$\operatorname{div}(\gamma \nabla w_j) \to \operatorname{div}(\gamma \nabla u) = 0 \text{ in } L^2(\Omega).$$

By the trace theorem $\partial_l w_j|_{\partial\Omega} \to \partial_l u|_{\partial\Omega}$ in $L^2(\partial\Omega)$ for each l. Thus, the lemma follows by applying the integral identity derived above to w_j and taking the limit as $j \to \infty$. □

Proof of Theorem 3.2.1. By Remark 3.1.5, the solutions u_M in Theorem 3.1.3 are in $H^2(\Omega)$ and their boundary values f_M are in $C^2(\partial\Omega)$. Applying

Lemma 3.2.3 to u_M, we have

$$\int_\Omega (\alpha \cdot \nabla\gamma) |\nabla u_M|^2 \, dx$$
$$= \int_{\partial\Omega} (\alpha \cdot \nu)\gamma |\nabla u_M|^2 \, dS - 2\int_{\partial\Omega} \mathrm{Re}((\alpha \cdot \nabla u_M)\gamma\overline{\partial_\nu u_M}) \, dS.$$

By the trace theorem $\partial_l u_M|_{\partial\Omega} \in L^2(\partial\Omega)$ for $1 \le l \le n$, and $\nu_l \in C^1(\partial\Omega)$ since $\partial\Omega$ is C^2. Denote the tangential part of ∇u_M on $\partial\Omega$ by

$$\nabla_T u_M = \nabla u_M - (\nabla u_M \cdot \nu)\nu\Big|_{\partial\Omega}.$$

Since $u_M|_{\partial\Omega} = f_M$, the tangential gradient of u_M on $\partial\Omega$ is just the tangential gradient of f_M. Using the pointwise expression for the DN map given in Exercise 2.7.5, we obtain

$$\nabla u_M|_{\partial\Omega} = \nabla_T u_M + (\partial_\nu u_M)\nu\Big|_{\partial\Omega} = \nabla_T f_M + \gamma^{-1}(\Lambda_\gamma f_M)\nu\Big|_{\partial\Omega}.$$

Thus we obtain

$$\int_\Omega (\alpha \cdot \nabla\gamma) |\nabla u_M|^2 \, dx = \int_{\partial\Omega} (\alpha \cdot \nu)\gamma |\nabla_T f_M|^2 \, dS + \int_{\partial\Omega} (\alpha \cdot \nu)\gamma^{-1} |\Lambda_\gamma f_M|^2 \, dS$$
$$- 2\int_{\partial\Omega} \mathrm{Re}((\alpha \cdot \nabla_T f_M)\overline{\Lambda_\gamma f_M}) \, dS - 2\int_{\partial\Omega} \mathrm{Re}((\alpha \cdot \nu)\gamma |\partial_\nu u_M|^2) \, dS.$$

This simplifies to

$$\int_\Omega (\alpha \cdot \nabla\gamma) |\nabla u_M|^2 \, dx = \int_{\partial\Omega} (\alpha \cdot \nu)\gamma |\nabla_T f_M|^2 \, dS - \int_{\partial\Omega} (\alpha \cdot \nu)\gamma^{-1} |\Lambda_\gamma f_M|^2 \, dS - 2\int_{\partial\Omega} \mathrm{Re}((\alpha \cdot \nabla_T f_M)\overline{\Lambda_\gamma f_M}) \, dS. \tag{3.12}$$

The right-hand side of (3.11) is therefore determined by the knowledge of $\Lambda_\gamma f_M$ and the restriction of γ to $\partial\Omega$. By Theorem 3.1.1 the boundary value $\gamma|_{\partial\Omega}$ can be reconstructed from the DN map Λ_γ, and it follows from (3.11) that $\alpha \cdot \nabla\gamma(x_0)$ can be reconstructed from the DN map Λ_γ for any constant vector α. The normal derivative is obtained by choosing $\alpha = \nu(x_0)$.

It remains to prove the local statement. Let $\Gamma \subset \partial\Omega$ be an open set containing x_0, and suppose that one knows $\Lambda_\gamma f|_\Gamma$ for all $f \in C^2(\partial\Omega)$ with $\mathrm{supp}(f) \subset \Gamma$. The functions f_M are supported near x_0 and (3.12) still holds, but the second term on the right-hand side of (3.12) is not fully known.

To remedy this, we suppose that $x_0 = 0$ and $B(0,r) \cap \partial\Omega \subset \Gamma$, and we choose $\psi \in C_c^\infty(\mathbb{R}^n)$ to be a function with $\psi = 1$ when $|x| \le r/2$ and $\mathrm{supp}(\psi) \subset B(0,r)$.

The argument in the proof of Lemma 3.2.3 gives

$$\begin{aligned}(3.13)\quad \int_\Omega (\alpha\cdot\nabla\gamma)\,|\nabla w|^2\,dx &= \int_{\partial\Omega}(\alpha\cdot\nu)\gamma\,|\nabla w|^2\,dS - 2\int_{\partial\Omega}\mathrm{Re}((\alpha\cdot\nabla w)\gamma\partial_\nu\bar{w})\,dS\\ &\quad + 2\int_\Omega \mathrm{Re}((\alpha\cdot\nabla w)\,\mathrm{div}(\gamma\nabla\bar{w}))\,dx\end{aligned}$$

for any $\gamma\in C^1(\overline{\Omega})$ and $w\in H^2(\Omega)$.

We apply (3.13) with γ replaced by $\psi\gamma$ and $w = u_M$. The first two terms on the right-hand side of (3.13) become integrals over Γ, and these are known since $u_M|_\Gamma$, $\partial_\nu u_M|_\Gamma$ and $\gamma|_\Gamma$ are known (the last one by Theorem 3.1.1). Since $\mathrm{div}(\gamma\nabla u_M) = 0$, for M large the last term on the right of (3.13) becomes

$$2\int_\Omega \mathrm{Re}((\alpha\cdot\nabla u_M)\gamma\nabla\psi\cdot\nabla(c_{M,N}\bar{v}_1))\,dx,$$

where $u_M = c_{M,N}(v_0 + v_1)$ as in the proof of Theorem 3.1.3. Here we used that v_0 is supported in $B(0,1/M)$ and $\nabla\psi = 0$ in this set. The correction term v_1 satisfies (3.7). By the choice of $c_{M,N}$ and by (3.2), the last term on the right of (3.13) goes to zero as $M\to\infty$. The left-hand side has limit $\alpha\cdot\nabla(\psi\gamma)(x_0) = \alpha\cdot\nabla\gamma(x_0)$ by (3.11). This proves the result. □

The proof of the stability result, Theorem 3.2.2, follows by comparing the expressions in the previous theorem for two different conductivities. To do this properly we will need certain further facts about the approximate solutions constructed in §3.1, and these facts will be proved in the exercises in the end of this section.

Proof of Theorem 3.2.2. In this proof, the constants C will only depend on Ω and E. Fix a point $x_0\in\partial\Omega$ (the constants C will also be independent of the choice of x_0), and for ease of notation assume that $x_0 = 0$. Let u_M and v_M be the solutions provided by Theorem 3.1.3 to the equations $\mathrm{div}(\gamma_1\nabla u_M) = 0$ and $\mathrm{div}(\gamma_2\nabla v_M) = 0$ in Ω. By Remark 3.1.5, these solutions are in $H^2(\Omega)$ and their boundary values satisfy $u_M|_{\partial\Omega} = v_M|_{\partial\Omega} = f_M\in C^2(\partial\Omega)$. Applying (3.12) to

u_M and v_M and subtracting the resulting expressions shows that

$$\int_\Omega (\alpha \cdot \nabla\gamma_1)\,|\nabla u_M|^2\,dx - \int_\Omega (\alpha\cdot\nabla\gamma_2)\,|\nabla v_M|^2\,dx$$
$$= \int_{\partial\Omega} (\alpha\cdot\nu)(\gamma_1-\gamma_2)\,|\nabla_T f_M|^2\,dS - \int_{\partial\Omega}(\alpha\cdot\nu)(\gamma_1^{-1}-\gamma_2^{-1})\,|\Lambda_{\gamma_1} f_M|^2\,dS$$
$$- \int_{\partial\Omega}(\alpha\cdot\nu)\gamma_2^{-1}\left[(\Lambda_{\gamma_1}f_M-\Lambda_{\gamma_2}f_M)\overline{\Lambda_{\gamma_1}f_M} + \Lambda_{\gamma_2}f_M\overline{(\Lambda_{\gamma_1}f_M-\Lambda_{\gamma_2}f_M)}\right] dS$$
$$-2\int_{\partial\Omega}\mathrm{Re}\left[(\alpha\cdot\nabla_T f_M)\overline{(\Lambda_{\gamma_1}f_M-\Lambda_{\gamma_2}f_M)}\right] dS.$$

Notice that

$$|\gamma_1^{-1}-\gamma_2^{-1}| = \left|\frac{\gamma_1-\gamma_2}{\gamma_1\gamma_2}\right| \le E^2\,|\gamma_1-\gamma_2|\,.$$

Assuming that α has unit length, we obtain

$$\left|\int_\Omega (\alpha\cdot\nabla\gamma_1)\,|\nabla u_M|^2\,dx - \int_\Omega(\alpha\cdot\nabla\gamma_2)\,|\nabla v_M|^2\,dx\right|$$
$$\le C\,\|\gamma_1-\gamma_2\|_{L^\infty(\partial\Omega)}\left[\|f_M\|_{H^1}^2 + \left\|\Lambda_{\gamma_1}f_M\right\|_{L^2}^2\right]$$
$$+ C\left\|(\Lambda_{\gamma_1}-\Lambda_{\gamma_2})f_M\right\|_{H^{-1/2}}\left[\left\|\Lambda_{\gamma_1}f_M\right\|_{H^{1/2}} + \left\|\Lambda_{\gamma_2}f_M\right\|_{H^{1/2}} + \|f_M\|_{H^{3/2}}\right].$$

Using the bounds for γ_j, we have (see Exercise 3.2.5)

$$\left\|\Lambda_{\gamma_j}f_M\right\|_{H^{1/2}} \le C\,\|f_M\|_{H^{3/2}}\,,\qquad \left\|\Lambda_{\gamma_j}f_M\right\|_{L^2} \le C\,\|f_M\|_{H^1}\,.$$

Using also Theorem 3.1.2, it follows that

$$\left|\int_\Omega (\alpha\cdot\nabla\gamma_1)\,|\nabla u_M|^2\,dx - \int_\Omega(\alpha\cdot\nabla\gamma_2)\,|\nabla v_M|^2\,dx\right|$$
$$\le C\left\|\Lambda_{\gamma_1}-\Lambda_{\gamma_2}\right\|_{H^{1/2}\to H^{-1/2}}\left[\|f_M\|_{H^1}^2 + \|f_M\|_{H^{1/2}}\,\|f_M\|_{H^{3/2}}\right].$$

Since f_M is an explicit function, we have the bounds (see Exercise 3.2.6)

$$\|f_M\|_{H^{1/2}} \le C,\quad \|f_M\|_{H^1} \le CN^{1/2},\quad \|f_M\|_{H^{3/2}} \le CN.$$

It follows that

$$\left|\int_\Omega (\alpha\cdot\nabla\gamma_1)\,|\nabla u_M|^2\,dx - \int_\Omega(\alpha\cdot\nabla\gamma_2)\,|\nabla v_M|^2\,dx\right| \le CN\left\|\Lambda_{\gamma_1}-\Lambda_{\gamma_2}\right\|_{H^{1/2}\to H^{-1/2}}.$$

Inspecting the construction in §3.1 carefully, for any $g\in C^1(\overline{\Omega})$ we have the quantitative estimate (see Exercise 3.2.7)

$$\left|\int_\Omega g\,|\nabla u_M|^2\,dx - g(0)\right| \le C(|g(0)|\,(M/N+M^{-1}) + \|\nabla g\|_{L^\infty(\Omega)}\,M^{-1}).$$

Applying this with $g = \alpha \cdot \nabla\gamma_j$, it follows that

$$|\alpha \cdot \nabla\gamma_1(0) - \alpha \cdot \nabla\gamma_2(0)| \leq C(N \left\|\Lambda_{\gamma_1} - \Lambda_{\gamma_2}\right\|_{H^{1/2}\to H^{-1/2}} + M/N + M^{-1}).$$

Let $\alpha = \nu(0)$. Choosing $N(M) = M^2$ we obtain

$$|\partial_\nu\gamma_1(0) - \partial_\nu\gamma_2(0)| \leq C(M^2 \left\|\Lambda_{\gamma_1} - \Lambda_{\gamma_2}\right\|_{H^{1/2}\to H^{-1/2}} + M^{-1}). \tag{3.14}$$

We now fix a suitable large number M, of the form

$$M = \left\|\Lambda_{\gamma_1} - \Lambda_{\gamma_2}\right\|_{H^{1/2}\to H^{-1/2}}^{s},$$

where the number $s > 0$ is optimized so that both terms on the right-hand side of (3.14) are comparable. Note that M is only large when Λ_{γ_1} and Λ_{γ_2} are sufficiently close. It is enough to consider this case, since when

$$\left\|\Lambda_{\gamma_1} - \Lambda_{\gamma_2}\right\|_{H^{1/2}\to H^{-1/2}} \geq c_0 > 0$$

the inequality in Theorem 3.2.2 already follows from the a priori bounds for γ_j.

Requiring that both terms on the right-hand side of (3.14) are comparable results in the equation

$$2s + 1 = -s,$$

or $s = -1/3$. With these choices, we have

$$|\partial_\nu\gamma_1(0) - \partial_\nu\gamma_2(0)| \leq C \left\|\Lambda_{\gamma_1} - \Lambda_{\gamma_2}\right\|_{H^{1/2}\to H^{-1/2}}^{1/3}.$$

Since this applies at any boundary point with a uniform constant C, we have proved the result. □

Remark 3.2.4. The proof above shows that under the assumptions of Theorem 3.2.2, for any unit vector $\alpha \in \mathbb{R}^n$ one has the estimate

$$\|\alpha \cdot \nabla\gamma_1 - \alpha \cdot \nabla\gamma_2\|_{L^\infty(\partial\Omega)} \leq C(E) \left\|\Lambda_{\gamma_1} - \Lambda_{\gamma_2}\right\|_{H^{1/2}(\partial\Omega)\to H^{-1/2}(\partial\Omega)}^{1/3}.$$

Exercise 3.2.5. Let $\Omega \subset \mathbb{R}^n$ be a bounded open set, and let $\gamma \in L^\infty(\Omega)$ be such that $1/E \leq \gamma \leq E$ a.e. in Ω for some $E > 0$. Show that for any $f \in H^{1/2}(\partial\Omega)$ the unique solution of $\operatorname{div}(\gamma\nabla u) = 0$ in Ω with $u|_{\partial\Omega} = f$ satisfies

$$\|u\|_{H^1(\Omega)} \leq C(E, \Omega) \|f\|_{H^{1/2}(\partial\Omega)}.$$

Show also that

$$\left\|\Lambda_\gamma f\right\|_{H^{-1/2}(\partial\Omega)} \leq C(E, \Omega) \|f\|_{H^{1/2}(\partial\Omega)}.$$

Moreover, if Ω has C^2 boundary and $\|\gamma\|_{C^1(\overline{\Omega})} \leq E$ and $f \in H^{3/2}(\partial\Omega)$, show that

$$\|u\|_{H^2(\Omega)} \leq C(E, \Omega) \|f\|_{H^{3/2}(\partial\Omega)}$$

and

$$\left\|\Lambda_\gamma f\right\|_{H^{1/2}(\partial\Omega)} \leq C(E,\Omega)\,\|f\|_{H^{3/2}(\partial\Omega)}.$$

Use also interpolation (as in Lemma C.3.2) to show that

$$\left\|\Lambda_\gamma f\right\|_{H^{s-1}(\partial\Omega)} \leq C(E,\Omega)\,\|f\|_{H^{s}(\partial\Omega)}, \quad 1/2 \leq s \leq 3/2.$$

Exercise 3.2.6. Let $\Omega \subset \mathbb{R}^n$ be a bounded open set with C^2 boundary. Show that the function f_M satisfies

$$\|f_M\|_{H^s(\partial\Omega)} \leq C(\Omega)N^{s-1/2}, \quad 1/2 \leq s \leq 3/2.$$

Exercise 3.2.7. Let $\Omega \subset \mathbb{R}^n$ be a bounded open set with C^2 boundary. Show that there is $C > 0$ such that for any $g \in C(\overline{\Omega})$ with modulus of continuity ω, meaning that

$$|g(x) - g(y)| \leq \omega(|x-y|), \quad x, y \in \overline{\Omega},$$

one has

$$\left|\int_\Omega g\,|\nabla u_M|^2\,dx - g(0)\right| \leq C(|g(0)|\,(M/N + M^{-1}) + \omega(M^{-1})).$$

3.3. Boundary normal coordinates

To recover higher-order derivatives of the conductivity, it will be useful to reduce to the case where the boundary is flat. Flattening the boundary can be carried out in different ways, and typically the scalar conductivity is transformed into a matrix conductivity in the process. Here it is convenient to choose the change of coordinates so that the new conductivity matrix has special form.

The relevant coordinates are called *boundary normal coordinates*, since (as may be seen from the proof) the point $F^{-1}(y', y_n)$ is obtained by choosing a boundary point according to y' and then moving y_n units in the direction of the inner unit normal.

Proposition 3.3.1 (Boundary normal coordinates). *Let Ω be a bounded domain with C^∞ boundary in $\mathbb{R}^n$, and let $p \in \partial\Omega$. There is a C^∞ diffeomorphism $F : U \to V$ between open sets of $\mathbb{R}^n$ where U is a neighborhood of p and V is a neighborhood of 0, such that*

$$F(p) = 0, \quad F(\Omega \cap U) = V \cap \{y_n > 0\}, \quad F(\partial\Omega \cap U) = V \cap \{y_n = 0\},$$

and further for any $y \in V \cap \{y_n \geq 0\}$,

$$(DF)(DF)^t|_{F^{-1}(y)} = \begin{bmatrix} h(y) & 0 \\ 0 & 1 \end{bmatrix}$$

for some C^∞ symmetric positive definite matrix $h(y) = \left[h^{\alpha,\beta}(y)\right]_{\alpha,\beta=1}^{n-1}$.

Proof. Since Ω has smooth boundary, there is a system of coordinates where $p = 0$ and where

$$\Omega \cap B(0,2r) = \{\, x \in B(0,2r) \mid x_n > h(x') \,\}$$

for some C^∞ map $h : \mathbb{R}^{n-1} \to \mathbb{R}$ with $h(0) = 0$. Write $q(y') = (y', h(y'))$, and define a map $\Phi : B(0,r) \to \mathbb{R}^n$ by

$$\Phi(y', y_n) = q(y') - y_n \nu(q(y')).$$

This is a C^∞ map in some neighborhood of 0, and its Jacobian matrix is given in terms of columns by

$$\begin{aligned} &D\Phi(y', y_n) \\ (3.15)\qquad &= \Big[\partial_1 q - y_n \partial_1\big(\nu(q(y'))\big), \; \dots, \; \partial_{n-1} q - y_n \partial_{n-1}\big(\nu(q(y'))\big), -\nu(q(y'))\Big]. \end{aligned}$$

Thus $D\Phi(0,0) = [\partial_1 q(0), \dots, \partial_{n-1} q(0), -\nu(0)]$. Since q parametrizes $\partial\Omega$, the vectors $\{\partial_1 q(0), \dots, \partial_{n-1} q(0)\}$ form a basis for all the tangent vectors to $\partial\Omega$ at 0. Consequently $D\Phi(0,0)$ is invertible. By the inverse function theorem, there is a neighborhood V of 0 and a neighborhood U of 0 such $\Phi : V \mapsto U$ is a diffeomorphism.

We set $F = \Phi^{-1} : U \to V$. This is a diffeomorphism with $F(0) = 0$, and $F(\Omega \cap U)$ coincides with $V \cap \{y_n > 0\}$ since ν was the unit outer normal. By (3.15) we have

$$D\Phi(y)^t D\Phi(y) = \begin{bmatrix} g(y) & \upsilon(y) \\ \upsilon(y)^t & 1 \end{bmatrix},$$

where $g(y)$ is some smooth $(n-1) \times (n-1)$ matrix. But $\upsilon(y) = 0$ since

$$\begin{aligned} \upsilon_j(y) &= \big[\partial_j q(y') - y_n \partial_j(\nu(q(y')))\big] \cdot \big[-\nu(q(y'))\big] \\ &= \frac{y_n}{2} \partial_j \big[\nu(q(y')) \cdot \nu(q(y'))\big] \\ &= 0 \end{aligned}$$

using that $\partial_j q(y')$ is tangent to $\partial\Omega$ and $|\nu| = 1$. This concludes the proof since $D\Phi(y) = DF(F^{-1}(y))^{-1}$, and thus $(DF)(DF)^t|_{F^{-1}(y)}$ has the required form where $h(y) = g(y)^{-1}$ is positive definite because $(DF)(DF)^t$ is. □

We now give a precise definition of higher-order normal derivatives $\partial_\nu^\ell \gamma$ at the boundary appearing in Theorem 3.0.1.

Definition 3.3.2 (Higher-order normal derivatives). Let Ω be a bounded domain with C^∞ boundary, let $\gamma \in C^\infty(\overline{\Omega})$, and let $p \in \partial\Omega$. For $\ell \geq 0$ we define

$$\partial_\nu^\ell \gamma(p) = \left(\frac{\partial}{\partial \nu}\right)^\ell \gamma(p) = \partial_{y_n}^\ell (\gamma(F^{-1}(y)))|_{y=0},$$

where F is as in Proposition 3.3.1.

An alternative interpretation of $\partial_\nu^\ell \gamma$, which shows that the above definition is independent of the choice of F, can be given as follows. Let γ be any C^∞ function that is defined in a neighborhood of $\partial\Omega$. For each point x, which is sufficiently close to $\partial\Omega$, there is a unique point $\pi(x) \in \partial\Omega$ that is nearest to x. The vector from x to $\pi(x)$ is normal to $\partial\Omega$ at $\pi(x)$. See Exercise 3.3.3. Let $\hat{n}(x)$ be a unit vector that is parallel to the vector from x to $\pi(x)$ and points from inside Ω to outside Ω. In the event that $x \in \partial\Omega$, so that $\pi(x) = x$, choose $\hat{n}(x)$ to be the unit outward normal to $\partial\Omega$ at x. The vector $\hat{n}(x) = \hat{n}\big(\pi(x)\big)$ is a C^∞ function of x in a neighborhood of $\partial\Omega$. Again, see Exercise 3.3.3. We define $\partial_\nu\gamma(x) = \hat{n}(x) \cdot \nabla\gamma(x)$, for all x in a neighborhood of $\partial\Omega$. Then we may define $\partial_\nu^\ell\gamma$ by ℓ applications of ∂_ν.

Exercise 3.3.3. Let $p \in \partial\Omega$. Let $x'(\xi')$ be a C^∞ parametrization of a neighborhood of p in $\partial\Omega$ with $x'(0) = p$. Denote by $\hat{n}(x')$ the unit outward normal to $\partial\Omega$ at $x' \in \partial\Omega$. Define $x(\xi', \xi_n) = x'(\xi') - \xi_n\hat{n}\big(x'(\xi')\big)$.

(a) Prove that $x(\xi', \xi_n)$ is a C^∞ diffeomorphism from a neighborhood of $0 \in \mathbb{R}^n$ to a neighborhood of $p \in \mathbb{R}^n$.

(b) Prove that, for all sufficiently small (ξ', ξ_n), $x'(\xi')$ is the point of $\partial\Omega$ that is nearest $x(\xi', \xi_n)$, so that the distance from $x(\xi', \xi_n)$ to $\partial\Omega$ is $|\xi_n|$.

Exercise 3.3.4. Show that Definition 3.3.2 is independent of the choice of F.

We will next determine how the conductivity operator

$$L_\gamma u = \nabla \cdot (\gamma \nabla u)$$

transforms under F. The idea is that solutions to $L_\gamma u = 0$ transform into solutions to $L_{\tilde{\gamma}}\tilde{u} = 0$ by the rule $\tilde{u} = u \circ F^{-1}$, where $\tilde{\gamma}$ is a certain matrix conductivity given below. However, since the change of coordinates is only defined near p we need to restrict our attention to functions defined in a neighborhood of p.

Lemma 3.3.5. *Let Ω, p, and $F : U \to V$ be as in Proposition* 3.3.1, *and let $\gamma \in C^\infty(\overline{\Omega})$ be a positive function. Assume that $v \in C^\infty(U \cap \overline{\Omega})$ and let $\tilde{v}(y) = v(F^{-1}(y))$ for $y \in V \cap \{y_n \geq 0\}$. If $\varphi \in C^\infty(\overline{\Omega})$ is supported in $U \cap \overline{\Omega}$ and if $\tilde{\varphi}(y) = \varphi(F^{-1}(y))$, then*

$$\int_\Omega \gamma \nabla v \cdot \nabla\varphi \, dx = \int_\Omega \tilde{\gamma} \nabla\tilde{v} \cdot \nabla\tilde{\varphi} \, dy,$$

where

$$\tilde{\gamma}(y) = \begin{bmatrix} ch & 0 \\ 0 & c \end{bmatrix}.$$

Here $h = \left[h^{\alpha,\beta}\right]_{\alpha,\beta=1}^{n-1}$ is the matrix in Proposition 3.3.1 *and*

$$c(y) = \frac{\gamma(F^{-1}(y))}{|\det DF(F^{-1}(y))|}.$$

Moreover, if $\varphi \in C_c^\infty(U \cap \Omega)$, *then*

$$\int_\Omega (L_\gamma v)\varphi\, dx = \int_\Omega (L_{\tilde\gamma}\tilde v)\tilde\varphi\, dy$$

and consequently

$$L_\gamma v(x) = |\det DF(x)|(L_{\tilde\gamma}\tilde v)(F(x)) \text{ in } U \cap \Omega.$$

Proof. Since φ is supported in $U \cap \overline{\Omega}$, we can make the change of coordinates $x = F^{-1}(y)$ to obtain

$$\begin{aligned}&\int_{U\cap\Omega} \gamma(x)\nabla v(x)\cdot\nabla\varphi(x)\, dx\\ &\quad= \int_{V\cap\{y_n>0\}} \gamma(F^{-1}(y))\nabla v(F^{-1}(y))\cdot\nabla\varphi(F^{-1}(y))|\det D(F^{-1})(y)|\, dy.\end{aligned}$$

The chain rule implies $Dv(x) = D(\tilde v \circ F)(x) = D\tilde v(F(x))DF(x)$, thus

$$\nabla v(x) = DF(x)^t \nabla\tilde v(F(x)). \tag{3.16}$$

Using the analogous result for φ, we have

$$\begin{aligned}&\int_{U\cap\Omega} \gamma(x)\nabla v(x)\cdot\nabla\varphi(x)\, dx\\ &= \int_{V\cap\{y_n>0\}} \frac{\gamma(F^{-1}(y))DF(F^{-1}(y))^t\nabla\tilde v(y)\cdot DF(F^{-1}(y))^t\nabla\tilde\varphi(y)}{|\det DF(F^{-1}(y))|}\, dy\\ &= \int_{V\cap\{y_n>0\}} \frac{(DF)\gamma(DF)^t}{|\det DF|}\Big|_{F^{-1}(y)} \nabla\tilde v(y)\cdot\nabla\tilde\varphi(y)\, dy\\ &= \int_{V\cap\{y_n>0\}} \tilde\gamma(y)\nabla\tilde v(y)\cdot\nabla\tilde\varphi(y)\, dy,\end{aligned}$$

where $\tilde\gamma$ is as required. Finally, if $\varphi \in C_c^\infty(U \cap \Omega)$, then

$$\int_\Omega (L_\gamma v)\varphi\, dx = -\int_{U\cap\Omega} \gamma\nabla v\cdot\nabla\varphi\, dx = -\int_{V\cap\{y_n>0\}} \tilde\gamma\nabla\tilde v\cdot\nabla\tilde\varphi\, dy = \int_\Omega (L_{\tilde\gamma}\tilde v)\tilde\varphi\, dy.$$

Changing variables $y = F(x)$ in the last integral and varying φ proves that $L_\gamma v(x) = |\det DF(x)|(L_{\tilde\gamma}\tilde v)(F(x))$. □

The same argument also proves the diffeomorphism invariance of the DN map for anisotropic conductivities discussed in §1.1.

Lemma 3.3.6 (Diffeomorphism invariance of DN map). *Let* $\Omega \subset \mathbb{R}^n$ *be a bounded open set with smooth boundary, and let* $F : \overline{\Omega} \to \overline{\Omega}$ *be a* C^∞ *diffeomorphism (i.e., F is bijective,* C^∞ *in* Ω*, and both F and* F^{-1} *extend as* C^∞ *maps* *to $\overline{\Omega}$) with $F|_{\partial\Omega} = \mathrm{Id}$. Let $\gamma = [\gamma^{j,k}]_{j,k=1}^n$ be a positive definite symmetric matrix with entries in $C^\infty(\overline{\Omega})$. Then*

$$\begin{cases} \nabla\cdot(\gamma\nabla u) = 0 & \text{in } \Omega, \\ u = f & \text{on } \partial\Omega \end{cases} \iff \begin{cases} \nabla\cdot(\tilde{\gamma}\nabla \tilde{u}) = 0 & \text{in } \Omega, \\ \tilde{u} = f & \text{on } \partial\Omega, \end{cases}$$

where $\tilde{u} = u\circ F^{-1}$, and $\tilde{\gamma} = F_*\gamma$ is the push-forward conductivity in $\overline{\Omega}$ given by

$$\tilde{\gamma}(y) = \frac{1}{|\det DF|}(DF)\gamma(DF)^t\Big|_{F^{-1}(y)}.$$

Moreover,

$$\Lambda_{\tilde{\gamma}} = \Lambda_\gamma.$$

Proof. Let $u, v \in H^1(\Omega)$ and $\tilde{u} = u\circ F^{-1}$, $\tilde{v} = v\circ F^{-1}$. One has $\tilde{u}, \tilde{v} \in H^1(\Omega)$ by Lemma C.1.48. The argument after (3.16) shows that

$$\int_\Omega \gamma\nabla u\cdot\nabla v\,dx = \int_\Omega \tilde{\gamma}\nabla\tilde{u}\cdot\nabla\tilde{v}\,dy. \tag{3.17}$$

Let $f \in H^{1/2}(\partial\Omega)$ and let $u_f \in H^1(\Omega)$ solve

$$\nabla\cdot(\gamma\nabla u_f) = 0 \text{ in } \Omega, \qquad u_f|_{\partial\Omega} = f.$$

Using (3.17) with $\tilde{v} \in H_0^1(\Omega)$, so $v = \tilde{v}\circ F \in H_0^1(\Omega)$, shows that the function $\tilde{u}_f = u_f\circ F^{-1}$ is a weak solution of $\nabla\cdot(\tilde{\gamma}\nabla\tilde{u}_f) = 0$ in Ω. We also have $\tilde{u}_f|_{\partial\Omega} = f$ since $F|_{\partial\Omega} = \mathrm{Id}$. The same argument works also in the converse direction, which proves the stated equivalence.

Now let $f, h \in H^{1/2}(\partial\Omega)$, let u_f be as above, and let $e_h \in H^1(\Omega)$ be any function with $e_h|_{\partial\Omega} = h$. From Theorem 2.7.3 (with $G = \gamma$) and (3.17) we obtain

$$\langle\Lambda_\gamma f, h\rangle_{\partial\Omega} = \int_\Omega \gamma\nabla u_f\cdot\nabla e_h\,dx = \int_\Omega \tilde{\gamma}\nabla\tilde{u}_f\cdot\nabla\tilde{e}_h\,dy.$$

Since $\nabla\cdot(\tilde{\gamma}\nabla\tilde{u}_f) = 0$ in Ω with $\tilde{u}_f|_{\partial\Omega} = f$, applying Theorem 2.7.3 again gives that

$$\langle\Lambda_\gamma f, h\rangle_{\partial\Omega} = \langle\Lambda_{\tilde{\gamma}} f, h\rangle_{\partial\Omega}.$$

The functions f and h were arbitrary, which proves that $\Lambda_\gamma = \Lambda_{\tilde{\gamma}}$. □

3.4. Concentrating solutions

In this section we will give a higher-order construction of concentrating solutions that can be used for determining higher-order normal derivatives from the DN map. We will give the construction in a form that can later be employed also for anisotropic conductivities in §3.6.

Using the change of coordinates in Proposition 3.3.1, we may assume that we are working in a domain $\tilde{\Omega}$ which is flat near 0 and for some $r > 0$ one has

$$B(0,2r) \cap \tilde{\Omega} = B(0,2r) \cap \{y_n > 0\}.$$

We will determine the conductivity in the set

$$\tilde{\Gamma} = B(0,r) \cap \{y_n = 0\}.$$

Below, we will also think of $\tilde{\Gamma}$ as a subset of $\mathbb{R}^{n-1}$. Motivated by Lemma 3.3.5, we consider a matrix conductivity $\tilde{\gamma}$ having the form

$$\tilde{\gamma}(y) = \begin{bmatrix} ch & 0 \\ 0 & c \end{bmatrix}, \tag{3.18}$$

where $h(y) = (h^{\alpha,\beta}(y))_{\alpha,\beta=1}^{n-1}$ is a symmetric positive definite matrix and $c(y)$ is a positive scalar function, both depending smoothly on y in the set $B(0,2r) \cap \{y_n \geq 0\}$. We assume that the matrix h is known, and that c is an unknown function which needs to be determined.

We construct approximate solutions to the conductivity equation in $\tilde{\Omega}$ which are supported in a small neighborhood of $\tilde{\Gamma}$ and oscillate rapidly on $\tilde{\Gamma}$, the speed of oscillation depending on a large parameter $s > 0$. To recover high order derivatives of the conductivity, we need to do an asymptotic construction in terms of powers of s. The proof is rather long but quite elementary.

We will use the following result several times below.

Exercise 3.4.1 (Leibniz rule). If $u(t)$ and $v(t)$ are C^m functions on some real interval I, then

$$\left(\frac{d}{dt}\right)^m (u(t)v(t)) = u^{(m)}(t)v(t) + u(t)v^{(m)}(t) + \sum_{j=1}^{m-1} \binom{m}{j} u^{(j)}(t)v^{(m-j)}(t), \tag{3.19}$$

where $\binom{m}{j} = \frac{m!}{j!(m-j)!}$ is the binomial coefficient.

Note that the approximate solutions in the next result are exponentially decaying in $\tilde{\Omega}$. They are related to the exponentially growing solutions which are used later in Chapter 4 for proving uniqueness results for the Calderón problem in the interior.

Proposition 3.4.2. *Let $N > 0$, let t' be a unit vector in $\mathbb{R}^{n-1}$, and assume that $\eta \in C^\infty(\mathbb{R}^{n-1})$ is a compactly supported function in $\tilde{\Gamma}$. For some small $\delta > 0$ and for any $s \geq 1$, there is a function $\tilde{v}_s \in C^\infty(\overline{\tilde{\Omega}})$ satisfying*

$$\tilde{v}_s(y', 0) = e^{isy' \cdot t'}\eta(y') \quad \text{on} \quad \tilde{\Gamma}, \qquad \mathrm{supp}(\tilde{v}_s) \subset \tilde{\Gamma} \times [0, \delta], \tag{3.20}$$

and

$$\|\tilde{v}_s\|_{H^1(\tilde{\Omega})} \leq Cs^{1/2}, \tag{3.21}$$

$$\left\|L_{\tilde{\gamma}}\tilde{v}_s\right\|_{L^2(\tilde{\Omega})} \leq Cs^{-N+3/2}, \tag{3.22}$$

where the constant C is independent of s. Further, this function has the form

$$\tilde{v}_s = e^{s\Phi}(a_0 + s^{-1}a_{-1} + \ldots + s^{-N}a_{-N}), \tag{3.23}$$

where Φ is a smooth complex function satisfying for some $\sigma > 0$

$$\begin{aligned} \Phi(y',0) &= iy' \cdot t' && \text{for } y' \in \tilde{\Gamma}, \\ \partial_n\Phi(y',0) &= f_1(y') && \text{for } y' \in \tilde{\Gamma}, \\ \mathrm{Re}(\Phi(y',y_n)) &\leq -\sigma y_n && \text{for } y \in \tilde{\Gamma} \times [0,\delta]. \end{aligned}$$

Here

$$f_1(y') = -\left(\sum_{\alpha,\beta=1}^{n-1} h^{\alpha,\beta}(y',0)t_\alpha t_\beta\right)^{1/2}.$$

Moreover, a_0, a_{-1}, …, a_{-N} are smooth complex functions independent of s and supported in the set $\tilde{\Gamma} \times [0,\delta]$ and they satisfy

$$\begin{aligned} a_0(y',0) &= \eta(y') && \text{for } y' \in \tilde{\Gamma}, \\ a_{-l}(y',0) &= 0 && \text{for } y' \in \tilde{\Gamma} \text{ and } l \geq 1. \end{aligned}$$

Proof. We try to find an approximate solution of the equation $L_{\tilde{\gamma}}\tilde{v} = 0$ having the form $\tilde{v}_s = e^{s\Phi}a$. One has the identity

$$\partial_j(e^{s\Phi}w) = e^{s\Phi}(\partial_j + s\partial_j\Phi)w.$$

Using this identity, the function $\tilde{v}_s$ satisfies

$$\begin{aligned} L_{\tilde{\gamma}}\tilde{v}_s &= \sum_{j,k=1}^{n} \partial_j(\tilde{\gamma}^{j,k}\partial_k\tilde{v}_s) \\ &= \sum_{j,k=1}^{n} e^{s\Phi}(\partial_j + s\partial_j\Phi)(\tilde{\gamma}^{j,k}(\partial_k + s\partial_k\Phi)a) \\ &= e^{s\Phi}\Big\{s^2\Big[\Big(\sum_{j,k=1}^{n} \tilde{\gamma}^{j,k}\partial_j\Phi\partial_k\Phi\Big)a\Big] \\ &\quad + s\Big[\sum_{j,k=1}^{n} \Big(2\tilde{\gamma}^{j,k}\partial_j\Phi\partial_k a + \partial_j(\tilde{\gamma}^{j,k}\partial_k\Phi)\Big)a\Big] + [L_{\tilde{\gamma}}a]\Big\}. \end{aligned} \tag{3.24}$$

Note that we have grouped the terms corresponding to different powers of s. The idea is to choose Φ and a so that the terms involving the largest powers of s are small, finally resulting in the estimate (3.22). Since we are only interested in finding approximate solutions in a small neighborhood of $\tilde{\Gamma}$, *it is sufficient* *to arrange that the terms vanish to high order on $\tilde{\Gamma}$ instead of vanishing in a full neighborhood of $\tilde{\Gamma}$.*

Finding Φ.

(1) Looking at the s^2 term in (3.24), the first task is to find a complex function Φ which satisfies

$$\partial_n^j\Big(\sum_{j,k=1}^{n} \tilde{\gamma}^{j,k}\partial_j\Phi\partial_k\Phi\Big)|_{\tilde{\Gamma}} = 0 \qquad \textit{for } j = 0, 1, \ldots, N-1. \tag{3.25}$$

We will look for Φ in the form

$$\Phi(y', y_n) = f_0(y') + y_n f_1(y') + \frac{y_n^2}{2} f_2(y') + \ldots + \frac{y_n^N}{N!} f_N(y'), \tag{3.26}$$

where $f_j(y') = \partial_n^j \Phi(y', 0)$ are functions to be determined. To ensure the boundary condition (3.20), we choose

$$f_0(y') = iy' \cdot t'.$$

(2) Next we find $f_1(y') = \partial_n \Phi(y', 0)$ so that (3.25) is satisfied for $j = 0$. Using the special form of $\tilde{\gamma}$ given in (3.18), we need that

$$c(\partial_n\Phi)^2 + c\sum_{\alpha,\beta=1}^{n-1} h^{\alpha,\beta}\partial_\alpha\Phi\partial_\beta\Phi\Big|_{\tilde{\Gamma}} = 0.$$

Since $\partial_\alpha\Phi|_{\tilde{\Gamma}} = it_\alpha$, this reduces to

$$f_1(y')^2 - \sum_{\alpha,\beta=1}^{n-1} h^{\alpha,\beta}(y')t_\alpha t_\beta = 0.$$

Recalling that we want $\mathrm{Re}(\Phi) \leq 0$ in $\tilde{\Gamma} \times [0, \delta]$, it makes sense to require that $f_1 = \partial_n\Phi|_{\tilde{\Gamma}}$ is negative. Thus, we make the choice

$$f_1(y') = -\left(\sum_{\alpha,\beta=1}^{n-1} h^{\alpha,\beta}(y', 0)t_\alpha t_\beta\right)^{1/2}.$$

Then f_1 is smooth and negative on $\tilde{\Gamma}$ since the matrix $(h^{\alpha,\beta}(y', 0))$ is positive definite and t' is a unit vector.

(3) We continue with (3.25) for $j = 1$ and try to find f_2 such that

$$\partial_n\Big(\sum_{j,k=1}^{n} \tilde{\gamma}^{j,k}\partial_j\Phi\partial_k\Phi\Big)\Big|_{\tilde{\Gamma}} = 0.$$

Since $(\partial_n\Phi)^2 + \sum_{\alpha,\beta=1}^{n-1} h^{\alpha,\beta}\partial_\alpha\Phi\partial_\beta\Phi|_{\tilde\Gamma} = 0$ by (3.25) for $j = 0$, we have

$$\partial_n\Big(\sum_{j,k=1}^{n} \tilde\gamma^{j,k}\partial_j\Phi\partial_k\Phi\Big)\Big|_{\tilde\Gamma} = c\partial_n\Big((\partial_n\Phi)^2 + \sum_{\alpha,\beta=1}^{n-1} h^{\alpha,\beta}\partial_\alpha\Phi\partial_\beta\Phi\Big)\Big|_{\tilde\Gamma}$$
$$= c\Big(2f_1\partial_n^2\Phi - \sum_{\alpha,\beta=1}^{n-1} (\partial_n h^{\alpha,\beta})t_\alpha t_\beta + T_0(h)\Big)\Big|_{\tilde\Gamma},$$

where $T_0(h)$ is a quantity only depending on t' and on y'-derivatives of $h^{\alpha,\beta}(y',0)$. The last expression vanishes if we choose $f_2 = \partial_n^2\Phi|_{\tilde\Gamma}$ as

$$f_2 = \frac{1}{2f_1}\sum_{\alpha,\beta=1}^{n-1} (\partial_n h^{\alpha,\beta})t_\alpha t_\beta - \frac{1}{2f_1}T_0(h).$$

(4) Let f_0 and f_1 be as above, and suppose that we have found $f_2, \dots, f_m$ such that[1]

$$\partial_n^j\Big(\sum_{j,k=1}^{n} \tilde\gamma^{j,k}\partial_j\Phi\partial_k\Phi\Big)\Big|_{\tilde\Gamma} = 0 \qquad \textit{for } 0 \le j \le m-1, \tag{3.27}$$

$$f_j = \frac{1}{2f_1}\sum_{\alpha,\beta=1}^{n-1} (\partial_n^{j-1} h^{\alpha,\beta})t_\alpha t_\beta + T_{j-2}(h) \qquad \textit{for } 2 \le j \le m, \tag{3.28}$$

where $T_l(h)$ is an expression only depending on t' and on y'-derivatives of $\partial_n^k h^{\alpha,\beta}(y',0)$ for $0 \le k \le l$. We wish to find f_{m+1} of the form (3.28) such that (3.27) is valid also for $j = m$. Since

$$\sum_{j,k=1}^{n} \tilde\gamma^{j,k}\partial_j\Phi\partial_k\Phi = c\Big((\partial_n\Phi)^2 + \sum_{\alpha,\beta=1}^{n-1} h^{\alpha,\beta}\partial_\alpha\Phi\partial_\beta\Phi\Big),$$

the identity (3.19) and the hypothesis (3.27) imply that

$$\partial_n^j\Big((\partial_n\Phi)^2 + \sum_{\alpha,\beta=1}^{n-1} h^{\alpha,\beta}\partial_\alpha\Phi\partial_\beta\Phi\Big)\Big|_{\tilde\Gamma} = 0 \qquad \textit{for } 0 \le j \le m-1.$$

Consequently, by (3.19) again and by (3.28),

$$\partial_n^m\Big(\sum_{j,k=1}^{n} \tilde\gamma^{j,k}\partial_j\Phi\partial_k\Phi\Big)\Big|_{\tilde\Gamma} = c\partial_n^m\Big((\partial_n\Phi)^2 + \sum_{\alpha,\beta=1}^{n-1} h^{\alpha,\beta}\partial_\alpha\Phi\partial_\beta\Phi\Big)\Big|_{\tilde\Gamma}$$
$$= c\Big(2f_1 f_{m+1} - \sum_{\alpha,\beta=1}^{n-1} (\partial_n^m h^{\alpha,\beta})t_\alpha t_\beta + T_{m-1}(h)\Big).$$

[1] The precise form of f_j is actually not needed. The only information that we need is that f_j exists and is independent of c.

We can make the last expression vanish by choosing f_{m+1} as

$$f_{m+1} = \frac{1}{2f_1} \sum_{\alpha,\beta=1}^{n-1} (\partial_n^m h^{\alpha,\beta}) t_\alpha t_\beta + T_{m-1}(h).$$

This is of the form (3.28) as required.

(5) We use the previous steps to inductively find f_0, f_1, ..., f_N so that Φ given by (3.26) satisfies the condition (3.25). Further, we have $\Phi(y', 0) = iy' \cdot t'$, and

$$\operatorname{Re}(\Phi(y',0)) = y_n \Big[f_1(y') + \operatorname{Re}\Big(\frac{y_n}{2} f_2(y') + \ldots + \frac{y_n^{N-1}}{N!} f_N(y') \Big) \Big]$$

If $\sigma > 0$ is such that $f_1|_{\tilde{\Gamma}} \leq -2\sigma$, we can find $\delta > 0$ such that

$$\left| \frac{y_n}{2} f_2(y') + \ldots + \frac{y_n^{N-1}}{N!} f_N(y') \right| \leq \sigma \quad \text{for } y \in \tilde{\Gamma} \times [0, \delta]$$

Then, for $y \in \tilde{\Gamma} \times [0, \delta]$ we have $\operatorname{Re}(\Phi(y)) \leq -\sigma y_n$. This completes the construction of Φ.

Finding a.

(1) We have constructed a function Φ so that the s^2 term in (3.24) vanishes to high order on $\tilde{\Gamma}$. The next task is to find a function a so that the s^1 and s^0 terms in (3.24) have the same property. We will in fact look for a in the form

$$a = a_0 + s^{-1} a_{-1} + \ldots + s^{-N} a_{-N}.$$

Here a_0, a_{-1}, ..., a_{-N} will be complex functions functions supported in $\tilde{\Gamma} \times [0, \delta]$ which are independent of s. For such a, the s^1 and s^0 terms in (3.24) become

$$\begin{aligned} s\Big[\sum_{j,k=1}^{n} &(2\tilde{\gamma}^{j,k} \partial_j \Phi \partial_k a + \partial_j(\tilde{\gamma}^{j,k} \partial_k \Phi) a \Big] + [L_{\tilde{\gamma}} a] \\ &= s[Ma_0] + [Ma_{-1} + L_{\tilde{\gamma}} a_0] + s^{-1}[Ma_{-2} + L_{\tilde{\gamma}} a_{-1}] + \ldots \\ &\quad + s^{-(N-1)}[Ma_{-N} + L_{\tilde{\gamma}} a_{-(N-1)}] + s^{-N}[L_{\tilde{\gamma}} a_{-N}], \end{aligned}$$

where M is the first-order differential operator given by

$$Mb = \sum_{j,k=1}^{n} \Big(2\tilde{\gamma}^{j,k} \partial_j \Phi \partial_k b + \partial_j(\tilde{\gamma}^{j,k} \partial_k \Phi) b \Big).$$

Note that the quantities in brackets are independent of s. We shall find a_0, $a_{-1}, \dots, a_{-N}$ successively so that one has

$$\partial_n^j(Ma_0)|_{\tilde{\Gamma}} = 0 \qquad \text{for } j = 0, 1, \dots, N-1, \tag{3.29}$$

$$\partial_n^j(Ma_{-1} + L_{\tilde{\gamma}}a_0)|_{\tilde{\Gamma}} = 0 \qquad \text{for } j = 0, 1, \dots, N-1, \tag{3.30}$$

$$\vdots$$

$$\partial_n^j(Ma_{-N} + L_{\tilde{\gamma}}a_{-(N-1)})|_{\tilde{\Gamma}} = 0 \qquad \text{for } j = 0, 1, \dots, N-1. \tag{3.31}$$

(2) The function a_0 is constructed in a similar way as Φ. We look for a_0 in the form

$$a_0(y', y_n) = \left(g_0(y') + y_n g_1(y') + \dots + \frac{y_n^N}{N!} g_N(y')\right)\zeta(y_n/\delta),$$

where $g_j(y') = \partial_n^j a_0(y', 0)$ are functions in $C_c^\infty(\tilde{\Gamma})$ to be determined, and $\zeta \in C^\infty(\mathbb{R})$ is a fixed cutoff function with $\zeta(t) = 1$ for $|t| \le 1/2$ and $\zeta(t) = 0$ for $|t| \ge 1$. It follows that a_0 is compactly supported in $\tilde{\Gamma} \times [0, \delta]$. Motivated by the boundary condition (3.20), we choose g_0 as

$$g_0(y') = \eta(y').$$

(3) Using the special form (3.18) for $\tilde{\gamma}$, it follows that

$$\begin{aligned} Mb &= \sum_{j,k=1}^{n} \left(2\tilde{\gamma}^{j,k}\partial_j\Phi\partial_k b + \tilde{\gamma}^{j,k}(\partial_j\partial_k\Phi)b + (\partial_j\tilde{\gamma}^{j,k})(\partial_k\Phi)b\right) \\ &= 2c\partial_n\Phi\partial_n b + 2c\sum_{\alpha,\beta=1}^{n-1} h^{\alpha,\beta}\partial_\alpha\Phi\partial_\beta b + c(\partial_n^2\Phi)b + c\sum_{\alpha,\beta=1}^{n-1} h^{\alpha,\beta}(\partial_\alpha\partial_\beta\Phi)b \\ &\quad + (\partial_n c)(\partial_n\Phi)b + \sum_{\alpha,\beta=1}^{n-1}(\partial_\alpha c)h^{\alpha,\beta}(\partial_\beta\Phi)b + c\sum_{\alpha,\beta=1}^{n-1}(\partial_\alpha h^{\alpha,\beta})(\partial_\beta\Phi)b. \end{aligned} \tag{3.32}$$

For the equation (3.29) with $j = 0$, using the choice $a_0(y', 0) = \eta(y')$, we have

$$Ma_0|_{\tilde{\Gamma}} = 2cf_1\partial_n a_0 + (\partial_n c)f_1\eta + c(\partial_n^2\Phi)\eta + T_0(c, h),$$

where $T_k(c, h)$ denotes an expression depending only on t', η, and on the y'-derivatives of $\partial_n^j c|_{\tilde{\Gamma}}$ and $\partial_n^j h^{\alpha,\beta}|_{\tilde{\Gamma}}$ for $0 \le j \le k$. Consequently, we have $Ma_0|_{\tilde{\Gamma}} = 0$ if $g_1 = \partial_n a_0|_{\tilde{\Gamma}}$ is chosen as

$$\begin{aligned} g_1(y') &= -\frac{\partial_n c}{2c}\eta - \frac{1}{2f_1}f_2\eta + T_0(c, h) \\ &= -\frac{\partial_n c}{2c}\eta - \left(\frac{1}{2f_1}\right)^2\left(\sum_{\alpha,\beta=1}^{n-1}(\partial_n h^{\alpha,\beta})t_\alpha t_\beta\right)\eta + T_0(c, h) \end{aligned}$$

by (3.28).

(4) Let g_0 be as above, and suppose that we have found $g_1, \dots, g_m$ such that (3.29) holds for $j = 0, 1, \dots, m-1$, and

$$(3.33)\quad g_j = -\frac{\partial_n^j c}{2c}\eta - \Big(\frac{1}{2f_1}\Big)^2 \Big(\sum_{\alpha,\beta=1}^{n-1} (\partial_n^j h^{\alpha,\beta}) t_\alpha t_\beta\Big)\eta + T_{j-1}(c,h),\ 1 \le j \le m.$$

We will find g_{m+1} having the form (3.33) so that (3.29) is satisfied also for $j = m$. By (3.32) and (3.19),

$$\partial_n^m (Ma_0)|_{\tilde{\Gamma}} = 2cf_1 \partial_n^{m+1} a_0 + (\partial_n^{m+1} c) f_1 \eta + c f_{m+2}\eta + T_m(c,h).$$

We can make this vanish by choosing $g_{m+1} = \partial_n^{m+1} a_0|_{\tilde{\Gamma}}$ having the required form (3.33), using the expression for f_{m+2} in (3.28).

(5) By the previous steps, we can find g_0, g_1, …, g_N inductively so that (3.29) is satisfied and each g_j for $j = 1, \dots, N$ has the form (3.33). This completes the construction of a_0.

(6) The construction of a_{-1} is similar to that of a_0. The function a_{-1} will have the form

$$a_{-1}(y', y_n) = \Big(g_{-1,0}(y') + y_n g_{-1,1}(y') + \dots + \frac{y_n^N}{N!} g_{-1,N}(y')\Big)\zeta(y_n/\delta)$$

for suitable $g_{-1,j} \in C_0^\infty(\tilde{\Gamma})$ and with ζ as above. Considering the boundary condition (3.20), we choose

$$g_{-1,0}(y') = 0.$$

Considering (3.30) for $j = 0$, since $a_{-1}|_{\tilde{\Gamma}} = 0$ we have by (3.32)

$$Ma_{-1} + L_{\tilde{\gamma}} a_0|_{\tilde{\Gamma}} = 2cf_1 \partial_n a_{-1} + L_{\tilde{\gamma}} a_0|_{\tilde{\Gamma}} = 2cf_1 \partial_n a_{-1}|_{\tilde{\Gamma}} + cg_2 + T_1(c,h).$$

This vanishes if $g_{-1,1}$ is chosen as

$$\begin{aligned} g_{-1,1} &= -\frac{1}{2f_1} g_2 + T_1(c,h) \\ &= -\Big(-\frac{1}{2f_1}\Big)\frac{\partial_n^2 c}{2c}\eta - \Big(-\frac{1}{2f_1}\Big)^3 \Big(\sum_{\alpha,\beta=1}^{n-1} (\partial_n^2 h^{\alpha,\beta}) t_\alpha t_\beta\Big)\eta + T_1(c,h). \end{aligned}$$

Here we used (3.33). The next case to consider is

$$\partial_n (Ma_{-1} + L_{\tilde{\gamma}} a_0)|_{\tilde{\Gamma}} = 2cf_1 g_{-1,2} + cg_3 + T_2(c,h).$$

This vanishes if one defines

$$\begin{aligned} g_{-1,2} &= -\frac{1}{2f_1} g_3 + T_2(c,h) \\ &= -\Big(-\frac{1}{2f_1}\Big)\frac{\partial_n^3 c}{2c}\eta - \Big(-\frac{1}{2f_1}\Big)^3 \Big(\sum_{\alpha,\beta=1}^{n-1} (\partial_n^3 h^{\alpha,\beta}) t_\alpha t_\beta\Big)\eta + T_2(c,h). \end{aligned}$$

Continuing, we obtain $g_{-1,0}, g_{-1,1}, \ldots, g_{-1,N}$ such that a_{-1} satisfies (3.30) and

$$g_{-1,j} = -\Big(-\frac{1}{2f_1}\Big)\frac{\partial_n^{1+j}c}{2c}\eta - \Big(-\frac{1}{2f_1}\Big)^3\Big(\sum_{\alpha,\beta=1}^{n-1}(\partial_n^{1+j}h^{\alpha,\beta})t_\alpha t_\beta\Big)\eta + T_{1+j-1}(c,h),$$
$$1 \le j \le N.$$

We have now constructed a_{-1}.

(7) The construction of $a_{-2}, \ldots, a_{-N}$ is completely analogous to that of a_{-1}. We leave it as an exercise to check that one can find $a_{-2}, \ldots, a_{-N}$ such that (3.29)–(3.31) are satisfied, and

$$\begin{aligned} a_{-l}|_{\tilde\Gamma} &= 0, \quad 1 \le l \le N,\\ \partial_n^j a_{-l}|_{\tilde\Gamma} &= -\frac{1}{2f_1}\partial_n^{j+1}a_{-(l-1)}|_{\tilde\Gamma} + T_{l+j-1}(c,h)\\ &= -\Big(-\frac{1}{2f_1}\Big)^l\frac{\partial_n^{l+j}c}{2c}\eta - \Big(-\frac{1}{2f_1}\Big)^{l+2}\Big(\sum_{\alpha,\beta=1}^{n-1}(\partial_n^{l+j}h^{\alpha,\beta})t_\alpha t_\beta\Big)\eta\\ &\qquad + T_{l+j-1}(c,h), \qquad 1 \le l \le N,\ 1 \le j \le N, \end{aligned}$$

where, as before, $T_k(c,h)$ denotes an expression only depending on t', η, and on y'-derivatives of $\partial_n^j c|_{\tilde\Gamma}$ and $\partial_n^j h^{\alpha,\beta}|_{\tilde\Gamma}$ for $0 \le j \le k$.

End of Proof. We have proved all the statements in the proposition except for the estimates (3.21) and (3.22). To verify (3.21), we first use that a is uniformly bounded in $\tilde\Gamma \times [0,\delta]$ and that $\mathrm{Re}(\Phi) \le -\sigma y_n$ to obtain

$$\begin{aligned} \|\tilde v_s\|_{L^2(\tilde\Gamma\times[0,\delta])} &\le C\left\|e^{s\,\mathrm{Re}(\Phi)}\right\|_{L^2(\tilde\Gamma\times[0,\delta])}\\ &\le C\Big(\int_{\tilde\Gamma}\int_0^\delta e^{-2s\sigma y_n}\,dy'\,dy_n\Big)^{1/2}\\ &\le Cs^{-1/2} \end{aligned}$$

by changing variables $y_n = t/s$. The derivatives of $\tilde v_s$ are given by

$$\partial_j\tilde v_s = e^{s\Phi}\big(s(\partial_j\Phi)a + \partial_j a\big),$$

and a similar argument as above shows that

$$\left\|\partial_j\tilde v_s\right\|_{L^2(\tilde\Gamma\times[0,\delta])} \le Cs^{1/2}$$

uniformly over $s \ge 1$. This implies (3.21).

To prove (3.22), we note that (3.24) and the construction of Φ and a imply that

$$L_{\tilde\gamma}\tilde v_s = e^{s\Phi}\big(s^2b_2 + sb_1 + b_0 + \ldots + s_{-(N-1)}b_{-(N-1)} + s^{-N}r_{-N}\big)$$

where each b_k satisfies

$$\partial_n^j b_k|_{\tilde\Gamma} = 0, \qquad 0 \le j \le N-1, \quad -(N-1) \le k \le 2,$$

and the remainder term r_{-N} has the form

$$r_{-N} = L_{\tilde{\gamma}} a_{-N}.$$

For $y = (y', y_n) \in \tilde{\Gamma} \times [0, \delta]$, we consider the Taylor expansion of b_k with respect to y_n as in Lemma 3.4.3. Since $\partial_n^j b_k(y', 0) = 0$ for $j \leq N-1$, it follows that for $y \in \tilde{\Gamma} \times [0, \delta]$ one has

$$|b_k(y', y_n)| \leq C y_n^N, \qquad -(N-1) \leq k \leq 2,$$

where the constant C depends on the Nth derivatives of b_k and is independent of s. Also $|r_{-N}(y', y_n)| \leq C$ with C independent of s. Since $s \geq 1$, we can estimate

$$\begin{aligned} \left|L_{\tilde{\gamma}} \tilde{v}_s(y', y_n)\right| &\leq C e^{s \operatorname{Re}(\Phi)} (s^2 y_n^N + s^{-N}) \\ &\leq C e^{-\sigma s y_n} (s^2 y_n^N + s^{-N}). \end{aligned}$$

The function $L_{\tilde{\gamma}} \tilde{v}_s$ is supported in $\tilde{\Gamma} \times [0, \delta]$ and satisfies

$$\left\| L_{\tilde{\gamma}} \tilde{v}_s \right\|_{L^2(\tilde{\Omega})}^2 \leq 2C^2 \int_{\tilde{\Gamma}} \int_0^\delta e^{-2\sigma s y_n} \left(s^4 y_n^{2N} + s^{-2N}\right) dy'\, dy_n.$$

We used the inequality $(a+b)^2 \leq 2(a^2 + b^2)$ for $a, b \geq 0$. The integrand is independent of y' and consequently

$$\begin{aligned} \left\| L_{\tilde{\gamma}} \tilde{v}_s \right\|_{L^2(\tilde{\Omega})}^2 &\leq C' \int_0^\infty e^{-2\sigma s y_n} \left(s^4 y_n^{2N} + s^{-2N}\right) dy_n \\ &= C' \int_0^\infty e^{-2\sigma t} \left(s^{4-2N} t^{2N} + s^{-2N}\right) s^{-1}\, dt \end{aligned}$$

by the change of variable $y_n = t/s$. Finally,

$$\left\| L_{\tilde{\gamma}} \tilde{v}_s \right\|_{L^2(\tilde{\Omega})}^2 \leq C' s^{3-2N} \int_0^\infty e^{-2\sigma t} t^{2N}\, dt + C' s^{-1-2N} \int_0^\infty e^{-2\sigma t}\, dt \leq C'' s^{3-2N}.$$

We have proved (3.22). □

Lemma 3.4.3 (Taylor expansion with remainder in integral form). *Let $b(y', y_n)$ be a C^k function in $W \times [0, \delta]$ where $W \subset \mathbb{R}^{n-1}$ is open. Then*

$$b(y', y_n) = \sum_{j=0}^{k-1} \frac{\partial_n^j b(y', 0)}{j!} y_n^j + \frac{\psi(y', y_n)}{k!} y_n^k,$$

where ψ is a C^0 function in $W \times [0, \delta]$ given by

$$\psi(y', y_n) = k \int_0^1 (1-t)^{k-1} \partial_n^k b(y', t y_n)\, dt.$$

In particular

$$\psi(y', 0) = \partial_n^k b(y', 0).$$

If $|\partial_n^k b(y', y_n)| \leq C$ *in* $W \times [0, \delta]$, *then also* $|\psi(y', y_n)| \leq C$ *in* $W \times [0, \delta]$.

Exercise 3.4.4. Prove Lemma 3.4.3.

3.5. Recovering higher-order derivatives

Let now Ω be a bounded domain with smooth boundary, and let p be a fixed point on $\partial\Omega$. Assume that $\gamma \in C^\infty(\overline{\Omega})$ is a positive function. We will prove that γ and its normal derivatives at p can be recovered from the DN map measured on a small neighborhood of p.

First, let $F : U \to V$ be the boundary flattening change of coordinates given in Proposition 3.3.1. We choose $r > 0$ so small that

$$B(0, 2r) \cap \{y_n > 0\} \subset V,$$

and define as in Lemma 3.3.5

$$\tilde{\gamma}(y) = \begin{bmatrix} ch & 0 \\ 0 & c \end{bmatrix}$$

where $h(y) = \left[h^{\alpha,\beta}(y)\right]_{\alpha,\beta=1}^{n-1}$ is the matrix in Proposition 3.3.1 and where

$$c(y) = \frac{\gamma(F^{-1}(y))}{|\det DF(F^{-1}(y))|}.$$

Note that c and h are only defined in $V \cap \{y_n \geq 0\}$. We also consider the flat boundary piece $\tilde{\Gamma} = B(0, r) \cap \{y_n = 0\}$, and $\Gamma = F^{-1}(\tilde{\Gamma})$ will be the corresponding neighborhood of p in $\partial\Omega$.

Let $t' \in \mathbb{R}^{n-1}$ be a unit vector, let $\eta \in C^\infty(\mathbb{R}^{n-1})$ be supported in $\tilde{\Gamma}$, and let N be large. We take $\tilde{v}_s$ to be the approximate solution given in Proposition 3.4.2, and we transport $\tilde{v}_s$ to the original domain Ω by

$$v_s(x) = \tilde{v}_s(F(x)), \quad x \in U \cap \overline{\Omega}.$$

Since $\tilde{v}_s$ is compactly supported in $\tilde{\Gamma} \times [0, \delta]$, it follows that we may extend v_s by zero to $\overline{\Omega}$ and obtain a new function, also denoted by v_s, in $C^\infty(\overline{\Omega})$ with $\mathrm{supp}(v_s)$ contained in a small neighborhood of p in $\overline{\Omega}$. We define the function ϕ_s on $\partial\Omega$ by

$$\phi_s = v_s|_{\partial\Omega}.$$

The next result shows that one can construct an exact solution of

$$L_\gamma u = 0$$

in Ω which looks like v_s when s is large.

Lemma 3.5.1. *There is a unique solution* $u_s \in H^1(\Omega)$ *of*

$$\begin{cases} L_\gamma u_s = 0 & \text{in } \Omega, \\ u_s = \phi_s & \text{on } \partial\Omega. \end{cases}$$

This solution has the form $u_s = v_s + r_s$ with

$$\|r_s\|_{H^1(\Omega)} \le Cs^{-N+3/2}. \tag{3.34}$$

Proof. The existence of a unique solution $u_s \in H^1(\Omega)$ follows from the well-posedness of the Dirichlet problem (see Theorem 2.5.6). To show the estimate (3.34) for the correction term r_s, we note that r_s solves

$$\begin{cases} L_\gamma r_s = -L_\gamma v_s & \text{in } \Omega, \\ \quad r_s = 0 & \text{on } \partial\Omega. \end{cases}$$

Again by well-posedness of the Dirichlet problem, we have

$$\|r_s\|_{H^1(\Omega)} \le C \left\| L_\gamma v_s \right\|_{H^{-1}(\Omega)} \le C \left\| L_\gamma v_s \right\|_{L^2(\Omega)}.$$

Now Lemma 3.3.5 together with a change of coordinates shows that for any $\varphi \in C_c^\infty(U \cap \Omega)$, with $\tilde{\varphi} = \varphi \circ F^{-1}$, one has

$$\begin{aligned} \left| \int_\Omega (L_\gamma v_s)\varphi \, dx \right| &= \left| \int_\Omega (L_{\tilde{\gamma}} \tilde{v}_s)\tilde{\varphi} \, dx \right| \\ &\le \left\| L_{\tilde{\gamma}} \tilde{v}_s \right\|_{L^2(V \cap \{y_n > 0\})} \|\tilde{\varphi}\|_{L^2(V \cap \{y_n > 0\})} \\ &\le C \left\| L_{\tilde{\gamma}} \tilde{v}_s \right\|_{L^2(V \cap \{y_n > 0\})} \|\varphi\|_{L^2(\Omega)}. \end{aligned}$$

Since $C_c^\infty(U \cap \Omega)$ is dense in $L^2(U \cap \Omega)$ by Lemma 2.1.6, the same result remains true for $\varphi \in L^2(U \cap \Omega)$. This implies that

$$\left\| L_\gamma v_s \right\|_{L^2(\Omega)} = \left\| L_\gamma v_s \right\|_{L^2(U \cap \Omega)} \le C \left\| L_{\tilde{\gamma}} \tilde{v}_s \right\|_{L^2(V \cap \{y_n > 0\})} \le Cs^{-N+3/2}$$

by (3.22). Thus

$$\|r_s\|_{H^1(\Omega)} \le C \left\| L_\gamma v_s \right\|_{L^2(\Omega)} \le Cs^{-N+3/2}. \qquad \square$$

Note that the boundary value ϕ_s is explicit and is given by

$$\phi_s(x) = \begin{cases} e^{isy'(x)\cdot t'} \eta(y'(x)) & \text{if } x \in \Gamma, \\ 0 & \text{otherwise,} \end{cases}$$

where $y'(x) = \big(F_1(x), \dots, F_{n-1}(x)\big)$ is the representation of Γ as a flat boundary piece. The next proposition proves that the boundary values of γ on Γ can be determined from the knowledge of Λ_γ.

Proposition 3.5.2. *We have*

$$\lim_{s \to \infty} s^{-1} \langle \Lambda_\gamma \phi_s, \bar{\phi}_s \rangle = - \int_{\tilde{\Gamma}} c(y', 0) f_1(y') \eta(y')^2 \, dy'.$$

Indeed, since $\eta \in C_c^\infty(\tilde{\Gamma})$ can be chosen arbitrarily and f_1 is known, it follows from the previous result that $c(y',0)$ is determined by Λ_γ on Γ. Finally, because $c(y) = |\det DF|^{-1}\gamma|_{F^{-1}(y)}$ where F only depends on p and Ω and can therefore be considered as known, we recover $\gamma(F^{-1}(y',0))$ for $y' \in \tilde{\Gamma}$. Thus we have determined $\gamma|_\Gamma$.

We first give a simple lemma needed for the proof.

Lemma 3.5.3. *If Φ is as in Proposition 3.4.2 and if b is a smooth function with $\operatorname{supp}(b) \subset \tilde{\Gamma} \times [0,\delta]$, then for any $k \geq 0$*

$$\lim_{s\to\infty} s^{k+1} \int_{\tilde{\Gamma}\times[0,\delta]} e^{2s\operatorname{Re}(\Phi(y))} y_n^k b(y)\,dy = k! \int_{\tilde{\Gamma}} \Big(-\frac{1}{2f_1(y')}\Big)^{k+1} b(y',0)\,dy'.$$

Proof. From Proposition 3.4.2 we have that $\operatorname{Re}(\Phi(y)) = y_n\psi(y',y_n)$ where $\psi \leq -\sigma$ in $\tilde{\Gamma}\times[0,\delta]$ and $\psi(y',0) = f_1(y')$. Consequently, the change of variable $y_n = t/s$ shows that

$$\begin{aligned} s^{k+1} \int_{\tilde{\Gamma}\times[0,\delta]} e^{2s\operatorname{Re}(\Phi(y))} y_n^k b(y)\,dy &= s^{k+1} \int_{\tilde{\Gamma}} \int_0^\delta e^{2sy_n\psi(y',y_n)} y_n^k b(y',y_n)\,dy'\,dy_n \\ &= \int_{\tilde{\Gamma}} \int_0^{s\delta} e^{2t\psi(y',t/s)} t^k b(y',t/s)\,dy'\,dt. \end{aligned}$$

Since $\psi \leq -\sigma$ the integral is absolutely convergent, and we may apply dominated convergence to obtain

$$\lim_{s\to\infty} s^{k+1} \int_{\tilde{\Gamma}\times[0,\delta]} e^{2s\operatorname{Re}(\Phi(y))} y_n^k b(y)\,dy = \int_{\tilde{\Gamma}} \int_0^\infty e^{2t\psi(y',0)} t^k b(y',0)\,dy'\,dt.$$

The result follows by noting that for any $\lambda > 0$,

$$\int_0^\infty e^{-\lambda t} t^k\,dt = \lambda^{-k-1} \int_0^\infty e^{-t} t^k\,dt = k!\,\lambda^{-k-1}.$$ □

Proof of Proposition 3.5.2. Assume that $N > 1$ in the construction of $\tilde{v}_s$, and let u_s be as in Lemma 3.5.1. Using that

$$u_s = v_s + r_s,$$

the definition of the DN map (see Theorem 2.7.3) gives

$$\langle \Lambda_\gamma \phi_s, \bar{\phi}_s \rangle = \int_\Omega \gamma \nabla u_s \cdot \nabla \bar{v}_s\,dx = \int_\Omega \gamma \nabla v_s \cdot \nabla \bar{v}_s\,dx + R_s$$

where $R_s = \int_\Omega \gamma \nabla r_s \cdot \nabla \bar{v}_s\,dx$ satisfies

$$|R_s| \leq C\,\|r_s\|_{H^1(\Omega)}\,\|v_s\|_{H^1(\Omega)} \leq Cs^{-N+2} \tag{3.35}$$

by (3.34) and the fact that $\|v_s\|_{H^1(\Omega)} \leq Cs^{1/2}$ which follows from (3.21). Lemma 3.3.5 implies that

$$\int_\Omega \gamma \nabla v_s \cdot \nabla \bar{v}_s \, dx = \int_{\tilde{\Gamma}\times[0,\delta]} \tilde{\gamma} \nabla \tilde{v}_s \cdot \overline{\nabla \tilde{v}_s} \, dy.$$

From the explicit form of $\tilde{v}_s$ in (3.23), we see that

$$\partial_j \tilde{v}_s = e^{s\Phi}\big(s(\partial_j \Phi)a_0 + r_{0,j}\big),$$

where $|r_{0,j}| \leq C$ uniformly in s. Combining all these facts, one has the identity

$$s^{-1}\langle \Lambda_\gamma \phi_s, \bar{\phi}_s\rangle = s \int_{\tilde{\Gamma}\times[0,\delta]} e^{2s\,\mathrm{Re}(\Phi)}(\tilde{\gamma}\nabla\Phi \cdot \nabla\bar{\Phi})|a_0|^2 \, dy$$
$$+ \int_{\tilde{\Gamma}\times[0,\delta]} e^{2s\,\mathrm{Re}(\Phi)} q \, dy + s^{-1}R_s,$$

where $|q| \leq C$ uniformly in s. By Lemma 3.5.3 for $k = 0$ the second term converges to zero as $s \to \infty$, and this is also true for the last term by (3.35) since $N > 1$. Thus by Lemma 3.5.3 again

$$\lim_{s\to\infty} s^{-1}\langle \Lambda_\gamma \phi_s, \bar{\phi}_s\rangle = -\frac{1}{2}\int_{\tilde{\Gamma}} \frac{(\tilde{\gamma}\nabla\Phi \cdot \nabla\bar{\Phi})|a_0|^2}{f_1} \, dy'.$$

It remains to observe that

$$(\tilde{\gamma}\nabla\Phi \cdot \nabla\bar{\Phi})|a_0|^2|_{\tilde{\Gamma}} = c(y',0)\Big[|\partial_n \Phi(y',0)|^2 + \sum_{\alpha,\beta=1}^{n-1} h^{\alpha,\beta}(y',0)t_\alpha t_\beta\Big]\eta(y')^2,$$

where the expression in brackets is equal to $2f_1(y')^2$. □

Proposition 3.5.2 recovered the boundary value of γ. The simplest way to recover the normal derivatives of γ is by comparing Λ_γ to Λ_{γ^k}, where γ^k is known and the normal derivatives of γ and γ^k agree up to order $k-1$.

Proposition 3.5.4. *Let $k \geq 1$ and let $\gamma^k \in C^\infty(\overline{\Omega})$ be any positive function such that for $0 \leq j \leq k-1$,*

$$\left(\frac{\partial}{\partial\nu}\right)^j \gamma^k = \left(\frac{\partial}{\partial\nu}\right)^j \gamma \quad \text{on } \Gamma \tag{3.36}$$

and

$$\left(\frac{\partial}{\partial\nu}\right)^k \gamma^k = 0 \quad \text{on } \Gamma. \tag{3.37}$$

Then

$$\lim_{s\to\infty} s^{k-1}\langle(\Lambda_\gamma - \Lambda_{\gamma^k})\phi_s, \bar{\phi}_s\rangle = \frac{1}{2}\int_{\tilde{\Gamma}} \left(-\frac{1}{2f_1(y')}\right)^{k-1} \frac{\partial_n^k(\gamma\circ F^{-1})(y',0)}{|\det DF(F^{-1}(y',0))|}\eta(y')^2 \, dy'.$$

Proof. Let $N > k+1$, let $u_s = v_s + r_s$ be the solution given in Lemma 3.5.1, and let $u_s^k = v_s^k + r_s^k$ be the corresponding solution in the case where γ is replaced by γ^k. Then we have $v_s^k = \tilde{v}_s^k \circ F$, where

$$\tilde{v}_s^k = e^{s\Phi}(a_0^k + s^{-1}a_{-1}^k + \ldots + s^{-N}a_{-N}^k)$$

is the function in Proposition 3.4.2 with c replaced by c^k, and

$$c^k(y) = \frac{\gamma^k(F^{-1}(y))}{|\det DF(F^{-1}(y))|}.$$

Note that Φ in Proposition 3.4.2 does not depend on c (see (3.26) and (3.28)), therefore Φ is the same both for $\tilde{v}_s$ and $\tilde{v}_s^k$. We also have $a_0|_{\tilde{\Gamma}} = a_0^k|_{\tilde{\Gamma}} = \eta$.

As in the proof of Proposition 3.5.2, but this time using the integral identity in Proposition 2.8.2, we have

$$\begin{aligned}\langle(\Lambda_\gamma - \Lambda_{\gamma^k})\phi_s, \bar{\phi}_s\rangle &= \int_\Omega (\gamma - \gamma^k)\nabla u_s \cdot \nabla \bar{u}_s^k \, dx \\ &= \int_{\tilde{\Gamma}\times[0,\delta]} (\tilde{\gamma} - \tilde{\gamma}^k)\nabla \tilde{v}_s \cdot \overline{\nabla \tilde{v}_s^k} \, dy + R_s,\end{aligned}$$

where R_s involves r_s and r_s^k and $|R_s| \leq Cs^{-N+2}$. Here

$$\tilde{\gamma} - \tilde{\gamma}^k = (c - c^k)\begin{bmatrix} h & 0 \\ 0 & 1 \end{bmatrix}.$$

By the condition (3.36) we have $\partial_n^j(c - c^k)|_{\tilde{\Gamma}} = 0$ for $0 \leq j \leq k-1$. Therefore by Taylor expansion (see Lemma 3.4.3)

$$(c - c^k)(y', y_n) = y_n^k \psi_k(y', y_n),$$

where ψ_k is a smooth function satisfying

$$\psi_k(y', 0) = \frac{\partial_n^k(c - c^k)(y', 0)}{k!} = \frac{1}{k!}\frac{\partial_n^k(\gamma \circ F^{-1})(y', 0)}{|\det DF(F^{-1}(y', 0))|} \tag{3.38}$$

by the Leibniz rule, (3.36), and (3.37). Using that

$$\begin{aligned}\partial_j \tilde{v}_s &= e^{s\Phi}(s(\partial_j\Phi)a_0 + r_{0,j}), \\ \partial_j \tilde{v}_s^k &= e^{s\Phi}(s(\partial_j\Phi)a_0^k + r_{0,j}^k),\end{aligned}$$

where $|r_{0,j}|$ and $|r_{0,j}^k|$ are bounded uniformly in s, it follows that

$$\begin{aligned}\langle(\Lambda_\gamma - \Lambda_{\gamma^k})\phi_s, \bar{\phi}_s\rangle \\ = s^2 \int_{\tilde{\Gamma}\times[0,\delta]} e^{2s\,\mathrm{Re}(\Phi)} y_n^k \psi_k \big[|\partial_n \Phi|^2 \\ + \sum_{\alpha,\beta=1}^{n-1} h^{\alpha,\beta} \partial_\alpha \Phi \partial_\beta \bar{\Phi}\big] a_0 \bar{a}_0^k \, dy \\ + s \int_{\tilde{\Gamma}\times[0,\delta]} e^{2s\,\mathrm{Re}(\Phi)} y_n^k q \, dy + R_s.\end{aligned}$$

Here $|q| \leq C$ uniformly in s, and Lemma 3.5.3 and the condition $N > k+1$ imply

$$\lim_{s\to\infty} s^{k-1} \left[s \int_{\tilde{\Gamma}\times[0,\delta]} e^{2s\,\mathrm{Re}(\Phi)} y_n^k q \, dy + R_s \right] = 0.$$

From Lemma 3.5.3 we conclude that

$$\begin{aligned}\lim_{s\to\infty} s^{k-1} \langle(\Lambda_\gamma - \Lambda_{\gamma^k})\phi_s, \bar{\phi}_s\rangle \\ = k! \int_{\tilde{\Gamma}} \left(-\frac{1}{2f_1(y')} \right)^{k+1} \psi_k(y',0)[2f_1(y')^2]\eta(y')^2 \, dy'.\end{aligned}$$

This finishes the proof by using (3.38). □

The main result of Chapter 3 now follows from the above propositions.

Proof of Theorem 3.0.1. If Γ is as stated and p is a fixed point on Γ, then it is enough to determine $\partial_\nu^\ell \gamma$ on Γ near p. By shrinking Γ if necessary, we can use boundary normal coordinates to flatten the boundary piece Γ. Thus we may assume that we are working in the setting described in the beginning of this section.

If $\ell \geq 0$ is given, we choose $N > \ell + 1$ and consider the solution u_s given in Lemma 3.5.1 such that the function $\phi_s = u_s|_{\partial\Omega}$ is supported in Γ. Since we have knowledge of $\Lambda_\gamma f|_\Gamma$ for f supported in Γ, we also know the quantities $\langle\Lambda_\gamma \phi_s, \bar{\phi}_s\rangle$ for $s \geq 1$. By Proposition 3.5.2 this determines γ on Γ by varying the function η.

From the knowledge of γ on Γ we can construct a conductivity γ^1 with $\gamma = \gamma^1$ and $\partial_\nu \gamma^1 = 0$ on $\partial\Omega$ near p (see the argument below). Since we know γ^1, we also know the map Λ_{γ^1}. Then Proposition 3.5.4 allows us to recover $\partial_\nu \gamma$ near p. Next we construct a conductivity γ^2 with $\gamma = \gamma^2$, $\partial_\nu \gamma = \partial_\nu \gamma^2$, and $\partial_\nu^2 \gamma^2 = 0$ on $\partial\Omega$ near p, and use Proposition 3.5.4 again to recover $\partial_\nu^2 \gamma$ near p. Continuing in this way, one finds the normal derivatives of γ up to order ℓ near p, and thus on all of Γ by varying the point p.

The previous argument used that if $\partial_\nu^j \gamma$ is known on Γ for $0 \le j \le k-1$, one can construct a smooth conductivity γ^k satisfying (3.36) and (3.37) near p. To see this, define the function

$$\tilde{\gamma}^k(y', y_n) = \sum_{j=0}^{k-1} \frac{\partial_{y_n}^j(\gamma \circ F^{-1})(y', 0)}{j!} y_n^j.$$

This function is known since it involves $\partial_\nu^j \gamma|_\Gamma$ for $0 \le j \le k-1$. Also, since γ is positive in $\overline{\Omega}$, we have $\tilde{\gamma}^k \ge \varepsilon_0 > 0$ in $\tilde{\Gamma} \times [0, \delta_0]$ for some sufficiently small $\delta_0 > 0$. Let $\chi_1(y')$ and $\chi_2(y_n)$ be smooth cutoff functions such that $\chi_1(y') = 1$ for $|y'|$ small and $\mathrm{supp}(\chi_1) \subset \tilde{\Gamma}$, and $\chi_2(y_n)$ is 1 for $|y_n| \le \delta_0/2$ and 0 for $|y_n| \ge \delta_0$. We then define $\chi(y', y_n) = \chi_1(y')\chi_2(y_n)$ and

$$\gamma^k(x) = \chi(F(x))\tilde{\gamma}^k(F(x)) + (1 - \chi(F(x))).$$

This gives a smooth positive function in $\overline{\Omega}$ for which (3.36) and and (3.37) hold near p, as desired. □

3.6. Anisotropic conductivities

In the final section of this chapter, we consider the anisotropic Calderón problem and employ the method of concentrating solutions to determine the Taylor series of a matrix conductivity $\gamma = [\gamma^{j,k}]_{j,k=1}^n$ at a boundary point from the DN map Λ_γ. As discussed in §1.1, one can only expect to determine a matrix conductivity up to a boundary fixing diffeomorphism that corresponds to a choice of coordinates. To find a suitable choice of coordinates, it is convenient to rewrite the conductivity operator

$$L_\gamma u = \sum_{j,k=1}^n \partial_j(\gamma^{j,k}\partial_k u)$$

in terms of a closely related operator of geometric nature. In this section we will assume that the reader is familiar with elements of Riemannian geometry.

Let $\Omega \subset \mathbb{R}^n$ be a bounded open set with C^∞ boundary. We say that a *Riemannian metric* on $\overline{\Omega}$ is a symmetric positive definite matrix function $g(x) = [g_{j,k}(x)]_{j,k=1}^n$ so that $g_{j,k} \in C^\infty(\overline{\Omega})$ for all j, k. The *Laplace–Beltrami operator* associated with the metric g is the operator

$$\Delta_g u = \sum_{j,k=1}^n |g|^{-1/2}\partial_j(|g|^{1/2}g^{j,k}\partial_k u),$$

where $[g^{j,k}] = g(x)^{-1}$ is the inverse matrix of $g(x)$, and $|g(x)| = \det(g(x))$. (Note that the indices in the entries of $g(x)$ are down, but in the entries of $g(x)^{-1}$ they are up.)

If $\overline{\Omega}$ is considered as a differentiable manifold with smooth boundary, the metric g induces an inner product for tangent vectors at each point x via $\langle v, w\rangle_{g(x)} = g(x)v \cdot w$. The Laplace–Beltrami operator is the natural elliptic operator associated with the Dirichlet energy for the metric g. If g is the Euclidean metric, i.e., $g_{j,k}(x) = \delta_{j,k}$, then Δ_g is just the standard Laplacian. See §3.7 for references on these and other facts on Riemannian geometry.

We record the following equivalence of the conductivity and Laplace–Beltrami operators in dimensions $n \geq 3$.

Lemma 3.6.1. *Let $\Omega \subset \mathbb{R}^n$ be a bounded open set with C^∞ boundary, and assume that $n \geq 3$.*

(a) *If $\gamma = [\gamma^{j,k}]$ is a C^∞ positive definite symmetric matrix function, there is a C^∞ positive definite symmetric matrix function $g = [g_{j,k}]$ such that*

$$L_\gamma = |g|^{1/2}\Delta_g.$$

(b) *If $g = [g_{j,k}]$ is a C^∞ positive definite symmetric matrix function, there is a C^∞ positive definite symmetric matrix function $\gamma = [\gamma^{j,k}]$ such that*

$$L_\gamma = |g|^{1/2}\Delta_g.$$

Proof. We note that $L_\gamma u = |g|^{1/2}\Delta_g u$ holds whenever

$$[\gamma^{j,k}] = |g|^{1/2}[g^{j,k}]. \tag{3.39}$$

If γ is given, taking determinants in (3.39) would give $\det(\gamma^{j,k}) = \det(g_{j,k})^{\frac{n-2}{2}}$ since $[g^{j,k}] = [g_{j,k}]^{-1}$. Since $n \geq 3$, this would give $|g| = \det(\gamma^{j,k})^{\frac{2}{n-2}}$, and we could choose

$$[g_{j,k}] = (|g|^{-1/2}[\gamma^{j,k}])^{-1} = \det(\gamma^{j,k})^{\frac{1}{n-2}}[\gamma^{j,k}]^{-1}.$$

Making this choice of g proves (a). The proof of (b) is similar. □

Lemma 3.6.1 shows that the equations $L_\gamma u = 0$ and $\Delta_g u = 0$ are equivalent in domains in $\mathbb{R}^n$ when $n \geq 3$. It is often convenient to work with the Laplace–Beltrami equation since this facilitates the use of ideas and methods from Riemannian geometry. We will take this point of view in this section.

Exercise 3.6.2. Let $n = 2$. Prove that the Laplace–Beltrami operator has the additional *conformal invariance*

$$\Delta_{\beta g} = \beta^{-1}\Delta_g$$

when β is a smooth and positive function. Derive a relation between $\Delta_{\beta g}$ and Δ_g when $n \geq 3$.

The DN map for Δ_g can be defined similarly as the DN map for a matrix conductivity, via the formula (with $f_1, f_2 \in H^{1/2}(\partial\Omega)$)

$$(3.40) \qquad \langle \Lambda_g f_1, f_2 \rangle_{\partial\Omega} = \int_\Omega \sum_{j,k=1}^n |g|^{1/2} g^{j,k} \partial_j u_{f_1} \partial_k e_{f_2}\, dx,$$

where u_{f_1} is the solution of $\Delta_g u = 0$ in Ω with $u|_{\partial\Omega} = f_1$, and e_{f_2} is any function in $H^1(\Omega)$ with $e_{f_2}|_{\partial\Omega} = f_2$.

We will prove the following boundary determination result for Λ_g.

Theorem 3.6.3. *Let Ω be a bounded domain in $\mathbb{R}^n$ with C^∞ boundary, and let g be a Riemannian metric in $\overline{\Omega}$. Assume that p is a point on $\partial\Omega$, let (y', y_n) be boundary normal coordinates for $(\overline{\Omega}, g)$ near p, and let $\tilde{g}(y)$ be the representation of g in these coordinates. If $n \geq 3$, then from the knowledge of Λ_g it is possible to determine $\partial_n^l \tilde{g}_{j,k}(y', 0)$ near p for any integer $l \geq 0$ and for all $1 \leq j, k \leq n$.*

The "boundary normal coordinates" in Theorem 3.6.3 refer to *boundary normal coordinates for the Riemannian metric* g (also-called *semigeodesic coordinates*). These coordinates are given by the diffeomorphism $F : U \to V$ in Proposition 3.6.4. Then the representation $\tilde{g}(y)$ is the pullback $\tilde{g}(y) = ((F^{-1})^* g)(y) = D(F^{-1})(y) g(F^{-1}(y)) D(F^{-1})(y)^t$ that satisfies

$$\tilde{g}(y)^{-1} = (DF) g (DF)^t|_{F^{-1}(y)}.$$

Note that the next result reduces to Proposition 3.3.1 in the Euclidean case $g_{j,k}(x) = \delta_{jk}$.

Proposition 3.6.4 (Riemannian boundary normal coordinates). *Let Ω be a bounded domain with C^∞ boundary in $\mathbb{R}^n$, let g be a Riemannian metric on $\overline{\Omega}$, and let $p \in \partial\Omega$. There is a C^∞ diffeomorphism $F : U \to V$ between open sets of $\mathbb{R}^n$, where U is a neighborhood of p and V is a neighborhood of 0, such that*

$$F(p) = 0, \quad F(\Omega \cap U) = V \cap \{y_n > 0\}, \quad F(\partial\Omega \cap U) = V \cap \{y_n = 0\},$$

and further for any $y \in V \cap \{y_n \geq 0\}$,

$$(3.41) \qquad (DF) g^{-1} (DF)^t|_{F^{-1}(y)} = \begin{bmatrix} [\tilde{g}^{\alpha,\beta}(y)] & 0 \\ 0 & 1 \end{bmatrix} =: \tilde{g}(y)^{-1}$$

for some C^∞ symmetric positive definite matrix $\left[\tilde{g}^{\alpha,\beta}(y)\right]_{\alpha,\beta=1}^{n-1}$.

The proof is analogous to that of Proposition 3.3.1, except that F is constructed by following the geodesic curves of the metric g in the normal direction from $\partial\Omega$ instead of following straight lines. The block diagonal form of $\tilde{g}(y)$ means that the y_n variable corresponds to moving along normal geodesics at unit speed, and that the infinitesimal variations in the y' direction are orthogonal to the y_n direction. For more details see [**KKL01**, §2.1].

Let us compute how Λ_g transforms under this change of coordinates. From the definition of Λ_g in (3.40), if f_2 and e_{f_2} are supported in $U \cap \overline{\Omega}$, the change of coordinates $x = F^{-1}(y)$ gives

$$\langle \Lambda_g f_1, f_2 \rangle_{\partial\Omega} = \int_{V \cap \{y_n > 0\}} \frac{|g|^{1/2} g^{-1}}{|\det DF|}\bigg|_{F^{-1}(y)} \nabla u_{f_1}(F^{-1}(y)) \cdot \nabla e_{f_2}(F^{-1}(y))\, dy.$$

Arguing as in Lemma 3.3.5 and writing $\tilde{u}_{f_1}(y) = u_{f_1}(F^{-1}(y))$ and $\tilde{e}_{f_2}(y) = e_{f_2}(F^{-1}(y))$, from (3.16) we obtain

$$\langle \Lambda_g f_1, f_2 \rangle_{\partial\Omega} = \int_{V \cap \{y_n > 0\}} \frac{|g|^{1/2}(DF) g^{-1} (DF)^t}{|\det DF|}\bigg|_{F^{-1}(y)} \nabla \tilde{u}_{f_1}(y) \cdot \nabla \tilde{e}_{f_2}(y)\, dy.$$

Taking determinants in (3.41) gives $|\det(DF)|^2 |g|^{-1}|_{F^{-1}(y)} = |\tilde{g}(y)|^{-1}$. Thus we have

$$\langle \Lambda_g f_1, f_2 \rangle_{\partial\Omega} = \int_{V \cap \{y_n > 0\}} |\tilde{g}(y)|^{1/2} \tilde{g}(y)^{-1} \nabla \tilde{u}_{f_1}(y) \cdot \nabla \tilde{e}_{f_2}(y)\, dy. \tag{3.42}$$

Note that this identity is almost the same as (3.40), but now the metric $\tilde{g}(y)$ has a simpler block diagonal form (and e_{f_2} is supported in $U \cap \overline{\Omega}$).

From the knowledge of Λ_g, we know the right-hand side of (3.42) for any f_1 and f_2 supported near p. We have now reduced matters to the situation in §3.4–3.5 and we can begin the proof of Theorem 3.6.3. Suppose that the boundary normal coordinates (y', y_n) are defined in

$$B(0, 2r) \cap \{y_n \geq 0\}.$$

Let $\tilde{\Gamma} = B(0, r) \cap \{y_n = 0\}$, which is identified with the corresponding set in $\mathbb{R}^{n-1}$. For simplicity we drop the tildes from $\tilde{g}(y)$ and write $g(y)$ for the representation of g in the y-coordinates.

To stay close to the notations in §3.5, we write

$$\tilde{\gamma}(y) = |g(y)|^{1/2} g(y)^{-1} = c(y) \begin{bmatrix} h(y) & 0 \\ 0 & 1 \end{bmatrix},$$

where $c(y) = |g(y)|^{1/2}$ and $h(y) = [g^{\alpha,\beta}(y)]_{\alpha,\beta=1}^{n-1}$.

Choose a unit vector $t' \in \mathbb{R}^{n-1}$ and a function $\eta \in C_c^\infty(\tilde{\Gamma})$, and let ϕ_s be the explicit boundary value supported in $\tilde{\Gamma}$ defined by

$$\phi_s(y', 0) = e^{isy' \cdot t'} \eta(y').$$

The next result states that certain quantities can be recovered from the boundary measurements Λ_g by using the method in §3.5.

Proposition 3.6.5. *In any dimension $n \geq 2$, one has*

$$\lim_{s \to \infty} s^{-1} \langle \Lambda_g \phi_s, \bar{\phi}_s \rangle_{\partial\Omega} = \int_{\tilde{\Gamma}} \Big[\sum_{\alpha,\beta=1}^{n-1} (|g|\, g^{\alpha,\beta})(y', 0) t_\alpha t_\beta \Big]^{1/2} \eta(y')^2\, dy'. \tag{3.43}$$

Further, if $l \geq 1$ and if g^l is any C^∞ Riemannian metric in $\overline{\Omega}$ so that in the (y', y_n) coordinates

$$g^l(y) = \begin{bmatrix} h^l(y) & 0 \\ 0 & 1 \end{bmatrix}$$

for some matrix h^l, and if for all $1 \leq j, k \leq n$

$$\partial_n^p(|g^l|^{1/2}(g^l)^{j,k}) = \partial_n^p(|g|^{1/2}\, g^{j,k}) \quad on \quad \tilde{\Gamma}\ for\ 0 \leq p \leq l-1, \tag{3.44}$$

$$\partial_n^l(|g^l|^{1/2}(g^l)^{j,k}) = 0 \qquad on \quad \tilde{\Gamma}, \tag{3.45}$$

then

$$\lim_{s\to\infty} s^{l-1}\langle(\Lambda_g - \Lambda_{g^l})\phi_s, \bar{\phi}_s\rangle = \int_{\tilde{\Gamma}} \Big(-\frac{1}{2f_1(y')}\Big)^{l+1} \cdot \sum_{\alpha,\beta=1}^{n} \Big[(\partial_n^l(|g|^{1/2})g^{\alpha,\beta} + \partial_n^l(|g|^{1/2}\, g^{\alpha,\beta}))(y',0)t_\alpha t_\beta\Big]\eta(y')^2\, dy'. \tag{3.46}$$

Proof. We first take $N > 1$ and let $\tilde{v}_s$ be the approximate solution in Proposition 3.4.2 with the choice $\tilde{\gamma}^{j,k} = |g|^{1/2}\, g^{j,k}$, satisfying

$$\Big\|\Delta_g \tilde{v}_s\Big\|_{L^2(\tilde{\Gamma}\times[0,\delta])} = \Big\||g|^{-1/2}\, L_{\tilde{\gamma}}\tilde{v}_s\Big\|_{L^2(\tilde{\Gamma}\times[0,\delta])} \leq Cs^{-N+3/2}.$$

Following the proof of Lemma 3.5.1, there is a solution $u_s = v_s + r_s$ of $\Delta_g u_s = 0$ in Ω with $u_s|_{\partial\Omega} = \phi_s$ and $\|r_s\|_{H^1(\Omega)} \leq Cs^{-N+3/2}$. Then, as in the proof of Proposition 3.5.2,

$$\begin{aligned}\langle\Lambda_g\phi_s, \bar{\phi}_s\rangle &= \int_\Omega \sum_{j,k=1}^{n} |g|^{1/2}\, g^{j,k}\partial_j u_s \partial_k \bar{v}_s\, dx \\ &= \int_{\tilde{\Gamma}\times[0,\delta]} \sum_{j,k=1}^{n} \tilde{\gamma}^{j,k}\partial_j\tilde{v}_s\partial_k\overline{\tilde{v}_s}\, dx + R_s,\end{aligned}$$

$$\langle\Lambda_g\phi_s, \bar{\phi}_s\rangle = \int_\Omega \sum_{j,k=1}^{n} |g|^{1/2}\, g^{j,k}\partial_j u_s \partial_k \bar{v}_s\, dx = \int_{\tilde{\Gamma}\times[0,\delta]} \sum_{j,k=1}^{n} \tilde{\gamma}^{j,k}\partial_j\tilde{v}_s\partial_k\overline{\tilde{v}_s}\, dx + R_s,$$

where $|R_s| \leq Cs^{-N+2}$. The computation in the proof of Proposition 3.5.2 implies that

$$\begin{aligned}\lim_{s\to\infty} s^{-1}\langle\Lambda_g\phi_s, \bar{\phi}_s\rangle &= -\frac{1}{2}\int_{\tilde{\Gamma}} \frac{(\tilde{\gamma}\nabla\Phi\cdot\nabla\bar{\Phi})\,|a_0|^2}{f_1}\bigg|_{\tilde{\Gamma}} dy' \\ &= -\int_{\tilde{\Gamma}} c(y',0)f_1(y')\eta(y')^2\, dy'.\end{aligned}$$

Since $c = |g|^{1/2}$, the expression for f_1 in Proposition 3.4.2 implies (3.43).

We move to the proof of (3.46), which proceeds in the same way as the proof of Proposition 3.5.4. Let g^l be the Riemannian metric in $\overline{\Omega}$ given above, and note that

$$\langle(\Lambda_g - \Lambda_{g^l})\phi_s, \bar{\phi}_s\rangle = \int_\Omega \sum_{j,k=1}^n |g|^{1/2} g^{j,k}\partial_j u_s \partial_k \bar{u}_s^l \, dx$$
$$- \int_\Omega \sum_{j,k=1}^n |g^l|^{1/2} (g^l)^{j,k}\partial_j u_s \partial_k \bar{u}_s^l \, dx,$$

where $\Delta_g u_s = 0$ and $\Delta_{g^l} u_s^l = 0$ in Ω and $u_s = u_s^l = \phi_s$ on $\partial\Omega$. Let $\tilde{v}_s$ be the approximate solution of Proposition 3.4.2 where $N > l+1$ and $\tilde{\gamma}^{j,k} = |g|^{1/2} g^{j,k}$, and let $\tilde{v}_s^l$ be the corresponding function where g is replaced by g^l. Then

$$\langle(\Lambda_g - \Lambda_{g^l})\phi_s, \bar{\phi}_s\rangle = \int_{\tilde{\Gamma}\times[0,\delta]} \sum_{j,k=1}^n (|g|^{1/2} g^{j,k} - |g^l|^{1/2}(g^l)^{j,k})\partial_j \tilde{v}_s \overline{\partial_k \tilde{v}_s^l} \, dy + R_s$$

with $|R_s| \le Cs^{-N+2}$. As a consequence of (3.44), we obtain that

$$\partial_n^p(|g|^{1/2} g^{j,k} - |g^l|^{1/2}(g^l)^{j,k})|_{\tilde{\Gamma}} = 0 \qquad \text{for } 0 \le p \le l-1.$$

By Taylor expansion (see Lemma 3.4.3), this implies

$$\langle(\Lambda_g - \Lambda_{g^l})\phi_s, \bar{\phi}_s\rangle = \int_{\tilde{\Gamma}\times[0,\delta]} \sum_{j,k=1}^n y_n^l \psi_l^{j,k}(y', y_n)\partial_j \tilde{v}_s \overline{\partial_k \tilde{v}_s^l} \, dy + R_s,$$

where $\psi_l^{j,k}$ is a smooth function satisfying

$$\psi_l^{j,k}(y', 0) = \partial_n^l(|g|^{1/2} g^{j,k})(y', 0)/l!\,.$$

Here we used (3.45). Then, as in Proposition 3.5.4,

$$\lim_{s\to\infty} s^{l-1}\langle(\Lambda_g - \Lambda_{g^l})\phi_s, \bar{\phi}_s\rangle$$
$$= l! \int_{\tilde{\Gamma}} \Big(-\frac{1}{2f_1(y')}\Big)^{l+1} \Big[\sum_{j,k=1}^n \psi_l^{j,k}(y', 0)\partial_j\Phi(y', 0)\partial_k\bar{\Phi}(y', 0)\Big]\eta(y')^2 \, dy'.$$

The quantity in brackets is equal to

$$\frac{1}{l!}\Big[\partial_n^l(|g|^{1/2})(y', 0)f_1(y')^2 + \sum_{\alpha,\beta=1}^{n-1} \partial_n^l(|g|^{1/2} g^{\alpha,\beta})(y', 0)t_\alpha t_\beta\Big].$$

The identity (3.46) follows upon substituting the expression for f_1. □

The next step is to show that the information contained in (3.43) and (3.46) is sufficient to determine all derivatives of $g^{\alpha,\beta}$ on $\tilde{\Gamma}$. To do this, we use a basic identity concerning the derivative of the logarithm of a determinant.

Exercise 3.6.6 (Jacobi's formula). If $A(t)$ is a symmetric $n\times n$ invertible matrix depending smoothly on t in some interval, then

$$\frac{d}{dt}\big(\log\det(A(t))\big)=\sum_{j,k=1}^{n}\big(A(t)^{-1}\big)_{j,k}\frac{d}{dt}A(t)_{j,k}$$

Proposition 3.6.7. *Let $n\geq 3$ and $l\geq 1$. From the knowledge of*

$$g^{\alpha,\beta}|_{\tilde{\Gamma}},\ \partial_n g^{\alpha,\beta}|_{\tilde{\Gamma}},\ \ldots,\ \partial_n^{l-1}g^{\alpha,\beta}|_{\tilde{\Gamma}},\ \textit{and}\ [\partial_n^l(|g|^{1/2})g^{\alpha,\beta}+\partial_n^l(|g|^{1/2}g^{\alpha,\beta})]|_{\tilde{\Gamma}}$$

it is possible to determine $\partial_n^l g^{\alpha,\beta}|_{\tilde{\Gamma}}$.

Proof. Fix l and assume that one has knowledge of the stated quantities. The Leibniz rule implies that

$$\partial_n^l\big(|g|^{1/2}\big)g^{\alpha,\beta}+\partial_n^l\big(|g|^{1/2}g^{\alpha,\beta}\big)=|g|^{1/2}\partial_n^l g^{\alpha,\beta}+2\partial_n^l\big(|g|^{1/2}\big)g^{\alpha,\beta}+T_{l-1},$$

where T_{l-1} denotes an expression depending on $\partial_n^j g^{\alpha,\beta}$ where $0\leq j\leq l-1$ and $1\leq\alpha,\beta\leq n-1$. Since

$$\partial_n(|g|^{1/2})=\frac{1}{2}|g|^{-1/2}\partial_n(|g|)=\frac{1}{2}|g|^{1/2}\partial_n(\log|g|)$$

we have again by the Leibniz rule

$$\begin{aligned}\partial_n^l(|g|^{1/2})g^{\alpha,\beta}&+\partial_n^l(|g|^{1/2}g^{\alpha,\beta})\\&=|g|^{1/2}\partial_n^l g^{\alpha,\beta}+\partial_n^{l-1}\big(|g|^{1/2}\partial_n(\log|g|)\big)g^{\alpha,\beta}+T_{l-1}\\&=|g|^{1/2}\big[\partial_n^l g^{\alpha,\beta}-\partial_n^l\big(\log|g^{-1}|\big)g^{\alpha,\beta}\big]+T_{l-1}.\end{aligned}$$

The last expression is then known on $\tilde{\Gamma}$. Denoting the expression in brackets by $k^{\alpha,\beta}$, the fact that $|g|^{1/2}|_{\tilde{\Gamma}}$ and $T_{l-1}|_{\tilde{\Gamma}}$ are known implies that $k^{\alpha,\beta}|_{\tilde{\Gamma}}$ is also known. By Exercise 3.6.6 (Jacobi's formula) we have in fact

$$\begin{aligned}k^{\alpha,\beta}&=\partial_n^l g^{\alpha,\beta}-\partial_n^{l-1}\Big(\sum_{\gamma,\delta=1}^{n-1}g_{\gamma,\delta}\partial_n g^{\gamma,\delta}\Big)g^{\alpha,\beta}\\&=\partial_n^l g^{\alpha,\beta}-\Big(\sum_{\gamma,\delta=1}^{n-1}g_{\gamma,\delta}\partial_n^l g^{\gamma,\delta}\Big)g^{\alpha,\beta}+T_{l-1}.\end{aligned}$$

Since $\sum_{\alpha,\beta=1}^{n-1}g_{\alpha,\beta}g^{\alpha,\beta}=n-1$, it follows that

$$\sum_{\alpha,\beta=1}^{n-1}g_{\alpha,\beta}k^{\alpha,\beta}=(2-n)\sum_{\gamma,\delta=1}^{n-1}g_{\gamma,\delta}\partial_n^l g^{\gamma,\delta}+T_{l-1}.$$

Therefore (using that $n\geq 3$)

$$\partial_n^l g^{\alpha,\beta}=k^{\alpha,\beta}+\frac{1}{2-n}\Big(\sum_{\gamma,\delta=1}^{n-1}g_{\gamma,\delta}k^{\gamma,\delta}\Big)g^{\alpha,\beta}+T_{l-1}.$$

Since $k^{\alpha,\beta}|_{\tilde{\Gamma}}$ was known, this determines $\partial_n^l g^{\alpha,\beta}|_{\tilde{\Gamma}}$ as required. □

The main result now follows from Propositions 3.6.5 and 3.6.7.

Proof of Theorem 3.6.3. We use boundary normal coordinates (y', y_n) to identify p with 0 as before, and denote by $\tilde{\Gamma}$ a flat neighborhood of 0 in $\partial\Omega$. Since $g_{\alpha,n} = 0$ for $1 \le \alpha \le n-1$ and $g_{n,n} = 1$, it is enough to determine $\partial_n^l g_{\alpha,\beta}|_{\tilde{\Gamma}}$ for all $l \ge 0$ and $1 \le \alpha, \beta \le n-1$. Further, to determine the matrix $(g_{\alpha,\beta})|_{\tilde{\Gamma}}$ it is enough to determine the inverse matrix $(g^{\alpha,\beta})|_{\tilde{\Gamma}}$.

For higher-order derivatives the identity

$$\sum_{\beta=1}^{n} g_{\alpha,\beta} g^{\beta,\gamma} = \delta_\alpha^\gamma,$$

where δ_α^γ is the Kronecker symbol, and the Leibniz rule show that

$$\sum_{\beta=1}^{n} (\partial_n^l g_{\alpha,\beta}) g^{\beta,\gamma} = T_\alpha^\gamma,$$

where T_α^γ contains terms depending on $\partial_n^j g_{\alpha,\beta}$ for $0 \le j \le l-1$ and $\partial_n^j g^{\beta,\gamma}$ for $0 \le j \le l$. Consequently

$$\partial_n^l g_{\alpha,\delta} = \sum_{\beta,\gamma=1}^{n} (\partial_n^l g_{\alpha,\beta}) g^{\beta,\gamma} g_{\gamma,\delta} = \sum_{\gamma=1}^{n} T_\alpha^\gamma g_{\gamma,\delta}.$$

This shows that it is sufficient to prove that Λ_g determines $\partial_n^l g^{\alpha,\beta}|_{\tilde{\Gamma}}$ for all l and α, β.

By (3.43), since the test function η can be chosen arbitrarily, it follows that Λ_g determines

$$\sum_{\alpha,\beta=1}^{n-1} (\,|g|\, g^{\alpha,\beta})(y', 0) t_\alpha t_\beta.$$

Also the unit vector $t' \in \mathbb{R}^{n-1}$ was arbitrary, and therefore we determine for all $1 \le \alpha, \beta \le n-1$

$$|g(y', 0)|\, g^{\alpha,\beta}(y', 0).$$

The $(n-1) \times (n-1)$ matrix $|g|\,[g^{\alpha,\beta}]$ has determinant $|g|^{n-1}\,|g|^{-1} = |g|^{n-2}$, which is also known on $\tilde{\Gamma}$. Since $n \ge 3$ we recover $|g|$ on $\tilde{\Gamma}$ and thus also

$$g^{\alpha,\beta}|_{\tilde{\Gamma}} = |g|^{-1}\,(|g|\, g^{\alpha,\beta})|_{\tilde{\Gamma}}.$$

This determines the boundary values of $g^{\alpha,\beta}$ on $\tilde{\Gamma}$.

For higher-order derivatives, we note that (3.46) shows upon varying η and t' that Λ_g determines on $\tilde{\Gamma}$ the quantity

$$\partial_n^l(|g|^{1/2}) g^{\alpha,\beta} + \partial_n^l(|g|^{1/2}\, g^{\alpha,\beta}).$$

Taking $l = 1$ and using that $g^{\alpha,\beta}|_{\tilde{\Gamma}}$ was known, Proposition 3.6.7 shows that $\partial_n g^{\alpha,\beta}|_{\tilde{\Gamma}}$ is determined by Λ_g. Proceeding inductively, we recover $\partial_n^l g^{\alpha,\beta}|_{\tilde{\Gamma}}$ for all $l \geq 0$ and $1 \leq \alpha, \beta \leq n-1$. □

3.7. Notes

The first boundary determination result in the Calderón problem was given in [**KV84a**], where it was shown that the Taylor series of a C^∞ conductivity at the boundary is uniquely determined by the DN map. The proof was based on solutions whose boundary values are rapidly oscillating and have vanishing moments. Another proof was given in [**SU88**] together with stability estimates. This proof was based on the fact that the DN map is a pseudodifferential operator whose symbol determines the Taylor series of the conductivity at the boundary. The direct construction of concentrating solutions in §3.1 is due to [**Bro01**] and it applies in low regularity settings. The Rellich identity argument for determining the normal derivative of a C^1 conductivity in §3.2 follows [**GZ16**, appendix with R. Brown]. Another method for reconstructing the boundary value and normal derivative in a low regularity setting is given in [**Nac96**].

The boundary determination result for a C^∞ anisotropic conductivity in §3.6 was proved in [**LU89**], who used boundary normal coordinates as in §3.3 and the fact that the DN map is a pseudodifferential operator. The direct construction of concentrating solutions to high order given in §3.4 and the recovery of derivatives of high order as in §3.5 appear in many places including [**NT01, KY02, Sal20**]. The argument for the anisotropic, or Riemannian, case in §3.6 follows these ideas. For basic facts on Riemannian geometry see [**Lee18**] or the work [**Sal13**] that includes an introduction from the point of view of the Calderón problem.

The methods based on concentrating solutions, either using the pseudodifferential approach or direct constructions, are constructive and local and they yield stability results. These methods have been extended to numerous other equations and systems. Selected results include [**NSU95, BS06**] for magnetic Schrödinger equation, [**DSFKSU09**] for Laplace–Beltrami equation with magnetic and electric potentials (with a correction in [**LLS22**]), [**McD97, JM00, CLLZ21**] for the Maxwell system, [**NU95**] for elasticity, [**HLW07, LW07**] for the Stokes and Navier-Stokes systems, [**NT00**] for a Dirac system, and [**OSSU24**] for general scalar equations.

Another approach to boundary determination, based on singular (Green function type) solutions instead of oscillatory solutions, was introduced in [**Isa88, Ale90**]. This method applies in low regularity settings and yields stability results. Boundary determination results can also be proved via monotonicity arguments and localized potentials [**Geb08**].

As stated in Theorem 3.0.2, the recovery of Taylor series of the conductivity at the boundary implies uniqueness for real-analytic conductivities. This was extended to the case of piecewise real-analytic conductivities in [**KV85**] and also in [**Dru82, Dru85**]. There have been many further works such as [**Har12, AdHG17**] where piecewise constant or piecewise real-analytic coefficients are recovered.

Chapter 4

The Calderón Problem in Dimensions ≥ 3

In this chapter, we prove interior uniqueness, stability, and reconstruction results for the Calderón problem in dimensions three and higher. To describe the contents of this chapter, we assume that Ω is a bounded open subset of $\mathbb{R}^n$, $n \geq 3$, with C^2 boundary, and all conductivities are positive functions in $C^2(\overline{\Omega})$. We will employ the Fourier transform whose basic properties are recalled in Appendix B.

The linearized version of the Calderón problem is considered in §4.1, and we present an argument showing uniqueness in the linearized problem. The proof is based on special harmonic functions given by the complex exponentials $e^{i\zeta \cdot x}$, where $\zeta \in \mathbb{C}^n$ is a vector satisfying $\zeta \cdot \zeta = 0$. In §4.2, we construct complex geometrical optics (CGO) solutions that resemble the harmonic exponentials by a Fourier series type argument. These solutions are used in §4.3 to prove that if two conductivities γ_1 and γ_2 give rise to the same boundary measurements, then $\gamma_1 = \gamma_2$ throughout Ω. This follows from the corresponding uniqueness result for Schrödinger operators. In §4.4 we give a quantitative version of the previous uniqueness theorem, stated in the form of a logarithmic stability result.

Another proof for the existence of CGO solutions, now based on the Fourier transform instead of Fourier series and including a uniqueness notion for these solutions, is given in §4.5. This is employed in §4.6 in a reconstruction procedure that allows one to construct a conductivity from the knowledge of the DN map.

4.1. The linearized Calderón problem

The "plane wave" exponential function

$$u = e^{i\zeta\cdot x}, \qquad \zeta \in \mathbb{C}^n,$$

is a solution to Laplace's equation,

$$\Delta u = 0,$$

if and only if

$$\zeta \cdot \zeta = 0.$$

If $\zeta = \eta + ik$ with $\eta, k \in \mathbb{R}^n$, then

$$0 = \zeta \cdot \zeta = |\eta|^2 - |k|^2 + 2i\eta \cdot k \iff |\eta| = |k| \text{ and } \eta \perp k. \tag{4.1}$$

Thus $\zeta \cdot \zeta = 0$ if and only if the real and imaginary parts of ζ are orthogonal and have the same length.

Any nonzero ζ obeying $\zeta \cdot \zeta = 0$ will have nonzero real and imaginary parts and the corresponding solution, $e^{i\zeta\cdot x}$, will grow or decay exponentially in some directions in $\mathbb{R}^n$ and will oscillate in some directions. The utility of exponentially growing solutions in solving the inverse conductivity problem was first observed by Calderón. We begin by exhibiting his uniqueness proof for the linearized inverse boundary value problem, when linearized at constant conductivity.

Theorem 4.1.1 (Uniqueness of the linearized Calderón problem). *The Fréchet derivative $D\Lambda_1$ of the map $\gamma \mapsto \Lambda_\gamma$ at $\gamma = 1$, $\delta\gamma \mapsto D\Lambda_1[\delta\gamma]$, is injective. That is, if*

$$D\Lambda_1[\delta\gamma] = 0$$

for some $\delta\gamma \in L^\infty(\Omega)$, then

$$\delta\gamma = 0.$$

Let us first show that the Fréchet derivative (see Definition A.2.28) exists.

Theorem 4.1.2. *Let $\Omega \subset \mathbb{R}^n$ be a bounded open set, and let $\gamma \in L^\infty(\Omega)$ satisfy $\gamma \geq c > 0$ a.e. in Ω. For $\delta\gamma \in L^\infty(\Omega)$, the identity*

$$\langle D\Lambda_\gamma[\delta\gamma]f, g\rangle_{\partial\Omega} = \int_\Omega \delta\gamma \nabla u \cdot \nabla v \, dx, \qquad f, g \in H^{1/2}(\partial\Omega),$$

where $u, v \in H^1(\Omega)$ satisfy $\operatorname{div}(\gamma\nabla u) = \operatorname{div}(\gamma\nabla v) = 0$ *in Ω with $u|_{\partial\Omega} = f$, $v|_{\partial\Omega} = g$, defines a bounded linear map*

$$D\Lambda_\gamma[\delta\gamma] : H^{1/2}(\partial\Omega) \to H^{-1/2}(\partial\Omega).$$

This map gives the Fréchet derivative of the map $\gamma \mapsto \Lambda_\gamma$ in the sense that

$$\lim_{\|\delta\gamma\|_{L^\infty(\Omega)}\to 0} \frac{\Lambda_{\gamma+\delta\gamma} - \Lambda_\gamma - D\Lambda_\gamma[\delta\gamma]}{\|\delta\gamma\|_{L^\infty(\Omega)}} = 0$$

in the space of bounded operators from $H^{1/2}(\partial\Omega)$ to $H^{-1/2}(\partial\Omega)$.

Proof. Let $\gamma(t) = \gamma_t$ be a C^1 curve of functions in $L^\infty(\Omega)$ that are uniformly bounded from below. This means that there is $E > 0$ such that $1/E \le \gamma(t) \le E$ a.e. in Ω for t near 0, and

$$\gamma(t) = \gamma(0) + t\dot{\gamma}(0) + t\varepsilon(t),$$

where $\dot{\gamma}(0) \in L^\infty(\Omega)$ and

$$\lim_{t\to 0} \|\varepsilon(t)\|_{L^\infty(\Omega)} = 0.$$

Let $f, g \in H^{1/2}(\partial\Omega)$, and let u_t and v_t satisfy

$$\operatorname{div}(\gamma_t \nabla u_t) = \operatorname{div}(\gamma_t \nabla v_t) = 0 \text{ in } \Omega, \quad u_t|_{\partial\Omega} = f, v_t|_{\partial\Omega} = g.$$

By the integral identity in Proposition 2.8.2,

$$\langle (\Lambda_{\gamma_t} - \Lambda_{\gamma_0}) f, g\rangle_{\partial\Omega} = \int_\Omega (\gamma_t - \gamma_0)\nabla u_t \cdot \nabla v_0 \, dx.$$

Consequently

$$\langle (\frac{1}{t}\Lambda_{\gamma_t} - \Lambda_{\gamma_0}) f, g\rangle_{\partial\Omega} = \int_\Omega (\dot{\gamma}(0) + \varepsilon(t))\nabla u_t \cdot \nabla v_0 \, dx. \tag{4.2}$$

Using the bounds for γ_t and Theorem 2.5.6, we have

$$\|u_t\|_{H^1(\Omega)} \le C(E, \Omega) \|f\|_{H^{1/2}(\partial\Omega)}.$$

Also, since $u_t - u_0$ solves

$$\begin{aligned} \operatorname{div}(\gamma_t \nabla(u_t - u_0)) &= -\operatorname{div}((\gamma_t - \gamma_0)\nabla u_0) \text{ in } \Omega, \\ u_t - u_0|_{\partial\Omega} &= 0, \end{aligned}$$

we have

$$\begin{aligned} \|u_t - u_0\|_{H^1(\Omega)} &\le C(E, \Omega) \|\operatorname{div}((\gamma_t - \gamma_0)\nabla u_0)\|_{H^{-1}(\Omega)} \\ &\le C(E, \Omega) \|\gamma_t - \gamma_0\|_{L^\infty(\Omega)} \|\nabla u_0\|_{L^2(\Omega)} \\ &\le C(E, \Omega) \|\gamma_t - \gamma_0\|_{L^\infty(\Omega)} \|f\|_{H^{1/2}(\Omega)}. \end{aligned}$$

Since $\|\gamma_t - \gamma_0\|_{L^\infty(\Omega)} \le Ct$, we may combine these facts and take the limit in (4.2) as $t \to 0$ to obtain

$$\langle D\Lambda_{\gamma(0)}[\dot{\gamma}(0)] f, g\rangle_{\partial\Omega} = \int_\Omega \dot{\gamma}(0)\nabla u_0 \cdot \nabla v_0 \, dx.$$

This defines $D\Lambda_{\gamma(0)}[\dot{\gamma}(0)] f$ weakly as an element of $H^{-1/2}(\partial\Omega)$, since

$$\begin{aligned} |\langle D\Lambda_{\gamma(0)}[\dot{\gamma}(0)] f, g\rangle_{\partial\Omega}| &\le \|\dot{\gamma}(0)\|_{L^\infty(\Omega)} \|\nabla u_0\|_{L^2(\Omega)} \|\nabla v_0\|_{L^2(\Omega)} \\ &\le \|\dot{\gamma}(0)\|_{L^\infty(\Omega)} \|f\|_{H^{1/2}(\partial\Omega)} \|g\|_{H^{1/2}(\partial\Omega)}. \end{aligned}$$

It also follows that $D\Lambda_{\gamma(0)}[\dot{\gamma}(0)]$ is bounded from $H^{1/2}(\partial\Omega)$ to $H^{-1/2}(\partial\Omega)$.

Finally, $D\Lambda_{\gamma(0)}[\dot{\gamma}(0)]$ is the Fréchet derivative since by (4.2) one has

$$\langle\Big[\frac{1}{t}(\Lambda_{\gamma_t}-\Lambda_{\gamma_0})-D\Lambda_{\gamma(0)}[\dot{\gamma}(0)]\Big]f,g\rangle_{\partial\Omega}$$
$$=\int_\Omega \dot{\gamma}(0)(\nabla u_t-\nabla u_0)\cdot\nabla v_0\,dx+\int_\Omega \varepsilon(t)\nabla u_t\cdot\nabla v_0\,dx.$$

Thus,

$$\Big|\langle\Big[(\frac{1}{t}\Lambda_{\gamma_t}-\Lambda_{\gamma_0})-D\Lambda_{\gamma(0)}[\dot{\gamma}(0)]\Big]f,g\rangle_{\partial\Omega}\Big|$$
$$\leq (\|\dot{\gamma}(0)\|_{L^\infty(\Omega)}\|\nabla u_t-\nabla u_0\|_{L^2(\Omega)}+\|\varepsilon(t)\|_{L^\infty(\Omega)}\|u_t\|_{H^1(\Omega)})\|v_0\|_{H^1(\Omega)}$$
$$\leq C(\|\dot{\gamma}(0)\|_{L^\infty(\Omega)}\|\gamma_t-\gamma_0\|_{L^\infty(\Omega)}+\|\varepsilon(t)\|_{L^\infty(\Omega)})\|f\|_{H^{1/2}(\partial\Omega)}\|g\|_{H^{1/2}(\partial\Omega)}.$$

The result follows by taking $\gamma(t)=\gamma+t\delta\gamma/\|\delta\gamma\|_{L^\infty(\Omega)}$ and letting $t=\|\delta\gamma\|_{L^\infty(\Omega)}\to 0$. □

Proof of Theorem 4.1.1. By Theorem 4.1.2, the equation

$$D\Lambda_\gamma[\delta\gamma]=0$$

is equivalent to

$$\int_\Omega \delta\gamma\nabla u_1\cdot\nabla u_2\,dx=0 \quad \text{for all } u_j\in H^1(\Omega) \text{ with } \operatorname{div}(\gamma\nabla u_j)=0. \tag{4.3}$$

If we further restrict to $\gamma=1$ then (4.3) must hold for every pair of harmonic functions u_1 and u_2. A natural set of choices for u_1 and u_2 are exponentials $e^{i\zeta\cdot x}$ with $\zeta\cdot\zeta=0$. Substituting

$$u_1=e^{i\zeta_1\cdot x},\qquad u_2=e^{i\zeta_2\cdot x}$$

with $\zeta_j\cdot\zeta_j=0$, into (4.3) gives

$$\zeta_1\cdot\zeta_2\int_\Omega e^{i(\zeta_1+\zeta_2)\cdot x}\delta\gamma\,dx=0$$

By (4.1), we may choose $\zeta_1=\frac{1}{2}(\eta+ik)$ and $\zeta_2=\frac{1}{2}(\eta-ik)$ with any $\eta,k\in\mathbb{R}^n$ for which $|\eta|=|k|$ and $\eta\perp k$. Then

$$\zeta_1+\zeta_2=\eta \qquad\text{and}\qquad \zeta_1\cdot\zeta_2=\tfrac{1}{2}|\eta|^2,$$

so $D\Lambda_1[\delta\gamma]=0$ implies that

$$|\eta|^2\int_\Omega e^{ix\cdot\eta}\delta\gamma\,dx=0.$$

This, in turn, implies that the Fourier transform $\hat{h}(-\eta)$ vanishes for every nonzero η, where h is the function with $h=\delta\gamma$ in Ω and $h=0$ outside Ω. However, h is an element of $L^1(\mathbb{R}^n)$, so $\hat{h}$ is continuous and therefore must vanish everywhere in $\mathbb{R}^n$. As a consequence

$$\delta\gamma=0,$$

which proves that $D\Lambda_1[\,\cdot\,]$ is injective. □

4.2. Complex geometrical optics solutions

Our approach for proving the uniqueness theorem in §4.3 is based on exponential solutions that are perturbations of those for the Laplacian. These solutions have many names, including exponentially growing solutions, Faddeev type solutions, and Sylvester–Uhlmann type solutions. We will call these solutions *complex geometrical optics (CGO) solutions*, since they are complex phase analogs of standard geometrical optics solutions for wave equations.

In this section, we will construct CGO solutions to the Schrödinger equation

$$(-\Delta + q)u = 0 \quad \text{in } \Omega.$$

The potential q is assumed to be in $L^\infty(\Omega)$. To motivate the construction, first let $q = 0$. We have seen that there are solutions to the equation $-\Delta u = 0$ having the form of a complex exponential at frequency $\zeta \in \mathbb{C}^n$,

$$u(x) = e^{i\zeta\cdot x}, \quad \zeta\cdot\zeta = 0.$$

Now suppose q is nonzero. The function $u = e^{i\zeta\cdot x}$ is not an exact solution of $(-\Delta + q)u = 0$ anymore, but we can find solutions which resemble complex exponentials. These are the CGO solutions, which have the form

$$u(x) = e^{i\zeta\cdot x}(1 + r(x,\zeta)). \tag{4.4}$$

Here r is a *correction term* which is needed to convert the approximate solution $e^{i\zeta\cdot x}$ to an exact solution.

In fact, we are interested in solutions in the *asymptotic limit* as $|\zeta| \to \infty$. This follows the principle that while it is usually not possible to obtain explicit formulas for solutions to variable coefficient equations, in suitable asymptotic limits explicit expressions for solutions may still exist.

The next theorem is the main result on the existence of CGO solutions. Note that the constant function $a \equiv 1$ always satisfies the transport equation $\zeta\cdot\nabla a = 0$, so as a special case one obtains the solutions $u = e^{i\zeta\cdot x}(1 + r)$ mentioned above.

Theorem 4.2.1 (CGO solutions). *Let $\Omega \subset \mathbb{R}^n$ be a bounded open set, and let $q \in L^\infty(\Omega)$. There is a constant C_0 depending only on Ω and n, such that for any $\zeta \in \mathbb{C}^n$ satisfying $\zeta\cdot\zeta = 0$ and $|\zeta| \geq \max(C_0\,\|q\|_{L^\infty(\Omega)}, 1)$, and for any function $a \in H^2(\Omega)$ satisfying*

$$\zeta\cdot\nabla a = 0 \quad \text{in } \Omega,$$

the equation $(-\Delta + q)u = 0$ in Ω has a solution $u \in H^2(\Omega)$ of the form

$$u(x) = e^{i\zeta\cdot x}(a + r), \tag{4.5}$$

where $r \in H^2(\Omega)$ satisfies

$$\|r\|_{H^k(\Omega)} \le C_0 |\zeta|^{k-1} \|(-\Delta + q)a\|_{L^2(\Omega)}, \qquad k = 0, 1, 2.$$

We note that (4.4) is a solution of $(-\Delta + q)u = 0$ if and only if

$$e^{-i\zeta\cdot x}(-\Delta + q)e^{i\zeta\cdot x}(1 + r) = 0. \tag{4.6}$$

It will be convenient to conjugate the exponentials $e^{i\zeta\cdot x}$ into the Laplacian. Writing $D_j = \frac{1}{i}\partial_{x_j}$ and $D = (D_1, \ldots, D_n)$, by this we mean that

$$\begin{aligned} e^{-i\zeta\cdot x}D_j(e^{i\zeta\cdot x}v) &= (D_j + \zeta_j)v, \\ e^{-i\zeta\cdot x}D^2(e^{i\zeta\cdot x}v) &= (D + \zeta)^2 v = (D^2 + 2\zeta \cdot D)v. \end{aligned}$$

Here we used the crucial condition $\zeta \cdot \zeta = 0$.

We can rewrite (4.6) as an equation for the *conjugated Laplacian*,

$$(D^2 + 2\zeta \cdot D + q)(1 + r) = 0.$$

This implies the following equation for r:

$$(D^2 + 2\zeta \cdot D + q)r = -q. \tag{4.7}$$

The solvability of (4.7) is the most important part in the construction of CGO solutions. We give the proof in several steps.

4.2.1. Basic estimate.

We first consider the free case in which there is no potential on the left-hand side of (4.7).

Theorem 4.2.2 (Estimates with no potential). *There is a constant C_0 depending only on Ω and n, such that for any $\zeta \in \mathbb{C}^n$ satisfying $\zeta \cdot \zeta = 0$ and $|\zeta| \ge 1$, and for any $f \in L^2(\Omega)$, the equation*

$$(D^2 + 2\zeta \cdot D)r = f \quad \text{in } \Omega \tag{4.8}$$

has a solution $r \in H^2(\Omega)$ satisfying

$$\|r\|_{H^k(\Omega)} \le C_0 |\zeta|^{k-1} \|f\|_{L^2(\Omega)}, \quad k = 0, 1, 2.$$

Note in particular that

$$\|r\|_{L^2(\Omega)} \le \frac{C_0}{|\zeta|} \|f\|_{L^2(\Omega)}.$$

If $\|f\|_{L^2(\Omega)}$ is uniformly bounded in ζ, this shows that $r \to 0$ in $L^2(\Omega)$ as $|\zeta| \to \infty$. Accordingly, the correction term r in the CGO solution (4.4) will be very small for large ζ, and the solution (4.4) will look like the complex exponential $e^{i\zeta\cdot x}$ then.

The idea of the proof is that (4.7) is a linear equation with constant coefficients, so one can try to solve it by the Fourier transform. Since $(D_j u)\hat{\ }(\xi) = \xi_j \hat{u}(\xi)$, the Fourier transformed equation is

$$(\xi^2 + 2\zeta \cdot \xi)\hat{r}(\xi) = \hat{f}(\xi).$$

We would like to divide by $\xi^2 + 2\zeta \cdot \xi$ and use the inverse Fourier transform to get a solution r. However, the symbol $\xi^2 + 2\zeta \cdot \xi$ vanishes for some $\xi \in \mathbb{R}^n$, and the division cannot be done directly.

It turns out that we can divide by the symbol if we use Fourier series in a large cube instead of the Fourier transform, and moreover take the Fourier coefficients in a shifted lattice instead of the usual integer coordinate lattice.

Proof of Theorem 4.2.2.

(1) Write $\zeta = s(\omega_1 + i\omega_2)$ where $s = |\zeta|/\sqrt{2}$ and ω_1 and ω_2 are orthogonal unit vectors in $\mathbb{R}^n$. By rotating coordinates in a suitable way, we can assume that $\omega_1 = e_1$ and $\omega_2 = e_2$ (the first and second coordinate vectors). Thus we need to solve the equation

$$(D^2 + 2s(D_1 + iD_2))r = f.$$

(2) We assume for simplicity that Ω is contained in the cube $Q = (-\pi, \pi)^n$. Extend f by zero outside Ω into Q, which gives a function in $L^2(Q)$ also denoted by f. We need to solve

$$(D^2 + 2s(D_1 + iD_2))r = f \quad \text{in } Q. \tag{4.9}$$

Let $w_k(x) = e^{i(k+\frac{1}{2}e_2)\cdot x}$ for $k \in \mathbb{Z}^n$. That is, we consider Fourier series in the lattice $\mathbb{Z}^n + \frac{1}{2}e_2$. Equipping $L^2(Q)$ with the inner product

$$(u, v) = (2\pi)^{-n} \int_Q u\bar{v}\, dx, \quad u, v \in L^2(Q),$$

we see that $(w_k, w_l) = 0$ if $k \neq l$ and $(w_k, w_k) = 1$, so $\{w_k\}$ is an orthonormal set in $L^2(Q)$. It is also a complete set: if $v \in L^2(Q)$ and $(v, w_k) = 0$ for all $k \in \mathbb{Z}^n$ then $(ve^{-\frac{1}{2}ix_2}, e^{ik\cdot x}) = 0$ for all $k \in \mathbb{Z}^n$. This implies $v = 0$ since $\{e^{ik\cdot x}\}_{k\in\mathbb{Z}^n}$ is an orthonormal basis of $L^2(Q)$.

(3) Hilbert space theory (see §A.1) gives that $\{w_k\}$ is an orthonormal basis of $L^2(Q)$ and f can be written as the series $f = \sum_{k\in\mathbb{Z}^n} f_k w_k$, where $f_k = (f, w_k)$ and $\|f\|_{L^2(Q)}^2 = \sum_{k\in\mathbb{Z}^n} |f_k|^2$. Seeking also r in the form $r = \sum_{k\in\mathbb{Z}^n} r_k w_k$, and using that

$$Dw_k = (k + \frac{1}{2}e_2)w_k,$$

the equation (4.9) implies that the Fourier coefficients should satisfy

$$p_k r_k = f_k, \quad k \in \mathbb{Z}^n,$$

where

$$p_k := \left(k + \tfrac{1}{2}e_2\right)^2 + 2s(k_1 + i(k_2 + \tfrac{1}{2})).$$

Note that $\mathrm{Im}(p_k) = 2s(k_2 + \frac{1}{2})$ is never zero for $k \in \mathbb{Z}^n$, which was the reason for considering the shifted lattice. We define

$$r_k := \frac{1}{p_k} f_k$$

and

$$r^{(N)} := \sum_{k \in \mathbb{Z}^n, |k| \leq N} r_k w_k.$$

The Fourier coefficients satisfy

$$|r_k| \leq \frac{1}{|p_k|} |f_k| \leq \frac{1}{|\mathrm{Im}(p_k)|} |f_k| \leq \frac{1}{\left|2s(k_2 + \frac{1}{2})\right|} |f_k| \leq \frac{1}{s} |f_k|.$$

Now $(r^{(N)})$ is a Cauchy sequence in $L^2(Q)$, since for $M \geq N$

$$\|r_M - r_N\|_{L^2(Q)} = \Big(\sum_{N<|k|\leq M} |r_k|^2 \Big)^{1/2} \leq \frac{1}{s} \Big(\sum_{N<|k|\leq M} |f_k|^2 \Big)^{1/2}$$

and the last quantity can be made arbitrarily small if M and N are large. This shows that $r^{(N)} \to r$ in $L^2(Q)$, where $r \in L^2(Q)$ satisfies

$$r = \sum_{k \in \mathbb{Z}^n} r_k w_k, \quad \|r\|_{L^2(Q)} \leq \frac{1}{s} \|f\|_{L^2(Q)}.$$

(4) We next show that $r \in H^2(Q)$. Note that

$$D_a r^{(N)} = \sum_{|k|\leq N} (k + \tfrac{1}{2}e_2)_a r_k w_k,$$

$$D_a D_b r^{(N)} = \sum_{|k|\leq N} (k + \tfrac{1}{2}e_2)_a (k + \tfrac{1}{2}e_2)_b r_k w_k.$$

We claim that for $a, b = 1, \ldots, n$ and for $k \in \mathbb{Z}^n$,

$$\left|(k + \tfrac{1}{2}e_2)_a r_k\right| \leq 4 |f_k|, \quad \left|(k + \tfrac{1}{2}e_2)_a (k + \tfrac{1}{2}e_2)_b r_k\right| \leq 16s |f_k|. \tag{4.10}$$

Consider two cases: if $\left|k + \frac{1}{2}e_2\right| \leq 4s$ (the small frequency case) we have

$$\left|(k + \tfrac{1}{2}e_2)_a r_k\right| \leq \frac{4s}{2s|k_2 + 1/2|} |f_k| \leq 4 |f_k|$$

and

$$\left|(k + \tfrac{1}{2}e_2)_a (k + \tfrac{1}{2}e_2)_b r_k\right| \leq \frac{(4s)^2}{2s|k_2 + 1/2|} \leq 16s |f_k|.$$

If $\left|k + \frac{1}{2}e_2\right| \geq 4s$ (the large frequency case), then

$$|\mathrm{Re}(p_k)| = \left|\left|k + \tfrac{1}{2}e_2\right|^2 + 2sk_1\right| \geq \left|k + \tfrac{1}{2}e_2\right|^2 - 2s\left|k + \tfrac{1}{2}e_2\right| \geq \tfrac{1}{2}\left|k + \tfrac{1}{2}e_2\right|^2$$

which implies

$$\left|(k+\tfrac{1}{2}e_2)_a r_k\right| \le \frac{\left|k+\frac{1}{2}e_2\right|}{\frac{1}{2}\left|k+\frac{1}{2}e_2\right|^2}\,|f_k| \le \frac{1}{2s}\,|f_k|$$

and

$$\left|(k+\tfrac{1}{2}e_2)_a(k+\tfrac{1}{2}e_2)_b r_k\right| \le \frac{\left|k+\frac{1}{2}e_2\right|^2}{\frac{1}{2}\left|k+\frac{1}{2}e_2\right|^2}\,|f_k| \le 2\,|f_k|\,.$$

This proves (4.10) since $s \ge 1$.

The estimates (4.10) imply that $D_a r^{(N)}$ and $D_a D_b r^{(N)}$ are Cauchy sequences in $L^2(Q)$, and thus converge to some v_a and v_{ab} in $L^2(Q)$. Further, the weak derivatives of r satisfy

$$D_a r = v_a, \quad D_a D_b r = v_{ab}.$$

To see this, let $\varphi \in C_c^\infty(Q)$ and note that

$$\begin{aligned} -\int_Q r D_a\varphi\,dx &= -\lim_{N\to\infty}\int_\Omega r^{(N)} D_a\varphi\,dx \\ &= \lim_{N\to\infty}\int_\Omega (D_a r^{(N)})\varphi\,dx \\ &= \int_\Omega v_a\varphi\,dx. \end{aligned}$$

The proof for $D_a D_b r$ is analogous.

(5) We have proved that $r \in H^2(Q)$ and that

$$D_a r = \sum_{k\in\mathbb{Z}^n}(k+\tfrac{1}{2}e_2)_a r_k w_k, \quad D_a D_b r = \sum_{k\in\mathbb{Z}^n}(k+\tfrac{1}{2}e_2)_a(k+\tfrac{1}{2}e_2)_b r_k w_k.$$

It immediately follows that r solves (4.9), and the norm estimates for r follow from (4.10) and the Parseval identity. □

4.2.2. Basic estimate with potential. Now we consider the solution of (4.7) in the presence of a nonzero potential. It will be convenient to give a name to the solution operator in the free case.

Definition 4.2.3 (Solution operator). Let $\zeta \in \mathbb{C}^n$ satisfy $\zeta\cdot\zeta = 0$ and $|\zeta| \ge 1$. We denote by G_ζ the solution operator

$$G_\zeta : L^2(\Omega) \to H^2(\Omega), \quad f \mapsto r,$$

where r is the solution to $(D^2 + 2\zeta\cdot D)r = f$ provided by Theorem 4.2.2.

Theorem 4.2.4 (Estimates with potential). *Let $q \in L^\infty(\Omega)$. There is a constant C_0 depending only on Ω and n, such that for any $\zeta \in \mathbb{C}^n$ satisfying $\zeta \cdot \zeta = 0$ and $|\zeta| \geq \max(C_0 \|q\|_{L^\infty(\Omega)}, 1)$, and for any $f \in L^2(\Omega)$, the equation*

$$(D^2 + 2\zeta \cdot D + q)r = f \quad \text{in } \Omega \tag{4.11}$$

has a solution $r \in H^2(\Omega)$ satisfying

$$\|r\|_{H^k(\Omega)} \leq C_0 |\zeta|^{k-1} \|f\|_{L^2(\Omega)}, \quad k = 0, 1, 2.$$

Proof. If one has $q = 0$, a solution would be given by $r = G_\zeta f$. Here q may be nonzero, so we try a solution of the form

$$r := G_\zeta \tilde{f}, \tag{4.12}$$

where $\tilde{f} \in L^2(\Omega)$ is a function to be determined. Inserting (4.12) in the equation (4.11), and using that $(D^2 + 2\zeta \cdot D)G_\zeta = \mathrm{Id}$, we see that $\tilde{f}$ should satisfy

$$(\mathrm{Id} + qG_\zeta)\tilde{f} = f \quad \text{in } \Omega. \tag{4.13}$$

By Theorem 4.2.2, we have the norm estimate

$$\left\|qG_\zeta\right\|_{L^2(\Omega)\to L^2(\Omega)} \leq \frac{C_0 \|q\|_{L^\infty(\Omega)}}{|\zeta|}.$$

If $|\zeta| \geq \max(2C_0 \|q\|_{L^\infty(\Omega)}, 1)$ then

$$\left\|qG_\zeta\right\|_{L^2(\Omega)\to L^2(\Omega)} \leq \frac{1}{2}.$$

It follows (see Theorem A.2.22) that $\mathrm{Id} + qG_\zeta$ is an invertible operator on $L^2(\Omega)$, and the inverse is given by a Neumann series

$$(\mathrm{Id} + qG_\zeta)^{-1} = \sum_{j=0}^\infty (-qG_\zeta)^j.$$

The equation (4.13) has a solution

$$\tilde{f} := (\mathrm{Id} + qG_\zeta)^{-1} f.$$

The definition (4.12) for r implies

$$(D^2 + 2\zeta \cdot D + q)r = \tilde{f} + qG_\zeta \tilde{f} = (\mathrm{Id} + qG_\zeta)\tilde{f} = f,$$

so r indeed solves the equation (4.11). Since $\left\|(\mathrm{Id} + qG_\zeta)^{-1}\right\|_{L^2(\Omega)\to L^2(\Omega)} \leq 2$, we have $\|\tilde{f}\|_{L^2(\Omega)} \leq 2\|f\|_{L^2(\Omega)}$. The norm estimates in Theorem 4.2.2 imply the desired estimates for r, if we replace C_0 by $2C_0$. □

It is now easy to prove the main result on the existence of CGO solutions.

Proof of Theorem 4.2.1. The function (4.5) solves $(-\Delta+q)u = 0$ if and only if

$$e^{-i\zeta\cdot x}(-\Delta + q)e^{i\zeta\cdot x}(a + r) = 0.$$

As in the beginning of this section, we conjugate the exponentials into the derivatives and rewrite (4.6) as

$$(D^2 + 2\zeta \cdot D + q)(a + r) = 0.$$

Since $\zeta \cdot Da = 0$, this implies the following equation for r:

$$(D^2 + 2\zeta \cdot D + q)r = -(D^2 + q)a.$$

Theorem 4.2.4 guarantees the existence of a solution r satisfying the given norm estimates. Then (4.5) is the required solution to $(-\Delta + q)u = 0$ in Ω. □

Exercise 4.2.5 (Small first-order perturbations). Prove the analog of Theorem 4.2.4 for the equation

$$(D^2 + 2\zeta \cdot D + 2A \cdot (D + \zeta) + q)r = f \quad \text{in } \Omega$$

where $A \in L^\infty(\Omega\,;\,\mathbb{C}^n)$, $q \in L^\infty(\Omega)$, and $\|A\|_{L^\infty(\Omega)}$ is sufficiently small (depending on Ω and n).

Exercise 4.2.6 (CGO solutions for small first-order perturbations). Prove the analog of Theorem 4.2.1 for the equation

$$(-\Delta + 2A \cdot D + q)u = 0 \quad \text{in } \Omega,$$

where $A \in L^\infty(\Omega\,;\,\mathbb{C}^n)$, $q \in L^\infty(\Omega)$, and $\|A\|_{L^\infty(\Omega)}$ is sufficiently small. In this case, the transport equation for a is

$$\zeta \cdot (D + A)a = 0 \quad \text{in } \Omega.$$

Exercise 4.2.7 (H^k to H^{k+2} mapping properties). Let $\Omega \subset \mathbb{R}^n$ be a bounded open set with C^k boundary. Show that there is a constant C_0 such that for any $\zeta \in \mathbb{C}^n$ with $\zeta \cdot \zeta = 0$ and $|\zeta| \geq 1$, there is a solution $r \in H^{k+2}(\Omega)$ of the equation

$$(D^2 + 2\zeta \cdot D)r = f \quad \text{in } \Omega$$

satisfying

$$\|r\|_{H^{k+l}(\Omega)} \leq C_0\, |\zeta|^{l-1}\, \|f\|_{H^k(\Omega)}, \quad l = 0, 1, 2.$$

(Hint: use the fact that there is a continuous extension operator $H^k(\Omega) \to H^k(\mathbb{R}^n)$.)

Exercise 4.2.8 (Additional decay for the H^1 norm). If f is a *fixed* function in $L^2(\Omega)$, show that for any $\zeta \in \mathbb{C}^n$ with $\zeta \cdot \zeta = 0$ and $|\zeta| \geq 1$ there is a solution $r = r(\,\cdot\,;\zeta) \in H^2(\Omega)$ of the equation

$$(D^2 + 2\zeta \cdot D)r = f \quad \text{in } \Omega$$

satisfying

$$\lim_{|\zeta|\to\infty} \|r(\,\cdot\,;\zeta)\|_{H^1(\Omega)} = 0.$$

(Hint: decompose f into a smooth part and a small part.)

4.3. Interior uniqueness

In this section, we use the special solutions constructed in §4.2 together with the boundary identifiability results in §3.1 and §3.2 to prove a global identifiability result for dimension $n \geq 3$. The case $n = 2$ will be considered in Chapter 5.

Theorem 4.3.1 (Interior uniqueness for Calderón problem). *Let $\Omega \subset \mathbb{R}^n$, $n \geq 3$, be a bounded open set with C^2 boundary, and let γ_1, γ_2 be positive functions in $C^2(\overline{\Omega})$. If*

$$\Lambda_{\gamma_1} = \Lambda_{\gamma_2},$$

then

$$\gamma_1 = \gamma_2.$$

We will obtain Theorem 4.3.1 as a consequence of an analogous theorem for the Schrödinger equation, via the reduction given in Proposition 2.8.5. We have seen in Example 2.5.8 that the Dirichlet problem for the Schrödinger equation need not always have a unique solution, and the DN map may not exist. It is quite natural to use the Cauchy data set, introduced already in Definition 2.7.8:

$$C_q = \{\, (u|_{\partial\Omega}, \partial_\nu u|_{\partial\Omega}) \mid u \in H^1(\Omega),\ (-\Delta + q)u = 0 \text{ in } \Omega \,\}.$$

Here, the normal derivative of a solution of $(-\Delta + q)u = 0$ in Ω is interpreted as an element of $H^{-1/2}(\partial\Omega)$ as in Exercise 2.7.7, and the Cauchy data set is a subset of $H^{1/2}(\partial\Omega) \times H^{-1/2}(\partial\Omega)$.

Theorem 4.3.2 (Interior uniqueness for the Schrödinger equation). *Let $\Omega \subset \mathbb{R}^n$, $n \geq 3$, be a bounded open set, and let $q_1, q_2 \in L^\infty(\Omega)$. If*

$$C_{q_1} = C_{q_2},$$

then

$$q_1 = q_2.$$

If 0 is not a Dirichlet eigenvalue of $-\Delta + q$ in Ω, the Cauchy data set C_q is just the graph of the DN map Λ_q (Exercise 2.7.9). Therefore, the previous theorem has an immediate corollary for the DN map:

Theorem 4.3.3 (Interior uniqueness for the Schrödinger equation). *Let $\Omega \subset \mathbb{R}^n$, $n \geq 3$, be a bounded open set, let $q_1, q_2 \in L^\infty(\Omega)$, and assume that 0 is not a Dirichlet eigenvalue of $-\Delta + q_1$ or $-\Delta + q_2$ in Ω. If*

$$\Lambda_{q_1} = \Lambda_{q_2},$$

then

$$q_1 = q_2.$$

Let us now give the proofs of these results.

Proof of Theorem 4.3.2. Since $C_{q_1} = C_{q_2}$, we know from Exercise 2.8.4 that

$$\int_\Omega (q_1 - q_2)u_1 u_2 \, dx = 0 \tag{4.14}$$

for *any* $H^1(\Omega)$ solutions u_j to the equations $(-\Delta + q_j)u_j = 0$, $j = 1, 2$. (In the case of DN maps, so that $\Lambda_{q_1} = \Lambda_{q_2}$, the same relation was proved in Proposition 2.8.3.) Thus, to prove that $q_1 = q_2$, it is enough to establish that products $u_1 u_2$ of such solutions are complete in $L^1(\Omega)$.

Fix $\xi \in \mathbb{R}^n$. We would like to choose the solutions in such a way that $u_1 u_2$ is close to $e^{ix\cdot\xi}$, since the functions $e^{ix\cdot\xi}$ form a complete set by invertibility of the Fourier transform. We begin by taking unit vectors ω_1 and ω_2 in $\mathbb{R}^n$ such that $\{\omega_1, \omega_2, \xi\}$ is an orthogonal set (here we need that $n \geq 3$). Let

$$\zeta = s(\omega_1 + i\omega_2),$$

so that $\zeta \cdot \zeta = 0$. By Theorem 4.2.1, if s is sufficiently large there exist H^1 solutions u_1 and u_2 which satisfy $(-\Delta + q_j)u_j = 0$, and which are of the form

$$u_1 = e^{i\zeta\cdot x}(e^{ix\cdot\xi} + r_1), \qquad u_2 = e^{-i\zeta\cdot x}(1 + r_2),$$

where

$$\left\| r_j \right\|_{L^2(\Omega)} \leq C/s \text{ for } j = 1, 2.$$

For the first solution we chose $a = e^{ix\cdot\xi}$ which satisfies $\zeta \cdot \nabla a = (\zeta \cdot \xi)e^{ix\cdot\xi} = 0$ by orthogonality, and for the second solution we chose a to be constant.

Inserting these solutions in (4.14), we obtain

$$\int_\Omega (q_1 - q_2)(e^{ix\cdot\xi} + r_1)(1 + r_2) \, dx = 0. \tag{4.15}$$

In this identity, only r_1 and r_2 depend on s. Since the L^2 norms of r_1 and r_2 are bounded by C/s, it is possible to take the limit as $s \to \infty$ in (4.15), and by Cauchy–Schwarz the terms involving r_1 and r_2 will vanish. This shows that

$$\int_\Omega (q_1 - q_2)e^{ix\cdot\xi} \, dx = 0.$$

This holds for every $\xi \in \mathbb{R}^n$. If $\tilde{q}$ is the function in $L^1(\mathbb{R}^n)$ which is equal to $q_1 - q_2$ in Ω and vanishes outside Ω, the last identity implies that the Fourier transform of $\tilde{q}$ vanishes for every frequency $\xi \in \mathbb{R}^n$. Consequently $\tilde{q} = 0$, and $q_1 = q_2$ in Ω. □

Proof of Theorem 4.3.1. Since $\Lambda_{\gamma_1} = \Lambda_{\gamma_2}$, Theorems 3.1.1 and 3.2.1 imply that $\gamma_1|_{\partial\Omega} = \gamma_2|_{\partial\Omega}$ and $\partial_\nu\gamma_1|_{\partial\Omega} = \partial_\nu\gamma_2|_{\partial\Omega}$. Also, part (b) of Proposition 2.8.5 guarantees that

$$\Lambda_{q_j} f = \gamma_j^{-1/2}\Lambda_{\gamma_j}(\gamma_j^{-1/2} f) + \tfrac{1}{2}\gamma_j^{-1}(\partial_\nu\gamma_j)f|_{\partial\Omega}$$

for all f, where q_1 and q_2 are defined by

$$q_j = \frac{\Delta\gamma_j^{1/2}}{\gamma_j^{1/2}}.$$

Thus we have

$$\Lambda_{q_1} = \Lambda_{q_2},$$

and Theorem 4.3.3 implies that $q_1 = q_2$.

We give two ways for showing that $\gamma_1 = \gamma_2$. First, writing $q = q_1 = q_2$, from the definition of q_j we see that

$$(-\Delta + q)(\gamma_j^{1/2}) = 0 \text{ in } \Omega$$

with $\gamma_1^{1/2}|_{\partial\Omega} = \gamma_2^{1/2}|_{\partial\Omega}$. Since q comes from a C^2 conductivity, the Dirichlet problem for $-\Delta + q$ is well-posed (see Proposition 2.8.5), so by uniqueness of solutions

$$\gamma_1 = \gamma_2 \text{ in } \Omega.$$

Alternatively, observe that the function

$$v = \log\left(\frac{\gamma_1}{\gamma_2}\right) = 2\left[\log\gamma_1^{1/2} - \log\gamma_2^{1/2}\right]$$

satisfies

$$\begin{aligned}\nabla\cdot\left((\gamma_1\gamma_2)^{1/2}\nabla v\right) &= 2\nabla\cdot\left[\gamma_2^{1/2}\nabla\gamma_1^{1/2} - \gamma_1^{1/2}\nabla\gamma_2^{1/2}\right]\\ &= 2(\gamma_1\gamma_2)^{1/2}(q_1 - q_2) = 0.\end{aligned}$$

Since also $v|_{\partial\Omega} = 0$, uniqueness of weak solutions implies that

$$v \equiv 0 \text{ in } \Omega,$$

i.e., $\gamma_1 = \gamma_2$ in Ω. □

4.4. Stability

In §4.3 we proved that if γ_1, γ_2 are two conductivities in $C^2(\overline{\Omega})$ such that Λ_{γ_1} is equal to Λ_{γ_2}, then $\gamma_1 = \gamma_2$. Here we address the stability question: if γ_1, γ_2 are two conductivities such that Λ_{γ_1} is close to Λ_{γ_2}, does this imply that γ_1 is close to γ_2?

More precisely, we are looking for an estimate of the form

$$\|\gamma_1 - \gamma_2\|_{L^\infty(\Omega)} \leq \omega\left(\left\|\Lambda_{\gamma_1} - \Lambda_{\gamma_2}\right\|_*\right), \tag{4.16}$$

where $\|\cdot\|_* = \|\cdot\|_{H^{1/2}(\partial\Omega)\to H^{-1/2}(\partial\Omega)}$ is the natural operator norm for the DN maps, and $\omega : [0,\infty) \to [0,\infty)$ is a *modulus of continuity*, that is, a continuous nondecreasing function satisfying $\omega(t) \to 0$ as $t \to 0^+$.

We begin with an example from [**Ale88**] showing that an estimate like (4.16) cannot hold for general conductivities in $L^\infty(\Omega)$. In this section we will employ the Sobolev spaces $H^s(\Omega)$ and $H^s(\partial\Omega)$ discussed in Appendix C.

Example 4.4.1. Let $\mathbb{D}$ be the unit disc in $\mathbb{R}^2$, and let $\gamma_1, \gamma_2 \in L^\infty(\mathbb{D})$ be two conductivities such that $\gamma_1 \equiv 1$ in $\mathbb{D}$, and

$$\gamma_2(x) = \begin{cases} 1 + A, & |x| < r_0, \\ 1, & r_0 < |x| < 1, \end{cases}$$

where A is a positive constant and $r_0 \in (0,1)$. If $f \in H^{1/2}(\partial\mathbb{D})$, then f may be written as Fourier series

$$f(e^{i\theta}) = \sum_{k=-\infty}^{\infty} \hat{f}(k) e^{ik\theta}.$$

The Sobolev norms may be expressed as $\|f\|^2_{H^s(\partial\mathbb{D})} = \sum_{k=-\infty}^{\infty} (1+k^2)^s |\hat{f}(k)|^2$. It can be shown that

$$\Lambda_{\gamma_1} f(e^{i\theta}) = \sum_{k=-\infty}^{\infty} |k|\, \hat{f}(k) e^{ik\theta},$$

$$\Lambda_{\gamma_2} f(e^{i\theta}) = \sum_{k=-\infty}^{\infty} |k| \frac{2 + A(1 + r_0^{2|k|})}{2 + A(1 - r_0^{2|k|})} \hat{f}(k) e^{ik\theta}.$$

Then

$$\begin{aligned} \left\|(\Lambda_{\gamma_1} - \Lambda_{\gamma_2}) f\right\|^2_{H^{-1/2}} &= \sum_{k=-\infty}^{\infty} (1+k^2)^{-1/2} k^2 \left| 1 - \frac{2 + A(1 + r_0^{2|k|})}{2 + A(1 - r_0^{2|k|})} \right|^2 |\hat{f}(k)|^2 \\ &= \sum_{k=-\infty}^{\infty} \frac{k^2}{1+k^2} \left(\frac{2A r_0^{2|k|}}{2 + A(1 - r_0^{2|k|})} \right)^2 (1+k^2)^{1/2} |\hat{f}(k)|^2. \end{aligned}$$

The sum is actually over $k \neq 0$, and then the expression in parentheses is $\leq A r_0^2$ since it attains its maximum value when $|k| = 1$. It follows that

$$\left\|\Lambda_{\gamma_1} - \Lambda_{\gamma_2}\right\|_* = \sup_{\|f\|_{H^{1/2}}=1} \left\|(\Lambda_{\gamma_1} - \Lambda_{\gamma_2}) f\right\|_{H^{-1/2}} \leq A r_0^2.$$

Now $\left\|\Lambda_{\gamma_1} - \Lambda_{\gamma_2}\right\|_* \to 0$ as $r_0 \to 0$, but

$$\|\gamma_1 - \gamma_2\|_{L^\infty(\mathbb{D})} = A.$$

Thus, an estimate of the form (4.16) cannot be valid if one only assumes that $\gamma_1, \gamma_2 \in L^\infty$.

Exercise 4.4.2. Let $f : \overline{\Omega} \to \mathbb{C}$ be continuous, where $\Omega \subset \mathbb{R}^n$ is bounded. Show that there is a modulus of continuity ω such that

$$|f(x) - f(y)| \leq \omega(|x-y|).$$

Exercise 4.4.3. Let $\gamma \equiv 1$ in the unit disc $\mathbb{D} \subset \mathbb{R}^2$. Show that the solution in $H^1(\mathbb{D})$ to $\nabla \cdot \gamma \nabla u = 0$ in $\mathbb{D}$, with $u|_{\partial \mathbb{D}} = f \in H^{1/2}(\partial \mathbb{D})$, is given by

$$u(re^{i\theta}) = \sum_{k=-\infty}^{\infty} r^{|k|} \hat{f}(k) e^{ik\theta}.$$

Exercise 4.4.4. Prove the formulas for $\Lambda_{\gamma_j} f(e^{i\theta})$ in Example 4.4.1.

It turns out that under certain *a priori* bounds on γ_1 and γ_2 in a higher Sobolev norm, a somewhat more carefully crafted version of the uniqueness proof in §4.3 will yield the stability estimate (4.16) with a logarithmic modulus of continuity.

As in the uniqueness proof, we will use the inverse problem for the Schrödinger equation to study the stability question. The following theorem gives a stability estimate for $q_1 - q_2$ in a weaker norm if one assumes a priori bounds in L^∞.

Theorem 4.4.5 (Stability for Schrödinger problem). *Let $\Omega \subset \mathbb{R}^n$, $n \geq 3$, be a bounded open set with C^∞ boundary, and let $q_j \in L^\infty(\Omega)$ be two potentials such that 0 is not a Dirichlet eigenvalue for $-\Delta + q_j$ in Ω. Further, assume that*

$$\left\| q_j \right\|_{L^\infty(\Omega)} \leq M.$$

There is a constant $C = C(\Omega, n, M)$ such that

$$\|q_1 - q_2\|_{H^{-1}(\Omega)} \leq \omega(\left\| \Lambda_{q_1} - \Lambda_{q_2} \right\|_*),$$

where ω is a modulus of continuity satisfying

$$\omega(t) \leq C \left|\log t\right|^{-\frac{2}{n+2}}, \quad 0 < t < 1/e.$$

Proof. Let $\xi \in \mathbb{R}^n$. We start from the integral identity in Proposition 2.8.3, which states that

$$\int_\Omega (q_1 - q_2) u_1 u_2 \, dx = \langle (\Lambda_{q_1} - \Lambda_{q_2})(u_1|_{\partial\Omega}), u_2|_{\partial\Omega} \rangle_{\partial\Omega}, \tag{4.17}$$

for any $u_j \in H^1(\Omega)$ which solve $(-\Delta + q_j)u_j = 0$ in Ω. As in the proof of Theorem 4.3.2, let ω_1 and ω_2 be unit vectors such that $\{\omega_1, \omega_2, \xi\}$ is an orthogonal set. We choose complex vectors

$$\zeta_1 = \frac{s}{\sqrt{2}}\Big(\sqrt{1 - \frac{|\xi|^2}{2s^2}}\omega_1 + \frac{1}{\sqrt{2}s}\xi + i\omega_2\Big),$$

$$\zeta_2 = -\frac{s}{\sqrt{2}}\Big(\sqrt{1 - \frac{|\xi|^2}{2s^2}}\omega_1 - \frac{1}{\sqrt{2}s}\xi + i\omega_2\Big).$$

By (4.1) and orthogonality, these satisfy $\zeta_j \cdot \zeta_j = 0$, $|\zeta_1| = |\zeta_2| = s$ as well as

$$\zeta_1 + \zeta_2 = \xi.$$

Note that our vector of interest ξ is "hidden" in ζ_1 and ζ_2 between large vectors in the direction of ω_1 and ω_2.

Theorem 4.2.1 ensures the existence of solutions u_j to $(-\Delta + q_j)u_j = 0$, provided that $s \geq \max(C_0 M, 1)$, of the form

$$u_1 = e^{i\zeta_1 \cdot x}(1 + r_1),$$
$$u_2 = e^{i\zeta_2 \cdot x}(1 + r_2),$$

with $\left\|r_j\right\|_{L^2(\Omega)} \leq \frac{C_0\left\|q_j\right\|_{L^\infty}}{s}$, and $C_0 = C_0(\Omega, n)$.

Inserting u_1 and u_2 in (4.17) and using that $e^{i(\zeta_1+\zeta_2)\cdot x} = e^{ix\cdot\xi}$, we obtain

$$\begin{aligned}\Big|\int_\Omega (q_1 - q_2)e^{ix\cdot\xi}\,dx\Big| &\leq \left\|\Lambda_{q_1} - \Lambda_{q_2}\right\|_* \|u_1\|_{H^{1/2}(\partial\Omega)} \|u_2\|_{H^{1/2}(\partial\Omega)} \\ &\quad + \Big|\int_\Omega (q_1 - q_2)e^{ix\cdot\xi}(r_1 + r_2 + r_1 r_2)\,dx\Big| \\ &\leq C\left\|\Lambda_{q_1} - \Lambda_{q_2}\right\|_* \|u_1\|_{H^1} \|u_2\|_{H^1} \\ &\quad + C(\|r_1\|_{L^2} + \|r_2\|_{L^2} + \|r_1\|_{L^2}\|r_2\|_{L^2})\end{aligned}$$

with $C = C(\Omega, n, M)$. If $\Omega \subset B(0, R)$, then

$$\left\|u_j\right\|_{H^1} \leq \left\|e^{i\zeta_j\cdot x}(1 + r_j)\right\|_{L^2} + \sum_{k=1}^n \left\|\partial_k(e^{i\zeta_j\cdot x})(1 + r_j) + e^{i\zeta_j\cdot x}\partial_k r_j\right\|_{L^2} \leq Cse^{Rs}.$$

We assume that s is so large that $s \leq e^{Rs}$, and have

$$|(\tilde{q}_1 - \tilde{q}_2)\hat{}\,(\xi)| \leq C\Big(e^{4Rs}\left\|\Lambda_{q_1} - \Lambda_{q_2}\right\|_* + \frac{1}{s}\Big), \tag{4.18}$$

where $\tilde{q}_j$ is the extension of q_j to $\mathbb{R}^n$ by zero (thus $\tilde{q}_j \in L^2(\mathbb{R}^n)$).

So far, we have proved that there are constants C and C', depending on Ω, n, and M, such that (4.18) holds whenever $s \geq C'$. It is possible to obtain a bound for $q_1 - q_2$ in H^{-1} by using (4.18) and the L^∞ bounds for q_j. If $\rho > 0$ is a constant which will be determined later, we have

$$\begin{aligned}
\|q_1 - q_2\|^2_{H^{-1}(\Omega)} &\leq \|\tilde{q}_1 - \tilde{q}_2\|^2_{H^{-1}(\mathbb{R}^n)} \\
&= \Big(\int_{|\xi|\leq\rho} + \int_{|\xi|>\rho}\Big)\frac{|(\tilde{q}_1 - \tilde{q}_2)\hat{\ }(\xi)|^2}{1+|\xi|^2}\,d\xi \\
&\leq C\rho^n\Big(e^{8Rs}\left\|\Lambda_{q_1} - \Lambda_{q_2}\right\|^2_* + \frac{1}{s^2}\Big) + (1+\rho^2)^{-1}\|q_1 - q_2\|^2_{L^2(\mathbb{R}^n)} \\
&\leq C\rho^n e^{8Rs}\left\|\Lambda_{q_1} - \Lambda_{q_2}\right\|^2_* + \frac{C\rho^n}{s^2} + \frac{C}{\rho^2}.
\end{aligned}$$

To make the last two terms of equal size, we choose

$$\rho = s^{\frac{2}{n+2}}.$$

Then

$$\|q_1 - q_2\|^2_{H^{-1}(\Omega)} \leq Ce^{16Rs}\left\|\Lambda_{q_1} - \Lambda_{q_2}\right\|^2_* + Cs^{-\frac{4}{n+2}} \tag{4.19}$$

for $s \geq C'(\Omega, n, M)$.

We make the final choice of the large parameter s so that the expression on the right of (4.19) becomes small when $\left\|\Lambda_{q_1} - \Lambda_{q_2}\right\|_*$ is small. We choose

$$s = \frac{1}{16R}\left|\log\left\|\Lambda_{q_1} - \Lambda_{q_2}\right\|_*\right|,$$

where we assume that

$$0 < \left\|\Lambda_{q_1} - \Lambda_{q_2}\right\|_* < c'(\Omega, n, M)$$

with c' chosen so that $s \geq C'$. With this assumption, it follows that

$$\|q_1 - q_2\|^2_{H^{-1}(\Omega)} \leq C\Big(\left\|\Lambda_{q_1} - \Lambda_{q_2}\right\|_* + \left|\log\left\|\Lambda_{q_1} - \Lambda_{q_2}\right\|_*\right|^{-\frac{4}{n+2}}\Big).$$

The required stability estimate is an easy consequence. □

In Theorem 4.4.5, one can remove the assumption that 0 is not a Dirichlet eigenvalue of $-\Delta + q_j$ if one replaces the distance between DN maps with the distance between Cauchy data sets (see Definition 2.7.8).

Exercise 4.4.6 (Stability in terms of the Cauchy data set). Define the distance between two Cauchy data sets by

$$\operatorname{dist}(C_{q_1}, C_{q_2}) = \max\left\{ \sup_{(f,g)\in C_{q_1}} \inf_{(\tilde{f},\tilde{g})\in C_{q_2}} \frac{\|(f,g)-(\tilde{f},\tilde{g})\|_{H^{1/2}\oplus H^{-1/2}}}{\|(f,g)\|_{H^{1/2}\oplus H^{-1/2}}}, \right.$$

$$\left. \sup_{(f,g)\in C_{q_2}} \inf_{(\tilde{f},\tilde{g})\in C_{q_1}} \frac{\|(f,g)-(\tilde{f},\tilde{g})\|_{H^{1/2}\oplus H^{-1/2}}}{\|(f,g)\|_{H^{1/2}\oplus H^{-1/2}}} \right\}.$$

The norm on the space $H^{1/2}(\Omega) \oplus H^{-1/2}(\Omega)$ is defined by the expression

$$\|(f,g)\|_{H^{1/2}\oplus H^{-1/2}} = \left(\|f\|^2_{H^{1/2}(\Omega)} + \|g\|^2_{H^{-1/2}(\Omega)}\right)^{1/2}.$$

Under the assumptions of Theorem 4.4.5, show the stability estimate

$$\|q_1 - q_2\|^2_{H^{-1}(\Omega)} \le C(\operatorname{dist}(C_{q_1}, C_{q_2}) + \left|\log \operatorname{dist}(C_{q_1}, C_{q_2})\right|^{-\frac{4}{n+2}}).$$

Exercise 4.4.7. Prove that if the spaces C_{q_j} are both graphs of corresponding DN maps Λ_{q_j}, then one has the estimates

$$\frac{\|\Lambda_{q_1} - \Lambda_{q_2}\|_*}{\sqrt{1+\|\Lambda_{q_1}\|_*^2}\sqrt{1+\|\Lambda_{q_2}\|_*^2}} \le \operatorname{dist}(C_{q_1}, C_{q_2}) \le \|\Lambda_{q_1} - \Lambda_{q_2}\|_*.$$

After having established stability for the Schrödinger problem, we proceed to prove the main stability result for the conductivity equation.

Theorem 4.4.8 (Stability for Calderón problem). *Let $\Omega \subset \mathbb{R}^n$ be a bounded open set with smooth boundary, where $n \ge 3$, and let γ_j, $j = 1, 2$, be two positive functions in $H^{s+2}(\Omega)$ with $s > n/2$, satisfying*

$$\frac{1}{M} \le \gamma_j \le M, \tag{4.20}$$

$$\left\|\gamma_j\right\|_{H^{s+2}(\Omega)} \le M. \tag{4.21}$$

There are constants $C = C(\Omega, n, M, s) > 0$ and $\sigma = \sigma(n,s) \in (0,1)$ such that

$$\|\gamma_1 - \gamma_2\|_{L^\infty(\Omega)} \le \omega(\left\|\Lambda_{\gamma_1} - \Lambda_{\gamma_2}\right\|_*),$$

where ω is a modulus of continuity satisfying

$$\omega(t) \le C\left|\log t\right|^{-\sigma}, \quad 0 < t < 1/e.$$

Note that $\gamma_j \in H^{s+2}(\Omega)$ with $s > n/2$ implies that $\gamma_j \in C^2(\overline{\Omega})$ by the Sobolev embedding theorem (see Appendix C). The logarithmic modulus of continuity is rather weak, in the sense that even small changes in γ can result in large changes in Λ_γ. However, it has been proved that the logarithmic modulus is optimal, and for instance Hölder type conditional stability cannot hold for conductivities of general form.

For the proof, we need to relate the difference of DN maps for the conductivity equation to a corresponding quantity in the Schrödinger case. To do this we will use the boundary stability estimates proved in Chapter 3.

Lemma 4.4.9. *Under the assumptions in Theorem* 4.4.8, *if*

$$q_j = \frac{\Delta(\gamma_j^{1/2})}{\gamma_j^{1/2}},$$

we have

$$\left\|\Lambda_{q_1} - \Lambda_{q_2}\right\|_* \leq C(\left\|\Lambda_{\gamma_1} - \Lambda_{\gamma_2}\right\|_* + \left\|\Lambda_{\gamma_1} - \Lambda_{\gamma_2}\right\|_*^{\frac{1}{3}}),$$

with $C = C(\Omega, n, M, s)$.

Proof. We use the identity in Proposition 2.8.5,

$$\Lambda_{q_j} f = \gamma_j^{-1/2} \Lambda_{\gamma_j}(\gamma_j^{-1/2} f) + \frac{1}{2}\gamma_j^{-1}\partial_\nu \gamma_j f\Big|_{\partial\Omega}.$$

If $f \in H^{1/2}(\partial\Omega)$, we obtain

$$\begin{aligned}(\Lambda_{q_1} - \Lambda_{q_2})f &= (\gamma_1^{-1/2} - \gamma_2^{-1/2})\Lambda_{\gamma_1}(\gamma_1^{-1/2} f) \\ &\quad + \gamma_2^{-1/2}(\Lambda_{\gamma_1} - \Lambda_{\gamma_2})(\gamma_1^{-1/2} f) + \gamma_2^{-1/2}\Lambda_{\gamma_2}((\gamma_1^{-1/2} - \gamma_2^{-1/2})f) \\ &\quad + \frac{1}{2}(\gamma_1^{-1} - \gamma_2^{-1})\frac{\partial \gamma_1}{\partial \nu} f + \frac{1}{2}\gamma_2^{-1}(\partial_\nu \gamma_1 - \partial_\nu \gamma_2) f.\end{aligned}$$

We estimate the $H^{-1/2}$ norm of this expression by the triangle inequality. For the first three terms we use the multiplier estimate (see Exercise C.3.7)

$$\|au\|_{H^{-1/2}(\partial\Omega)} \leq \|a\|_{C^1(\partial\Omega)} \|u\|_{H^{-1/2}(\partial\Omega)},$$

and for the last two terms we use the easy estimate

$$\|au\|_{H^{-1/2}(\partial\Omega)} \leq \|au\|_{L^2(\partial\Omega)} \leq \|a\|_{L^\infty(\partial\Omega)} \|u\|_{H^{1/2}(\partial\Omega)}.$$

The a priori estimates for γ_j imply (for $p = -\frac{1}{2}$ or $p = -1$)

$$\begin{aligned}\left\|\gamma_1^p - \gamma_2^p\right\|_{C^1(\partial\Omega)} &\leq C\|\gamma_1 - \gamma_2\|_{C^1(\partial\Omega)}, \\ \|\gamma_j^p\|_{C^1(\partial\Omega)} &\leq C, \\ \|\Lambda_{\gamma_j}\|_* &\leq C,\end{aligned}$$

where $C = C(\Omega, n, M)$. Consequently

$$\left\|\Lambda_{q_1} - \Lambda_{q_2}\right\|_* \leq C(\left\|\Lambda_{\gamma_1} - \Lambda_{\gamma_2}\right\|_* + \|\gamma_1 - \gamma_2\|_{C^1(\partial\Omega)} + \|\partial_\nu\gamma_1 - \partial_\nu\gamma_2\|_{L^\infty(\partial\Omega)}).$$

Using Theorem 3.1.2 and Remark 3.2.4 for the last two terms, we obtain

$$\left\|\Lambda_{q_1} - \Lambda_{q_2}\right\|_* \leq C(\left\|\Lambda_{\gamma_1} - \Lambda_{\gamma_2}\right\|_* + \left\|\Lambda_{\gamma_1} - \Lambda_{\gamma_2}\right\|_*^{1/3}).$$ □

We can now give the proof of the main stability result for the conductivity equation. We will follow the method in the end of proof of Theorem 4.3.1.

Proof of Theorem 4.4.8. We introduce the function

$$v = \log \frac{\gamma_1^{1/2}}{\gamma_2^{1/2}} = \frac{1}{2}(\log \gamma_1 - \log \gamma_2).$$

This is a C^2 function in $\overline{\Omega}$, and satisfies

$$\begin{cases} \nabla \cdot (\gamma_1\gamma_2)^{1/2} \nabla v = (\gamma_1\gamma_2)^{1/2}(q_1 - q_2) & \text{in } \Omega, \\ v = \frac{1}{2}(\log \gamma_1 - \log \gamma_2) & \text{on } \partial\Omega, \end{cases}$$

where $q_j = \Delta\sqrt{\gamma_j}/\sqrt{\gamma_j}$. Therefore

$$\begin{aligned} \frac{1}{2} \|\log \gamma_1 - \log \gamma_2\|_{H^1(\Omega)} &= \|v\|_{H^1(\Omega)} \\ &\leq C\|(\gamma_1\gamma_2)^{1/2}(q_1 - q_2)\|_{H^{-1}(\Omega)} + C \|\log \gamma_1 - \log \gamma_2\|_{H^{1/2}(\partial\Omega)} \\ &\leq C \|q_1 - q_2\|_{H^{-1}(\Omega)} + C \|\log \gamma_1 - \log \gamma_2\|_{H^{1/2}(\partial\Omega)}. \end{aligned}$$

By Theorem 4.4.5 and Lemma 4.4.9, if $\left\|\Lambda_{\gamma_1} - \Lambda_{\gamma_2}\right\|_*$ is small we have

$$\begin{aligned} \|q_1 - q_2\|_{H^{-1}(\Omega)} &\leq C \left|\log \left\|\Lambda_{q_1} - \Lambda_{q_2}\right\|_*\right|^{-\frac{2}{n+2}} \\ &\leq C \left|\log \left\|\Lambda_{\gamma_1} - \Lambda_{\gamma_2}\right\|_*^{\frac{1}{3}}\right|^{-\frac{2}{n+2}} \\ &\leq C \left|\log \left\|\Lambda_{\gamma_1} - \Lambda_{\gamma_2}\right\|_*\right|^{-\frac{2}{n+2}}. \end{aligned}$$

We obtain

$$\|\log \gamma_1 - \log \gamma_2\|_{H^1(\Omega)} \leq C \left|\log \left\|\Lambda_{\gamma_1} - \Lambda_{\gamma_2}\right\|_*\right|^{-\frac{2}{n+2}} + C \|\log \gamma_1 - \log \gamma_2\|_{H^{1/2}(\partial\Omega)}.$$

We would like to change the norms of $\log \gamma_1 - \log \gamma_2$ on both sides to L^∞ norms. This can be done by Sobolev embedding (Exercise C.1.16), logarithmic convexity of Sobolev norms (Exercise C.1.17), and the a priori bounds on γ_j. Since $s > n/2$, we have

$$\begin{aligned} \|\log \gamma_1 - \log \gamma_2\|_{L^\infty(\Omega)} &\leq C \|\log \gamma_1 - \log \gamma_2\|_{H^s(\Omega)} \\ &\leq C \|\log \gamma_1 - \log \gamma_2\|_{H^1(\Omega)}^{\frac{2}{s+1}} \|\log \gamma_1 - \log \gamma_2\|_{H^{s+2}(\Omega)}^{\frac{s-1}{s+1}} \\ &\leq C \|\log \gamma_1 - \log \gamma_2\|_{H^1(\Omega)}^{\frac{2}{s+1}}. \end{aligned}$$

Note that Exercises C.1.16–C.1.17 are given for $H^s(\mathbb{R}^n)$, but the same results are valid for $H^s(\Omega)$ when Ω has smooth boundary. Alternatively, we could use the results in $\mathbb{R}^n$ by working with positive functions $\tilde{\gamma}_j \in H^{s+2}(\mathbb{R}^n)$ such that $\tilde{\gamma}_j|_\Omega = \gamma_j$ and $\left\|\tilde{\gamma}_j\right\|_{H^{s+2}} \leq 2M$, which is possible by the definition of $H^{s+2}(\Omega)$ in Definition C.1.19.

Next, using logarithmic convexity of Sobolev norms (Exercise C.1.47) and the trace theorem on $H^{s+2}(\Omega)$ (Remark C.2.2),

$$\begin{aligned}\|\log\gamma_1 - \log\gamma_2\|_{H^{1/2}(\partial\Omega)} &\leq C\,\|\log\gamma_1 - \log\gamma_2\|_{L^2(\partial\Omega)}^{\frac{2s}{2s+3}}\,\|\log\gamma_1 - \log\gamma_2\|_{H^{s+\frac{3}{2}}(\partial\Omega)}^{\frac{1}{2s+3}}\\ &\leq C\,\|\log\gamma_1 - \log\gamma_2\|_{L^\infty(\partial\Omega)}^{\frac{2s}{2s+3}}.\end{aligned}$$

It follows that

$$\|\log\gamma_1 - \log\gamma_2\|_{L^\infty(\Omega)}^{\frac{s+1}{2}} \leq C\left|\log\left\|\Lambda_{\gamma_1} - \Lambda_{\gamma_2}\right\|_*\right|^{-\frac{2}{n+2}} + C\,\|\log\gamma_1 - \log\gamma_2\|_{L^\infty(\partial\Omega)}^{\frac{2s}{2s+3}}.$$

Finally, we obtain bounds in terms of $\gamma_1 - \gamma_2$ by using that

$$\begin{aligned}\log\gamma_1 - \log\gamma_2 &= \int_0^1 \frac{d}{dt}\log\left((1-t)\gamma_2 + t\gamma_1\right)dt\\ &= \Big(\int_0^1 \frac{1}{(1-t)\gamma_2 + t\gamma_1}\,dt\Big)(\gamma_1 - \gamma_2).\end{aligned}$$

The a priori bounds on γ_j imply that

$$\begin{aligned}\|\gamma_1 - \gamma_2\|_{L^\infty(\Omega)} &\leq C\,\|\log\gamma_1 - \log\gamma_2\|_{L^\infty(\Omega)},\\ \|\log\gamma_1 - \log\gamma_2\|_{L^\infty(\partial\Omega)} &\leq C\,\|\gamma_1 - \gamma_2\|_{L^\infty(\partial\Omega)}.\end{aligned}$$

This shows that

$$\|\gamma_1 - \gamma_2\|_{L^\infty(\Omega)}^{\frac{s+1}{2}} \leq C\left|\log\left\|\Lambda_{\gamma_1} - \Lambda_{\gamma_2}\right\|_*\right|^{-\frac{2}{n+2}} + C\,\|\gamma_1 - \gamma_2\|_{L^\infty(\partial\Omega)}^{\frac{2s}{2s+3}}.$$

The result follows by Theorem 3.1.2. □

4.5. CGO solutions: Second proof

The construction of complex geometrical optics solutions in Theorem 4.2.1 was based on considering Fourier series in a slightly shifted lattice, which avoided the problem of dividing by symbols having zeros. This construction was sufficient for the interior uniqueness and stability results in the previous sections.

However, to obtain a reconstruction procedure for determining a conductivity from the DN map, it is useful to give another proof of the existence of CGO solutions. This proof is valid in $\mathbb{R}^n$ instead of just bounded domains, and it comes with a reasonable uniqueness notion for the CGO solutions upon fixing a decay condition at infinity. Similar arguments will appear later in the context of inverse scattering problems in Chapter 7.

To construct the solutions we shall make use of polynomially weighted L^2 norms, defined for any $u \in C_c^\infty(\mathbb{R}^n)$ and any $\delta \in \mathbb{R}$ by

$$\|u\|_{L^2_\delta} = \|\langle x\rangle^\delta u\|_{L^2} = \left(\int_{\mathbb{R}^n} \langle x\rangle^{2\delta}|u|^2\,dx\right)^{1/2},$$

where $\langle x\rangle$ is the Japanese bracket

$$\langle x\rangle = (1+|x|^2)^{1/2}.$$

Similarly, for $k \geq 0$ we define weighted Sobolev norms

$$\|u\|_{H^k_\delta} = \left(\sum_{|\alpha|\leq k} \|\partial^\alpha u\|_{L^2_\delta}\right)^{1/2}.$$

The space L^2_δ is defined as the completion of $C_c^\infty(\mathbb{R}^n)$ with respect to the norm $\|\cdot\|_{L^2_\delta}$, and similarly for H^k_δ.

The main theorem in this section is the following.

Theorem 4.5.1 (CGO solutions). *Let $-1 < \delta < 0$. There exists $C = C(\delta)$ such that, for every $q \in L^2_{\delta+1}$ with $\langle x\rangle q \in L^\infty$ and every $\zeta \in \mathbb{C}^n$ satisfying*

$$\zeta\cdot\zeta = 0 \qquad \text{and} \qquad |\zeta| \geq \max\{C\,\|\langle x\rangle q\|_{L^\infty}, 1\}$$

there exists a unique solution to

$$-\Delta u + qu = 0 \quad \text{in } \mathbb{R}^n$$

of the form

$$u = e^{i\zeta\cdot x}(1 + r(x,\zeta))$$

with $r(\,\cdot\,,\zeta) \in H^2_\delta$. Furthermore,

$$\|r\|_{H^k_\delta} \leq C|\zeta|^{k-1}\|q\|_{L^2_{\delta+1}}, \qquad k = 0,1,2.$$

This theorem has a counterpart for the conductivity problem. This is obtained by invoking the correspondence of Proposition 2.8.5 between the Schrödinger equation and the conductivity equation.

Theorem 4.5.2. *Let $-1 < \delta < 0$ and let γ be a positive function with $\frac{\Delta\gamma^{1/2}}{\gamma^{1/2}} \in L^2_{\delta+1}$ and $\langle x\rangle\frac{\Delta\gamma^{1/2}}{\gamma^{1/2}} \in L^\infty$. If $\zeta \in \mathbb{C}^n$ satisfies $\zeta\cdot\zeta = 0$ and $|\zeta|$ is sufficiently large, there exists a unique solution to*

$$\nabla\cdot(\gamma\nabla u) = 0 \text{ in } \mathbb{R}^n$$

of the form

$$u = \gamma^{-1/2}e^{i\zeta\cdot x}(1 + r(x,\zeta))$$

with $r(\,\cdot\,,\zeta) \in H^2_\delta$. Furthermore,

$$\|r(x,\zeta)\|_{H^k_\delta} \leq C|\zeta|^{k-1}\left\|\frac{\Delta\gamma^{1/2}}{\gamma^{1/2}}\right\|_{L^2_{\delta+1}}, \qquad k = 0,1,2.$$

Exercise 4.5.3. Prove Theorem 4.5.2 assuming Theorem 4.5.1 is known.

As in §4.2, we will consider the conjugated operator

$$e^{-i\zeta\cdot x}(-\Delta)(e^{i\zeta\cdot x}v) = (D^2 + 2\zeta \cdot D)v,$$

where $D = (D_1, \dots, D_n)$ and $D_j = \frac{1}{i}\partial_{x_j}$. Most of the work necessary for the proof of Theorem 4.5.1 is associated with establishing the following proposition.

Proposition 4.5.4 (Solution operator). *Let $-1 < \delta < 0$ and let $\zeta \in \mathbb{C}^n$ satisfy $\zeta \cdot \zeta = 0$ and $|\zeta| \geq 1$. For any $f \in L^2_{\delta+1}$, there exists a unique $w \in H^2_\delta$ such that*

$$(D^2 + 2\zeta \cdot D)w = f \text{ in } \mathbb{R}^n. \tag{4.22}$$

Moreover, $G_\zeta : L^2_{\delta+1} \to H^2_\delta$, $f \mapsto w$ is bounded with norm estimates

$$\|G_\zeta f\|_{H^k_\delta} \leq C(\delta)|\zeta|^{k-1}\|f\|_{L^2_{\delta+1}}, \qquad k = 0, 1, 2.$$

We postpone the proof of this proposition to the end of this section. Instead we first show how it may be applied to prove Theorem 4.5.1.

Proof of Theorem 4.5.1. We seek u of the form

$$u = e^{i\zeta\cdot x}(1 + r)$$

satisfying

$$(-\Delta + q)\{e^{i\zeta\cdot x}(1 + r)\} = 0$$

or equivalently

$$(D^2 + 2\zeta \cdot D + q)r = -q. \tag{4.23}$$

To solve (4.23), we look for a solution

$$r = G_\zeta f$$

for a suitable $f \in L^2_{\delta+1}$. Proposition 4.5.4 gives $(D^2 + 2\zeta \cdot D)G_\zeta f = f$, and therefore f should solve

$$(\mathrm{Id} + qG_\zeta)f = -q. \tag{4.24}$$

When $|\zeta| \geq 1$, Proposition 4.5.4 implies

$$\left\|qG_\zeta f\right\|_{L^2_{\delta+1}} \leq \|\langle x\rangle q\|_{L^\infty} \left\|G_\zeta f\right\|_{L^2_\delta} \leq \frac{C(\delta)\,\|\langle x\rangle q\|_{L^\infty}}{|\zeta|}\,\|f\|_{L^2_{\delta+1}}.$$

Thus if $|\zeta| \geq \max\{2C(\delta)\,\|\langle x\rangle q\|_{L^\infty}, 1\}$, we have

$$\left\|qG_\zeta\right\|_{L^2_{\delta+1}\to L^2_{\delta+1}} \leq \frac{1}{2}.$$

It follows that (4.24) has a solution $f \in L^2_{\delta+1}$ given by the Neumann series

$$f = \sum_{j=0}^{\infty}(-qG_\zeta)^j(-q)$$

with norm estimate

$$\|f\|_{L^2_{\delta+1}} \leq 2\,\|q\|_{L^2_{\delta+1}}.$$

Then $r = G_\zeta f$ will solve (4.23) and u will solve $(-\Delta + q)u = 0$. The estimates for r follow from Proposition 4.5.4.

It remains to prove uniqueness. Suppose that $u = e^{i\zeta\cdot x}(1+r)$ and $\tilde{u} = e^{i\zeta\cdot x}(1+\tilde{r})$ solve $(-\Delta+q)u = (-\Delta+q)\tilde{u} = 0$ with $r, \tilde{r} \in H^2_\delta$. It follows that $w = r - \tilde{r}$ solves

$$(D^2 + 2\zeta\cdot D)w = -qw.$$

Since $w \in H^2_\delta$ and $\langle x\rangle q \in L^\infty$, one has $qw \in L^2_{\delta+1}$. Then the uniqueness part of Proposition 4.5.4 gives

$$w = -G_\zeta(qw)$$

and consequently

$$\|w\|_{L^2_\delta} \leq \frac{C(\delta)}{|\zeta|}\,\|qw\|_{L^2_{\delta+1}} \leq \frac{C(\delta)\,\|\langle x\rangle q\|_{L^\infty}}{|\zeta|}\,\|w\|_{L^2_\delta}.$$

But $|\zeta| \geq 2C(\delta)\,\|\langle x\rangle q\|_{L^\infty}$, so

$$\|w\|_{L^2_\delta} \leq \frac{1}{2}\,\|w\|_{L^2_\delta}$$

which can only occur if $w = 0$. This proves that $u = \tilde{u}$. □

We now return to Proposition 4.5.4. The idea of the proof is to Fourier transform the equation

$$(D^2 + 2\zeta\cdot D)w = f,$$

which gives

$$(|\xi|^2 + 2\zeta\cdot\xi)\hat{w} = \hat{f}.$$

We would like to obtain a solution by dividing by the symbol $|\xi|^2 + 2\zeta\cdot\xi$, and most of the work goes to showing that this division can be done rigorously. Since $\zeta\cdot\zeta = 0$, looking at the real and imaginary parts of the symbol shows that it vanishes on a codimension 2 sphere in $\mathbb{R}^n$. Near a point in the zero set, one can choose new coordinates η so that the symbol becomes a multiple of $\eta_1 + i\eta_2$. Dividing by such a symbol corresponds to applying the inverse of a Cauchy–Riemann operator $D_1 + iD_2$.

We will now give the details of the proof. Below, we will write $\langle u, \varphi\rangle$ for the pairing of a distribution u and a test function φ in $\mathbb{R}^n$.

Proof of Proposition 4.5.4. We first prove uniqueness. If $w \in \mathscr{S}(\mathbb{R}^n)$ and

$$(D^2 + 2\zeta\cdot D)w = 0,$$

Fourier transformation gives

$$(|\xi|^2 + 2\zeta\cdot\xi)\,\hat{w}(\xi) = 0. \tag{4.25}$$

As this equation is invariant under rotations, we may assume, without loss of generality, that the imaginary part of ζ is in the positive e_1 direction and the real part of ζ is in the span of e_1 and e_2 with negative e_2 component. By (4.1), the real and imaginary parts of ζ must be mutually perpendicular and of the same length, so that

$$\zeta = -se_2 + ise_1 \qquad \text{with} \qquad s = \frac{|\zeta|}{\sqrt{2}},$$

in which case (4.25) is equivalent to

$$\big[(\xi_1^2 + (\xi_2 - s)^2 + \xi_3^2 \cdots + \xi_n^2 - s^2) + 2is\xi_1\big]\,\hat{w}(\xi) = 0. \tag{4.26}$$

Let

$$\mathcal{M}(s) = \{\, \xi \in \mathbb{R}^n \mid \xi_1 = 0,\ \xi_1^2 + (\xi_2 - s)^2 + \xi_3^2 \cdots + \xi_n^2 = s^2 \,\}$$

denote the codimension 2 sphere which arises as the intersection of the plane $\xi_1 = 0$ and the $(n-1)$-dimensional sphere with center se_2 and radius s. The content of (4.26) is that $\hat{w}$ is supported on $\mathcal{M}(s)$, and we will prove that w must vanish.

Now let $w \in L^2_\delta$ be any weak solution to $(D^2 + 2\zeta \cdot D)w = 0$. To show that $w = 0$, it suffices to show that $\langle w, \bar{\varphi}\rangle = 0$ for all $\varphi \in \mathcal{S}(\mathbb{R}^n)$. To do so, we approximate w by $w_\varepsilon \in L^2(\mathbb{R}^n)$. Let

$$\chi \in C^\infty([0,\infty)) \quad \text{with} \quad \operatorname{supp}\chi \subset [0,1) \quad \text{and} \qquad \int_{\mathbb{R}^n} \chi(|\xi|^2)\,\frac{d\xi}{(2\pi)^n} = 1$$

and

$$\beta(x) = \int_{\mathbb{R}^n} e^{i\xi\cdot x}\chi(|\xi|^2)\,\frac{d\xi}{(2\pi)^n}.$$

Then $\beta = \mathcal{F}^{-1}\{\chi(|\xi|^2)\} \in \mathcal{S}(\mathbb{R}^n)$ and

$$w_\varepsilon(x) = \beta(\varepsilon x)w(x) \in L^2(\mathbb{R}^n).$$

As $\varphi \in \mathcal{S}(\mathbb{R}^n)$, $\varphi_\varepsilon(x) = \beta(\varepsilon x)\varphi(x) \in \mathcal{S}(\mathbb{R}^n)$ and

$$\lim_{\varepsilon\searrow 0}\langle x\rangle^{2-\delta/2}\varphi_\varepsilon(x) = \langle x\rangle^{2-\delta/2}\varphi(x) \qquad \text{in } L^2(\mathbb{R}^n)$$

by the Lebesgue dominated convergence theorem. Consequently

$$\langle w, \bar{\varphi}\rangle = \lim_{\varepsilon\searrow 0}\langle w, \bar{\varphi}_\varepsilon\rangle = \lim_{\varepsilon\searrow 0}\langle w_\varepsilon, \bar{\varphi}\rangle = \lim_{\varepsilon\searrow 0}\int_{\mathbb{R}^n} \hat{w}_\varepsilon(\xi)\overline{\hat{\varphi}(\xi)}\,\frac{d\xi}{(2\pi)^n}.$$

Observe that $\big(\frac{1}{\varepsilon^n}\hat{\beta}(\frac{\xi}{\varepsilon})\big)^\vee(x) = \beta(\varepsilon x)$. By Exercise 4.5.8,

$$\operatorname{supp}\hat{w}_\varepsilon \subset N_\varepsilon\big(\mathcal{M}(s)\big) = \{\,\xi \mid \operatorname{dist}(\xi, \mathcal{M}(s)) \leq \varepsilon\,\}$$

so that

$$|\langle w, \bar{\varphi}\rangle| \leq \limsup_{\varepsilon\searrow 0}\varepsilon\Big(\int_{N_\varepsilon} |\hat{w}_\varepsilon(\xi)|^2\frac{d\xi}{(2\pi)^n}\Big)^{1/2}\Big(\frac{1}{\varepsilon^2}\int_{N_\varepsilon} |\hat{\varphi}(\xi)|^2\frac{d\xi}{(2\pi)^n}\Big)^{1/2}.$$

As $\hat{\varphi}$ is smooth and $\frac{1}{\varepsilon^2}$(volume of N_ε) converges to a constant times the surface area of $\mathcal{M}(s)$,

$$|\langle w, \bar{\varphi}\rangle| \leq C\Big(\limsup_{\varepsilon\searrow 0} \varepsilon\|\hat{w}_\varepsilon\|_{L^2}\Big)\Big(\int_{\mathcal{M}(s)} |\hat{\varphi}(\xi)|^2\, dS(\xi)\Big)^{1/2}. \tag{4.27}$$

Moreover,

$$\begin{aligned}\frac{1}{(2\pi)^n}\|\hat{w}_\varepsilon\|_{L^2}^2 = \|w_\varepsilon\|_{L^2}^2 &= \int_{\mathbb{R}^n} |\beta(\varepsilon x)|^2 |w(x)|^2\, dx \\ &\leq (\sup_x \langle x\rangle^{-2\delta}\beta(\varepsilon x)^2)\, \|w\|_{L^2_\delta}^2.\end{aligned}$$

$$\frac{1}{(2\pi)^n}\|\hat{w}_\varepsilon\|_{L^2}^2 = \|w_\varepsilon\|_{L^2}^2 = \int_{\mathbb{R}^n} |\beta(\varepsilon x)|^2 |w(x)|^2\, dx \leq (\sup_x \langle x\rangle^{-2\delta}\beta(\varepsilon x)^2)\, \|w\|_{L^2_\delta}^2.$$

As $\beta \in \mathcal{S}(\mathbb{R}^n)$ and $\delta < 0$, we have

$$\begin{aligned}\|\hat{w}_\varepsilon\|_{L^2}^2 &\leq C\,\|w\|_{L^2_\delta}^2 \sup_x \left(1+|x|^2\right)^{-\delta}\left(1+\varepsilon^2|x|^2\right)^{\delta} \\ &= C\,\|w\|_{L^2_\delta}^2 \sup_x \left[\frac{1+|x|^2}{1+\varepsilon^2|x|^2}\right]^{|\delta|} \\ &\leq C\,\varepsilon^{2\delta}\,\|w\|_{L^2_\delta}^2.\end{aligned}$$

Returning to (4.27) we have

$$|\langle w, \bar{\varphi}\rangle| \leq C \limsup_{\varepsilon\searrow 0} (\varepsilon\cdot\varepsilon^{\delta})\,\|w\|_{L^2_\delta}\left(\int_{\mathcal{M}(s)} |\hat{\varphi}(\xi)|^2\, dS(\xi)\right)^{1/2}.$$

Since $\delta > -1$, it therefore follows that

$$\langle w, \bar{\varphi}\rangle = 0$$

for every $\varphi \in \mathcal{S}$, so that $w = 0$.

We turn to proving existence of a solution to (4.22). Suppose for now that $f \in \mathcal{S}(\mathbb{R}^n)$ and define

$$\hat{w}(\xi) = \frac{\hat{f}(\xi)}{|\xi|^2 + 2\zeta\cdot\xi}.$$

We shall prove that w is well defined and satisfies the estimate

$$\|w\|_{L^2_\delta} \leq \frac{C}{|\zeta|}\|f\|_{L^2_{\delta+1}}. \tag{4.28}$$

Once this estimate is established, we will prove H^k_δ estimates and dispense with the assumption that $f \in \mathcal{S}(\mathbb{R}^n)$ by continuity.

As we did in the uniqueness proof, we may assume that

$$\zeta = -se_2 + ise_1 \qquad \text{with} \qquad s = \frac{|\zeta|}{\sqrt{2}}$$

and therefore

$$|\xi|^2 + 2\zeta \cdot \xi = \left[\xi_1^2 + (\xi_2 - s)^2 + \xi_3^2 + \cdots + \xi_n^2 - s^2\right] + 2is\xi_1 = P(\xi, s).$$

Since the polynomial $P(\xi, s)$ is homogeneous of degree two,

$$P(\xi, s) = s^2 P(\xi/s, 1).$$

As before we denote

$$N_\rho\big(\mathcal{M}(s)\big) = \{\, \xi \in \mathbb{R}^n \mid \operatorname{dist}(\xi, \mathcal{M}(s)) \leq \rho \,\}.$$

Every point $p \in \mathcal{M}(s)$ obeys $p_1 = 0$ and $|p - se_2| = s$. Hence for all $\xi \in \mathbb{R}^n$ and $p \in \mathcal{M}(s)$ we have that

$$|\xi - p| \geq |\xi_1|, \qquad |\xi - p| \geq \big||\xi - se_2| - |p - se_2|\big| = \big||\xi - se_2| - s\big|,$$

which implies that $\operatorname{dist}(\xi, \mathcal{M}(s)) \geq |\xi_1|$ and $\operatorname{dist}(\xi, \mathcal{M}(s)) \geq \big||\xi - se_2| - s\big|$. As a result, if $\xi \in N_{s/2n}\big(\mathcal{M}(s)\big)$, then $|\xi_1| \leq \frac{s}{2n}$ and $|\xi - se_2| \geq s - \frac{s}{2n}$ so that at least one component of $\xi - se_2$ must be at least $\frac{1}{\sqrt{n}}\big(s - \frac{s}{2n}\big) > \frac{s}{2n}$.

The above discussion implies that the sets

$$\mathcal{O}_1(s) = \mathbb{R}^n \setminus N_{s/2n}\big(\mathcal{M}(s)\big),$$

$$\mathcal{O}_2(s) = \Big\{|\xi_2 - s| > \frac{s}{2n}\Big\} \cap N_s\big(\mathcal{M}(s)\big)^\circ,$$

$$\mathcal{O}_j(s) = \Big\{|\xi_j| > \frac{s}{2n}\Big\} \cap N_s\big(\mathcal{M}(s)\big)^\circ \quad \text{for } 3 \leq j \leq n,$$

form an open cover of $\mathbb{R}^n$. The singularity of $\frac{\hat{f}(\xi)}{P(\xi,s)}$ on $\mathcal{M}(s)$ has been excluded from $\mathcal{O}_1(s)$. The remaining sets $\mathcal{O}_2(s), \cdots, \mathcal{O}_n(s)$ cover the set $N_{s/2n}\big(\mathcal{M}(s)\big) \subset N_s\big(\mathcal{M}(s)\big)^\circ$ with the jth component of $|\xi - se_2|$ being relatively large on $\mathcal{O}_j$. It is useful to note that $\mathcal{M}(s) = s\mathcal{M}(1)$ and that $\mathcal{O}_j(s) = s\mathcal{O}_j(1)$.

Let $\chi_j(\xi)$ be a partition of unity subordinate to this open cover. We wish to define $\hat{w}$ as

$$\hat{w}(\xi) = \sum_{j=1}^{n} \hat{w}_j(\xi), \qquad \hat{w}_j(\xi) = \frac{\chi_j(\xi)\hat{f}(\xi)}{P(\xi, s)}.$$

Since $\mathcal{O}_1(1)$ is bounded away from $\mathcal{M}(1)$ and since $P(\xi, 1) \to \infty$ as $|\xi| \to \infty$, there exists a constant c such that

$$|P(\xi, 1)| \geq c > 0, \qquad \xi \in \mathcal{O}_1(1).$$

For $\xi \in \mathcal{O}_1(s)$ this leads to the estimate

$$|P(\xi, s)| = s^2 |P(\xi/s, 1)| \geq cs^2,$$

so from the definition of $\hat{w}_1$ we obtain

$$\|\hat{w}_1\|_{L^2} \leq \frac{1}{cs^2}\|\hat{f}\|_{L^2}.$$

The Plancherel identity gives

$$\|w_1\|_{L^2_\delta} \leq \|w_1\|_{L^2} \leq \frac{1}{cs^2}\|f\|_{L^2} \leq \frac{1}{cs^2}\|f\|_{L^2_{\delta+1}}. \tag{4.29}$$

Here we have used the assumptions that $\delta < 0$ and $\delta + 1 > 0$. Since our hypothesis guarantees that $|\zeta| = \sqrt{2}s \geq 1$, (4.29) gives the desired estimate (4.28) for w_1.

To estimate each w_j, with $j = 2, \cdots, n$, we first introduce new coordinates in $\mathcal{O}_j(s)$ by

$$\begin{aligned} \eta_1 &= 2\xi_1, \\ \eta_\ell &= \xi_\ell \qquad \text{for } \ell \neq 1, j, \\ \eta_j &= \frac{\xi_1^2 + (\xi_2 - s)^2 + \xi_3^2 + \cdots + \xi_n^2 - s^2}{s}. \end{aligned} \tag{4.30}$$

In terms of these new coordinates

$$\hat{w}_j(\eta) = \frac{\chi_j(\xi)\hat{f}(\xi)}{s(\eta_j + i\eta_1)}.$$

Since

$$\nabla_\xi \eta_\ell = \begin{cases} 2e_1 & \text{if } \ell = 1 \\ e_\ell & \text{if } \ell \neq 1, j \\ \frac{2\xi - 2se_2}{s} & \text{if } \ell = j \end{cases}$$

the Jacobian of this coordinate transformation on $\mathcal{O}_j(s)$ is

$$|\det[\frac{\partial \eta}{\partial \xi}]| = \begin{cases} \frac{4|\xi_j|}{s} & \text{if } j \neq 2 \\ \frac{4|\xi_2 - s|}{s} & \text{if } j = 2 \end{cases}$$

which is bounded above by 8 and below by $\frac{2}{n}$ on $\mathcal{O}_j(s)$, $j = 2, \ldots, n$ for all s. At this point we shall make use of the following three results, the proofs of which will be given later.

Lemma 4.5.5. *For each $j = 2, \ldots, n$, the map Z_j defined by*

$$Z_j f = \mathcal{F}^{-1}\left\{\frac{1}{\eta_j + i\eta_1}\hat{f}(\eta)\right\}, \qquad f \in \mathcal{S}(\mathbb{R}^n),$$

has a unique continuous extension to a bounded linear operator from $L^2_{\delta+1}$ to L^2_δ. For each $f \in L^2_{\delta+1}$, $Z_j f$ is a weak solution to $(D_{x_j} + iD_{x_1})w = f$. That is,

$$-\left\langle Z_j f, (D_{x_j} + iD_{x_1})\varphi\right\rangle = \langle \varphi, f\rangle$$

for all $\varphi \in \mathcal{S}(\mathbb{R}^n)$ and $f \in L^2_{\delta+1}$.

Lemma 4.5.6. *For any $\chi \in C_c^\infty(\mathbb{R}^n)$ and any $f \in \mathcal{S}(\mathbb{R}^n)$*

$$\left\|\mathcal{F}^{-1}\{\chi(\xi)\hat{f}(\xi)\}\right\|_{L^2_{\delta+1}} \le C\|f\|_{L^2_{\delta+1}},$$

where the constant C depends on χ, but is independent of f.

Lemma 4.5.7. *Let $\mathcal{O}$ and $\mathcal{O}'$ be open subsets of $\mathbb{R}^n$ and $\mathcal{C}$ be a compact subset of $\mathcal{O}$. Let $\hat{f} \in C_c^\infty(\mathcal{C})$ and let Ψ be a smooth diffeomorphism from $\mathcal{O}$ to $\mathcal{O}'$. Then,*

$$\left\|\mathcal{F}^{-1}\{\hat{f}\circ\Psi^{-1}\}\right\|_{L^2_\delta} \le C\|f\|_{L^2_\delta}.$$

The constant C depends on Ψ, $\mathcal{C}$ and δ, but is independent of f.

The proof of Proposition 4.5.4 now proceeds as follows. If $\Psi(\eta)$ is the inverse map of the change of coordinates (4.5), then

$$\hat{w}_j(\xi) = \frac{\chi_j(\xi)\hat{f}(\xi)}{P(\xi,s)} = \left[\frac{(\chi_j\hat{f})\circ\Psi}{s(\eta_j + i\eta_1)}\right]\circ\Psi^{-1}(\xi).$$

Set

$$g_j = \mathcal{F}^{-1}\{(\chi_j\hat{f}\,)\circ\Psi\}, \qquad h_j = \mathcal{F}^{-1}\{\chi_j\hat{f}\}.$$

In this notation

$$\hat{w}_j = \frac{1}{s}\widehat{Z_jg_j}\circ\Psi^{-1}.$$

Using, in order, Lemma 4.5.7, Lemma 4.5.5, Lemma 4.5.7 and Lemma 4.5.6, we obtain that

$$\|w_j\|_{L^2_\delta} \le \frac{c_1}{s}\|Z_jg_j\|_{L^2_\delta} \le \frac{c_2}{s}\|g_j\|_{L^2_{\delta+1}} \le \frac{c_3}{s}\|h_j\|_{L^2_{\delta+1}} \le \frac{c_4}{s}\|f\|_{L^2_{\delta+1}}.$$

Recalling that $s = \frac{|\zeta|}{\sqrt{2}}$ and invoking the formula $w = \sum_{j=1}^n w_j$ completes the proof of (4.28) for $f \in \mathcal{S}(\mathbb{R}^n)$. Writing $G_\zeta : \mathcal{S} \to L^2_\delta$, $G_\zeta f = w$, noting that $\mathcal{S}$ is dense in $L^2_{\delta+1}$ and using the B.L.T. theorem (Theorem A.2.13) proves the estimate

$$\left\|G_\zeta f\right\|_{L^2_\delta} \le \frac{C}{s}\,\|f\|_{L^2_{\delta+1}}$$

for $f \in L^2_{\delta+1}$.

It remains to establish the estimates in H^k_δ for $k = 1, 2$. The Plancherel theorem yields

$$\left\|D_jG_\zeta f\right\|_{L^2_\delta} = \|\langle x\rangle^\delta D_jG_\zeta f\|_{L^2} = (2\pi)^{-n/2}\left\|\frac{\xi_j}{|\xi|^2 + 2\zeta\cdot\xi}\hat{f}\right\|_{H^\delta}.$$

Let $\chi \in C_c^\infty(\mathbb{R}^n)$ satisfy $\chi = 1$ for $|\xi| \le 4$ and $\chi = 0$ for $|\xi| \ge 8$. We write $\xi_j = \chi(\xi/s)\xi_j + (1-\chi(\xi/s))\xi_j$. Since $\left\|\chi(\xi/s)\xi_j\right\|_{C^1(\mathbb{R}^n)} \le Cs$, we have

$$\left\|\frac{\chi(\xi/s)\xi_j}{|\xi|^2 + 2\zeta\cdot\xi}\hat{f}\right\|_{H^\delta} \le Cs\left\|\frac{1}{|\xi|^2 + 2\zeta\cdot\xi}\hat{f}\right\|_{H^\delta} \le Cs\left\|G_\zeta f\right\|_{L^2_\delta} \le C\,\|f\|_{L^2_{\delta+1}}.$$

On the other hand, $1 - \chi(\xi/s)$ is supported in $\{|\xi| \geq 4s\}$, and in this set one has $||\xi|^2 + 2\zeta \cdot \xi| \geq c|\xi|^2$. Arguing as in (4.29), using $-1 < \delta < 0$, yields

$$\begin{aligned} \left\| \frac{(1-\chi(\xi/s))\xi_j}{|\xi|^2 + 2\zeta \cdot \xi} \hat{f} \right\|_{H^\delta} &\leq \left\| \frac{(1-\chi(\xi/s))\xi_j}{|\xi|^2 + 2\zeta \cdot \xi} \hat{f} \right\|_{L^2} \\ &\leq \left(\sup_{|\xi|\geq 4s} \frac{|\xi_j|}{c|\xi|^2} \right) \|\hat{f}\|_{L^2} \\ &\leq \frac{C}{s} \|f\|_{L^2_{\delta+1}} . \end{aligned}$$

Combining these estimates gives

$$\left\| D_j G_\zeta f \right\|_{L^2_\delta} \leq C \|f\|_{L^2_{\delta+1}} .$$

A similar argument yields

$$\left\| D_j D_k G_\zeta f \right\|_{L^2_\delta} \leq C s \|f\|_{L^2_{\delta+1}} .$$

This concludes the proof of the proposition. □

Exercise 4.5.8. Let $w \in L^2_\delta$ be any weak solution to $(D^2 + 2\zeta \cdot D)w = 0$. Let

$$\chi \in C^\infty([0,\infty)) \quad \text{with} \quad \operatorname{supp}\chi \subset [0,1) \quad \text{and} \quad \int_{\mathbb{R}^n} \chi(|\xi|^2) \, \frac{d\xi}{(2\pi)^n} = 1$$

and

$$w_\varepsilon(x) = \beta(\varepsilon x) w(x) \qquad \text{where} \qquad \beta(x) = \int_{\mathbb{R}^n} e^{ix\cdot\xi} \chi(|\xi|^2) \, \frac{d\xi}{(2\pi)^n}.$$

(a) Prove that if the Fourier transform of $\varphi \in \mathcal{S}(\mathbb{R}^n)$ vanishes in

$$N_\varepsilon\big(\mathcal{M}(s)\big) = \{ \, \xi \mid \operatorname{dist}(\xi, \mathcal{M}(s)) \leq \varepsilon \, \}$$

then there is a $r \in \mathcal{S}(\mathbb{R}^n)$ such that

$$\beta(\varepsilon x)\varphi(x) = (D^2 + 2\bar{\zeta} \cdot D) r.$$

(b) Prove that $\hat{w}_\varepsilon(\xi)$ is supported in $N_\varepsilon\big(\mathcal{M}(s)\big)$.

It still remains to prove the three auxiliary Lemmas 4.5.5–4.5.7. If we note that $\|f\|_{H^s(\mathbb{R}^n)} = \frac{1}{(2\pi)^{n/2}} \|\hat{f}\|_{L^2_s(\mathbb{R}^n)}$, then Lemmas 4.5.6 and 4.5.7 merely state the well–known facts that multiplication by smooth, compactly supported functions and composition with smooth diffeomorphisms are bounded operators on $H^s(\mathbb{R}^n)$. The former is Lemma C.1.32. The latter is the following exercise.

Exercise 4.5.9. Let $s \in \mathbb{R}$. Let $\mathcal{O}$ and $\mathcal{O}'$ be open subsets of $\mathbb{R}^n$ and $\mathcal{C}$ be a compact subset of $\mathcal{O}$. Let Ψ be a smooth diffeomorphism from $\mathcal{O}$ to $\mathcal{O}'$. Prove that there is a constant C, depending only on Ψ, s, $\mathcal{O}$ and $\mathcal{C}$, such that

$$\left\| u \circ \Psi^{-1} \right\|_{H^s(\mathbb{R}^n)} \leq C \|u\|_{H^s(\mathbb{R}^n)}$$

for all $u \in C_c^\infty(\mathcal{C})$.

It thus only remains to establish the properties of Z_j.

Proof of Lemma 4.5.5. To prove the lemma, it suffices to consider a single value of the index j, like $j = 2$. We furthermore claim that it suffices to prove the estimate $\|Z_2 f\|_{L^2_\delta} \leq C\|f\|_{L^2_{\delta+1}}$ in $\mathbb{R}^2$. To see this we note that

$$\begin{aligned}\|u\|^2_{L^2_\delta(\mathbb{R}^n)} &= \int_{\mathbb{R}^n} (1+|x|^2)^\delta |u(x)|^2\, dx \\ &\leq \int_{\mathbb{R}^n} \left(1+x_1^2+x_2^2\right)^\delta |u(x)|^2\, dx\end{aligned}$$

since $\delta < 0$. Therefore, writing $x = (x_1, x_2, x'')$ with $x'' \in \mathbb{R}^{n-2}$,

$$\|Z_2 f\|^2_{L^2_\delta(\mathbb{R}^n)} \leq \int_{\mathbb{R}^{n-2}} \|Z_2 f(\cdot\,,\cdot\,,x'')\|^2_{L^2_\delta(\mathbb{R}^2)}\, dx''. \tag{4.31}$$

Here we use the fact that $(Z_2 f)(x_1, x_2, x'') = [Z_2 f(\cdot, x'')](x_1, x_2)$, i.e., we use that x'' may be treated as parameters untouched by Z_2. At the same time

$$\begin{aligned}\|f\|^2_{L^2_{\delta+1}(\mathbb{R}^n)} &= \int_{\mathbb{R}^n} (1+|x|^2)^{\delta+1} |f(x)|^2\, dx \\ &\geq \int_{\mathbb{R}^n} (1+x_1^2+x_2^2)^{\delta+1} |f(x)|^2\, dx\end{aligned}$$

since $\delta + 1 > 0$. Therefore

$$\|f\|^2_{L^2_{\delta+1}(\mathbb{R}^n)} \geq \int_{\mathbb{R}^{n-2}} \|f(\cdot\,,\cdot\,,x'')\|^2_{L^2_\delta(\mathbb{R}^2)}\, dx''. \tag{4.32}$$

The estimates (4.31) and (4.32) immediately imply that it suffices to prove the estimate $\|Z_2 f\|_{L^2_\delta} \leq C\|f\|_{L^2_{\delta+1}}$ in two dimensions. This latter estimate is a consequence of Lemma 4.5.10 with $p = 2$.

We now prove

$$-\left\langle Z_j f, (D_{x_j} + iD_{x_1})\bar{\varphi}\right\rangle = \langle f, \bar{\varphi}\rangle$$

for all $\varphi \in \mathcal{S}(\mathbb{R}^n)$ and $f \in L^2_{\delta+1}$, assuming the boundedness of the map $Z_j : L^2_{\delta+1} \to L^2_\delta$. For any $\varphi \in \mathcal{S}(\mathbb{R}^n)$, we have $\left(D_{x_j} + iD_{x_1}\right)\varphi \in L^2_{-\delta}$ and $\varphi \in L^2_{-\delta-1}$.

So, by continuity, it suffices to consider $f \in \mathcal{S}(\mathbb{R}^n)$. For $f, \varphi \in \mathcal{S}(\mathbb{R}^n)$

$$\begin{aligned}
\langle f, \bar{\varphi} \rangle &= \frac{1}{(2\pi)^n} \langle \hat{f}(\xi), \overline{\hat{\varphi}(\xi)} \rangle \\
&= \frac{1}{(2\pi)^n} \left\langle \frac{\hat{f}(\xi)}{\eta_j + i\eta_1}, \overline{(\eta_j - i\eta_1)\hat{\varphi}(\xi)} \right\rangle \\
&= \frac{1}{(2\pi)^n} \left\langle \frac{\hat{f}(\xi)}{\eta_j + i\eta_1}, \overline{((D_{x_j} - iD_{x_1})\varphi)\hat{\ }(\xi)} \right\rangle \\
&= \langle Z_j f, (D_{x_j} + iD_{x_1})\bar{\varphi} \rangle.
\end{aligned}$$

The formula $\frac{1}{(2\pi)^n} \langle \hat{g}(\xi), \overline{\hat{r}(\xi)} \rangle = \langle g, \bar{r} \rangle$ is usually first proven for $g, r \in \mathcal{S}(\mathbb{R}^n)$. But, by the Lebesgue dominated convergence theorem, it extends to $\hat{g} \in L^1(\mathbb{R}^n)$, since $\mathcal{S}(\mathbb{R}^n)$ is dense in $L^1(\mathbb{R}^n)$ and $\hat{g} \in L^1(\mathbb{R}^n)$ implies $g \in L^\infty(\mathbb{R}^n)$. □

The following lemma gives weighted L^p estimates for the operator Z corresponding to the Fourier multiplier $\frac{1}{\xi_1 + i\xi_2}$ in $\mathbb{R}^2$. This operator is called the *Cauchy transform* and it is an inverse of the Cauchy-Riemann operator $D_{x_1} + iD_{x_2}$. We will encounter the Cauchy transform again in §5.3.

Lemma 4.5.10 (Weighted estimates). *Define Z by*

$$(Zf)(u_1, u_2) := \int_{\mathbb{R}^2} \frac{1}{(u_1 - v_1) + i(u_2 - v_2)} f(v_1, v_2)\, dv, \qquad f \in \mathcal{S}(\mathbb{R}^2).$$

(a) *The map Z is bounded from $L^p_{\delta+1}(\mathbb{R}^2)$ to $L^p_\delta(\mathbb{R}^2)$ provided $1 < p < \infty$ and $-\frac{2}{p} < \delta < 1 - \frac{2}{p}$. Here $L^p_\delta = \{ u \mid \langle x \rangle^\delta u \in L^p(\mathbb{R}^n) \}$ equipped with the norm $\|u\|_{L^p_\delta} = \|\langle x \rangle^\delta u\|_{L^p}$.*

(b) *Furthermore*

$$Zf = -2\pi i \mathcal{F}^{-1} \left\{ \frac{1}{\xi_1 + i\xi_2} \hat{f}(\xi) \right\}, \qquad f \in \mathcal{S}(\mathbb{R}^2),$$

and $(D_{x_1} + iD_{x_2})Zf = -2\pi i f$ for $f \in \mathcal{S}(\mathbb{R}^2)$.

Proof.

(a) Since the spaces $L^q(\mathbb{R}^n)$ and $L^p(\mathbb{R}^n)$ are dual provided $\frac{1}{p} + \frac{1}{q} = 1$ and $1 < p < \infty$, the same is true for $L^q_{-\delta}$ and L^p_δ. As a result, it suffices to verify the estimate $|\langle Zf,g\rangle| \le C\|f\|_{L^p_{\delta+1}}\|g\|_{L^q_{-\delta}}$ for all $g \in L^q_{-\delta}$. We have

$$\begin{aligned}|\langle Zf,g\rangle| &= \left|\int_{\mathbb{R}^2}\int_{\mathbb{R}^2}\frac{f(v)\overline{g(u)}}{(u_1-v_1)+i(u_2-v_2)}\,du\,dv\right| \\ &\le \int_{\mathbb{R}^2}\int_{\mathbb{R}^2}\frac{|f(v)|(1+|u|^2)^{-\beta/2}(1+|v|^2)^{\alpha/2}}{|u-v|^{1/p}} \\ &\quad\times\frac{|g(u)|(1+|u|^2)^{\beta/2}(1+|v|^2)^{-\alpha/2}}{|u-v|^{1/q}}\,du\,dv,\end{aligned}$$

where $\alpha > 0$ and $\beta > 0$ will be chosen later. Employing Hölder's inequality,

$$\begin{aligned}|\langle Zf,g\rangle| \le{}& \left(\int_{\mathbb{R}^2}\left\{\int_{\mathbb{R}^2}\frac{(1+|u|^2)^{-p\beta/2}(1+|v|^2)^{p(\alpha-\delta-1)/2}}{|u-v|}\,du\right\}\right. \\ &\qquad\left.(1+|v|^2)^{p(\delta+1)/2}|f(v)|^p\,dv\right)^{1/p} \\ &\times\left(\int_{\mathbb{R}^2}\left\{\int_{\mathbb{R}^2}\frac{(1+|u|^2)^{q(\beta+\delta)/2}(1+|v|^2)^{-q\alpha/2}}{|u-v|}\,dv\right\}\right. \\ &\qquad\left.(1+|u|^2)^{-q\delta/2}|g(u)|^q\,du\right)^{1/q} \\ \le{}& C\|f\|_{L^p_{\delta+1}}\|g\|_{L^q_{-\delta}}\end{aligned}$$

provided that the constant

$$\begin{aligned}C ={}& \left(\sup_v\int_{\mathbb{R}^2}\frac{(1+|u|^2)^{-p\beta/2}(1+|v|^2)^{p(\alpha-\delta-1)/2}}{|u-v|}\,du\right)^{1/p} \\ &\times\left(\sup_u\int_{\mathbb{R}^2}\frac{(1+|u|^2)^{q(\beta+\delta)/2}(1+|v|^2)^{-q\alpha/2}}{|u-v|}\,dv\right)^{1/q}\end{aligned}$$

is finite.

Since $1+|x|^2 \le (1+|x|)^2 \le 2(1+|x|^2)$ it suffices to check that

$$\sup_v\int_{\mathbb{R}^2}\frac{(1+|u|)^{-p\beta}(1+|v|)^{p(\alpha-\delta-1)}}{|u-v|}\,du$$

and

$$\sup_u\int_{\mathbb{R}^2}\frac{(1+|u|)^{q(\beta+\delta)}(1+|v|)^{-q\alpha}}{|u-v|}\,dv$$

are finite for appropriate choices of α, β, p, q and δ, which we do next.

We impose the constraint that $\delta = \alpha - \beta - \frac{1}{q}$, which implies that

$$\begin{aligned}p(\alpha-\delta-1) &= p(\beta+\tfrac{1}{q}-1) = p(\beta-\tfrac{1}{p}) = p\beta-1, \\ q(\beta+\delta) &= q(\alpha-\tfrac{1}{q}) = q\alpha-1.\end{aligned}$$

For each fixed v with $|v| \geq 1$, let R be a rotation chosen so that $v = |v|Re_1$. Making the change of variables $u = |v|Rw$, we see that the integral

$$\begin{aligned}\int_{\mathbb{R}^2} &\frac{(1+|u|)^{-p\beta}(1+|v|)^{p(\alpha-\delta-1)}}{|u-v|}\,du \\ &= (1+|v|)^{p(\alpha-\delta-1)}|v| \int_{\mathbb{R}^2} \frac{(1+|v||w|)^{-p\beta}}{|w-e_1|}\,dw \\ &\leq (1+|v|)^{p(\alpha-\delta-1)}|v| \int_{\mathbb{R}^2} \frac{(|v||w|)^{-p\beta}}{|w-e_1|}\,dw \\ &= (1+|v|)^{p\beta-1}|v|^{1-p\beta} \int_{\mathbb{R}^2} \frac{1}{|w-e_1|\,|w|^{p\beta}}\,dw\end{aligned}$$

converges and is bounded uniformly for $|v| \geq 1$ if $1 < p\beta < 2$. For each fixed v with $|v| < 1$, we make the change of variables $u = w + v$ and use the bound $1 + |w + v| \geq c(1 + |w|)$ to see that the integral

$$\begin{aligned}\int_{\mathbb{R}^2} &\frac{(1+|u|)^{-p\beta}(1+|v|)^{p(\alpha-\delta-1)}}{|u-v|}\,du \\ &= (1+|v|)^{p(\alpha-\delta-1)} \int_{\mathbb{R}^2} \frac{(1+|w+v|)^{-p\beta}}{|w|}\,dw \\ &\leq c^{-p\beta}(1+|v|)^{p(\alpha-\delta-1)} \int_{\mathbb{R}^2} \frac{1}{|w|(1+|w|)^{p\beta}}\,dw\end{aligned}$$

again converges and is bounded uniformly for $|v| \leq 1$ if $p\beta > 1$.

Similarly, for each fixed u with $|u| \geq 1$, let R be a rotation chosen so that $u = |u|Re_1$. Making the change of variables $v = |u|Rw$, we see that the integral

$$\begin{aligned}\int_{\mathbb{R}^2} &\frac{(1+|u|)^{q(\beta+\delta)}(1+|v|)^{-q\alpha}}{|u-v|}\,dv \\ &= (1+|u|)^{q(\beta+\delta)}|u| \int_{\mathbb{R}^2} \frac{(1+|u||w|)^{-q\alpha}}{|w-e_1|}\,dw \\ &\leq (1+|u|)^{q(\beta+\delta)}|u| \int_{\mathbb{R}^2} \frac{(|u||w|)^{-q\alpha}}{|w-e_1|}\,dw \\ &= (1+|u|)^{q\alpha-1}|u|^{1-q\alpha} \int_{\mathbb{R}^2} \frac{1}{|w-e_1|\,|w|^{q\alpha}}\,dw\end{aligned}$$

converges and is bounded uniformly for $|u| \geq 1$ if $1 < q\alpha < 2$. For each fixed u with $|u| < 1$, we make the change of variables $v = w + u$ and use the bound

$1 + |w + u| \geq c(1 + |w|)$ to see that the integral

$$\int_{\mathbb{R}^2} \frac{(1+|u|)^{q(\beta+\delta)}(1+|v|)^{-q\alpha}}{|u-v|}\,dv = (1+|u|)^{q(\beta+\delta)} \int_{\mathbb{R}^2} \frac{(1+|w+u|)^{-q\alpha}}{|w|}\,dw$$
$$\leq c^{-q\alpha}(1+|u|)^{q(\beta+\delta)} \int_{\mathbb{R}^2} \frac{1}{|w|(1+|w|)^{q\alpha}}\,dw$$

again converges and is bounded uniformly for $|u| \leq 1$ if $q\alpha > 1$.
Thus, in order to guarantee that C is finite, it suffices to require that

$$\frac{1}{p} < \beta < \frac{2}{p} \quad \text{and} \quad \frac{1}{q} < \alpha < \frac{2}{q} \tag{4.33}$$

with

$$\delta = \alpha - \beta - \frac{1}{q}. \tag{4.34}$$

As α and β run over the region (4.33), $-\beta$ runs over $-\frac{2}{p} < -\beta < -\frac{1}{p}$ and $\alpha-\beta-\frac{1}{q}$ runs over

$$-\frac{2}{p} = \frac{1}{q} - \frac{2}{p} - \frac{1}{q} < \alpha - \beta - \frac{1}{q} < \frac{2}{q} - \frac{1}{p} - \frac{1}{q} = 1 - \frac{2}{p}.$$

Thus, if $1 < p < \infty$ and δ satisfies

$$-\frac{2}{p} < \delta < 1 - \frac{2}{p},$$

then it is always possible to select α and β such that (4.33) and (4.34) are satisfied. This completes the proof of part (a).

(b) For $f \in \mathcal{S}(\mathbb{R}^n)$, the fact that the Fourier transform takes convolutions to products implies that

$$\mathcal{F}^{-1}\left\{\frac{1}{\xi_1 + i\xi_2}\hat{f}(\xi)\right\} = \mathcal{F}^{-1}\left\{\frac{1}{\xi_1 + i\xi_2}\right\} * f.$$

The inverse Fourier transform is computed in Exercise 4.5.11. This gives the required formula for Zf, and the identity $(D_{x_1} + iD_{x_2})Zf = -2\pi f$ for $f \in \mathcal{S}(\mathbb{R}^2)$ follows via the Fourier transform. □

Exercise 4.5.11 (Inverse Fourier transform of $\frac{1}{\xi_1+i\xi_2}$). For $R > 0$, define the function f_R in $\mathbb{R}^2$ by

$$f_R(x) = \int_{|\xi|\leq R} \frac{e^{ix\cdot\xi}}{\xi_1 + i\xi_2}\,d\xi.$$

(a) Apply polar coordinates to prove that

$$f_R(x) = \int_0^{2\pi} e^{-i\theta}\frac{e^{iRx\cdot(\cos\theta,\sin\theta)} - 1}{ix\cdot(\cos\theta,\sin\theta)}\,d\theta.$$

(b) Write $e^{-i\theta} = \cos\theta - i\sin\theta$ and use parity to show that

$$f_R(|x|e_1) = \frac{1}{i|x|}\int_0^{2\pi}(e^{iR|x|\cos\theta} - 1)\,d\theta.$$

(c) Use rotation invariance to show that

$$f_R(x) = \frac{1}{i(x_1 + ix_2)}\int_0^{2\pi}(e^{iR|x|\cos\theta} - 1)\,d\theta.$$

(d) Take the limit of f_R as $R \to \infty$ in $\mathcal{S}'(\mathbb{R}^2)$ to obtain

$$\mathcal{F}^{-1}\left\{\frac{1}{\xi_1 + i\xi_2}\right\} = -\frac{1}{2\pi i}\frac{1}{x_1 + ix_2}.$$

Remark 4.5.12 (Faddeev–Green function). For later purposes we observe that G_ζ can be written as a convolution operator. The proof of Proposition 4.5.4 shows that for $f \in \mathcal{S}(\mathbb{R}^n)$, one has

$$G_\zeta f = \mathcal{F}^{-1}\left\{\frac{1}{|\xi|^2 + 2\zeta\cdot\xi}\hat{f}(\xi)\right\}.$$

The proof was based on the decomposition

$$\frac{1}{|\xi|^2 + 2\zeta\cdot\xi} = \sum_{j=1}^{n}\frac{\chi_j(\xi)}{|\xi|^2 + 2\zeta\cdot\xi}.$$

The term involving χ_1 is a smooth bounded function in $\mathbb{R}^n$, and the terms involving χ_j for $j \geq 2$ were transformed by changes of coordinates to terms of the form

$$\left[\frac{\chi_j \circ \Psi}{s(\eta_j + i\eta_1)}\right] \circ \Psi^{-1}$$

that are functions in $L^1(\mathbb{R}^n)$ supported in the compact set $\overline{\mathcal{O}_j(s)}$. This proves that $\frac{1}{|\xi|^2+2\zeta\cdot\xi} \in L^1(\mathbb{R}^n) + L^\infty(\mathbb{R}^n)$, hence it is a tempered distribution. The formula for $G_\zeta f$ then gives that

$$G_\zeta f = g_\zeta * f,$$

where g_ζ is the tempered distribution

$$g_\zeta = \mathcal{F}^{-1}\left\{\frac{1}{|\xi|^2 + 2\zeta\cdot\xi}\right\}.$$

The distribution g_ζ is sometimes called a *Faddeev–Green function.*

4.6. Reconstruction

In §4.3 we proved a uniqueness result in the Calderón problem, stating that if two positive conductivities $\gamma_1, \gamma_2 \in C^2(\overline{\Omega})$ satisfy $\Lambda_{\gamma_1} = \Lambda_{\gamma_2}$, then necessarily $\gamma_1 = \gamma_2$ in Ω. The proof was not constructive and did not give an algorithm to determine γ from Λ_γ. In this section we will give a constructive proof that results in a reconstruction procedure for this inverse problem.

Theorem 4.6.1. *Let Ω be a bounded domain in $\mathbb{R}^n$, $n \geq 3$, with smooth boundary and suppose that $\gamma \in C^2(\overline{\Omega})$ is a positive function. From the knowledge of the map*

$$\Lambda_\gamma : H^{3/2}(\partial\Omega) \to H^{1/2}(\partial\Omega)$$

it is possible to determine γ in Ω in a constructive way.

As before, this result will be obtained as a consequence of a reconstruction procedure for the inverse problem for a Schrödinger equation.

Theorem 4.6.2. *Let Ω be a bounded domain in $\mathbb{R}^n$, $n \geq 3$, with smooth boundary and suppose that $q \in L^\infty(\Omega)$. Assume that 0 is not a Dirichlet eigenvalue of $-\Delta + q$ in Ω. From the knowledge of the map*

$$\Lambda_q : H^{3/2}(\partial\Omega) \to H^{1/2}(\partial\Omega)$$

it is possible to determine q in Ω in a constructive way.

Let us give an outline of the proof. It is very similar to the uniqueness proof earlier, and relies on CGO solutions $u_\zeta(x) = e^{i\zeta\cdot x}(1 + r(x,\zeta))$ for $\zeta \in \mathbb{C}^n$ such that $\zeta \cdot \zeta = 0$ and $|\zeta|$ is large. Even though the original problem is stated in the domain Ω, we extend q by zero outside of Ω and consider the solutions u_ζ in $\mathbb{R}^n$. It is important that the solution u_ζ is unique as long as $r(x,\zeta)$ satisfies a decay condition as $|x| \to \infty$.

It will be possible to characterize the boundary value $u_\zeta|_{\partial\Omega}$ as the unique solution $f \in H^{3/2}(\partial\Omega)$ of the following integral equation on the boundary:

$$(\mathrm{Id} + S_\zeta(\Lambda_q - \Lambda_0))f = e^{i\zeta\cdot x} \qquad \text{on } \partial\Omega.$$

Here S_ζ is a modified single layer potential depending on the complex vector ζ. The point is that the operator on the left-hand side only depends on the data Λ_q and other known quantities, so one can compute $u_\zeta|_{\partial\Omega}$ from the boundary data by solving this integral equation. By using these functions in a suitable integral identity and taking the limit as $|\zeta| \to \infty$, we are able to recover the Fourier transform of q.

Most of the work in the proof of Theorem 4.6.2 boils down to establishing basic properties for the single layer potential. Before starting the proof, let us see how Theorem 4.6.1 follows from it.

Proof of Theorem 4.6.1. Suppose we know the map

$$\Lambda_\gamma : H^{3/2}(\partial\Omega) \to H^{1/2}(\partial\Omega).$$

If q is defined by $q = \Delta\gamma^{1/2}/\gamma^{1/2}$, it was proved in Proposition 2.8.5 that

$$\Lambda_q f = \gamma^{-1/2}\Lambda_\gamma(\gamma^{-1/2} f) + \frac{1}{2}\gamma^{-1}(\partial_\nu \gamma) f|_{\partial\Omega}.$$

From Theorems 3.1.1 and 3.2.1, we know that from the knowledge of Λ_γ it is possible to reconstruct the boundary value $\gamma|_{\partial\Omega}$ and the normal derivative $\partial_\nu\gamma|_{\partial\Omega}$. Thus, we have access to the map $\Lambda_q : H^{3/2}(\partial\Omega) \to H^{1/2}(\partial\Omega)$.

By Theorem 4.6.2, we can determine q constructively from this information. Consider the unique weak solution $v \in H^1(\Omega)$ of the equation

$$(-\Delta + q)v = 0 \qquad \text{in } \Omega$$

with boundary value $v|_{\partial\Omega} = \gamma^{1/2}|_{\partial\Omega}$. Since the coefficient q and the boundary value are known, we can compute the solution v. But the function $\gamma^{1/2}$ solves this Dirichlet problem, so we also know $v = \gamma^{1/2}$. This determines γ in Ω. □

Assume that $q \in L^\infty(\Omega)$ is such that 0 is not a Dirichlet eigenvalue of $-\Delta + q$ in Ω. The objective is to reconstruct q in Ω from the knowledge of Λ_q. The first step is to extend q by zero into $\mathbb{R}^n$, with the extension also denoted by q. We have the following CGO solutions from Theorem 4.5.1.

Proposition 4.6.3. *Let $q \in L^\infty_c(\mathbb{R}^n)$, and fix δ with $-1 < \delta < 0$. There exists $C = C(\delta, q) > 0$ such that for any $\zeta \in \mathbb{C}^n$ satisfying $\zeta \cdot \zeta = 0$ and $|\zeta| \geq C$, there exists a unique solution*

$$u(x) = u_\zeta(x) = e^{i\zeta\cdot x}(1 + r(x, \zeta))$$

of the equation $(-\Delta + q)u = 0$ in $\mathbb{R}^n$ where $r(\,\cdot\,, \zeta) \in H^2_\delta(\mathbb{R}^n)$. Moreover,

$$\|r\|_{L^2_\delta(\mathbb{R}^n)} \leq \frac{C}{|\zeta|}.$$

In the uniqueness proof of the inverse problem for the Schrödinger equation, we began from the assumption $\Lambda_{q_1} = \Lambda_{q_2}$ and used the integral identity

$$\langle(\Lambda_{q_1} - \Lambda_{q_2})(u_1|_{\partial\Omega}), u_2|_{\partial\Omega}\rangle_{\partial\Omega} = \int_\Omega (q_1 - q_2) u_1 u_2 \, dx,$$

where u_j are solutions of $(-\Delta + q_j)u_j = 0$ in Ω. We then chose complex geometrical optics solutions u_j such that $u_1 u_2 \approx e^{ix\cdot\xi}$ for large values of the complex vectors. Since the left-hand side of the identity is zero, this resulted in the vanishing of the Fourier transform of $q_1 - q_2$.

In the reconstruction problem we are instead given the DN map Λ_q for an unknown potential q, and the objective is to determine the Fourier transform of q from this information. This will be achieved by comparing Λ_q to Λ_0, the DN map with zero potential. The next result shows how this is precisely done.

Proposition 4.6.4. *Let $q \in L^\infty(\Omega)$ be such that 0 is not a Dirichlet eigenvalue of $-\Delta + q$ in Ω. Let also $\xi \in \mathbb{R}^n$ with $\xi \neq 0$. If $n \geq 3$ and $s > 0$ is sufficiently large, there exist $\zeta_j = \zeta_j(s,\xi) \in \mathbb{C}^n$ with $\zeta_j \cdot \zeta_j = 0$ and $|\zeta_j| = s$ for $j = 1, 2$, such that*

$$\lim_{s\to\infty} \langle (\Lambda_q - \Lambda_0)(u_{\zeta_1}|_{\partial\Omega}), e^{i\zeta_2\cdot x}|_{\partial\Omega}\rangle_{\partial\Omega} = \int_\Omega q e^{ix\cdot\xi}\,dx.$$

Here u_{ζ_1} is the solution of $(-\Delta + q)u = 0$ in $\mathbb{R}^n$ given in Proposition 4.6.3.

Proof. Let $\alpha, \beta \in \mathbb{R}^n$ be unit vectors such that $\{\alpha, \beta, \xi/|\xi|\}$ is an orthonormal set in $\mathbb{R}^n$ (here we need $n \geq 3$). We define complex vectors

$$\zeta_1 = \sqrt{\frac{s^2}{2} - \frac{|\xi|^2}{4}}\alpha + \frac{\xi}{2} + i\frac{s}{\sqrt{2}}\beta,$$

$$\zeta_2 = -\sqrt{\frac{s^2}{2} - \frac{|\xi|^2}{4}}\alpha + \frac{\xi}{2} - i\frac{s}{\sqrt{2}}\beta.$$

By using the fact that $\alpha, \beta, \xi/|\xi|$ are orthonormal, it follows that $\zeta_j \cdot \zeta_j = 0$ and $|\zeta_j| = s$. The main point in the choice of ζ_1 and ζ_2 is that

$$\zeta_1 + \zeta_2 = \xi.$$

Let u_{ζ_1} be the solution of $(-\Delta+q)u = 0$ in $\mathbb{R}^n$ in Proposition 4.6.3, and note that $e^{i\zeta_2\cdot x}$ solves the same equation with zero potential, that is, $\Delta(e^{i\zeta_2\cdot x}) = 0$. The integral identity in Proposition 2.8.3 implies that

$$\begin{aligned}\langle (\Lambda_q - \Lambda_0)(u_{\zeta_1}|_{\partial\Omega}), e^{i\zeta_2\cdot x}|_{\partial\Omega}\rangle_{\partial\Omega} &= \int_\Omega q u_{\zeta_1} e^{i\zeta_2\cdot x}\,dx \\ &= \int_\Omega q e^{i(\zeta_1+\zeta_2)\cdot x}(1 + r(x,\zeta_1))\,dx \\ &= \int_\Omega q e^{ix\cdot\xi}(1 + r(x,\zeta_1))\,dx.\end{aligned}$$

Since $\|r(\,\cdot\,,\zeta_1)\|_{L^2_\delta(\mathbb{R}^n)} \leq \frac{C}{s}$, the result follows by taking the limit as $s \to \infty$. □

From Proposition 4.6.4 we see that the Fourier transform of q at nonzero frequencies ξ can be recovered from the map Λ_q, as long as the boundary value $u_{\zeta_1}|_{\partial\Omega}$ of the solution in Proposition 4.6.3 can somehow be determined from Λ_q. Since q is compactly supported, its Fourier transform is continuous and would therefore be determined also at $\xi = 0$.

The determination of $u_\zeta|_{\partial\Omega}$ from Λ_q will require certain facts on layer potentials. This may be motivated as follows.

Motivation 4.6.5. Since q is extended by zero outside of Ω, the equation $(-\Delta+q)u_\zeta = 0$ in $\mathbb{R}^n$ implies

$$-\Delta u_\zeta = 0 \quad \text{in } \mathbb{R}^n \setminus \overline{\Omega}.$$

Thus u_ζ is harmonic in the exterior domain $\mathbb{R}^n \setminus \overline{\Omega}$. Note that this equation does not involve the unknown potential q. Writing $u_\zeta = e^{i\zeta\cdot x} + v_\zeta$, we see that also the correction term v_ζ is harmonic in the exterior domain. Further, the function $e^{-i\zeta\cdot x}v_\zeta$ satisfies a decay condition at infinity by construction.

It is known that harmonic functions in the exterior domain satisfying certain decay conditions can be represented in terms of single layer potentials, which are integral operators mapping functions on $\partial\Omega$ to functions in $\mathbb{R}^n$. In our case we will have

$$v_\zeta|_{\mathbb{R}^n\setminus\overline{\Omega}} = S_\zeta h|_{\mathbb{R}^n\setminus\overline{\Omega}}$$

for some function $h \in H^{1/2}(\partial\Omega)$, where S_ζ is a modified (or Faddeev type) single layer potential that differs from the standard one by factors of $e^{i\zeta\cdot x}$ in its integral kernel. It will turn out that $h = -(\Lambda_q - \Lambda_0)(u_\zeta|_{\partial\Omega})$, and collecting these facts gives that

$$u_\zeta = e^{i\zeta\cdot x} - S_\zeta(\Lambda_q - \Lambda_0)(u_\zeta|_{\partial\Omega}) \quad \text{in } \mathbb{R}^n \setminus \overline{\Omega}.$$

The integral equation characterizing $u_\zeta|_{\partial\Omega}$ in terms of Λ_q follows by restricting the last identity on $\partial\Omega$.

Let us begin the development of the required layer potentials. We decompose $\mathbb{R}^n$ in three disjoint parts,

$$\mathbb{R}^n = \Omega_- \cup \Gamma \cup \Omega_+,$$

where $\Omega_- = \Omega$ is the interior domain, $\Gamma = \partial\Omega$ is the boundary of Ω, and $\Omega_+ = \mathbb{R}^n \setminus \overline{\Omega}$ is the exterior domain.

Consider the trace operator on Γ,

$$\gamma : H^1(\mathbb{R}^n) \to H^{1/2}(\Gamma), \quad \gamma u = u|_{\partial\Omega},$$

and the corresponding trace operators from the interior and exterior,

$$\gamma_- : H^1(\Omega) \to H^{1/2}(\Gamma), \quad \gamma_- u = u|_\Gamma,$$

$$\gamma_+ : H^1(\Omega_+) \to H^{1/2}(\Gamma), \quad \gamma_+ u = u|_\Gamma.$$

Since taking traces is a local operation near Γ, the operators $\gamma_\pm$ can also be applied to functions that are in $H^1(U\cap\Omega_\pm)$ where U is some open set containing Γ. If u is a function that is H^1 in a full neighborhood U of Γ, it follows that

$$\gamma_- u = \gamma_+ u = \gamma u.$$

If u is a H^2 function in $U \cap \Omega_\pm$ where U is some neighborhood of Γ, we denote by $(\partial_\nu u)_- = \nabla u\cdot\nu|_{\partial\Omega}$ the normal derivative of u from the interior and $(\partial_\nu u)_+ = \nabla u\cdot\nu|_{\partial\Omega}$ the normal derivative from the exterior, where ν denotes the *unit outer normal of* Ω_-. Note that in this case $(\partial_\nu u)_\pm$ is in $H^{1/2}(\Gamma)$.

The standard single layer potential on $\mathbb{R}^n$ will be obtained from the fundamental solution of the Laplacian given by the Newtonian potential.

Proposition 4.6.6 (Newtonian potential). *Let $n \geq 3$, and let*

$$k_0(x) = \frac{1}{n(n-2)\alpha(n)} |x|^{2-n}, \quad x \in \mathbb{R}^n,$$

where $\alpha(n)$ is the volume of the unit ball in $\mathbb{R}^n$. This function gives rise to a convolution operator

$$K_0 : L^2_c(\mathbb{R}^n) \to H^2_{\mathrm{loc}}(\mathbb{R}^n), \quad K_0 f(x) = \int_{\mathbb{R}^n} k_0(x-y) f(y)\, dy.$$

Furthermore, K_0 extends as an operator $H^{-1}_c(\mathbb{R}^n) \to H^1_{\mathrm{loc}}(\mathbb{R}^n)$, and

$$-\Delta K_0 f = f, \qquad f \in H^{-1}_c(\mathbb{R}^n).$$

One has the symmetry property $\langle K_0 f, g\rangle = \langle f, K_0 g\rangle$ for $f, g \in H^{-1}_c(\mathbb{R}^n)$.

Proof. Let $\chi, \psi \in C^\infty_c(\mathbb{R}^n)$. We will show that

$$\|\chi K_0(\psi f)\|_{L^2(\mathbb{R}^n)} \leq C_{\chi,\psi} \|f\|_{L^2(\mathbb{R}^n)}.$$

Define

$$L = \{\, x - y \mid x \in \operatorname{supp}(\chi),\ y \in \operatorname{supp}(\psi) \,\}.$$

Then L is a compact set, and for $x \in \operatorname{supp}(\chi)$ we have

$$\chi K_0(\psi f)(x) = \int_{\mathbb{R}^n} \chi(x)(\chi_L k_0)(x-y)\psi(y) f(y)\, dy.$$

Since $\chi_L k_0 \in L^1(\mathbb{R}^n)$, Young's inequality for convolutions (Lemma C.3.10) implies that

$$\|\chi K_0(\psi f)\|_{L^2} \leq C_\chi \|\chi_L k_0\|_{L^1} \|\psi f\|_{L^2} \leq C_{\chi,\psi} \|f\|_{L^2}.$$

This proves that $K_0 f \in L^2_{\mathrm{loc}}$ whenever $f \in L^2_c$. The same argument applied to $\partial_j(\chi K_0 \psi f)$ for $1 \leq j \leq n$ implies that

$$\chi K_0 \psi : L^2(\mathbb{R}^n) \to H^1(\mathbb{R}^n) \text{ is bounded for any } \chi, \psi \in C^\infty_c(\mathbb{R}^n). \tag{4.35}$$

Thus K_0 maps L^2_c to H^1_{loc}.

The fact that k_0 is a fundamental solution, i.e.,

$$-\Delta K_0 f = f, \qquad f \in C^\infty_c(\mathbb{R}^n),$$

is standard (see [**Eva10**, §2.2]). We wish to show that

$$-\Delta K_0 f = f, \qquad f \in L^2_c(\mathbb{R}^n). \tag{4.36}$$

Let $f \in L^2(\mathbb{R}^n)$ with $\operatorname{supp}(f)$ compact. By density (see Lemma 2.1.6) there is a sequence $(f_j) \subset C^\infty_c(\mathbb{R}^n)$ so that $f_j \to f$ in $L^2(\mathbb{R}^n)$ and $\operatorname{supp}(f_j)$ lies in a fixed compact set for each j. By (4.35), we have

$$K_0 f_j \to K_0 f \qquad \text{in } H^1_{\mathrm{loc}}(\mathbb{R}^n).$$

Then also

$$-\Delta K_0 f_j \to -\Delta K_0 f \qquad \text{in } H^{-1}_{\mathrm{loc}}(\mathbb{R}^n).$$

Moreover, we have already seen that $-\Delta K_0 f_j = f_j$, so also

$$-\Delta K_0 f_j \to f \qquad \text{in } L^2(\mathbb{R}^n).$$

Uniqueness of limits implies (4.36).

We next show that $K_0 f \in H^2_{\mathrm{loc}}$ whenever $f \in L^2_c$. By the previous arguments, $u = K_0 f$ satisfies

$$-\Delta u = f \in L^2_c.$$

We have seen above that $u \in H^1_{\mathrm{loc}}$. Interior elliptic regularity (Theorem 2.6.2) and (4.35) readily imply that $u \in H^2_{\mathrm{loc}}$ and that

$$\chi K_0 \psi : L^2(\mathbb{R}^n) \to H^2(\mathbb{R}^n) \text{ is bounded for any } \chi, \psi \in C_c^\infty(\mathbb{R}^n). \tag{4.37}$$

Finally, we consider the formal adjoint $K_0^* : H_c^{-2} \to L^2_{\mathrm{loc}}$ defined via

$$\langle K_0^* F, \varphi \rangle = \langle F, K_0 \varphi \rangle, \qquad F \in H_c^{-2},\ \varphi \in L^2_c.$$

For $f, g \in C_c^\infty(\mathbb{R}^n)$ one has the symmetry property $\langle K_0 f, g \rangle = \langle f, K_0 g \rangle$ by the Fubini theorem since $k_0(-x) = k_0(x)$. This extends to $f, g \in L^2_c$ by (4.37) and density. Thus K_0^* agrees with K_0 on $L^2_c(\mathbb{R}^n)$, so K_0^* is indeed an extension of K_0 and we may write K_0 instead of K_0^*. By (4.37) we have bounded operators

$$\chi K_0 \psi : L^2(\mathbb{R}^n) \to H^2(\mathbb{R}^n),$$
$$\chi K_0 \psi : H^{-2}(\mathbb{R}^n) \to L^2(\mathbb{R}^n).$$

Interpolation (Lemma C.3.2) shows that $\chi K_0 \psi$ is bounded H^{-1} to H^1, so K_0 maps H_c^{-1} to H^1_{loc}. Then (4.36) follows since $C_c^\infty(\mathbb{R}^n)$ is dense in $H^{-1}(\mathbb{R}^n)$, and the symmetry property extends to $f, g \in H_c^{-1}$. □

The single layer potential is the composition of K_0 and γ^*, the adjoint of the trace operator, defined as the map $\gamma^* : H^{-1/2}(\Gamma) \to H^{-1}(\mathbb{R}^n)$ via

$$\langle \gamma^* f, \varphi \rangle = \langle f, \gamma \varphi \rangle_\Gamma$$

for $\varphi \in H^1(\mathbb{R}^n)$. Thus formally $\gamma^* f = f\, dS$ where dS is the surface measure of Γ.

Definition 4.6.7 (Single layer potential). The standard single layer potential S_0 on $\mathbb{R}^n$ is defined as the operator

$$S_0 = K_0 \gamma^* : H^{-1/2}(\Gamma) \to H^1_{\mathrm{loc}}(\mathbb{R}^n).$$

We will next discuss the basic jump and mapping properties of S_0. Let u be a function on $U \setminus \Gamma$, where U is an open set containing Γ, such that $u|_{U \cap \Omega_\mp}$ are H^1 in $U \cap \Omega_\mp$ and satisfy $-\Delta(u|_{U \cap \Omega_\mp}) = 0$ in $U \cap \Omega_\mp$. We can then define the normal derivatives from the interior or exterior weakly as elements of $H^{-1/2}(\Gamma)$ by

$$\langle (\partial_\nu u)_\mp, h \rangle_\Gamma = \pm \int_{U \cap \Omega_\mp} \nabla u \cdot \nabla e_h \, dx, \quad h \in H^{1/2}(\Gamma),$$

where $e_h \in H^1_c(U)$ is any function with $e_h|_\Gamma = h$.

Definition 4.6.8 (Jumps). If u is as above, the jumps of u on Γ are defined by

$$[u]_\Gamma = \gamma_- u - \gamma_+ u, \qquad [\partial_\nu u]_\Gamma = (\partial_\nu u)_- - (\partial_\nu u)_+.$$

Lemma 4.6.9 (Jump relations). *If $f \in H^{-1/2}(\Gamma)$, then $u = S_0 f \in H^1_{\mathrm{loc}}(\mathbb{R}^n)$ satisfies*

$$\begin{aligned} -\Delta u &= 0 \quad \textit{in} \quad \Omega_\pm, \\ [u]_\Gamma &= 0, \\ [\partial_\nu u]_\Gamma &= f. \end{aligned}$$

If $f \in H^s(\Gamma)$ for $s \geq -1/2$, then $u|_{\Omega_-} \in H^{s+3/2}(\Omega_-)$ and $u|_{\Omega_+} = \tilde{u}|_{\Omega_+}$ for some $\tilde{u} \in H^{s+3/2}_{\mathrm{loc}}(\mathbb{R}^n)$.

Proof. Harmonicity and the first jump property are direct consequences of the properties of K_0. The definitions also imply that for $h \in H^{1/2}(\Gamma)$,

$$\langle [\partial_\nu u] |_\Gamma, h\rangle_\Gamma = \int_{\mathbb{R}^n} \nabla u \cdot \nabla e_h \, dx = \langle \gamma^* f, e_h\rangle = \langle f, h\rangle_\Gamma.$$

This shows the second jump property. If $f \in H^{k+1/2}(\Gamma)$ with $k \geq 0$, then u satisfies

$$\begin{aligned} -\Delta u &= 0 \quad \text{in} \quad \Omega_\pm, \\ [u]_\Gamma &= 0, \\ [\partial_\nu u]_\Gamma &= f \in H^{k+1/2}(\Gamma). \end{aligned}$$

The transmission property [**McL00**, Theorem 4.20] implies that $u|_{\Omega_\pm}$ are H^{k+2} near Γ. Then the properties for u follow from interior elliptic regularity for $-\Delta$ and from interpolation. ☐

We now move to the modified single layer potential S_ζ, which will be obtained from the following inverse operator K_ζ of the Laplacian. This operator involves the inverse G_ζ of the conjugated operator $e^{-i\zeta\cdot x}(-\Delta)(e^{i\zeta\cdot x}\,\cdot\,)$ studied in Proposition 4.5.4.

Proposition 4.6.10. *Let $\zeta \in \mathbb{C}^n$ satisfy $\zeta \cdot \zeta = 0$ and $|\zeta| \geq 1$. The operator*

$$K_\zeta : L^2_c(\mathbb{R}^n) \to H^2_{\mathrm{loc}}(\mathbb{R}^n), \quad K_\zeta f = e^{i\zeta\cdot x} G_\zeta(e^{-i\zeta\cdot x} f)$$

satisfies

$$-\Delta K_\zeta f = f, \qquad f \in L^2_c(\mathbb{R}^n).$$

Proof. This follows from Proposition 4.5.4. If $f \in L^2_c$, then $e^{-i\zeta\cdot x} f \in L^2_c$ and thus $G_\zeta(e^{-i\zeta\cdot x} f) \in H^2_\delta$ whenever $-1 < \delta < 0$. It follows that $K_\zeta f \in H^2_{\mathrm{loc}}$ for any $f \in L^2_c$. Also, if $u = G_\zeta(e^{-i\zeta\cdot x} f)$, we have $e^{-i\zeta\cdot x}(-\Delta)(e^{i\zeta\cdot x} u) = e^{-i\zeta\cdot x} f$, showing that $-\Delta K_\zeta f = f$. ☐

The next result shows that the operator K_ζ differs from the usual fundamental solution K_0 by a smoothing operator.

Lemma 4.6.11 (Properties of K_ζ). *Let $\zeta \in \mathbb{C}^n$ satisfy $\zeta \cdot \zeta = 0$ and $|\zeta| \geq 1$. Then*

$$K_\zeta = K_0 + R_\zeta$$

where R_ζ is an operator

$$R_\zeta : L^2_c(\mathbb{R}^n) \to C^\infty(\mathbb{R}^n).$$

There is $k_\zeta \in L^1_{\mathrm{loc}}(\mathbb{R}^n)$ that is smooth in $\mathbb{R}^n \setminus \{0\}$ and $r_\zeta \in C^\infty(\mathbb{R}^n)$ such that $-\Delta k_\zeta = \delta_0$, $-\Delta r_\zeta = 0$, and for $f \in L^2_c(\mathbb{R}^n)$,

$$K_\zeta f = k_\zeta * f, \qquad R_\zeta f = r_\zeta * f.$$

Proof. By Remark 4.5.12, G_ζ can be written as a convolution operator

$$G_\zeta f = g_\zeta * f,$$

where $g_\zeta = \mathscr{F}^{-1}\{\frac{1}{|\xi|^2 + 2\zeta\cdot\xi}\}$ is a tempered distribution. Let $g_{\zeta,j} \in C^\infty_c(\mathbb{R}^n)$ satisfy $g_{\zeta,j} \to g_\zeta$ as $j \to \infty$ in $\mathcal{S}'(\mathbb{R}^n)$. If $f \in C^\infty_c(\mathbb{R}^n)$, we have

$$\begin{aligned} K_\zeta f(x) &= e^{i\zeta\cdot x}(g_\zeta * e^{-i\langle\zeta,\cdot\rangle} f)(x) \\ &= \lim e^{i\zeta\cdot x}(g_{\zeta,j} * e^{-i\langle\zeta,\cdot\rangle} f)(x) \\ &= \lim \int_{\mathbb{R}^n} e^{i\zeta\cdot x} g_{\zeta,j}(x-y) e^{-i\zeta\cdot y} f(y)\, dy \\ &= \lim\, (e^{i\langle\zeta,\cdot\rangle} g_{\zeta,j} * f)(x) \\ &= (e^{i\langle\zeta,\cdot\rangle} g_\zeta * f)(x). \end{aligned}$$

Thus, writing $k_\zeta = e^{i\langle\zeta,\cdot\rangle} g_\zeta$ which is now a distribution in $\mathcal{D}'(\mathbb{R}^n)$ instead of $\mathcal{S}'(\mathbb{R}^n)$, we have

$$K_\zeta f = k_\zeta * f, \qquad f \in C^\infty_c(\mathbb{R}^n).$$

On the other hand, $K_0 f = k_0 * f$ where k_0 is the Newtonian potential. Thus

$$R_\zeta f = r_\zeta * f, \qquad r_\zeta = k_\zeta - k_0.$$

Since $-\Delta R_\zeta f = 0$ for all $f \in C^\infty_c(\mathbb{R}^n)$, it follows that $-\Delta r_\zeta = 0$, i.e., r_ζ is a harmonic distribution. From elliptic regularity (see, e.g., [**Hör03a**, Theorem 7.9.7] for the case of distributional solutions) it follows that r_ζ is a C^∞ function and $R_\zeta f$ is smooth whenever $f \in L^2_c$. It also follows that $k_\zeta = k_0 + r_\zeta \in L^1_{\mathrm{loc}}(\mathbb{R}^n)$ satisfies $-\Delta k_\zeta = \delta_0$ and is smooth outside the origin. □

We can now define the modified single layer potential

$$S_\zeta = K_\zeta \gamma^*.$$

Since $K_\zeta = K_0 + R_\zeta$, we have

$$S_\zeta f = S_0 f + R_\zeta \gamma^* f$$

where the function $R_\zeta \gamma^* f$ is smooth and harmonic for any $f \in H^{-1/2}(\Gamma)$. It follows that S_ζ has the same jump and mapping properties as those stated in Lemma 4.6.9 for S_0.

Lemma 4.6.12 (Jump relations for S_ζ). *Let $\zeta \in \mathbb{C}^n$ satisfy $\zeta \cdot \zeta = 0$ and $|\zeta| \geq 1$. Then the conclusions of Lemma 4.6.9 hold with S_0 replaced by S_ζ.*

The following result is the final component in the reconstruction procedure. It considers four equivalent characterizations of solutions of the equation $(-\Delta + q)u = 0$ in $\mathbb{R}^n$, where q vanishes outside Ω, that have the form

$$u = u_0 + e^{i\zeta \cdot x} r,$$

where u_0 is any harmonic function in $\mathbb{R}^n$, and $r \in H^1_\delta(\mathbb{R}^n)$. We will later take $u_0 = e^{i\zeta \cdot x}$, but the following equivalences work for any harmonic function u_0. The characterizations are given in terms of a differential equation (DE), integral equation (IE), exterior problem (EP) and boundary integral equation (BE).

Proposition 4.6.13 (Equivalent problems). *Let $q \in L^\infty(\Omega)$, extended by zero outside Ω, be such that 0 is not a Dirichlet eigenvalue of $-\Delta + q$ in Ω. Let $\zeta \in \mathbb{C}^n$ satisfy $\zeta \cdot \zeta = 0$ with $|\zeta|$ sufficiently large, and let $-1 < \delta < 0$. Further, let $u_0 \in H^2_{\mathrm{loc}}(\mathbb{R}^n)$ be such that $\Delta u_0 = 0$ in $\mathbb{R}^n$. Consider the following problems:*

(DE) $$\begin{cases} (-\Delta + q)u = 0 \text{ in } \mathbb{R}^n \\ e^{-i\zeta \cdot x}(u - u_0) \in H^2_\delta(\mathbb{R}^n), \end{cases}$$

(IE) $$\begin{cases} u + K_\zeta(qu) = u_0 \text{ in } \mathbb{R}^n \\ u \in H^2_{\mathrm{loc}}(\mathbb{R}^n), \end{cases}$$

(EP) $$\begin{cases} \text{(i)} & \Delta u = 0 \text{ in } \Omega_+ \\ \text{(ii)} & u = \tilde{u}|_{\Omega_+} \text{ for some } \tilde{u} \in H^2_{\mathrm{loc}}(\mathbb{R}^n) \\ \text{(iii)} & e^{-i\zeta \cdot x}(u - u_0)|_{\Omega_+} = \tilde{r}|_{\Omega_+} \text{ for some } \tilde{r} \in H^2_\delta(\mathbb{R}^n) \\ \text{(iv)} & (\partial_\nu u)_+ = \Lambda_q(\gamma_+ u) \text{ on } \Gamma, \end{cases}$$

(BE) $$\begin{cases} (\mathrm{Id} + \gamma S_\zeta(\Lambda_q - \Lambda_0))f = u_0 \text{ on } \Gamma \\ f \in H^{3/2}(\Gamma). \end{cases}$$

Each of these problems has a unique solution. Further, these problems are equivalent in the sense that u solves (DE) iff u solves (IE), if u solves (DE) then $u|_{\Omega_+}$ solves (EP), if u solves (EP) then there is a solution $\tilde{u}$ of (DE) with $\tilde{u}|_{\Omega_+} = u$, if u solves (DE) then $f = u|_\Gamma$ solves (BE), and finally if f solves (BE) then there is a solution u of (DE) with $u|_\Gamma = f$.

Proof. The function $u = u_0 + e^{i\zeta \cdot x} r$ solves $(-\Delta + q)u = 0$ in $\mathbb{R}^n$ if and only if

$$e^{-i\zeta \cdot x}(-\Delta + q)e^{i\zeta \cdot x} r = -e^{-i\zeta \cdot x} q u_0 \quad \text{in } \mathbb{R}^n.$$

The right-hand side is in $L^2_c(\mathbb{R}^n)$, so by Proposition 4.6.3 there is a unique solution $r \in H^1_\delta(\mathbb{R}^n)$ where $-1 < \delta < 0$ if $|\zeta|$ is sufficiently large. This proves that (DE) has a unique solution. It remains to prove that all four problems are equivalent in the sense described above.

(DE) $\Longrightarrow$ (IE): Assume u solves (DE). Then $u = u_0 + e^{i\zeta\cdot x}r$ where $r \in H^2_\delta(\mathbb{R}^n)$, and

$$e^{-i\zeta\cdot x}(-\Delta + q)e^{i\zeta\cdot x}r = -e^{-i\zeta\cdot x}qu_0 \quad \text{in } \mathbb{R}^n.$$

By Proposition 4.5.4, after adding the potential q and choosing $|\zeta|$ sufficiently large, we have $r = G_\zeta v \in H^2_{\mathrm{loc}}(\mathbb{R}^n)$ where v satisfies

$$v + qr = -e^{-i\zeta\cdot x}qu_0.$$

Since q is compactly supported in $\mathbb{R}^n$ also $v = -q(r + e^{-i\zeta\cdot x}u_0)$ is compactly supported. Thus we may apply G_ζ to both sides of the last identity to obtain

$$r + G_\zeta(qr) = -G_\zeta(e^{-i\zeta\cdot x}qu_0).$$

Multiplying by $e^{i\zeta\cdot x}$ and adding u_0 to both sides gives (IE).

(IE) $\Longrightarrow$ (DE): Assume u solves (IE). Then $r = e^{-i\zeta\cdot x}(u - u_0)$ satisfies

$$r = -G_\zeta(e^{-i\zeta\cdot x}qu). \tag{4.38}$$

This shows that $r \in H^2_\delta(\mathbb{R}^n)$, and (DE) follows by applying $-\Delta$ to both sides of (IE).

(DE) $\Longrightarrow$ (EP): Let $\tilde{u}$ solve (DE), and define $u = \tilde{u}|_{\Omega_+}$. Since q vanishes outside Ω, properties (i), (ii), and (iii) of (EP) are valid. We need to show (iv). Since $\tilde{u}$ solves the equation $(-\Delta + q)\tilde{u} = 0$ in Ω, we have

$$(\partial_\nu u)_+ = \partial_\nu \tilde{u}|_\Gamma = \Lambda_q(\tilde{u}|_\Gamma) = \Lambda_q(\gamma_+ u).$$

(EP) $\Longrightarrow$ (DE): Suppose u solves (EP). Define $v \in H^2(\Omega)$ as the unique solution of the equation $(-\Delta + q)v = 0$ in Ω with $v|_\Gamma = \gamma_+ u|_\Gamma$, and define

$$\tilde{u}(x) = \begin{cases} v(x), & x \in \Omega, \\ u(x), & x \in \Omega_+. \end{cases}$$

Then $\gamma_- \tilde{u}|_\Gamma = \gamma_+ \tilde{u}|_\Gamma$ and

$$(\partial_\nu \tilde{u})_-|_\Gamma = \Lambda_q(\gamma_+ u|_\Gamma) = (\partial_\nu \tilde{u})_+|_\Gamma$$

by (EP)(iv). It follows that $\tilde{u} \in H^2_{\mathrm{loc}}(\mathbb{R}^n)$ and $(-\Delta + q)\tilde{u} = 0$ in $\mathbb{R}^n$. Further, $e^{-i\zeta\cdot x}(\tilde{u} - u_0) \in H^2_\delta(\mathbb{R}^n)$ by (EP)(iii).

(DE) $\Longrightarrow$ (BE): Let u solve (DE), and let $f = u|_\Gamma$. We fix a point $x \in \Omega_+$ and let $v(y) = k_\zeta(x - y)$ where $y \in \Omega$. Then v is a smooth harmonic function in Ω by Lemma 4.6.11.

Now Green's theorem implies

$$\int_\Gamma (u\partial_\nu v - v\partial_\nu u)\,dS = \int_\Omega (u\Delta v - v\Delta u)\,dy.$$

By (DE) we have $\Delta u = qu$ and $\partial_\nu u|_\Gamma = \Lambda_q f$. Using the properties in Lemma 4.6.11 we obtain

$$\int_\Gamma u\partial_\nu v\,dS - S_\zeta\Lambda_q f(x) = -K_\zeta(qu)(x),$$

which is valid for $x \in \Omega_+$. The function v is harmonic in Ω, hence $\partial_\nu v|_\Gamma = \Lambda_0(v|_\Gamma)$. The symmetry of Λ_0 implies

$$\int_\Gamma u\partial_\nu v\,dS = \int_\Gamma u\Lambda_0(v|_\Gamma)\,dS = \int_\Gamma \Lambda_0(u|_\Gamma)v\,dS = S_\zeta\Lambda_0 f(x).$$

We obtain

$$S_\zeta(\Lambda_q - \Lambda_0)f = K_\zeta(qu) \quad \text{in} \quad \Omega_+. \tag{4.39}$$

Adding u to both sides, using the fact that u solves (IE), and taking traces on Γ gives (BE).

(BE) $\Longrightarrow$ (EP): Let f solve (BE). We define a function $\tilde{u}$ by

$$\tilde{u} = u_0 - S_\zeta(\Lambda_q - \Lambda_0)f.$$

Here $f \in H^{3/2}(\Gamma)$, and $(\Lambda_q - \Lambda_0)f = \partial_\nu v_q - \partial_\nu v_0\big|_\Gamma \in H^{1/2}(\Gamma)$, where $v_q \in H^2(\Omega)$ solves $(-\Delta + q)v_q = 0$ in Ω with $v_q|_\Gamma = f$ and similarly for v_0. By Lemma 4.6.12, we have $\tilde{u} \in H^2_{\mathrm{loc}}(\mathbb{R}^n)$.

The function $\tilde{u}$ is harmonic in $\mathbb{R}^n \setminus \Gamma$ by Lemma 4.6.12, and $\tilde{u}|_\Gamma = f$ by using (BE). The jump relation for S_ζ implies that on Γ

$$(\partial_\nu\tilde{u})_- - (\partial_\nu\tilde{u})_+ = -(\Lambda_q - \Lambda_0)f.$$

But $(\partial_\nu\tilde{u})_- = \Lambda_0 f$, so we have $(\partial_\nu\tilde{u})_+ = \Lambda_q(\gamma_+\tilde{u})$. Therefore $\tilde{u}|_{\Omega_+}$ satisfies (EP)(i) and (iv). Also (EP)(ii) is valid by mapping properties of S_ζ.

To prove (EP)(iii) it is sufficient to show that for any $h \in H^{1/2}(\Gamma)$,

$$e^{-i\zeta\cdot x}S_\zeta h|_{\Omega_+} = w|_{\Omega_+} \quad \text{for some} \quad w \in H^2_\delta(\mathbb{R}^n).$$

Formally, one has $e^{-i\zeta\cdot x}S_\zeta h = G_\zeta e^{-i\zeta\cdot x}\gamma^* h$, where G_ζ maps $L^2_{\delta+1}(\mathbb{R}^n)$ to $H^2_\delta(\mathbb{R}^n)$. Here we only have $\gamma^* h \in H^{-1}_c(\mathbb{R}^n)$, and we have so far not discussed mapping properties of G_ζ in negative Sobolev spaces. We will instead construct an extension of $e^{-i\zeta\cdot x}S_\zeta h|_{\Omega_+}$ to $\mathbb{R}^n$ by

$$w(x) = \begin{cases} e^{-i\zeta\cdot x}(P_0(\gamma S_\zeta h) + F), & x \in \Omega, \\ e^{-i\zeta\cdot x}S_\zeta h, & x \in \Omega_+. \end{cases}$$

Here $P_0 : H^{3/2}(\Gamma) \to H^2(\Omega)$ is the Poisson operator mapping f to the function $v \in H^2(\Omega)$ solving $-\Delta v = 0$ in Ω with $v|_{\partial\Omega} = f$, and $F \in H^2(\Omega)$ is a function that will be chosen so that $e^{i\zeta\cdot x}w \in H^2_{\mathrm{loc}}(\mathbb{R}^n)$. Thus we need $F|_\Gamma = 0$, and since

$$[\partial_\nu(e^{i\zeta\cdot x}w)]_\Gamma = \Lambda_0(\gamma S_\zeta h) + \partial_\nu F|_\Gamma - (\partial_\nu S_\zeta h)_+$$

we also require that $\partial_\nu F|_\Gamma = (\partial_\nu S_\zeta h)_+ - \Lambda_0(\gamma S_\zeta h)$. Choosing F to be any function in $H^2(\Omega)$ with this Cauchy data yields $e^{i\zeta\cdot x}w \in H^2_{\mathrm{loc}}(\mathbb{R}^n)$, so also $w \in H^2_{\mathrm{loc}}(\mathbb{R}^n)$.

Next we note that

$$e^{-i\zeta\cdot x}(-\Delta)(e^{i\zeta\cdot x}w) = \begin{cases} -e^{-i\zeta\cdot x}\Delta F, & x \in \Omega, \\ 0, & x \in \Omega_+. \end{cases}$$

Since $w \in H^2_{\mathrm{loc}}$ this implies that $e^{-i\zeta\cdot x}(-\Delta)(e^{i\zeta\cdot x}w) \in L^2_c(\mathbb{R}^n)$. The uniqueness part of Proposition 4.5.4 shows that $w \in H^2_\delta(\mathbb{R}^n)$, which concludes the proof of (EP)(iii). □

The reconstruction procedure now follows.

Proof of Theorem 4.6.2. By Proposition 4.6.4, we can recover the Fourier transform of q from Λ_q if we can reconstruct the boundary values $u_\zeta|_{\partial\Omega}$ of the CGO solutions u_ζ in Proposition 4.6.3. Consider the integral equation

$$(\mathrm{Id} + \gamma S_\zeta(\Lambda_q - \Lambda_0))f = e^{i\zeta\cdot x} \text{ on } \Gamma.$$

Since Λ_q is known, the operator on the left-hand side is also known. Proposition 4.6.13 shows that the boundary integral equation has a unique solution $f \in H^{3/2}(\partial\Omega)$ and that $f = u_\zeta|_{\partial\Omega}$. Thus solving the boundary integral equation for different values of ζ allows us to recover the boundary values $u_\zeta|_{\partial\Omega}$, and in this way we can reconstruct q. □

The reconstruction procedure above involves solving the boundary integral equation (BE) in Proposition 4.6.13 for many choices of ζ. Finally, let us verify that (BE) is a Fredholm integral equation (so it satisfies the Fredholm alternative in Proposition A.3.15).

Proposition 4.6.14. *The operator*

$$\gamma S_\zeta(\Lambda_q - \Lambda_0) : H^{3/2}(\Gamma) \to H^{3/2}(\Gamma)$$

is compact.

Proof. Let $f \in H^{3/2}(\Gamma)$, and let $u = P_q f$, where $P_q : H^{3/2}(\Gamma) \to H^2(\Omega)$ is the Poisson operator mapping h_0 to v_0. where $(-\Delta + q)v_0 = 0$ in Ω and $v_0|_\Gamma = h_0$. The exact same argument leading to (4.39) in the proof of Proposition 4.6.13 shows that

$$S_\zeta(\Lambda_q - \Lambda_0)f = K_\zeta(qEJu) \quad \text{in } \Omega_+,$$

where $E : L^2(\Omega) \to L^2(\mathbb{R}^n)$ is extension by zero and $J : H^2(\Omega) \to L^2(\Omega)$ is the natural inclusion. Taking traces on Γ, we obtain the factorization

$$\gamma S_\zeta(\Lambda_q - \Lambda_0) = \gamma K_\zeta q E J P_q.$$

The result follows since on the right-hand side J is compact by Theorem 2.4.8 and all other operators are bounded (here we use Proposition A.3.8). □

4.7. Notes

The uniqueness result for the linearized Calderón problem via harmonic exponentials in §4.1 follows the pioneering work [**Cal80**], which also gave an approximate reconstruction procedure for conductivities close to constant (for implementations see, e.g., [**II89, BM08, CJSZ23**]). Complex geometrical optics (CGO) solutions, which generalize the harmonic exponentials, have been a central component in many further developments in the Calderón problem. These solutions were constructed in the general case in [**SU87**], although such solutions had appeared before in [**Fad16, NA84, BC85, New85, NK87**] in relation to scattering problems.

Our first approach to CGO solutions in §4.2 follows [**Häh96**] and it involves shifted Fourier series. The global uniqueness theorem for C^2 conductivities in dimensions $n \geq 3$ proved in §4.3 is from [**SU87**]. The logarithmic stability result in §4.4 is due to [**Ale88**]. It has been proved that this type of stability is optimal [**Man01, KRS21**], though one has Lipschitz stability if the conductivity lies in a finite-dimensional space (see, e.g., [**AV05**]). For further stability results see the survey [**Ale07**]. In §4.5 we present a second approach to CGO solutions based on the Fourier transform, which follows the original work [**SU87**] and involves a notion of uniqueness of CGO solutions when one prescribes a decay condition at infinity. The uniqueness part is related to Agmon-Hörmander estimates [**Hör03a**, §7.1], and the weighted estimates are similar to those in [**NW73**]. The CGO solutions are used in §4.6 to give a reconstruction procedure due to [**Nac88**]. Another procedure motivated by inverse scattering problems has been given in [**Nov88**].

There is a vast literature on different variations of the Calderón problem and CGO solutions in dimension $n \geq 3$. Here we only discuss a selection of results related to Chapter 4, with a few selected references for each topic and with a particular emphasis on uniqueness results. Further references may be found in the survey article [**Uhl14b**]. The two-dimensional case, partial data problems and inverse scattering problems will be discussed separately in subsequent chapters.

(1) **Low regularity.** The methods discussed in §4.3 can be extended to prove uniqueness for conductivities with 3/2 derivatives [**BT03**]. A significant advance was made in [**HT13**] who proved uniqueness for C^1 conductivities

by introducing ideas such as Bourgain spaces and averaging methods from nonlinear dispersive PDEs. This was extended to Lipschitz conductivities in any dimension $n \geq 3$ in [**CR16**]. Even better results are available in low dimensions. For example, for $n = 3, 4$ uniqueness was proved for positive conductivities in $W^{1,n} \cap L^\infty$ in [**Hab15b**] by using methods related to L^p Carleman estimates. L^p Carleman estimates have also been employed for determining a potential $q \in L^{n/2}$ from the DN map Λ_q, see [**Cha90, Nac92, DSFKS13**].

At the time of writing this it is an open problem whether uniqueness holds for $W^{1,n} \cap L^\infty$ conductivities in high dimensions, or for piecewise smooth conductivities. The case of conductivities with less than one derivative is not understood either. However, there is a counterexample for the partial data problem with anisotropic Hölder continuous conductivities [**DKN20**] based on a counterexample to the unique continuation principle in dimensions $n \geq 3$. Connections between the Calderón problem and unique continuation methods are discussed further in Chapter 6.

(2) **The anisotropic problem.** As discussed in §1.1, if the conductivity is a positive definite matrix one can only expect to determine the conductivity up to a boundary fixing diffeomorphism (this was observed by L. Tartar, see [**KV84b**]). In §3.6 we proved that the Taylor series of a C^∞ matrix conductivity, when written in boundary normal coordinates, can be determined from the DN map. This was a key component in [**LU89**] where it was proved that a real-analytic conductivity matrix, or a real-analytic Riemannian metric, can be determined up to diffeomorphism in dimensions $n \geq 3$. The result of [**LU89**] involved extra conditions, which were later removed in [**LU01, LTU03**] via a method based on Green functions (an alternative approach using Poisson kernels is given in [**LLS20**]). A similar result for Einstein metrics was given in [**GSB08**].

It is an outstanding open problem whether a C^∞ conductivity or metric is determined up to diffeomorphism by the DN map in dimensions $n \geq 3$. There has been progress for metrics in a fixed conformal class, or for determining a potential in a Schrödinger equation with a known metric, if the metric is *conformally transversally anisotropic* (i.e., has suitable conformal product structure) and certain additional conditions hold. See [**DSFKSU09, DSFKLS16, CFO23**].

(3) **Invisibility.** One does get counterexamples to uniqueness in the Calderón problem if one allows degenerate conductivity matrices or Riemannian metrics. Moreover, one can create singular Riemannian metrics $\tilde{g}$ on $\Omega = B(0,2)$ so that $\tilde{g}$ has suitable structure in $B(0,2) \setminus \overline{B(0,1)}$, has zero and/or

infinite eigenvalues on $\partial B(0,1)$, and $\tilde{g}|_{B(0,1)}$ can be an arbitrary smooth Riemannian metric, such that

$$\Lambda_{\tilde{g}} = \Lambda_{g_0},$$

where g_0 is the Euclidean metric on $B(0,2)$. Here, one needs to define the DN map and solutions to the singular equation $\Delta_{\tilde{g}} u = 0$ in an appropriate way. The exotic structure in

$$B(0,2) \setminus \overline{B(0,1)}$$

functions as an *invisibility cloak* that can be used to hide any object in $B(0,1)$, so that the electric measurements at the boundary are the same as those for the Euclidean metric.

The result described above first appeared in [**GLU03a, GLU03b**] as a theoretical counterexample in the Calderón problem with singular metrics. Similar constructions appeared later in the physics literature [**Leo06**], [**PSS06**], where it was proposed that it might be possible to construct such invisibility cloaks by using metamaterials. Research on invisibility cloaks and transformation optics has become an active area, and further results may be found in the survey articles [**GKLU09, Uhl14b**].

(4) **First-order terms.** For elliptic equations with a first-order term (magnetic potential or convection term), even the construction of CGO solutions requires additional arguments. Small first-order terms were considered in [**Sun93**] and one can proceed as in Exercises 4.2.5–4.2.6. For large first-order terms a construction via pseudodifferential conjugation was given in [**NSU95**] together with a uniqueness result for recovering a smooth magnetic field and electric potential. Another approach for constructing CGO solutions involves convexification and Carleman estimates, see [**DSFKSU07**]. Results for low regularity coefficients may be found in [**Sal04, Sal06, KU14, Hab18**].

(5) **Higher-order equations.** One can also study Calderón type problems for polyharmonic operators such as $(-\Delta)^m + A(x,D)$ where $m \geq 1$ is an integer and $A(x,D)$ is a differential operator of order $\leq 2m-1$. See, e.g., [**Isa91, KLU12a, KU16a**] and the references in [**BKSU25**].

(6) **Systems.** There are uniqueness results for various elliptic systems such as the Maxwell equations [**OPS93, OS96, Ola94, KSU11a**], elasticity [**NU94, ER02, ER03**], Dirac [**NT00, Li07**], and Stokes and Navier-Stokes systems [**HLW07, LW07**].

(7) **Finite measurements.** The results above consider the idealized case where one has knowledge of the full DN map involving infinitely many measurements. Of course in practice one can only measure $\Lambda_\gamma f_j$ for finitely many Dirichlet data f_j, and in this case one could consider γ belonging to a suitable finite-dimensional space. We refer the reader to [**AS19, AS22a,**

AS22b, AAS23, Har19, Har23] and references therein for results and Lipschitz stability estimates in the finite measurement case.

(8) **Fractional equations.** A nonlocal version of the Calderón problem, known as the fractional Calderón problem, arises when we replace the Schrödinger equation by its fractional version

$$((-\Delta)^s + q)u = 0 \text{ in } \Omega.$$

Here $(-\Delta)^s$ is the fractional Laplacian defined by the Fourier multiplier $|\xi|^{2s}$. Since the equation is nonlocal, it is natural to consider the exterior problem where one prescribes $u|_{\mathbb{R}^n\setminus\overline{\Omega}} = f$ instead of prescribing Dirichlet data. Then the inverse problem is to determine the potential q in Ω from a suitable exterior Dirichlet-to-Neumann map.

Uniqueness for this inverse problem was established in [**GSU20**]. For this nonlocal equation it is not known if CGO solutions are available. However, it turns out that solutions of the fractional Schrödinger equation are much more flexible than those of the standard Schrödinger equation, and there are strong uniqueness and approximation properties for solutions that are helpful in the inverse problem. There is a large literature on such problems including low regularity and stability results [**RS20**], reconstruction even in the single measurement case [**GRSU20**], and uniqueness in the anisotropic case [**FGKU21**]. The surveys [**Sal17a, Cov25**] contain further information, and the textbooks [**KR23, LL24**] discuss inverse problems for various fractional equations.

(9) **Nonlinear equations.** An important variant of the Calderón problem arises in the case of nonlinear equations, such as the semilinear equation

$$\Delta u + a(x, u) = 0$$

or the quasilinear equation

$$\nabla \cdot (a(x, u)\nabla u) = 0.$$

Under suitable assumptions on the nonlinearity $a(x, u)$, one obtains a well-posed Dirichlet problem (possibly for small data) and a nonlinear DN map Λ_a. The inverse problem is to determine the nonlinearity $a(x, u)$ from the knowledge of the nonlinear DN map Λ_a.

The *first-order linearization method* proceeds by showing that the Fréchet derivative $D\Lambda_a$, evaluated at suitable boundary data, is the DN map of a linear equation with suitable potential or conductivity. One can then solve the inverse problem for the linear equation by the methods in Chapter 4 and attempt to recover the full nonlinearity $a(x, u)$ via additional arguments. This idea goes back to [**Isa93**] and it has been implemented for semilinear [**IS94**] and quasilinear [**Sun96, SU97**] equations. Further references may be found within [**LLLS21**].

The *higher-order linearization method* due to [**FO20, LLLS21**] builds on earlier work for hyperbolic equations [**KLU18**]. In this method one exploits higher-order Fréchet derivatives $D^k\Lambda_a$ and observes that the nonlinear interactions can in fact be helpful and yield more information than in corresponding inverse problems for linear equations. The works [**FO20, LLLS21**] study semilinear equations and [**CFK$^+$21**] studies quasilinear equations, and the nonlinearity is assumed to be real-analytic with respect to the solution. Results in the absence of real-analyticity and further references are given in [**JNS23**].

Another case that has been studied is the inverse problem for p-Laplace type conductivity equations, see [**SZ12, Bra16, CF24**].

(10) **CGO solutions.** The CGO solutions are an interesting special class of solutions to elliptic equations. They have been used in other inverse problems such as determining inclusions via the enclosure method [**Ike00, Ike01, Ike10**], hybrid imaging problems [**KLQ$^+$17**] and corner scattering [**BPS14**]. They have also found applications in the spectral theory of periodic Schrödinger operators [**Tho73, Sob99, KU17**].

We have given two constructions of CGO solutions in §4.2 and §4.5. Several further constructions are available. The early work [**Isa91**] considers operators of the form $P(D) + V$, constructs CGO solutions and proves results regarding density of products of solutions. In [**KSU11a**] there is a construction that involves a mixture of the Fourier transform and Fourier series approaches and is valid also in transversally anisotropic geometries. Another very important approach, originally arising in partial data problems [**BU02, KSU07**], is based on integration by parts and Carleman estimate techniques. This approach may avoid Fourier analysis altogether, and it also gives rise to more general CGO solutions with nonlinear phase functions related to so-called limiting Carleman weights. We will discuss this approach in more detail in Chapter 6 and the Notes section therein.

A more geometric point of view to CGO solutions is provided by *microlocal analysis*. This gives further insight to the nature of CGO solutions and therefore we conclude this Notes section with a brief explanation. For the following microlocal terminology we refer the reader to the textbooks [**GS94, Zwo12**], the comprehensive treatment [**Hör85a, Hör85b**], or to the notes [**Sal20, Sal24**] that discuss the connection between microlocal analysis and inverse problems.

In the Calderón problem it is natural to try to find special boundary data that could capture some properties of the unknown conductivity in the interior. These special boundary data could be taken as boundary values of suitable special solutions. For real principal type equations such as the wave equation, *propagation of singularities* provides useful special solutions that concentrate along spatial projections of null bicharacteristic curves. One challenge in the

Calderón problem is that the underlying equation is elliptic, so singularities do not propagate and one does not obtain useful special solutions in this way. However, it turns out that after conjugating the equation by exponentials, one obtains a new equation for which singularities do propagate.

More precisely, if we consider a large parameter $s = 1/h$, where $h > 0$ is a small (semiclassical) parameter, the conjugated operator

$$P(x, hD) = e^{-x_1/h}(-h^2\Delta)e^{x_1/h}$$

is a semiclassical *complex principal type* operator whose characteristic set is involutive. This means that the semiclassical principal symbol $p(x, \xi)$ satisfies the Poisson bracket condition $\{\bar{p}, p\} = 0$ when $p = 0$. This holds since the phase function x_1 is a limiting Carleman weight (see the Notes section in Chapter 6 for more information). Of course any linear phase function $\alpha \cdot x$ would work just as well. For such an operator, (semiclassical) singularities propagate along two-dimensional manifolds, and correspondingly there exist special solutions that concentrate near two-planes. If we write $x = (x_1, x_2, x'') \in \mathbb{R}^n$ and the two-plane is the set $\{x'' = 0\}$, the special solution has the form

$$u = e^{\frac{x_1+ix_2}{h}}(a(x'') + r(x)),$$

where $a(x'')$ could be chosen as a smooth function supported near 0.

The CGO solutions described above are very close to the ones constructed in Theorems 4.2.1 or 4.5.1. Solutions concentrating near two-planes were constructed in the Euclidean setting in [**GU01**] and in a more geometric setting in [**DSFKSU09, DSFKLS16**]. The complex principal type structure has also been exploited in the detection of singularities [**GLS**$^+$**18**]. The microlocal point of view to CGO solutions is further explained in [**Sal17b**].

Chapter 5

The Calderón Problem in the Plane

In this chapter we discuss the Calderón problem in two dimensions. The arguments presented here have a somewhat different flavor compared to the case $n \geq 3$, and will rely heavily on complex analysis. We will give the proof of Astala and Päivärinta [**AP06a**] which allows us to treat bounded measurable conductivities. This will involve the theory of quasiconformal mappings, a generalization of the standard theory of analytic functions.

In this chapter we will denote the conductivity by σ instead of γ. This is customary in the two-dimensional results and emphasizes the fact that the conductivity has to be real valued. Let us state more precisely the result that will be proved.

Theorem 5.0.1. *Let Ω be a bounded domain in $\mathbb{R}^2$ with smooth boundary and suppose that $\sigma_1, \sigma_2 \in L^\infty(\Omega)$ satisfy $\sigma_j \geq c > 0$ a.e. in Ω. If*

$$\Lambda_{\sigma_1} = \Lambda_{\sigma_2}$$

then

$$\sigma_1 = \sigma_2.$$

We will begin with a reduction. If $B \subset \mathbb{R}^2$ is an open ball centered at the origin such that $\overline{\Omega} \subset B$, we define new conductivities

$$\tilde{\sigma}_j(x) = \begin{cases} \sigma_j(x) & \text{if } x \in \Omega, \\ 1 & \text{if } x \in B \setminus \Omega. \end{cases}$$

Then $\tilde{\sigma}_j$ are positive functions in $L^\infty(B)$, and the condition $\Lambda_{\sigma_1} = \Lambda_{\sigma_2}$ on $\partial\Omega$ implies that $\Lambda_{\tilde{\sigma}_1} = \Lambda_{\tilde{\sigma}_2}$ on ∂B by Proposition 2.8.7. (Indeed, working with bounded measurable conductivities avoids the need for boundary determination results when doing the extension.) By a simple rescaling, it is sufficient to prove Theorem 5.0.1 when the domain is the unit disc $\mathbb{D}$. We will assume throughout this chapter that $\Omega = \mathbb{D}$.

Notation. In this chapter we will work with L^p based Sobolev spaces. If $\Omega \subset \mathbb{R}^2$ is a bounded open set and $1 \leq p \leq \infty$, recall from Definition 2.3.5 the space

$$W^{1,p}(\Omega) = \{\, u \in L^p(\Omega) \mid \nabla u \in L^p(\Omega) \,\}.$$

Here $\nabla u = (\partial_1 u, \partial_2 u)$ denotes the weak gradient. We will frequently perform computations in the weak sense. For instance, if $F_1, F_2 \in L^2(\Omega)$, the curl free condition

$$\partial_1 F_2 - \partial_2 F_1 = 0 \text{ in } \Omega$$

means that for all $\varphi \in C_c^\infty(\Omega)$ we have

$$\int_\Omega (F_1 \partial_2 \varphi - F_2 \partial_1 \varphi)\, dx = 0.$$

We will use the spaces

$$L^p_c(\Omega) = \{\, u \in L^p(\Omega) \mid \mathrm{supp}(u) \text{ is a compact subset of } \Omega \,\},$$
$$L^p_{\mathrm{loc}}(\Omega) = \{\, u \mid u|_{\Omega'} \in L^p(\Omega') \text{ whenever } \Omega' \text{ is open with } \overline{\Omega'} \subset \Omega \,\}.$$

The spaces $W_c^{1,p}(\Omega)$ and $W_{\mathrm{loc}}^{1,p}(\Omega)$ are defined in a similar way.

We also define the spaces $L^{2\pm}(\mathbb{R}^2)$ and $C_0(\mathbb{R}^2)$ by

$$L^{2\pm} = \{\, f \mid f \in L^{2+\varepsilon} \cap L^{2-\varepsilon} \text{ for some } \varepsilon > 0 \,\}, \tag{5.1}$$
$$C_0 = \{\, f : \mathbb{R}^2 \to \mathbb{C} \text{ continuous} \mid f(z) \to 0 \text{ as } |z| \to \infty \,\}. \tag{5.2}$$

It follows from the definition that any function in $C_0(\mathbb{R}^2)$ is uniformly bounded on $\mathbb{R}^2$.

We will also frequently use the "big O" and "small o" notations

$$f(z) = O(h(z)) \text{ as } |z| \to \infty,$$
$$g(z) = o(h(z)) \text{ as } |z| \to \infty,$$

to mean that

$$|f(z)| \leq C|h(z)| \text{ when } |z| \text{ is sufficiently large},$$
$$\frac{g(z)}{h(z)} \to 0 \text{ as } |z| \to \infty.$$

5.1. Complex derivatives

This section contains a brief review of the complex derivatives ∂ and $\bar{\partial}$. If $x = (x_1, x_2) \in \mathbb{R}^2$ we write $z = x_1 + ix_2$ and identify $\mathbb{R}^2$ with $\mathbb{C}$ in this way. The *complex derivatives* (or Wirtinger derivatives) are defined by

$$\bar{\partial} = \partial_{\bar{z}} := \frac{1}{2}\left(\frac{\partial}{\partial x_1} + i\frac{\partial}{\partial x_2}\right),$$

$$\partial = \partial_z := \frac{1}{2}\left(\frac{\partial}{\partial x_1} - i\frac{\partial}{\partial x_2}\right).$$

Let $\Omega \subset \mathbb{R}^2$ be an open set, and let $f : \Omega \to \mathbb{C}$ be a C^1 function. Write $f = u + iv$ where $u = \mathrm{Re}(f)$, $v = \mathrm{Im}(f)$. Recall that f is said to be *analytic*, or *holomorphic*, in Ω if it satisfies the Cauchy-Riemann equations

$$\partial_1 u = \partial_2 v, \quad \partial_2 u = -\partial_1 v.$$

Recall also that any analytic function is C^∞. The $\bar{\partial}$ operator satisfies

$$\bar{\partial} f = \frac{1}{2}(\partial_1 + i\partial_2)(u + iv) = \frac{1}{2}(\partial_1 u - \partial_2 v + i(\partial_1 v + \partial_2 u)).$$

This immediately implies that the $\bar{\partial}$ operator characterizes analytic functions.

Lemma 5.1.1. *f is analytic in Ω if and only if $\bar{\partial} f = 0$ in Ω.*

Another fact to note is that the Laplacian $\Delta = \partial_1^2 + \partial_2^2$ factors in terms of the complex derivatives as

$$\Delta = 4\partial\bar{\partial}.$$

A function u is called harmonic if $\Delta u = 0$. We obtain:

Lemma 5.1.2. *If f is analytic, then the real and imaginary parts of f are harmonic functions.*

Proof. If $f = u + iv$ is analytic, then $\Delta f = 4\partial(\bar{\partial} f) = 0$. Since u and v are real valued it follows that $\Delta u = \Delta v = 0$. □

We can combine the above result with *Weyl's lemma* (see, e.g., [**Hör03a**, Theorem 7.9.7]), so that we can talk about analytic functions in $L^1_{\mathrm{loc}}(\Omega)$.

Lemma 5.1.3 (Weyl's lemma). *If $u \in L^1_{\mathrm{loc}}(\Omega)$ and $\Delta u = 0$ in Ω in the weak sense, then $u \in C^\infty(\Omega)$. Consequently, if $f \in L^1_{\mathrm{loc}}(\Omega)$ and $\bar{\partial} f = 0$ in Ω in the weak sense, then $f \in C^\infty(\Omega)$ and f is analytic in Ω.*

The next result establishes the existence of conjugate harmonic functions. Recall that an open set Ω is *simply connected* if Ω is connected and every closed curve in Ω can be continuously deformed to a point, or equivalently if Ω and $S^2 \setminus \Omega$ are connected where $S^2 = \mathbb{C} \cup \{\infty\}$ is the Riemann sphere.

Lemma 5.1.4 (Conjugate harmonic functions). *Let $\Omega \subset \mathbb{R}^2$ be a simply connected open set, and let $u : \Omega \to \mathbb{R}$ be a C^∞ function with $\Delta u = 0$. There exists a C^∞ function $v : \Omega \to \mathbb{R}$, unique up to an additive constant, such that $\Delta v = 0$ and $f = u + iv$ is analytic.*

The following exercises outline some basic properties of the ∂ and $\bar{\partial}$ operators and include a proof of Lemma 5.1.4.

Exercise 5.1.5 (Properties of ∂ and $\bar{\partial}$). Let $f, g : \mathbb{R}^2 \to \mathbb{C}$ be C^1 functions.

(a) Suppose that $f(x_1, x_2) = F(x_1 + ix_2)$ with $F : \mathbb{C} \to \mathbb{C}$ analytic and $g(x_1, x_2) = G(x_1 - ix_2)$ with $G : \mathbb{C} \to \mathbb{C}$ analytic. Prove that

$$\partial f(x_1, x_2) = F'(x_1 + ix_2), \qquad \bar{\partial} f(x_1, x_2) = 0,$$
$$\partial g(x_1, x_2) = 0, \qquad \bar{\partial} g(x_1, x_2) = G'(x_1 - ix_2).$$

Prove conversely that, if $\bar{\partial} f = 0$, then there is an analytic function $F(z)$ such that $f(x_1, x_2) = F(x_1 + ix_2)$ and if $\partial g = 0$, then there is an analytic function $G(z)$ such that $g(x_1, x_2) = G(x_1 - ix_2)$.

(b) Prove that

$$\partial f \bar{\partial} g + \bar{\partial} f \partial g = \tfrac{1}{2} \nabla f \cdot \nabla g.$$

Prove that, if f is C^2, then

$$4\partial\bar{\partial} f = 4\bar{\partial}\partial f = \Delta f.$$

Prove that, if f is C^2, then $\Delta f = 0$ if and only if there are analytic functions F and G such that $f(x_1, x_2) = F(x_1 + ix_2) + G(x_1 - ix_2)$.

(c) Prove that

$$\partial(fg) = f\partial g + g\partial f, \qquad \partial(f \circ g) = (\partial f) \circ g \, \partial g + (\bar{\partial} f) \circ g \, \partial \bar{g},$$
$$\bar{\partial}(fg) = f\bar{\partial} g + g\bar{\partial} f, \qquad \bar{\partial}(f \circ g) = (\partial f) \circ g \, \bar{\partial} g + (\bar{\partial} f) \circ g \, \bar{\partial} \bar{g}.$$

(d) Prove that

$$\overline{(\bar{\partial} f)(z)} = \partial\big(\overline{f(z)}\big), \qquad \partial(f(\bar{z})) = (\bar{\partial} f)(\bar{z}), \qquad \partial\big(\overline{f(\bar{z})}\big) = \overline{(\partial f)(\bar{z})},$$
$$\overline{(\partial f)(z)} = \bar{\partial}\big(\overline{f(z)}\big), \qquad \bar{\partial}(f(\bar{z})) = (\partial f)(\bar{z}), \qquad \bar{\partial}\big(\overline{f(\bar{z})}\big) = \overline{(\bar{\partial} f)(\bar{z})}.$$

Exercise 5.1.6 (Poincaré lemma). Let $\Omega \subset \mathbb{R}^2$ be a simply connected open set. Let $F : \Omega \to \mathbb{R}^2$ be a C^∞ vector field whose curl vanishes,

$$\partial_1 F_2 - \partial_2 F_1 = 0 \quad \text{in } \Omega.$$

Show that $F = \nabla p$ for some C^∞ function $p : \Omega \to \mathbb{R}$. (Hint: try to find p in terms of a line integral of F over curves.)

Exercise 5.1.7. Prove Lemma 5.1.4 by using Exercise 5.1.6.

Exercise 5.1.8 (Integration by parts). Let Ω be a bounded open subset of $\mathbb{R}^2$ with smooth boundary. Denote by (ν_1, ν_2) the unit outer normal to $\partial\Omega$. Give $\partial\Omega$ the standard orientation. That is, when you walk along $\partial\Omega$ in the positive direction, ν is on your right-hand side.

(a) Let each component of the vector field (f_1, f_2) be in $C^1(\overline{\Omega})$. Prove that

$$\int_\Omega \left[\frac{\partial f_1}{\partial x_1} + \frac{\partial f_2}{\partial x_2}\right] dx = \int_{\partial\Omega} [\nu_1 f_1 + \nu_2 f_2]\, dS$$

where dS is the surface (or arc length) measure of $\partial\Omega$.

(b) Let $f \in C^1(\overline{\Omega})$. Prove that

$$\int_\Omega \partial f\, dx = \int_{\partial\Omega} \nu f\, dS \qquad \int_\Omega \bar\partial f\, dx = \int_{\partial\Omega} \bar\nu f\, dS,$$

where $\nu = \frac{1}{2}(\nu_1 - i\nu_2)$ and $\bar\nu = \frac{1}{2}(\nu_1 + i\nu_2)$.

Exercise 5.1.9. Let Ω be a bounded, open, simply connected subset of $\mathbb{R}^2$ with smooth boundary. Let each component of the vector field (f_1, f_2) be in $C^1(\overline{\Omega})$. Use Exercise 5.1.6 to prove that if $\bar\partial f_1 = \partial f_2$, then there is a function $g \in C^2(\overline{\Omega})$ such that $f_1 = \partial g$ and $f_2 = \bar\partial g$.

5.2. Reduction to Beltrami equation

In the first step of the proof of Theorem 5.0.1, we reduce the conductivity equation $\nabla \cdot \sigma \nabla u = 0$ to a certain first-order equation involving the complex derivatives, namely the *Beltrami equation*

$$\bar\partial f = \mu \overline{\partial f}.$$

In this section we will also show that the DN map Λ_σ for the conductivity equation uniquely determines a corresponding boundary map for this first-order equation, namely the *μ-Hilbert transform $\mathcal{H}_\mu$*.

Proposition 5.2.1 (Reduction to Beltrami equation). *Let $\sigma \in L^\infty(\Omega)$ with $\sigma \geq c > 0$ a.e. in $\mathbb{D}$, and let $u \in H^1(\mathbb{D})$ be a real valued solution of*

$$\nabla \cdot \sigma \nabla u = 0 \quad \text{in } \mathbb{D}.$$

There exists a real valued $v \in H^1(\mathbb{D})$, unique up to an additive constant, such that $f = u + iv$ satisfies the equation

$$\bar\partial f = \mu \overline{\partial f} \quad \text{in } \mathbb{D}, \tag{5.3}$$

where

$$\mu = \frac{1-\sigma}{1+\sigma}.$$

Conversely, if $f = u + iv \in H^1(\mathbb{D})$ *satisfies* (5.3) *where* $\mu \in L^\infty(\mathbb{D})$ *with* $\|\mu\|_{L^\infty(\mathbb{D})} < 1$, *then*

$$\nabla \cdot \sigma \nabla u = 0 \quad \textit{in } \mathbb{D},$$
$$\nabla \cdot \sigma^{-1} \nabla v = 0 \quad \textit{in } \mathbb{D},$$

where

$$\sigma = \frac{1-\mu}{1+\mu}.$$

Proof. Suppose that $\nabla \cdot \sigma \nabla u = 0$, and define the vector field

$$F = (-\sigma \partial_2 u, \sigma \partial_1 u)$$

Then in the weak sense

$$\partial_1 F_2 - \partial_2 F_1 = 0.$$

It follows from Exercise 5.2.2 that there is a real valued $v \in H^1(\mathbb{D})$, unique up to an additive constant, such that

$$(\partial_1 v, \partial_2 v) = (-\sigma \partial_2 u, \sigma \partial_1 u).$$

Let $f = u + iv$. Then

$$\bar\partial f = \frac{1}{2}(\partial_1 u - \partial_2 v + i(\partial_2 u + \partial_1 v)) = \frac{1}{2}(1-\sigma)(\partial_1 u + i\partial_2 u),$$
$$\mu \overline{\partial f} = \frac{1-\sigma}{1+\sigma}\frac{1}{2}(\partial_1 u + \partial_2 v + i(\partial_2 u - \partial_1 v)) = \frac{1}{2}(1-\sigma)(\partial_1 u + i\partial_2 u).$$

This shows (5.3).

For the converse direction, if $f = u + iv$ solves (5.3) and if $\sigma = \frac{1-\mu}{1+\mu}$, then

$$\bar\partial u + i\bar\partial v = \bar\partial f = \tfrac{1-\sigma}{1+\sigma}\overline{\partial f} = \tfrac{1-\sigma}{1+\sigma}(\bar\partial u - i\bar\partial v)$$

and thus

$$(1+\sigma)(\bar\partial u + i\bar\partial v) = (1-\sigma)(\bar\partial u - i\bar\partial v).$$

This shows that $2i\bar\partial v + 2\sigma\bar\partial u = 0$, and therefore

$$(\partial_1 v, \partial_2 v) = (-\sigma \partial_2 u, \sigma \partial_1 u).$$

It follows that

$$\partial_1(\sigma \partial_1 u) + \partial_2(\sigma \partial_2 u) = \partial_1 \partial_2 v - \partial_2 \partial_1 v = 0$$

and

$$\partial_1(\sigma^{-1}\partial_1 v) + \partial_2(\sigma^{-1}\partial_2 v) = -\partial_1\partial_2 u + \partial_2 \partial_1 u = 0.$$

The last two computations are easy to justify in the weak sense. This proves the result. □

Exercise 5.2.2 (Poincaré lemma for L^2 vector fields)**.** Let $\Omega \subset \mathbb{R}^2$ be a bounded simply connected open set, and let $F : \Omega \to \mathbb{R}^2$ be a vector field with components in $L^2(\Omega)$ whose curl vanishes in the weak sense,

$$\partial_1 F_2 - \partial_2 F_1 = 0 \quad \text{in } \Omega.$$

Show that $F = \nabla p$ for some real valued function $p \in H^1(\Omega)$.

You may use the following outline:

(a) Find $p_0 \in H^1_0(\Omega)$ with $\Delta p_0 = \nabla \cdot F$ in Ω.

(b) Define $\tilde{F} = F - \nabla p_0 \in L^2(\Omega)^2$ and show that

$$\partial_1 \tilde{F}_2 - \partial_2 \tilde{F}_1 = 0, \qquad \partial_1 \tilde{F}_1 + \partial_2 \tilde{F}_2 = 0.$$

(c) Show that

$$\Delta \tilde{F}_1 = \Delta \tilde{F}_2 = 0.$$

Use Weyl's lemma, Lemma 5.1.3, to conclude that $\tilde{F}_1$ and $\tilde{F}_2$ are C^∞.

(d) Use the standard Poincaré lemma (Exercise 5.1.6) to $\tilde{F}$.

Exercise 5.2.3. Justify in the weak sense the computations in the end of the proof in Proposition 5.2.1.

Several remarks are in order. The first-order equation (5.3) is called a *Beltrami equation.* There is a one-to-one correspondence between solutions of $\nabla \cdot \sigma \nabla u = 0$ and of $\bar{\partial} f = \mu \overline{\partial f}$, where σ and μ are related by the formulas in Proposition 5.2.1. It is very important that the complex coefficient μ satisfies

$$\|\mu\|_{L^\infty} < 1. \tag{5.4}$$

This is the condition for uniform ellipticity of the Beltrami equation and ensures that the corresponding conductivity stays positive and bounded.

Solutions of $\nabla \cdot \sigma \nabla u = 0$ are called *σ-harmonic functions.* The function v constructed in Proposition 5.2.1 is called a *σ-harmonic conjugate* of u. The situation is symmetric so that the σ-harmonic conjugate v of u is itself a $1/\sigma$-harmonic function.

In a sense, a σ-harmonic conjugate "completes" the original σ-harmonic function u into a "μ-analytic" function $f = u + iv$. This generalizes the fact that a harmonic function u can be "completed" into an analytic function $f = u + iv$ by using the harmonic conjugate v. This is also obtained from Proposition 5.2.1 in the special case where $\sigma = 1$, so that $\mu = 0$ and the Beltrami equation is just $\bar{\partial} f = 0$.

The theory of analytic functions (solutions of $\bar{\partial} f = 0$) provides strong complex analysis tools to study harmonic functions (solutions of $\Delta u = 0$). In the same way, the point of view of the Beltrami equation provides powerful tools (now based on the theory of quasiconformal mappings) to the analysis of σ-harmonic functions and also the Calderón problem. For completeness, we will give the definition of quasiconformal mappings in the end of the section.

Above we have reduced the study of solutions of $\nabla\cdot\sigma\nabla u = 0$ to the Beltrami equation $\bar{\partial} f = \mu\overline{\partial f}$. Now we make a similar reduction on the level of boundary measurements. Given a σ-harmonic function $u \in H^1(\mathbb{D})$, we specify a unique σ-harmonic conjugate $v \in H^1(\mathbb{D})$ by requiring that

$$\int_{\partial\mathbb{D}} v\, dS = 0.$$

If $\mu \in L^\infty(\mathbb{D})$ with $\|\mu\|_{L^\infty(\mathbb{D})} < 1$, we define the μ-Hilbert transform

$$\mathcal{H}_\mu : H^{1/2}(\partial\mathbb{D};\mathbb{R}) \to H^{1/2}(\partial\mathbb{D};\mathbb{R}), \quad u|_{\partial\mathbb{D}} \mapsto v|_{\partial\mathbb{D}},$$

where $f = u + iv$ solves $\bar{\partial} f = \mu\overline{\partial f}$ in $\mathbb{D}$ and $\int_{\partial\mathbb{D}} v\, dS = 0$.

Exercise 5.2.4. Show that $\mathcal{H}_\mu$ is a well-defined bounded linear map on $H^{1/2}(\partial\mathbb{D};\mathbb{R})$.

Proposition 5.2.5. *Knowledge of Λ_σ determines the operators $\mathcal{H}_\mu$, $\mathcal{H}_{-\mu}$, and $\Lambda_{1/\sigma}$. Further, one has the identity*

$$\mathcal{H}_{-\mu}\mathcal{H}_\mu g = \mathcal{H}_\mu\mathcal{H}_{-\mu} g = -g \tag{5.5}$$

for any $g \in H^{1/2}(\partial\mathbb{D})$ with $\int_{\partial\mathbb{D}} g\, dS = 0$.

Proof.

(1) Let $g \in H^{1/2}(\partial\mathbb{D})$, let $u \in H^1(\mathbb{D})$ satisfy $\nabla\cdot\sigma\nabla u = 0$ with $u|_{\partial\mathbb{D}} = g$, and let $\varphi \in H^1(\mathbb{D})$. We have

$$\langle \Lambda_\sigma g, \varphi\rangle_{\partial\Omega} = \int_{\mathbb{D}} \sigma\nabla u \cdot \nabla\varphi\, dx.$$

Letting v be the σ-harmonic conjugate of u with $\int_{\partial\mathbb{D}} v\, dS = 0$, so that $(\partial_1 v, \partial_2 v) = (-\sigma\partial_2 u, \sigma\partial_1 u)$, we have

$$\langle \Lambda_\sigma g, \varphi\rangle_{\partial\Omega} = \int_{\mathbb{D}} (\partial_2 v, -\partial_1 v)\cdot\nabla\varphi\, dx = \int_{\partial\mathbb{D}} ((-\nu_2,\nu_1)\cdot\nabla v)\varphi\, dS.$$

This shows that

$$\Lambda_\sigma = \partial_T \mathcal{H}_\mu$$

in the weak sense, where $\partial_T = (-\nu_2, \nu_1)\cdot\nabla$ is the tangential derivative along $\partial\mathbb{D}$.

(2) We have recovered the tangential derivative of $\mathcal{H}_\mu$ from the DN map. It is now enough to prove that any function $h \in H^{1/2}(\mathbb{D})$ whose integral over

$\partial\mathbb{D}$ vanishes is uniquely determined by $\partial_T h$. This can be done by using Fourier coefficients (see Exercise 5.2.6): one expands h in Fourier series

$$h(e^{i\theta}) = \sum_{m=-\infty}^{\infty} \hat{h}(m)e^{im\theta},$$

where the Fourier coefficients are given by

$$\hat{h}(m) = \frac{1}{2\pi}\int_0^{2\pi} e^{-im\theta} h(e^{i\theta})\, d\theta.$$

The Fourier coefficients of $\partial_T h$ are given by

$$(\partial_T h)\hat{}\,(m) = im\hat{h}(m).$$

One has $\hat{h}(0) = 0$ since the integral of h over $\partial\mathbb{D}$ vanishes. Consequently

$$h(e^{i\theta}) = \sum_{m=-\infty, m\neq 0}^{\infty} \frac{(\partial_T h)\hat{}\,(m)}{im} e^{im\theta}.$$

We have proved that $\Lambda_\sigma g = \partial_T \mathcal{H}_\mu g$ determines $\mathcal{H}_\mu g$.

(3) The identity (5.5) is proved in Exercise 5.2.7. Assuming this identity, we see that $\mathcal{H}_{\pm\mu}$ are bijective operators on the space of functions in $H^{1/2}(\partial D)$ whose integral vanishes. Therefore

$$\mathcal{H}_{-\mu} = -(\mathcal{H}_\mu)^{-1}$$

for such functions. Since $\mathcal{H}_{\pm\mu}(c) = 0$ for any constant c, the operator $\mathcal{H}_\mu$ determines $\mathcal{H}_{-\mu}$. Noting that

$$-\mu = \frac{\sigma - 1}{\sigma + 1} = \frac{1 - 1/\sigma}{1 + 1/\sigma}$$

we have $\Lambda_{1/\sigma} = \partial_T \mathcal{H}_{-\mu}$. Thus the operator Λ_σ indeed determines $\mathcal{H}_\mu$, $\mathcal{H}_{-\mu}$ and $\Lambda_{1/\sigma}$. □

Exercise 5.2.6. If $s \geq 0$, show that

$$H^s(\partial\mathbb{D}) = \{ \sum_{m=-\infty}^{\infty} a_m e^{im\theta} \mid \sum_{m=-\infty}^{\infty} (1 + |m|)^{2s} |a_m|^2 < \infty \}.$$

Extend this characterization to all $s \in \mathbb{R}$, and show that any element $h \in H^s(\partial\mathbb{D})$ can be written as the Fourier series

$$h = \sum \hat{h}(m)e^{im\theta}, \quad \hat{h}(m) = \frac{1}{2\pi}\int_{\partial\mathbb{D}} e^{-im\theta} h(e^{i\theta})\, d\theta$$

with suitable interpretations for the sum and the integral. Show that

$$(\partial_T h)\hat{}\,(m) = im\hat{h}(m).$$

Exercise 5.2.7. Prove the identity (5.5) for any $g \in H^{1/2}(\partial\mathbb{D})$ whose integral over $\partial\mathbb{D}$ vanishes.

We now explain how the previous reduction is related to quasiconformal mappings. To do this, we first need to give a definition of quasiconformal mappings. A comprehensive reference to this theory is [**AIM09**].

Let $\Omega \subset \mathbb{C}$ be a bounded open set, and let $f \in W^{1,1}_{\mathrm{loc}}(\Omega)$. Then the first partial derivatives of f exist at almost every point in Ω, and at these points we can define the directional derivative

$$\partial_\alpha f(z) = \lim_{r\to 0} \frac{f(z+re^{i\alpha}) - f(z)}{r} = \cos(\alpha)\partial_1 f(z) + \sin(\alpha)\partial_2 f(z).$$

For small r, we consider the image of the circle $\alpha \mapsto z + re^{i\alpha}$ under the map f. The value $|\partial_\alpha f(z)|$ measures how much f distorts an infinitesimal circle at z in direction α. The next exercise expresses this in terms of the complex derivatives.

Exercise 5.2.8. Show that

$$\partial_\alpha f(z) = \bar\partial f(z) e^{-2i\alpha} + \partial f(z).$$

Show that $\partial_\alpha f(z)$ is independent of α if and only if $\bar\partial f(z) = 0$.

Recall that a map $f : \Omega \to \Omega'$ between two open sets in $\mathbb{C}$ is called *conformal* if it is analytic and bijective onto its image (this implies that the derivative of f is nonvanishing in Ω). The previous exercise can be interpreted so that conformal functions map infinitesimal circles to infinitesimal circles: if $\bar\partial f = 0$ then

$$f(z+re^{i\alpha}) \approx f(z) + \partial_\alpha f(z) re^{i\alpha} = f(z) + \partial f(z) re^{i\alpha}.$$

This is, in a sense, a very strong requirement. For instance, it follows from the Schwarz lemma in complex analysis that

$$f : \mathbb{D} \to \mathbb{D} \text{ conformal} \implies f \text{ is a Möbius map, } f(z) = \frac{az+b}{cz+d}.$$

Quasiconformal mappings relax this requirement, and require that f maps infinitesimal circles to infinitesimal ellipses with uniformly bounded eccentricity. Thus, instead of the condition that $|\partial_\alpha f(z)|$ is independent of α, the condition is

$$\max_\alpha |\partial_\alpha f(z)| \leq K \min_\alpha |\partial_\alpha f(z)| \quad \text{for a.e. } z \in \Omega, \tag{5.6}$$

where K is a uniform constant with $1 \leq K < \infty$. The precise definition is as follows.

Definition 5.2.9. Let $1 \leq K < \infty$. A mapping $f \in W^{1,2}_{\mathrm{loc}}(\Omega)$ is called *K-quasiregular* if it is orientation preserving (in the sense that its Jacobian $|\partial f|^2 - |\bar\partial f|^2$ is nonnegative almost everywhere) and if the condition (5.6) holds. If in addition f is a homeomorphism onto its image, then f is called *K-quasiconformal.*

Quasiconformal mappings form a much larger class of mappings than the conformal ones. For instance, compare the result of the next exercise to the fact mentioned above that any conformal map $\mathbb{D} \to \mathbb{D}$ must be a Möbius map.

Exercise 5.2.10. Show that any C^1 orientation preserving diffeomorphism $f : \Omega \to \Omega'$, where Ω and Ω' are open sets containing $\overline{\mathbb{D}}$, restricts to a quasiconformal map $\mathbb{D} \to \mathbb{D}$.

Despite being a much larger class than conformal mappings, quasiconformal mappings still have a powerful and well established theory with many applications in elliptic PDE (both linear and nonlinear), conformal geometry, complex dynamics, and inverse problems. It is remarkable that quasiconformal maps can also be characterized as the solutions of a PDE, the Beltrami equation, showing that the inequality (5.6) is in effect equivalent with a partial differential equation. The next exercise discusses this equivalence for mappings with C^1 regularity.

Exercise 5.2.11. Let $f : \Omega \to \Omega'$ be a C^1 orientation preserving diffeomorphism between two open subsets of $\mathbb{C}$. Show that f is quasiconformal if and only if

$$\bar{\partial} f = \mu \partial f \quad \text{in } \Omega$$

for some $\mu \in L^\infty(\Omega)$ with $\|\mu\|_{L^\infty(\Omega)} < 1$.

Note that the Beltrami equation in (5.3) differs from the Beltrami equation in Exercise 5.2.11 by having $\overline{\partial f}$ instead of ∂f on the right-hand side. However, one can sometimes make a reduction: (5.3) can be written formally as

$$\bar{\partial} f = \tilde{\mu} \partial f, \quad \tilde{\mu} = \mu \frac{\overline{\partial f}}{\partial f}.$$

The last step can often be made rigorous, since in many cases of interest one has $\partial f \neq 0$ almost everywhere (see Proposition 5.9.2).

5.3. Cauchy and Beurling transforms

In Proposition 5.2.1, we made a reduction from the conductivity equation

$$\nabla \cdot \sigma \nabla u = 0$$

into the Beltrami equation

$$\bar{\partial} f = \mu \overline{\partial f}.$$

In the study of the Calderón problem in the plane, it will be very useful to understand the properties of solutions of various Beltrami equations. The simplest such equation is the $\bar{\partial}$-equation

$$\bar{\partial} u = f \quad \text{in } \mathbb{R}^2.$$

We will construct a solution operator P, called the *Cauchy transform* (sometimes also *solid Cauchy transform*), such that $u = Pf$ will be a unique solution of this equation in suitable function classes. In fact, P is a constant multiple of the operator Z already encountered in Lemma 4.5.10.

We will prove the following result. If $K \subset \mathbb{R}^2$ is a compact set write

$$L^p_K(\mathbb{R}^2) := \{\, f \in L^p(\mathbb{R}^2) \mid \operatorname{supp}(f) \subset K \,\}.$$

This is a closed subspace of $L^p(\mathbb{R}^2)$.

Proposition 5.3.1 (Cauchy transform). *Let $p > 2$. There is a linear operator*

$$P : L^p_c(\mathbb{R}^2) \to W^{1,p}(\mathbb{R}^2)$$

such that for any $f \in L^p_c(\mathbb{R}^2)$ the function $u = Pf$ is the unique solution in $W^{1,p}(\mathbb{R}^2)$ of the equation

$$\bar{\partial} u = f \quad \text{in } \mathbb{R}^2.$$

If $K \subset \mathbb{R}^2$ is a compact set then there is a constant $C > 0$ such that

$$\|Pf\|_{W^{1,p}(\mathbb{R}^2)} \leq C \, \|f\|_{L^p(\mathbb{R}^2)}, \quad f \in L^p_K(\mathbb{R}^2).$$

Moreover, P extends as a linear operator

$$P : L^{2\pm}(\mathbb{R}^2) \to C_0(\mathbb{R}^2),$$

where $L^{2\pm}$ and C_0 are as in (5.1)–(5.2).

Remark 5.3.2. The Cauchy transform has many other useful mapping properties besides those in Proposition 5.3.1. If $1 < p < 2$, we define the *Sobolev conjugate exponent*

$$p^* = \frac{2p}{2-p}.$$

This appears in the *Sobolev embedding theorem* (see, e.g., [**Eva10**, §5])

$$W^{1,p}(\mathbb{R}^2) \subset \begin{cases} L^{p^*}(\mathbb{R}^2), & 1 < p < 2, \\ \cap_{q<\infty} L^q_{\mathrm{loc}}(\mathbb{R}^2), & p = 2, \\ C_0(\mathbb{R}^2), & 2 < p < \infty. \end{cases} \tag{5.7}$$

We will prove later in §5.7 that P is a bounded linear operator

$$P : L^p(\mathbb{R}^2) \to L^{p^*}(\mathbb{R}^2).$$

We also mention here the weighted mapping property proved in Lemma 4.5.10,

$$P : L^p_{\delta+1}(\mathbb{R}^2) \to L^p_\delta(\mathbb{R}^2),$$

where $1 < p < \infty$ and $-\frac{2}{p} < \delta < 1 - \frac{2}{p}$. We refer the reader to [**AIM09**, §4] for further properties of the Cauchy transform.

The other main result in this section concerns a somewhat more general Beltrami type equation, given by

$$\bar{\partial} u - \nu \partial u = f \quad \text{in } \mathbb{R}^2.$$

Here we assume that $\nu \in L^\infty_c(\mathbb{R}^2)$ satisfies the ellipticity condition

$$\|\nu\|_{L^\infty} < 1.$$

We can try to solve this equation by treating it as a perturbation of the $\bar{\partial}$-equation and by looking for a solution in the form Pw. Since one should have $\bar{\partial} Pw = w$, this equation reduces to

$$(I - \nu S)w = f \quad \text{in } \mathbb{R}^2,$$

where S is the *Beurling operator* (or *Beurling–Ahlfors operator*)

$$S = \partial P.$$

This operator turns out to be bounded on L^p spaces, and the ellipticity condition $\|\nu\|_{L^\infty} < 1$ allows one to solve the equation $(I - \nu S)w = f$ at least for p close to 2. The Beurling operator has the important property that it intertwines the ∂ and $\bar{\partial}$ operators. The precise properties that will be needed are stated in the following result.

Proposition 5.3.3 (Beurling operator). *Let $1 < p < \infty$. There is a bounded linear operator*

$$S : L^p(\mathbb{R}^2) \to L^p(\mathbb{R}^2)$$

such that for any $u \in L^1_{\mathrm{loc}}(\mathbb{R}^2)$ with $\nabla u \in L^p(\mathbb{R}^2)$ one has

$$S\bar{\partial} u = \partial u.$$

If $f \in L^p(\mathbb{R}^2)$ with $1 < p < \infty$ one also has

$$Sf = \partial P f.$$

The norm of this operator satisfies

$$\|S\|_{L^2 \to L^2} = \lim_{p \to 2} \|S\|_{L^p \to L^p} = 1.$$

The mapping properties of the Beurling transform are a consequence of the Calderón–Zygmund theory of singular integrals. We will not give a proof here, except to mention that formally one may write $S = \partial \bar{\partial}^{-1}$, so that on the Fourier side the Beurling transform is given by

$$Sf = \mathcal{F}^{-1}\{m(\xi)\hat{f}(\xi)\}, \qquad m(\xi) = \frac{\xi_1 - i\xi_2}{\xi_1 + i\xi_2}.$$

The Fourier multiplier $m(\xi)$ is unimodular, i.e., $|m(\xi)| = 1$, and hence by the Plancherel theorem one indeed has $\|S\|_{L^2 \to L^2} = 1$. A proof of Proposition 5.3.3 together with further information on the Beurling transform may be found in [**AIM09**, §4].

The rest of this section is devoted to proving Proposition 5.3.1.

Motivation for the Proof of Proposition 5.3.1. Let us try to solve the equation

$$\bar{\partial} u = f \quad \text{in } \mathbb{R}^2.$$

This is a linear partial differential equation with constant coefficients, and we can formally solve it by Fourier analysis. Fourier transforming this equation leads to

$$\frac{i}{2}(\xi_1 + i\xi_2)\hat{u}(\xi) = \hat{f}(\xi),$$

and formally dividing by the symbol shows that the solution u should be given by

$$u = \mathcal{F}^{-1}\left\{\frac{2}{i}\frac{1}{\xi_1 + i\xi_2}\hat{f}\right\}.$$

Since the Fourier transform maps convolutions to products, this formally implies that u should be given by the convolution $u = k * f$, where

$$k := \mathcal{F}^{-1}\left\{\frac{2}{i}\frac{1}{\xi_1 + i\xi_2}\right\}. \tag{5.8}$$

We have already computed this inverse Fourier transform in Exercise 4.5.11. It is given by

$$k(z) = \frac{1}{\pi z}.$$

Motivated by the above discussion, we define the operator P as follows.

Definition 5.3.4. The *Cauchy transform* is the map

$$Pf = \frac{1}{\pi z} * f,$$

whenever f is a measurable function on $\mathbb{R}^2$ and the convolution is well defined.

We can now prove the proposition on the Cauchy transform.

Proof of Proposition 5.3.1. Let first $f \in L^p_K(\mathbb{R}^2)$ where $2 < p < \infty$ and $K \subset \mathbb{R}^2$ is compact. We can write

$$\frac{1}{z} = k_1(z) + k_2(z), \qquad k_1 = \chi\frac{1}{z}, \quad k_2 = (1-\chi)\frac{1}{z},$$

where $\chi = \chi_{\{|z|\leq 1\}}$ is the characteristic function of the unit disc. Then $k_1 \in L^1(\mathbb{R}^2)$ and $k_2 \in L^p(\mathbb{R}^2)$, so Young's inequality for convolutions (Lemma C.3.10) yields

$$\|k_1 * f\|_{L^p} \leq \|k_1\|_{L^1}\|f\|_{L^p}, \quad \|k_2 * f\|_{L^p} \leq \|k_2\|_{L^p}\|f\|_{L^1}.$$

We have $\|f\|_{L^1} \leq C_{K,p}\|f\|_{L^p}$ using that $f \in L^p_K$. This shows that P is a well-defined and bounded linear map $L^p_K(\mathbb{R}^2) \to L^p(\mathbb{R}^2)$.

Now $u = Pf$ satisfies $\bar\partial u = f$ and, by Proposition 5.3.3,

$$\partial u = \partial P f = S f.$$

Thus the L^p estimate for the Beurling transform yields

$$\left\|\bar\partial u\right\|_{L^p} + \|\partial u\|_{L^p} \le C_p \|f\|_{L^p}.$$

It follows that P is bounded $L^p_K(\mathbb{R}^2) \to W^{1,p}(\mathbb{R}^2)$.

Finally, let us show that the Cauchy transform Pf is also well defined when $f \in L^{2\pm}(\mathbb{R}^2)$. One has $f \in L^q \cap L^{q'}$ for some $q < 2$ where $\frac{1}{q} + \frac{1}{q'} = 1$. Writing $\frac{1}{z} = k_1 + k_2$, where $k_1 = \chi_R \frac{1}{z}$ and $k_2 = (1 - \chi_R)\frac{1}{z}$ with $\chi_R = \chi_{\{|z| \le R\}}$, Hölder's inequality and polar coordinates give

$$\|k_1 * f\|_{L^\infty} \le \|k_1\|_{L^q} \|f\|_{L^{q'}} = \left(\frac{2\pi R^{2-q}}{2-q}\right)^{1/q} \|f\|_{L^{q'}},$$

$$\|k_2 * f\|_{L^\infty} \le \|k_2\|_{L^{q'}} \|f\|_{L^q} = \left(\frac{2\pi R^{2-q'}}{2-q'}\right)^{1/q'} \|f\|_{L^q}.$$

We choose R so that the terms on the right are equal, i.e.,

$$R = c_q \left(\frac{\|f\|_{L^q}}{\|f\|_{L^{q'}}}\right)^{\frac{q}{2(2-q)}}.$$

With this choice of R, we get

$$\left\|\frac{1}{z} * f\right\|_{L^\infty} \le C_q \sqrt{\|f\|_{L^q} \|f\|_{L^{q'}}} \le 2C_q(\|f\|_{L^q} + \|f\|_{L^{q'}}). \tag{5.9}$$

Thus P is a bounded operator from $L^q \cap L^{q'}$, which is a Banach space with norm $\|f\|_{L^q} + \|f\|_{L^{q'}}$, to L^∞. If $f \in C_c^\infty(\mathbb{R}^2)$, then $Pf \in W^{1,p}(\mathbb{R}^2) \subset C_0(\mathbb{R}^2)$ by the first part of the theorem and Sobolev embedding (5.7). By density and the B.L.T. theorem, P is bounded from $L^q \cap L^{q'}$ to C_0. □

5.4. Existence and uniqueness of CGO solutions

We are ready to introduce the complex geometrical optics (CGO) solutions that will be used to resolve the Calderón problem in the plane. Recall that in dimensions $n \ge 3$, we employed exponentially growing solutions to the Schrödinger equation $(-\Delta + q)u = 0$ which resembled the harmonic exponentials depending on a vector $\zeta \in \mathbb{C}^n$ with $\zeta \cdot \zeta = 0$,

$$u_0(x) = e^{i\zeta\cdot x}, \quad \Delta u_0 = 0.$$

In the case $n = 2$ we have reduced the conductivity equation to a Beltrami equation. The CGO solutions to the Beltrami equation will be based on the analytic exponentials depending on a complex number $k \in \mathbb{C}$,

$$f_0(z) = e^{ikz}, \quad \bar\partial f_0 = 0.$$

The next proposition is the basic existence and uniqueness result for such solutions. Their behavior for large values of k is a subtle issue and will be considered only later.

Proposition 5.4.1 (CGO solutions). *Let $\mu \in L^\infty_c(\mathbb{R}^2)$ with $\|\mu\|_{L^\infty} < 1$. For any $k \in \mathbb{C}$, there is a unique solution $f_\mu = f_\mu(\,\cdot\,,k)$ of the equation*

$$\bar\partial f = \mu \overline{\partial f} \quad \text{in } \mathbb{R}^2$$

having the form

$$f_\mu(z,k) = e^{ikz}(1 + \eta_\mu(z,k)),$$

where

$$\eta_\mu(\,\cdot\,,k) \in W^{1,2}_{\mathrm{loc}}(\mathbb{R}^2),$$
$$\eta_\mu(z,k) = O(\frac{1}{z}) \text{ as } |z| \to \infty.$$

Further, $\eta_\mu(\,\cdot\,,k) \in W^{1,p}(\mathbb{R}^2)$ for some $p > 2$. In the case $k = 0$ one has $f_\mu(z,0) = 1$.

This result will ultimately be a consequence of a strong form of *Liouville's theorem*. The usual form of this theorem states that any analytic function which is bounded on $\mathbb{R}^2$ is constant. The following exercise gives a few other versions. Recall that $C_0(\mathbb{R}^2)$ is the space of continuous functions vanishing at infinity in (5.2), and also recall Weyl's lemma (Lemma 5.1.3).

Exercise 5.4.2 (Liouville theorem). Prove the following statements:

(a) Any bounded analytic function on $\mathbb{R}^2$ is constant.
(b) Any analytic function in $C_0(\mathbb{R}^2)$ is identically zero.
(c) Any function $f \in L^1_{\mathrm{loc}}(\mathbb{R}^2)$ satisfying $\bar\partial f = 0$ and $f(z) = o(1)$ as $|z| \to \infty$ is identically zero. (The last condition means that

$$\lim_{R\to\infty} \operatorname{ess\,sup}_{|z|\geq R} |f(z)| = 0.)$$

(d) Any analytic function in $L^p(\mathbb{R}^2)$, $1 \leq p < \infty$, is identically zero. (Hint: use the mean value property, or the distribution theory fact that any tempered distribution solving $\bar\partial f = 0$ is a polynomial.)

The following is the strong form that we will use.

Lemma 5.4.3 (Strong Liouville theorem). *Let $\nu \in L^\infty_c(\mathbb{R}^2)$ with $\|\nu\|_{L^\infty} < 1$, and assume that $\alpha \in L^\infty_c(\mathbb{R}^2)$. Let also $p > 2$ be such that*

$$\|\nu\|_{L^\infty} \|S\|_{L^p \to L^p} < 1.$$

Then for any $f \in L^p_c(\mathbb{R}^2)$, the equation

$$\bar\partial g - \nu \overline{\partial g} + \alpha \bar g = f \quad \text{in } \mathbb{R}^2$$

has a unique solution $g \in W^{1,2}_{\text{loc}}(\mathbb{R}^2)$ *with* $g(z) = o(1)$ *as* $|z| \to \infty$*. Moreover,* g *is in* $W^{1,p}(\mathbb{R}^2) \cap C_0(\mathbb{R}^2)$ *and* $g(z) = O(1/z)$ *as* $|z| \to \infty$*. If* f *is supported in a fixed compact set* K*, then*

$$\|g\|_{W^{1,p}(\mathbb{R}^2)} \leq C\,\|f\|_{L^p(\mathbb{R}^2)}$$

for some constant C *independent of* f*.*

At this point, Proposition 5.4.1 follows easily. In the proof and at many points below, we will use the notation

$$e_k(z) := e^{i(kz+\overline{kz})}.$$

Proof of Proposition 5.4.1. Note that

$$\bar\partial(e^{ikz}v) = e^{ikz}\bar\partial v, \qquad \partial(e^{ikz}v) = e^{ikz}(\partial + ik)v.$$

Inserting the form $f = e^{ikz}(1+\eta)$ into $\bar\partial f = \mu\overline{\partial f}$, we have the equivalences

$$\begin{aligned}\bar\partial f = \mu\overline{\partial f} &\iff e^{ikz}\bar\partial\eta = \mu e^{-i\overline{kz}}(\overline{\partial\eta} - i\bar k\bar\eta - i\bar k)\\ &\iff \bar\partial\eta - \mu e_{-k}\overline{\partial\eta} + i\bar k\mu e_{-k}\bar\eta = -i\bar k\mu e_{-k}.\end{aligned}$$

Notice that $|e_k(z)| = 1$. Therefore the coefficients of the last equation are in $L^\infty_c(\mathbb{R}^2)$ with $\|\mu e_{-k}\|_{L^\infty} < 1$, and the right-hand side is in $L^q_c(\mathbb{R}^2)$ for any $q \geq 1$. By Proposition 5.3.3, we may choose $p > 2$ such that

$$\|\mu e_{-k}\|_{L^\infty}\,\|S\|_{L^p\to L^p} < 1.$$

By Lemma 5.4.3 the last equation for η has a unique solution with $\eta \in W^{1,2}_{\text{loc}}(\mathbb{R}^2)$ and $\eta(z) = O(\frac{1}{z})$ as $|z| \to \infty$, and further $\eta \in W^{1,p}(\mathbb{R}^2)$. If $k = 0$ the right-hand side is 0, and then the unique solution is $\eta(z,0) = 0$ showing that $f(z,0) = 1$. □

To warm up for the proof of Lemma 5.4.3, we first give a slightly simpler result for a related equation.

Lemma 5.4.4 (Strong Liouville theorem). *Let* $\nu \in L^\infty_c(\mathbb{R}^2)$ *with* $\|\nu\|_{L^\infty} < 1$*, and assume that* $\alpha \in L^\infty_c(\mathbb{R}^2)$*. Let also* $p > 2$ *be such that*

$$\|\nu\|_{L^\infty}\,\|S\|_{L^p\to L^p} < 1.$$

Then for any $f \in L^p_c(\mathbb{R}^2)$*, the equation*

$$\bar\partial g - \nu\partial g + \alpha g = f \quad \text{in } \mathbb{R}^2 \tag{5.10}$$

has a unique solution $g \in W^{1,p} \cap C_0(\mathbb{R}^2)$*.*

Proof.

(1) The first step is to reduce (5.10) to the case $\alpha = 0$. To do this, we find a function $\beta \in W^{1,p}(\mathbb{R}^2)$ (*integrating factor*) satisfying

$$\bar{\partial}\beta - \nu\partial\beta = \alpha.$$

A solution is given by $\beta = Pw$, provided that $w \in L^p_c(\mathbb{R}^2)$ satisfies

$$w - \nu S w = \alpha.$$

This has the solution $w = (\mathrm{Id} - \nu S)^{-1}\alpha$, where $\mathrm{Id} - \nu S$ is invertible on L^p via Neumann series using that $\|\nu S\|_{L^p \to L^p} < 1$ by the assumption. By writing $w = \alpha + \nu S(\mathrm{Id} - \nu S)^{-1}\alpha$ we see that $w \in L^p_c$, and consequently $\beta \in W^{1,p}(\mathbb{R}^2)$ as required.

(2) To show the existence of a solution, we use the integrating factor β and note that the equation (5.10) is equivalent with

$$(\bar{\partial} - \nu\partial)(e^\beta g) = e^\beta f.$$

We try $e^\beta g = Pw$. This solves the last equation provided that w satisfies

$$w - \nu S w = e^\beta f.$$

Since $\beta \in W^{1,p} \subset C_0$ by Sobolev embedding (5.7), it follows that $e^\beta f \in L^p_c$ and the function $w = (\mathrm{Id} - \nu S)^{-1}(e^\beta f) \in L^p_c$ satisfies the equation required of w. Then

$$g = e^{-\beta} P w$$

is a solution of (5.10). Since $\beta \in W^{1,p} \cap C_0$ and $Pw \in W^{1,p} \cap C_0$ by Proposition 5.3.1, we have $g \in W^{1,p} \cap C_0$.

(3) For uniqueness, assume that $g \in W^{1,p} \cap C_0(\mathbb{R}^2)$ solves

$$\bar{\partial} g - \nu \partial g + \alpha g = 0.$$

Choosing β as in Step 2, this is equivalent with

$$(\bar{\partial} - \nu\partial)(e^\beta g) = 0.$$

Since $e^\beta g \in W^{1,p}$, the function $w := \bar{\partial}(e^\beta g) = \nu\partial(e^\beta g) \in L^p_c$ satisfies $e^\beta g = Pw$ by Liouville's theorem (Exercise 5.4.2). Consequently

$$w - \nu S w = 0.$$

Now $\mathrm{Id} - \nu S$ is invertible on L^p so $w = 0$. Thus $\bar{\partial}(e^\beta g) = 0$, and since $e^\beta g \in C_0$ we obtain $e^\beta g = 0$ from Exercise 5.4.2. $\square$

The proof of Lemma 5.4.3 is similar to the previous proof, the main difference being the appearance of complex conjugates of g and ∂g.

Proof of Lemma 5.4.3.

(1) We begin by proving uniqueness of solutions. Suppose that $g \in W^{1,2}_{\mathrm{loc}}(\mathbb{R}^2)$, $g(z) = o(1)$ as $|z| \to \infty$, and

$$\bar\partial g - \nu \overline{\partial g} + \alpha \bar g = 0.$$

Define

$$\tilde\nu(z) = \begin{cases} \nu \frac{\overline{\partial g}}{\partial g}, & \partial g(z) \neq 0, \\ 0, & \partial g(z) = 0, \end{cases} \qquad \tilde\alpha(z) = \begin{cases} \alpha \frac{\bar g}{g}, & g(z) \neq 0, \\ 0, & g(z) = 0. \end{cases}$$

Then $\tilde\nu, \tilde\alpha \in L^\infty_c(\mathbb{R}^2)$. These functions are measurable since g and ∂g are. This may be seen for instance by writing

$$\frac{\bar g}{g} \chi_{\{g \neq 0\}} = \lim_{\varepsilon \to 0} \frac{\bar g^2}{|g|^2 + \varepsilon} \chi_{\{g \neq 0\}}.$$

Note also that

$$\|\tilde\nu S\|_{L^p \to L^p} \leq \|\nu S\|_{L^p \to L^p} < 1.$$

Then g is a solution of

$$\bar\partial g - \tilde\nu \partial g + \tilde\alpha g = 0.$$

Choosing $\beta \in W^{1,p}(\mathbb{R}^2)$ such that $\bar\partial \beta - \tilde\nu \partial \beta = \tilde\alpha$ as in Step 1 of the proof of Lemma 5.4.4 and writing $h = e^\beta g$, we reduce matters to the equation

$$\bar\partial h - \tilde\nu \partial h = 0. \tag{5.11}$$

(2) It is sufficient to prove that any solution $h \in W^{1,2}_{\mathrm{loc}}(\mathbb{R}^2)$ of (5.11) satisfying $h(z) = o(1)$ as $|z| \to \infty$ is identically zero. From (5.11) we see that $\bar\partial h \in L^2_c(\mathbb{R}^2)$ and consequently $\partial h = S \bar\partial h$. It follows that $w = \bar\partial h$ satisfies

$$w - \tilde\nu S w = 0.$$

Since $\mathrm{Id} - \tilde\nu S$ is invertible on L^2, it follows that $w = 0$. Thus $\bar\partial h = 0$, and Liouville's theorem (Exercise 5.4.2) shows that $h = 0$.

(3) It remains to prove existence of solutions. Because of the $\bar g$ term the reduction to the case $\alpha = 0$ is not immediately available. However, we can use the uniqueness of solutions combined with the Fredholm alternative (Proposition A.3.15) to get the desired result. Write $C : g \mapsto \bar g$ for the complex conjugation operator. We need to solve

$$(\bar\partial - \nu C \partial + \alpha C) g = f.$$

Looking for a solution in the form $g = Pw$ for $w \in L^p_c$, it follows that w should satisfy

$$(\mathrm{Id} - \nu C S + \alpha C P) w = f. \tag{5.12}$$

Since $\|\nu CS\|_{L^p \to L^p} < 1$, the real-linear operator $\mathrm{Id} - \nu CS$ has the bounded inverse

$$(\mathrm{Id} - \nu CS)^{-1} = \sum_{j=0}^{\infty} (\nu CS)^j.$$

The equation in (5.12) for w is equivalent with

(5.13) $$(\mathrm{Id} + R)w = (\mathrm{Id} - \nu CS)^{-1} f,$$

where

$$R = (\mathrm{Id} - \nu CS)^{-1} \alpha CP.$$

(4) Let $K \subset \mathbb{R}^2$ be a compact set containing the supports of ν, α, and f. We note that P is bounded $L^p_K(\mathbb{R}^2) \to W^{1,p}(\mathbb{R}^2)$ and that multiplication by the compactly supported function α is a compact operator from $W^{1,p}(\mathbb{R}^2)$ to $L^p(\mathbb{R}^2)$. The latter fact is a consequence of the compact Sobolev embedding, Theorem 2.4.12. It follows that R is a compact real-linear operator on $L^p_K(\mathbb{R}^2)$. By the Fredholm alternative, (5.13) has a solution $w \in L^p_K(\mathbb{R}^2)$ if and only if $\mathrm{Id} + R$ is injective on $L^p_K(\mathbb{R}^2)$. But by tracing back the steps above, any solution $w \in L^p_K(\mathbb{R}^2)$ of $(\mathrm{Id} + R)w = 0$ gives rise to $h = Pw \in W^{1,p} \cap C_0$ satisfying

$$\bar\partial h - \nu \overline{\partial h} + \alpha \bar h = 0.$$

Thus $h = 0$, and also $w = \bar\partial h = 0$. This shows injectivity of $\mathrm{Id} + R$ and solvability of (5.13), which implies existence of a solution $g \in W^{1,p}(\mathbb{R}^2) \cap C_0(\mathbb{R}^2)$ to the original equation. The Fredholm alternative also implies that the inverse of $\mathrm{Id} + R$ is bounded on $L^p_K(\mathbb{R}^2)$, and the norm bound for g follows from (5.13) since

$$g = P(\mathrm{Id} + R)^{-1}(\mathrm{Id} - \nu CS)^{-1} f.$$

(5) The final step is to prove that any solution $g \in C_0(\mathbb{R}^2)$ satisfies $g(z) = O(1/z)$ as $|z| \to \infty$. Since $\bar\partial g = 0$ outside some large disk with radius R, the function $h(z) = g(1/z)$ is analytic in $B \setminus \{0\}$, where $B = B(0, 1/R)$. By the condition $g \in C_0(\mathbb{R}^2)$, we have that $h(z) \to 0$ as $z \to 0$. The Riemann removable singularities theorem in complex analysis implies that h is analytic in B with $h(0) = 0$, and consequently we may write $h(z) = zv(z)$ for some function v analytic in B. Thus $g(z) = v(1/z)/z$ for $|z|$ large, showing that $g(z) = O(1/z)$ as $|z| \to \infty$. □

Exercise 5.4.5. Prove the Riemann removable singularities theorem used in the previous proof: if w is analytic and uniformly bounded in $\mathbb{D} \setminus \{0\}$, then w has a unique extension as an analytic function into $\mathbb{D}$.

We also give a variant of Lemma 5.4.3 that will be used later.

Lemma 5.4.6. *Let $\nu \in L_c^\infty(\mathbb{R}^2)$ with $\|\nu\|_{L^\infty} < 1$, and assume that $\alpha \in L_c^\infty(\mathbb{R}^2)$. Let also $p > 2$ be such that*

$$\|\nu\|_{L^\infty} \|S\|_{L^p \to L^p} < 1.$$

If $g \in W^{1,2}_{\mathrm{loc}}(\mathbb{R}^2)$ solves the equation

$$\bar\partial g - \nu \overline{\partial g} + \alpha \bar g = 0 \quad \text{in } \mathbb{R}^2,$$

then g is continuous. If additionally g is bounded, then $g = Ce^\beta$ where C is a constant and $\beta \in W^{1,p}(\mathbb{R}^2)$.

Exercise 5.4.7. Prove Lemma 5.4.6 by modifying the proof of Lemma 5.4.3.

We conclude this section with an important consequence of Lemma 5.4.3 on the higher integrability of the gradient of solutions.

Proposition 5.4.8 (Higher integrability). *Let $\Omega \subset \mathbb{R}^2$ be open and let $\sigma \in L^\infty(\Omega)$ with $\sigma \geq c > 0$ a.e. in Ω. If $u \in H^1_{\mathrm{loc}}(\Omega)$ solves*

$$\nabla \cdot (\sigma \nabla u) = 0 \text{ in } \Omega,$$

then $u \in W^{1,p}_{\mathrm{loc}}(\Omega)$ for some $p > 2$.

Proof. By splitting u in real and imaginary parts, we may assume that u is real valued. Let B_1 and B_2 be two concentric balls with $\overline{B}_1 \subset B_2$ and $\overline{B}_2 \subset \Omega$. Then $u|_{B_2} \in H^1(B_2)$ solves

$$\nabla \cdot \sigma \nabla u = 0 \text{ in } B_2.$$

By Proposition 5.2.1 there is $v \in H^1(B_2)$ such that $f := u + iv \in H^1(B_2)$ solves

$$\bar\partial f = \mu \overline{\partial f} \text{ in } B_2, \qquad \mu = \frac{1-\sigma}{1+\sigma}.$$

Let $\psi \in C_c^\infty(B_2)$ satisfy $\psi \geq 0$ and $\psi = 1$ near $\overline{B}_1$. Then $f_1 := \psi f \in H^1_c(\mathbb{R}^2)$ satisfies

$$\bar\partial f_1 - \mu \overline{\partial f_1} = (\bar\partial \psi) f - \mu \overline{(\partial \psi) f} \text{ in } \mathbb{R}^2.$$

By Sobolev embedding (5.7), $(\partial\psi) f \in L^q_c(\mathbb{R}^2)$ for any $q > 2$, which implies that the right-hand side of the last equation is in $L^q_c(\mathbb{R}^2)$ for any $q > 2$. Lemma 5.4.3 implies that $f_1 \in W^{1,p}(\mathbb{R}^2)$ for some $p > 2$. This gives that $u|_{B_1} \in W^{1,p}(B_1)$. □

5.5. Basic properties of CGO solutions

Using Proposition 5.4.1, we can easily show one basic property of the solutions f_μ: their values in $\mathbb{R}^2 \setminus \mathbb{D}$ are uniquely determined by the DN map. This will be an important step in the solution of the Calderón problem.

Proposition 5.5.1 (CGO solutions outside $\mathbb{D}$). *If* $\Lambda_{\sigma_1} = \Lambda_{\sigma_2}$, *then*

$$f_{\pm\mu_1}(z,k) = f_{\pm\mu_2}(z,k) \quad \textit{for } z \in \mathbb{R}^2 \setminus \mathbb{D},$$

where $\mu_j = \frac{1-\sigma_j}{1+\sigma_j}$.

Proof. Since $\Lambda_{\sigma_1} = \Lambda_{\sigma_2}$, we obtain from Proposition 5.2.5 that

$$\mathcal{H}_{\pm\mu_1} = \mathcal{H}_{\pm\mu_2}.$$

It is enough to consider the solutions f_{μ_j}. Decompose $f_{\mu_j} = f_{\mu_j}(\,\cdot\,,k)$ into real and imaginary parts as

$$f_{\mu_j} = u_j + iv_j.$$

We wish to define a function

$$\tilde{f}(z) = \begin{cases} \tilde{u}(z) + i\tilde{v}(z), & z \in \mathbb{D}, \\ u_2(z) + iv_2(z), & z \in \mathbb{R}^2 \setminus \mathbb{D}, \end{cases} \tag{5.14}$$

such that $\tilde{f}$ solves $\bar{\partial}\tilde{f} = \mu_1 \overline{\partial \tilde{f}}$ in $\mathbb{R}^2$.

Let first $\tilde{u}$ be the unique $W^{1,2}(\mathbb{D})$ solution of

$$\nabla \cdot \sigma_1 \nabla \tilde{u} = 0 \text{ in } \mathbb{D}, \quad \tilde{u}|_{\partial\mathbb{D}} = u_2|_{\partial\mathbb{D}}.$$

Since $v_2|_{\mathbb{D}}$ is a σ-harmonic conjugate of $u_2|_{\mathbb{D}}$ in $\mathbb{D}$, it follows that for some constant c_0 one has

$$v_2|_{\partial\mathbb{D}} = \mathcal{H}_{\mu_2}(u_2|_{\partial\mathbb{D}}) + c_0.$$

Let $\tilde{v} \in W^{1,2}(\mathbb{D})$ be a σ_1-harmonic conjugate of $\tilde{u}$ in $\mathbb{D}$. This is unique up to an additive constant, and we fix this constant by requiring that

$$\int_{\partial\mathbb{D}} (\tilde{v} - c_0)\, dS = 0.$$

Then

$$\begin{aligned} \tilde{v}|_{\partial\mathbb{D}} &= (\tilde{v} - c_0)|_{\partial\mathbb{D}} + c_0 \\ &= \mathcal{H}_{\mu_1}(\tilde{u}|_{\partial\mathbb{D}}) + c_0 \\ &= \mathcal{H}_{\mu_2}(\tilde{u}|_{\partial\mathbb{D}}) + c_0 \\ &= \mathcal{H}_{\mu_2}(u_2|_{\partial\mathbb{D}}) + c_0 \\ &= v_2|_{\partial\mathbb{D}}. \end{aligned}$$

Now define $\tilde{f}$ by (5.14). Since $\tilde{u} + i\tilde{v}|_{\partial\mathbb{D}} = u_2 + iv_2|_{\partial\mathbb{D}}$, the function $\tilde{f}$ is in $W^{1,2}_{\mathrm{loc}}(\mathbb{R}^2)$. Then

$$\bar{\partial}\tilde{f} = \mu_1 \overline{\partial \tilde{f}} \quad \text{in } \mathbb{R}^2$$

since $\tilde{f}$ satisfies this equation both in $\mathbb{D}$ and $\mathbb{R}^2 \setminus \mathbb{D}$. Also, one has

$$\tilde{f} = e^{ikz}(1 + \tilde{\eta}),$$

where $\tilde{\eta} = e^{-ikz}\tilde{f} - 1$ is in $W^{1,2}_{\mathrm{loc}}(\mathbb{R}^2)$. But since $\tilde{f} = f_{\mu_2} = e^{ikz}(1+\eta_2)$ in $\mathbb{R}^2 \setminus \mathbb{D}$, where $\eta_2(z) = O(1/z)$ for $|z|$ large, it follows that $\tilde{\eta}(z) = O(1/z)$ for $|z|$ large. We can now invoke the uniqueness part of Proposition 5.4.1 to conclude that

$$\tilde{f} = f_{\mu_1} \quad \text{in } \mathbb{R}^2.$$

This shows that $f_{\mu_1} = f_{\mu_2}$ in $\mathbb{R}^2 \setminus \mathbb{D}$. □

Another property of the solutions f_μ that follows quite easily is their smoothness with respect to the parameter k. This will be used in the next section when deriving the $\partial_{\bar{k}}$ equation for the scattering transform. Recall that we write $f_\mu(z,k) = e^{ikz}(1 + \eta_\mu(z,k))$ where $\eta_\mu(\,\cdot\,,k) \in W^{1,p}(\mathbb{R}^2)$ for some $p > 2$.

Proposition 5.5.2 (Smoothness in k). *For some $p > 2$, the maps*

$$k \mapsto \eta_\mu(\,\cdot\,,k),$$
$$k \mapsto \chi(\,\cdot\,)f_\mu(\,\cdot\,,k)$$

are C^∞ maps from $\mathbb{C}$ into $W^{1,p}(\mathbb{R}^2)$ for any fixed $\chi \in C_c^\infty(\mathbb{R}^2)$.

The statement for $F : k \mapsto \eta_\mu(\,\cdot\,,k)$ means that all Fréchet derivatives $(D^m F)_k$ exist at each $k \in \mathbb{C}$, and the maps $k \mapsto (D^m F)_k$ are continuous $\mathbb{C} \to \mathcal{L}(\mathbb{C}^m, W^{1,p}(\mathbb{R}^2))$ where $\mathcal{L}(X,Y)$ denotes the space of bounded linear operators from X to Y (see Definition A.2.29). The proof is an easy consequence of the next result.

Lemma 5.5.3. *If $k \in \mathbb{C}$ define*

$$L_\mu(k) : W^{1,p}(\mathbb{R}^2) \to W^{1,p}(\mathbb{R}^2), \quad L_\mu(k)g = P(\mu\overline{\partial(e_k g)}).$$

(a) *The map $k \mapsto e_{\pm k}|_B$ is a C^∞ map from $\mathbb{C}$ to $W^{1,\infty}(B)$ for any bounded open set $B \subset \mathbb{R}^2$.*
(b) *The map $k \mapsto L_\mu(k)$ is a C^∞ map from $\mathbb{C}$ to $\mathcal{L}(W^{1,p}(\mathbb{R}^2))$.*
(c) *For each k the map $\mathrm{Id} - L_\mu$ is bounded and invertible on $W^{1,p}(\mathbb{R}^2)$, and $k \mapsto (\mathrm{Id} - L_\mu)^{-1}$ is a C^∞ map from $\mathbb{C}$ to $\mathcal{L}(W^{1,p}(\mathbb{R}^2))$.*

Exercise 5.5.4. If X, Y, Z are normed spaces and $F : U \to Y$ is C^k (resp. C^∞) in some open set $U \subset X$, and if $A : Y \to Z$ is a bounded linear map, then $A \circ F$ is C^k (resp. C^∞) and $D^m(A \circ F) = A \circ D^m F$.

Exercise 5.5.5. If X, Y, Z are normed spaces and $F : U \to V$, $G : V \to Z$ are C^k (resp. C^∞) maps where $U \subset X$, $V \subset Y$ are open sets, then $G \circ F : U \to Z$ is C^k (resp. C^∞) and

$$D(G \circ F)_x = DG_{F(x)} \circ DF_x.$$

Exercise 5.5.6. If X is a Banach space and U is the set of invertible elements of $\mathcal{L}(X)$, show that the map $U \to \mathcal{L}(X), A \mapsto A^{-1}$ is C^∞.

Proof of Proposition 5.5.2. Note that the solutions $f = e^{ikz}(1+\eta)$ given in Proposition 5.4.1 are characterized by

$$\begin{aligned}\bar{\partial} f = \mu \overline{\partial f} \quad &\Longleftrightarrow \quad e^{ikz}\bar{\partial}\eta = \mu\overline{\partial(e^{ikz}\eta)} - \mu i\bar{k}e^{-i\overline{kz}} \\ &\Longleftrightarrow \quad \bar{\partial}\eta = \mu\overline{\partial(e_k\eta)} - \mu i\bar{k}e_{-k}.\end{aligned}$$

Since $\eta \in W^{1,p}(\mathbb{R}^2)$ and the right-hand side is in $L^p_c(\mathbb{R}^2)$ for some $p > 2$, it follows that

$$\eta = P(\mu\overline{\partial(e_k\eta)}) - i\bar{k}P(\mu e_{-k}).$$

Equivalently, $(\mathrm{Id} - L_\mu(k))\eta = -i\bar{k}P(\mu e_{-k})$. By Lemma 5.5.3, we have

$$\eta_\mu(\,\cdot\,,k) = -i\bar{k}(\mathrm{Id} - L_\mu(k))^{-1}(P(\mu e_{-k})).$$

Using Exercises 5.5.4 and 5.5.5 with Lemma 5.5.3, we see that $k \mapsto \eta_\mu(\,\cdot\,,k)$ is a C^∞ map from $\mathbb{C}$ to $W^{1,p}(\mathbb{R}^2)$.

The smoothness of $k \mapsto \chi(\,\cdot\,)f_\mu(\,\cdot\,,k)$ follows since $k \mapsto \chi(\,\cdot\,)e^{ik\,\cdot}$ is smooth by the same argument as in Lemma 5.5.3(a). □

Proof of Lemma 5.5.3.

(a) Let $F : k \mapsto e_k|_B$ from $\mathbb{C}$ to $W^{1,\infty}(B)$. If $t \in \mathbb{R}$, the Taylor expansion of $f(t) = e^{it}$ implies that

$$e^{it} = 1 + it + \ldots + \frac{i^m}{m!}t^m + R_{m+1}(t), \quad R_{m+1}(t) = \frac{i^{m+1}}{m!}\int_0^t e^{is}(t-s)^m\,ds.$$

Since $e_{k+h} = e_k e_h$ for $k, h \in \mathbb{C}$, we have

$$e_{k+h} = e_k\left[1 + i(hz + \overline{hz}) + \ldots + \frac{i^m}{m!}(hz+\overline{hz})^m\right] + e_k R_{m+1}(hz + \overline{hz}).$$

The remainder term satisfies

$$\left\|e_k R_{m+1}(hz+\overline{hz})\right\|_{W^{1,\infty}(B)} \le C_{m,B,k}\,|h|^{m+1}.$$

This proves that F is C^m for all m, hence C^∞.

(b) Let $B = B(0,R)$ be a ball containing the support of μ. Define

$$(DL_\mu)_k[h]g = P(\mu C\partial(e_k(i(hz+\overline{hz}))g)).$$

By the argument in (a) we have

$$(L_\mu(k+h) - L_\mu(k) - (DL_\mu)_k[h])g = P(\mu C\partial(e_k R_2(hz+\overline{hz})g)).$$

The $W^{1,p}$ norm of the right-hand side is $O(|h|^2)\,\|g\|_{W^{1,p}}$, which shows that $k \mapsto L_\mu(k)$ is C^1. We leave it as an exercise to check that the higher derivatives also exist and $k \mapsto L_\mu(k)$ is a C^∞ map.

(c) It follows from (a) that $\mathrm{Id} - L_\mu$ is bounded on $W^{1,p}(\mathbb{R}^2)$. To show that it is invertible, we need to show that the equation $(\mathrm{Id} - L_\mu)g = f$ has a unique solution $g \in W^{1,p}(\mathbb{R}^2)$ for any $f \in W^{1,p}(\mathbb{R}^2)$, and that $\|g\|_{W^{1,p}} \leq C \|f\|_{W^{1,p}}$ for some constant C independent of f.

Writing $g = f + h$ where $h \in W^{1,p}(\mathbb{R}^2)$, the equation is equivalent with

$$(\mathrm{Id} - L_\mu)h = L_\mu f.$$

We write $L_\mu w = P(\mu e_k \overline{\partial w} + \mu(\bar{\partial} e_k)\bar{w})$. Since μ is compactly supported and $h, f \in W^{1,p}(\mathbb{R}^2)$, we may take $\bar{\partial}$ of the earlier equation and obtain the equivalent equation

$$\bar{\partial} h - \mu e_k \overline{\partial h} - \mu(\bar{\partial} e_k)\bar{h} = \mu e_k \overline{\partial f} + \mu(\bar{\partial} e_k)\bar{f}.$$

The right-hand side is in $L^p_c(\mathbb{R}^2)$ and the coefficients μe_k, $\mu(\bar{\partial} e_k)$ are in $L^\infty_c(\mathbb{R}^2)$ with $\|\mu e_k\|_{L^\infty} < 1$. Thus Lemma 5.4.3 applies, and it follows that the last equation has a unique solution $h \in W^{1,p}(\mathbb{R}^2)$ for any $f \in W^{1,p}(\mathbb{R}^2)$ with the norm bound

$$\|h\|_{W^{1,p}} \leq C \left\| \mu e_k \overline{\partial f} + \mu(\bar{\partial} e_k)\bar{f} \right\|_{L^p} \leq C \|f\|_{W^{1,p}}.$$

The invertibility of $\mathrm{Id} - L_\mu$ on $W^{1,p}(\mathbb{R}^2)$ follows. The fact that the map $k \mapsto (\mathrm{Id} - L_\mu(k))^{-1}$ is C^∞ is a consequence of (b) and Exercise 5.5.6. □

Exercise 5.5.7. Complete the details of the proof of part (b) in Lemma 5.5.3.

5.6. Scattering transform

In the previous sections we have made a reduction from the conductivity equation $\nabla \cdot \sigma \nabla u = 0$ into a Beltrami equation $\bar{\partial} f = \mu \overline{\partial f}$, where μ and σ are related by

$$\mu = \frac{1-\sigma}{1+\sigma}, \qquad -\mu = \frac{1 - 1/\sigma}{1 + 1/\sigma}. \tag{5.15}$$

Note the symmetry between $\pm\mu$ and $\sigma^{\pm 1}$. From this point on it is convenient to consider both μ and $-\mu$ (or σ and $1/\sigma$) simultaneously.

We have also defined the μ-Hilbert transform $\mathcal{H}_\mu$, constructed CGO solutions f_μ to the Beltrami equation, and proved the following implications:

$$\Lambda_{\sigma_1} = \Lambda_{\sigma_2} \implies \mathcal{H}_{\pm\mu_1} = \mathcal{H}_{\pm\mu_2} \implies f_{\pm\mu_1}|_{\mathbb{C}\setminus\mathbb{D}} = f_{\pm\mu_2}|_{\mathbb{C}\setminus\mathbb{D}}.$$

The final step in the proof is to show that $f_{\pm\mu_1} = f_{\pm\mu_2}$ also in $\mathbb{D}$. From this fact, one can conclude (at least formally) that

$$\mu_1|_{\mathbb{D}} = \left.\frac{\bar{\partial} f_{\mu_1}}{\overline{\partial f_{\mu_1}}}\right|_{\mathbb{D}} = \left.\frac{\bar{\partial} f_{\mu_2}}{\overline{\partial f_{\mu_2}}}\right|_{\mathbb{D}} = \mu_2|_{\mathbb{D}}.$$

Using the relationship between σ and μ, this would immediately imply that $\sigma_1 = \sigma_2$ in $\mathbb{D}$.

The proof that $f_{\pm\mu_1}|_{\mathbb{D}} = f_{\pm\mu_2}|_{\mathbb{D}}$ is involved, and it relies on the fact that the solutions $f_{\pm\mu} = f_{\pm\mu}(z,k)$ satisfy a $\bar\partial$-type equation also with respect to the complex variable k. This is an important feature of the $\bar\partial$-scattering method that underlies this part of the proof. In fact, the $\partial_{\bar k}$ equation for these solutions turns out to be in a sense simpler than the $\partial_{\bar z}$ equation $\partial_{\bar z} f_{\pm\mu} = \pm\mu\overline{\partial_z f_{\pm\mu}}$ that we have studied before.

To state the $\partial_{\bar k}$ equations, it is convenient to switch from solutions of the Beltrami equation back to the original conductivity equation. From Proposition 5.2.1 and the relation (5.15) we know that $\mathrm{Re}(f_\mu)$ and $\mathrm{Im}(f_{-\mu})$ are solutions of $\nabla\cdot\sigma\nabla u = 0$ in $\mathbb{R}^2$, and $\mathrm{Im}(f_\mu)$ and $\mathrm{Re}(f_{-\mu})$ are solutions of $\nabla\cdot(1/\sigma)\nabla u = 0$ in $\mathbb{R}^2$. We define two complex valued functions

$$u_\sigma := \mathrm{Re}(f_\mu) + i\,\mathrm{Im}(f_{-\mu}), \quad u_{1/\sigma} := \mathrm{Re}(f_{-\mu}) + i\,\mathrm{Im}(f_\mu).$$

Note that the pair $(u_\sigma, u_{1/\sigma})$ uniquely determines the pair $(f_\mu, f_{-\mu})$, and vice versa.

Proposition 5.6.1 (CGO solutions u_σ and $\partial_{\bar k}$ equation). *Let a denote either σ or $1/\sigma$. For any $k\in\mathbb{C}$, the function u_a is the unique complex solution of*

$$\nabla\cdot(a\nabla u_a(\,\cdot\,,k)) = 0 \text{ in } \mathbb{R}^2$$

having the form

$$u_a(z,k) = e^{ikz}(1 + r_a(z,k)),$$

where $r_a(\,\cdot\,,k)\in W^{1,2}_{\mathrm{loc}}(\mathbb{R}^2)$ and $r_a(z,k) = O(1/z)$ as $z\to\infty$. Furthermore, $k\mapsto r_a$ is a C^∞ map $\mathbb{C}\to W^{1,p}(\mathbb{R}^2)$ for some $p>2$. Thus u_a is C^∞ with respect to k, and for any fixed $z\in\mathbb{C}$ it satisfies the $\partial_{\bar k}$-equation

$$\partial_{\bar k} u_a(z,\,\cdot\,) = -i\tau_a(\,\cdot\,)\overline{u_a(z,\,\cdot\,)} \text{ in } \mathbb{R}^2,$$

where $\tau_a(k)$ is a complex function in $\mathbb{R}^2$.

The coefficient $\tau_a(k)$ deserves a special name. We will explain the reason for this terminology in the end of this section.

Definition 5.6.2. The function $\tau_\sigma(k)$ is called the *scattering transform* or *nonlinear Fourier transform* of a positive L^∞ conductivity σ.

A number of basic properties of $\tau_\sigma(k)$ for a positive L^∞ conductivity σ, with $\sigma = 1$ in $\mathbb{R}^2\setminus\mathbb{D}$, are given in the next proposition.

Proposition 5.6.3. *The scattering transform is a C^∞ function with respect to k, and it satisfies*

$$|\tau_\sigma(k)| \le 1 \text{ for any } k\in\mathbb{C}, \tag{5.16}$$

$$\tau_{1/\sigma}(k) = -\tau_\sigma(k). \tag{5.17}$$

It is determined from the special solutions $f_{\pm\mu}$ by

$$\overline{\tau_\sigma(k)} = \frac{1}{2}(b_1^{\mu}(k) - b_1^{\ \mu}(k)), \tag{5.18}$$

$$b_1^{\pm\mu}(k) := \lim_{z\to\infty} z(e^{-ikz} f_{\pm\mu}(z,k) - 1), \quad k \in \mathbb{C}. \tag{5.19}$$

In particular, if $\Lambda_{\sigma_1} = \Lambda_{\sigma_2}$, then $\tau_{\sigma_1} = \tau_{\sigma_2}$.

We will begin the proof of Propositions 5.6.1 and 5.6.3. Recall that

$$\eta_\mu(z,k) = e^{-ikz} f_\mu(z,k) - 1.$$

The equation $\bar\partial f_\mu = \mu \overline{\partial f_\mu}$, together with the fact that $\mu = 0$ in $\mathbb{R}^2 \setminus \mathbb{D}$, implies that

$$\partial_{\bar z}\eta_\mu(\,\cdot\,,k) = 0 \quad \text{in } \mathbb{R}^2 \setminus \mathbb{D}.$$

Since also $\eta_\mu(z,k) = O(1/z)$ as $z \to \infty$, the function $z \mapsto \eta_\mu(1/z,k)$ is an analytic function in $\mathbb{D}$ vanishing at 0 and continuous up to $\partial\mathbb{D}$. It follows that

$$\eta_\mu(1/z,k) = \sum_{m=1}^{\infty} b_m^{\mu}(k) z^m, \quad |z| \le 1,$$

where $b_m^{\mu}(k)$ is obtained as the Fourier coefficient

$$b_m^{\mu}(k) := \frac{1}{2\pi}\int_0^{2\pi} e^{-im\theta}\eta_\mu(e^{-i\theta},k)\,d\theta. \tag{5.20}$$

A similar argument works for $\eta_{-\mu}$ and we have

$$\eta_{\pm\mu}(z,k) = \sum_{m=1}^{\infty} b_m^{\pm\mu}(k) z^{-m}, \quad |z| \ge 1. \tag{5.21}$$

Proof of Proposition 5.6.1. Since $f_\mu = e^{ikz}(1+\eta_\mu)$, we have

$$\begin{aligned}
\operatorname{Re}(f_\mu) &= \frac{1}{2}e^{ikz}(1 + \eta_\mu + e_{-k}(1+\overline{\eta_\mu})),\\
\operatorname{Im}(f_{-\mu}) &= \frac{1}{2i}e^{ikz}(1 + \eta_{-\mu} - e_{-k}(1+\overline{\eta_{-\mu}})).
\end{aligned}$$

Consequently

$$u_\sigma = \operatorname{Re}(f_\mu) + i\operatorname{Im}(f_{-\mu}) = e^{ikz}\left(1 + \frac{\eta_\mu + \eta_{-\mu}}{2} + e_{-k}\frac{\bar\eta_\mu - \bar\eta_{-\mu}}{2}\right). \tag{5.22}$$

Thus u_σ is a solution of $\nabla\cdot\sigma\nabla u = 0$ in $\mathbb{R}^2$ of the form $e^{ikz}(1+r_\sigma)$, where $r_\sigma(z,k)$ has the required properties. A similar statement is valid for $u_{1/\sigma}$. This proves the existence of solutions. Uniqueness will be shown in the end of this proof.

Proposition 5.5.2 and (5.22) show that $k \mapsto r_\sigma$ is a C^∞ map from $\mathbb{C}$ to $W^{1,p}(\mathbb{R}^2)$ for some $p > 2$. Computing the $\partial_{\bar{k}}$ derivative gives

$$\partial_{\bar{k}} u_\sigma = e^{ikz}\left[-i\bar{z}e_{-k}\frac{\bar{\eta}_\mu - \bar{\eta}_{-\mu}}{2} + \partial_{\bar{k}}\left(\frac{\eta_\mu + \eta_{-\mu}}{2}\right) + e_{-k}\partial_{\bar{k}}\left(\frac{\bar{\eta}_\mu - \bar{\eta}_{-\mu}}{2}\right)\right].$$

The expression (5.21) implies that

$$\bar{z}\frac{\bar{\eta}_\mu - \bar{\eta}_{-\mu}}{2} = \frac{\overline{b_1^\mu(k) - b_1^{-\mu}(k)}}{2} + \tilde{r}(z,k),$$

where $\tilde{r}(\,\cdot\,,k) \in W^{1,2}_{\mathrm{loc}}(\mathbb{R}^2)$ and $\tilde{r} = O(1/z)$ as $z \to \infty$. Defining $\tau_\sigma(k)$ by

$$\overline{\tau_\sigma(k)} = \frac{1}{2}(b_1^\mu(k) - b_1^{-\mu}(k)),$$

we obtain

$$\partial_{\bar{k}} u_\sigma = e^{-i\overline{kz}}\left[-i\tau_\sigma(k) + \hat{r}(z,k)\right]$$

with $\hat{r}(\,\cdot\,,k) \in W^{1,2}_{\mathrm{loc}}(\mathbb{R}^2)$ and $\hat{r} = O(1/z)$ as $z \to \infty$.

On the other hand, since u_σ solves $\nabla \cdot \sigma\nabla u_\sigma = 0$ in $\mathbb{R}^2$ and since it is smooth in k, also the function $\partial_{\bar{k}} u_\sigma$ solves the same equation. Thus we have two solutions of the equation $\nabla \cdot \sigma\nabla u = 0$ given by

$$\overline{\partial_{\bar{k}} u_\sigma} = e^{ikz}\left[i\overline{\tau_\sigma(k)} + O(1/z)\right],$$
$$i\overline{\tau_\sigma(k)}u_\sigma = e^{ikz}\left[i\overline{\tau_\sigma(k)} + O(1/z)\right],$$

as $z \to \infty$. The uniqueness statement for these solutions implies that they have to be equal, thus resulting in the $\partial_{\bar{k}}$ equation

$$\partial_{\bar{k}} u_\sigma(z,\,\cdot\,) = -i\tau_\sigma(\,\cdot\,)\overline{u_\sigma(z,\,\cdot\,)}.$$

It remains to prove uniqueness. Let $u_1 = e^{ikz}(1 + r_1)$ and $u_2 = e^{ikz}(1 + r_2)$ be two solutions of $\nabla \cdot (\sigma\nabla u_j) = 0$ with the given properties. By the higher integrability result in Proposition 5.4.8, we have $u_j \in W^{1,p}_{\mathrm{loc}}(\mathbb{R}^2)$ for some $p > 2$. Since $r_j = O(1/z)$ as $z \to \infty$, it follows that $r_j \in W^{1,p}(\mathbb{R}^2)$. Then $u := u_1 - u_2 = e^{ikz}r$ is another solution where $r := r_1 - r_2 \in W^{1,p}$. We wish to show that $u \equiv 0$. Since σ is real, also $\mathrm{Re}(u)$ and $\mathrm{Im}(u)$ are solutions and they are of the form $e^{ikz}r$ where $r \in W^{1,p}$. Thus we may assume that u is real and $u = e^{ikz}r$, where $r \in W^{1,p}$.

Since $\nabla \cdot (\sigma\nabla u) = 0$, we can use the argument in Proposition 5.2.1 together with the version of Poincaré lemma given in Lemma 5.6.5 to find a real valued conjugate function $v \in W^{1,p}_{\mathrm{loc}}(\mathbb{R}^2)$ solving $\nabla \cdot (\sigma^{-1}\nabla v) = 0$ such that

$$\nabla v = (-\sigma\partial_2 u, \sigma\partial_1 u).$$

Note that v is unique up to an additive constant. As in Proposition 5.2.1, $f := u + iv \in W^{1,p}_{\mathrm{loc}}(\mathbb{R}^2)$ will solve

(5.23) $$\bar\partial f = \mu\overline{\partial f} \text{ in } \mathbb{R}^2, \qquad \mu = \frac{1-\sigma}{1+\sigma}.$$

We write $m := e^{-ikz} f$ and $s := e^{-ikz} v$. Then $m = r + is$ solves

$$\bar\partial m = \mu e_{-k}\overline{(\partial + ik)m}.$$

Since the right-hand side is in $L^p_c(\mathbb{R}^2)$, we may write

$$m = m_0 + h_0,$$

where $m_0 = P(\mu e_{-k}\overline{(\partial + ik)m}) \in W^{1,p}(\mathbb{R}^2)$ by Proposition 5.3.1, and h_0 is holomorphic ($\bar\partial h_0 = 0$). We have $m = r + is = m_0 + h_0$ with $m_0, r \in W^{1,p}$, so

(5.24) $$s = s_0 + h,$$

where $s_0 \in W^{1,p}(\mathbb{R}^2)$ and $\bar\partial h = 0$.

Now, since v is real valued and $v = e^{ikz} s$, we obtain the important property

$$\bar s = e_k s.$$

Taking complex conjugates in (5.24) gives $e_k s = \bar s_0 + \bar h$, or in other words

(5.25) $$s = s_1 + e_{-k}\bar h,$$

where $s_1 \in W^{1,p}(\mathbb{R}^2)$. We write

$$w := h - e_{-k}\bar h.$$

From (5.24) and (5.25) we get $w \in W^{1,p}(\mathbb{R}^2)$, and since $\bar\partial h = 0$ we also have

$$\bar\partial\,[(\partial + ik)w] = 0.$$

Since $(\partial + ik)w \in L^p(\mathbb{R}^2)$, the Liouville theorem (Exercise 5.4.2) implies that $(\partial + ik)w = 0$. Now $(\partial + ik)w = e_{-k}\partial(e_k w) = 0$, so another application of the Liouville theorem gives $w = 0$. Thus $h = e_{-k}\bar h$, or in other words

$$e^{ikz} h = \overline{e^{-ikz}}\bar h.$$

The left-hand side is holomorphic and the right-hand side is antiholomorphic, so $e^{ikz}h$ has vanishing gradient and one must have $e^{ikz}h = c_0$ for some constant $c_0 \in \mathbb{C}$.

Going back to (5.24) gives that

$$v = e^{ikz} s = e^{ikz} s_0 + c_0$$

for some $s_0 \in W^{1,p}(\mathbb{R}^2)$. Replacing v by $v - c_0$, we see that $v = e^{ikz} s_0$ and thus $f = e^{ikz}(r + is_0)$ with $r, s_0 \in W^{1,p}(\mathbb{R}^2)$. Since $\bar\partial(r + is_0) = \bar\partial(e^{-ikz} f) = 0$ outside $\overline{\mathbb{D}}$, the argument in the end of proof of Lemma 5.4.3 gives that $r + is_0 = O(1/z)$ as $z \to \infty$. Now the uniqueness part of Proposition 5.4.1 shows that $f \equiv 0$, so also $u \equiv 0$. This concludes the uniqueness proof. □

Remark 5.6.4. The uniqueness argument in Proposition 5.6.1 is related to the absence of exceptional points $k \in \mathbb{C}$ of a related integral equation (compare with [**Nac96**]). Another possible proof of uniqueness would follow by applying Stoilow factorization and writing the solution f of (5.23) as $f = h \circ \varphi$, where h is an entire holomorphic function and φ is a principal solution of the Beltrami equation (see [**AIM09**, §5.3 and §5.5]). The form $u = e^{ikz} r$ with $r \in W^{1,p}$ implies $|\operatorname{Re}(h(z))| = o(|e^{ikz}|)$ as $z \to \infty$, and the Poisson integral formula for $\operatorname{Re}(h(z))$ implies $|h'(z)| = o(|e^{ikz}|)$ as $z \to \infty$. Then $e^{-ikz}h'(z)$ is an analytic function in C_0, hence zero by the Liouville theorem, which implies that f is constant.

The following version of the Poincaré lemma was used in the uniqueness proof above.

Lemma 5.6.5 (Poincaré lemma for L^p_{loc} vector fields). *Let $F \in L^p_{\mathrm{loc}}(\mathbb{R}^2, \mathbb{R}^2)$ for some p with $2 < p < \infty$. Suppose that*

$$\partial_1 F_2 - \partial_2 F_1 = 0.$$

Then there is $v \in W^{1,p}_{\mathrm{loc}}(\mathbb{R}^2, \mathbb{R})$ such that

$$F = \nabla v.$$

One possible proof is based on the next Poincaré inequality.

Lemma 5.6.6 (Poincaré inequality). *Let $n = 2$, $2 < p \le \infty$ and $R > 0$. There is $C_p > 0$ such that*

$$\|v - v(0)\|_{L^p(B(0,R))} \le C_p R \, \|\nabla v\|_{L^p(B(0,R))}, \qquad v \in C^\infty(\overline{B(0,R)}).$$

Exercise 5.6.7. Prove Lemma 5.6.6 by using the identity

$$v(x) - v(0) = \int_0^1 \frac{\partial}{\partial t} v(tx)\, dt.$$

It is enough to give the proof for $R = 1$ and then use scaling.

Exercise 5.6.8. Prove Lemma 5.6.5. (Hint: use mollifications $F^\varepsilon = F * \varphi_\varepsilon$ as in Theorem 2.1.4, the Poincaré lemma for C^∞ vector fields in Exercise 5.1.6 to find v_ε with $\nabla v_\varepsilon = F^\varepsilon$ and $v_\varepsilon(0) = 0$, and the Poincaré inequality in Lemma 5.6.6 to show that $v_\varepsilon|_{B(0,R)}$ forms a Cauchy sequence in $W^{1,p}(B(0,R))$ for any $R > 0$.)

For the proof of the bound (5.16) for the scattering transform, we first record a number of useful properties of the functions

$$M_{\pm\mu}(z,k) := e^{-ikz} f_{\pm\mu}.$$

We will make use of the next elementary fact and also the classical Schwarz lemma from complex analysis.

Exercise 5.6.9. If $z, w \in \mathbb{C}$, $w \neq 0$, and $\operatorname{Re}(z/w) > 0$, then one has $z + w \neq 0$ and

$$\left|\frac{z-w}{z+w}\right| < 1.$$

Exercise 5.6.10 (Schwarz lemma)**.** Prove that if h is analytic in $\mathbb{D}$ with $h(0) = 0$ and $|h(z)| \leq 1$ in $\mathbb{D}$, then $|h(z)| \leq |z|$ in $\mathbb{D}$.

Lemma 5.6.11. *The function $M_{\pm\mu}$ is nonvanishing. Moreover,*

$$M_{\pm\mu}(z,k) = e^{\beta_\pm(z,k)},$$

where, for some $p > 2$, $\beta_\pm(\,\cdot\,,k) \in W^{1,p}(\mathbb{R}^2)$ for each k. One also has

$$\operatorname{Re}\left(\frac{M_\mu}{M_{-\mu}}\right) > 0,$$

so that $M_\mu + M_{-\mu} \neq 0$ everywhere and

$$\left|\frac{M_\mu - M_{-\mu}}{M_\mu + M_{-\mu}}\right| < 1, \qquad \left|\frac{f_\mu - f_{-\mu}}{f_\mu + f_{-\mu}}\right| < 1. \tag{5.26}$$

If $|z| \geq 1$ then

$$\left|\frac{M_\mu - M_{-\mu}}{M_\mu + M_{-\mu}}\right| \leq \frac{1}{|z|}, \qquad \left|\frac{f_\mu - f_{-\mu}}{f_\mu + f_{-\mu}}\right| \leq \frac{1}{|z|}. \tag{5.27}$$

Proof. The Beltrami equation for f_μ immediately implies that

$$\bar\partial M_\mu = \mu e_{-k}\overline{\partial M_\mu} - i\bar k \mu e_{-k}\overline{M_\mu} \quad \text{in } \mathbb{R}^2.$$

Since $M_\mu = 1 + \eta_\mu$, where $\eta_\mu(\,\cdot\,,k) \in W^{1,p}(\mathbb{R}^2)$, the function $M_\mu(\,\cdot\,,k)$ is in $W^{1,2}_{\mathrm{loc}}$ and bounded for any $k \in \mathbb{C}$. Lemma 5.4.6 shows that $M_\mu = Ce^\beta$ for some constant C and some $\beta \in W^{1,p}(\mathbb{R}^2)$ for some $p > 2$. Since $M_\mu \to 1$ as $z \to \infty$, we have $C = 1$. The same proof works for $M_{-\mu}$.

To show the positivity of $\operatorname{Re}(M_\mu/M_{-\mu})$, assume on the contrary that the real part is nonpositive at some point in $\mathbb{C}$. Since $M_\mu/M_{-\mu}$ is continuous and has limit 1 as $z \to \infty$, it is not possible that $\operatorname{Re}(M_\mu/M_{-\mu}) < 0$ everywhere. Thus, $\operatorname{Re}(M_\mu/M_{-\mu}) = 0$ for some $z_0 \in \mathbb{C}$, so there is $t \in \mathbb{R}$ with

$$M_\mu(z_0,k) = itM_{-\mu}(z_0,k).$$

Writing $h(z,k) = M_\mu(z,k) - itM_{-\mu}(z,k)$, the equations for $M_{\pm\mu}$ show that

$$\bar\partial h = \mu e_k \overline{\partial h} - i\bar k \mu e_{-k}\bar h \quad \text{in } \mathbb{R}^2.$$

Now h is bounded, so by Lemma 5.4.6 we have $h = Ce^\beta$. Using that $h(z_0) = 0$ we must have $C = 0$, so $M_\mu = itM_{-\mu}$. This contradicts the fact that the limit of $M_{\pm\mu}$ as $z \to \infty$ is 1.

We have proved that $\mathrm{Re}(M_\mu/M_{-\mu}) > 0$. Consequently $M_\mu + M_{-\mu}$ is nonvanishing, and the conditions (5.26) follow from Exercise 5.6.9 and the fact that $f_{\pm\mu} = e^{ikz}M_{\pm\mu}$. If $|z| \geq 1$, define

$$m(z,k) := \frac{M_\mu - M_{-\mu}}{M_\mu + M_{-\mu}}.$$

By (5.26) we have

$$|m(z,k)| < 1.$$

The function $h(z) := m(1/z,k)$ is then analytic in $\mathbb{D}$, satisfies $h(0) = 0$ by the asymptotics for $M_\mu = 1+\eta_\mu$, and $|h(z)| < 1$ in $\overline{\mathbb{D}}$. The Schwarz lemma (Exercise 5.6.10) implies that $|h(z)| \leq |z|$. Consequently $|m(z,k)| \leq 1/|z|$, which shows the bound (5.27). □

The previous result implies the useful fact that the solution u_σ never vanishes.

Proposition 5.6.12. *The function $u_\sigma(z,k)$ in Proposition 5.6.1 satisfies $u_\sigma(z,k) \neq 0$ for all $z, k \in \mathbb{C}$.*

Proof. We argue by contradiction and assume that $u_\sigma(z_0,k_0) = 0$ for some $z_0, k_0 \in \mathbb{C}$. Since

$$u_\sigma(z_0,k_0) = \mathrm{Re}(f_\mu(z_0,k_0)) + i\,\mathrm{Im}(f_{-\mu}(z_0,k_0))$$

it follows that $f_\mu(z_0,k_0) = it$ and $f_{-\mu}(z_0,k_0) = s$ for some $t, s \in \mathbb{R}$. But then

$$1 = \left|\frac{it-s}{it+s}\right| = \left|\frac{f_\mu(z_0,k_0) - f_{-\mu}(z_0,k_0)}{f_\mu(z_0,k_0) + f_{-\mu}(z_0,k_0)}\right|.$$

The last quantity is < 1 by (5.26). This is not possible, so u_σ must be nonvanishing. □

We can now prove the basic properties of the scattering transform.

Proof of Proposition 5.6.3. In the proof of Proposition 5.6.1 we defined $\tau_\sigma(k)$ so that (5.18) is satisfied, and the expression (5.19) follows immediately from (5.21) and the definition of $f_{\pm\mu}$. Since Proposition 5.5.1 shows that Λ_σ determines $f_{\pm\mu}$ in $\mathbb{R}^2 \setminus \mathbb{D}$, we see from (5.19) that Λ_σ determines $b_1^{\pm\mu}(k)$ and thus $\tau_\sigma(k)$ for all k. The property (5.17) is a direct consequence of (5.18) since the conductivity $1/\sigma$ corresponds to Beltrami coefficient $-\mu$. Also, the fact that τ_σ is smooth with respect to k follows since $b_1^{\pm\mu}$ are smooth, using the Fourier coefficient definition (5.20) and Proposition 5.5.2.

It remains to prove the bound (5.16). By (5.18) and (5.19) the scattering transform can be expressed in terms of the functions $M_{\pm\mu}$ as

$$\overline{\tau_\sigma(k)} = \lim_{z\to\infty} z\frac{M_\mu - M_{-\mu}}{2} = \lim_{z\to\infty} z\frac{M_\mu - M_{-\mu}}{M_\mu + M_{-\mu}}.$$

Here we used that $M_{\pm\mu} \to 1$ as $z \to \infty$. Now the bound (5.16) follows from (5.27). $\square$

To explain the origin of the terminology for $\tau_\sigma(k)$, we describe briefly and in a formal way a one-dimensional scattering problem where similar concepts appear. The multidimensional case will be discussed in more detail in Chapter 7.

Consider an infinite string whose displacement at time t is given by the function $u(x,t)$. If the string is homogeneous, the displacement function solves the wave equation

$$(\partial_t^2 - \partial_x^2)u = 0.$$

A particular solution is given by the one-dimensional plane wave

$$u_0(t,x) = \delta(t-x).$$

This is a Dirac delta function moving to the right, whose peak at time t is at the point $x = t$.

Consider now an inhomogeneous string, whose displacement function solves the perturbed wave equation

$$(\partial_t^2 - \partial_x^2 + q)u = 0.$$

The inhomogeneity is given by the potential $q = q(x) \in C_c^\infty(\mathbb{R})$, and we assume $\mathrm{supp}(q) \subset [-R,R]$. The free plane wave $u_0(x,t) = \delta(t-x)$ solves this equation if $t < -R$, and uniqueness in the Cauchy problem (see, e.g., [**Hör85a**, §23.2]) implies that there is a unique solution $u_q(x,t)$ satisfying

$$(\partial_t^2 - \partial_x^2 + q)u_q = 0, \quad u_q = u_0 \text{ when } t < -R. \tag{5.28}$$

The function u_q is called the (outgoing) distorted plane wave corresponding to u_0.

To describe the distorted plane waves in more detail, we take Fourier transforms formally with respect to t in (5.28). For a precise justification of the following steps see, e.g., [**RS20b**]. Writing

$$\psi_q(x,k) := \int_{-\infty}^{\infty} e^{ikt} u_q(x,t)\,dt,$$

$$\psi_0(x,k) := \int_{-\infty}^{\infty} e^{ikt} u_0(x,t)\,dt = e^{ikx},$$

this results in the equation

$$(-\partial_x^2 - k^2 + q)\psi_q = 0 \text{ in } \mathbb{R}.$$

The condition $u_q - u_0 = 0$ for $t < -R$ is transformed into analyticity of $\psi_q - \psi_0$ for $\mathrm{Im}(k) > 0$. The unique solution $\psi_q(x,k)$ satisfying this analyticity condition can be explicitly written in terms of the resolvent operator $(-\partial_x^2-(k+i0)^2+q)^{-1}$ as $\psi_q = \psi_0 + r_q$, where

$$r_q = (-\partial_x^2 - (k+i0)^2 + q)^{-1}(-q\psi_0). \tag{5.29}$$

Resolvents are studied in more detail in Chapter 7. In the present one-dimensional case, solving the equation $-u'' - k^2 u = f$, where $\mathrm{Im}(k) \geq 0$, $k \neq 0$, and f is compactly supported yields the formula

$$(-\partial_x^2 - (k+i0)^2)^{-1} f(x) = \frac{i}{2k}\int_{-\infty}^{\infty} e^{ik|x-y|} f(y)\,dy. \tag{5.30}$$

The functions ψ_0 and ψ_q are also-called plane waves (even though they are the time Fourier transforms of actual solutions of the wave equation). They are of fundamental importance in scattering theory, and can be used to parametrize the generalized eigenfunctions of the operators $-\partial_x^2$ and $-\partial_x^2 + q$ and also to study various scattering measurements. The family of exponentials $\psi_0(x,k) = e^{ikx}$ is related to the usual Fourier transform

$$\hat{f}(k) = \int_{-\infty}^{\infty} f(x)\overline{\psi_0(x,k)}\,dx.$$

Similarly, the family $\psi_q(x,k)$ gives rise to a *distorted Fourier transform*

$$\mathcal{F}_q\{f\}(k) := \int_{-\infty}^{\infty} f(x)\overline{\psi_q(x,k)}\,dx.$$

Counterparts of the inversion and Plancherel formulas are valid also for the distorted transform $\mathcal{F}_q$ in any dimension (see [**Agm75**, §6]).

Finally, we discuss an analogy between the scattering transform $\tau_\sigma(k)$ and the above concepts. Consider the inverse backscattering problem of determining the potential q from the measurements

$$\lim_{x\to-\infty} u_q(x,t), \quad t \in \mathbb{R}.$$

This corresponds to sending a free plane wave from $x = -\infty$ and then measuring the response also at $x = -\infty$. From these measurements one can formally determine $\lim_{x\to-\infty}\psi_q(x,k)$ for any k by taking the Fourier transform in time. The equation for r_q in (5.29) implies that

$$(-\partial_x^2 - k^2) r_q = -q\psi_0 - q r_q = -q\psi_q,$$

which yields the *Lippmann-Schwinger equation*

$$\psi_q = \psi_0 + (-\partial_x^2 - (k+i0)^2)^{-1}(-q\psi_q).$$

Applying (5.30) and the formula

$$|x-y| = |x|\sqrt{1 - 2xy/|x|^2 + |y|^2/|x|^2} = |x|(1 - xy/|x|^2) + O(|x|^{-1}),$$

which holds uniformly for $|y| \leq R$ as $|x| \to \infty$, gives the asymptotics as $x \to -\infty$,

$$\psi_q(x,k) = \psi_0(x,k) - \frac{i}{2k} e^{-ikx} \int_{-\infty}^{\infty} e^{iky} q(y) \psi_q(y,k)\, dy + O(|x|^{-1}).$$

Consequently, from the measurements of ψ_q as $x \to -\infty$ one determines the *backscattering transform* of q,

$$B_q(k) := \int_{-\infty}^{\infty} e^{iky} q(y) \psi_q(y,k)\, dy, \quad k \in \mathbb{R}.$$

Note that in the Born approximation $\psi_q \approx \psi_0$, one recovers the usual Fourier transform

$$B_q(k) \approx \int_{-\infty}^{\infty} e^{2iky} q(y)\, dy.$$

We have shown that the backscattering measurements for the equation

$$(-\partial_x^2 - k^2 + q)\psi = 0$$

determine a nonlinear Fourier transform of q, via the special solutions ψ_q, by

$$B_q(k) = \lim_{x \to -\infty} 2ike^{ikx}(\psi_q(x,k) - \psi_0(x,k)), \quad k \in \mathbb{R}.$$

Analogously, boundary measurements for the two-dimensional conductivity equation

$$\nabla \cdot \sigma \nabla u = 0$$

(or equivalently the Beltrami equation $\bar{\partial} f = \mu \overline{\partial f}$) determine, via the special CGO solutions $f_{\pm\mu}$, the nonlinear Fourier transform of σ,

$$b_1^{\pm\mu}(k) = \lim_{z \to \infty} ze^{-ikz}(f_{\pm\mu}(z,k) - f_0(z,k)), \quad k \in \mathbb{C},$$

$$\tau_\sigma(k) = \frac{1}{2}\overline{(b_1^{\mu}(k) - b_1^{-\mu}(k))}.$$

This analogy motivates calling τ_σ the scattering transform or nonlinear Fourier transform of the conductivity σ.

5.7. Uniqueness for C^2 conductivities

In this section we temporarily add the assumption that the conductivity is C^2 and finish the uniqueness proof of the Calderón problem. The more difficult case of bounded measurable conductivities will be dealt with in the following sections.

Theorem 5.7.1 (Uniqueness for C^2 conductivities, $n = 2$). *Let σ_1, σ_2 be two positive functions in $C^2(\overline{\mathbb{D}})$ such that $\sigma_1 = \sigma_2 = 1$ near $\partial\mathbb{D}$. If*

$$\Lambda_{\sigma_1} = \Lambda_{\sigma_2}$$

then

$$\sigma_1 = \sigma_2.$$

In fact the argument below works for $\sigma_j \in W^{2,\infty}(\mathbb{D})$ positive. After a reduction involving boundary determination and Proposition 2.8.7, this implies an analogous result on any C^2 domain.

Exercise 5.7.2. If Ω is a bounded domain in $\mathbb{R}^2$ with C^2 boundary and $\sigma_1, \sigma_2 \in W^{2,\infty}(\Omega)$ are positive functions with $\Lambda_{\sigma_1} = \Lambda_{\sigma_2}$, show that $\sigma_1 = \sigma_2$.

Assume that σ_j are extended as functions in $C^2(\mathbb{R}^2)$ so that $\sigma_j = 1$ outside of $\mathbb{D}$. Even for bounded measurable conductivities, up to this point we have shown that

$$\Lambda_{\sigma_1} = \Lambda_{\sigma_2} \implies \mathcal{H}_{\pm\mu_1} = \mathcal{H}_{\pm\mu_2} \implies f_{\pm\mu_1}|_{\mathbb{C}\setminus\mathbb{D}} = f_{\pm\mu_2}|_{\mathbb{C}\setminus\mathbb{D}} \implies \tau_{\sigma_1} = \tau_{\sigma_2},$$

where $\tau_\sigma(k)$ is the scattering transform appearing as a coefficient in the $\partial_{\bar{k}}$ equation in Proposition 5.6.1 (where $z \in \mathbb{C}$ is fixed):

$$\partial_{\bar{k}} u_\sigma(z, \cdot\,) = -i\tau_\sigma(\,\cdot\,)\overline{u_\sigma(z, \cdot\,)} \quad \text{in } \mathbb{R}^2.$$

Here u_σ is the unique solution of $\nabla \cdot \sigma \nabla u = 0$ in $\mathbb{R}^2$ of the form $u = e^{ikz}(1 + r)$, where $r \in W^{1,2}_{\mathrm{loc}}(\mathbb{R}^2)$ and $r = O(1/z)$ as $z \to \infty$.

In particular, writing

$$\tau(k) = \tau_{\sigma_1}(k) = \tau_{\sigma_2}(k),$$

the functions u_{σ_1} and u_{σ_2} both solve the same $\partial_{\bar{k}}$ equation for any fixed z:

$$\partial_{\bar{k}} u_{\sigma_1}(z, \cdot\,) = -i\tau(\,\cdot\,)\overline{u_{\sigma_1}(z, \cdot\,)} \quad \text{in } \mathbb{R}^2,$$

$$\partial_{\bar{k}} u_{\sigma_2}(z, \cdot\,) = -i\tau(\,\cdot\,)\overline{u_{\sigma_2}(z, \cdot\,)} \quad \text{in } \mathbb{R}^2.$$

Using that $u_\sigma = e^{ikz}(1 + r_\sigma)$, the equations become

$$\partial_{\bar{k}} r_{\sigma_1}(z, k) = -i\tau(k)e_{-k}(z)\overline{r_{\sigma_1}(z, k)} - i\tau(k)e_{-k}(z) \quad \text{for } k \in \mathbb{R}^2,$$

$$\partial_{\bar{k}} r_{\sigma_2}(z, k) = -i\tau(k)e_{-k}(z)\overline{r_{\sigma_2}(z, k)} - i\tau(k)e_{-k}(z) \quad \text{for } k \in \mathbb{R}^2.$$

Under the assumption that σ_j are C^2 functions, we will show that for any fixed $z \in \mathbb{C}$ solutions to the above equation are unique. The key point is that for C^2 conductivities, the remainder terms in CGO solutions and the scattering transform are bounded or decay as $k \to \infty$.

Proposition 5.7.3 (Decay in k). *Let $\sigma \in C^2(\mathbb{R}^2)$ with $\sigma = 1$ for $|z| > 1$. Then, for any fixed $z \in \mathbb{C}$,*

$$|r_\sigma(z, \cdot\,)| \le C_z,$$

$$\lim_{k\to\infty} r_\sigma(z, k) = \sigma^{-1/2}(z) - 1.$$

Furthermore, $\tau_\sigma \in L^{2\pm}(\mathbb{R}^2)$, and one has

$$|\tau_\sigma(k)| \le C(1 + |k|)^{-1}, \quad k \in \mathbb{C}.$$

Uniqueness will follow from the next Liouville type result, which differs from Lemma 5.4.3 (the case $\nu = 0$) by involving a coefficient α that is not compactly supported but lies in the space $L^{2\pm}$.

Proposition 5.7.4 (Liouville theorem). *Let $\alpha \in L^{2\pm}(\mathbb{R}^2)$. If $g \in W^{1,2}_{\mathrm{loc}}(\mathbb{R}^2)$ is a bounded solution of*

$$\bar\partial g = \alpha \bar g \quad \text{in } \mathbb{R}^2,$$

then $g = Ce^{\beta}$ for some constant C and some $\beta \in C_0(\mathbb{R}^2)$. If further $g(z) \to 0$ as $z \to \infty$, then $g = 0$.

Proof. We make the same reduction as in the proof of Lemma 5.4.3: define

$$\tilde\alpha(z) = \begin{cases} \alpha\frac{\bar g}{g}, & g(z) \neq 0, \\ 0, & g(z) = 0, \end{cases}$$

so $\tilde\alpha \in L^{2\pm}(\mathbb{R}^2)$ and g solves

$$\bar\partial g - \tilde\alpha g = 0 \quad \text{in } \mathbb{R}^2.$$

We use Proposition 5.3.1 to find an integrating factor $\beta \in C_0(\mathbb{R}^2)$ such that

$$\bar\partial \beta = \tilde\alpha \quad \text{in } \mathbb{R}^2.$$

Then

$$\bar\partial(e^{-\beta} g) = 0$$

in the weak sense. Since $e^{-\beta}g$ is bounded, the Liouville theorem implies that $g = Ce^{\beta}$. Further, if $g \to 0$ as $z \to \infty$, it follows that $g = 0$. □

Given these results, we can complete the proof of uniqueness for C^2 conductivities.

Proof of Theorem 5.7.1. We have seen that $\Lambda_{\sigma_1} = \Lambda_{\sigma_2}$ implies $\tau_{\sigma_1} = \tau_{\sigma_2}$. By the discussion in the beginning of this section, and writing $\tau = \tau_{\sigma_1} = \tau_{\sigma_2}$, for any fixed $z \in \mathbb{C}$ the function $r = r_{\sigma_1} - r_{\sigma_2}$ solves the equation

$$\partial_{\bar k} r(z,k) = -i\tau(k) e_{-k}(z)\overline{r(z,k)} \quad \text{for } k \in \mathbb{R}^2.$$

By Proposition 5.6.1, the function $r(z,\,\cdot\,)$ is C^∞. Then by Proposition 5.7.3 we have $r(z,\,\cdot\,) \in W^{1,2}_{\mathrm{loc}}(\mathbb{R}^2) \cap L^\infty(\mathbb{R}^2)$, and $\tau \in L^{2\pm}(\mathbb{R}^2)$. The Liouville theorem, Proposition 5.7.4, shows that

$$r(z,k) = C(z)e^{\beta(z,k)},$$

where $\beta(z,\,\cdot\,) \in C_0(\mathbb{R}^2)$.

We will evaluate the last condition at $k = 0$. Recall that

$$f_{\pm\mu_j}(z,0) = 1$$

and thus $u_{\sigma_j}(z,0) = \mathrm{Re}(f_\mu) + i\,\mathrm{Im}(f_{-\mu})|_{k=0} = 1$, so $r_{\sigma_j}(z,0) = 0$. Thus $C(z) = 0$ for all $z \in \mathbb{C}$. But also

$$0 = \lim_{k\to\infty} C(z)e^{\beta(z,k)} = \lim_{k\to\infty}(r_{\sigma_1}(z,k) - r_{\sigma_2}(z,k)) = \sigma_1^{-1/2}(z) - \sigma_2^{-1/2}(z).$$

This shows that $\sigma_1 = \sigma_2$. □

It remains to prove the estimates in Proposition 5.7.3. This is most conveniently done by reducing the conductivity equation to the Schrödinger equation, as we did in the uniqueness proof for the Calderón problem in three and higher dimensions in Chapter 4. Recall from Proposition 2.8.5 that

$$\nabla \cdot \sigma \nabla u = 0 \quad \Longleftrightarrow \quad (-\Delta + q_\sigma)\psi = 0,$$

where $u = \sigma^{-1/2}\psi$, and the potential q_σ is given by

$$q_\sigma = \frac{\Delta \sigma^{1/2}}{\sigma^{1/2}}.$$

The following result discusses CGO solutions for the Schrödinger equation.

Proposition 5.7.5 (CGO solutions for $-\Delta + q_\sigma$). *Let $\sigma \in C^2(\mathbb{R}^2)$ be positive with $\sigma = 1$ for $|z| > 1$. For any $k \in \mathbb{C}$, the equation*

$$(-\Delta + q_\sigma)\psi = 0 \quad \text{in } \mathbb{R}^2$$

has a unique solution $\psi_\sigma(z,k) = e^{ikz}(1 + s_\sigma(z,k))$ with $s_\sigma \in W^{1,2}_{\mathrm{loc}}(\mathbb{R}^2)$ and $s_\sigma(z,k) = O(1/z)$ as $z \to \infty$. The solution ψ_σ is related to u_σ by

$$\psi_\sigma = \sigma^{1/2}u_\sigma, \quad s_\sigma = \sigma^{1/2} - 1 + \sigma^{1/2}r_\sigma.$$

The map $k \mapsto s_\sigma$ is C^∞ from $\mathbb{C}$ to $W^{1,p}(\mathbb{R}^2)$ for some $p > 2$.

Proof. Existence follows from Proposition 5.6.1: if u_σ is the solution of $\nabla \cdot (\sigma \nabla u) = 0$ in $\mathbb{R}^2$ with $u_\sigma = e^{ikz}(1+r_\sigma)$, then $\psi_\sigma = \sigma^{1/2}u_\sigma$ solves $(-\Delta+q_\sigma)\psi_\sigma = 0$ in $\mathbb{R}^2$ and

$$\psi_\sigma = e^{ikz}(\sigma^{1/2} + \sigma^{1/2}r_\sigma) = e^{ikz}(1 + s_\sigma),$$

where s_σ is as above. For uniqueness, if $\psi = e^{ikz}(1+s)$ and $\tilde{\psi} = e^{ikz}(1+\tilde{s})$ are two solutions of $(-\Delta + q_\sigma)v = 0$ with $s, \tilde{s} = O(1/z)$ as $z \to \infty$, then $u = \sigma^{-1/2}\psi$ and $\tilde{u} = \sigma^{-1/2}\tilde{\psi}$ are two solutions of $\nabla \cdot \sigma \nabla v = 0$ and

$$u = e^{ikz}(1 + (\sigma^{-1/2} - 1 + \sigma^{-1/2}s)),$$
$$\tilde{u} = e^{ikz}(1 + (\sigma^{-1/2} - 1 + \sigma^{-1/2}\tilde{s})).$$

The uniqueness part of Proposition 5.6.1 implies that $s = \tilde{s}$. The smoothness of the map $k \mapsto s_\sigma$ follows from Proposition 5.5.2. □

Remark 5.7.6. One could also consider CGO solutions to $(-\Delta + q)u = 0$ of the same form as in Chapter 4, namely

$$u = e^{i\zeta\cdot x}(1 + r), \qquad \zeta \in \mathbb{C}^2, \quad \zeta \cdot \zeta = 0.$$

But if $\zeta \in \mathbb{C}^2$ satisfies $\zeta \cdot \zeta = 0$, then $\zeta_2^2 = -\zeta_1^2$ and consequently $\zeta_2 = \pm i\zeta_1$. Thus

$$e^{i\zeta\cdot x} = e^{i\zeta_1(x_1 \pm ix_2)}.$$

Writing $k = \zeta_1$ we see that $e^{i\zeta\cdot x}$ is equal to e^{ikz} or $e^{ik\bar{z}}$, and the latter case can be reduced to the case of e^{ikz} by complex conjugation. This shows that the CGO solutions in Proposition 5.7.5 are in fact the same as those considered in Chapter 4.

We will next show that the correction term $s_\sigma(z,k)$ in the Schrödinger solution goes to zero as $k \to \infty$, when z is kept fixed. To do this, we give a representation of s_σ in terms an inverse of the conjugated Laplacian. Note that

$$e^{-ikz}\Delta(e^{ikz}v) = 4\bar{\partial}e^{-ikz}\partial(e^{ikz}v) = 4\bar{\partial}(\partial + ik)v.$$

We will give suitable inverses for $\bar{\partial}$ and $\partial + ik$. The first result shows in particular that for $1 < p < 2$ the Cauchy transform P maps L^p to L^{p^*}, where p^* is the Sobolev conjugate exponent

$$p^* = \frac{2p}{2-p}.$$

Lemma 5.7.7 (The operators $\bar{\partial}^{-1}$ and ∂^{-1} on L^p). *Let $1 < p < 2$. For any $f \in L^p(\mathbb{R}^2)$, the equation*

$$\bar{\partial}u = f$$

has a unique solution $u = \bar{\partial}^{-1}f \in L^{p^*}(\mathbb{R}^2)$. Further, $\nabla u \in L^p(\mathbb{R}^2)$ and

$$\|u\|_{L^{p*}} + \|\nabla u\|_{L^p} \le C_p \|f\|_{L^p}.$$

The same result holds with $\bar{\partial}$ replaced by ∂.

Proof. Recall from §5.3 that $Pf = \frac{1}{\pi z} * f$. By the Hardy-Littlewood-Sobolev inequality [**Hör03a**, Theorem 4.5.3], the map $f \mapsto \frac{1}{|z|} * f$ is bounded $L^p(\mathbb{R}^2) \to L^{p^*}(\mathbb{R}^2)$. This implies that for any $f \in L^p$, $Pf(z)$ is well defined for a.e. $z \in \mathbb{R}^2$ and satisfies $\|Pf\|_{L^{p*}} \le C\|f\|_{L^p}$. For any $f \in C_c^\infty(\mathbb{R}^2)$, Propositions 5.3.1 and 5.3.3 yield the identities

$$\bar{\partial}Pf = f,$$
$$\partial Pf = Sf.$$

Since S is bounded on L^p and C_c^∞ is dense in L^p, the B.L.T. theorem implies that these identities remain true for $f \in L^p$.

We have proved that $u = Pf$ solves $\bar{\partial}u = f$ and satisfies the required norm estimates. If u and $\tilde{u}$ are two solutions in L^{p^*}, then $\bar{\partial}(u - \tilde{u}) = 0$ in $\mathbb{R}^2$ and $u - \tilde{u} \in L^{p^*}$. The Liouville theorem (Exercise 5.4.2) implies that $u = \tilde{u}$. Thus $\bar{\partial}^{-1} = P$ satisfies the required properties, and a corresponding result for ∂^{-1} follows by taking complex conjugates. □

Lemma 5.7.8 (The operator $(\partial + ik)^{-1}$). *Let $k \in \mathbb{C}$ and let $1 < p < 2$.*

(a) *For any $f \in L^p(\mathbb{R}^2)$, the equation*

$$(\partial + ik)u = f$$

has a unique solution $u = (\partial + ik)^{-1} f \in L^{p^*}(\mathbb{R}^2)$. Further, $\nabla u \in L^p(\mathbb{R}^2)$ and

$$\|u\|_{L^{p*}} \leq C_p \|f\|_{L^p}.$$

(b) *If $v \in L^{p^*}$, $\nabla v \in L^p$ and $k \neq 0$, the equation $(\partial + ik)u = v$ has a unique solution $u \in W^{1,p^*}$ of the form*

$$u = \frac{1}{ik}(v - (\partial + ik)^{-1}\partial v).$$

Proof.

(a) We observe that $\partial + ik = e^{-ikz}\partial e^{ikz} = e_{-k}(z)\partial e_k(z)$, where $e_k(z) = e^{i(kz+\overline{kz})}$ is unimodular. Then existence, uniqueness and norm estimates for u reduce to Lemma 5.7.7.

(b) If $v \in L^{p^*}$, $\nabla v \in L^p$ and $k \neq 0$, the function $u = \frac{1}{ik}(v - (\partial + ik)^{-1}\partial v)$ is in L^{p^*} with $\nabla u \in L^p$ by (a). A direct computation yields $(\partial + ik)u = v$. Then also $\partial u = v - iku \in L^{p^*}$, and one has $\bar{\partial} u = \bar{S}\partial u \in L^{p^*}$ by the mapping properties of the Beurling transform in Proposition 5.3.3. □

The following result gives a solution operator G_k to the Laplace equation conjugated by complex exponentials, somewhat similar to the one in Proposition 4.5.4 for $n \geq 3$.

Proposition 5.7.9 (Solution operator G_k). *Let $k \in \mathbb{C} \setminus \{0\}$ and $1 < p < 2$. For any $f \in L^p(\mathbb{R}^2)$ the equation*

$$e^{-ikz}\Delta(e^{ikz}u) = f$$

has a unique solution $u = G_k f \in W^{1,p^*}(\mathbb{R}^2)$ given by

$$G_k f = \frac{1}{4ik}(\bar{\partial}^{-1} f - (\partial + ik)^{-1}(\partial\bar{\partial}^{-1} f)).$$

Further,

$$\|u\|_{L^{p*}} \leq \frac{C_p}{|k|} \|f\|_{L^p}, \quad \|\nabla u\|_{L^{p*}} \leq C_p \|f\|_{L^p}.$$

Proof. The equation reads

$$4\bar{\partial}(\partial + ik)u = f.$$

For existence, we first use Lemma 5.7.7 and choose

$$v = \bar{\partial}^{-1}(f/4).$$

Then $\|v\|_{L^{p^*}} + \|\nabla v\|_{L^p} \le C_p \|f\|_{L^p}$. Now use Lemma 5.7.8(b) to choose a solution $u \in W^{1,p^*}$ of

$$(\partial + ik)u = v$$

having the form

$$u = \frac{1}{ik}(v - (\partial + ik)^{-1}\partial v).$$

It follows that

$$\|u\|_{L^{p^*}} \le \frac{C_p}{|k|}(\|v\|_{L^{p^*}} + \|\partial v\|_{L^p}) \le \frac{C_p}{|k|}\|f\|_{L^p}.$$

Moreover, $\partial u = v - iku \in L^{p^*}$ with

$$\|\partial u\|_{L^{p^*}} = \|v - iku\|_{L^{p^*}} \le C_p \|f\|_{L^p}$$

and by Proposition 5.3.3

$$\left\|\bar\partial u\right\|_{L^{p^*}} = \left\|\bar S \partial u\right\|_{L^{p^*}} \le C_p \|\partial u\|_{L^{p^*}} \le C_p \|f\|_{L^p}.$$

This shows the existence of a solution u with the required properties.

For uniqueness, suppose that $u \in W^{1,p^*}$ solves

$$4\bar\partial(\partial + ik)u = 0.$$

Since $(\partial+ik)u \in L^{p^*}$, the uniqueness part of Lemma 5.7.7 shows that $(\partial+ik)u = 0$. Uniqueness in Lemma 5.7.7 again gives $u = 0$. □

Proposition 5.7.10 (CGO solutions for large k). *Let $2 < r < \infty$. There is $C > 0$ such that for $|k|$ sufficiently large, the solution $\psi = e^{ikz}(1+s)$ in Proposition 5.7.5 satisfies*

$$\|s\|_{L^r} \le \frac{C}{|k|}, \quad \|s\|_{W^{1,r}} \le C$$

and

$$\|s\|_{L^\infty} \le \frac{C}{|k|}.$$

Proof. Let $1 < p < 2$. Then $\psi = e^{ikz}(1+s)$ with $s \in W^{1,p^*}$ solves $(-\Delta+q_\sigma)\psi = 0$ if and only if

$$e^{-ikz}\Delta(e^{ikz}s) - q_\sigma s = q_\sigma.$$

By Proposition 5.7.9 this is equivalent with the integral equation

$$s - G_k(q_\sigma s) = G_k q_\sigma. \tag{5.31}$$

Here we used that q_σ is supported in $\overline{\mathbb{D}}$ and $s \in L^{p^*}$, so $q_\sigma s \in L^p$. The estimates for G_k imply that

$$\|G_k q_\sigma f\|_{L^{p^*}} \le \frac{C_p}{|k|}\|q_\sigma f\|_{L^p} \le \frac{C_p\|q_\sigma\|_{L^\infty}}{|k|}\|f\|_{L^{p^*}}.$$

If $|k| \geq 2C_p\|q_\sigma\|_{L^\infty}$, a Neumann series argument shows that $\mathrm{Id} - G_k(q_\sigma\,\cdot\,)$ is invertible on L^{p^*} and its inverse has norm ≤ 2. It follows that for $|k|$ large the equation (5.31) has a unique solution s, and

$$\|s\|_{L^{p^*}} \leq 2\|G_k q_\sigma\|_{L^{p^*}} \leq \frac{C}{|k|}.$$

To obtain the remaining estimates, we observe that $s = G_k f$ where $f = q_\sigma(1+s)$. The W^{1,p^*} estimate for G_k gives

$$\|\nabla s\|_{L^{p^*}} \leq C.$$

For the L^∞ estimate we use the expression for G_k in Proposition 5.7.9, so that

$$s = G_k f = \frac{1}{4ik}(\bar{\partial}^{-1} f - (\partial + ik)^{-1}(\partial\bar{\partial}^{-1} f)).$$

Fix some p with $1 < p < 2$. Now $f = q_\sigma(1+s)$ where $s \in L^r$ whenever $2 < r < \infty$, which implies that $f \in L^p \cap L^{p'}$. The L^p bounds for the Beurling transform in Proposition 5.3.1 imply that also $\partial\bar{\partial}^{-1} f \in L^p \cap L^{p'}$. By Proposition 5.3.1 (see (5.9)) we have

$$\|\bar{\partial}^{-1} f\|_{C_0} \leq C_p(\|f\|_{L^p} + \|f\|_{L^{p'}}).$$

Since $(\partial + ik)^{-1} g = e_{-k}\partial^{-1} e_k g$ with $g = \partial\bar{\partial}^{-1} f$, also the second term in $G_k f$ satisfies the same estimate. This proves that

$$\|s(\,\cdot\,, k)\|_{L^\infty} \leq C|k|^{-1}$$

whenever $|k| \geq 2C_p\|q_\sigma\|_{L^\infty}$. □

The estimates for r_σ in Proposition 5.7.3 now follow from Proposition 5.7.10, since $s_\sigma = \sigma^{1/2} - 1 + \sigma^{1/2} r_\sigma$ and $r_\sigma(z, \,\cdot\,)$ is C^∞ for fixed z. It remains to prove the decay estimates for the scattering transform τ_σ.

Recall from Proposition 5.6.12 that u_σ is never zero. The formula

$$\tau_\sigma(k) = i\frac{\partial_{\bar{k}} u_\sigma(z,k)}{\overline{u_\sigma(z,k)}},$$

which is valid for any fixed z, implies that

$$\tau_\sigma(k) = ie_k(z)\frac{\partial_{\bar{k}} r_\sigma(z,k)}{\overline{1 + r_\sigma(z,k)}}.$$

The relation between r_σ and s_σ further yields

$$\tau_\sigma(k) = ie_k(z)\frac{\partial_{\bar{k}} s_\sigma(z,k)}{\overline{1 + s_\sigma(z,k)}}. \tag{5.32}$$

Since we know that $s_\sigma(z,k) \leq C/|k|$ for large $|k|$, it is sufficient to determine the asymptotics of $\partial_{\bar{k}} s_\sigma(z,k)$ as $|k| \to \infty$. From (5.31) we have

$$s_\sigma = G_k(q_\sigma(1 + s_\sigma)),$$

where G_k has the formula in Proposition 5.7.9.

The discussion above shows that the key point is to obtain estimates for $\partial_{\bar{k}}((\partial+ik)^{-1}f)$ with suitable f. This will be done in Lemma 5.7.13. The lemma involves the Fourier transform with a slightly different notation.

Definition 5.7.11 (Fourier transform $\tilde{f}$). If $f \in L^p(\mathbb{R}^2)$ with $1 \leq p \leq 2$, we define

$$\tilde{f}(k) = \hat{f}(-2k_1, 2k_2), \qquad k = k_1 + ik_2 \in \mathbb{C}.$$

Remark 5.7.12. If $f \in L^p(\mathbb{R}^2)$ with $1 \leq p \leq 2$, then $\hat{f} \in L^{p'}(\mathbb{R}^2)$ where $\frac{1}{p} + \frac{1}{p'} = 1$ by the Hausdorff–Young theorem (Exercise C.3.8). Thus $\tilde{f}(k)$ is defined for almost every k. Writing $k = k_1 + ik_2$ and $w = w_1 + iw_2$, we have

$$e_k(w) = e^{2i\,\mathrm{Re}(kw)} = e^{2i(k_1w_1 - k_2w_2)}.$$

Therefore, if $f \in L^1(\mathbb{R}^2)$, the Fourier transform $\tilde{f}$ may be written also as

$$\int_{\mathbb{R}^2} e_k(w)f(w)\,dA(w) = \tilde{f}(k). \tag{5.33}$$

Here and below, dA denotes the Lebesgue measure in $\mathbb{R}^2$.

Lemma 5.7.13 (Derivative of $(\partial+ik)^{-1}$). *Let $f \in L^p \cap L^{p'}(\mathbb{R}^2)$ where $1 < p < 2$ and $\frac{1}{p} + \frac{1}{p'} = 1$, and let $z_0 \in \mathbb{C}$. Then the function*

$$F(k) = (\partial + ik)^{-1}f(z_0)$$

is in $W^{1,p'}_{\mathrm{loc}}(\mathbb{R}^2)$ and its $\partial_{\bar{k}}$-derivative is given by

$$\partial_{\bar{k}}\left[(\partial + ik)^{-1}f(z_0)\right] = \frac{1}{\pi i}\tilde{f}(k)e_{-k}(z_0) \quad \textit{for a.e. } k \in \mathbb{C}.$$

Proof. It is sufficient to consider $z_0 = 0$ (the argument for general z_0 is analogous). Recall that $(\partial+ik)^{-1} = e_{-k}\partial^{-1}e_k$. Since $f \in L^{2\pm}$, we have $(\partial+ik)^{-1}f \in C_0$ by Proposition 5.3.1 applied to $\bar{P}$. In particular, $F(k)$ is a well defined complex function on $\mathbb{C}$. Using the convolution form $\partial^{-1}f = \frac{1}{\pi\bar{z}} * f$, we obtain

$$F(k) = (\partial + ik)^{-1}f(0) = -\frac{1}{\pi}\int_{\mathbb{R}^2} \frac{e_k(w)}{\bar{w}}f(w)\,dA(w).$$

By (5.33), $F(k)$ may be interpreted as a Fourier transform of $\frac{f(w)}{\bar{w}}$. Let χ be the characteristic function of the unit disc. Since $\frac{\chi(w)}{\bar{w}} \in L^r$ for $r < 2$, $\frac{1-\chi(w)}{\bar{w}} \in L^\infty$, and $f \in L^p \cap L^{p'}$, we have $\frac{f(w)}{\bar{w}} \in L^1 + L^p$ and therefore $F \in C_0 + L^{p'}$ by the Riemann–Lebesgue lemma (Lemma B.2.8) and Hausdorff–Young theorem (Exercise C.3.8). Moreover, since $w_j\frac{f(w)}{\bar{w}} \in L^p$, the Hausdorff–Young theorem implies that $\partial_{k_j}F \in L^{p'}$. It follows that $F \in W^{1,p'}_{\mathrm{loc}}$.

Let us compute $\partial_{k_1} F$. For $h \in \mathbb{C}$, Taylor expansion gives

$$e_h = 1 + \int_0^1 \frac{d}{dt} e_{th}\, dt = 1 + i(hz + \overline{hz}) \int_0^1 e^{th}\, dt = 1 + 2i\,\mathrm{Re}(hz) + R_h,$$

where

$$R_h(z) = 2i\,\mathrm{Re}(hz) \int_0^1 (e_{th} - 1)(z)\, dt.$$

For $h > 0$, this yields

$$\frac{F(k+h) - F(k)}{h} + \frac{2i}{\pi} \int e_k(w) \frac{\mathrm{Re}(w)}{\bar{w}} f(w)\, dA(w)$$
$$= -\frac{1}{\pi} \int_{\mathbb{R}^2} \frac{e_k(w)}{\bar{w}} \frac{R_h(w)}{h} f(w)\, dA(w).$$

Writing $g(w) = \frac{\mathrm{Re}(w)}{\bar{w}} f(w)$, the remainder term is a constant multiple of

$$\int_{\mathbb{R}^2} \int_0^1 (e_{k+th}(w) - e_k(w)) g(w)\, dA(w) = \int_0^1 (\tilde{g}(k+th) - \tilde{g}(k))\, dt.$$

Since $g \in L^p$, one has $\tilde{g} \in L^{p'}$, and the $L^{p'}$ norm of the remainder term is

$$\leq \int_0^1 \|\tilde{g}(\,\cdot\, + th) - \tilde{g}(\,\cdot\,)\|_{L^{p'}}\, dt.$$

The last quantity goes to zero as $h \to 0$ since translation is a continuous operation on $L^{p'}$ (this follows from approximation by C_c functions as in Theorem 2.1.4(d)). Thus the remainder term goes to zero and $\partial_{k_1} F$ exists for a.e. k as $h \to 0$.

Repeating the above argument for h replaced by ih with $h > 0$, we see that also $\partial_{k_2} F$ exists a.e. and

$$\partial_{k_1}((\partial + ik)^{-1} f)(0) = -\frac{2i}{\pi} \int e_k(w) \frac{\mathrm{Re}(w)}{\bar{w}} (w)\, dA(w),$$
$$\partial_{k_2}((\partial + ik)^{-1} f)(0) = \frac{2i}{\pi} \int e_k(w) \frac{\mathrm{Im}(w)}{\bar{w}} f(w)\, dA(w).$$

Now $\partial_{\bar{k}} = \frac{1}{2}(\partial_{k_1} + i\partial_{k_2})$, so it follows that

$$\partial_{\bar{k}}((\partial + ik)^{-1} f)(0) = \frac{1}{\pi i} \int_{\mathbb{R}^2} e_k(w) f(w)\, dA(w). \qquad \square$$

From the previous results we obtain a useful expression for the scattering transform τ_σ in terms of the Fourier transform of $q_\sigma(1 + s_\sigma)$.

Lemma 5.7.14 (Scattering transform for C^2 conductivities). *Let $\sigma \in C^2(\mathbb{R}^2)$ be positive with $\sigma = 1$ for $|z| > 1$. The scattering transform τ_σ has the formula*

$$\tau_\sigma(k) = \frac{i}{4\pi\bar{k}} \left(q_\sigma[1+s_\sigma]\right)^{\tilde{}}(k).$$

Proof. From (5.32) and (5.31), we have

$$\tau_\sigma(k) = ie_k(z)\frac{\partial_{\bar{k}}(G_k f)(z)}{\overline{1+s_\sigma(z,k)}},$$

where $f = f_k = q_\sigma(1+s_\sigma) \in L^p$ for $p \in (1,\infty)$ since $q \in L_c^\infty$ and $s_\sigma \in L^\infty$. Note that $k \mapsto f_k$ is a C^∞ map from $\mathbb{C}$ to $W_c^{1,r}$ for some $r > 2$ by Proposition 5.7.5. By the L^p estimates for Beurling transform, we also have $\partial\bar{\partial}^{-1} f \in L^p$ for any $p \in (1,\infty)$. Proposition 5.7.9 and Lemma 5.7.13 give that

$$\begin{aligned}\partial_{\bar{k}}(G_k f)(z) &= -\frac{1}{4ik}\partial_{\bar{k}}((\partial+ik)^{-1}(\partial\bar{\partial}^{-1}f_k)))(z) + G_k(\partial_{\bar{k}}f_k)(z)\\ &= \frac{1}{4\pi k}(\partial\bar{\partial}^{-1}f)^{\tilde{}}(k)e_{-k}(z) + G_k(\partial_{\bar{k}}f_k).\end{aligned}$$

We further have

$$(\partial\bar{\partial}^{-1}f)^{\tilde{}}(k) = (\partial\bar{\partial}^{-1}f)^{\hat{}}(-2k_1,2k_2) = \frac{k}{\bar{k}}\tilde{f}(k).$$

Combining these facts yields

$$\tau_\sigma(k) = \frac{i}{4\pi\bar{k}}\frac{\tilde{f}(k)}{\overline{1+s_\sigma(z,k)}} + ie_k(z)\frac{G_k(\partial_{\bar{k}}f_k)(z)}{\overline{1+s_\sigma(z,k)}}.$$

The last identity is valid for any $z \in \mathbb{C}$. Now $G_k(\partial_{\bar{k}}f_k) \in W^{1,p^*}$ for any $p < 2$ and $s_\sigma(\,\cdot\,,k) \in W^{1,r}$ for some $r > 2$, so both these functions go to zero as $|z| \to \infty$. Thus taking the limit as $|z| \to \infty$ proves the result. □

Remark 5.7.15. We can rewrite the above results in the notation of [**Nac96**]. Let $\mathbf{t}(k)$ be the scattering transform in [**Nac96**] given by

$$\mathbf{t}(k) = (q_\sigma m_\sigma)^{\tilde{}}(k) = \int_{\mathbb{R}^2} q_\sigma(w)m_\sigma(w,k)e_k(w)\,dA(w),$$

where $m_\sigma(z,k) = e^{-ikz}\psi_\sigma = 1 + s_\sigma$. The conclusion of Lemma 5.7.14 may be rewritten as

$$\tau_\sigma(k) = \frac{i}{4\pi\bar{k}}\mathbf{t}(k).$$

Further, (5.32) gives

$$\partial_{\bar{k}}m_\sigma = -ie_{-k}\tau_\sigma\overline{m_\sigma} = \frac{1}{4\pi\bar{k}}\mathbf{t}(k)e_{-k}\overline{m_\sigma}.$$

This is the $\partial_{\bar{k}}$-equation appearing in [**Nac96**, equation (0.12)].

We can now conclude the uniqueness proof for C^2 conductivities by establishing Proposition 5.7.3.

Proof of Proposition 5.7.3. As discussed above, the estimates for r_σ in Proposition 5.7.3 follow from Proposition 5.7.10 since $s_\sigma = \sigma^{1/2} - 1 + \sigma^{1/2} r_\sigma$ and $r_\sigma(z, \,\cdot\,)$ is C^∞ for fixed z.

To prove the estimates for τ_σ, we use Lemma 5.7.14 to conclude that

$$\tau_\sigma(k) = \frac{1}{\bar{k}} \tilde{f}(k),$$

where $f = \frac{i}{4\pi} q_\sigma(1 + s_\sigma) \in L^\infty_c(\mathbb{R}^2)$. Let $\chi(k)$ be the characteristic function of $\mathbb{D}$. Since $\tau_\sigma \in C^\infty$, we have $\chi\tau_\sigma \in L^r$ for any r. On the other hand, if $r > 1$ we have

$$\|(1-\chi)\tau_\sigma\|_{L^r} \leq \left\| \frac{1-\chi(k)}{\bar{k}} \right\|_{L^{2r}} \|\tilde{f}\|_{L^{2r}}.$$

The last expression is finite by the Hausdorff–Young theorem, so $\tau_\sigma \in L^r(\mathbb{R}^2)$ for any $r > 1$. Finally, the Riemann–Lebesgue lemma implies the estimate $|\tau_\sigma(k)| \leq C/|k|$. □

5.8. Topological methods

We return to the proof of uniqueness for L^∞ conductivities. Recall the $\partial_{\bar{k}}$ equation for the solutions u_σ,

$$\partial_{\bar{k}} u_\sigma(z,k) = -i\tau_\sigma(k)\overline{u_\sigma(z,k)} \quad \text{in } \mathbb{R}^2.$$

We would like to conclude that solutions to this equation are unique. However, unlike in the case where $\sigma \in C^2$, for $\sigma \in L^\infty$ the solutions u_σ or the scattering transform τ_σ will not have sufficient decay properties as $k \to \infty$ to guarantee uniqueness. The logarithm of u_σ will have some decay properties instead. We will eventually use nonlinear $\partial_{\bar{k}}$ equations for the logarithms to obtain a uniqueness statement.

Topological methods will be the main tool for dealing with the nonlinear partial differential equations that arise in this process. The first result is a simple surjectivity statement.

Lemma 5.8.1. *Let $\lambda \in \mathbb{C}\setminus\{0\}$, and let $F : \mathbb{C} \to \mathbb{C}$ be a continuous function such that*

$$F(z) = \lambda z + z\varepsilon(z) \quad \text{when } |z| > 1, \quad \varepsilon(z) \to 0 \text{ as } z \to \infty.$$

Then F is surjective.

Intuitively, Lemma 5.8.1 should be true since for large $r > 0$ the curves $t \mapsto F(re^{it})$ look like circles with large radius, and for very small $r > 0$ the same curves are close to the point $F(0)$. By continuously deforming these curves with large r into the curves for small r, one should pass through any given point of $\mathbb{C}$ (the point $F(0)$ is obtained as the limit when $r \to 0$).

The second result is a version of the argument principle for certain solutions of $\bar{\partial}F + \kappa F = 0$ where $\kappa \in L^\infty(\mathbb{R}^2)$. It shows that one has some control of the zeros of F even if there is no decay for κ at infinity.

Lemma 5.8.2. *Let $\lambda \in \mathbb{C}\setminus\{0\}$, and let $F : \mathbb{C} \to \mathbb{C}$ be a continuous function such that*

$$F(z) = \lambda z + z\varepsilon(z) \quad \text{when } |z| > 1, \quad \varepsilon(z) \to 0 \text{ as } z \to \infty.$$

If, additionally, $F \in W^{1,p}_{\mathrm{loc}}(\mathbb{C})$ for some $p > 2$ and for some $C > 0$

$$|\bar{\partial}F| \leq C|F| \quad \text{almost everywhere in } \mathbb{C},$$

then F has exactly one zero in $\mathbb{C}$.

Let us give some definitions to prepare for the proofs.

Definition 5.8.3 (Closed curves). Let Ω be a connected open subset of $\mathbb{C}$. A *curve* in Ω is a continuous map $\gamma : [0,1] \to \Omega$. The curve is closed if $\gamma(0) = \gamma(1)$, and the image of the curve is

$$\gamma^* := \gamma([0,1]).$$

Definition 5.8.4 (Homotopy). Let Ω be a connected open subset of $\mathbb{C}$. Two closed curves γ_0 and γ_1 in Ω are *homotopic* in Ω if there is a continuous map $H : [0,1] \times [0,1] \to \Omega$ satisfying for all $s, t \in [0,1]$

$$H(0,t) = \gamma_0(t), \quad H(1,t) = \gamma_1(t),$$
$$H(s,0) = H(s,1).$$

Writing $\gamma_s(t) := H(s,t)$, we say that curve γ_0 is continuously deformed into γ_1 through the family $\{\gamma_s\}_{s\in[0,1]}$.

We also need a way of detecting if a given point is "inside" or "outside" a closed curve. This is given by the index of a point.

Definition 5.8.5 (Index). Let γ be a closed C^1 curve in $\mathbb{C}$. The *index* (or *winding number*) of a point $z \in \mathbb{C} \setminus \gamma^*$ relative to the curve γ is defined as

$$\mathrm{Ind}_\gamma(z) := \frac{1}{2\pi i}\int_\gamma \frac{1}{w - z}\,dw.$$

The next set of exercises contains some basic facts about the index, which will be needed for the proof of Lemma 5.8.1. These facts are also discussed in [**Rud87**, §10].

Exercise 5.8.6. If $\gamma(t) = Re^{2\pi i t}$ for $t \in [0,1]$, show that $\mathrm{Ind}_\gamma(z) = 1$ when $|z| < R$ and $\mathrm{Ind}_\gamma(z) = 0$ for $|z| > R$.

Exercise 5.8.7. Prove that the map $z \mapsto \mathrm{Ind}_\gamma(z)$ assumes only integer values, is constant on the connected components of $\mathbb{C} \setminus \gamma^*$, and is equal to 0 on the unbounded component.

Exercise 5.8.8. If $\gamma_0, \gamma_1 : [0,1] \to \mathbb{C}$ are closed C^1 curves and if $z \in \mathbb{C}$ is such that

$$|\gamma_0(t) - \gamma_1(t)| < |\gamma_0(t) - z|, \quad t \in [0,1],$$

show that $\mathrm{Ind}_{\gamma_0}(z) = \mathrm{Ind}_{\gamma_1}(z)$.

Exercise 5.8.9. Prove that if Ω is a connected open subset of $\mathbb{C}$ and if γ_0 and γ_1 are closed C^1 curves homotopic to each other in Ω, then

$$\mathrm{Ind}_{\gamma_0}(z) = \mathrm{Ind}_{\gamma_1}(z), \quad z \notin \Omega.$$

Exercise 5.8.10. Show that $\mathrm{Ind}_\gamma(z)$ is well defined for any closed curve γ (which may not be C^1) by

$$\mathrm{Ind}_\gamma(z) := \lim_{j\to\infty} \mathrm{Ind}_{\gamma_j}(z),$$

where γ_j are trigonometric polynomials that approximate γ in the $L^\infty([0,1])$ norm. Show also that the results of Exercises 5.8.7 and 5.8.9 remain valid for continuous curves.

Proof of Lemma 5.8.1. Given $w_0 \in \mathbb{C}$, it is required to show that $F(z_0) = w_0$ for some $z_0 \in \mathbb{C}$. Replacing F by $F - w_0$, we may assume that $w_0 = 0$. We argue by contradiction and suppose that $F(z) \neq 0$ for all $z \in \mathbb{C}$. Since F is continuous, the curves

$$\gamma_0(t) = F(\delta e^{it}), \quad \gamma_1(t) = F(Re^{it})$$

are homotopic in $\mathbb{C} \setminus \{0\}$ for any $\delta, R > 0$. It follows that

$$\mathrm{Ind}_{\gamma_0}(0) = \mathrm{Ind}_{\gamma_1}(0).$$

Now, using that $F(0) \neq 0$, we may find some small $\delta > 0$ so that

$$F(\partial B(0,\delta)) \subset B(F(0), |F(0)|/2).$$

Then 0 is in the unbounded component of $\mathbb{C} \setminus \gamma_0^*$, and

$$\mathrm{Ind}_{\gamma_0}(0) = 0.$$

However, using the assumption on F we also see that $F(\partial B(0,R))$ is homotopic to the circle $\partial B(0,R)$ in $\mathbb{C} \setminus \{0\}$ for some $R > 0$ sufficiently large. This implies that

$$\mathrm{Ind}_{\gamma_1}(0) = \mathrm{Ind}_{\partial B(0,R)}(0) = 1.$$

We have reached a contradiction. □

The next exercises contain the proof of Lemma 5.8.2.

Exercise 5.8.11. Prove a standard version of the argument principle: if F is a holomorphic function in a ball $B(0,R)$, if $\gamma(t) = re^{it}$ for some $r < R$, and if F has no zeros on γ^*, then the number of zeros N of F in $B(0,r)$ (counted with multiplicities) is equal to

$$N = \frac{1}{2\pi i}\int_\gamma \frac{F'(z)}{F(z)}\,dz = \mathrm{Ind}_{F(\gamma)}(0).$$

Exercise 5.8.12. Prove another version of the argument principle: if F is a holomorphic function in $\{|z| < R+\varepsilon\}$ for some $\varepsilon > 0$, and if the curve $t \mapsto F(Re^{2\pi it})$ is homotopic to $t \mapsto Re^{2\pi it}$ in $\mathbb{C}\setminus\{0\}$, then the function F has exactly one zero in $\{|z| < R\}$.

Exercise 5.8.13. Prove Lemma 5.8.2 by filling in the details of the following outline: reduce the inequality $|\bar\partial F| \leq C|F|$ to the equation $\bar\partial F + \kappa F = 0$ for some $\kappa \in L^\infty$, let $\beta \in W^{1,p}(\mathbb{R}^2)$ with $p > 2$ satisfy $\bar\partial\beta = \kappa\chi_B$, where χ_B is the characteristic function of a suitable large disc B, and show that the previous exercise can be applied to the function $e^\beta F$ that is holomorphic in B.

5.9. Uniqueness for bounded measurable conductivities

In this section we consider the case of bounded measurable conductivities. We will give the proof of Theorem 5.0.1 modulo the fundamental subexponential growth estimate for the CGO solutions. Assume that $\sigma_1, \sigma_2 \in L^\infty(\mathbb{D})$ are positive functions such that $\Lambda_{\sigma_1} = \Lambda_{\sigma_2}$. In Proposition 5.6.1 we have constructed CGO solutions u_{σ_j} to the corresponding conductivity equations. Since these solutions are uniquely determined by $f_{\pm\mu_j}$ (where μ_j is the complex coefficient related to σ_j), Proposition 5.5.1 implies that

$$u_{\sigma_1}(z,k) = u_{\sigma_2}(z,k), \quad z \in \mathbb{R}^2\setminus\mathbb{D}, \quad k \in \mathbb{C}.$$

The next proposition gives a similar result in the interior of $\mathbb{D}$.

Proposition 5.9.1 (CGO solutions agree everywhere). *If $\Lambda_{\sigma_1} = \Lambda_{\sigma_2}$, then*

$$u_{\sigma_1}(z,k) = u_{\sigma_2}(z,k), \quad z,k \in \mathbb{C}.$$

The solution of the Calderón problem is an immediate consequence of this result and a basic fact on quasiregular mappings.

Proposition 5.9.2 (Jacobian of quasiregular maps). *Let $\Omega \subset \mathbb{R}^2$ be a connected open set. If $f \in W^{1,2}_{\mathrm{loc}}(\mathbb{R}^2)$ satisfies*

$$\bar\partial f = \mu\overline{\partial f} \quad \text{in } \Omega,$$

where $\mu \in L^\infty(\Omega)$ and $\|\mu\|_{L^\infty(\Omega)} < 1$, then $\partial f \neq 0$ almost everywhere in Ω unless f is constant.

Note that if $\mu = 0$, the function f in Proposition 5.9.2 is analytic. Then $\partial f = f'$ is also analytic, and thus ∂f is either constant or its zero set has no accumulation point in Ω. In the latter case the zero set of ∂f is countable and thus it has measure zero.

Proposition 5.9.2 is a generalization of the above fact to quasiregular mappings. It is a consequence of the Stoilow factorization and the details may be found in [**AIM09**, Corollary 5.5.2]. However, in the case where $\mu \in W^{1,\infty}(\Omega)$ is real valued (corresponding to a Lipschitz continuous conductivity), we can give an alternative proof of a slightly weaker result. This result would still be sufficient for the uniqueness proof for C^1 conductivities.

Proposition 5.9.3 (Jacobian of quasiregular maps with $\mu \in W^{1,\infty}$). *Let $\Omega \subset \mathbb{R}^2$ be a connected open set. If $f \in W^{1,2}_{\mathrm{loc}}(\Omega)$ satisfies*

$$\bar\partial f = \mu \overline{\partial f} \quad \text{in } \Omega,$$

where $\mu \in W^{1,\infty}(\Omega)$ is real valued and $\|\mu\|_{L^\infty(\Omega)} < 1$, then $\partial f \neq 0$ in a dense subset of Ω unless f is constant.

Proof. Since $\partial f \in L^2_{\mathrm{loc}}(\Omega)$, there is a set $E \subset \Omega$ with measure zero such that $\partial f(x)$ is defined for all $x \in \tilde\Omega$ where $\tilde\Omega = \Omega \setminus E$. Define

$$S = \{\, x \in \tilde\Omega \mid \partial f(x) = 0 \,\}.$$

If $\tilde\Omega \setminus S$ is not dense in Ω, there is a point $x_0 \in \Omega$ and a ball B centered at x_0 such that $\partial f = 0$ a.e. on B. Using the equation for f, we see that also $\bar\partial f = 0$ a.e. on B, so the gradients of $u = \mathrm{Re}(f)$ and $v = \mathrm{Im}(f)$ vanish a.e. on B. Then u and v are constant on B (see Exercise 2.4.5).

But u and v satisfy the conductivity equations

$$\nabla \cdot \sigma \nabla u = \nabla \cdot \sigma^{-1} \nabla v = 0 \text{ in } \Omega,$$

where

$$\sigma = \frac{1-\mu}{1+\mu}.$$

Since $\sigma \in W^{1,\infty}$, we have $\Delta u = -\sigma^{-1}\nabla\sigma \cdot \nabla u \in L^2_{\mathrm{loc}}$. Interior elliptic regularity, Theorem 2.6.2, gives $u \in H^2_{\mathrm{loc}}$ and similarly $v \in H^2_{\mathrm{loc}}$. From the version of the unique continuation principle proved in Theorem 6.4.2, it follows that u and v must be constant everywhere in Ω. Consequently, f is a constant map in Ω. □

Proof of Theorem 5.0.1. Since $\Lambda_{\sigma_1} = \Lambda_{\sigma_2}$ implies $\Lambda_{1/\sigma_1} = \Lambda_{1/\sigma_2}$, Proposition 5.9.1 is valid also with σ_j replaced by $1/\sigma_j$. The pair $(f_{\mu_j}, f_{-\mu_j})$ is uniquely determined by the pair $(u_{\sigma_j}, u_{1/\sigma_j})$, so it follows that

$$f_{\mu_1}(z,k) = f_{\mu_2}(z,k), \quad z, k \in \mathbb{C}.$$

The Beltrami equations $\bar{\partial} f_{\mu_j} = \mu_j \overline{\partial f_{\mu_j}}$ imply that

$$\mu_1 = \frac{\bar{\partial} f_{\mu_1}}{\overline{\partial f_{\mu_1}}} = \frac{\bar{\partial} f_{\mu_2}}{\overline{\partial f_{\mu_2}}} = \mu_2,$$

at all those points of $\mathbb{D}$, where ∂f_{μ_1} and ∂f_{μ_2} are nonzero. By Proposition 5.9.2 this is true almost everywhere. It follows that $\mu_1 = \mu_2$ almost everywhere in $\mathbb{D}$, and since

$$\sigma_1 = \frac{1-\mu_1}{1+\mu_1}, \qquad \sigma_2 = \frac{1-\mu_2}{1+\mu_2}$$

we see that also $\sigma_1 = \sigma_2$ almost everywhere in $\mathbb{D}$. □

We now focus on proving Proposition 5.9.1. Due to Proposition 5.6.1, the solutions u_{σ_j} satisfy the $\partial_{\bar{k}}$ equations

$$\partial_{\bar{k}} u_{\sigma_j}(z,\cdot) = -i\tau_{\sigma_j}(\cdot)\overline{u_{\sigma_j}(z,\cdot)}. \tag{5.34}$$

Also, from the fact that $\Lambda_{\sigma_1} = \Lambda_{\sigma_2}$ and from Proposition 5.6.3 we know that

$$\tau_{\sigma_1}(k) = \tau_{\sigma_2}(k), \quad k \in \mathbb{C}.$$

Let $\tau(k) := \tau_{\sigma_1}(k) = \tau_{\sigma_2}(k)$. It follows that for fixed $z \in \mathbb{C}$, both $u_{\sigma_1}(z,\cdot)$ and $u_{\sigma_2}(z,\cdot)$ are solutions of the equation

$$\partial_{\bar{k}} u(k) = -i\tau(k)\overline{u(k)}.$$

We proved in §5.7 that for C^2 conductivities, both the coefficient τ and the solution u (or the function $e^{-ikz}u - 1$) have suitable decay properties as $k \to \infty$, and therefore the solutions to the above equation are unique and one obtains that $u_{\sigma_1} = u_{\sigma_2}$ as required.

It turns out that when the conductivities are only bounded and measurable, one cannot expect any decay with respect to k in the solutions u_σ. However, one can prove that at least $e^{-ikz}u_\sigma(z,k)$ does not grow exponentially in k. The proof will be discussed in the next section.

Proposition 5.9.4 (Subexponential growth). *We have*

$$u_\sigma(z,k) = e^{\delta_\sigma(z,k)},$$

where, for fixed $k \in \mathbb{C}$, one has $\delta_\sigma(\cdot,k) \in W^{1,p}_{\mathrm{loc}}(\mathbb{R}^2)$ for some $p > 2$ and

$$\delta_\sigma(z,k) = ikz + v_k(z), \quad v_k(z) \to 0 \text{ as } z \to \infty.$$

Also, for fixed $z \in \mathbb{C}$, the function $\delta_\sigma(z,\cdot)$ is C^∞ and

$$\delta_\sigma(z,k) = ikz + k\varepsilon_z(k), \quad \varepsilon_z(k) \to 0 \text{ as } k \to \infty.$$

Not discouraged by the lack of decay with respect to k in the solution u_σ, we will try to exploit the decay in its logarithm δ_σ. The $\partial_{\bar{k}}$ equations (5.34) for u_{σ_j} imply the following nonlinear equations for δ_{σ_j}:

$$\begin{aligned}\partial_{\bar{k}}\delta_{\sigma_j}(z,k) &= -i\tau(k)e^{\overline{\delta_{\sigma_j}(z,k)}-\delta_{\sigma_j}(z,k)} \\ &= -i\tau(k)e^{-2i\,\mathrm{Im}(\delta_{\sigma_j}(z,k))}.\end{aligned}$$

This shows that we have two families of solutions $\{\delta_{\sigma_j}(z,\,\cdot\,)\}_{z\in\mathbb{C}}$ to the same nonlinear $\partial_{\bar{k}}$ equation. Moreover, the two families have the same asymptotics as $k\to\infty$ by Proposition 5.9.4. This turns out to be sufficient for uniqueness.

Proof of Proposition 5.9.1. To have $u_{\sigma_1}(z,k) = u_{\sigma_2}(z,k)$ for all $z,k\in\mathbb{C}$, it is enough to show that

$$\delta_{\sigma_1}(z,k) = \delta_{\sigma_2}(z,k), \quad z,k\in\mathbb{C}.$$

We generalize the setup a little bit and consider the function

$$F_k(z,w) := \delta_{\sigma_1}(z,k) - \delta_{\sigma_2}(w,k), \quad z,w,k\in\mathbb{C}.$$

It is enough to prove that F_k vanishes on the diagonal for any k.

If $k=0$, we know by Proposition 5.4.1 that $f_{\pm\mu}(z,0)=1$. This implies that $u_\sigma(z,0)=1$, so the logarithm satisfies $\delta_\sigma(z,0)=0$. This proves that

$$F_0(z,w)=0, \quad z,w\in\mathbb{C}.$$

Let now $k_0\neq 0$, and suppose that $z_0\in\mathbb{C}$ is fixed. The result will follow if we can show that

$$F_{k_0}(z_0,z_0)=0.$$

We do the proof in two steps. The first step is to show that

$$F_{k_0}(z_0,w_0)=0 \quad \text{for some } w_0\in\mathbb{C}. \tag{5.35}$$

To see this, it is enough to observe that by Proposition 5.9.4 and Lemma 5.8.1 the map $w\mapsto\delta_{\sigma_2}(w,k_0)$ is a surjective map from $\mathbb{C}$ to $\mathbb{C}$. This shows that there exists some $w_0\in\mathbb{C}$ such that $\delta_{\sigma_2}(w_0,k_0)=\delta_{\sigma_1}(z_0,k_0)$, which gives (5.35).

The second step is to show that for any $z,w\in\mathbb{C}$, one has

$$F_{k_0}(z,w)=0 \implies z=w. \tag{5.36}$$

Assume that $F_{k_0}(z,w)=0$, and note that the map $k\mapsto F_k(z,w)$ satisfies the $\partial_{\bar{k}}$ equation

$$\partial_{\bar{k}}F_k(z,w) = -i\tau(k)\left[e^{-2i\,\mathrm{Im}(\delta_{\sigma_1}(z,k))} - e^{-2i\,\mathrm{Im}(\delta_{\sigma_2}(w,k))}\right].$$

We now use the bound $|\tau(k)|\leq 1$ from Proposition 5.6.3 and the elementary fact that

$$\left|e^{is}-e^{it}\right| \leq |s-t|, \quad s,t\in\mathbb{R}. \tag{5.37}$$

Taking absolute values in the previous equation for $F_k(z,w)$ yields

$$\begin{aligned}|\partial_{\bar{k}} F_k(z,w)| &\leq 2\left|\mathrm{Im}(\delta_{\sigma_1}(z,k)) - \mathrm{Im}(\delta_{\sigma_2}(w,k))\right| \\ &\leq 2\,|F_k(z,w)|\,.\end{aligned}$$

We also know from Proposition 5.9.4 that

$$F_k(z,w) = i(z-w)k + k\tilde{\varepsilon}(k), \quad \tilde{\varepsilon}(k) \to 0 \text{ as } k \to \infty.$$

If $z \neq w$, it follows from Lemma 5.8.2 that the map $k \mapsto F_k(z,w)$ has only one zero. However, we always have $F_0(z,w) = 0$, which contradicts the assumption that $F_{k_0}(z,w) = 0$, where $k_0 \neq 0$. It must follow that $z = w$, finishing the proof of (5.36).

Finally, combining (5.35) and (5.36) yields $F_{k_0}(z_0, z_0) = 0$. Since $k_0 \neq 0$ and $z_0 \in \mathbb{C}$ were arbitrary, this concludes the proof that F_k vanishes on the diagonal. □

5.10. Subexponential growth

To complete the uniqueness result for bounded measurable conductivities, it remains to prove Proposition 5.9.4 from the previous section concerning subexponential growth of $e^{-ikz}u_\sigma$ with respect to k. This will follow from a corresponding result for the solutions f_μ of the Beltrami equation. Given the subexponential growth for f_μ, we will conclude with the proof of Proposition 5.9.4 for the solutions u_σ.

In order to pass from growth properties of solutions of Beltrami equations back to the conductivity equation, it is useful to generalize the setup slightly. If $\lambda \in \partial\mathbb{D}$, we denote by $f_{\lambda\mu}$ the solution of the Beltrami equation

$$\bar{\partial} f_{\lambda\mu} = \lambda\mu\overline{\partial f_\mu} \quad \text{in } \mathbb{R}^2$$

satisfying $f_{\lambda\mu}(\,\cdot\,,k) \in W^{1,2}_{\mathrm{loc}}(\mathbb{R}^2)$ and

$$\begin{aligned}f_{\lambda\mu}(z,k) &= e^{ikz}(1+\eta_{\lambda\mu}(z,k)), \\ \eta_{\lambda\mu}(z,k) &= O(1/z) \text{ as } z \to \infty.\end{aligned}$$

Here $\mu = (1-\sigma)/(1+\sigma)$ as before. Precisely the same proof as in Proposition 5.4.1 shows that such solutions exist, are unique, and satisfy $\eta_{\lambda\mu}(\,\cdot\,,k) \in W^{1,p}(\mathbb{R}^2)$ for some $p > 2$. The following result states that the functions $e^{-ikz}f_{\lambda\mu}(z,k)$ have subexponential growth with respect to k.

Proposition 5.10.1 (Subexponential growth). *For fixed $z \in \mathbb{C}$, we have*

$$f_{\lambda\mu}(z,k) = e^{ik\phi_{\lambda\mu}(z,k)},$$

where $\phi_{\lambda\mu}$ is C^∞ in k and, for any $k \in \mathbb{C}$,

$$\phi_{\lambda\mu}(z,k) = z + v_1(z,k), \quad v_1(\,\cdot\,,k) \in W^{1,p}(\mathbb{R}^2)$$

for some $p > 2$. For any $z \in \mathbb{C}$ the function $\phi_{\lambda\mu}(z, \cdot\,)$ satisfies

$$\phi_{\lambda\mu}(z,k) = z + \varepsilon_z(k), \quad \varepsilon_z(k) \to 0 \text{ as } k \to \infty.$$

The proof of Proposition 5.10.1 requires several facts from the theory of quasiconformal mappings and it will only be sketched here in the case $\lambda = 1$. A detailed proof may be found in [**AIM09**, §18.6] or in the original work [**AP06a**].

Sketch of proof of Proposition 5.10.1 for $\lambda = 1$. One first notes that f_μ solves the Beltrami equation $\bar\partial f_\mu = \mu_1 \partial f_\mu$ with a modified Beltrami coefficient

$$\mu_1 = \mu \frac{\overline{\partial f_\mu}}{\partial f_\mu}.$$

By the Stoilow factorization, $f_\mu = h \circ \phi$, where ϕ is a quasiconformal homeomorphism solving the same Beltrami equation with asymptotics $\phi = z + O(1/z)$ as $z \to \infty$ and where h is an entire analytic function. Since h/e^{ikz} is entire and has limit 1 as $z \to \infty$, the Liouville theorem gives that $h(z) = e^{ikz}$ and thus

$$f_\mu = e^{ik\phi}.$$

This proves that f_μ becomes the function e^{ikz} in the quasiconformal change of variables $z \mapsto \phi(z)$.

From the Beltrami equation for f_μ, it follows that ϕ satisfies the nonlinear Beltrami equation

$$\bar\partial \phi = -\frac{\bar k}{k} \mu (e_{-k} \circ \phi) \overline{\partial \phi}.$$

Since ϕ is a quasiconformal homeomorphism $\mathbb{C} \to \mathbb{C}$, it is convenient to work with its inverse ψ which is also quasiconformal. The inverse function satisfies the nonlinear Beltrami equation

$$\bar\partial \psi = -\frac{\bar k}{k} (\mu \circ \psi) e_{-k} \partial \phi$$

with $\psi = z + O(1/z)$ as $z \to \infty$. Note that there is no complex conjugate on the right, so this equation is very close to the standard Beltrami equation $\bar\partial \Psi = \nu \partial \Psi$ appearing in the study of quasiconformal maps.

By compactness arguments, one can extract a uniform limit

$$\psi_\infty = \lim_{k_n \to \infty} \psi(\,\cdot\,, k_n)$$

which is also a quasiconformal homeomorphism. One then compares ψ above with the solution of the *linear* Beltrami equation

$$\bar\partial \Psi = -\frac{\bar k}{k} (\mu \circ \psi_\infty) e_{-k} \partial \Psi$$

with asymptotics $\Psi = z + O(1/z)$ as $z \to \infty$. One can prove that $\Psi \to z$ uniformly as $k \to \infty$.

Now $\psi(\cdot, k_n) - \Psi$ can be estimated by $\|\mu \circ \psi(\cdot, k_n) - \mu \circ \psi_\infty\|_{L^p}$, which converges to zero as $n \to \infty$ by higher integrability of the Jacobian of $\psi(\cdot, k_n)$ and ψ_∞. This shows that $\psi(\cdot, k_n) \to z$ uniformly as $k \to \infty$. Passing to the inverse function proves the proposition. □

After some work, Proposition 5.9.4 will now follow.

Proof of Proposition 5.9.4. We need to show that

$$u_\sigma = e^{ikz+v(z,k)},$$

where

$$v(z,k) \to 0 \text{ as } z \to \infty, \text{ when } k \text{ is fixed,}$$

and

$$v(z,k) = |k|\, o(1) \text{ as } k \to \infty, \text{ when } z \text{ is fixed.}$$

In Proposition 5.6.12 we proved that u_σ never vanishes. More generally, we show that in fact u_σ has a well-defined logarithm. Write

$$\begin{aligned} u_\sigma &= \frac{1}{2}(f_\mu + \overline{f_\mu} + f_{-\mu} - \overline{f_{-\mu}}) \\ &= \frac{1}{2}(f_\mu + f_{-\mu})\left(1 + \frac{\overline{f_\mu} - \overline{f_{-\mu}}}{f_\mu + f_{-\mu}}\right) \\ &= f_\mu \frac{f_\mu + f_{-\mu}}{2f_\mu}\left(1 + \frac{\overline{f_\mu} - \overline{f_{-\mu}}}{f_\mu + f_{-\mu}}\right) \\ &= f_\mu \left(1 + \frac{f_\mu - f_{-\mu}}{f_\mu + f_{-\mu}}\right)^{-1}\left(1 + \frac{\overline{f_\mu} - \overline{f_{-\mu}}}{f_\mu + f_{-\mu}}\right). \end{aligned}$$

If $\mathbb{D}(z,r)$ is the open disc centered at z with radius r, then by (5.26)

$$1 + \frac{f_\mu - f_{-\mu}}{f_\mu + f_{-\mu}} \in \mathbb{D}(1,1), \quad 1 + \frac{\overline{f_\mu} - \overline{f_{-\mu}}}{f_\mu + f_{-\mu}} \in \mathbb{D}(1,1).$$

Consider the principal branch of the complex logarithm,

$$\operatorname{Log} : \mathbb{C} \setminus (-\infty, 0] \to \{\, z \in \mathbb{C} \mid |\operatorname{Im}(z)| < \pi \,\}.$$

Since $z/w \in \mathbb{C} \setminus (-\infty, 0]$ whenever $\operatorname{Re}(z), \operatorname{Re}(w) > 0$, we may define

$$g := \operatorname{Log}\left[\left(1 + \frac{f_\mu - f_{-\mu}}{f_\mu + f_{-\mu}}\right)^{-1}\left(1 + \frac{\overline{f_\mu} - \overline{f_{-\mu}}}{f_\mu + f_{-\mu}}\right)\right].$$

Proposition 5.10.1 shows that

$$f_\mu(z,k) = e^{ikz+\beta(z,k)},$$

where $\beta(\cdot, k) \in W^{1,p}(\mathbb{R}^2)$ for some $p > 2$, when $k \in \mathbb{C}$ is fixed. We choose

$$\delta_\sigma(z,k) := ikz + \beta(z,k) + g(z,k).$$

It follows that $u_\sigma = e^{\delta_\sigma}$. It is also clear that δ_σ is C^∞ in k, since this is true for β and g by Propositions 5.10.1 and 5.5.2.

Let now $k \in \mathbb{C}$ be fixed, and consider the asymptotics as $z \to \infty$. Define

$$v(z,k) := \beta(z,k) + g(z,k).$$

Here $\beta(\,\cdot\,,k) \in W^{1,p}(\mathbb{R}^2)$, and $g(\,\cdot\,,k)$ is in $W^{1,p}_{\mathrm{loc}}(\mathbb{R}^2)$ since it is the logarithm of a $W^{1,p}_{\mathrm{loc}}(\mathbb{R}^2)$ function. By (5.27) we have

$$g(z,k) = o(1) \quad \text{as } z \to \infty.$$

This shows that $v(\,\cdot\,,k) \in W^{1,p}_{\mathrm{loc}}(\mathbb{R}^2)$ with $v(z,k) = o(1)$ as $z \to \infty$.

Finally, let $z \in \mathbb{C}$ be fixed and consider the asymptotics as $k \to \infty$. We have

$$u_\sigma(z,k) = e^{ikz + v(z,k)}$$

and we need to show that

$$v(z,k) = |k|\, o_z(1) \quad \text{as } k \to \infty.$$

Here we write $o_z(1)$ for any quantity converging to 0 as $k \to \infty$.

Using that $v(z,k) = \beta(z,k) + g(z,k)$, we have

$$|\mathrm{Im}(v)| \le |\beta| + |g| \le |k|\, o_z(1) + \pi = |k|\, o_z(1).$$

To bound $\mathrm{Re}(v)$, we note that

$$u_\sigma = \frac{1}{2}(f_\mu + \overline{f_\mu} + f_{-\mu} - \overline{f_{-\mu}}),$$

which implies

$$e^v = e^{-ikz} u_\sigma = \frac{1}{2}(M_\mu + e_{-k}\overline{M_\mu} + M_{-\mu} - e_{-k}\overline{M_{-\mu}})$$

and consequently, by the triangle inequality,

$$e^{\mathrm{Re}(v)} = |e^v| \le |M_\mu| + |M_{-\mu}| \le e^{|k| o_z(1)}.$$

This shows that

$$\mathrm{Re}(v) \le |k|\, o_z(1).$$

It remains to prove that

$$\mathrm{Re}(v) \ge -\,|k|\, o_z(1),$$

or equivalently,

$$|e^{-ikz} u_\sigma| \ge e^{-|k| o_z(1)}.$$

This is the only point where we need to use the more general solutions $f_{\lambda\mu}$ with λ not equal to 1. We write

$$u_\sigma = \frac{f_\mu + f_{-\mu}}{2}\left(1 + \frac{\overline{f_\mu} - \overline{f_{-\mu}}}{f_\mu + f_{-\mu}}\right) = \frac{f_\mu + f_{-\mu}}{2}\overline{\left[1 + e^{it}\frac{f_\mu - f_{-\mu}}{f_\mu + f_{-\mu}}\right]},$$

where $t \in \mathbb{R}$ is chosen so that

$$e^{-it} = \frac{\overline{f_\mu} + \overline{f_{-\mu}}}{f_\mu + f_{-\mu}}.$$

Therefore

$$|e^{-ikz} u_\sigma| = \left|\frac{M_\mu + M_{-\mu}}{2}\right| \left|1 + e^{it} \frac{f_\mu - f_{-\mu}}{f_\mu + f_{-\mu}}\right|.$$

Here

$$\left|\frac{M_\mu + M_{-\mu}}{2}\right| = \left|\frac{1 + M_{-\mu}/M_\mu}{2M_\mu}\right| \geq e^{-|k| o_z(1)}$$

since $\mathrm{Re}(M_{-\mu}/M_\mu) > 0$ by Lemma 5.6.11 and by Proposition 5.10.1. The result will follow if we can prove that

$$\inf_{t \in \mathbb{R}} \left|1 + e^{it} \frac{f_\mu - f_{-\mu}}{f_\mu + f_{-\mu}}\right| \geq e^{-|k| o_z(1)}.$$

To show the last inequality, we note that

$$1 + e^{it} \frac{f_\mu - f_{-\mu}}{f_\mu + f_{-\mu}} = \frac{2\tilde{f}_t}{f_\mu + f_{-\mu}},$$

where the function

$$\tilde{f}_t := \frac{1}{2}\left[(1 + e^{it}) f_\mu + (1 - e^{it}) f_{-\mu}\right]$$

satisfies, by a direct computation, the equation

$$\bar{\partial} \tilde{f}_t = e^{it} \mu \overline{\partial \tilde{f}_t}.$$

The properties of $f_{\pm\mu}$ imply that $\tilde{f}_t$ is a $W^{1,2}_{\mathrm{loc}}(\mathbb{R}^2)$ function satisfying $\tilde{f}_t = e^{ikz}(1 + O(1/z))$ as $z \to \infty$. The uniqueness of CGO solutions, as discussed in the beginning of this section, shows that $\tilde{f}_t = f_{\lambda\mu}$ for $\lambda = e^{it}$. Consequently, by Proposition 5.10.1,

$$\left|1 + e^{it} \frac{f_\mu - f_{-\mu}}{f_\mu + f_{-\mu}}\right| = \left|\frac{2M_{\lambda\mu}}{M_\mu + M_{-\mu}}\right| \geq \frac{2|M_{\lambda\mu}|}{|M_\mu| + |M_{-\mu}|} \geq e^{-|k| o_z(1)}.$$

This concludes the proof. □

5.11. Notes

We recall from §1.1 the heuristic dimension count argument showing that the Calderón problem for $n = 2$ is formally well determined as opposed to being formally overdetermined when $n \geq 3$. This suggests that the two-dimensional case could be different, and indeed the uniqueness, stability, and reconstruction results in Chapter 4 fail when $n = 2$ essentially since one cannot find three orthonormal vectors in $\mathbb{R}^2$. Thus, for fixed $\xi \in \mathbb{R}^2$ it is not possible to find $\zeta_1, \zeta_2 \in \mathbb{C}^2$ with $\zeta_j \cdot \zeta_j = 0$ such that $\zeta_1 + \zeta_2 = \xi$ as $|\zeta_j| \to \infty$. Instead of using CGO solutions in the high frequency limit as $|\zeta| \to \infty$, the arguments in this chapter employ CGO solutions for all values $\zeta \in \mathbb{C}^2$ with $\zeta \cdot \zeta = 0$.

Uniqueness for $W^{2,p}$ conductivities in two dimensions for $p > 1$ was proved in **[Nac96]**, with an early result for conductivities close to constant given in **[SU86]**. The method of **[Nac96]** was based on a reduction to the Schrödinger equation $(-\Delta + q)u = 0$ and on a version of the $\bar\partial$-scattering method developed in **[NA84, BC85]**. The uniqueness result was improved to $W^{1,p}$ conductivities with $p > 2$ in **[BU97]** by using a different reduction involving a matrix $\bar\partial$-type equation

$$\left[\begin{pmatrix} \bar\partial & 0 \\ 0 & \partial \end{pmatrix} - \begin{pmatrix} 0 & q \\ \bar q & 0 \end{pmatrix}\right]\begin{pmatrix} v_1 \\ v_2 \end{pmatrix} = 0,$$

as well as the $\bar\partial$-scattering method and theory of pseudoanalytic functions (i.e., solutions of $\bar\partial f = af + b\bar f$).

The definitive result of **[AP06a]** established uniqueness for L^∞ conductivities, this time via a reduction to a Beltrami equation $\bar\partial f = \mu \overline{\partial f}$. Solutions of such an equation are quasiregular maps, and the theory of these maps plays a central role in the argument. We have presented the approach of **[AP06a]** in this chapter, except that in §5.7 we temporarily added the assumption that the conductivity is C^2 and gave a shortcut to prove uniqueness by the method of **[Nac96]**. We have partly followed another textbook treatment of the Astala-Päivärinta approach given in **[AIM09]**, which also contains a comprehensive account of the theory of quasiregular maps.

One can go even beyond L^∞ conductivities in two dimensions. There are uniqueness results for positive measurable conductivities σ in H^1 with $\|\nabla \log \sigma\|_{L^2}$ small in **[CW18]**, and for σ with $\log \sigma \in \dot H^1$ (homogeneous Sobolev space) in **[NRT20]**. The work **[ALP16]** studies the borderline between visibility and invisibility, and gives both uniqueness results and counterexamples (cloaking constructions) involving exponential integrability conditions.

Stability and reconstruction results are available for Nachman's method **[Nac96, Liu97]**, the Brown-Uhlmann method **[AP06b, BBR01, KT04]**, and the Astala-Päivärinta method **[BFR07, CFR10, FR13, FP18]**. Further reconstruction schemes may be found in **[Nov88, HM08, HN11]** also on Riemann surfaces.

The anisotropic Calderón problem for matrix conductivities or Riemannian metrics is rather well understood in two dimensions. A matrix conductivity can only be determined up to a change of coordinates, and for $n = 2$ there exist special *isothermal coordinates* that reduce the anisotropic problem to an isotropic one involving scalar conductivities. This observation was made in **[Syl90]**. Consequently the isotropic results **[Nac96, BU97, AP06a]** lead to corresponding results in **[Nac96, SU03, APL05]** for the anisotropic Calderón problem when $n = 2$. Results for conductivity tensors on domains with more complicated topology and on Riemann surfaces are given in [**HS10, HS12**].

In two dimensions, the inverse problem for the Laplace–Beltrami equation $\Delta_g u = 0$ is not equivalent to the problem for matrix conductivities due to an additional conformal invariance of the Laplace–Beltrami operator (see Exercise 3.6.2). The inverse problem for Δ_g in two dimensions was studied in [**LU01**], where it was shown that a smooth Riemannian metric g is uniquely determined by the DN map up to conformal scaling and diffeomorphism. The method was based on boundary determination and identifying interior points with corresponding Green functions. Alternative methods using maximal ideals and Banach algebra theory [**Bel03, PSU23**], embeddings into $\mathbb{C}^n$ [**HM07**], and Poisson kernels [**LLS20**] are available.

The scattering transforms discussed in this chapter appear in solvability results for completely integrable nonlinear equations based on inverse scattering methods. The basic example is the KdV equation (a one-dimensional nonlinear evolution equation) that can be solved by taking a scattering transform of the initial data, solving a *linear* evolution equation, and then taking the inverse scattering transform [**GGKM67**]. Other completely integrable equations, such as certain Novikov-Veselov or Davey-Stewartson equations, can be transformed into linear evolution equations by the same scheme. The scattering transform can be understood as a nonlinear Fourier transform, and one can ask for its mapping properties or analogs of the Fourier inversion and Plancherel theorems. Further information on inverse scattering methods related to the material in this chapter may be found in [**FA84, BC89, AC91, Per14, NRT20**].

The methods in Chapter 5 apply to the Schrödinger equation $(-\Delta+q)u = 0$ only when the potential q is of *conductivity type*, i.e., of the form

$$q = \frac{\Delta\sigma^{1/2}}{\sigma^{1/2}}$$

for some positive conductivity function γ. The case of general potentials had to wait until the work of Bukhgeim [**Buk08**] who introduced a new method in two-dimensional inverse problems.

In the method of Bukhgeim, the CGO solutions $u(z) = e^{ikz}(1 + r)$ with linear phase are replaced by CGO solutions with quadratic phase, of the form

$$u_1 = e^{is(z-z_0)^2}(1 + r_1), \qquad u_2 = e^{is(\bar{z}-\bar{z}_0)^2}(1 + r_2),$$

where z_0 is a fixed point, $s > 0$ is a large parameter, and the remainder terms r_j have certain decay as $s \to \infty$. Then the product $u_1 u_2$ takes the form

$$u_1 u_2 = e^{2is\mathrm{Re}((z-z_0)^2)}(1 + r_1 + r_2 + r_1 r_2).$$

The phase function $\mathrm{Re}((z - z_0)^2)$ is real and quadratic with a nondegenerate critical point at z_0. By the *stationary phase method*, one has the limit

$$\lim_{s\to\infty} \frac{2s}{\pi} \int_\Omega (q_1 - q_2) e^{2is\mathrm{Re}((z-z_0)^2)}\, dx = (q_1 - q_2)(z_0).$$

By additional arguments, one can show that the corresponding part involving remainder terms converges to zero if q_j have certain regularity. Thus, if $\Lambda_{q_1} = \Lambda_{q_2}$, this argument proves that $q_1 = q_2$ pointwise in Ω. We note that this method works for complex potentials and therefore also for complex conductivities with positive real part and sufficient regularity, unlike the methods in Chapter 5 that require the conductivity to be real valued.

The method of Bukhgeim has been extended to low regularity settings with L^p potentials for $p > 4/3$ [**BIY15, BTW20**], there are stability results [**BIY15**] and reconstruction results [**AFR16, Tej17**]. The method has been extended to compact Riemann surfaces with boundary, and in this case the quadratic phase functions $(z - z_0)^2$ are replaced by Morse holomorphic functions [**GT10, GT11a**]. One can also determine several coefficients (metric, magnetic field, potential) up to natural gauges [**GT11b, IUY12, CLT24**]. Inverse problems for systems have been treated as well [**AGTU13, IY12b, IY15, IY16, IY17**]. Many of the previous results apply in the case of partial boundary measurements.

Survey articles on the Calderón problem in two dimensions may be found in [**Uhl03, GT13, ALP13, IY13b**].

Chapter 6

Partial Data

In Chapters 4 and 5 we proved that if the DN maps for two conductivities coincide on the whole boundary, then the conductivities are equal. In this chapter we consider the case where measurements are made only on part of the boundary.

The first result that we will prove is due to Isakov [**Isa07**]. It states that if one knows the DN map on a open subset $\Gamma \subset \partial\Omega$ for any Dirichlet data supported in the same set Γ, and if the inaccessible part $\Gamma_0 = \partial\Omega \setminus \Gamma$ is part of a hyperplane, then this data determines the conductivity.

Theorem 6.0.1. *Let $\Omega \subset \mathbb{R}^n$ be a bounded domain with smooth boundary, where $n \geq 3$, and let γ_1 and γ_2 be two positive functions in $C^2(\overline{\Omega})$. Assume that $\Omega \subset \{x_n > 0\}$, let $\Gamma_0 = \partial\Omega \cap \{x_n = 0\}$, and let $\Gamma = \partial\Omega \setminus \Gamma_0$. If*

$$\Lambda_{\gamma_1} f|_\Gamma = \Lambda_{\gamma_2} f|_\Gamma \quad \textit{for all } f \in H^{1/2}(\partial\Omega) \textit{ with } \operatorname{supp}(f) \subset \Gamma,$$

then $\gamma_1 = \gamma_2$ in Ω.

The result follows quite easily from the full data results in Chapter 4 by using a reflection argument. We will also prove a similar theorem where the inaccessible set Γ_0 is part of a sphere.

For more general domains, the first partial data result was proved by Bukhgeim and Uhlmann [**BU02**]. It involves a unit vector $\alpha \in \mathbb{R}^n$ and the subsets of the boundary

$$\partial\Omega_{-,\varepsilon} = \{\, x \in \partial\Omega \mid \alpha \cdot \nu(x) < \varepsilon \,\},$$
$$\partial\Omega_{+,\varepsilon} = \{\, x \in \partial\Omega \mid \alpha \cdot \nu(x) > \varepsilon \,\},$$

where ν is the unit outer normal of $\partial\Omega$. The theorem is as follows.

Theorem 6.0.2. *Let $\Omega \subset \mathbb{R}^n$ be a bounded open set with smooth boundary, where $n \geq 3$, and let γ_1 and γ_2 be two positive functions in $C^2(\overline{\Omega})$. If $\alpha \in \mathbb{R}^n$ is a unit vector, and if for some $\varepsilon > 0$ one has $\gamma_1|_{\partial\Omega_{+,\varepsilon}} = \gamma_2|_{\partial\Omega_{+,\varepsilon}}$ and*

$$\Lambda_{\gamma_1} f|_{\partial\Omega_{-,\varepsilon}} = \Lambda_{\gamma_2} f|_{\partial\Omega_{-,\varepsilon}} \quad \text{for all } f \in H^{1/2}(\partial\Omega),$$

then $\gamma_1 = \gamma_2$ in Ω.

The proof is based on CGO solutions, but it requires new elements since we need some control of the solutions on parts of the boundary. The main tool is a weighted norm estimate known as a *Carleman estimate*. This estimate also gives rise to a new construction of CGO solutions that does not require Fourier analysis.

6.1. Reflection approach

As before, we will obtain the partial data result for the conductivity equation by proving a uniqueness result for the Schrödinger equation.

Theorem 6.1.1 (Partial data for Schrödinger). *Let $\Omega \subset \{x_n > 0\}$ be a bounded open set with smooth boundary, let $\Gamma_0 = \partial\Omega \cap \{x_n = 0\}$, and let $\Gamma = \partial\Omega \setminus \Gamma_0$. Let $q_1, q_2 \in L^\infty(\Omega)$ be such that 0 is not a Dirichlet eigenvalue for $-\Delta + q_j$ in Ω. If one has*

$$\Lambda_{q_1} f|_\Gamma = \Lambda_{q_2} f|_\Gamma \quad \text{for all } f \in H^{1/2}(\partial\Omega) \text{ with } \operatorname{supp}(f) \subset \Gamma,$$

then $q_1 = q_2$ in Ω.

Theorem 6.0.1 is an immediate consequence of Theorem 6.1.1 and the unique continuation result proved later in Theorem 6.4.1.

Proof of Theorem 6.0.1. Define $q_j = \Delta(\gamma_j^{1/2})/\gamma_j^{1/2}$. By Proposition 2.8.5 the Dirichlet problem for $-\Delta + q_j$ in Ω is well-posed, and the DN maps are related by

$$\Lambda_{q_j} f = \gamma_j^{-1/2} \Lambda_{\gamma_j}(\gamma_j^{-1/2} f) + \frac{1}{2}\gamma_j^{-1}(\partial_\nu \gamma_j) f|_{\partial\Omega}.$$

Recall from the boundary determination results, Theorems 3.1.1 and 3.2.1, that the knowledge of Λ_γ on Γ for any f supported in Γ determines $\gamma|_\Gamma$ and $\partial_\nu \gamma|_\Gamma$ uniquely. The assumption

$$\Lambda_{\gamma_1} f|_\Gamma = \Lambda_{\gamma_2} f|_\Gamma \quad \text{for all } f \in H^{1/2}(\partial\Omega) \text{ with } \operatorname{supp}(f) \subset \Gamma,$$

therefore implies that

$$\gamma_1|_\Gamma = \gamma_2|_\Gamma, \quad \partial_\nu \gamma_1|_\Gamma = \partial_\nu \gamma_2|_\Gamma.$$

The expression for Λ_{q_j} above shows that

$$\Lambda_{q_1} f|_\Gamma = \Lambda_{q_2} f|_\Gamma \quad \text{for all } f \in H^{1/2}(\partial\Omega) \text{ with } \operatorname{supp}(f) \subset \Gamma.$$

By Theorem 6.1.1, we have $q_1 = q_2$ in Ω. Define $q = q_1 = q_2$. Since $q_j = \Delta\gamma_j^{1/2}/\gamma_j^{1/2}$, we have

$$(-\Delta + q)\gamma_j^{1/2} = 0 \qquad \text{in } \Omega.$$

We also have the boundary conditions

$$\gamma_1^{1/2}|_\Gamma = \gamma_2^{1/2}|_\Gamma, \qquad \partial_\nu\gamma_1^{1/2}|_\Gamma = \partial_\nu\gamma_2^{1/2}|_\Gamma.$$

By the unique continuation principle (Theorem 6.4.1), any two solutions of $(-\Delta+q)u = 0$ having the same Cauchy data on an open subset of the boundary must be equal in the domain Ω which is connected. This proves that $\gamma_1 = \gamma_2$. □

We move to the proof of Theorem 6.1.1. In the present setting where the inaccessible part of the boundary is part of the hyperplane $\{x_n = 0\}$, it is natural to use the reflection which takes a complex vector

$$z = (z', z_n) \in \mathbb{C}^n,$$

where $z' = (z_1, \ldots, z_{n-1}) \in \mathbb{C}^{n-1}$, to the vector

$$z^* = (z', -z_n).$$

Proof of Theorem 6.1.1. Recall from Proposition 2.8.3 the integral identity

$$\langle(\Lambda_{q_1} - \Lambda_{q_2})(u_1|_{\partial\Omega}), u_2|_{\partial\Omega}\rangle_{\partial\Omega} = \int_\Omega (q_1 - q_2)u_1u_2\,dx$$

valid for any $u_j \in H^1(\Omega)$ with $(-\Delta + q_j)u_j = 0$ in Ω. Assume that $u_1|_{\partial\Omega}$ is supported in Γ. In this case we know that

$$\Lambda_{q_1}(u_1|_{\partial\Omega})|_\Gamma = \Lambda_{q_2}(u_1|_{\partial\Omega})|_\Gamma,$$

and consequently $(\Lambda_{q_1} - \Lambda_{q_2})(u_1|_{\partial\Omega})$ vanishes on Γ. If also $u_2|_{\partial\Omega}$ is supported in Γ, the whole boundary integral is zero. Therefore, our partial data assumption implies that

$$\int_\Omega (q_1 - q_2)u_1u_2\,dx = 0 \tag{6.1}$$

for any $u_j \in H^1(\Omega)$ with $(-\Delta + q_j)u_j = 0$ in Ω and $\operatorname{supp}(u_j|_{\partial\Omega}) \subset \Gamma$.

We now use the reflection idea and consider the open set

$$U = \Omega \cup \operatorname{int}(\Gamma_0) \cup \Omega^*,$$

where the interior of Γ_0 is relative to $\{x_n = 0\}$, and

$$\Omega^* = \{\, x^* \mid x \in \Omega \,\}.$$

We also define the even extensions of the potentials,

$$q_j^e(x) = \begin{cases} q_j(x) & \text{if } x \in \Omega, \\ q_j(x^*) & \text{if } x \in \Omega^*. \end{cases}$$

These are L^∞ functions in U.

If $\zeta_1, \zeta_2 \in \mathbb{C}^n$ satisfy $\zeta_j \cdot \zeta_j = 0$ with $|\zeta_j|$ large enough, Theorem 4.2.1 shows that there exist $w_j \in H^2(U)$ satisfying

$$(-\Delta + q_j)w_j = 0 \quad \text{in } U,$$

having the form

$$w_j = e^{\zeta_j \cdot x}(1 + \psi_j),$$

where

$$\left\|\psi_j\right\|_{L^2(U)} \leq \frac{C}{|\zeta_j|}.$$

We have made a small change in notation and chosen solutions that resemble $e^{\zeta \cdot x}$ instead of $e^{i\zeta \cdot x}$.

We now define functions $u_j(x) = w_j(x) - w_j^*(x)$ in the original domain Ω, where we write

$$f^*(x) = f(x^*).$$

An easy computation shows that $u_j \in H^2(\Omega)$ satisfies $(-\Delta + q_j)u_j = 0$ in Ω. Since the set $\{x_n = 0\}$ is invariant under reflection, we have $u_j|_{\Gamma_0} = 0$. Thus (6.1) is valid for these choices of u_j.

We wish to compute the product $u_1 u_2$. Since

$$u_j = e^{\zeta_j \cdot x}(1 + \psi_j) - e^{\zeta_j \cdot x^*}(1 + \psi_j^*),$$

we have

$$\begin{aligned} u_1 u_2 = {} & e^{(\zeta_1 + \zeta_2)\cdot x}(1 + \psi_1)(1 + \psi_2) - e^{(\zeta_1^* + \zeta_2)\cdot x}(1 + \psi_1^*)(1 + \psi_2) \\ & - e^{(\zeta_1 + \zeta_2^*)\cdot x}(1 + \psi_1)(1 + \psi_2^*) + e^{(\zeta_1 + \zeta_2)\cdot x^*}(1 + \psi_1^*)(1 + \psi_2^*). \end{aligned}$$

We would like to arrange $u_1 u_2$ to look like $e^{ix\cdot\xi}$ for given $\xi \in \mathbb{R}^n$ when $|\zeta_j|$ are large. In particular, we do not want $u_1 u_2$ to grow exponentially with respect to $|\zeta_j|$. It will be useful to choose ζ_j so that

$$\zeta_j \cdot \zeta_j = 0, \quad \zeta_1 + \zeta_2 = i\xi, \quad \mathrm{Re}(\zeta_1^* + \zeta_2) = \mathrm{Re}(\zeta_1 + \zeta_2^*) = 0.$$

Write $\xi = (\xi', \xi_n)$, where $\xi' = (\xi_1, \ldots, \xi_{n-1})$. We assume for the moment that $|\xi'| > 0$. It will be convenient to use the reflection invariant unit vector

$$v_1 = \frac{1}{|\xi'|}(\xi', 0).$$

Let $v_3 = e_n$ be the nth coordinate vector, and choose some unit vector $v_2 \in \mathbb{R}^n$ such that

$$v_2 \cdot v_1 = v_2 \cdot v_3 = 0.$$

Then $\{v_1, v_2, v_3\}$ are orthogonal unit vectors (here we used that $n \geq 3$). If $\tau > 0$ is large, define the complex vectors

$$\zeta_1 = |\xi| \sqrt{\tau^2 + \frac{1}{4}} v_2 + i\left[\left(\frac{1}{2}|\xi'| + \tau\xi_n\right) v_1 + \left(\frac{1}{2}\xi_n - \tau|\xi'|\right) v_3\right],$$

$$\zeta_2 = -|\xi| \sqrt{\tau^2 + \frac{1}{4}} v_2 + i\left[\left(\frac{1}{2}|\xi'| - \tau\xi_n\right) v_1 + \left(\frac{1}{2}\xi_n + \tau|\xi'|\right) v_3\right].$$

It is easy to check that $\zeta_1 \cdot \zeta_1 = \zeta_2 \cdot \zeta_2 = 0$ and

$$|\zeta_1| = |\zeta_2| = \sqrt{2}\,|\xi| \sqrt{\tau^2 + 1/4}.$$

We also have

$$\zeta_1 + \zeta_2 = i(|\xi'|\, v_1 + \xi_n v_3) = i\xi$$

and, since v_1 and v_2 are reflection invariant,

$$\zeta_1^* + \zeta_2 = i(|\xi'|\, v_1 + 2\tau\,|\xi'|\, v_3),$$

$$\zeta_1 + \zeta_2^* = i(|\xi'|\, v_1 - 2\tau\,|\xi'|\, v_3).$$

Thus ζ_1 and ζ_2 satisfy all the properties mentioned above.

Keeping ξ fixed and using that $\left\|\psi_j\right\|_{L^2(\Omega)} \leq C/\tau$ and $\|\psi_j^*\|_{L^2(\Omega)} \leq C/\tau$, all terms involving the remainder terms ψ_j will be small as $\tau \to \infty$. By (6.1) we have

$$\begin{aligned} 0 &= \int_\Omega (q_1 - q_2) u_1 u_2 \, dx \\ &= \int_\Omega (q_1 - q_2)(e^{ix\cdot\xi} - e^{ix\cdot\eta_1(\tau)} - e^{ix\cdot\eta_2(\tau)} + e^{ix^*\cdot\xi}) \, dx + o(1) \end{aligned}$$

as $\tau \to \infty$, where $\eta_1(\tau) = |\xi'|\, v_1 + 2\tau\,|\xi'|\, v_3$ and $\eta_2(\tau) = |\xi'|\, v_1 - 2\tau\,|\xi'|\, v_3$. Defining $q(x) = q_1(x) - q_2(x)$ when $x \in \Omega$ and $q(x) = 0$ otherwise, we have

$$\int_\Omega q e^{ix\cdot\eta_j(\tau)} \, dx = \hat{q}(-\eta_j(\tau)).$$

Now $q \in L^1(\mathbb{R}^n)$, so $\hat{q}(\eta) \to 0$ as $|\eta| \to \infty$ by the Riemann–Lebesgue lemma (Lemma B.2.8). Since $|\eta_j(\tau)| \to \infty$ as $\tau \to \infty$, we have

$$\lim_{\tau\to\infty} \int_\Omega q e^{ix\cdot\eta_j(\tau)} \, dx = 0.$$

Therefore,

$$\begin{aligned} &\int_\Omega (q_1 - q_2)(e^{ix\cdot\xi} + e^{ix^*\cdot\xi}) \, dx \\ &\qquad = \lim_{\tau\to\infty} \int_\Omega (q_1 - q_2)(e^{ix\cdot\xi} - e^{ix\cdot\eta_1(\tau)} - e^{ix\cdot\eta_2(\tau)} + e^{ix^*\cdot\xi}) \, dx \\ &\qquad = 0. \end{aligned}$$

In the integral involving $e^{ix^*\cdot\xi}$, we can make the change of variables $x \to x^*$ to obtain the following statement for even extensions of q_j in the double domain U:

$$\int_U (q_1^e - q_2^e)e^{ix\cdot\xi}\,dx = 0.$$

Thus $(q_1^e - q_2^e)\hat{\ }(-\xi) = 0$. This is true for any fixed $\xi \in \mathbb{R}^n$ with $|\xi'| > 0$. But since $q_1^e - q_2^e$ is compactly supported in $\mathbb{R}^n$, its Fourier transform is continuous, and we have $(q_1^e - q_2^e)\hat{\ }(-\xi) = 0$ for any $\xi \in \mathbb{R}^n$. This shows that $q_1 = q_2$. □

Note that in the previous proof, one of the crucial points was the fact that a solution of the Schrödinger equation in the double domain $\Omega \cup \mathrm{int}(\Gamma_0) \cup \Omega^*$ can be reflected from Ω^* to produce a solution of the Schrödinger equation back in Ω. That is, the Schrödinger operator $-\Delta + q$ should be preserved under reflection. This is quite clear for the reflection $(x', x_n) \mapsto (x', -x_n)$, but one can ask if there are other reflection operators that have this property.

This question can be answered by using some differential (or Riemannian) geometry facts. However, to keep things simple, we will describe the argument in a self-contained way. Let $F : U \to V$ be a C^∞ bijective map between two open subsets of $\mathbb{R}^n$ (the map F is our reflection operator). We denote the Euclidean Laplacian in V by

$$\Delta_e = \sum_{j=1}^n \frac{\partial^2}{\partial x_j^2}.$$

Here e is the Euclidean metric, corresponding to the identity matrix $e(x) = I = [\delta_{j,k}]_{j,k=1}^n$. We want to compute how Δ_e transforms under the reflection F. To do this, we will "pull back" quantities on V by the map F into quantities on U.

First, define the pullback of a function $v \in C^\infty(V)$ as the function

$$F^*v(x) = (v \circ F)(x), \quad x \in U.$$

Next, define the pullback of the Euclidean metric e on V as the matrix function

$$F^*e(x) = (DF(x))^t DF(x), \quad x \in U.$$

Also, if $g = [g_{j,k}(x)]_{j,k=1}^n$ is a positive definite matrix function whose entries are C^∞ functions on U, as in §3.6 we define the Laplace–Beltrami operator

$$\Delta_g u = |g|^{-1/2} \sum_{j,k=1}^n \frac{\partial}{\partial x_j}\left[|g|^{1/2} g^{j,k} \frac{\partial u}{\partial x_k}\right], \quad u \in C^\infty(U).$$

Here $[g^{j,k}]_{j,k=1}^n$ is the inverse matrix of $[g_{j,k}]_{j,k=1}^n$, and $|g| = \det[g_{j,k}]$.

The next exercise shows that under a map F, the Euclidean Laplacian Δ_e transforms into the Laplace–Beltrami operator Δ_{F^*e}.

Exercise 6.1.2. Let $F : U \to V$ be a bijective C^∞ map between open sets of F. If $v \in C^\infty(V)$, show that

$$F^*(\Delta_e v)(x) = (\Delta_{F^*e} F^* v)(x), \quad x \in U.$$

Our question was to find those reflection operators F such that the Euclidean Schrödinger operator $-\Delta_e + q$ transforms into another Euclidean Schrödinger operator $-\Delta_e + \tilde{q}$ under F. This is true when $F^*e = e$, or equivalently when $(DF)^t DF$ is the identity matrix. The reflection $F(x', x_n) = (x', -x_n)$ satisfies this property.

However, we also know that it is possible to convert conductivity operators $v \mapsto \operatorname{div}(\gamma \nabla v)$ into Schrödinger operators by choosing $v = \gamma^{-1/2} u$. Thus we could ask to find reflections F such that $F^*e = ce$ for some positive scalar function c. Using the definition of F^*e and taking determinants, the last condition is equivalent with

$$(DF(x))^t DF(x) = \det(DF(x))^{2/n} \operatorname{Id}.$$

Any C^∞ bijective map satisfying this equation is called a *conformal transformation.*

It is a theorem of Liouville (see [**IM01**]) that any conformal transformation between two open subsets of $\mathbb{R}^n$, $n \geq 3$, is obtained by composing rotations, translations, scalings, reflections $(x', x_n) \mapsto (x', -x_n)$, and *Kelvin transforms*

$$F(x) = \frac{x}{|x|^2}, \quad x \in \mathbb{R}^n \setminus \{0\}.$$

In the same way that $(x', x_n) \mapsto (x', -x_n)$ reflects across the hyperplane $\{x_n = 0\}$, the Kelvin transform acts as a reflection across the unit sphere $\{|x| = 1\}$. It also has the property that it maps hyperplanes and spheres in $\mathbb{R}^n \setminus \{0\}$ to hyperplanes and spheres.

Exercise 6.1.3. Show that F maps the set $\{x \in \mathbb{R}^n\,;\, 0 < |x| < 1\}$ onto the set $\{x \in \mathbb{R}^n\,;\, |x| > 1\}$ and preserves the set $\{|x| = 1\}$. Show also that $F(F(x)) = x$, and that

$$F^*e(x) = |x|^{-4} \operatorname{Id}.$$

Exercise 6.1.4. Show that F maps the spherical set

$$\{\, x = (x', x_n) \in \mathbb{R}^n \mid |x'|^2 + (x_n - 1/2)^2 = (1/2)^2,\ x \neq 0 \,\}$$

onto the hyperplane $\{\, (x', x_n) \mid x_n = 1 \,\}$.

We now give the partial data results for the conductivity and Schrödinger equations for the case where the inaccessible part of the boundary is part of a sphere. The first theorem follows from the second one exactly as in the case where part of the boundary is flat, so we will only prove the Schrödinger case.

Theorem 6.1.5. *Let $\Omega \subset \mathbb{R}^n$ be a bounded domain with smooth boundary, where $n \geq 3$, and let γ_1 and γ_2 be two positive functions in $C^2(\overline{\Omega})$. Assume that $\Omega \subset B$ for some open ball B in $\mathbb{R}^n$, let $\Gamma_0 = \partial\Omega \cap \partial B$, and $\Gamma = \partial\Omega \setminus \Gamma_0$. Assume also that $\partial B \setminus \partial\Omega \neq \emptyset$. If*

$$\Lambda_{\gamma_1} f|_\Gamma = \Lambda_{\gamma_2} f|_\Gamma \quad \textit{for all } f \in H^{1/2}(\partial\Omega) \textit{ with } \operatorname{supp}(f) \subset \Gamma,$$

then $\gamma_1 = \gamma_2$ in Ω.

Theorem 6.1.6. *Let $\Omega \subset \mathbb{R}^n$ be a bounded open set with smooth boundary, where $n \geq 3$, and let $q_1, q_2 \in L^\infty(\Omega)$ be such that 0 is not a Dirichlet eigenvalue of $-\Delta + q_1$ or $-\Delta + q_2$ in Ω. Assume that $\Omega \subset B$ for some open ball B in $\mathbb{R}^n$, let $\Gamma_0 = \partial\Omega \cap \partial B$, and let $\Gamma = \partial\Omega \setminus \Gamma_0$. Assume also that $\partial B \setminus \partial\Omega \neq \emptyset$. If*

$$\Lambda_{q_1} f|_\Gamma = \Lambda_{q_2} f|_\Gamma \quad \textit{for all } f \in H^{1/2}(\partial\Omega) \textit{ with } \operatorname{supp}(f) \subset \Gamma,$$

then $q_1 = q_2$ in Ω.

Proof. Since $\Omega \subset B$ and $\partial B \setminus \partial\Omega \neq \emptyset$, there is some $x_0 \in \partial B$ with $x_0 \notin \overline{\Omega}$. Thus there is a small ball centered at x_0 which does not intersect Ω. We may choose coordinates so that $x_0 = 0$ and $B \subset \{x_n > 0\}$. If B has radius r, it follows that

$$B = \left\{ x = (x', x_n) \mid |x'|^2 + |x_n - r|^2 < r^2 \right\}.$$

For simplicity we will assume that $r = 1/2$ (the general case can be reduced to this by scaling).

Let $F(x) = x/|x|^2$ be the Kelvin transform as above. Define $\tilde{\Omega} = F^{-1}(\Omega)$, and note that $\tilde{\Omega}$ is a bounded domain with C^∞ boundary such that $F^{-1}(\Gamma_0)$ is contained in the hyperplane $\{ (x', x_n) \mid x_n = 1 \}$ by Exercise 6.1.4. If $u \in H^1(\Omega)$ solves $(-\Delta_e + q)u = 0$ in Ω, then by Exercise 6.1.2, $\tilde{u} = F^* u \in H^1(\tilde{\Omega})$ solves

$$(-\Delta_{F^*e} + F^*q)\tilde{u} = 0 \qquad \text{in } \tilde{\Omega}.$$

Using Exercise 6.1.3, we have

$$\Delta_{F^*e} v = \Delta_{|x|^{-4}e} v = |x|^{2n} \sum_{j=1}^{n} \partial_j(|x|^{-2n+4} \partial_j v).$$

This looks like the conductivity operator $\operatorname{div}(\gamma \nabla v)$ with $\gamma(x) = |x|^{-2n+4}$. The substitution $\tilde{u} = \gamma^{-1/2}\tilde{v} = |x|^{n-2}\, \tilde{v}$ gives, as in Proposition 2.8.5, that

$$\Delta_{F^*e}(|x|^{n-2}\, \tilde{v}) = |x|^{n+2} (\Delta_e - c)\tilde{v}, \tag{6.2}$$

where

$$c(x) = \frac{\Delta(|x|^{2-n})}{|x|^{2-n}}.$$

However, since $|x|^{2-n}$ is harmonic when $n \geq 3$ and $x \neq 0$, we have $c = 0$. Combining these facts, we have seen that if $u \in H^1(\Omega)$ solves $(-\Delta_e + q)u = 0$ in Ω, then $\tilde{v} = |x|^{2-n} F^* u \in H^1(\tilde{\Omega})$ solves

$$(-\Delta_e + |x|^{-4} F^* q)\tilde{v} = 0 \qquad \text{in } \tilde{\Omega}.$$

Tracing back the steps, we see that if $\tilde{v}$ solves the above equation, then $u = F^*(|x|^{n-2}\,\tilde{v})$ solves $(-\Delta_e + q)u = 0$ in Ω (recall that $F(F(x)) = x$). It also follows that the Dirichlet problem for $(-\Delta_e + q)u = 0$ in Ω is well-posed if and only if the Dirichlet problem for the corresponding equation for $\tilde{v}$ is well-posed in $\tilde{\Omega}$.

Denote by Λ_q the DN map for the equation $(-\Delta_e + q)u = 0$ in Ω, and by $\tilde{\Lambda}_{\tilde{q}}$ the DN map for the equation $(-\Delta_e + \tilde{q})\tilde{v} = 0$ in $\tilde{\Omega}$. We will prove that if

$$\Lambda_{q_1} f|_\Gamma = \Lambda_{q_2} f|_\Gamma \quad \text{for all } f \in H^{1/2}(\partial\Omega) \text{ with } \operatorname{supp}(f) \subset \Gamma,$$

then

$$\tilde{\Lambda}_{|x|^{-4}F^*q_1} \tilde{f}|_{\tilde{\Gamma}} = \tilde{\Lambda}_{|x|^{-4}F^*q_2} \tilde{f}|_{\tilde{\Gamma}} \quad \text{for all } \tilde{f} \in H^{1/2}(\partial\tilde{\Omega}) \text{ with } \operatorname{supp}(\tilde{f}) \subset \tilde{\Gamma}.$$

Here $\tilde{\Gamma} = F^{-1}(\Gamma)$. Since $\partial\tilde{\Omega} \setminus \tilde{\Gamma} = F^{-1}(\Gamma_0)$ is contained in the hyperplane $\{x_n = 1\}$ and since $\tilde{\Omega}$ is contained in $\{x_n > 1\}$, this would imply by Theorem 6.1.1 that $|x|^{-4} F^* q_1 = |x|^{-4} F^* q_2$ in $\tilde{\Omega}$, and therefore $q_1 = q_2$ in Ω.

Let $\tilde{f}_j \in H^{1/2}(\partial\tilde{\Omega})$ with $\operatorname{supp}(\tilde{f}_j) \subset \tilde{\Gamma}$, and let $\tilde{v}_j \in H^1(\tilde{\Omega})$ solve

$$(-\Delta_e + |x|^{-4} F^* q_j)\tilde{v}_j = 0 \quad \text{in } \tilde{\Omega}$$

with boundary condition $\tilde{v}_j|_{\partial\tilde{\Omega}} = \tilde{f}_j$. Define $u_j = F^*(|x|^{n-2}\,\tilde{v}_j) \in H^1(\Omega)$, so that $(-\Delta_e + q_j)u_j = 0$ in Ω with $u_j|_{\partial\Omega} = F^*(|x|^{n-2}\,\tilde{f}_j)$. We now note that, by Proposition 2.8.3,

$$\langle (\tilde{\Lambda}_{|x|^{-4}F^*q_1} - \tilde{\Lambda}_{|x|^{-4}F^*q_2})\tilde{f}_1, \tilde{f}_2\rangle_{\partial\tilde{\Omega}} = \int_{\tilde{\Omega}} |x|^{-4} F^*(q_1 - q_2)\tilde{v}_1\tilde{v}_2\,dx.$$

Changing variables $x = F^{-1}(y)$ and noting that $|\det D(F^{-1})(y)| = |y|^{-2n}$, the previous expression becomes

$$\int_\Omega |y|^4 (q_1 - q_2)(y)\,|y|^{n-2} u_1(y)\,|y|^{n-2} u_2(y)\,|y|^{-2n}\,dy$$
$$= \int_\Omega (q_1 - q_2)u_1u_2\,dy = \langle (\Lambda_{q_1} - \Lambda_{q_2})f_1, f_2\rangle_{\partial\Omega},$$

where $f_j = F^*(|x|^{n-2}\,\tilde{f}_j)$. Since

$$\operatorname{supp}(f_j) \subset \Gamma,$$

the assumption on the DN maps Λ_{q_1} and Λ_{q_2} implies that

$$\langle (\Lambda_{q_1} - \Lambda_{q_2})f_1, f_2\rangle_{\partial\Omega} = 0.$$

This concludes the proof that $\tilde{\Lambda}_{|x|^{-4}F^*q_1}\tilde{f}|_{\tilde{\Gamma}} = \tilde{\Lambda}_{|x|^{-4}F^*q_2}\tilde{f}|_{\tilde{\Gamma}}$ whenever we have $\mathrm{supp}(\tilde{f}) \subset \tilde{\Gamma}$. □

6.2. Carleman estimate approach

We proceed to the proof of Theorem 6.0.2. Again we first consider the Schrödinger equation $(-\Delta+q)u = 0$ in Ω, where $q \in L^\infty(\Omega)$ and $\Omega \subset \mathbb{R}^n$ is a bounded open set with smooth boundary.

Motivation 6.2.1. Recall from Theorem 4.2.4 that in the construction of complex geometrical optics solutions, which depend on a large vector $\zeta \in \mathbb{C}^n$ satisfying $\zeta \cdot \zeta = 0$, we needed to solve equations of the form

$$(D^2 + 2\zeta \cdot D + q)r = f \qquad \text{in } \Omega,$$

or written in another way,

$$e^{-i\zeta\cdot x}(-\Delta + q)e^{i\zeta\cdot x}r = f \qquad \text{in } \Omega.$$

In particular, Theorem 4.2.4 shows the existence of a solution and implies the estimate

$$\|r\|_{L^2(\Omega)} \leq \frac{C_0}{|\zeta|}\|f\|_{L^2(\Omega)}.$$

We write

$$\zeta = \frac{1}{h}(\beta + i\alpha),$$

where α and β are orthogonal unit vectors in $\mathbb{R}^n$, and $h > 0$ is a *small parameter*. Replacing r by $e^{-\frac{i\beta\cdot x}{h}}r$, the L^2 estimate for r may be written as

$$\|r\|_{L^2(\Omega)} \leq Ch\|e^{\frac{1}{h}\alpha\cdot x}(-\Delta + q)e^{-\frac{1}{h}\alpha\cdot x}r\|_{L^2(\Omega)}.$$

It is possible to view this as a uniqueness result: if the right-hand side is zero, then a solution r with suitable properties also vanishes. It turns out that such a uniqueness result can be proved directly without Fourier analysis, and this is sufficient to imply also the existence of a solution.

Remark 6.2.2. We will systematically use a small parameter h instead of a large parameter $|\zeta|$ (these are related by $h = \frac{\sqrt{2}}{|\zeta|}$). This is of course just a matter of convention, but has the benefit of being consistent with *semiclassical calculus*, which is a well-developed theory for the analysis of certain asymptotic limits [**Zwo12**]. We will also arrange so that our basic partial derivatives will be hD_j instead of $\frac{\partial}{\partial x_j}$. The usefulness of these choices will hopefully be evident below.

6.2.1. Carleman estimates for test functions. We begin with the simplest Carleman estimate, which is valid for test functions and does not involve boundary terms.

Theorem 6.2.3 (Carleman estimate). *Let $q \in L^\infty(\Omega)$, let α be a unit vector in $\mathbb{R}^n$, and let $\varphi(x) = \alpha \cdot x$. There exist constants $C > 0$ and $h_0 > 0$ such that whenever $0 < h \leq h_0$, we have*

$$\|u\|_{L^2(\Omega)} \leq Ch\|e^{\varphi/h}(-\Delta + q)e^{-\varphi/h}u\|_{L^2(\Omega)}, \quad u \in C_c^\infty(\Omega).$$

We introduce some notation which will be used in the proof and also later. If $u, v \in L^2(\Omega)$ we write

$$(u, v) = \int_\Omega u\bar{v}\,dx, \quad \|u\| = (u, u)^{1/2} = \|u\|_{L^2(\Omega)}.$$

Consider the semiclassical Laplacian

$$P_0 = -h^2\Delta = (hD)^2$$

and the corresponding Schrödinger operator

$$P = h^2(-\Delta + q) = P_0 + h^2 q.$$

The operators conjugated with exponential weights will be denoted by

$$P_{0,\varphi} = e^{\varphi/h}P_0 e^{-\varphi/h}, \quad P_\varphi = e^{\varphi/h}Pe^{-\varphi/h} = P_{0,\varphi} + h^2 q.$$

We will also need the concept of adjoints of differential operators. If

$$L = \sum_{|\alpha| \leq m} a_\alpha(x) D^\alpha$$

is a differential operator in Ω, with $a_\alpha \in W^{|\alpha|,\infty}(\Omega)$ (that is, all partial derivatives up to order $|\alpha|$ are in $L^\infty(\Omega)$), then L^* (the formal adjoint of L) is the differential operator which satisfies

$$(Lu, v) = (u, L^* v), \quad u, v \in C_c^\infty(\Omega).$$

For L of the above form, an integration by parts shows that

$$L^* v = \sum_{|\alpha| \leq m} D^\alpha(\overline{a_\alpha(x)} v).$$

Proof of Theorem 6.2.3. Using the notation above, the desired estimate can be written as

$$h\|u\| \leq C\left\|P_\varphi u\right\|, \quad u \in C_c^\infty(\Omega).$$

First consider the case $q = 0$, that is, the estimate

$$h\|u\| \leq C\left\|P_{0,\varphi} u\right\|, \quad u \in C_c^\infty(\Omega).$$

We need an explicit expression for $P_{0,\varphi}$. On the level of operators, one has

$$e^{\varphi/h} hD_j e^{-\varphi/h} = hD_j + i\partial_j\varphi.$$

Since $\varphi(x) = \alpha \cdot x$, where α is a unit vector, we obtain

$$\begin{aligned} P_{0,\varphi} &= \sum_{j=1}^{n} (e^{\varphi/h} hD_j e^{-\varphi/h})(e^{\varphi/h} hD_j e^{-\varphi/h}) \\ &= \sum_{j=1}^{n} (hD_j + i\alpha_j)^2 \\ &= (hD)^2 - 1 + 2i\alpha \cdot hD. \end{aligned}$$

The objective is to prove a positive lower bound for

$$\left\| P_{0,\varphi} u \right\|^2 = (P_{0,\varphi} u, P_{0,\varphi} u).$$

To this end, we decompose $P_{0,\varphi}$ in a way which is useful for determining which parts in the inner product are positive and which may be negative. Write

$$P_{0,\varphi} = A + iB,$$

where $A^* = A$ and $B^* = B$. Here, A and iB are the self-adjoint and skew-adjoint parts of $P_{0,\varphi}$. Since

$$\begin{aligned} P_{0,\varphi}^* &= (e^{\varphi/h} P_0 e^{-\varphi/h})^* = e^{-\varphi/h} P_0 e^{\varphi/h} = P_{0,-\varphi} \\ &= (hD)^2 - 1 - 2i\alpha \cdot hD, \end{aligned}$$

we obtain A and B from the formulas (cf. the real and imaginary parts of a complex number)

$$\begin{aligned} A &= \frac{P_{0,\varphi} + P_{0,\varphi}^*}{2} = (hD)^2 - 1, \\ B &= \frac{P_{0,\varphi} - P_{0,\varphi}^*}{2i} = 2\alpha \cdot hD. \end{aligned}$$

Now we have

$$\begin{aligned} \left\| P_{0,\varphi} u \right\|^2 &= (P_{0,\varphi} u, P_{0,\varphi} u) \\ &= ((A + iB)u, (A + iB)u) \\ &= (Au, Au) + (Bu, Bu) + i(Bu, Au) - i(Au, Bu) \\ &= \|Au\|^2 + \|Bu\|^2 + (i[A, B]u, u), \end{aligned}$$

where $[A, B] = AB - BA$ is the *commutator* of A and B. This argument used integration by parts and the fact that $A^* = A$ and $B^* = B$. There are no boundary terms since $u \in C_c^\infty(\Omega)$.

The terms $\|Au\|^2$ and $\|Bu\|^2$ are nonnegative, so the only negative contributions could come from the commutator term. But in our case A and B are constant coefficient differential operators, and these operators always satisfy

$$[A, B] \equiv 0.$$

Therefore,

$$\left\|P_{0,\varphi} u\right\|^2 = \|Au\|^2 + \|Bu\|^2. \tag{6.3}$$

By the Poincaré inequality (see Proposition 2.4.7),

$$\|Bu\| = 2h\,\|\alpha \cdot Du\| \geq ch\,\|u\|,$$

where c depends on Ω. This proves that for any $h > 0$, one has

$$h\,\|u\| \leq C\left\|P_{0,\varphi} u\right\|, \qquad u \in C_c^\infty(\Omega).$$

Finally, consider the case where q may be nonzero. The last estimate implies that for $u \in C_c^\infty(\Omega)$, one has

$$\begin{aligned}
h\,\|u\| &\leq C\left\|P_{0,\varphi} u\right\| \\
&\leq C\left\|(P_{0,\varphi} + h^2 q)u\right\| + C\left\|h^2 q u\right\| \\
&\leq C\left\|P_\varphi u\right\| + Ch^2\,\|q\|_{L^\infty(\Omega)}\,\|u\|.
\end{aligned}$$

Choose h_0 so that $C\,\|q\|_{L^\infty(\Omega)}\,h_0 = \frac{1}{2}$, that is,

$$h_0 = \frac{1}{2C\,\|q\|_{L^\infty(\Omega)}}.$$

Then, if $0 < h \leq h_0$,

$$h\,\|u\| \leq C\left\|P_\varphi u\right\| + \frac{1}{2}h\,\|u\|.$$

The last term may be absorbed in the left-hand side, which completes the proof.

□

Exercise 6.2.4 (H^1 Carleman estimate). Let $\varphi(x) = \alpha \cdot x$ and $q \in L^\infty(\Omega)$. Show that there are $C > 0$ and $h_0 > 0$ such that for any h with $0 < h \leq h_0$, one has

$$\|u\| + \|hDu\| \leq Ch\|e^{\varphi/h}(-\Delta + q)e^{-\varphi/h}u\|, \quad u \in C_c^\infty(\Omega).$$

(Hint: write $\|hDu\|^2 = ((hD)^2 u, u) = (Au, u) + \|u\|^2$ and use the $\|Au\|^2$ term in (6.3).)

Exercise 6.2.5 (H^2 Carleman estimate). Improve the estimate in Exercise 6.2.4 and show that

$$\|u\| + \|hDu\| + \left\|h^2 \nabla^2 u\right\| \leq Ch\|e^{\varphi/h}(-\Delta + q)e^{-\varphi/h}u\|, \quad u \in C_c^\infty(\Omega),$$

where $\nabla^2 u = [\partial_{x_j x_k} u]_{j,k=1}^n$ encodes the second order derivatives of u.

(Hint: write $\sum_{j,k} \left\|h^2 D_{jk} u\right\|^2 = \sum_{j,k}(h^2 D_{jj} u, h^2 D_{kk} u) = (Au + u, Au + u)$ and use (6.3).)

Exercise 6.2.6 (Large first-order perturbations). Let $\varphi(x) = \alpha \cdot x$, let $W = (W_1, \dots, W_n) \in L^\infty(\Omega; \mathbb{R}^n)$ be a vector field, and let $q \in L^\infty(\Omega)$. Show that there are $C > 0$ and $h_0 > 0$ such that for any h with $0 < h \le h_0$, one has

$$\|u\| + \|hDu\| \le Ch\|e^{\varphi/h}(-\Delta + W \cdot \nabla + q)e^{-\varphi/h}u\|, \quad u \in C_c^\infty(\Omega).$$

(Hint: use the convexified weight $\varphi_\varepsilon = \varphi + \frac{h}{\varepsilon}\frac{\varphi^2}{2}$, where $\varepsilon > 0$ is small but fixed.)

6.2.2. Complex geometrical optics solutions. Here, we show how the Carleman estimate gives a new method for constructing complex geometrical optics solutions (CGO solutions). We first establish an existence result for an inhomogeneous equation, analogous to Theorem 4.2.4.

Proposition 6.2.7 (Solvability). *Let $q \in L^\infty(\Omega)$, let α be a unit vector in $\mathbb{R}^n$, and let $\varphi(x) = \alpha \cdot x$. There exist constants $C > 0$ and $h_0 > 0$ such that whenever $0 < h \le h_0$, the equation*

$$e^{\varphi/h}(-\Delta + q)e^{-\varphi/h} r = f \quad \text{in } \Omega$$

has a solution $r \in L^2(\Omega)$ for any $f \in L^2(\Omega)$, satisfying

$$\|r\|_{L^2(\Omega)} \le Ch\,\|f\|_{L^2(\Omega)}\,.$$

Remark 6.2.8. With some knowledge of unbounded operators on Hilbert space, the proof is immediate. Consider $P_\varphi^* : L^2(\Omega) \to L^2(\Omega)$ with domain $C_c^\infty(\Omega)$. It is a general fact that

$$\left.\begin{array}{c} T \text{ injective} \\ \text{range of } T \text{ closed} \end{array}\right\} \implies T^* \text{ surjective.}$$

Since the Carleman estimate is valid for P_φ^* one obtains injectivity and closed range for P_φ^*, and thus solvability for P_φ. Below we give a direct proof based on duality and the Hahn–Banach theorem, and also obtain the norm bound.

Proof of Proposition 6.2.7. Note that $P_\varphi^* = P_{0,-\varphi} + h^2\bar{q}$. If h_0 is as in Theorem 6.2.3, for $h \le h_0$ we have

$$\|u\| \le \frac{C}{h}\left\|P_\varphi^* u\right\|, \quad u \in C_c^\infty(\Omega). \tag{6.4}$$

Let $D = P_\varphi^* C_c^\infty(\Omega)$ be a subspace of $L^2(\Omega)$, and consider the linear functional

$$L : D \to \mathbb{C}, \quad L(P_\varphi^* v) = (v, f), \quad \text{for } v \in C_c^\infty(\Omega).$$

This is well defined since any element of D has a unique representation as $P_\varphi^* v$ with $v \in C_c^\infty(\Omega)$, by the Carleman estimate (6.4). Also, the Carleman estimate implies

$$\left|L(P_\varphi^* v)\right| \le \|v\|\,\|f\| \le \frac{C}{h}\|f\|\left\|P_\varphi^* v\right\|.$$

Thus L is a bounded linear functional on D.

The Hahn–Banach theorem ensures that there is a bounded linear functional $\hat{L} : L^2(\Omega) \to \mathbb{C}$ satisfying $\hat{L}|_D = L$ and $\|\hat{L}\| \leq Ch^{-1}\|f\|$. By the Riesz representation theorem, there is $\tilde{r} \in L^2(\Omega)$ such that

$$\hat{L}(w) = (w, \tilde{r}), \quad w \in L^2(\Omega),$$

and $\|\tilde{r}\| \leq Ch^{-1}\|f\|$. Then, for $v \in C_c^\infty(\Omega)$, by the definition of weak derivatives we have

$$(v, P_\varphi \tilde{r}) = (P_\varphi^* v, \tilde{r}) = \hat{L}(P_\varphi^* v) = L(P_\varphi^* v) = (v, f).$$

This proves that $P_\varphi \tilde{r} = f$ in the weak sense.

Finally, set $r = h^2 \tilde{r}$. This satisfies $e^{\varphi/h}(-\Delta + q)e^{-\varphi/h} r = f$ in Ω, and $\|r\| \leq Ch\|f\|$. □

We now give a construction of CGO solutions to the Schrödinger equation $(-\Delta + q)u = 0$ in Ω based on Proposition 6.2.7. This is a complex phase version of the *WKB construction* used for finding geometrical optics solutions for the wave equation. We already used such a construction in Chapter 4.

Our CGO solutions are of the form

$$u = e^{-\frac{1}{h}(\varphi + i\psi)}(a + r). \tag{6.5}$$

Here $h > 0$ is small and $\varphi(x) = \alpha \cdot x$ as before, ψ is a real valued phase function, a is a complex amplitude, and r is a correction term which will be small when h is small.

Writing $\rho = \varphi + i\psi$ for the complex phase function, using the formula

$$e^{\rho/h} hD_j(e^{-\rho/h} u) = (hD_j + i\partial_j \rho)u$$

and inserting (6.5) in the equation, we have

$$\begin{aligned}(-\Delta + q)u = 0 &\Longleftrightarrow e^{\rho/h}((hD)^2 + h^2 q)e^{-\rho/h}(a + r) = 0 \\ &\Longleftrightarrow e^{\rho/h}((hD)^2 + h^2 q)e^{-\rho/h} r = -((hD + i\nabla\rho)^2 + h^2 q)a.\end{aligned}$$

The last equation may be written as

$$e^{\varphi/h}(-\Delta + q)e^{-\varphi/h}(e^{-i\psi/h} r) = f,$$

where

$$f = -e^{-i\psi/h}\Big(- h^{-2}(\nabla\rho)^2 + h^{-1}[2\nabla\rho \cdot \nabla + \Delta\rho] + (-\Delta + q)\Big)a.$$

Now Proposition 6.2.7 ensures that one can find a correction term r that satisfies $\|r\| \leq Ch$, thus showing the existence of CGO solutions, provided that

$$\|f\| \leq C$$

with C independent of h. Looking at the expression for f, we see that it is enough to choose ψ and a in such a way that

$$(\nabla\rho)^2 = 0, \quad 2\nabla\rho \cdot \nabla a + (\Delta\rho)a = 0.$$

Since $\varphi(x) = \alpha\cdot x$ with α a unit vector, expanding the square in the equation $(\nabla\rho)^2 = 0$ gives the following equations for ψ:

$$|\nabla\psi|^2 = 1, \qquad \alpha \cdot \nabla\psi = 0.$$

The first equation is an *eikonal equation* (a certain nonlinear first-order PDE) for ψ. We obtain one solution by choosing $\psi(x) = \beta \cdot x$, where $\beta \in \mathbb{R}^n$ is a unit vector satisfying $\alpha \cdot \beta = 0$. It would be possible to use other solutions ψ, but this choice is close to the discussion in Chapter 4.

If $\psi(x) = \beta \cdot x$, then the second equation becomes

$$(\alpha + i\beta) \cdot \nabla a = 0.$$

This is a *complex transport equation* (a first-order linear equation) for a, analogous to the equation for a in Theorem 4.2.1. One solution is given by $a \equiv 1$. Again, other choices are possible.

This concludes the construction of CGO solutions based on Carleman estimates. There is a difference with the analogous result in Theorem 4.2.1: the correction term r given by this argument is so far only in $L^2(\Omega)$, not in $H^2(\Omega)$. The same is true for the solution u. One can in fact obtain r and u in $H^2(\Omega)$ and prove the following restatement of Theorem 4.2.1. We will give a sketch of the proof, since it involves the useful notion of Carleman estimates that are shifted to a different smoothness index.

Theorem 6.2.9 (CGO solutions). *Let $\Omega \subset \mathbb{R}^n$ be a bounded open set and let $q \in L^\infty(\Omega)$. Suppose that $\alpha, \beta \in \mathbb{R}^n$ are orthogonal unit vectors. There are $C, h_0 > 0$ such that for any $a \in H^2(\Omega)$ satisfying*

$$(\alpha + i\beta) \cdot \nabla a = 0 \text{ in } \Omega,$$

and for any h with $0 < h \leq h_0$, there is a solution $u \in H^2(\Omega)$ of

$$(-\Delta + q)u = 0 \text{ in } \Omega$$

having the form

$$u = e^{-\frac{1}{h}(\alpha+i\beta)\cdot x}(a + r).$$

The function $r \in H^2(\Omega)$ satisfies

$$\|r\|_{L^2(\Omega)} + h\|\nabla r\|_{L^2(\Omega)} + h^2\left\|\nabla^2 r\right\|_{L^2(\Omega)} \leq Ch\|(-\Delta + q)a\|_{L^2(\Omega)},$$

where $\nabla^2 r = [\partial_{x_j x_k} r]_{j,k=1}^n$ encodes the second order derivatives of r.

Sketch of Proof. As discussed above, we have already proved this result in the case where we obtain $u, r \in L^2(\Omega)$ with an L^2 estimate for r. By the argument above, the theorem follows if we can show an improved version of Proposition 6.2.7 where one obtains a solution $r \in H^2(\Omega)$ with estimates

$$\|r\| + h\|\nabla r\| + h^2\|\nabla^2 r\| \leq Ch\,\|f\|.$$

We consider the semiclassical Sobolev norm

$$\|u\|_{H^k_{\mathrm{scl}}} = \sum_{|\alpha|\leq k} h^{|\alpha|}\|\partial^\alpha u\|_{L^2(\Omega)}.$$

Note that $H^0_{\mathrm{scl}} = L^2$.

Our starting point is the H^2 Sobolev estimate from Exercise 6.2.5,

$$\|u\|_{H^2_{\mathrm{scl}}} \leq Ch^{-1}\left\|P_{0,\varphi}u\right\|_{L^2} \tag{6.6}$$

for $0 < h \leq h_0$ and $u \in C_c^\infty(\Omega)$. Applying the Hahn–Banach argument in the proof of Proposition 6.2.7 with this estimate would lead to a solvability result for right-hand sides f in the dual of $H^2_{\mathrm{scl}}(\Omega)$, which is a H^{-2} type space, and yielding solutions in $L^2(\Omega)$.

In order to obtain solutions in $H^2(\Omega)$, we need to "shift" the estimate (6.6) to a different Sobolev index. This means that we need to prove an estimate

$$\|u\|_{H^{s+2}_{\mathrm{scl}}} \leq Ch^{-1}\left\|P_{0,\varphi}u\right\|_{H^s_{\mathrm{scl}}} \tag{6.7}$$

for $0 < h \leq h_0$ and $u \in C_c^\infty(\Omega)$, where $s \in \mathbb{R}$ is the smoothness index. Of course, this requires an appropriate definition of the H^s_{scl} norm. To this end we choose $R > 0$ so that $\overline{\Omega} \subset (-R,R)^n$ and consider $\overline{\Omega}$ as a subset of the torus $T^n = [-R,R]^n$ with opposite sides identified. The torus T^n is a compact manifold, and for $f \in C^\infty(T^n)$ we can define the semiclassical Sobolev norm

$$\|f\|_{H^s_{\mathrm{scl}}} = \|\langle hD\rangle^s f\|_{L^2}.$$

The operator $\langle hD\rangle^s := (1 - h^2\Delta)^{s/2}$ can be defined either in terms of Fourier coefficients, or by the functional calculus obtained from the spectral theorem for unbounded operators on $L^2(T^n)$. Now any function $u \in C_c^\infty(\Omega)$ can be identified with a smooth function on T^n via zero extension, and this defines the H^s_{scl} norm of u for any $s \in \mathbb{R}$.

To prove (6.7), we would like to apply the H^2 estimate in (6.6) to the function $\langle hD\rangle^s u$. This function is not compactly supported in Ω for general s though, so one needs to choose a cutoff function $\chi \in C_c^\infty(\mathbb{R}^n)$ such that $\chi = 1$ near $\overline{\Omega}$ and $\mathrm{supp}(\chi)$ is slightly larger than $\overline{\Omega}$. Then (6.7) can be proved by applying (6.6) (in a slightly larger domain $\tilde{\Omega}$) to $\chi\langle hD\rangle^s u$. The errors resulting from the cutoff χ can be dealt with by *pseudolocal estimates*. These estimates imply that operators of the type $(1-\chi)\langle hD\rangle^s\psi$, where ψ is a cutoff function with $\psi = 1$

near $\overline{\Omega}$ and $\chi = 1$ near $\operatorname{supp}(\psi)$, are semiclassically smoothing (i.e., very small when h is small).

After proving (6.7) (for $P_{0,\varphi}^*$) we can apply this estimate with $s = -2$, use the Hahn–Banach argument in the proof of Proposition 6.2.7, and employ duality between H^2 and H^{-2} spaces to prove a version of Proposition 6.2.7 for $P_{0,\varphi}$ that yields a right inverse for $P_{0,\varphi}$ with the required estimates. If $h > 0$ is sufficiently small, one can add a potential $q \in L^\infty(\Omega)$ as a small perturbation and prove a solvability result for P_φ. □

Exercise 6.2.10. Complete the details of the proof of Proposition 6.2.9. You may follow the exposition in [**DSFKSU09**] or [**ST09**].

6.2.3. Carleman estimates with boundary terms. The next step in the proof of Theorem 6.0.2 is to derive a Carleman estimate for functions which vanish at the boundary but are not compactly supported. The estimate will include terms involving the normal derivative. We will use the notation

$$(u, v)_{\partial\Omega} = \int_{\partial\Omega} u\bar{v}\, dS,$$
$$\partial_\nu u = \nabla u \cdot \nu|_{\partial\Omega},$$

and

$$\partial\Omega_\pm = \partial\Omega_\pm(\alpha) = \{\, x \in \partial\Omega \mid \pm\alpha \cdot \nu(x) \geq 0 \,\}.$$

Theorem 6.2.11 (Carleman estimate with boundary terms). *Let $q \in L^\infty(\Omega)$, let α be a unit vector in $\mathbb{R}^n$, and let $\varphi(x) = \alpha \cdot x$. There exist constants $C > 0$ and $h_0 > 0$ such that whenever $0 < h \leq h_0$, we have*

$$\begin{aligned} &- h((\alpha \cdot \nu)\partial_\nu u, \partial_\nu u)_{\partial\Omega_-} + \|u\|^2_{L^2(\Omega)} \\ &\qquad \leq Ch^2\|e^{\varphi/h}(-\Delta + q)e^{-\varphi/h}u\|^2_{L^2(\Omega)} + Ch((\alpha \cdot \nu)\partial_\nu u, \partial_\nu u)_{\partial\Omega_+} \end{aligned}$$

for any $u \in C^\infty(\overline{\Omega})$ with $u|_{\partial\Omega} = 0$.

Note that the sign of $\alpha \cdot \nu$ on $\partial\Omega_\pm$ ensures that all terms in the Carleman estimate are nonnegative.

Proof. We first claim that

$$ch^2\|u\|^2 - 2h^3((\alpha \cdot \nu)\partial_\nu u, \partial_\nu u)_{\partial\Omega} \leq \left\|P_{0,\varphi} u\right\|^2 \tag{6.8}$$

for $u \in C^\infty(\overline{\Omega})$ with $u|_{\partial\Omega} = 0$. It is easy to see that this implies the desired estimate in the case $q = 0$.

As in the proof of Theorem 6.2.3, we decompose

$$P_{0,\varphi} = A + iB,$$

where $A = (hD)^2 - 1$ and $B = 2\alpha \cdot hD$, and $A^* = A$, $B^* = B$. Then

$$\begin{aligned}\left\|P_{0,\varphi}u\right\|^2 &= (P_{0,\varphi}u, P_{0,\varphi}u)\\ &= ((A+iB)u, (A+iB)u)\\ &= \|Au\|^2 + \|Bu\|^2 + i(Bu, Au) - i(Au, Bu).\end{aligned}$$

We wish to integrate by parts to obtain the commutator term involving $i[A,B]$, but this time boundary terms will arise. We have

$$\begin{aligned}&i(Bu, (hD)^2u)\\ &\quad= \sum_{j=1}^n i(Bu, (hD_j)^2u)\\ &\quad= \sum_{j=1}^n \Big[i(Bu, \frac{h}{i}\nu_j hD_j u)_{\partial\Omega} + i(hD_jBu, hD_ju)\Big]\\ &\quad= -2h^3(\alpha\cdot\nabla u, \partial_\nu u)_{\partial\Omega} + \sum_{j=1}^n \Big[i(hD_jBu, \frac{h}{i}\nu_j u)_{\partial\Omega} + i((hD_j)^2Bu, u)\Big].\end{aligned}$$

But $u|_{\partial\Omega} = 0$, so the boundary term involving $\frac{h}{i}\nu_j u$ is zero. For the first boundary term we use the decomposition

$$\nabla u|_{\partial\Omega} = (\partial_\nu u)\nu + (\nabla u)_{\tan},$$

where $(\nabla u)_{\tan} := \nabla u - (\nabla u \cdot \nu)\nu|_{\partial\Omega}$ is the tangential part of ∇u, which vanishes since $u|_{\partial\Omega} = 0$. By these facts, we obtain

$$i(Bu, Au) = i(ABu, u) - 2h^3((\alpha\cdot\nu)\partial_\nu u, \partial_\nu u)_{\partial\Omega}.$$

Similarly, using that $u|_{\partial\Omega} = 0$,

$$i(Au, Bu) = i(Au, 2\alpha\cdot\frac{h}{i}\nu u)_{\partial\Omega} + i(BAu, u) = i(BAu, u).$$

We have proved that

$$\left\|P_{0,\varphi}u\right\|^2 = \|Au\|^2 + \|Bu\|^2 + (i[A,B]u, u) - 2h^3((\alpha\cdot\nu)\partial_\nu u, \partial_\nu u)_{\partial\Omega}.$$

Again, since A and B are constant coefficient operators, we have $[A,B] = AB - BA \equiv 0$. The Poincaré inequality gives $\|Bu\| \geq ch\,\|u\|$, which proves (6.8).

Writing (6.8) in a different form, we have

$$\begin{aligned}&-2h((\alpha\cdot\nu)\partial_\nu u, \partial_\nu u)_{\partial\Omega_-} + c\,\|u\|^2\\ &\qquad\qquad \leq h^2\|e^{\varphi/h}(-\Delta)e^{-\varphi/h}u\|^2 + 2h((\alpha\cdot\nu)\partial_\nu u, \partial_\nu u)_{\partial\Omega_+}.\end{aligned}$$

We can add and subtract the potential q to obtain

$$\begin{aligned}-2h((\alpha\cdot\nu)\partial_\nu u,\partial_\nu u)_{\partial\Omega_-}+c\,\|u\|^2&\\ \le h^2\|e^{\varphi/h}(-\Delta+q)e^{-\varphi/h}u\|^2+h^2\,\|q\|^2_{L^\infty(\Omega)}\,\|u\|^2&\\ +2h((\alpha\cdot\nu)\partial_\nu u,\partial_\nu u)_{\partial\Omega_+}.&\end{aligned}$$

Choosing h small enough (depending on $\|q\|_{L^\infty(\Omega)}$), the term involving $\|u\|^2$ on the right can be absorbed to the left-hand side. This concludes the proof. □

6.3. Uniqueness with partial data

Let Ω be a bounded open set in $\mathbb{R}^n$ with smooth boundary, where $n \geq 3$. If $\alpha \in \mathbb{R}^n$, recall the subsets of the boundary

$$\begin{aligned}\partial\Omega_\pm &= \{\, x\in\partial\Omega \mid \pm\alpha\cdot\nu(x)>0 \,\},\\ \partial\Omega_{-,\varepsilon} &= \{\, x\in\partial\Omega \mid \alpha\cdot\nu(x)<\varepsilon \,\}.\end{aligned}$$

Also, let $\partial\Omega_{+,\varepsilon} = \partial\Omega_{+,\varepsilon}(\alpha) = \{\, x\in\partial\Omega \mid \alpha\cdot\nu(x)>\varepsilon \,\}$. We first consider a partial data uniqueness result for the Schrödinger equation.

Theorem 6.3.1. *Let q_1 and q_2 be two functions in $L^\infty(\Omega)$ such that the Dirichlet problems for $-\Delta+q_1$ and $-\Delta+q_2$ are well-posed. If α is a unit vector in $\mathbb{R}^n$ and if*

$$\Lambda_{q_1} f|_{\partial\Omega_{-,\varepsilon}} = \Lambda_{q_2} f|_{\partial\Omega_{-,\varepsilon}} \quad \textit{for all } f\in H^{1/2}(\partial\Omega),$$

then $q_1 = q_2$ in Ω.

Given this result, it is easy to prove the corresponding theorem for the conductivity equation.

Proof of Theorem 6.0.2. Define

$$q_j = \frac{\Delta\gamma_j^{1/2}}{\gamma_j^{1/2}}.$$

By Proposition 2.8.5, we have the relation

$$\Lambda_{q_j} f = \gamma_j^{-1/2}\Lambda_{\gamma_j}(\gamma_j^{-1/2} f) + \frac{1}{2}\gamma_j^{-1}(\partial_\nu\gamma_j) f|_{\partial\Omega}.$$

Since $\Lambda_{\gamma_1} f|_{\partial\Omega_{-,\varepsilon}} = \Lambda_{\gamma_2} f|_{\partial\Omega_{-,\varepsilon}}$ for all f, the boundary determination results in Theorems 3.1.1 and 3.2.1 imply that

$$\gamma_1|_{\partial\Omega_{-,\varepsilon}} = \gamma_2|_{\partial\Omega_{-,\varepsilon}} \quad\text{and}\quad \partial_\nu\gamma_1|_{\partial\Omega_{-,\varepsilon}} = \partial_\nu\gamma_2|_{\partial\Omega_{-,\varepsilon}}.$$

Using the additional assumption $\gamma_1|_{\partial\Omega_{+,\varepsilon}} = \gamma_2|_{\partial\Omega_{+,\varepsilon}}$, we see that the DN maps for the Schrödinger equation satisfy $\Lambda_{q_1} f|_{\partial\Omega_{-,\varepsilon}} = \Lambda_{q_2} f|_{\partial\Omega_{-,\varepsilon}}$ for all $f\in H^{1/2}(\partial\Omega)$.

From Theorem 6.3.1 we obtain $q_1 = q_2$, or

$$\frac{\Delta\gamma_1^{1/2}}{\gamma_1^{1/2}} = \frac{\Delta\gamma_2^{1/2}}{\gamma_2^{1/2}} \quad \text{in} \quad \Omega.$$

Writing $q = q_1 = q_2$, both $\gamma_1^{1/2}$ and $\gamma_2^{1/2}$ solve the equation

$$(-\Delta + q)u = 0 \quad \text{in} \quad \Omega.$$

Since $\gamma_1|_{\partial\Omega} = \gamma_2|_{\partial\Omega}$, well-posedness of the Dirichlet problem implies that $\gamma_1 = \gamma_2$ in Ω. □

We proceed to the proof of Theorem 6.3.1. The main tool is the Carleman estimate in Theorem 6.2.11, which will be applied with the weight $-\varphi$ instead of φ. The estimate then has the form

$$\begin{aligned} &h((\alpha\cdot\nu)\partial_\nu u, \partial_\nu u)_{\partial\Omega_+} + \|u\|_{L^2(\Omega)}^2 \\ &\qquad \leq Ch^2\|e^{-\varphi/h}(-\Delta+q)e^{\varphi/h}u\|_{L^2(\Omega)}^2 - Ch((\alpha\cdot\nu)\partial_\nu u, \partial_\nu u)_{\partial\Omega_-} \end{aligned}$$

for any $u \in C^\infty(\overline{\Omega})$ with $u|_{\partial\Omega} = 0$. Choosing $v = e^{\varphi/h}u$ and noting that $v|_{\partial\Omega} = 0$, this may be written as

$$\begin{aligned} (6.9)\quad &h\big((\alpha\cdot\nu)e^{-\varphi/h}\partial_\nu v, e^{-\varphi/h}\partial_\nu v\big)_{\partial\Omega_+} + \|e^{-\varphi/h}v\|_{L^2(\Omega)}^2 \\ &\qquad \leq Ch^2\|e^{-\varphi/h}(-\Delta+q)v\|_{L^2(\Omega)}^2 - Ch\big((\alpha\cdot\nu)e^{-\varphi/h}\partial_\nu v, e^{-\varphi/h}\partial_\nu v\big)_{\partial\Omega_-}. \end{aligned}$$

If $v \in H^2 \cap H_0^1(\Omega)$, we can find $v_j \in C^\infty(\overline{\Omega})$ with $v_j|_{\partial\Omega} = 0$ such that $v_j \to v$ in $H^2(\Omega)$. (This can be done e.g. by choosing $F_j \in C^\infty(\overline{\Omega})$ with $F_j \to \Delta v$ in $L^2(\Omega)$ and choosing v_j as the solution of $\Delta v_j = F_j$ with $v_j|_{\partial\Omega} = 0$.) Applying (6.9) to v_j and taking the limit as $j \to \infty$ proves that (6.9) remains true for $v \in H^2 \cap H_0^1(\Omega)$.

Proof of Theorem 6.3.1. Recall from Proposition 2.8.3 that

$$(6.10)\qquad \int_\Omega (q_1 - q_2)u_1u_2\,dx = \big\langle(\Lambda_{q_1} - \Lambda_{q_2})(u_1|_{\partial\Omega}), u_2|_{\partial\Omega}\big\rangle_{\partial\Omega}$$

whenever $u_j \in H^1(\Omega)$ are solutions of $(-\Delta+q_j)u_j = 0$ in Ω. By the assumption on the DN maps, the boundary integral is really over $\partial\Omega_{+,\varepsilon}$. If further $u_1 \in H^2(\Omega)$, then

$$\Lambda_{q_1}(u_1|_{\partial\Omega}) = \partial_\nu u_1|_{\partial\Omega}$$

since $\nabla u_1 \in H^1(\Omega)$ and $\partial_\nu u_1|_{\partial\Omega} \in H^{1/2}(\partial\Omega)$. Also,

$$\Lambda_{q_2}(u_1|_{\partial\Omega}) = \partial_\nu \tilde{u}_2|_{\partial\Omega},$$

where $\tilde{u}_2$ solves

$$\begin{cases} (-\Delta + q_2)\,\tilde{u}_2 = 0 & \text{in } \Omega \\ \qquad\qquad\ \tilde{u}_2 = u_1 & \text{on } \partial\Omega. \end{cases}$$

We have $\tilde{u}_2 \in H^2(\Omega)$ since $u_1|_{\partial\Omega} \in H^{3/2}(\partial\Omega)$. Therefore, (6.10) implies

$$\int_\Omega (q_1 - q_2)u_1 u_2 \, dx = \int_{\partial\Omega_{+,\varepsilon}} \partial_\nu (u_1 - \tilde{u}_2) u_2 \, dS$$

for any $u_j \in H^2(\Omega)$ which solve $(-\Delta + q_j)u_j = 0$ in Ω.

Given the unit vector $\alpha \in \mathbb{R}^n$, let $\xi \in \mathbb{R}^n$ be a vector orthogonal to α, and let $\beta \in \mathbb{R}^n$ be a unit vector such that $\{\alpha, \beta, \xi\}$ is an orthogonal triplet (here we use that $n \geq 3$). Write $\varphi(x) = \alpha \cdot x$ and $\psi(x) = \beta \cdot x$. Theorem 6.2.9 ensures that there exist CGO solutions to $(-\Delta + q_j)u_j = 0$ of the form

$$u_1 = e^{\frac{1}{h}(\varphi + i\psi)} e^{ix\cdot\xi}(1 + r_1), \quad u_2 = e^{-\frac{1}{h}(\varphi + i\psi)}(1 + r_2),$$

where $\left\|r_j\right\| \leq Ch$, $\left\|\nabla r_j\right\| \leq C$, and $u_j \in H^2(\Omega)$. Then, writing

$$u := u_1 - \tilde{u}_2 \in H^2 \cap H^1_0(\Omega),$$

we have

$$\int_\Omega e^{ix\cdot\xi}(q_1 - q_2)(1 + r_1 + r_2 + r_1 r_2) \, dx = \int_{\partial\Omega_{+,\varepsilon}} (\partial_\nu u) u_2 \, dS. \tag{6.11}$$

By the estimates for r_j, the limit as $h \to 0$ of the left-hand side is

$$\int_\Omega e^{ix\cdot\xi}(q_1 - q_2) \, dx.$$

We wish to show that the right-hand side converges to zero as $h \to 0$.

By Cauchy–Schwarz, one has

$$\begin{aligned} \Big|\int_{\partial\Omega_{+,\varepsilon}} (\partial_\nu u) u_2 \, dS\Big|^2 &= \Big|\int_{\partial\Omega_{+,\varepsilon}} e^{-\varphi/h}(\partial_\nu u) e^{\varphi/h} u_2 \, dS\Big|^2 \\ &\leq \Big(\int_{\partial\Omega_{+,\varepsilon}} |e^{-\varphi/h}\partial_\nu u|^2 \, dS\Big)\Big(\int_{\partial\Omega_{+,\varepsilon}} |e^{\varphi/h} u_2|^2 \, dS\Big). \end{aligned} \tag{6.12}$$

To use the Carleman estimate, we note that $\varepsilon \leq \alpha \cdot \nu$ on $\partial\Omega_{+,\varepsilon}$. By (6.9) applied to u with potential q_2, and using that $\partial_\nu u|_{\partial\Omega_{-,\varepsilon}} = 0$ by the assumption on DN maps, we obtain for small h that

$$\begin{aligned} \int_{\partial\Omega_{+,\varepsilon}} |e^{-\varphi/h}\partial_\nu u|^2 \, dS &\leq \frac{1}{\varepsilon} \int_{\partial\Omega_{+,\varepsilon}} (\alpha \cdot \nu)|e^{-\varphi/h}\partial_\nu u|^2 \, dS \\ &\leq \frac{1}{\varepsilon} Ch \|e^{-\varphi/h}(-\Delta + q_2)u\|^2_{L^2(\Omega)}. \end{aligned}$$

The reason for choosing the potential q_2 is that

$$(-\Delta + q_2)u = (-\Delta + q_2)u_1 = (q_2 - q_1)u_1.$$

Thus, the solution $\tilde{u}_2$ goes away, and we are left with an expression involving only u_1 for which we know exact asymptotics. We obtain

$$\int_{\partial\Omega_{+,\varepsilon}} |e^{-\varphi/h}\partial_\nu u|^2\,dS \leq \frac{1}{\varepsilon}Ch\|(q_2-q_1)e^{i\psi/h}e^{ix\cdot\xi}(1+r_1)\|^2_{L^2(\Omega)} \leq Ch.$$

This takes care of the first term on the right-hand side of (6.12). For the other term we compute

$$\begin{aligned}\int_{\partial\Omega_{+,\varepsilon}} |e^{\varphi/h}u_2|^2\,dS &= \int_{\partial\Omega_{+,\varepsilon}} |1+r_2|^2\,dS\\ &\leq 2\int_{\partial\Omega_{+,\varepsilon}} (1+|r_2|^2)\,dS \leq C(1+\|r_2\|^2_{L^2(\partial\Omega)}).\end{aligned}$$

By the trace theorem, we have $\|r_2\|_{L^2(\partial\Omega)} \leq C\,\|r_2\|_{H^1(\Omega)} \leq C$. Combining these estimates, from (6.12) we obtain that, for small h,

$$\Big|\int_{\partial\Omega_{+,\varepsilon}} (\partial_\nu u)u_2\,dS\Big| \leq Ch^{1/2}.$$

Taking the limit as $h \to 0$ in (6.11), we are left with

$$\int_\Omega e^{ix\cdot\xi}(q_1-q_2)\,dx = 0. \tag{6.13}$$

This is true for all $\xi \in \mathbb{R}^n$ orthogonal to α. Next we write

$$\partial\Omega_{-,\varepsilon}(\tilde{\alpha}) = \{\, x \in \partial\Omega \mid \tilde{\alpha}\cdot\nu(x) < \varepsilon \,\}.$$

Since the DN maps agree on $\partial\Omega_{-,\varepsilon}(\alpha)$ for a fixed constant $\varepsilon > 0$, they also agree on $\partial\Omega_{-,\tilde{\varepsilon}}(\tilde{\alpha})$ for $\tilde{\alpha}$ sufficiently close to α on the unit sphere and for some smaller constant $\tilde{\varepsilon}$. Thus, in particular, (6.13) holds for ξ in an open cone in $\mathbb{R}^n$. Writing $\tilde{q}$ for the function which is equal to $q_1 - q_2$ in Ω and which is zero outside of Ω, this implies that the Fourier transform of $\tilde{q}$ vanishes in an open set. But since $\tilde{q}$ is compactly supported, its Fourier transform is analytic, and this implies that $\tilde{q} \equiv 0$. We have proved that $q_1 \equiv q_2$. □

6.4. Unique continuation

In this section we prove the unique continuation result required in the proof of Theorem 6.0.1. The uniqueness results below are true for more general elliptic equations, but for our purposes it is sufficient to consider operators

$$Lu = -\operatorname{div}(\gamma\nabla u) + W\cdot\nabla u + qu,$$

where the coefficients satisfy the following conditions:

(6.14) $$\gamma \in W^{1,\infty}(\Omega) \text{ satisfies } \gamma \geq c > 0 \text{ in } \Omega,$$

(6.15) $$W \in L^\infty(\Omega, \mathbb{C}^n),$$

(6.16) $$q \in L^\infty(\Omega, \mathbb{C}).$$

Here $W^{1,\infty}(\Omega)$ can be identified with the set of Lipschitz continuous functions in Ω, see [**Eva10**, §5.8].

Theorem 6.4.1 (Unique continuation from local Cauchy data). *Let $\Omega \subset \mathbb{R}^n$ be a bounded connected open set with smooth boundary, and suppose that* (6.14)–(6.16) *hold. Assume that Γ is a nonempty open subset of $\partial\Omega$. If $u \in H^2(\Omega)$ satisfies*

$$Lu = 0 \quad \text{in } \Omega$$

and

$$u|_\Gamma = \partial_\nu u|_\Gamma = 0,$$

then $u = 0$ in Ω.

As an immediate corollary, if two solutions to the equation $Lu = 0$ have the same Cauchy data on an open subset of the boundary, then the solutions are identical in the whole domain. This is an instance of the *unique continuation principle* for elliptic equations.

We also state two closely related variants. The first one is called *weak unique continuation* and it concerns uniqueness of solutions which vanish in some ball. One also has *strong unique continuation* meaning that any solution vanishing to infinite order at a point in a suitable sense must vanish everywhere, but we will not need this.

Theorem 6.4.2 (Weak unique continuation). *Let $\Omega \subset \mathbb{R}^n$ be a connected bounded open set, and suppose that* (6.14)–(6.16) *hold. If $u \in H^2(\Omega)$ satisfies*

$$Lu = 0 \quad \text{in } \Omega$$

and

$$u = 0 \text{ in some ball } B \text{ contained in } \Omega,$$

then $u = 0$ in Ω.

The next variant states that a solution that vanishes on one side of a hypersurface must vanish also on the other side.

Theorem 6.4.3 (Unique continuation across a hypersurface). *Let $\Omega \subset \mathbb{R}^n$ be an open set, and suppose that* (6.14)–(6.16) *hold. Assume that S is a C^∞ hypersurface such that $\Omega = S_+ \cup S \cup S_-$ where S_+ and S_- denote the two sides of S. If $x_0 \in S$ and if V is an open neighborhood of x_0 in Ω, and if $u \in H^2(V)$ satisfies*

$$Lu = 0 \quad \text{in } V$$

and

$$u = 0 \quad in\, V \cap S_+,$$

then $u = 0$ in some neighborhood of x_0.

We will prove Theorem 6.4.3 and deduce the other unique continuation results from that. In fact we will give two proofs, one based on the Carleman estimate with a linear weight and another one based on Carleman estimates with more general weights.

Remark 6.4.4 (Unique continuation for differential inequalities). Theorems 6.4.1 and 6.4.2 remain valid if one replaces the condition $Lu = 0$ in Ω with the differential inequality

$$|\operatorname{div}(\gamma \nabla u)| \leq C(|u| + |\nabla u|) \quad \text{a.e. in } \Omega.$$

Similarly, Theorem 6.4.3 remains valid if instead of the equation $Lu = 0$ in V one assumes the above differential inequality in V. Moreover, the results are true if one only assumes that $u \in H^1(\Omega)$ or $u \in H^1(V)$. See Remark 6.4.12 for further details on these extensions.

6.4.1. Unique continuation across a hyperplane. We will warm up by considering a very simple case, which illustrates the main ideas with minimal technicalities. This case is related to unique continuation across a hyperplane for solutions in an infinite strip.

Proposition 6.4.5. *Let $\Omega = \{ x \in \mathbb{R}^n \mid a < x_n < b \}$, let $q \in L^\infty(\Omega)$, and assume that $u \in H^2(\Omega)$ solves*

$$(-\Delta + q)u = 0 \text{ in } \Omega$$

and

$$u|_{\{x_n = b\}} = \partial_\nu u|_{\{x_n = b\}} = 0.$$

Then $u = 0$ in Ω.

Since the domain Ω is unbounded, a few remarks are in order. The Sobolev spaces on Ω are defined for $k \geq 0$ by

$$H^k(\Omega) = \{ u \in L^2(\Omega) \mid \partial^\alpha u \in L^2(\Omega) \text{ for } |\alpha| \leq k \}.$$

We define $H_0^k(\Omega)$ to be the closure of $C_c^\infty(\Omega)$ in $H^k(\Omega)$. The trace theorem is valid on Ω (see Proposition C.2.17), and as in Theorem C.2.1 one can show that

$$H_0^1(\Omega) = \{ u \in H^1(\Omega) \mid u|_{\partial\Omega} = 0 \},$$
$$H_0^2(\Omega) = \{ u \in H^2(\Omega) \mid u|_{\partial\Omega} = \partial_\nu u|_{\partial\Omega} = 0 \}.$$

Proposition 6.4.5 states that any solution whose Cauchy data vanishes on one part of the boundary ($\{x_n = b\}$) has to be identically zero, with no a priori conditions on the other part of the boundary ($\{x_n = a\}$). The result will follow from the next Carleman estimate. It involves a linear weight that is well adapted to unique continuation across hyperplanes.

Proposition 6.4.6 (Carleman estimate). *Let* $\Omega = \{\, x \in \mathbb{R}^n \mid a < x_n < b \,\}$ *and let* $q \in L^\infty(\Omega)$. *There are* $C, h_0 > 0$ *such that when* $0 < |h| \le h_0$ *one has*

$$\|w\|_{L^2(\Omega)} \le C|h|\,\|e^{x_n/h}(-\Delta + q)(e^{-x_n/h}w)\|_{L^2(\Omega)}, \quad w \in H_0^2(\Omega).$$

Proof. This result is proved in exactly the same way as Theorem 6.2.3. It is sufficient to note that the Poincaré inequality, Proposition 2.4.7, is valid in the infinite strip Ω and yields

$$\left\|P_{0,\varphi}w\right\| \ge \|Bw\| = 2|h|\,\|D_n w\| \ge \frac{2\sqrt{2}}{b-a}|h|\,\|w\|. \qquad \square$$

We can now prove the unique continuation result. We wish to show that $u = 0$ near $\{x_n = b\}$, and for this we use the exponential weight $e^{x_n/h}$, where x_n is large near the set $\{x_n = b\}$ where we want to prove vanishing. Applying the Carleman estimate to χu where χ is a cutoff function with $\chi = 1$ near $\{x_n = b\}$, using that the function

$$(-\Delta + q)(\chi u) = -[\Delta, \chi]u$$

is supported away from $\{x_n = b\}$ and hence x_n is smaller in its support, and taking the limit as $h \to 0$ yields the result.

Proof of Proposition 6.4.5. Suppose that $u \in H^2(\Omega)$ solves $(-\Delta + q)u = 0$ in Ω with $u = \partial_\nu u = 0$ on $\{x_n = b\}$. Let c_0 be any number satisfying $a < c_0 < b$. We want to show that $u = 0$ in $\{c_0 < x_n < b\}$.

We let $0 < h \le h_0$ and rewrite the estimate from Proposition 6.4.6 as

$$\|e^{x_n/h}w\|_{L^2(\Omega)} \le Ch\|e^{x_n/h}(-\Delta + q)w\|_{L^2(\Omega)},$$

which holds for all $w \in H_0^2(\Omega)$. Now choose

$$w = \chi u,$$

where $\chi(x', x_n) = \zeta(x_n)$ where $\zeta \in C^\infty(\mathbb{R})$ satisfies $\zeta = 1$ for $t \ge c_0$ and $\zeta = 0$ near $t \le a$. Using that $\chi = 0$ near $\{x_n = a\}$ and $u = \partial_\nu u = 0$ on $\{x_n = b\}$, we have $w \in H_0^2(\Omega)$. Therefore

$$\begin{aligned}
\|e^{x_n/h}u\|_{L^2(\{c_0<x_n<b\})} &\le \|e^{x_n/h}\chi u\|_{L^2(\Omega)} \\
&\le Ch\|e^{x_n/h}(-\Delta + q)(\chi u)\|_{L^2(\Omega)} \\
&\le Ch(\|e^{x_n/h}\chi(-\Delta + q)u\|_{L^2(\Omega)} + \|e^{x_n/h}[\Delta, \chi]u\|_{L^2(\Omega)}),
\end{aligned}$$

where $[\Delta, \chi]v = 2\nabla\chi \cdot \nabla v + (\Delta\chi)v$. In particular $[\Delta, \chi]u$ is supported in $\operatorname{supp}(\nabla\chi) \subset \{a \le x_n \le c_0\}$.

Using that $(-\Delta + q)u = 0$, the previous inequality implies

$$\|e^{x_n/h}u\|_{L^2(\{c_0<x_n<b\})} \leq Ch\|e^{x_n/h}[\Delta,\chi]u\|_{L^2(\{a<x_n<c_0\})}.$$

Now $e^{x_n/h}$ is $\geq e^{c_0/h}$ when $x_n \geq c_0$ and $\leq e^{c_0/h}$ when $x_n \leq c_0$. This implies that

$$\begin{aligned} e^{c_0/h}\|u\|_{L^2(\{c_0<x_n<b\})} &\leq \|e^{x_n/h}u\|_{L^2(\{c_0<x_n<b\})} \\ &\leq Ch\|e^{x_n/h}[\Delta,\chi]u\|_{L^2(\{a<x_n<c_0\})} \\ &\leq Che^{c_0/h}\|[\Delta,\chi]u\|_{L^2(\{a<x_n<c_0\})}. \end{aligned}$$

But $[\Delta,\chi]u$ is a fixed function, so dividing by $e^{c_0/h}$ and taking $h \to 0$ shows that $\|u\|_{L^2(\{c_0<x_n<b\})} = 0$. Thus $u|_{\{c_0<x_n<b\}}$ as required. □

6.4.2. Unique continuation via linear weights. We proceed to prove the unique continuation theorems. First observe that since $\gamma \in W^{1,\infty}(\Omega)$, we may write

$$Lu = -\gamma\Delta u - \nabla\gamma\cdot\nabla u + W\cdot\nabla u + qu.$$

Dividing by the positive function γ, we see that $Lu = 0$ is equivalent with the equation

$$(-\Delta + W\cdot\nabla + q)u = 0$$

for new L^∞ coefficients W and q. Thus it is enough to consider operators of the above form.

We have seen in Exercise 6.2.6 that the Carleman estimate with linear weight holds also for these operators, i.e., one has

$$\|e^{x_n/h}w\|_{L^2(\Omega)} \leq Ch\|e^{x_n/h}(-\Delta + W\cdot\nabla + q)w\|_{L^2(\Omega)} \tag{6.17}$$

whenever $0 < h \leq h_0$ and $w \in H_0^2(\Omega)$. A shorter proof, valid for functions with sufficiently small support, could be obtained upon scaling the H^1 Carleman estimate in Exercise 6.2.4 from the unit ball into a ball of radius ε, and by observing that the first-order perturbation can be absorbed in the $\|hDu\|$ term when ε is small enough.

Now, if in Theorem 6.4.1 the geometry is such that $\Omega \subset \{a < x_n < b\}$ and $\Gamma \cap \{x_n = b\} \neq \emptyset$, we can use the argument in Proposition 6.4.6 to show that $u = 0$ near $\{x_n = b\}$. The general case can be reduced to this geometry by noting that Ω satisfies the *exterior ball property* given in the next lemma, and by using the Kelvin transform that was already used in Theorem 6.1.6 to map Ω into the exterior ball.

Lemma 6.4.7 (Exterior ball property). *Let $\Omega \subset \mathbb{R}^n$ be an open set with C^2 boundary. Given any $x_0 \in \partial\Omega$, there is an open ball $B \subset \mathbb{R}^n$ such that $\overline{B} \subset \mathbb{R}^n \setminus \Omega$ and $\partial B \cap \partial\Omega = \{x_0\}$.*

Proof. As in Theorem 2.2.2 we can choose coordinates so that $x_0 = 0$ and there is a C^2 function $h : \mathbb{R}^{n-1} \to \mathbb{R}$ with $h(0) = 0$, $\nabla h(0) = 0$, and

$$\Omega \cap B(0,r) = \{x_n > h(x')\} \cap B(0,r).$$

By Taylor expanding the C^2 function h at 0, there is $c > 0$ such that

$$h(x') \geq -c|x'|^2$$

for small x'. We now choose $\varepsilon > 0$ small and consider the ball $B(-\varepsilon e_n, \varepsilon)$. Near 0 the boundary of this ball is of the form $\{(x', g(x'))\}$ where

$$g(x') = -\varepsilon + \sqrt{\varepsilon^2 - |x'|^2}.$$

Since $f(t) = \sqrt{\varepsilon^2 - t}$ is smooth near 0 and $f''(t) < 0$, Taylor expanding the square root gives

$$g(x') \leq -\frac{1}{2\varepsilon}|x'|^2.$$

This is $\leq -c|x'|^2 \leq h(x')$ when ε is chosen sufficiently small. It follows that the ball $B(-\varepsilon e_n, \varepsilon)$ satisfies the conditions in the lemma. □

We first prove a local version of Theorem 6.4.1.

Proposition 6.4.8. *Let $\Omega \subset \mathbb{R}^n$ be a bounded open set with C^2 boundary and let L satisfy* (6.14)–(6.16)*. Suppose that $u \in H^2(\Omega)$ solves*

$$Lu = 0 \text{ in } \Omega$$

and, for some $x_0 \in \partial\Omega$, one has

$$u = \partial_\nu u = 0 \text{ near } x_0 \text{ on } \partial\Omega.$$

Then $u = 0$ near x_0.

Proof. As discussed above it is enough to consider $L = -\Delta + W \cdot \nabla + q$. Given $x_0 \in \partial\Omega$, we choose an exterior ball B as in Lemma 6.4.7. After a translation, rotation and scaling, we may assume that $B = B(0,1)$ and $x_0 = e_n$. Let $F(x) = x/|x|^2$ be the Kelvin transform, so that $\tilde{\Omega} := F(\Omega)$ is contained in B and $e_n \in \partial\tilde{\Omega}$. A short computation using Exercise 6.1.2 gives that

$$0 = F^*(-\Delta_e u + W \cdot \nabla u + qu) = -\Delta_{F^*e}(F^*u) + \tilde{W} \cdot \nabla(F^*u) + \tilde{q}F^*u$$

for some L^∞ coefficients $\tilde{W}$ and $\tilde{q}$. Writing $F^*u = |x|^{n-2}\tilde{v}$ and using (6.2) shows that $\tilde{v}$ solves

$$(-\Delta + \tilde{W}' \cdot \nabla + \tilde{q}')\tilde{v} = 0 \text{ in } \tilde{\Omega}$$

for new L^∞ coefficients $\tilde{W}'$ and $\tilde{q}'$. Moreover, $\tilde{v} = \partial_\nu \tilde{v} = 0$ near e_n on $\partial\tilde{\Omega}$.

Changing notation, we have reduced to the case where $\Omega \subset B(0,1)$ with $e_n \in \partial\Omega$, $u \in H^2(\Omega)$ solves

$$(-\Delta + W \cdot \nabla + q)u = 0 \text{ in } \Omega,$$

and
$$u = \partial_\nu u = 0 \text{ on } \partial\Omega \text{ near } e_n.$$
We can now use the argument in Proposition 6.4.5. Let $\chi \in C^\infty(\mathbb{R})$ satisfy $\chi(t) = 1$ for $t \geq 1 - \varepsilon$ and $\chi(t) = 0$ for $t \leq 1 - 2\varepsilon$, where $\varepsilon > 0$ is small but fixed so that $w = \chi u$ is in $H_0^2(\Omega)$ and has small support. Applying the Carleman estimate in (6.17) to w, and using that u is a solution, gives
$$\begin{aligned} e^{(1-\varepsilon)/h} \|u\|_{L^2(\Omega\cap\{1-\varepsilon<x_n<1\})} &\leq \|e^{x_n/h}\chi u\|_{L^2(\Omega)} \\ &\leq Ch\|e^{x_n/h}(-\Delta + W\cdot\nabla + q)(\chi u)\|_{L^2(\Omega)} \\ &\leq Ch\|e^{x_n/h}([-\Delta,\chi]u + W\cdot(\nabla\chi)u\|_{L^2(\Omega)} \\ &\leq Che^{(1-\varepsilon)/h}\|[-\Delta,\chi]u + W\cdot(\nabla\chi)u\|_{L^2(\Omega)}. \end{aligned}$$
Dividing by $e^{(1-\varepsilon)/h}$ and letting $h \to 0$, we obtain that $u = 0$ in the set $\Omega\cap\{1-\varepsilon < x_n < 1\}$. □

We can now easily prove the unique continuation statements.

Proof of Theorem 6.4.3. Since $u = 0$ in $V \cap S_+$, we have $u = \partial_\nu u = 0$ on $V \cap S$. Proposition 6.4.8 implies that $u = 0$ near S. □

Weak unique continuation follows by using a connectedness argument. We first prove a special case.

Proposition 6.4.9 (Weak unique continuation for concentric balls). *Let $\Omega = B(x_0, R)$ be an open ball in $\mathbb{R}^n$, and let L satisfy* (6.14)–(6.16). *If $u \in H^2(\Omega)$ satisfies*
$$Lu = 0 \quad \text{in} \quad \Omega$$
and
$$u = 0 \text{ in some ball } B(x_0, r_0) \text{ with } r_0 < R,$$
then $u = 0$ in Ω.

Proof. Let
$$I = \{r \in (0,R)\,;\, u = 0 \text{ in } B(x_0,r)\}.$$
By assumption, I is nonempty. It is closed in $(0,R)$ since whenever u vanishes on $B(x_0,r_j)$ and $r_j \to r$, then u vanishes on $B(x_0,r)$. We will show that I is open, which implies $I = (0,R)$ by connectedness and therefore proves the result.

Suppose $r \in I$, so $u = 0$ in $B(x_0,r)$. Let S be the hypersurface $\partial B(x_0,r)$. We know that $u = 0$ on one side of this hypersurface. Now Theorem 6.4.3 implies that for any $z \in S$, there is some open ball $B(z,r_z)$ contained in Ω so that u vanishes in $B(z,r_z)$. Define the open set
$$U = B(x_0,r) \cup \left(\bigcup_{z\in S} B(z,r_z) \right).$$

The distance between the compact set S and the closed set $\overline{B(x_0, R)} \setminus U$ is positive. In particular, there is some $\varepsilon > 0$ such that $u = 0$ in $B(x_0, r + \varepsilon)$. This shows that I is open. □

Proof of Theorem 6.4.2. Suppose $u \in H^2(\Omega)$ satisfies $Lu = 0$ in Ω and $u = 0$ in some open ball contained in Ω. Set

$$A = \{ x \in \Omega \mid u = 0 \text{ in some neighborhood of } x \text{ in } \Omega \}.$$

By assumption, A is a nonempty open subset of Ω. We will show that it is also closed. This implies by connectedness that $A = \Omega$, so indeed u vanishes in Ω as required.

Suppose on the contrary that A is not closed as a subset of Ω. Then there is some point x_0 on the boundary of A relative to Ω, for which $x_0 \notin A$. Choose $r_0 > 0$ so that $B(x_0, r_0) \subset \Omega$, and choose some point $y \in B(x_0, r_0/4)$ with $y \in A$. Since $y \in A$, we know that u vanishes on some ball $B(y, s_0)$ with $s_0 < r_0/2$. By Proposition 6.4.9, we see that u vanishes in the ball $B(y, r_0/2) \subset \Omega$. But $x_0 \in B(y, r_0/2)$, so u vanishes near x_0. This contradicts the fact that $x_0 \notin A$. □

Proof of Theorem 6.4.1. Fix some $x_0 \in \Gamma$. By Proposition 6.4.8, u vanishes near x_0 in Ω. Then u vanishes in some small ball contained in Ω, so Theorem 6.4.2 implies that $u = 0$ in Ω. □

6.4.3. Unique continuation via general weights. We have proved the main unique continuation statements by using Carleman estimates with linear weights and easy geometric arguments. The linear Carleman weight is very special, and typically unique continuation results are proved via stronger weights that give better powers of h in the Carleman estimate. For completeness we give another proof of Theorem 6.4.3 using this method.

We begin by recalling some notation from the proof of Theorem 6.2.3. Let (u, v) be the inner product in $L^2(\Omega)$ and $\|u\|$ the corresponding norm, and let $P_0 = (hD)^2$ be the semiclassical Laplacian. If $\psi \in C^\infty(\overline{\Omega})$ is a real valued function, we define the conjugated Laplacian

$$P_{0,\psi} = e^{\psi/h} P_0 e^{-\psi/h}.$$

We also write ψ'' for the Hessian matrix

$$\psi''(x) = [\partial_{x_j x_k} \psi(x)]_{j,k=1}^n.$$

The following is an analog of Theorem 6.2.3 for a general weight function. The point is that a Carleman estimate of the type $\left\|P_{0,\psi} u\right\| \geq ch \|u\|$ may follow if the weight ψ is chosen so that $(i[A, B]u, u)$ is at least nonnegative. In the case when ψ was a linear function, both A and B were constant coefficient operators and the commutator $i[A, B]$ was identically zero. However, if ψ is convex (meaning that the Hessian ψ'' is positive definite) one obtains a better lower bound.

Proposition 6.4.10 (Carleman estimate with general weight). *Let $\Omega \subset \mathbb{R}^n$ be a bounded open set and let $\psi \in C^\infty(\overline{\Omega})$. Then*

$$P_{0,\psi} = A + iB,$$

where A and B are the formally self-adjoint operators

$$A = (hD)^2 - |\nabla\psi|^2,$$
$$B = \nabla\psi \circ hD + hD \circ \nabla\psi.$$

If $u \in C_c^\infty(\Omega)$, one has

$$\left\|P_{0,\psi}u\right\|^2 = \|Au\|^2 + \|Bu\|^2 + (i[A,B]u,u),$$

where the commutator $i[A,B]$ satisfies

$$(i[A,B]u,u) = 4h(\psi'' hDu, hDu) + 4h((\psi''\nabla\psi \cdot \nabla\psi)u,u) - h^3((\Delta^2\psi)u,u).$$

Proof. The first step is to decompose $P_{0,\psi}$ into self-adjoint and skew-adjoint parts as

$$P_{0,\psi} = A + iB,$$

where A and B are the formally self-adjoint operators

$$A = \frac{P_{0,\psi} + P_{0,\psi}^*}{2}, \quad B = \frac{P_{0,\psi} - P_{0,\psi}^*}{2i}.$$

We have

$$\begin{aligned} P_{0,\psi} &= \sum_{j=1}^n (e^{\psi/h} hD_j e^{-\psi/h})^2 \\ &= \sum_{j=1}^n (hD_j + i\partial_j\psi)^2 \\ &= (hD)^2 - |\nabla\psi|^2 + i\nabla\psi \circ hD + ihD \circ \nabla\psi, \end{aligned}$$

and

$$\begin{aligned} P_{0,\psi}^* &= (e^{\psi/h} P_0 e^{-\psi/h})^* \\ &= e^{-\psi/h} P_0 e^{\psi/h} \\ &= (hD)^2 - |\nabla\psi|^2 - i\nabla\psi \circ hD - ihD \circ \nabla\psi. \end{aligned}$$

The required expressions for A and B follow.

If $u \in C_c^\infty(\Omega)$ we compute

$$\left\|P_{0,\psi}u\right\|^2 = ((A+iB)u,(A+iB)u) = \|Au\|^2 + \|Bu\|^2 + (i[A,B]u,u).$$

It remains to compute the commutator:

$$\begin{aligned} i[A,B]u &= h\big[((hD)^2 - |\nabla\psi|^2)(2\nabla\psi\cdot\nabla u + (\Delta\psi)u) \\ &\quad - (2\nabla\psi\cdot\nabla + \Delta\psi)((hD)^2u - |\nabla\psi|^2 u)\big] \\ &= h\big[2\nabla(hD)^2\psi\cdot\nabla u + 4hD\partial_k\psi\cdot hD\partial_k u + ((hD)^2\Delta\psi)u \\ &\quad + 2hD\Delta\psi\cdot hDu + 2\nabla\psi\cdot\nabla(|\nabla\psi|^2)u\big] \\ &= h\big[4(\psi''\nabla\psi\cdot\nabla\psi)u - 4h^2\partial_{jk}\psi\partial_{jk}u - 4h^2\nabla\Delta\psi\cdot\nabla u \\ &\quad - h^2(\Delta^2\psi)u\big]. \end{aligned}$$

Integrating by parts once yields

$$(i[A,B]u,u) = 4h^3(\psi''\nabla u,\nabla u) + 4h((\psi''\nabla\psi\cdot\nabla\psi)u,u) - h^3((\Delta^2\psi)u,u). \qquad \square$$

The next result shows that if one starts with any function φ with nonvanishing gradient, the convexified weight function $\psi = e^{\lambda\varphi}$ for λ sufficiently large will have a good Carleman estimate. Note that the power $h^{3/2}$ on the right is better than the power h in the estimate (6.17) for the linear weight, and that one has a second large parameter λ.

Proposition 6.4.11 (Carleman estimate with weight $e^{\lambda\varphi}$). *Let $\Omega \subset \mathbb{R}^n$ be a bounded open set, let $W \in L^\infty(\Omega,\mathbb{C}^n)$ and $q \in L^\infty(\Omega)$, and assume that $\varphi \in C^\infty(\overline{\Omega})$ satisfies $\varphi \geq 0$ and $\nabla\varphi \neq 0$ in $\overline{\Omega}$. Let*

$$\psi = e^{\lambda\varphi}.$$

There exist $C_0, \lambda_0, h_0 > 0$ such that whenever $\lambda \geq \lambda_0$ and $0 < h \leq h_0$, one has

$$\lambda^2\|u\| + \lambda\|hDu\| \leq C_0 h^{3/2}\|e^{\psi/h}(-\Delta + W\cdot\nabla + q)e^{-\psi/h}u\|$$

for any $u \in H_0^2(\Omega)$.

Proof. In the following, the positive constants c and C are always independent of λ, h and u and they may change from line to line. (We understand that c is small and C may be large.) Since $\psi = e^{\lambda\varphi}$, we have

$$\nabla\psi = \lambda e^{\lambda\varphi}\nabla\varphi, \quad \psi'' = \lambda^2 e^{\lambda\varphi}\nabla\varphi\otimes\nabla\varphi + \lambda e^{\lambda\varphi}\varphi'',$$

where $\nabla\varphi\otimes\nabla\varphi$ denotes the matrix $[\partial_j\varphi\partial_k\varphi]_{j,k=1}^n$. Assuming that $\lambda \geq 1$, we also have

$$|\Delta^2\psi| \leq C\lambda^4 e^{\lambda\varphi}.$$

By Proposition 6.4.10, for $u \in C_c^\infty(\Omega)$ we have

$$\|P_{0,\psi}u\|^2 = \|Au\|^2 + \|Bu\|^2 + (i[A,B]u,u),$$

where

$$\begin{aligned}(i[A,B]u,u) &= 4h(\psi'' hDu, hDu) + 4h((\psi''\nabla\psi\cdot\nabla\psi)u,u) - h^3((\Delta^2\psi)u,u)\\ &= 4h\lambda^4(e^{3\lambda\varphi}|\nabla\varphi|^4 u,u) + 4h\lambda^3(e^{3\lambda\varphi}(\varphi''\nabla\varphi\cdot\nabla\varphi)u,u)\\ &\quad - h^3((\Delta^2\psi)u,u)\\ &\quad + 4h\lambda^2(e^{\lambda\varphi}\nabla\varphi\cdot hDu, \nabla\varphi\cdot hDu)\\ &\quad + 4h\lambda(e^{\lambda\varphi}\varphi'' hDu, hDu).\end{aligned}$$

Consequently,

$$\begin{aligned}(i[A,B]u,u) \geq 4h\lambda^3(e^{3\lambda\varphi}[\lambda|\nabla\varphi|^4 + \varphi''\nabla\varphi\cdot\nabla\varphi]u,u) - Ch^3\lambda^4(e^{3\lambda\varphi}u,u)\\ - Ch\lambda(e^{\lambda\varphi}hDu, hDu).\end{aligned}$$

We used that $1 \leq e^{\lambda\varphi}$ and that $(e^{\lambda\varphi}\nabla\varphi\cdot hDu, \nabla\varphi\cdot hDu) \geq 0$. Now choose $\lambda \geq 1$ so large that $\lambda|\nabla\varphi|^4 + \varphi''\nabla\varphi\cdot\nabla\varphi \geq \lambda|\nabla\varphi|^4/2$ in $\overline{\Omega}$, i.e.

$$\lambda \geq \max\left\{1,\ 2\sup_{x\in\overline{\Omega}} \frac{|\varphi''\nabla\varphi\cdot\nabla\varphi|}{|\nabla\varphi|^4}\right\}.$$

This is possible since $\nabla\varphi$ is nonvanishing in $\overline{\Omega}$. If h is chosen sufficiently small (independent of λ), it follows that

$$(i[A,B]u,u) \geq ch\lambda^4(e^{3\lambda\varphi}u,u) - Ch\lambda(e^{\lambda\varphi}hDu, hDu).$$

We have proved the inequality

$$\|P_{0,\psi}u\|^2 \geq \|Au\|^2 + \|Bu\|^2 + ch\lambda^4(e^{3\lambda\varphi}u,u) - Ch\lambda(e^{\lambda\varphi}hDu, hDu).$$

The last negative term can be absorbed in the positive term $\|Au\|^2$ as follows. The argument is elementary but slightly tricky. Integration by parts gives

$$\begin{aligned}(e^{\lambda\varphi}hDu, hDu) &= (e^{\lambda\varphi}(hD)^2u,u) + (hD(e^{\lambda\varphi})\cdot hDu, u)\\ &= (e^{\lambda\varphi}Au,u) + (e^{\lambda\varphi}|\nabla\psi|^2 u,u) - ih\lambda(e^{\lambda\varphi}\nabla\varphi\cdot hDu, u)\\ &= (Au, e^{\lambda\varphi}u) + \lambda^2(e^{3\lambda\varphi}|\nabla\varphi|^2 u,u) - ih\lambda(e^{\lambda\varphi}\nabla\varphi\cdot hDu, u).\end{aligned}$$

By Young's inequality we have $(Au, e^{\lambda\varphi}u) \leq \frac{1}{\delta}\|Au\|^2 + \frac{\delta}{4}\|e^{\lambda\varphi}u\|^2$ where $\delta > 0$ is a number to be determined later. We obtain

$$\begin{aligned}(e^{\lambda\varphi}hDu, hDu) \leq \frac{1}{\delta}\|Au\|^2 + \frac{\delta}{4}\|e^{\lambda\varphi}u\|^2 + C\lambda^2(e^{3\lambda\varphi}u,u)\\ + Ch\lambda\|e^{\lambda\varphi/2}hDu\|\,\|e^{\lambda\varphi/2}u\|.\end{aligned}$$

Multiplying by δ and rearranging, we have

$$\begin{aligned}\|Au\|^2 \geq \delta(e^{\lambda\varphi}hDu, hDu) - \frac{\delta^2}{4}\|e^{\lambda\varphi}u\|^2 - C\delta\lambda^2(e^{3\lambda\varphi}u,u)\\ - Ch\lambda\delta\|e^{\lambda\varphi/2}hDu\|\,\|e^{\lambda\varphi/2}u\|.\end{aligned}$$

Combining the above inequalities gives that

$$\begin{aligned}\|P_{0,\psi}u\|^2 \geq \delta(e^{\lambda\varphi}hDu, hDu) - \frac{\delta^2}{4}(e^{3\lambda\varphi}u, u) - C\delta\lambda^2(e^{3\lambda\varphi}u, u) \\ - Ch\lambda\delta\|e^{\lambda\varphi/2}hDu\|\,\|e^{\lambda\varphi/2}u\| + ch\lambda^4(e^{3\lambda\varphi}u, u) \\ - Ch\lambda(e^{\lambda\varphi}hDu, hDu).\end{aligned}$$

We used that $1 \leq e^{\lambda\varphi}$ and $\|Bu\|^2 \geq 0$.

The idea is to choose δ so that the last expression is positive. By inspection, we arrive at the choice

$$\delta = \varepsilon h\lambda^2,$$

where ε is a fixed constant independent of h and λ. If ε is chosen sufficiently small (depending of c and C), we have

$$\begin{aligned}\|P_{0,\psi}u\|^2 \geq ch\lambda^4(e^{3\lambda\varphi}u, u) + (\varepsilon\lambda - C)h\lambda(e^{\lambda\varphi}hDu, hDu) \\ - C\varepsilon h^2\lambda^3\|e^{\lambda\varphi/2}hDu\|\,\|e^{\lambda\varphi/2}u\|.\end{aligned}$$

Choosing $\lambda \geq \lambda_0$ with λ_0 large enough (depending on ε and C) gives

$$\begin{aligned}\|P_{0,\psi}u\|^2 \geq ch\lambda^4(e^{3\lambda\varphi}u, u) + ch\lambda^2(e^{\lambda\varphi}hDu, hDu) \\ - C\varepsilon h^2\lambda^3\|e^{\lambda\varphi/2}hDu\|\,\|e^{\lambda\varphi/2}u\|.\end{aligned}$$

Now

$$2\lambda^3\|e^{\lambda\varphi/2}hDu\|\,\|e^{\lambda\varphi/2}u\| \leq \lambda^4(e^{3\lambda\varphi}u, u) + \lambda^2(e^{\lambda\varphi}hDu, hDu).$$

If $h \leq h_0$ with h_0 sufficiently small depending on C, c and ε, we have

$$\|P_{0,\psi}u\|^2 \geq ch\lambda^4(e^{3\lambda\varphi}u, u) + ch\lambda^2(e^{\lambda\varphi}hDu, hDu).$$

This implies

$$h\lambda^4\|e^{3\lambda\varphi/2}u\|^2 + h\lambda^2\|e^{\lambda\varphi/2}hDu\|^2 \leq C\|P_{0,\psi}u\|^2.$$

Consequently

$$\lambda^2\|e^{3\lambda\varphi/2}u\| + \lambda\|e^{\lambda\varphi/2}hDu\| \leq Ch^{3/2}\|e^{\psi/h}(-\Delta)e^{-\psi/h}u\|.$$

Adding and subtracting $W \cdot \nabla + q$ gives

$$\begin{aligned}\lambda^2\|e^{3\lambda\varphi/2}u\| + \lambda\|e^{\lambda\varphi/2}hDu\| \leq Ch^{3/2}\|e^{\psi/h}(-\Delta + W \cdot \nabla + q)e^{-\psi/h}u\| \\ + Ch^{3/2}(\|u\| + \|Du\| + h^{-1}\,\|(W \cdot \nabla\psi)u\|).\end{aligned}$$

Since $1 \leq e^{\lambda\varphi/2}$, choosing h so small that $Ch^{1/2} \leq 1/2$ and using that $\lambda \geq 1$ gives the required estimate for $u \in C_c^\infty(\Omega)$. The same estimate is true for $u \in H_0^2(\Omega)$ by density. □

We give the second proof of unique continuation across a hypersurface. In this proof we start with a weight φ such that $S = \{\varphi = \varphi(x_0)\}$ is a level set of φ and $u = 0$ in $\{\varphi > \varphi(x_0)\}$. We then force the solution u to vanish near x_0 by using a slightly bent weight $\tilde{\varphi}$, so that the sets $\{\varphi \leq \varphi(x_0)\}$ and $\{\tilde{\varphi} \geq \tilde{\varphi}(x_0) - \varepsilon\}$ have small intersection when ε is small. The convexified weights in Lemma 6.4.11 will also be used.

Proof of Theorem 6.4.3. As discussed before (6.17), it is enough to consider the case $\gamma \equiv 1$. Let $x_0 \in S$ and let V be a neighborhood of x_0 in Ω. The statement is local, so we may assume that V is a small ball centered at x_0 and that there is a real valued function $\varphi \in C^\infty(\overline{V})$, with $\varphi \neq 0$ in $\overline{V}$, such that

$$\begin{aligned} S \cap V &= \{\, x \in V \mid \varphi(x) = \varphi(x_0) \,\}, \\ S_+ \cap V &= \{\, x \in V \mid \varphi(x) > \varphi(x_0) \,\}, \\ S_- \cap V &= \{\, x \in V \mid \varphi(x) < \varphi(x_0) \,\}. \end{aligned}$$

In fact, φ is just a defining function for the hypersurface S, given near x_0 by $\varphi(x', x_n) = x_n - g(x')$ if S is locally the graph $(x', g(x'))$. By adding a constant we may assume that $\varphi \geq 1$ in $\overline{V}$. Also, write $\{\varphi = \varphi(x_0)\}$ for $S \cap V$, $\{\varphi > \varphi(x_0)\}$ for $S \cap V_+$, and $\{\varphi < \varphi(x_0)\}$ for $S \cap V_-$.

Assume that $u \in H^2(V)$ satisfies $(-\Delta + W \cdot \nabla + q)u = 0$ in V and $u = 0$ in $\{\varphi > \varphi(x_0)\}$. We need to show that u vanishes in some neighborhood of x_0. This will be done by applying a Carleman estimate to χu for a suitable cutoff function $\chi \in C_c^\infty(V)$. Given φ, we have seen that the function

$$\psi = e^{\lambda\varphi}$$

for λ large admits a good Carleman estimate and also satisfies

$$\begin{aligned} \{\psi = \psi(x_0)\} &= \{\varphi = \varphi(x_0)\}, \\ \{\psi > \psi(x_0)\} &= \{\varphi > \varphi(x_0)\}, \\ \{\psi < \psi(x_0)\} &= \{\varphi < \varphi(x_0)\}. \end{aligned}$$

However, to obtain the conclusion that $u = 0$ near x_0 we need to bend the weights a little bit. Define

$$\tilde{\varphi}(x) = \varphi(x) - \frac{1}{2}|x - x_0|^2.$$

Then $\nabla\tilde{\varphi}(x) = \nabla\varphi(x) - (x - x_0)$, and by shrinking V if necessary we have $\nabla\tilde{\varphi} \neq 0$ and $\tilde{\varphi} \geq 0$ in $\overline{V}$. By Proposition 6.4.11 the weight

$$\tilde{\psi} = e^{\lambda\tilde{\varphi}}$$

admits for h small (and for λ fixed but sufficiently large) the Carleman estimate

$$\|v\| + \|hDv\| \leq Ch^{3/2}\|e^{\tilde{\psi}/h}(-\Delta + W \cdot \nabla + q)e^{-\tilde{\psi}/h}v\|, \quad v \in H_0^2(V).$$

We will eventually show that $u = 0$ in $\{\psi \le \psi(x_0)\} \cap \{\tilde{\psi} \ge \tilde{\psi}(x_0) - \varepsilon\}$. The new weight $\tilde{\varphi}$ was chosen so that this set will be small whenever ε is small (see (6.20)).

Let V' be an open ball centered at x_0 and strictly contained inside V, and choose $\chi \in C_c^\infty(V)$ so that $\chi = 1$ near $\overline{V'}$. The Carleman estimate applied to $v = e^{\tilde{\psi}/h}\chi u$ implies

$$
\begin{aligned}
(6.18) \qquad & \|e^{\tilde{\psi}/h}\chi u\| + \|e^{\tilde{\psi}/h}\chi hDu\| \\
&= \|e^{\tilde{\psi}/h}\chi u\| + \|hD(e^{\tilde{\psi}/h}\chi u) - hD(e^{\tilde{\psi}/h}\chi)u\| \\
&\le \|e^{\tilde{\psi}/h}\chi u\| + \|hD(e^{\tilde{\psi}/h}\chi u)\| + C\|e^{\tilde{\psi}/h}\chi u\| + h\|e^{\tilde{\psi}/h}(\nabla\chi)u\| \\
&\le Ch^{3/2}\|e^{\tilde{\psi}/h}(-\Delta + W\cdot\nabla + q)(\chi u)\| + h\|e^{\tilde{\psi}/h}(\nabla\chi)u\| \\
&\le Ch^{3/2}\|e^{\tilde{\psi}/h}\chi(-\Delta + W\cdot\nabla + q)u\| \\
&\quad + Ch^{3/2}\|e^{\tilde{\psi}/h}([-\Delta,\chi]u + (W\cdot\nabla\chi)u\| + h\|e^{\tilde{\psi}/h}(\nabla\chi)u\|.
\end{aligned}
$$

Since $(-\Delta + W\cdot\nabla + q)u = 0$ in V, the first term on the right vanishes. Also, the properties of χ and the fact that $u = 0$ in $\{\psi > \psi(x_0)\}$ show that the functions $[-\Delta,\chi]u$ and $(\nabla\chi)u$ are supported in the set

$$U := (V \setminus V') \cap \{\psi \le \psi(x_0)\}.$$

Using that $\tilde{\varphi} < \tilde{\varphi}(x_0)$ in $(V \setminus \{x_0\}) \cap \{\varphi \le \varphi(x_0)\}$, we also have $\tilde{\psi} < \tilde{\psi}(x_0)$ in $(V \setminus \{x_0\}) \cap \{\psi \le \psi(x_0)\}$. Thus we can find $\varepsilon > 0$ so that

$$U \subset \{\, x \in V \mid \tilde{\psi}(x) < \tilde{\psi}(x_0) - \varepsilon \,\}.$$

The inequality (6.18) thus implies that

$$
\begin{aligned}
(6.19) \qquad \|e^{\tilde{\psi}/h}\chi u\| &\le Ch(\|e^{\tilde{\psi}/h}u\|_{L^2(U)} + \|e^{\tilde{\psi}/h}\nabla u\|_{L^2(U)}) \\
&\le Che^{(\tilde{\psi}(x_0)-\varepsilon)/h}(\|u\|_{L^2(U)} + \|\nabla u\|_{L^2(U)}).
\end{aligned}
$$

We claim that by decreasing ε if necessary, we may arrange that

$$(6.20) \qquad \{\psi \le \psi(x_0)\} \cap \{\tilde{\psi} \ge \tilde{\psi}(x_0) - \varepsilon\} \subset V'.$$

Indeed, if $x \in \{\psi \le \psi(x_0)\} \cap \{\tilde{\psi} \ge \tilde{\psi}(x_0) - \varepsilon\}$ then $\varphi(x) \le \varphi(x_0)$. Furthermore, by the definition of $\tilde{\psi}$ and $\tilde{\varphi}$ we have

$$e^{\lambda\varphi(x_0)} - \varepsilon \le e^{\lambda\varphi(x)}e^{-\frac{1}{2}|x-x_0|^2} \le e^{\lambda\varphi(x_0)}e^{-\frac{1}{2}|x-x_0|^2}.$$

Here λ is fixed, so taking ε sufficiently small we can make $|x - x_0|$ as small as we wish. This proves (6.20).

Now since $\chi = 1$ near $\overline{V'}$, (6.20) and (6.19) imply

$$
\begin{aligned}
e^{(\tilde{\psi}(x_0)-\varepsilon)/h}&\|u\|_{L^2(\{\psi\le\psi(x_0)\}\cap\{\tilde{\psi}\ge\tilde{\psi}(x_0)-\varepsilon\})} \\
&\le \|e^{\tilde{\psi}/h}u\|_{L^2(\{\psi\le\psi(x_0)\}\cap\{\tilde{\psi}\ge\tilde{\psi}(x_0)-\varepsilon\})} \\
&\le \|e^{\tilde{\psi}/h}\chi u\| \\
&\le Che^{(\tilde{\psi}(x_0)-\varepsilon)/h}(\|u\|_{L^2(U)} + \|\nabla u\|_{L^2(U)}).
\end{aligned}
$$

Canceling the exponentials and letting $h \to 0$, we obtain that $u = 0$ near x_0 as required. □

Remark 6.4.12. We now give the proof of Theorem 6.4.3 when the condition $Lu = 0$ in V is replaced by the inequality

$$|\operatorname{div}(\gamma\nabla u)| \le C(|u| + |\nabla u|) \quad \text{a.e. in } V.$$

Using the identity $\operatorname{div}(\gamma\nabla u) = \gamma\Delta u + \nabla\gamma \cdot \nabla u$ and the fact that $\gamma \in W^{1,\infty}$ has a positive lower bound, we obtain that

$$|\Delta u| \le C(|u| + |\nabla u|) \quad \text{a.e. in } V. \tag{6.21}$$

Next we apply (6.18) with $W = 0$ and $q = 0$ to obtain the inequality

$$
\begin{aligned}
&\|e^{\tilde{\psi}/h}\chi u\| + \|e^{\tilde{\psi}/h}\chi hDu\| \\
&\qquad \le Ch^{3/2}\|e^{\tilde{\psi}/h}\chi\Delta u\| + Ch^{3/2}\|e^{\tilde{\psi}/h}([-\Delta,\chi]u\| + h\|e^{\tilde{\psi}/h}(\nabla\chi)u\|.
\end{aligned}
$$

Combining this with (6.21) and choosing h small enough allows us to absorb the first term on the right to the left hand side. This gives

$$\|e^{\tilde{\psi}/h}\chi u\| + \|e^{\tilde{\psi}/h}\chi hDu\| \le Ch^{3/2}\|e^{\tilde{\psi}/h}([-\Delta,\chi]u\| + h\|e^{\tilde{\psi}/h}(\nabla\chi)u\|.$$

We are now in the same setting as on the paragraph after (6.18). Thus repeating the end of the proof establishes Theorem 6.4.3 if one assumes the differential inequality above. The other unique continuation theorems assuming the differential inequality follow from this version of Theorem 6.4.3 as before.

The unique continuation theorems for differential inequalities remain valid when u is only assumed to be H^1 instead of H^2 (the differential inequality then implies the further condition that $\Delta u \in L^2$). This follows by replacing u by suitable mollifications of u in the proofs and by applying a lemma due to Friedrichs. We refer the reader to [**Ler19**, Lemma B.16] for details.

6.5. Notes

There are many variants of the Calderón problem with partial data. Let $\Omega \subset \mathbb{R}^n$ be a bounded open set with smooth boundary and let $\gamma \in L^\infty(\Omega)$ satisfy

$\gamma \geq c > 0$ a.e. in Ω. If Γ_D and Γ_N are nonempty open subsets of $\partial\Omega$, one can consider the partial DN map

$$\Lambda_\gamma^{\Gamma_D,\Gamma_N} : H_c^{1/2}(\Gamma_D) \to H^{-1/2}(\Gamma_N), \quad f \mapsto \Lambda_\gamma f|_{\Gamma_N}.$$

Knowing $\Lambda_\gamma^{\Gamma_D,\Gamma_N}$ corresponds to the case where one can input Dirichlet data only on Γ_D (with the potential being grounded on the rest of $\partial\Omega$), and one can measure Neumann data only on Γ_N.

Similarly, if $q \in L^\infty(\Omega)$ one can consider the partial Cauchy data set for the Schrödinger equation,

$$C_q^{\Gamma_D,\Gamma_N} = \{ (u|_{\Gamma_D}, \partial_\nu u|_{\Gamma_N}) \mid u \in H^1(\Omega) \text{ solves } (-\Delta + q)u = 0 \text{ in } \Omega \\ \text{with } \operatorname{supp}(u|_{\partial\Omega}) \subset \Gamma_D \}.$$

If $\Gamma_D = \Gamma_N = \partial\Omega$, this is just the (full) Cauchy data set in Definition 2.7.8. As we have seen in this chapter, the partial data problem for C^2 conductivities can be reduced to the case of L^∞ potentials. We note that for technical reasons, the space $H_\Delta(\Omega) = \{ u \in L^2(\Omega) \mid \Delta u \in L^2(\Omega) \}$ is sometimes used instead of $H^1(\Omega)$ above.

In the *Calderón problem with partial data* one would like to determine information on γ or q from the knowledge of $\Lambda_\gamma^{\Gamma_D,\Gamma_N}$ or $C_q^{\Gamma_D,\Gamma_N}$, with various choices of Γ_D and Γ_N. If $\Gamma_D = \Gamma_N = \Gamma$, we write $\Lambda_\gamma^\Gamma = \Lambda_\gamma^{\Gamma,\Gamma}$ and similarly C_q^Γ. We call Λ_γ^Γ the *local DN map*.

In the following we will discuss a number of results that are available for the Calderón problem with partial data, together with selected references. Survey articles on partial data problems include [**KS14**] for $n \geq 3$ and [**GT13, IY13b**] for $n = 2$.

(1) **Boundary determination.** The boundary determination results in Chapter 3 work with local data, and they show that knowledge of the local DN map on Γ determines the Taylor series of a smooth conductivity on Γ.

Furthermore, if the conductivity γ is piecewise real-analytic, or if γ is *a priori* known in a neighborhood of $\partial\Omega$, then γ can be determined also in the interior. This can be done by recovering the full DN map in a smaller domain via a Runge approximation argument. See [**KV85, Isa88, AU04, Fat07, AK12, RS19**].

(2) **Carleman estimates with linear weights.** In [**BU02**] it was shown, in the notation of §6.3, that knowledge of $C_q^{\partial\Omega,\partial\Omega_{-,\varepsilon}}$ for some $\varepsilon > 0$ determines a potential $q \in L^\infty(\Omega)$ when $n \geq 3$. The starting point was the integral identity

$$(6.22) \qquad \langle (\Lambda_{q_1} - \Lambda_{q_2})(u_1|_{\partial\Omega}), u_2|_{\partial\Omega} \rangle_{\partial\Omega} = \int_\Omega (q_1 - q_2) u_1 u_2 \, dx$$

for any u_j solving $(-\Delta + q_j)u_j = 0$ in Ω.

If one only knows that $\Lambda_{q_1} = \Lambda_{q_2}$ on $\partial\Omega_{-,\varepsilon}$, then the left-hand side of (6.22) becomes an integral over $\partial\Omega_{+,\varepsilon}$. This unknown boundary term was handled by a Carleman estimate involving a linear weight and boundary terms. Here u_1 and u_2 are CGO solutions of the form

$$u = e^{(\varphi+i\psi)/h}(a + r), \tag{6.23}$$

where φ and ψ are linear functions. We have presented the construction of such CGO solutions via Carleman estimates in §6.2 and the result of [**BU02**] in §6.3.

(3) **Carleman estimates with limiting weights.** The result of [**BU02**] was extended significantly in [**KSU07**]. In this work the authors constructed CGO solutions of the form (6.23) for more general Carleman weights φ. In the argument involving completeness of products of solutions, one typically needs CGO solutions $u_1 = e^{\varphi/h}m_1$ and $u_2 = e^{-\varphi/h}m_2$ so that the exponential growth will be cancelled in the product u_1u_2. Thus the Carleman estimate should be valid both for φ and $-\varphi$.

The notion of *limiting Carleman weights* (LCWs) was introduced for this purpose. By definition, a LCW is a smooth real function φ with nonvanishing gradient such that

$$\{\bar{p}, p\} = 0 \quad \text{when} \quad p = 0,$$

where $p(x, \xi) = |\xi|^2 - |\nabla\varphi|^2 + 2i\nabla\varphi \cdot \xi$ is the principal symbol of the conjugated Laplacian, and $\{a, b\} = \nabla_\xi a \cdot \nabla_x b - \nabla_x a \cdot \nabla_\xi b$ is the Poisson bracket. In domains in $\mathbb{R}^2$ the LCWs are precisely the harmonic functions whereas for $n \geq 3$ any LCW is (up to a conformal transformation, scaling and adding a constant) of one of the following three forms [**DSFKSU09, AFGS20**]:

$$x_1, \qquad \log|x|, \qquad \arctan\left(\frac{x_1}{x_2}\right).$$

By using the weight $\varphi(x) = \log|x - x_0|$ with $x_0 \notin \overline{\Omega}$, it was proved in [**KSU07**] that q is uniquely determined by $C_q^{\tilde{F},\tilde{B}}$, where $\tilde{F}$ is a neighborhood of the front face $\{\, x \in \partial\Omega \mid (x - x_0) \cdot \nu \leq 0 \,\}$ in $\partial\Omega$, and $\tilde{B}$ is a neighborhood of the back face $\{(x - x_0) \cdot \nu \geq 0\}$ in $\partial\Omega$. Note that if Ω is strictly convex, then $\tilde{F}$ can be made arbitrarily small, but then $\tilde{B}$ has to be almost all of $\partial\Omega$.

Further contributions of [**KSU07**] include a convexification argument as in Exercise 6.2.6, and a way of using Carleman estimates to construct CGO solutions that vanish on suitable parts of the boundary. Such solutions are useful in identities like (6.22) when only partial information on the DN map is available. The determination of the coefficient was done by using a microlocal argument. This was replaced by the microlocal Helgason support theorem for the two-plane transform in the subsequent work [**DSFKSU07**] that also included first-order terms.

(4) **Reflection approach.** Another important approach to partial data problems was introduced in [**Isa07**]. The idea was to construct CGO solutions that vanish on parts of the boundary by applying reflections to standard CGO solutions. The corresponding results were presented in §6.1 and they state that C_q^Γ determines q provided that the *inaccessible part* $\Gamma_0 = \partial\Omega \setminus \Gamma$ is contained in a hyperplane or in a sphere.

(5) **A synthesis.** Even though the Carleman estimate and reflection approaches seem different, it was shown in [**KS13**] that they can be placed within a common framework. For a given LCW φ the set $\partial\Omega_{\mathrm{tan}} = \{ x \in \partial\Omega \mid \nabla\varphi(x) \cdot \nu = 0 \}$ plays a special role, and parts of hyperplanes and spheres are contained in $\partial\Omega_{\mathrm{tan}}$ for suitable φ. Either Carleman estimates or reflection arguments can be employed to construct CGO solutions vanishing on part of $\partial\Omega_{\mathrm{tan}}$, with additional control on either the front face or back face. The work [**KS13**] also generalizes the earlier approaches, for example by requiring the inaccessible part to satisfy only a conformal flatness condition in one direction and allowing certain non-Euclidean geometries. The method boils down to inverting a (broken) geodesic X-ray transform with partial data, which is a question of independent interest. A related result appears in [**IY13a**].

(6) **Stability, reconstruction etc.** For the reflection approach there are logarithmic stability results [**HW16**]. However, for the Carleman estimate approach the known stability results involve a log log modulus [**HW06**], [**CDSFR14, CDSFR16, PMRT22**]. A reconstruction procedure is given in [**NS10**]. There are various low regularity results [**Knu06, KS07, KU16b, CT20**], results for other equations [**Chu14a**], systems [**COS09, ST10, Car11, IY14, CST17, COST18**] and slab geometries [**LU10, KLU12b, CM16**]. We mention that in the partial data case there is no direct reduction from the partial ND map to a DN map, and one needs to deal with the ND map separately [**Chu14b**].

(7) **The linearized problem with partial data.** One can also study the linearized version (at zero potential) of the partial data problem. Uniqueness for this problem was proved in [**DSFKSU09**], where it was shown that products of harmonic functions that vanish on an arbitrary fixed part of the boundary form a complete set. The method was based on analytic microlocal analysis. This result has since found applications in inverse problems for nonlinear equations, with extensions given in [**KU20, CGU23, KKU23**].

(8) **The two-dimensional case.** The results discussed above were restricted to $n \geq 3$. Again, the two-dimensional case requires separate arguments. A

result involving knowledge of both a local DN map and ND map was given in [**APL05**]. In [**IUY10**] it was proved that the local DN map measured on an arbitrary open subset of the boundary determines the coefficient everywhere. The proof was based on an adaptation of Bukhgeim's method with Carleman weights given by holomorphic Morse functions.

A similar result valid on Riemann surfaces with boundary was proved in [**GT11a**] with extensions in [**GT11b, AGTU13, Tzo17, AT18**]. The method has been extended to more general equations [**IUY12, CLT24**], less regular coefficients [**IY12a**], the ND map [**IUY15**] and systems [**IY12b, IY15, IY16, IY17**].

(9) **Disjoint data and counterexamples.** For the case where the measurement sets Γ_D and Γ_N are disjoint, there are only few results. See [**KS13**] for a result where Γ_D and Γ_N cover all but a measure zero subset of $\partial\Omega$, and [**IUY11**] for a two-dimensional result with certain configurations.

There is an interesting series of counterexamples with [**DKN19a**], [**DKN19b**] for disjoint data, [**DKN20**] for the local data problem with Hölder continuous coefficients, and [**DHKN24**] for the anisotropic problem with nonzero frequency. The work [**DKN20**] is based on a counterexample to the unique continuation principle.

(10) **Unique continuation.** In §6.4 we have discussed unique continuation results for the operator

$$L = -\operatorname{div}(\gamma \nabla u) + W \cdot \nabla u + qu.$$

The work [**Car39**] introduced the method of exponentially weighted estimates, nowadays called Carleman estimates, for such problems. An earlier uniqueness result of Holmgren was given for real-analytic coefficients [**Hör03a**, Theorem 8.6.5]. Carleman estimates have since found various applications in unique continuation results, control theory and inverse problems.

The unique continuation results in §6.4 were proved by L^2 Carleman estimates, and the presentation involving linear weights follows [**Hab15a**]. The method of L^2 Carleman estimates extends to cover differential inequalities for H^1 solutions with Lipschitz continuous matrix valued γ, $W \in L^\infty$ and $q \in L^n$ in dimensions $n \geq 3$ (see, e.g., [**Hör85a**, §17.2]). With L^p Carleman estimates and additional ideas one can also deal with $W \in L^n$ and $q \in L^{n/2}$. Counterexamples involving Hölder continuous matrix valued γ exist, but for $n = 2$ the unique continuation result holds with $\gamma \in L^\infty$. See [**ARRV09, Ler19, LRLR22a, LRLR22b**] for further references and various improvements.

Chapter 7

Scattering Theory

This chapter has two main parts. In the first, up to and including §7.3, we give a brief description of direct scattering theory for the Schrödinger operator. In the second, §7.4, we consider an inverse problem and show that the scattering amplitude of a Schrödinger operator at fixed frequency uniquely determines its potential.

First, fix $\lambda > 0$ and consider the free Hamiltonian

$$H_0 u = (-\Delta - \lambda)u.$$

Note that, for each $\omega \in S^{n-1}$, the function

$$\psi_0(x, \omega) = e^{i\sqrt{\lambda}\, x\cdot\omega}$$

obeys

$$H_0 \psi_0 = 0.$$

The functions $\psi_0(\,\cdot\,, \omega)$ are "eigenfunctions" of $-\Delta$ with eigenvalue λ, and they are parametrized by the unit sphere S^{n-1} in $\mathbb{R}^n$. The corresponding solution of the time-dependent Schrödinger equation $(i\frac{\partial}{\partial t} + \Delta)\psi = 0$, namely $\psi(x,t) = e^{-i\lambda t} e^{i\sqrt{\lambda}\, x\cdot\omega}$, has phase velocity $\sqrt{\lambda}\,\omega$. That is, if you move with $x(t) = x(0) + \sqrt{\lambda}\,\omega\, t$, you always see the same value of $\psi(x,t)$. We think of $\psi_0(x,\omega)$ as an incoming wave in direction ω.

Next consider the perturbed operator

$$H_q u = (-\Delta - \lambda + q)u,$$

where q is bounded and compactly supported in $\mathbb{R}^n$. Then, as we will demonstrate in §7.2, the eigenfunctions $\psi_q(x,\omega)$ are also parametrized by the sphere.

They are unique if we insist that they have the form

$$\psi_q(x,\omega) = e^{i\sqrt{\lambda}\,x\cdot\omega} + \psi_q^s(x,\omega)$$

with the condition that the *scattered wave* $\psi_q^s(\,\cdot\,,\omega)$ is *outgoing.*

There are several equivalent definitions of outgoing. The one we shall use is that ψ_q^s is in the range of a certain linear operator, $R_0(\lambda)$, which we will introduce in the next section. In Corollary 7.3.2, we shall show that, asymptotically for large $|x|$,

$$\psi_q(\lambda,x,\omega) = e^{i\sqrt{\lambda}\,x\cdot\omega} + \frac{a_q(\sqrt{\lambda},\theta,\omega)}{|x|^{\frac{n-1}{2}}}e^{i\sqrt{\lambda}|x|} + O(|x|^{-\frac{n-1}{2}-1}), \tag{7.1}$$

where $\theta = \frac{x}{|x|}$. We think of the first term as an incoming wave in direction ω and of the second term as an outgoing radially expanding wave that arises when the incoming wave scatters off of the potential q. The amplitude, $a_q(\sqrt{\lambda},\theta,\omega)$, of this outgoing wave depends on the direction of view, θ, and is called the *scattering amplitude.* We saw a similar setup in the context of acoustic scattering in §1.3 and in the one-dimensional case in the end of §5.6.

7.1. The free resolvent

Our first step is to solve the constant coefficient equation

$$(-\Delta - \lambda)u = f \text{ in } \mathbb{R}^n. \tag{7.2}$$

Here $f \in L^2_\delta(\mathbb{R}^n)$ for some $\delta > \frac{1}{2}$, where L^2_δ is the space introduced in §4.5 with norm

$$\|f\|_{L^2_\delta} = \left\|\langle x\rangle^\delta f\right\|_{L^2}.$$

We will solve (7.2) by employing the Fourier transform and simply writing

$$\hat{u}(\xi) = \frac{\hat{f}(\xi)}{|\xi|^2 - \lambda}.$$

If $\operatorname{Im}\lambda \neq 0$, the denominator does not vanish and $\hat{u}$ is an unambiguously defined element of $L^2(\mathbb{R}^n)$ with

$$\|u\|_{L^2} = \frac{1}{(2\pi)^{n/2}}\|\hat{u}\|_{L^2} \le \frac{1}{(2\pi)^{n/2}|\operatorname{Im}\lambda|}\|\hat{f}\|_{L^2} = \frac{1}{|\operatorname{Im}\lambda|}\|f\|_{L^2}.$$

For real nonzero λ we shall define the unique *outgoing* solution to (7.2) by

$$\hat{u}(\xi) = \lim_{\varepsilon\downarrow 0}\frac{\hat{f}(\xi)}{|\xi|^2 - \lambda - i\varepsilon}. \tag{7.3}$$

Letting ε increase to zero in (7.3) would define the unique *incoming* solution to (7.2). We will only consider outgoing solutions below.

We begin with the following solvability result.

Theorem 7.1.1 (Agmon estimates). *Let $\lambda \in \mathbb{C} \setminus [0, \infty)$, $|\lambda| \geq \varepsilon > 0$, $\delta > \frac{1}{2}$, and $n \geq 2$. Given any $f \in L^2_\delta(\mathbb{R}^n)$, there exists a unique weak solution $u \in L^2{}_{-\delta}(\mathbb{R}^n)$ to*

$$(-\Delta - \lambda)u = f \text{ in } \mathbb{R}^n. \tag{7.4}$$

Moreover, there is a constant $C = C(n, \delta, \varepsilon)$ such that

$$\|u\|_{H^k_{-\delta}} \leq C|\lambda|^{\frac{k-1}{2}} \|f\|_{L^2_\delta}, \qquad k = 0, 1, 2. \tag{7.5}$$

We shall denote the solution operator to (7.4) by

$$u = R_0(\lambda)f.$$

The operator $R_0(\lambda)$ is called the *free resolvent* at frequency $\lambda \in \mathbb{C} \setminus [0, \infty)$. In order to extend $R_0(\lambda)$ to positive real values of λ we shall prove the following. Recall that $\langle u, \varphi \rangle$ denotes the pairing of a function or distribution u with a function φ in $\mathbb{R}^n$.

Proposition 7.1.2 (Limiting absorption principle). *For every $f \in L^2_\delta(\mathbb{R}^n)$ and $g \in L^2_\delta(\mathbb{R}^n)$, the map*

$$\lambda \mapsto \langle R_0(\lambda)f, g \rangle$$

defined in $\{\operatorname{Im} \lambda > 0\}$ has unique limits on $\mathbb{R} \setminus \{0\}$. For $\lambda \in \mathbb{R} \setminus \{0\}$, the formula

$$\langle R_0(\lambda)f, g \rangle = \lim_{\varepsilon \downarrow 0} \langle R_0(\lambda + i\varepsilon)f, g \rangle$$

defines a bounded linear map $R_0(\lambda) : L^2_\delta \to H^2_{-\delta}$.

Definition 7.1.3 (Outgoing resolvent). For $\lambda > 0$, the operator

$$R_0(\lambda) = \lim_{\varepsilon \downarrow 0} R_0(\lambda + i\varepsilon)$$

is called the *outgoing resolvent*. A function u is said to be *outgoing* if it is of the form $R_0(\lambda)f$ for some $f \in L^2_\delta$ where $\delta > \frac{1}{2}$.

The following argument is rather similar to the proof of Proposition 4.5.4 related to norm estimates for CGO solutions.

Proof of Theorem 7.1.1. We will prove the estimate in $L^2_{-\delta}$. The $H^k_{-\delta}$ estimates for $k = 1, 2$ will follow from this (see Exercise 7.1.6).

Let u and f obey (7.4). Then

$$(|\xi|^2 - \lambda)\hat{u}(\xi) = \hat{f}(\xi).$$

The quantity $|\xi|^2 - \lambda$ does not vanish since $\lambda \in \mathbb{C} \setminus [0, \infty)$, so

$$\hat{u}(\xi) = \frac{\hat{f}(\xi)}{|\xi|^2 - \lambda}.$$

The solution u is therefore unique. If $\lambda \leq -\varepsilon$, using that $\delta > 1/2$ we immediately obtain the estimate

$$\|u\|_{L^2_{-\delta}} \leq \|u\|_{L^2} \leq \frac{1}{|\lambda|}\|f\|_{L^2} \leq \frac{C_\varepsilon}{|\lambda|^{1/2}}\|f\|_{L^2_\delta}.$$

We may therefore suppose that $\operatorname{Im}\lambda \neq 0$.

We define a smooth partition of unity $\chi_0(\xi,\lambda)^2, \chi_1(\xi,\lambda)^2, \ldots, \chi_n(\xi,\lambda)^2$ with χ_0 supported away from $|\xi| = \sqrt{|\lambda|}$, the other χ_j's supported near $|\xi| = \sqrt{|\lambda|}$ and, for $1 \leq j \leq n$, $\frac{|\xi_j|^2}{|\xi|^2} \geq \frac{1}{2n}$ on the support of χ_j. To do so, let $\approx{\chi}_0(\xi), \ldots, \approx{\chi}_n(\xi)$ be nonnegative smooth functions in $\mathbb{R}^n \setminus \{0\}$ which take values in $[0,1]$ and satisfy

$$\overset{\approx}{\chi}_0(\xi) = \begin{cases} 0, & \frac{1}{2} \leq |\xi| \leq \frac{3}{2}, \\ 1, & |\xi| \leq \frac{1}{4} \text{ or } |\xi| \geq 2, \end{cases}$$

and, for $j = 1, \ldots, n$,

$$\overset{\approx}{\chi}_j(\xi) = \begin{cases} 1, & |\xi_j|^2 \geq \frac{1}{n}|\xi|^2, \\ 0, & |\xi_j|^2 \leq \frac{1}{2n}|\xi|^2. \end{cases}$$

Let

$$\widetilde{\chi}_0(\xi,\lambda) = \overset{\approx}{\chi}_0(\xi/\sqrt{|\lambda|}),$$
$$\widetilde{\chi}_j(\xi,\lambda) = \big(1 - \widetilde{\chi}_0(\xi,\lambda)\big)\overset{\approx}{\chi}_j(\xi) \qquad \text{for } j = 1, \ldots, n,$$

and, finally, define

$$\chi_k(\xi,\lambda) = \frac{\widetilde{\chi}_k(\xi,\lambda)}{\sqrt{\sum_{j=0}^n \widetilde{\chi}_j(\xi,\lambda)^2}} \qquad \text{for } k = 0, \ldots, n,$$

so that

$$\sum_{j=0}^n \chi_j(\xi,\lambda)^2 = 1.$$

The functions χ_k are well defined and smooth in $\mathbb{R}^n$ since for any $0 \neq \xi \in \mathbb{R}^n$, we must have $|\xi_j|^2 \geq \frac{1}{n}|\xi|^2$ for at least one $1 \leq j \leq n$ so that $\sum_{j=1}^n \overset{\approx}{\chi}_j(\xi)^2 \geq 1$ for all $\xi \neq 0$ and

$$\sum_{j=0}^n \widetilde{\chi}_j(\xi,\lambda)^2 \geq \widetilde{\chi}_0(\xi,\lambda)^2 + \big(1 - \widetilde{\chi}_0(\xi,\lambda)\big)^2 \geq \frac{1}{4}.$$

Combining this with the estimate

$$\sup_{\substack{|\lambda|\geq\varepsilon \\ \xi\in\mathbb{R}^n \\ 0\leq j\leq n}} |D_\lambda^m D_\xi^\alpha \widetilde{\chi}_j(\xi,\lambda)| < \infty$$

yields that, for $|\lambda| \geq \varepsilon$,

$$|D_\lambda^m D_\xi^\alpha \chi_j(\xi, \lambda)| \leq C(m, \alpha, \varepsilon). \tag{7.6}$$

We now write

$$R_0(\lambda) f = \mathcal{F}^{-1}\Big\{q_0 \hat{f} + \sum_{j=1}^{n} q_j m_j \hat{f}\Big\} \tag{7.7}$$

with

$$q_0(\xi) = \frac{\chi_0(\xi, \lambda)}{|\xi|^2 - \lambda}, \tag{7.8}$$

$$q_j(\xi) = \frac{\chi_j(\xi, \lambda)}{|\xi| + \sqrt{\lambda}} \qquad \text{for} \quad j = 1, \ldots, n, \tag{7.9}$$

$$m_j(\xi) = \frac{\chi_j(\xi, \lambda)}{|\xi| - \sqrt{\lambda}} \qquad \text{for} \quad j = 1, \ldots, n. \tag{7.10}$$

Above we choose $\sqrt{\lambda}$ to have positive imaginary part, which is possible since $\operatorname{Im} \lambda \neq 0$. In the remainder of the proof, we shall assume that $\operatorname{Im} \lambda > 0$, so $\operatorname{Re}(\sqrt{\lambda})$ is also nonnegative and the magnitude of the denominator in (7.9) is bounded below by $\sqrt{|\lambda|} \geq \sqrt{\varepsilon}$ for all $\xi \in \mathbb{R}^n$ and $|\lambda| \geq \varepsilon$. In this case the estimates for q_j are relatively easy, but the estimates for m_j are more involved since the denominator in (7.10) is not bounded away from zero. On the other hand, the argument for the case where $\operatorname{Im} \lambda < 0$ (so $\operatorname{Re}(\sqrt{\lambda})$ is negative) is analogous to the case $\operatorname{Re}(\sqrt{\lambda}) > 0$ after interchanging the definitions of the q_j and the m_j.

We now study the terms involving q_j. It follows from (7.6), (7.8), and (7.9) that, for $|\lambda| \geq \varepsilon$,

$$|D_\lambda^m D_\xi^\alpha q_j(\xi, \lambda)| \leq \frac{C(m, \alpha, \varepsilon)}{\sqrt{|\lambda|}}$$

with a new constant $C(m, \alpha, \varepsilon)$. By Lemmas C.1.29 and (when δ is not an integer) C.3.2,

$$\|q_j \hat{f}\|_{H^\delta(\mathbb{R}^n)} \leq \frac{C(\delta, \varepsilon)}{\sqrt{|\lambda|}} \|\hat{f}\|_{H^\delta(\mathbb{R}^n)}$$

for any δ and $0 \leq j \leq n$. Therefore the operators

$$Q_j : f \mapsto \mathcal{F}^{-1}\{q_j(\xi) \hat{f}(\xi)\}$$

satisfy the estimate

$$\|Q_j f\|_{L^2_\delta} \leq \frac{C(\delta, \varepsilon)}{\sqrt{|\lambda|}} \|f\|_{L^2_\delta}. \tag{7.11}$$

Moreover, the norms of the derivatives of the operators Q_j with respect to λ in $\{\operatorname{Im}\lambda \geq 0\} \cap \{|\lambda| \geq \varepsilon\}$ are also bounded, so that the Q_j are continuous functions of λ in the uniform operator topology.

From (7.7) we see that when $\operatorname{Im}\lambda > 0$, we have

$$R_0(\lambda)f = Q_0(\lambda)f + \sum_{j=1}^{n} Q_j(\lambda)M_j(\lambda)f, \tag{7.12}$$

where M_j are the operators

$$M_j : f \mapsto \mathcal{F}^{-1}\{m_j(\xi)\hat{f}(\xi)\}.$$

The estimate (7.5), as well as Theorem 7.1.1, will then follow from (7.11), the fact that $\delta > \frac{1}{2}$, and the next estimate for the operators M_j. □

Lemma 7.1.4. *Let $\delta > \frac{1}{2}$, $\operatorname{Im}\lambda > 0$ and $|\lambda| > \varepsilon > 0$. For all $1 \leq j \leq n$,*

$$\|M_j f\|_{L^2_{-\delta}} \leq C(\varepsilon, \delta)\|f\|_{L^2_\delta}.$$

Moreover, for $f, g \in L^2_\delta$, the map

$$\lambda \mapsto \langle M_j f, g\rangle$$

defined in $\{\operatorname{Im}\lambda > 0\}$ has unique limits on $\mathbb{R}$.

Proof. We will prove the lemma for M_1. Let $\eta = \psi(\xi)$ define the change of coordinates

$$\eta_1 = |\xi|, \quad \eta_j = \xi_j \quad \text{for} \quad j = 2, \ldots, n.$$

By construction the Jacobian, $\det(D\psi) = \frac{\xi_1}{|\xi|}$, is bounded and bounded away from zero on an open set Ω containing the support of χ_1. Let $\tilde{\chi}_1$ be a C^∞ function, taking values in $[0,1]$, that is supported in Ω, is identically one on the support of χ_1 and has similar form as χ_1. By Lemma C.1.48 and Lemma C.3.2 (when δ is not an integer) $u \mapsto \Phi u = (\chi_1 u) \circ \psi^{-1}$ and $u \mapsto \tilde{\Phi} u = \left(\frac{\tilde{\chi}_1 u}{|D\psi|}\right) \circ \psi^{-1}$ are bounded maps on $H^\delta(\mathbb{R}^n)$ with operator norms depending only on δ and ε. Furthermore these operators are norm continuous in λ in $\{\operatorname{Im}\lambda \geq 0\} \cap \{|\lambda| \geq \varepsilon\}$.

(They depend on λ only through χ_1 and $\tilde{\chi}_1$.) Since

$$\begin{aligned}
(M_1 f, g) &= \int \frac{\chi_1(\xi)}{|\xi| - \sqrt{\lambda}} \hat{f}(\xi)\, \overline{\hat{g}(\xi)}\, \frac{d\xi}{(2\pi)^n} \\
&= \int \frac{1}{|\xi| - \sqrt{\lambda}} \tilde{\chi}_1(\xi)\hat{f}(\xi)\, \overline{\chi_1(\xi)\hat{g}(\xi)}\, \frac{d\xi}{(2\pi)^n} \\
&= \int \frac{1}{\eta_1 - \sqrt{\lambda}} \tilde{\chi}_1(\psi^{-1}(\eta))\hat{f}(\psi^{-1}(\eta))\, \overline{\chi_1(\psi^{-1}(\eta))\hat{g}(\psi^{-1}(\eta))} \\
&\qquad\qquad \times \frac{1}{(D\psi)(\psi^{-1}(\eta))} \frac{d\eta}{(2\pi)^n} \\
&= \int \frac{1}{\eta_1 - \sqrt{\lambda}} (\tilde{\Phi}\hat{f})(\eta)\, \overline{(\Phi\hat{g})(\eta)}\, \frac{d\eta}{(2\pi)^n} \\
&= (N_1((\tilde{\Phi}\hat{f})^\vee), (\Phi\hat{g})^\vee) \quad \text{where} \quad N_1 f = \mathcal{F}^{-1}\left\{\frac{\hat{f}}{\eta_1 - \sqrt{\lambda}}\right\},
\end{aligned}$$

it suffices to prove that

$$|(N_1 f, g)| \le \widetilde{C}(\delta, \varepsilon)\|f\|_{L^2_\delta}\|g\|_{L^2_\delta} \tag{7.13}$$

and that $(N_1 f, g)$ converges as λ approaches $\mathbb{R}$ from above.

Now, by Exercise 7.1.5, N_1 can be also expressed as

$$N_1 f(x_1, x') = i \int_{-\infty}^{x_1} e^{i\sqrt{\lambda}(x_1 - y_1)} f(y_1, x')\, dy_1. \tag{7.14}$$

Because $\operatorname{Im}\sqrt{\lambda} \ge 0$ and $x_1 - y_1 \ge 0$ on the domain of integration, the exponential in (7.14) is uniformly bounded. Fixing x' for the moment and applying the Cauchy–Schwarz inequality in $\mathbb{R}$, we see that, since $\delta > \frac{1}{2}$,

$$\begin{aligned}
|N_1 f(x_1, x')|^2 &\le \int_{-\infty}^{x_1} (1 + y_1^2)^{-\delta}\, dy_1 \int_{-\infty}^{\infty} |f(y_1, x')|^2 (1 + y_1^2)^{\delta}\, dy_1 \\
&\le C(\delta)\|f(\cdot, x')\|^2_{L^2_\delta(\mathbb{R})}
\end{aligned}$$

so that

$$\|N_1 f(\cdot, x')\|^2_{L^2_{-\delta}(\mathbb{R})} \le C(\delta) \int_{-\infty}^{\infty} (1 + x_1^2)^{-\delta}\, dx_1\, \|f(\cdot, x')\|^2_{L^2_\delta(\mathbb{R})}.$$

The integral $\tilde{C}(\delta) = C(\delta)\int_{-\infty}^{\infty} (1 + x_1^2)^{-\delta}\, dx_1$ is finite as long as $\delta > \frac{1}{2}$. Since, for $\delta > 0$,

$$\begin{aligned}
(1 + x_1^2 + |x'|^2)^{-\delta} &\le (1 + x_1^2)^{-\delta}, \\
(1 + x_1^2 + |x'|^2)^{\delta} &\ge (1 + x_1^2)^{\delta}
\end{aligned}$$

we have that

$$
\begin{aligned}
\|N_1 f\|^2_{L^2_{-\delta}(\mathbb{R}^n)} &= \int_{-\infty}^{\infty}\int_{\mathbb{R}^{n-1}} (1+x_1^2+|x'|^2)^{-\delta} |N_1 f(x_1,x')|^2\,dx'\,dx_1 \\
&\le \int_{\mathbb{R}^{n-1}}\int_{-\infty}^{\infty} (1+x_1^2)^{-\delta} |N_1 f(x_1,x')|^2\,dx_1\,dx' \\
&= \int_{\mathbb{R}^{n-1}} \|N_1 f(\cdot,x')\|^2_{L^2_{-\delta}(\mathbb{R})}\,dx' \\
&\le \tilde{C}(\delta)\int_{\mathbb{R}^{n-1}}\int_{-\infty}^{\infty} (1+x_1^2)^{\delta} |f(x_1,x')|^2\,dx_1\,dx' \\
&\le \tilde{C}(\delta)\int_{\mathbb{R}^{n-1}}\int_{-\infty}^{\infty} (1+x_1^2+|x'|^2)^{\delta} |f(x_1,x')|^2\,dx_1\,dx'.
\end{aligned}
$$

This implies (7.13). The existence of unique limiting values of

$$(N_1 f, g) = i\int_{\mathbb{R}^n}\int_{-\infty}^{x_1} e^{i\sqrt{\lambda}(x_1-y_1)} f(y_1,x')\,\overline{g(x)}\,dy_1\,dx$$

as $\operatorname{Im}\lambda$ decreases to zero follows from the dominated convergence theorem and the existence of pointwise limits for the function $e^{i\sqrt{\lambda}(x_1-y_1)}$. □

Exercise 7.1.5. Let $\operatorname{Im}\lambda > 0$ and choose $\operatorname{Im}\sqrt{\lambda} > 0$. Define the map

$$N_1 : L^2(\mathbb{R}^n) \to L^2(\mathbb{R}^n), \quad N_1 f = \mathcal{F}^{-1}\left\{\frac{\hat{f}(\eta)}{\eta_1 - \sqrt{\lambda}}\right\}.$$

Prove that

$$N_1 f(x_1,x') = i\int_{-\infty}^{x_1} e^{i\sqrt{\lambda}(x_1-y_1)} f(y_1,x')\,dy_1.$$

Exercise 7.1.6. Prove the $H^k_{-\delta}$ estimate for $k = 1, 2$ in Theorem 7.1.1. You may adapt the corresponding argument in the proof of Proposition 4.5.4.

Exercise 7.1.7. Let $\lambda > 0$, $\delta > \frac{1}{2}$ and $f \in L^2_\delta(\mathbb{R}^n)$.

(a) Prove that $R_0(\lambda)f$ is a weak solution of the differential equation

$$(-\Delta - \lambda)u = f.$$

In other words, prove that $u = R_0(\lambda)f$ obeys $\langle u, (-\Delta-\lambda)\varphi\rangle = \langle f, \varphi\rangle$ for all $\varphi \in \mathcal{S}(\mathbb{R}^n)$.

(b) Prove that if u is a weak solution of the differential equation

$$(-\Delta - \lambda)u = f$$

and if u is in the range of $R_0(\lambda)$, then $u = R_0(\lambda)f$.

We have now completed the proof of Theorem 7.1.1. The proof of the limiting absorption principle follows easily from the arguments above.

Proof of Proposition 7.1.2. The limits for λ on the negative real axis follow from the formula $\hat{u}(\xi) = \frac{\hat{f}(\xi)}{|\xi|^2-\lambda}$. Thus it is enough to consider the case of the positive real axis. Let $\lambda_0 > 0$ and suppose that $(\lambda_k) \subset \{\operatorname{Im}\lambda > 0\}$ with $\lambda_k \to \lambda_0$. We recall (7.12) and claim that

$$\langle R_0(\lambda_k)f, g\rangle \to \langle Q_0(\lambda_0)f + \sum_{j=1}^{n} Q_j(\lambda_0)M_j(\lambda_0)f, g\rangle.$$

The operators $Q_j(\lambda)$ were continuous in λ in $\{\operatorname{Im}\lambda \geq 0\}\setminus\{0\}$, so $Q_0(\lambda_k)f \to Q_0(\lambda_0)f$ and

$$\|(Q_j(\lambda_k) - Q_j(\lambda_0))M_j(\lambda_k)f\|_{L^2_{-\delta}} \leq \left\|Q_j(\lambda_k) - Q_j(\lambda_0)\right\|_{L^2_{-\delta}\to L^2_{-\delta}} \left\|M_j(\lambda_k)f\right\|_{L^2_{-\delta}}.$$

This converges to zero as $k \to \infty$ since

$$\left\|M_j(\lambda_k)f\right\|_{L^2_{-\delta}} \leq C\,\|f\|_{L^2_\delta}$$

uniformly over k large.

It remains to show that, as $k \to \infty$,

$$\langle Q_j(\lambda_0)\left[M_j(\lambda_k) - M_j(\lambda_0)\right]f, g\rangle \to 0.$$

This follows from Lemma 7.1.4 upon writing

$$\langle Q_j(\lambda_0)\left[M_j(\lambda_k) - M_j(\lambda_0)\right]f, g\rangle = \langle (M_j(\lambda_k) - M_j(\lambda_0))f, Q_j(\lambda_0)'g\rangle,$$

where the adjoint $Q_j(\lambda_0)'$ has the same form as $Q_j(\lambda_0)$ since $\lambda_0 > 0$. □

In §7.2 we shall make use of the following lemma. Before stating the lemma we define the codimension 1 sphere

$$M_\lambda = \{\,\xi \in \mathbb{R}^n \mid \xi\cdot\xi - \lambda = 0\,\}.$$

Lemma 7.1.8. *Let $\lambda > 0$ and $f \in L^2_\delta(\mathbb{R}^n)$ with $\delta > \frac{1}{2}$. Then*

$$\operatorname{Im}\int_{\mathbb{R}^n} (R_0(\lambda)f)(x)\overline{f(x)}\,dx = \frac{1}{2^{n+1}\pi^{n-1}\sqrt{\lambda}}\int_{M_\lambda} |\hat{f}(\omega)|^2\,dS(\omega), \tag{7.15}$$

where dS is the surface measure on the sphere M_λ of radius $\sqrt{\lambda}$.

Proof. We apply the Parseval identity to obtain

$$(R_0(\lambda+i\varepsilon)f, f)_{L^2(\mathbb{R}^n,dx)} = \Big(\frac{1}{\xi\cdot\xi - \lambda - i\varepsilon}\hat{f}, \hat{f}\Big)_{L^2(\mathbb{R}^n,\frac{d\xi}{(2\pi)^n})}$$

so that

$$\begin{aligned}\operatorname{Im}(R_0(\lambda+i\varepsilon)f,f) &= \int_{\mathbb{R}^n} \frac{|\hat f(\xi)|^2\varepsilon}{(\xi\cdot\xi-\lambda)^2+\varepsilon^2}\frac{d\xi}{(2\pi)^n}\\ &= \frac{1}{(2\pi)^n}\int_0^\infty\int_{M_{\rho^2}} \frac{|\hat f(\omega)|^2\varepsilon}{(\rho^2-\lambda)^2+\varepsilon^2}\, dS(\omega)\, d\rho\\ &= \int_0^\infty \frac{\tilde F(\rho)\varepsilon}{(\rho^2-\lambda)^2+\varepsilon^2}\, d\rho,\end{aligned}$$

where

$$\tilde F(\rho) = \frac{1}{(2\pi)^n}\int_{M_{\rho^2}} |\hat f(\omega)|^2\, dS(\omega).$$

Now make the change of variables $\rho^2-\lambda=\varepsilon t$. This gives

$$\operatorname{Im}(R_0(\lambda+i\varepsilon)f,f) = \int_{-\lambda/\varepsilon}^\infty \frac{F_\varepsilon(t)}{t^2+1}\, dt,$$

where

$$F_\varepsilon(t) = \frac{1}{2\rho}\tilde F(\rho)\big|_{\rho=\sqrt{\lambda+\varepsilon t}}.$$

Since $f\in L^2_\delta(\mathbb{R}^n)$, we have that $\hat f\in H^\delta(\mathbb{R}^n)$. Applying Lemma C.2.4 and Exercise C.2.7, in ρ,ω coordinates, yields that $\tilde F(\rho)$ is locally bounded and continuous. Hence,

$$\lim_{\varepsilon\downarrow 0}\int_{-\lambda/\varepsilon}^{-\lambda/(2\varepsilon)} \frac{F_\varepsilon(t)}{t^2+1}\, dt = \lim_{\varepsilon\downarrow 0}\int_0^{\sqrt{\lambda/2}} \frac{\tilde F(\rho)\varepsilon}{(\rho^2-\lambda)^2+\varepsilon^2}\, d\rho = 0$$

and, by the Lebesgue dominated convergence theorem,

$$\lim_{\varepsilon\downarrow 0}\int_{-\lambda/(2\varepsilon)}^\infty \frac{F_\varepsilon(t)}{t^2+1}\, dt = \int_{-\infty}^\infty \frac{F_0(t)}{t^2+1}\, dt.$$

Hence,

$$\operatorname{Im}(R_0(\lambda)f,f) = \frac{\tilde F(\sqrt\lambda)}{2\sqrt\lambda}\int_{-\infty}^\infty \frac{1}{t^2+1}\, dt = \frac{\pi}{2\sqrt\lambda(2\pi)^n}\int_{M_\lambda} |\hat f(\omega)|^2\, dS(\omega). \qquad \square$$

Remark 7.1.9. Lemma 7.1.8 may be understood as an instance of an integral identity ("boundary pairing") for the scattering problem, analogous to the integral identities in the Calderón problem given in §2.8. See [**Mel95**, §2.3].

The following theorem, originally due to Rellich, will play an important role both in proving the existence of eigenfunctions of H_q and in our treatment of the inverse scattering problem in §7.4.

Theorem 7.1.10 (Rellich uniqueness theorem). *Let $\lambda > 0$, $r > 0$ and $f \in L^2$, with* $\operatorname{supp} f \subset \bar{B}_r$, *the closed ball of radius r. Write $u = R_0(\lambda)f$ for the outgoing solution of $(-\Delta - \lambda)u = f$. The following are equivalent.*

(i) $u \in L^2_\mu$ *for some* $\mu > -\frac{1}{2}$.

(ii) $\hat{f}|_{\partial B(0,\sqrt{\lambda})} = 0$.

(iii) $\operatorname{supp}(u) \subset \bar{B}_r$.

Proof. We shall show that (i) ⇒ (ii) ⇒ (iii) ⇒ (i).

Proof of (i) ⇒ (ii). Suppose first that (ii) is false. That is, there exists $\xi_0 \in M_\lambda$ such that $\hat{f}(\xi_0) \neq 0$. As $R_0(\lambda)$ commutes with rotations, we may assume that $\xi_0 = (\sqrt{\lambda}, 0, \cdots, 0)$. Let $\mathcal{U}$ be an open neighborhood of ξ_0 in $\mathbb{R}^n$ that is sufficiently small that it does not intersect the support of any $\chi_j(\xi)$ with $0 \leq j \leq n$, $j \neq 1$. The χ_j's were defined in the proof of Theorem 7.1.1. Let $\chi_{\mathcal{U}}$ be a C^∞ function that takes values in $[0,1]$, is one on a neighborhood of ξ_0 and is supported in $\mathcal{U}$. In the notation of (7.7)

$$\begin{aligned}\chi_{\mathcal{U}}(\xi)(R_0(\lambda)f)^\wedge(\xi) &= \chi_{\mathcal{U}}(\xi)q_0(\xi)\hat{f}(\xi) + \sum_{j=1}^{n} \chi_{\mathcal{U}}(\xi)q_j(\xi)m_j(\xi)\hat{f}(\xi)\\ &= \chi_{\mathcal{U}}(\xi)q_1(\xi)\hat{f}(\xi)m_1(\xi).\end{aligned}$$

Now, if (i) is true, $(R_0(\lambda)f)^\wedge(\xi) \in H^\mu(\mathbb{R}^n)$ so that,

$$\chi_{\mathcal{U}}(\xi)q_1(\xi)\hat{f}(\xi)m_1(\xi) \in H^\mu(\mathbb{R}^n).$$

Since we are assuming that f has compact support, $\hat{f}$ and hence $q_1\hat{f}$ is C^∞ in $\mathcal{U}$. Furthermore we are free to choose $\mathcal{U}$ small enough that $q_1\hat{f}$ is bounded away from zero on $\mathcal{U}$. Hence $(\chi_{\mathcal{U}} m_1)\circ\psi^{-1}$, which is the product of $\frac{1}{q_1\hat{f}}$ and $\chi_{\mathcal{U}} q_1 \hat{f} m_1$ composed with ψ^{-1}, is still in $H^\mu(\mathbb{R}^n)$. Denote by ζ the inverse Fourier transform of $\chi_{\mathcal{U}} \circ \psi^{-1}$ and observe that the inverse Fourier transform of $m_1 \circ \psi^{-1} = \frac{1}{\eta_1 - \sqrt{\lambda}}$ is $-i\delta(x')\, e^{i\sqrt{\lambda}x_1} H(x_1)$ where H is the one-dimensional Heaviside function and $\delta(x')$ is the $n-1$-dimensional delta function. We have now shown that, under the assumptions that (ii) is false and (i) is true, the inverse Fourier transform of $(\chi_{\mathcal{U}} m_1) \circ \psi^{-1}$, namely

$$-i \int e^{i\sqrt{\lambda}(x_1-y_1)} H(x_1 - y_1)\zeta(y_1, x')\, dy_1 \tag{7.16}$$

is in L^2_μ. But $(1 + x_1^2)^s$ is not in $L^1(\mathbb{R})$ for any $s \geq -\frac{1}{2}$. So, by Exercise 7.1.11, the function of (7.16) is not in L^2_s for any $s \geq -\frac{1}{2}$, contradicting our initial hypothesis and proving the implication (i) ⇒ (ii).

Proof of (ii) ⇒ (iii). For any function f, which is supported inside the ball of radius r, and any $\xi \in \mathbb{C}^n$,

$$|\hat{f}(\xi)| = \left| \int_{\bar{B}_r} e^{-ix\cdot\xi} f(x)\, dx \right| \le e^{r|\operatorname{Im}\xi|} \int_{\bar{B}_r} |f(x)|\, dx.$$

If f is C^∞, then writing $e^{-ix\cdot\xi} = \langle\xi\rangle^{-2N}(1-\Delta_x)^N(e^{-ix\cdot\xi})$ and integrating by parts gives that

$$|\hat{f}(\xi)| \le \frac{C_N}{(1+|\xi|^2)^N} e^{r|\operatorname{Im}\xi|} \tag{7.17}$$

for all $N \in \mathbb{N}$. In fact, according to the Paley-Wiener theorem, Exercise B.2.10, a function f is C^∞ and supported in $\bar{B}_r$ if and only if $\hat{f}(\xi)$ extends to a holomorphic function on $\mathbb{C}^n$ which obeys (7.17).

Start by assuming that f is C^∞. We claim first that $\hat{f}$ restricted to the complex manifold $\mathcal{M}_\lambda^{\mathbb{C}} = \{\, \zeta \mid \zeta\cdot\zeta = \lambda \,\}$ is identically zero. To see this consider the power series expansion of $f|_{\mathcal{M}_\lambda^{\mathbb{C}}}$ near $\zeta = (\sqrt{\lambda}, 0, \ldots, 0)$. Let $\tilde{\zeta} = (\zeta_2, \ldots, \zeta_n)$. Now, $g(\tilde{\zeta}) = \hat{f}\left(\sqrt{\lambda - \tilde{\zeta}\cdot\tilde{\zeta}}, \tilde{\zeta}\right)$ is holomorphic in a neighborhood of zero and, by hypothesis (ii), vanishes whenever $\tilde{\zeta}$ is real. Hence all the coefficients in the power series expansion of g are zero. Thus, g and, since $\mathcal{M}_\lambda^{\mathbb{C}}$ is connected by Exercise 7.1.12, $f|_{\mathcal{M}_\lambda^{\mathbb{C}}}$, are identically zero.

Now consider $\hat{u} = \big(R_0(\lambda) f\big)\hat{}$, which automatically extends to $\zeta \in \mathbb{C}^n$ via

$$\hat{u}(\zeta) = \frac{\hat{f}(\zeta)}{\zeta\cdot\zeta - \lambda}. \tag{7.18}$$

By the Cauchy integral formula

$$|\hat{f}(\zeta)| \le \frac{C_N'}{1+|\zeta|^{2N}} e^{r|\operatorname{Im}\zeta|} \operatorname{dist}(\zeta, \mathcal{M}_\lambda^{\mathbb{C}}) \tag{7.19}$$

within distance one, for example, of $\mathcal{M}_\lambda^{\mathbb{C}}$. If $|\zeta\cdot\zeta| \ge \frac{\lambda}{2}$, then, choosing $\sqrt{\zeta\cdot\zeta}$ to have non-negative real part, we have that $\frac{\sqrt{\lambda}}{\sqrt{\zeta\cdot\zeta}}\zeta$ is on $\mathcal{M}_\lambda^{\mathbb{C}}$ so that

$$\begin{aligned}
\operatorname{dist}(\zeta, \mathcal{M}_\lambda^{\mathbb{C}}) &\le \left|\zeta - \frac{\sqrt{\lambda}}{\sqrt{\zeta\cdot\zeta}}\zeta\right| = \left|\frac{\sqrt{\zeta\cdot\zeta}-\sqrt{\lambda}}{\sqrt{\zeta\cdot\zeta}}\right| |\zeta| \\
&= \left|\frac{\zeta\cdot\zeta - \lambda}{\sqrt{\zeta\cdot\zeta}(\sqrt{\zeta\cdot\zeta}+\sqrt{\lambda})}\right| |\zeta| \\
&\le \frac{\sqrt{2}}{\lambda} |\zeta|\, |\zeta\cdot\zeta - \lambda|.
\end{aligned}$$

On the other hand, if $|\zeta\cdot\zeta|\le\frac{\lambda}{2}$, then

$$|\zeta\cdot\zeta-\lambda|\ge\frac{\lambda}{2}\ge\frac{\lambda}{2}\frac{\operatorname{dist}(\zeta,\mathcal{M}_\lambda^{\mathbb{C}})}{\sqrt{\lambda}+|\zeta|}$$

since $(\sqrt{\lambda},0,\cdots,0)\in\mathcal{M}_\lambda^{\mathbb{C}}$. In both cases there is a λ–dependent constant D so that

$$\frac{1}{|\zeta\cdot\zeta-\lambda|}\le D\frac{1+|\zeta|^2}{\operatorname{dist}(\zeta,\mathcal{M}_\lambda^{\mathbb{C}})}. \tag{7.20}$$

Hence (7.18) can have no poles, so $\hat{u}(\zeta)$ is holomorphic and satisfies a bound of the form (7.17). The Paley-Wiener theorem implies (iii).

Now we extend to a general $f\in L^2$ that is supported in $\bar{B}_r$. Let φ be a nonnegative C_c^∞ function that is supported in the unit ball $\bar{B}_1$ and has total mass one. That is, $\int\varphi(x)\,dx=1$. Set, for each $\varepsilon>0$, $\varphi_\varepsilon(x)=\frac{1}{\varepsilon^n}\varphi(\frac{x}{\varepsilon})$ and $f_\varepsilon=\varphi_\varepsilon*f$. Then f_ε is C^∞ and is supported in $\bar{B}_{r+\varepsilon}$. Since $\hat{f}_\varepsilon(\xi)=\hat{\varphi}_\varepsilon(\xi)\hat{f}(\xi)$, f_ε satisfies condition (ii). So by the part of (ii) $\Rightarrow$ (iii) that we have already proven, $u_\varepsilon\equiv R_0(\lambda)f_\varepsilon$ also vanishes outside of $\bar{B}_{r+\varepsilon}$. Since convolution with φ_ε commutes with application of the Laplacian, $u_\varepsilon=\varphi_\varepsilon*u$, where $u=R_0(\lambda)f$. Since u_ε converges to u, locally in L^2, as $\varepsilon\to0$, u vanishes outside of $\bar{B}_{r+\varepsilon}$ for all $\varepsilon>0$. This is all we need.

Proof of (iii) $\Rightarrow$ (i). Since f vanishes outside $\bar{B}_r$ it is in L^2_δ for any $\delta>\frac{1}{2}$. By Proposition 7.1.2, $R_0(\lambda)f\in L^2_{-\delta}$ so that the restriction of $R_0(\lambda)f$ to any bounded set is in L^2. Since we are assuming that $R_0(\lambda)f$ vanishes outside of $\bar{B}_r$, it is itself in $L^2(\mathbb{R}^n)$. □

Exercise 7.1.11. Let

$$H(x)=\begin{cases}1 & \text{if } x\ge0\\ 0 & \text{if } x<0\end{cases}$$

be the Heaviside step function.

(a) Let $\zeta\in\mathcal{S}(\mathbb{R})$ obey $\hat{\zeta}(\alpha)=\int e^{-i\alpha x}\zeta(x)\,dx\ne0$ for some $\alpha\in\mathbb{R}$. Prove that $z(x)=\int e^{-i\alpha y}H(x-y)\zeta(y)\,dy$ is not in $L^2_s(\mathbb{R})$ for any $s\ge-\frac{1}{2}$.

(b) Let $\zeta\in\mathcal{S}(\mathbb{R}^n)$ obey $\int e^{-i\alpha x_1}\zeta(x_1,0)\,dx=\int_{\mathbb{R}^{n-1}}\hat{\zeta}(\alpha,\xi')\,\frac{d\xi'}{(2\pi)^{n-1}}\ne0$ for some $\alpha\in\mathbb{R}$. Prove that $z(x_1,x')=\int e^{-i\alpha y_1}H(x_1-y_1)\zeta(y_1,x')\,dy_1$ is not in $L^2_s(\mathbb{R}^n)$ for any $s\ge-\frac{1}{2}$.

Exercise 7.1.12. Prove that if $\lambda\ne0$, then $\mathcal{M}_\lambda^{\mathbb{C}}$ is a smooth connected manifold.

We shall need one more uniqueness result in the next section.

Proposition 7.1.13 (Uniqueness of outgoing solutions). *Let $\lambda > 0$ and let $q \in L^\infty(\mathbb{R}^n)$ be compactly supported. If $u \in L^2_\mu(\mathbb{R}^n)$ for some $\mu > -\frac{1}{2}$, u is outgoing, and u solves*

$$(-\Delta - \lambda + q)u = 0 \text{ in } \mathbb{R}^n,$$

then $u \equiv 0$.

Proof. Write $f = -qu \in L^2_c(\mathbb{R}^n)$ and note that u solves

$$(-\Delta - \lambda)u = f.$$

Since u is outgoing, $u = R_0(\lambda)f$ by Exercise 7.1.7 and thus $u \in H^2_{-\delta}(\mathbb{R}^n)$ for any $\delta > \frac{1}{2}$. The Rellich uniqueness theorem, Theorem 7.1.10, ensures that u is compactly supported since f is.

We have proved that $u \in H^2(\mathbb{R}^n)$ solves

$$(-\Delta - \lambda + q)u = 0 \text{ in } \mathbb{R}^n$$

and u vanishes outside a large ball. The unique continuation principle, Theorem 6.4.2, implies that $u \equiv 0$. □

7.2. Outgoing eigenfunctions

In the previous section we studied the free resolvent $R_0(\lambda)$ related to outgoing solutions of $(-\Delta - \lambda)u = f$. Now we include a compactly supported potential q.

Theorem 7.2.1 (Resolvent $R_q(\lambda)$). *Let $\lambda > 0$ and $\delta > \frac{1}{2}$. Let $q \in L^\infty(\mathbb{R}^n)$ be real and compactly supported. Then for any $f \in L^2_\delta$ there exists a unique outgoing solution $u = R_q(\lambda)f \in H^2_{-\delta}$ to*

$$(-\Delta - \lambda + q)u = f \text{ in } \mathbb{R}^n. \tag{7.21}$$

Moreover, we have the formulas

$$R_q(\lambda) = R_0(\lambda)\big(\mathrm{Id} + qR_0(\lambda)\big)^{-1}, \tag{7.22}$$

$$R_q(\lambda) = \big(\mathrm{Id} + R_0(\lambda)q\big)^{-1}R_0(\lambda). \tag{7.23}$$

Proof. A function $u \in L^2_{-\delta}$ is outgoing if and only if there is a $v \in L^2_\delta$ such that $u = R_0(\lambda)v$. By Exercise 7.1.7, the (weak) equation (7.21) for outgoing u is thus equivalent to the equation

$$\big(\mathrm{Id} + qR_0(\lambda)\big)v = f$$

for v. So we must establish the invertibility of

$$\big(\mathrm{Id} + qR_0(\lambda)\big) : L^2_\delta \to L^2_\delta \tag{7.24}$$

to prove (7.22). Then (7.23) will follow from writing (7.21) as

$$(-\Delta - \lambda)\big(\mathrm{Id} + R_0(\lambda)q\big)u = f$$

and observing that $(\mathrm{Id} + R_0(\lambda)q) : L^2_{-\delta} \to L^2_{-\delta}$ has a bounded inverse because it is the Banach space adjoint[1] of $(\mathrm{Id} + qR_0(\lambda))$.

It remains to establish the invertibility of (7.24). For this we need the property that

$$qR_0(\lambda) : L^2_\delta(\mathbb{R}^n) \to L^2_\delta(\mathbb{R}^n)$$

is a compact operator (see Definition A.3.1). To see this, let χ be a smooth cutoff function which is one on the support of q and is supported inside a ball of radius r. Since $R_0(\lambda)$ is bounded $L^2_\delta \to H^2_{-\delta}$, the operator $\chi R_0(\lambda)$ is bounded $L^2_\delta \to H^1_0(B_r)$. By compact Sobolev embedding (Theorem 2.4.3) the inclusion $H^1_0(B_r) \to L^2(B_r)$ is compact. Applying a zero extension, we see that $qR_0(\lambda) = q(\chi R_0(\lambda))$ is compact $L^2_\delta \to L^2_\delta$ by Proposition A.3.8.

For large λ, it follows from (7.5) that

$$\begin{aligned} \|qR_0(\lambda)\| &\le \left\|(1+|x|^2)^\delta q(x)\right\|_{L^\infty} \|R_0(\lambda)\| \\ &\le \|(1+|x|^2)^\delta q(x)\|_{L^\infty} \frac{C}{\sqrt{\lambda}} \end{aligned}$$

so that $\mathrm{Id} + qR_0(\lambda)$ is invertible for λ sufficiently large. By the Fredholm alternative (Proposition A.3.15), to prove the bounded invertibility of the operator $\mathrm{Id} + qR_0(\lambda)$ at any λ we need only prove that the kernel is empty. This can be seen as follows.

Suppose that

$$f + qR_0(\lambda)f = 0.$$

Then

$$(f, R_0(\lambda)f) = -(qR_0(\lambda)f, R_0(\lambda)f) \tag{7.25}$$

and, because the right-hand side of (7.25) is real, we may conclude via (7.15) that $\hat{f}|_{M_\lambda} = 0$. According to Theorem 7.1.10, $u = R_0(\lambda)f$ is a compactly supported outgoing solution to $(-\Delta - \lambda)u = f = -qu$. Hence $u \equiv 0$ by Proposition 7.1.13. By Exercise 7.1.7, $f = 0$ as well and the theorem is proved. □

The existence of the generalized eigenfunctions for $H_q = -\Delta - \lambda + q$ now follows easily.

Corollary 7.2.2. *Let $\lambda > 0$ and $\omega \in S^{n-1}$. Let $q \in L^\infty$ be of compact support. There exists a unique λ–outgoing eigenfunction for H_q in the direction ω. That is, there exists a unique $\psi_q(\lambda, \cdot, \omega) \in H^2_{\mathrm{loc}}(\mathbb{R}^n)$ satisfying*

$$(-\Delta - \lambda + q)\psi_q = 0$$

[1] Let X and Y be Banach spaces and X^* and Y^* be their respective dual spaces. If T is a bounded linear operator from X to Y, then the Banach space adjoint of T, denoted T' is the bounded linear operator from Y^* to X^* defined by $(T'\ell)(x) = \ell(Tx)$ for all $\ell \in Y^*$ and $x \in X$. In the current application, we have $X = Y = L^2_\delta(\mathbb{R}^n)$, $X^* = Y^* = L^2_{-\delta}(\mathbb{R}^n)$ and $\ell(x) = \int_{\mathbb{R}^n} \ell(y)x(y)\,dy$. In this case, the Banach space adjoint is the "L^2–adjoint without the complex conjugate."

of the form

$$\psi_q = \psi_0(\lambda, x, \omega) + \psi_q^s(\lambda, x, \omega)$$

such that $\psi_q^s(\lambda, \cdot, \omega) \in L^2_\mu$, *for all* $\mu < -\frac{1}{2}$, *and is* λ*–outgoing. Moreover,*

$$\psi_q(\lambda, \cdot, \omega) = \big(\mathrm{Id} - R_q(\lambda)q\big)\psi_0(\lambda, \cdot, \omega) \tag{7.26}$$

and

$$\psi_q(\lambda, \cdot, \omega) = \psi_0(\lambda, \cdot, \omega) - R_0(\lambda)(q\psi_q(\lambda, \cdot, \omega)).$$

Proof. In terms of ψ_q^s, the equation $(-\Delta - \lambda + q)\psi_q = 0$ is

$$(-\Delta - \lambda + q)\psi_q^s = -q\psi_0. \tag{7.27}$$

By Theorem 7.2.1, this equation has a unique outgoing solution and that solution is

$$\psi_q^s = -R_q(\lambda)(q\psi_0(\lambda, \cdot, \omega)). \tag{7.28}$$

Rewriting (7.27) as

$$(-\Delta - \lambda)\psi_q^s = -q\psi_0 - q\psi_q^s = -q\psi_q$$

yields the last claim. □

Our next step is to extend Theorem 7.1.10 to H_q. In this extension, we need to replace the Fourier transform

$$\hat{f}(-\sqrt{\lambda}\omega) = \int_{\mathbb{R}^n} f(x)\psi_0(\lambda, x, \omega)\, dx$$

with the *distorted Fourier transform*

$$\mathcal{F}_q f(-\sqrt{\lambda}\omega) = \int_{\mathbb{R}^n} f(x)\psi_q(\lambda, x, \omega)\, dx.$$

Theorem 7.2.3. *Let* $\lambda > 0$, $r > 0$ *and* $f \in L^2$, *with* $\operatorname{supp} f \subset B_r$ *and* $\operatorname{supp} q \subset B_r$. *The following are equivalent.*

(i) $R_q(\lambda)f \in L^2_\mu$ *for some* $\mu > -\frac{1}{2}$.

(ii) $\int_{\mathbb{R}^n} f(x)\psi_q(\lambda, x, \omega)\, dx = 0$ *for all* $\omega \in S^{n-1}$.

(iii) $\operatorname{supp}(R_q(\lambda)f) \subset B_r$.

Proof. We first note, from (7.23), that

$$\big(\mathrm{Id} + R_0(\lambda)q\big)R_q(\lambda)f = R_0(\lambda)f$$

so that

$$R_q(\lambda)f = R_0(\lambda)\big(\mathrm{Id} - qR_q(\lambda)\big)f = R_0(\lambda)F, \tag{7.29}$$

where $F = \big(\mathrm{Id} - qR_q(\lambda)\big)f$. If (i) is satisfied, then Theorem 7.1.10 yields

$$\begin{aligned} 0 &= \int_{\mathbb{R}^n} F(x)\psi_0(x,\omega)\,dx \\ &= \int_{\mathbb{R}^n} \big([\mathrm{Id} - qR_q(\lambda)]f\big)(x)\psi_0(x,\omega)\,dx \\ &= \int_{\mathbb{R}^n} f(x)\big([\mathrm{Id} - R_q(\lambda)q]\psi_0\big)(x,\omega)\,dx \\ &= \int_{\mathbb{R}^n} f(x)\psi_q(\lambda,x,\omega)\,dx \end{aligned}$$

which proves that (i) ⇒ (ii).

The previous calculation along with Theorem 7.1.10 implies that

$$\operatorname{supp}(R_0(\lambda)F) \subset B_r$$

which, in view of (7.29), shows that (ii) ⇒ (iii). As before, (iii) ⇒ (i) is trivial. □

7.3. Asymptotics and the scattering amplitude

Let $R_q(\lambda, x, y)$ denote the outgoing Green's kernel defined by

$$(R_q(\lambda)f)(x) = \int_{\mathbb{R}^n} R_q(\lambda, x, y) f(y)\,dy.$$

It is, for each fixed y, the solution of

$$(-\Delta_x + q(x) - \lambda)R_q(\lambda, x, y) = \delta(x - y) \tag{7.30}$$

subject to the Sommerfeld radiation condition (see (1.17))

$$\frac{\partial}{\partial r}R_q - i\sqrt{\lambda}R_q = o\left(\frac{1}{|x|^{\frac{n-1}{2}}}\right) \qquad \text{as} \qquad |x| \to \infty.$$

The "unperturbed" or "free" outgoing Green's kernel $R_0(\lambda, x, y)$ can be computed using

$$R_0(\lambda, x, y) = \lim_{\varepsilon \downarrow 0} \int_{\mathbb{R}^n} \frac{e^{i(x-y)\cdot\xi}}{|\xi|^2 - (\lambda + i\varepsilon)} \frac{d\xi}{(2\pi)^n}.$$

Lemma 7.3.1 (Free resolvent kernel). *If $\lambda > 0$, then*

$$R_0(\lambda, x, y) = \frac{i}{4}\left(\frac{\sqrt{\lambda}}{2\pi|x-y|}\right)^{\frac{n-2}{2}} H^{(1)}_{\frac{n-2}{2}}(\sqrt{\lambda}\,|x-y|\,),$$

where $H^{(1)}_\nu(z) = J_\nu(z) + iY_\nu(z)$ is a Hankel function. In particular, for $n = 3$,

$$R_0(\lambda, x, y) = \frac{e^{i\sqrt{\lambda}\,|x-y|}}{4\pi|x-y|}.$$

Asymptotically, for $|x-y|$ large,

$$R_0(\lambda,x,y)=\frac{1}{2\sqrt{\lambda}}\left(\frac{\sqrt{\lambda}}{2\pi|x-y|}\right)^{\frac{n-1}{2}}e^{i(\sqrt{\lambda}|x-y|-\frac{n-3}{4}\pi)}+O\left(\frac{1}{|x-y|^{\frac{n-1}{2}-1}}\right).$$

Proof. By rotation invariance, we may assume, without loss of generality, that $x-y=(|x-y|,0,\cdots,0)$. By scaling $\xi\to\sqrt{\lambda}\,\xi$,

$$\begin{aligned}R_0(\lambda,x,y)&=\lim_{\varepsilon\downarrow 0}\lambda^{\frac{n}{2}-1}\int_{\mathbb{R}^n}\frac{e^{i\xi_1\sqrt{\lambda}|x-y|}}{|\xi|^2-1-i\varepsilon/\lambda}\frac{d\xi}{(2\pi)^n}\\&=\lim_{\varepsilon\downarrow 0}\lambda^{\frac{n}{2}-1}\int_{\mathbb{R}^n}\frac{e^{i\xi_1 r}}{|\xi|^2-1-i\varepsilon}\frac{d\xi}{(2\pi)^n},\end{aligned}$$

where $r=\sqrt{\lambda}\,|x-y|$. An argument based on the residue theorem yields

$$\int_{\mathbb{R}}\frac{e^{i\xi_1 r}}{\xi_1^2+p^2-1-i\varepsilon}\frac{d\xi_1}{2\pi}=2\pi i\frac{1}{2\pi}\frac{e^{-r\sqrt{p^2-1-i\varepsilon}}}{2i\sqrt{p^2-1-i\varepsilon}}=\frac{e^{-r\sqrt{p^2-1-i\varepsilon}}}{2\sqrt{p^2-1-i\varepsilon}},$$

where $p=(\xi_2,\cdots,\xi_n)$ and $\sqrt{p^2-1-i\varepsilon}$ denotes the square root with positive real part. The corresponding imaginary part is negative. As $\sqrt{p^2-1}$ has an integrable, square root, singularity at $|p|=1$ and $e^{-r\sqrt{p^2-1}}$ decays exponentially quickly for large $|p|$, the Lebesgue dominated convergence theorem gives

$$R_0(\lambda,x,y)=\tfrac{1}{2}\lambda^{\frac{n}{2}-1}\int_{\mathbb{R}^{n-1}}\frac{e^{-r\sqrt{p^2-1}}}{\sqrt{p^2-1}}\frac{dp}{(2\pi)^{n-1}}.$$

Now $\sqrt{p^2-1}$ is positive for $|p|>1$ and is i times a negative number when $|p|<1$. Going to spherical coordinates in $\mathbb{R}^{n-1}$

$$R_0(\lambda,x,y)=\frac{\Omega_{n-1}}{2^n\pi^{n-1}}\lambda^{\frac{n}{2}-1}\int_0^\infty\frac{e^{-r\sqrt{\rho^2-1}}}{\sqrt{\rho^2-1}}\rho^{n-2}\,d\rho,$$

where $\Omega_m=\frac{2\pi^{m/2}}{\Gamma(m/2)}$ denotes the surface area of a unit sphere in $\mathbb{R}^m$. When $\rho\ge 1$, make the change of variables $\rho=\cosh\tau$, $\frac{1}{\sqrt{\rho^2-1}}d\rho=\frac{1}{\sinh\tau}\sinh\tau\,d\tau=d\tau$ to obtain

$$\int_1^\infty\frac{e^{-r\sqrt{\rho^2-1}}}{\sqrt{\rho^2-1}}\rho^{n-2}\,d\rho=\int_0^\infty e^{-r\sinh\tau}\cosh^{n-2}\tau\,d\tau$$

For $0\le\rho\le 1$, make the change of variables $\rho=\sin\alpha$, $\sqrt{\rho^2-1}=-i\cos\alpha$, $d\rho=\cos\alpha\,d\alpha$ to obtain

$$\int_0^1\frac{e^{-r\sqrt{\rho^2-1}}}{\sqrt{\rho^2-1}}\rho^{n-2}\,d\rho=i\int_0^{\frac{\pi}{2}}e^{ir\cos\alpha}\sin^{n-2}\alpha\,d\alpha.$$

Poisson's integral representation [**MOS66**, Third Chapter, §5] for the Bessel functions of the first and second kind are (for $\nu > -\frac{1}{2}$ and $\operatorname{Re} z > 0$)

$$J_\nu(z) = \frac{2(\frac{z}{2})^\nu}{\sqrt{\pi}\,\Gamma(\nu+\frac{1}{2})} \int_0^{\frac{\pi}{2}} \cos\big(z\cos\alpha\big)\sin^{2\nu}\alpha\, d\alpha,$$

$$Y_\nu(z) = \frac{2(\frac{z}{2})^\nu}{\sqrt{\pi}\,\Gamma(\nu+\frac{1}{2})}\Big[\int_0^{\frac{\pi}{2}} \sin\big(z\cos\alpha\big)\sin^{2\nu}\alpha\, d\alpha - \int_0^\infty e^{-z\sinh\tau}\cosh^{2\nu}\tau\, d\tau\Big].$$

The Hankel function $H_\nu^{(1)}(r)$ is defined by

$$\begin{aligned} H_\nu^{(1)}(r) &= J_\nu(r) + iY_\nu(r) \\ &= \frac{2(\frac{r}{2})^\nu}{\sqrt{\pi}\,\Gamma(\nu+\frac{1}{2})}\Big[\int_0^{\frac{\pi}{2}} e^{ir\cos\alpha}\sin^{2\nu}\alpha\, d\alpha - i\int_0^\infty e^{-r\sinh\tau}\cosh^{2\nu}\tau\, d\tau\Big] \\ &= \frac{2(\frac{r}{2})^\nu}{\sqrt{\pi}\,\Gamma(\nu+\frac{1}{2})}\Big[-i\int_0^\infty \frac{e^{-r\sqrt{\rho^2-1}}}{\sqrt{\rho^2-1}}\,\rho^{n-2}d\rho\Big] \end{aligned}$$

with $\nu = \frac{n-2}{2}$. Thus

$$\begin{aligned} R_0(\lambda, x, y) &= \frac{\Omega_{n-1}}{2^n\pi^{n-1}}\lambda^{\frac{n}{2}-1} i \frac{\sqrt{\pi}\,\Gamma(\frac{n-1}{2})}{2(\frac{r}{2})^{\frac{n-2}{2}}} H^{(1)}_{\frac{n-2}{2}}(r) \\ &= \frac{i}{4}\left(\frac{\sqrt{\lambda}}{2\pi|x-y|}\right)^{\frac{n-2}{2}} H^{(1)}_{\frac{n-2}{2}}(\sqrt{\lambda}\,|x-y|\,). \end{aligned}$$

In particular

$$H^{(1)}_{\frac{1}{2}}(z) = -i\sqrt{\frac{2}{\pi z}}e^{iz}$$

so that, for $n = 3$,

$$R_0(\lambda, x, y) = \frac{e^{i\sqrt{\lambda}|x-y|}}{4\pi|x-y|}.$$

We already saw the $n = 3$ case in Exercise 1.3.3.

Asymptotically, for large r,

$$H_\nu^{(1)}(\sqrt{\lambda}\,r) = \sqrt{\frac{2}{\pi\sqrt{\lambda}\,r}}e^{i(\sqrt{\lambda}r-\frac{1}{2}\nu\pi-\frac{1}{4}\pi)} + O(\frac{1}{r^{3/2}}).$$

This is formula (9.2.3) in [**AS64**]. So

$$
\begin{aligned}
&R_0(\lambda, x, y)\\
&\quad = \frac{i}{4}\Big(\frac{\sqrt{\lambda}}{2\pi|x-y|}\Big)^{\frac{n-2}{2}} \sqrt{\frac{2}{\pi\sqrt{\lambda}\,|x-y|}}\, e^{i(\sqrt{\lambda}\,|x-y|-\frac{n}{4}\pi+\frac{1}{4}\pi)} + O\Big(\frac{1}{|x-y|^{\frac{n+1}{2}}}\Big)\\
&\quad = \frac{1}{2\sqrt{\lambda}}\Big(\frac{\sqrt{\lambda}}{2\pi|x-y|}\Big)^{\frac{n-1}{2}} e^{i(\sqrt{\lambda}\,|x-y|-\frac{n-3}{4}\pi)} + O\Big(\frac{1}{|x-y|^{\frac{n+1}{2}}}\Big). \qquad \square
\end{aligned}
$$

Corollary 7.3.2 (Asymptotics). *Let $\lambda > 0$ and $\omega \in S^{n-1}$. Let $q \in L^\infty$ be of compact support.*

(a) *There is a C^∞ function $a_q(\sqrt{\lambda}, \cdot, \omega) : S^{n-1} \to \mathbb{C}$ such that, asymptotically for large $|x|$,*

$$\psi_q(\lambda, x, \omega) = e^{i\sqrt{\lambda}\,x\cdot\omega} + \frac{a_q(\sqrt{\lambda}, \theta, \omega)}{|x|^{\frac{n-1}{2}}} e^{i\sqrt{\lambda}\,|x|} + O(|x|^{-\frac{n-1}{2}-1}),$$

where $\theta = \frac{x}{|x|}$.

(b) *There is a constant $C_{\sqrt{\lambda},n}$ such that*

$$R_q(\lambda, x, y) = C_{\sqrt{\lambda},n} \frac{e^{i\sqrt{\lambda}\,|x|}}{|x|^{\frac{n-1}{2}}} \psi_q(\lambda, y, -\frac{x}{|x|}) + O\Big(|x|^{-\frac{n-1}{2}-1}\Big)$$

Proof.

(a) Rewriting the last conclusion of Corollary 7.2.2 in terms of the Green's kernel gives the Lippmann–Schwinger equation (compare with (1.18))

$$\psi_q(\lambda, x, \omega) = e^{i\sqrt{\lambda}\,x\cdot\omega} - \int R_0(\lambda, x, y)\, q(y) \psi_q(\lambda, y, \omega)\, dy. \tag{7.31}$$

For y in the support of q and large $|x|$, Taylor expanding the square root gives

$$
\begin{aligned}
|x-y|^2 = |x|^2 - 2x\cdot y + |y|^2 = |x|^2\big(1 - 2\frac{\theta\cdot y}{|x|} + \frac{|y|^2}{|x|^2}\big)\\
\Longrightarrow\ |x-y| = |x| - \theta\cdot y + O\big(\frac{1}{|x|}\big)
\end{aligned}
$$

so that

$$R_0(\lambda, x, y) = C_{\sqrt{\lambda},n} \frac{1}{|x|^{\frac{n-1}{2}}}\, e^{i\sqrt{\lambda}\,(|x|-\theta\cdot y)} + O\Big(\frac{1}{|x|^{\frac{n+1}{2}}}\Big), \tag{7.32}$$

where

$$C_{\sqrt{\lambda},n} = \frac{1}{2\sqrt{\lambda}}\left(\frac{\sqrt{\lambda}}{2\pi}\right)^{\frac{n-1}{2}} e^{-i\frac{n-3}{4}\pi}.$$

Substituting this into (7.31), defining

$$a_q(\sqrt{\lambda},\theta,\omega) = -C_{\sqrt{\lambda},n}\int e^{-i\sqrt{\lambda}\theta\cdot y}\, q(y)\psi_q(\lambda,y,\omega)\,dy$$

and noting that ψ_q is locally L^2 and hence locally L^1 gives part (a).

(b) By (7.22),

$$R_q = (\mathrm{Id} - R_q q)R_0.$$

From (7.32)

$$R_0(\lambda,y,x) = R_0(\lambda,x,y) = C_{\sqrt{\lambda},n}\frac{e^{i\sqrt{\lambda}|x|}}{|x|^{\frac{n-1}{2}}}\psi_0(\lambda,y,-\theta) + O\Big(\frac{1}{|x|^{\frac{n+1}{2}}}\Big)$$

so that, by (7.26)

$$\begin{aligned} R_q(\lambda,x,y) &= R_q(\lambda,y,x) \\ &= C_{\sqrt{\lambda},n}\frac{e^{i\sqrt{\lambda}|x|}}{|x|^{\frac{n-1}{2}}}(\mathrm{Id} - R_q q)\psi_0(\lambda,\,\cdot\,,-\theta) + O\Big(\frac{1}{|x|^{\frac{n+1}{2}}}\Big) \\ &= C_{\sqrt{\lambda},n}\frac{e^{i\sqrt{\lambda}|x|}}{|x|^{\frac{n-1}{2}}}\psi_q(\lambda,y,-\theta) + O\Big(\frac{1}{|x|^{\frac{n+1}{2}}}\Big). \end{aligned}$$

□

7.4. Inverse scattering

Let $n \geq 2$ and $q \in L^\infty(\mathbb{R}^n)$ be supported in $B_R = \{\, x \in \mathbb{R}^n \mid |x| < R \,\}$. Suppose that $\lambda > 0$ and ω is a direction in $S^{n-1} = \{\, x \in \mathbb{R}^n \mid |x| = 1 \,\}$. In §7.1–§7.3 we have developed the stationary approach to scattering theory by a potential. Replacing λ by λ^2, we have proved that there exists a unique outgoing solution $\psi_q(\lambda,x,\omega)$ of

$$(-\Delta + q - \lambda^2)\psi(\lambda,x,\omega) = 0$$

of the form

$$\psi_q(\lambda,x,\omega) = e^{i\lambda x\cdot\omega} + \psi_q^s(\lambda,x,\omega),$$

where

$$\begin{aligned} \psi_q^s(\lambda,x,\omega) &= -R_q(\lambda^2)(qe^{i\lambda x\cdot\omega}) \\ &= \frac{a_q(\lambda,\theta,\omega)}{|x|^{\frac{n-1}{2}}}e^{i\lambda|x|} + O\big(|x|^{-\frac{n-1}{2}-1}\big) \end{aligned} \tag{7.33}$$

with $\theta = \frac{x}{|x|}$. See Corollaries 7.2.2 and 7.3.2.

Definition 7.4.1. The function $a_q(\lambda,\theta,\omega)$ in (7.33) is called the *scattering amplitude*. The function $\psi_q(\lambda,x,\omega)$ is called an *outgoing eigenfunction*.

From the proof of Corollary 7.3.2 we have

$$a_q(\lambda,\theta,\omega) = c_n\lambda^{\frac{n-3}{2}} \int e^{-i\lambda\theta\cdot y} q(y)\psi_q(\lambda,y,\omega)\,dy. \tag{7.34}$$

The scattering amplitude measures, roughly speaking, the amplitude, measured in the direction

$$\theta = \frac{x}{|x|},$$

of a spherical wave produced by the potential q when it interacts with plane waves of energy λ^2 moving in the direction ω. We discussed a classical analog of this picture in §1.3. In particular (7.33) is the analog for the Schrödinger equation of (1.19).

Our first result shows that the Fourier transform of q, and therefore q itself, can be recovered from the high frequency behavior of the scattering amplitude a_q.

Theorem 7.4.2 (Uniqueness in full data inverse scattering). *Let $n \geq 2$ and let $q_1, q_2 \in L^\infty(\mathbb{R}^n)$ be supported in B_R. If*

$$a_{q_1} = a_{q_2},$$

then

$$q_1 = q_2.$$

Proof. Fix $\xi \in \mathbb{R}^n$ and let $\eta \in \mathbb{R}^n$ be a unit vector with $\eta\cdot\xi = 0$. For $\lambda \geq |\xi|$, define

$$\omega = \omega(\lambda) = \frac{1}{\lambda}(-\xi + \sqrt{\lambda^2 - |\xi|^2}\eta), \tag{7.35}$$

$$\theta = \theta(\lambda) = \frac{1}{\lambda}(\xi + \sqrt{\lambda^2 - |\xi|^2}\eta). \tag{7.36}$$

Note that $\lambda(\theta-\omega) = 2\xi$. From (7.34) and (7.33) we have

$$\lambda^{\frac{3-n}{2}} a_q(\lambda,\theta,\omega) = c_n\hat{q}(2\xi) + c_n \int e^{-i\lambda\theta\cdot y} q(y)\psi_q^s(\lambda,y,\omega)\,dy, \tag{7.37}$$

where

$$\psi_q^s = -R_q(\lambda^2)(q\psi_0) = -R_0(\lambda^2)(\mathrm{Id} + qR_0(\lambda^2))^{-1}(q\psi_0).$$

The estimate $\left\|R_0(\lambda^2)\right\|_{L^2_\delta\to L^2_{-\delta}} \leq C\lambda^{-1}$ from Theorem 7.1.1 implies

$$\left\|qR_0(\lambda^2)\right\|_{L^2_\delta\to L^2_\delta} \leq \frac{1}{2}$$

when $\lambda \geq \lambda_0$ with λ_0 large enough. A Neumann series argument yields

$$\left\|(\mathrm{Id} + qR_0(\lambda^2))^{-1}\right\|_{L^2_\delta\to L^2_\delta} \leq 2$$

for $\lambda \geq \lambda_0$, and therefore

$$\left\|\psi_q^s(\lambda, \cdot, \omega)\right\|_{L^2_{-\delta}} \leq C\lambda^{-1}$$

uniformly over $\lambda \geq \lambda_0$ and $\omega \in S^{n-1}$.

Since q is compactly supported, (7.37) gives that

$$\lim_{\lambda\to\infty} \lambda^{\frac{3-n}{2}} a_q(\lambda, \theta(\lambda), \omega(\lambda)) = c_n \hat{q}(2\xi).$$

This proves that $\hat{q}$, and hence q, can be recovered from a_q. □

We now fix $\lambda_0 > 0$. The subject of *inverse scattering at fixed energy* is the study of the invertibility of the map

$$q \mapsto a_q(\lambda_0, \cdot, \cdot).$$

We will prove uniqueness when $n \geq 3$.

Theorem 7.4.3 (Uniqueness in fixed energy inverse scattering). *Let $n \geq 3$ and let $q_1, q_2 \in L^\infty(\mathbb{R}^n)$ be supported in B_R. Fix $\lambda_0 > 0$, and assume that*

$$a_{q_1}(\lambda_0, \cdot, \cdot) = a_{q_2}(\lambda_0, \cdot, \cdot).$$

Then

$$q_1 = q_2.$$

We will give a proof that essentially reduces this problem to proving the injectivity of the DN map $\Lambda_{q-\lambda_0^2}$ acting on a large ball. Define the bilinear form

$$E(u_1, u_2) := \int_{B_R} (q_1 - q_2) u_1 \bar{u}_2 \, dx.$$

Proposition 7.4.4. *Let $q_1, q_2 \in L^\infty(\mathbb{R}^n)$ be supported in B_R and assume that $a_{q_1}(\lambda_0, \cdot, \cdot) = a_{q_2}(\lambda_0, \cdot, \cdot)$. Then*

$$E(u_1, u_2) = 0 \tag{7.38}$$

for all $u_j \in H^2(B_R)$ obeying

$$(-\Delta + q_1 - \lambda_0^2)u_1 = 0, \qquad (-\Delta + q_2 - \lambda_0^2)u_2 = 0. \tag{7.39}$$

The proof of the proposition is based on two lemmas.

Lemma 7.4.5. *Under the hypotheses of Theorem* 7.4.3,

$$E(\psi_{q_1}(\lambda_0, \cdot, \omega), \psi_{q_2}(\lambda_0, \cdot, \omega)) = 0 \qquad \textit{for all } \omega \in S^{n-1}.$$

Lemma 7.4.6 (Runge approximation). *Let $q \in L^\infty(\mathbb{R}^n)$ be supported in the closed ball $\bar{B}_{R'} \subset B_R$ and define*

$$\mathcal{E} = \operatorname{span}\{\, \psi_q(\lambda_0, \cdot, \omega) \mid \omega \in S^{n-1} \,\}.$$

Then $\mathcal{E}$ is dense in

$$N_q = \{\, u \in H^2(B_R) \mid (-\Delta + q - \lambda_0^2)u = 0 \,\}$$

with respect to the $H^2(B_{R'})$ topology. In other words, for any $u \in N_q$ there are $u_j \in \mathcal{E}$ such that $\left\|u_j - u\right\|_{H^2(B_{R'})} \to 0$ as $j \to \infty$.

Proof of Proposition 7.4.4. By Lemma 7.4.5 we have

$$\int_{B_R} (q_1 - q_2)\psi_{q_1}(\lambda_0, \cdot, \omega)\overline{\psi_{q_2}(\lambda_0, \cdot, \omega)}\, dx = 0. \tag{7.40}$$

Since q_1 and q_2 are supported in B_R, they are also supported in $\bar{B}_{R'}$ for some $R' < R$ and thus the integral is over $B_{R'}$. Now let $u_1, u_2 \in H^2(B_R)$ satisfy (7.39). Lemma 7.4.6 ensures that we can find

$$u_{1,j} \in \operatorname{span}\{\, \psi_{q_1}(\lambda_0, \cdot, \omega) \mid \omega \in S^{n-1} \,\},$$
$$u_{2,j} \in \operatorname{span}\{\, \psi_{q_2}(\lambda_0, \cdot, \omega) \mid \omega \in S^{n-1} \,\},$$

such that $\left\|u_{1,j} - u_1\right\|_{L^2(B_{R'})} \to 0$ and $\left\|u_{2,j} - u_2\right\|_{L^2(B_{R'})} \to 0$ as $j \to \infty$. From (7.40) we have

$$\int_{B_R} (q_1 - q_2)u_{1,j}\bar{u}_{2,j}\, dx = 0.$$

Taking the limit as $j \to \infty$ proves the proposition. □

Proof of Theorem 7.4.3. Proposition 7.4.4 ensures that

$$\int_{B_R} (q_1 - q_2)u_1\bar{u}_2\, dx = 0$$

for all solutions $u_j \in H^2(B_R)$ of $(-\Delta + q_j - \lambda_0^2)u_j = 0$ in B_R. The conclusion $q_1 = q_2$ follows by applying CGO solutions as in the portion of the proof of Theorem 4.3.2 that starts with (4.14). □

It remains to prove the two lemmas.

Proof of Lemma 7.4.5. Integrating by parts as in Theorem 2.2.11, for all u_j obeying (7.39),

$$\begin{aligned} E(u_1, u_2) &= \int_{B_R} ((\Delta u_1)\bar{u}_2 - u_1\Delta\bar{u}_2)\, dx \\ &= \int_{\partial B_R} \left(\frac{\partial u_1}{\partial r}\bar{u}_2 - u_1\overline{\frac{\partial u_2}{\partial r}} \right) dS, \end{aligned} \tag{7.41}$$

where dS denotes the surface measure on ∂B_R. Note, in particular, that the outgoing eigenfunctions ψ_{q_j} satisfy (7.39).

Since $a_{q_1}(\lambda_0, \theta, \omega) = a_{q_2}(\lambda_0, \theta, \omega)$ for all θ and ω, it follows from (7.33) that

$$\psi_{q_1}(\lambda_0, x, \omega) - \psi_{q_2}(\lambda_0, x, \omega) = O\big(|x|^{-\frac{n-1}{2}-1}\big)$$

is in $L^2(\mathbb{R}^n)$. We also have that

$$\big(-\Delta - \lambda_0^2\big)\big(\psi_{q_1}(\lambda_0, \cdot, \omega) - \psi_{q_2}(\lambda_0, \cdot, \omega)\big) = -q_1\psi_{q_1}(\lambda_0, \cdot, \omega) + q_2\psi_{q_2}(\lambda_0, \cdot, \omega)$$

is supported in B_R. By (7.28) and (7.22), both $\psi_{q_j}(\lambda_0, \cdot, \omega) - e^{i\lambda_0 x\cdot\omega}$, $j = 1, 2$ are outgoing so that $\psi_{q_1}(\lambda_0, \cdot, \omega) - \psi_{q_2}(\lambda_0, \cdot, \omega)$ is outgoing as well. Hence, by the Rellich uniqueness theorem (Theorem 7.1.10) with $f = -q_1\psi_{q_1} + q_2\psi_{q_2}$,

$$\psi_{q_1}(\lambda_0, x, \omega) - \psi_{q_2}(\lambda_0, x, \omega) = 0 \qquad \text{for all } |x| > R.$$

Combining this with the fact that $\psi_{q_j}(\lambda_0, \cdot, \omega) \in H^2_{\text{loc}}(\mathbb{R}^n)$, the trace theorem yields

$$\begin{aligned} \psi_{q_1}(\lambda_0, \cdot, \omega)\big|_{\partial B_R} &= \psi_{q_2}(\lambda_0, \cdot, \omega)\big|_{\partial B_R}, \\ \frac{\partial}{\partial r}\psi_{q_1}(\lambda_0, \cdot, \omega)\big|_{\partial B_R} &= \frac{\partial}{\partial r}\psi_{q_2}(\lambda_0, \cdot, \omega)\big|_{\partial B_R}, \end{aligned}$$

as elements of $H^{1/2}(\partial B_R) \subset L^2(\partial B_R)$. The lemma follows from (7.41). □

Proof of Lemma 7.4.6. First we prove that $\mathcal{E}$ is dense in N_q with respect to the $L^2(B_R)$ topology. To do so it suffices to prove that if the function $f \in N_q$ obeys

$$\int_{B_R} f(x)\,\psi_q(\lambda_0, x, \omega)\,dx = 0 \qquad \text{for all } \omega \in S^{n-1} \tag{7.42}$$

then f must vanish. Define

$$w(x) = \int_{B_R} R_q(\lambda_0, x, y) f(y)\,dy. \tag{7.43}$$

Then, by (7.30),

$$-\Delta w - \lambda_0^2 w = \chi_{B_R} f - qw$$

vanishes outside B_R and, by Corollary 7.3.2(b) and (7.42),

$$w(x) = O\Big(|x|^{-\frac{n-1}{2}-1}\Big)$$

is in $L^2(\mathbb{R}^n)$. As $\chi_{B_R} f - qw$ is also in $L^2(\mathbb{R}^n)$, we have that, as in the proof of Theorem 7.2.1, $w \in H^2(B_{2R})$. By the Rellich uniqueness theorem, we obtain that w vanishes outside B_R and therefore

$$w\big|_{\partial B_R} = \frac{\partial w}{\partial r}\Big|_{B_R} = 0. \tag{7.44}$$

Now using the divergence theorem and $(-\Delta + q - \lambda_0^2)w = \chi_{B_R} f$ we have

$$\begin{aligned}\int_{B_R} |f|^2\,dx &= \int_{B_R} \bar{f}(x)\,(-\Delta + q - \lambda_0^2)w\,dx \\ &= \int_{B_R} w(x)\,(-\Delta + q - \lambda_0^2)\bar{f}(x)\,dx - \int_{\partial B_R} \Big(\frac{\partial w}{\partial r}\bar{f} - w\frac{\partial \bar{f}}{\partial r}\Big)\,dS.\end{aligned}$$

Since q is real–valued and $f \in N_q$, we have that $(-\Delta + q - \lambda_0^2)\bar{f} = 0$ in B_R and conclude that $f = 0$.

Now we finish the proof by showing density in $H^2(B_{R'})$. By interior elliptic regularity, Proposition D.3.5, with $\gamma = \mathrm{Id}$ and $\ell = 2$, there is a constant $C = C(R, R')$ such that

(7.45)

$$\begin{aligned}\|u\|_{H^2(B_{R'})} &\le C\Big(\|-\Delta u\|_{L^2(B_R)} + \|u\|_{L^2(B_R)}\Big) \\ &\le C\Big(\|(-\Delta + q - \lambda_0^2)u\|_{L^2(B_R)} + \|(q - \lambda_0^2)u\|_{L^2(B_R)} + \|u\|_{L^2(B_R)}\Big) \\ &\le C'\Big(\|(-\Delta + q - \lambda_0^2)u\|_{L^2(B_R)} + \|u\|_{L^2(B_R)}\Big)\end{aligned}$$

for all $u \in H^2(B_R)$. Now let $f \in N_q$. We already know that there exists a sequence $\{f_j\}_{j\in\mathbb{N}} \subset \mathcal{E}$ such that

$$\lim_{j\to\infty} \|f - f_j\|_{L^2(B_R)} = 0$$

Then from (7.45) we conclude that

$$\lim_{j\to\infty} \Big\|f - f_j\Big\|_{H^2(B_{R'})} = 0$$

concluding the proof of Lemma 7.4.6. □

7.5. Notes

In §7.1–§7.3 we presented elements of the stationary, or time-independent, approach to *direct scattering theory* in the presence of a compactly supported L^∞ potential. We have followed the development in [**Agm75**] based on estimates in weighted L^2 spaces. Some of the arguments extend readily to short range potentials such as those satisfying $|q(x)| \le C(1 + |x|)^{-1-\varepsilon}$ for some $\varepsilon > 0$. More detailed presentations of this theory may be found in [**RS75, RS79**] (see in particular §IX.9, §XI.6, §XIII.8), in [**Hör03b**, §14] or in [**Yaf10**], with many historical references given therein. Alternative time-dependent approaches to scattering theory work either with the time-dependent Schrödinger equation or with the wave equation. See [**RS79**, §XI.17], [**Mel95**], [**DZ19**] or [**RS20b**] for related presentations.

In *inverse scattering problems* for a potential q, the objective is to determine information on q from (partial) knowledge of the scattering amplitude a_q. Recall that knowing $a_q(\lambda, \theta, \omega)$ roughly corresponds to sending a plane wave with frequency λ propagating in direction ω, and measuring the scattered spherical wave far away in direction θ. Textbook references on various aspects of inverse scattering theory include [**CCPR97, Isa17, CK19, Iso20**] and survey articles may be found in [**Uhl92, Nov08, Nov22**].

In the following we formulate four classical inverse problems in potential scattering.

(a) **Full data:** determine q from $a_q(\,\cdot\,,\,\cdot\,,\,\cdot\,)$.

(b) **Fixed frequency:** determine q from $a_q(\lambda_0,\,\cdot\,,\,\cdot\,)$ for fixed λ_0.

(c) **Fixed angle:** determine q from $a_q(\,\cdot\,,\,\cdot\,,\omega_0)$ for fixed ω_0.

(d) **Backscattering:** determine q from $a_q(\lambda, -\omega, \omega)$ for $\lambda > 0$, $\omega \in S^{n-1}$.

There are many other relevant variants, for example the important phaseless data problem where one has information on the absolute value $|a_q|^2$.

When setting up inverse problems it is useful to do a heuristic dimension count as we did in §1.1. Our unknown q is a function of n variables, whereas a_q is a function with $2n-1$ degrees of freedom (one for the frequency λ, $n-1$ for the outgoing direction θ, and $n-1$ for the incident direction ω). It follows that the full data problem is formally overdetermined when $n \geq 2$. The fixed frequency problem is overdetermined when $n \geq 3$ and well determined when $n = 2$. The fixed angle and backscattering problems are formally well determined in all dimensions.

An early milestone in inverse scattering is the one-dimensional theory developed by Borg, Gelfand, Levinson, Levitan, Marchenko, and others in the 1950s. More details and references may be found in the recent textbook [**Iso20**]. We will focus on the case $n \geq 2$ below.

The full data problem can be studied in several ways. In Theorem 7.4.2 we proved that one can recover the Fourier transform of the potential q from the high frequency limit of a_q as $\lambda \to \infty$ (this goes back to [**Fad56**]). Another possibility is to determine the hyperbolic DN map associated with the wave equation $(\partial_t^2 - \Delta + q)u = 0$ in $B_R \times \mathbb{R}$ from a_q, and to solve the wave equation inverse problem via the boundary control method to recover q (see [**KKL01**]). Alternatively, using high frequency or singular solutions one can determine the X-ray transform of q, i.e., the integrals of q along straight lines, and invert the X-ray transform [**Sal20**]. Of course the full data problem is a special case of the fixed frequency problem, and the uniqueness result in Theorem 7.4.3 also applies.

Uniqueness in the fixed angle and backscattering problems is open even for potentials in $C_c^\infty(\mathbb{R}^n)$ at the time of writing this. However, in the fixed angle problem uniqueness is known with measurements from two opposite directions ω_0 and $-\omega_0$ [**RS20a**], as well as for potentials with generalized reflection symmetries [**RS20b**]. For the backscattering problem there are results for small, generic or angularly controlled potentials [**ER92, Ste92, MU08, RU14**] as well as recovery of singularities results [**GU93, OPS01, Mer18**].

The fixed frequency problem is closest to the topic of this book. In §7.4 we proved uniqueness in this problem for compactly supported potentials when $n \geq 3$. In fact, by the methods in §7.4 it is not hard to show that knowing $a_q(\lambda_0, \cdot, \cdot)$ is equivalent to knowing the DN map $\Lambda_{q-\lambda_0^2}$ on the boundary of a suitable large ball, see [**Ber58, Nac88, Uhl92**]. As in the Calderón problem, CGO solutions have been crucial in many developments in fixed frequency inverse scattering. We will next discuss selected results related to the fixed frequency problem.

Uniqueness in the fixed frequency problem for $n \geq 3$ was proved first for small potentials in [**NK87**] and in general in [**Nov88, Ram88, Nac88**] (with [**Nac88**] considering a near field setting). These works follow the approach developed in [**Fad16, Fad74**] and the $\bar{\partial}$-method in [**NA84, BC85**]. Alternative arguments may be found in [**Wed91, Mel95**]. The case of exponentially decaying potentials was considered in [**Nov94, UV02**], and uniqueness results for the magnetic Schrödinger equation may be found in [**ER95, ER97, PSU10**]. We also mention [**Iso97**] on inverse scattering for Dirac systems.

In two dimensions, uniqueness in the fixed frequency problem for compactly supported potentials follows from the reduction to the DN map problem and the result of [**Buk08**]. This was extended to potentials with super-Gaussian decay in [**GST11**], which also considered scattering on Riemann surfaces with Euclidean ends. An earlier result for exponentially decaying potentials with small norm is in [**Nov92**]. The magnetic case and Dirac systems are considered in [**PT17, Tzo20**].

In the absence of sufficient decay for the potentials, there are counterexamples to uniqueness in the fixed frequency problem. In two dimensions [**GN95**] gives a counterexample involving potentials in Schwartz space, and in dimension three there are counterexamples with potentials decaying like $|x|^{-3/2}$ [**New62, Sab66**]. However, if the potentials have regular behavior at infinity (outside a ball they are given by convergent asymptotic sums of homogeneous functions in the radial variable), one still has uniqueness even in the magnetic case by the results of [**WY05, Wed07**] (see also [**Jos00, JSB98, JSB99**]).

Appendix A

Functional Analysis

In this appendix we provide a summary (mostly without proofs) of the most basic definitions and results concerning Banach and Hilbert spaces (§A.1) and bounded operators (§A.2). We also develop (with proofs) the most basic results concerning compact operators (§A.3).

A.1. Banach and Hilbert spaces

Definition A.1.1 (Vector space). A vector space over $\mathbb{C}$ is a set $\mathcal{V}$ equipped with two operations,

$$(\mathbf{v}, \mathbf{w}) \in \mathcal{V} \times \mathcal{V} \mapsto \mathbf{v} + \mathbf{w} \in \mathcal{V}, \qquad (\alpha, \mathbf{v}) \in \mathbb{C} \times \mathcal{V} \mapsto \alpha\mathbf{v} \in \mathcal{V},$$

called addition and scalar multiplication, respectively, that obey the following axioms.

Additive axioms: There is an element $\mathbf{0} \in \mathcal{V}$ and, for each $\mathbf{x} \in \mathcal{V}$ there is an element $-\mathbf{x} \in \mathcal{V}$ such that, for all $\mathbf{x}, \mathbf{y}, \mathbf{z} \in \mathcal{V}$,

(1) $\mathbf{x} + \mathbf{y} = \mathbf{y} + \mathbf{x}$
(2) $(\mathbf{x} + \mathbf{y}) + \mathbf{z} = \mathbf{x} + (\mathbf{y} + \mathbf{z})$
(3) $\mathbf{0} + \mathbf{x} = \mathbf{x} + \mathbf{0} = \mathbf{x}$
(4) $(-\mathbf{x}) + \mathbf{x} = \mathbf{x} + (-\mathbf{x}) = \mathbf{0}$

Multiplicative axioms: For every $\mathbf{x} \in \mathcal{V}$ and $\alpha, \beta \in \mathbb{C}$,

(5) $0\mathbf{x} = \mathbf{0}$
(6) $1\mathbf{x} = \mathbf{x}$
(7) $(\alpha\beta)\mathbf{x} = \alpha(\beta\mathbf{x})$

Distributive axioms: For every $\mathbf{x}, \mathbf{y} \in \mathcal{V}$ and $\alpha, \beta \in \mathbb{C}$,

(8) $\alpha(\mathbf{x} + \mathbf{y}) = \alpha\mathbf{x} + \alpha\mathbf{y}$
(9) $(\alpha + \beta)\mathbf{x} = \alpha\mathbf{x} + \beta\mathbf{x}$

Definition A.1.2 (Subspace). A subset $\mathcal{W}$ of a vector space $\mathcal{V}$ is called a *linear subspace* of $\mathcal{V}$ if it is closed under addition and scalar multiplication. That is, if $\mathbf{x}+\mathbf{y} \in \mathcal{W}$ and $\alpha\mathbf{x} \in \mathcal{W}$ for all $\mathbf{x},\mathbf{y} \in \mathcal{W}$ and all $\alpha \in \mathbb{C}$. Then $\mathcal{W}$ is itself a vector space over $\mathbb{C}$.

Definition A.1.3 (Inner product).

(a) An inner product on a vector space $\mathcal{V}$ is a function

$$(\mathbf{x},\mathbf{y}) \in \mathcal{V} \times \mathcal{V} \mapsto \langle \mathbf{x},\mathbf{y}\rangle \in \mathbb{C}$$

that obeys
(1) $\langle \alpha\mathbf{x},\mathbf{z}\rangle = \alpha\langle \mathbf{x},\mathbf{z}\rangle$, $\langle \mathbf{x}+\mathbf{y},\mathbf{z}\rangle = \langle \mathbf{x},\mathbf{z}\rangle + \langle \mathbf{y},\mathbf{z}\rangle$ (linearity)
(2) $\langle \mathbf{x},\mathbf{y}\rangle = \overline{\langle \mathbf{y},\mathbf{x}\rangle}$ (conjugate symmetry)
(3) $\langle \mathbf{x},\mathbf{x}\rangle > 0$ if $\mathbf{x} \neq \mathbf{0}$ (positive–definiteness)
for all $\mathbf{x},\mathbf{y},\mathbf{z} \in \mathcal{V}$ and $\alpha \in \mathbb{C}$.

(b) Two vectors $\mathbf{x}$ and $\mathbf{y}$ are said to be *orthogonal* with respect to the inner product $\langle\,\cdot\,,\,\cdot\,\rangle$ if $\langle \mathbf{x},\mathbf{y}\rangle = 0$.

(c) We will use the terms "inner product space" or "pre–Hilbert space" to mean a vector space over $\mathbb{C}$ equipped with an inner product.

Definition A.1.4 (Norm).

(a) A norm on a vector space $\mathcal{V}$ is a function $\mathbf{x} \in \mathcal{V} \mapsto \|\mathbf{x}\| \in [0,\infty)$ that obeys
(1) $\|\mathbf{x}\| = 0$ if and only if $\mathbf{x} = \mathbf{0}$.
(2) $\|\alpha\mathbf{x}\| = |\alpha|\|\mathbf{x}\|$
(3) $\|\mathbf{x}+\mathbf{y}\| \le \|\mathbf{x}\| + \|\mathbf{y}\|$
for all $\mathbf{x},\mathbf{y} \in \mathcal{V}$ and $\alpha \in \mathbb{C}$.

(b) A sequence $\{\mathbf{v}_n\}_{n\in\mathbb{N}} \subset \mathcal{V}$ is said to be *Cauchy* with respect to the norm $\|\cdot\|$ if

$$\forall\varepsilon > 0 \quad \exists N \in \mathbb{N} \quad \text{s.t.} \quad m,n > N \implies \|\mathbf{v}_n - \mathbf{v}_m\| < \varepsilon$$

(c) A sequence $\{\mathbf{v}_n\}_{n\in\mathbb{N}} \subset \mathcal{V}$ is said to *converge* to $\mathbf{v}$ in the norm $\|\cdot\|$ if

$$\forall\varepsilon > 0 \quad \exists N \in \mathbb{N} \quad \text{s.t.} \quad n > N \implies \|\mathbf{v}_n - \mathbf{v}\| < \varepsilon$$

(d) A normed vector space is said to be *complete* if every Cauchy sequence converges.

(e) A subset $\mathcal{D}$ of a normed vector space $\mathcal{V}$ is said to be *dense* in $\mathcal{V}$ if $\overline{\mathcal{D}} = \mathcal{V}$, where $\overline{\mathcal{D}}$ is the closure of $\mathcal{D}$. That is, if every element of $\mathcal{V}$ is a limit of a sequence of elements of $\mathcal{D}$.

Theorem A.1.5. *Let $\langle\,\cdot\,,\,\cdot\,\rangle$ be an inner product on a vector space $\mathcal{V}$ and set $\|\mathbf{x}\| = \sqrt{\langle \mathbf{x},\mathbf{x}\rangle}$ for all $\mathbf{x} \in \mathcal{V}$.*

(a) *The inner product is sesquilinear. That is,*

$$\langle \alpha\mathbf{x}+\beta\mathbf{y},\mathbf{z}\rangle = \alpha\langle \mathbf{x},\mathbf{z}\rangle + \beta\langle \mathbf{y},\mathbf{z}\rangle$$

$$\langle \mathbf{x}, \alpha\mathbf{y} + \beta\mathbf{z}\rangle = \overline{\alpha}\,\langle \mathbf{x}, \mathbf{y}\rangle + \overline{\beta}\,\langle \mathbf{x}, \mathbf{z}\rangle$$

for all $\mathbf{x}, \mathbf{y}, \mathbf{z} \in \mathcal{V}$ *and* $\alpha, \beta \in \mathbb{C}$.[1]

(b) $\|\mathbf{x}\|$ *is a norm.*

(c) *The inner product and associated norm obeys*

(1) (*Cauchy–Schwarz inequality*) $|\langle \mathbf{x}, \mathbf{y}\rangle| \le \|\mathbf{x}\|\,\|\mathbf{y}\|$

(2) (*Parallelogram law*) $\|\mathbf{x} + \mathbf{y}\|^2 + \|\mathbf{x} - \mathbf{y}\|^2 = 2\|\mathbf{x}\|^2 + 2\|\mathbf{y}\|^2$

(3) (*Polarization identities*)

$$\begin{aligned}\langle \mathbf{x}, \mathbf{y}\rangle &= \tfrac{1}{2}\{\|\mathbf{x} + \mathbf{y}\|^2 - \|\mathbf{x}\|^2 - \|\mathbf{y}\|^2\} + \tfrac{1}{2i}\{\|\mathbf{x} + i\mathbf{y}\|^2 - \|\mathbf{x}\|^2 - \|\mathbf{y}\|^2\}\\ &= \tfrac{1}{4}\{\|\mathbf{x} + \mathbf{y}\|^2 - \|\mathbf{x} - \mathbf{y}\|^2\} + \tfrac{1}{4i}\{\|\mathbf{x} + i\mathbf{y}\|^2 - \|\mathbf{x} - i\mathbf{y}\|^2\}\end{aligned}$$

for all $\mathbf{x}, \mathbf{y} \in \mathcal{V}$.

Proof.

(a) is obvious.

(b) See [**RS80**, Theorem II.2] or [**Con90**, Corollary 1.5].

(c) (1) See [**RS80**, Corollary to Theorem II.1] or [**Con90**, paragraph 1.4].

(c) (2) and (3) are obvious. □

Lemma A.1.6. *Let* $\|\ \cdot\ \|$ *be a norm on a vector space* $\mathcal{V}$. *There exists an inner product* $\langle\ \cdot\ ,\ \cdot\ \rangle$ *on* $\mathcal{V}$ *such that*

$$\langle \mathbf{x}, \mathbf{x}\rangle = \|\mathbf{x}\|^2 \qquad \textit{for all} \quad \mathbf{x} \in \mathcal{V}$$

if and only if $\|\ \cdot\ \|$ *obeys the parallelogram law.*

Proof. This is Exercise A.1.7. □

Exercise A.1.7. Let $\|\ \cdot\ \|$ be a norm on a vector space $\mathcal{V}$. Prove that there exists an inner product $\langle\ \cdot\ ,\ \cdot\ \rangle$ on $\mathcal{V}$ such that

$$\langle \mathbf{x}, \mathbf{x}\rangle = \|\mathbf{x}\|^2 \qquad \text{for all} \quad \mathbf{x} \in \mathcal{V}$$

if and only if $\|\ \cdot\ \|$ obeys the parallelogram law.

Definition A.1.8 (Banach space).

(a) A Banach space is a complete normed vector space.

(b) Two Banach spaces $\mathcal{B}_1$ and $\mathcal{B}_2$ are said to be isometric if there exists a map $U : \mathcal{B}_1 \to \mathcal{B}_2$ that is

(1) linear (meaning that $U(\alpha\mathbf{x} + \beta\mathbf{y}) = \alpha U(\mathbf{x}) + \beta U(\mathbf{y})$ for all $\mathbf{x}, \mathbf{y} \in \mathcal{B}_1$ and $\alpha, \beta \in \mathbb{C}$)

(2) surjective (also-called onto)

(3) isometric (meaning that $\|U\mathbf{x}\|_{\mathcal{B}_2} = \|\mathbf{x}\|_{\mathcal{B}_1}$ for all $\mathbf{x} \in \mathcal{B}_1$). This implies that U is injective (also-called 1–1).

[1] Physicists and mathematical physicists tend to use the convention that inner products are linear in the second argument and conjugate linear in the first.

Definition A.1.9 (Hilbert space).

(a) A Hilbert space $\mathcal{H}$ is a complex inner product space that is complete under the associated norm.

(b) Two Hilbert spaces $\mathcal{H}_1, \mathcal{H}_2$ are said to be isomorphic (denoted $\mathcal{H}_1 \cong \mathcal{H}_2$) if there exists a map $U : \mathcal{H}_1 \to \mathcal{H}_2$ that is
(1) linear
(2) onto
(3) inner product preserving (meaning that $\langle U\mathbf{x}, U\mathbf{y}\rangle_{\mathcal{H}_2} = \langle \mathbf{x}, \mathbf{y}\rangle_{\mathcal{H}_1}$ for all $\mathbf{x}, \mathbf{y} \in \mathcal{H}_1$).
Such a map, U, is called *unitary*.

Theorem A.1.10 (Completion). *If $(\mathcal{V}, \langle \cdot, \cdot \rangle_{\mathcal{V}})$ is any inner product space, then there exists a Hilbert space $(\mathcal{H}, \langle \cdot, , \cdot \rangle_{\mathcal{H}})$ and a map $U : \mathcal{V} \to \mathcal{H}$ such that*

(1) *U is 1–1,*

(2) *U is linear,*

(3) *$\langle U\mathbf{x}, U\mathbf{y}\rangle_{\mathcal{H}} = \langle \mathbf{x}, \mathbf{y}\rangle_{\mathcal{V}}$ for all $\mathbf{x}, \mathbf{y} \in \mathcal{V}$ and*

(4) *$U(\mathcal{V}) = \{ U\mathbf{x} \mid \mathbf{x} \in \mathcal{V} \}$ is dense in $\mathcal{H}$. If $\mathcal{V}$ is complete, then $U(\mathcal{V}) = \mathcal{H}$.*

$\mathcal{H}$ is called the completion of $\mathcal{V}$.

Proof. See [**RS80**, Theorem I.3 and Problem 1 of Chapter II]. □

Example A.1.11.

(a) $\mathbb{C}^n = \{ \mathbf{x} = (x_1, \cdots x_n) \mid x_1, \cdots x_n \in \mathbb{C} \}$ together with the inner product $\langle \mathbf{x}, \mathbf{y}\rangle = \sum_{\ell=1}^{n} x_\ell \bar{y}_\ell$ is a Hilbert space.

(b) If $1 \le p < \infty$, then $\ell^p = \{ (x_n)_{n\in\mathbb{N}} \mid \sum_{n=1}^{\infty} |x_n|^p < \infty \}$ together with the norm

$$\left\| (x_n)_{n\in\mathbb{N}} \right\|_p = \Big[\sum_{n=1}^{\infty} |x_n|^p \Big]^{1/p}$$

is a Banach space. Here, and in the next two examples, each entry x_n in the sequence $(x_n)_{n\in\mathbb{N}}$ is to be a complex number.

(c) $\ell^2 = \{ (x_n)_{n\in\mathbb{N}} \mid \sum_{n=1}^{\infty} |x_n|^2 < \infty \}$ is a Hilbert space with the inner product $\langle (x_n)_{n\in\mathbb{N}}, (y_n)_{n\in\mathbb{N}} \rangle = \sum_{n=1}^{\infty} x_n \overline{y_n}$.

(d) $\ell^\infty = \{ (x_n)_{n\in\mathbb{N}} \mid \sup_n |x_n| < \infty \}$ and $c_0 = \{ (x_n)_{n\in\mathbb{N}} \mid \lim_{n\to\infty} x_n = 0 \}$ are both Banach spaces with the norm

$$\left\| (x_n)_{n\in\mathbb{N}} \right\|_\infty = \sup_n |x_n|.$$

(e) Let $\mathcal{X}$ be a metric space (or more generally a topological space) and

$$C(\mathcal{X}) = \{ f : \mathcal{X} \to \mathbb{C} \mid f \text{ continuous, bounded} \},$$
$$C_c(\mathcal{X}) = \{ f : \mathcal{X} \to \mathbb{C} \mid f \text{ continuous and compactly supported} \}.$$

If $\mathcal{X}$ is a subset of $\mathbb{R}^n$ or $\mathbb{C}^n$ for some $n \in \mathbb{N}$, let

$$C_0(\mathcal{X}) = \{ f : \mathcal{X} \to \mathbb{C} \mid f \text{ continuous}, \lim_{|x|\to\infty} f(x) = 0 \}.$$

Then $C(\mathcal{X})$ and $C_0(\mathcal{X})$ are Banach spaces with the norm $\|f\| = \sup_{x\in\mathcal{X}} |f(x)|$. $C_c(\mathcal{X})$ is a normed vector space, but need not be complete.

(f) Let $1 \le p \le \infty$. Let $(X, \mathcal{M}, \mu)$ be a measure space, with X a set, $\mathcal{M}$ a σ–algebra and μ a measure. For $p < \infty$, set

$$\mathcal{L}^p(X, \mathcal{M}, \mu) = \{ \varphi : X \to \mathbb{C} \mid \varphi\ \mathcal{M}\text{–measurable}, \textstyle\int |\varphi(x)|^p \, d\mu(x) < \infty \},$$
$$\|\varphi\|_p = \Big[\int |\varphi(x)|^p \, d\mu(x) \Big]^{1/p}.$$

For $p = \infty$, set[2]

$$\mathcal{L}^\infty(X, \mathcal{M}, \mu) = \{ \varphi : X \to \mathbb{C} \mid \varphi\ \mathcal{M}\text{–measurable, ess sup} |\varphi(x)| < \infty \},$$
$$\|\varphi\|_\infty = \text{ess sup} |\varphi(x)|.$$

This is not quite a Banach space because any function φ that is zero almost everywhere has "norm" zero. So we define an equivalence relation on $\mathcal{L}^p(X, \mathcal{M}, \mu)$ by

$$\varphi \sim \psi \iff \varphi = \psi \text{ a.e.}$$

As usual, the equivalence class of $\varphi \in \mathcal{L}^p(X, \mathcal{M}, \mu)$ is

$$[\varphi] = \{ \psi \in \mathcal{L}^p(X, \mathcal{M}, \mu) \mid \psi \sim \varphi \}.$$

Then

$$L^p(X, \mathcal{M}, \mu) = \{ [\varphi] \mid \varphi \in \mathcal{L}^p(X, \mathcal{M}, \mu) \}$$

is a Banach space with

$$[\varphi] + [\psi] = [\varphi + \psi], \qquad a[\varphi] = [a\varphi], \qquad \big\|[\varphi]\big\|_p = \|\varphi\|_p,$$

for all $\varphi, \psi \in \mathcal{L}^p(X, \mathcal{M}, \mu)$ and $a \in \mathbb{C}$, and $L^2(X, \mathcal{M}, \mu)$ is a Hilbert space with inner product

$$\langle [\varphi], [\psi] \rangle = \int \varphi(x) \overline{\psi(x)} \, d\mu(x)$$

for all $\varphi, \psi \in \mathcal{L}^2(X, \mathcal{M}, \mu)$. It is standard to write φ in place of $[\varphi]$.

---

[2] The essential supremum of $|\varphi|$, with respect to the measure μ, is denoted $\text{ess sup}_{x\in X} |\varphi(x)|$ and is defined to be $\inf\{ a \ge 0 \mid |\varphi(x)| \le a \text{ almost everywhere with respect to } \mu \}$.

(g) Let D be an open subset of $\mathbb{C}$. Then

$$A^2(D) = \{ \varphi : D \to \mathbb{C} \mid \varphi \text{ analytic}, \quad \textstyle\int_D |\varphi(x+iy)|^2 \, dx \, dy < \infty \}$$

is a Hilbert space with the inner product

$$\langle \varphi, \psi \rangle = \int_D \varphi(x+iy) \, \overline{\psi(x+iy)} \, dx \, dy.$$

(h) Let $\ell \geq 0$ be an integer and Ω be an open subset of $\mathbb{R}^n$ for some $n \in \mathbb{N}$. If $\alpha = (\alpha_1, \cdots, \alpha_n) \in \mathbb{N}_0^n$, where $\mathbb{N}_0 = \{0\} \cup \mathbb{N}$, we shall use $\partial^\alpha \varphi(x)$ to denote the partial derivative $\frac{\partial^{\alpha_1}}{\partial x_1^{\alpha_1}} \cdots \frac{\partial^{\alpha_n}}{\partial x_n^{\alpha_n}} \varphi(x)$. The order of this partial derivative is $|\alpha| = \alpha_1 + \cdots + \alpha_n$. Define

$$\|\varphi\|_{H^\ell(\Omega)} = \Big\{ \sum_{|\alpha| \leq \ell} \int_\Omega |\partial^\alpha \varphi(x)|^2 \, dx \Big\}^{1/2}$$

for each $\varphi \in C^\ell(\Omega)$ for which the right-hand side is finite. The Sobolev space $H^\ell(\Omega)$ is the completion of the vector space

$$\{ \varphi \in C^\ell(\Omega) \mid \|\varphi\|_{H^\ell(\Omega)} < \infty \}$$

equipped with the inner product

$$\langle \varphi, \psi \rangle_{H^\ell(\Omega)} = \sum_{|\alpha| \leq \ell} \int_\Omega \partial^\alpha \varphi(x) \, \overline{\partial^\alpha \psi(x)} \, dx.$$

Similarly, $H_0^\ell(\Omega)$ is the completion of $C_c^\infty(\Omega)$.

Theorem A.1.12. *Let $-\infty < a < b < \infty$ and $1 \leq p < \infty$. The following sets of functions are dense in $L^p([a,b])$.*

(a) *Simple functions (functions of the form $\sum_{j=1}^n a_j \chi_{E_j}(x)$ with $n \in \mathbb{N}$ and the sets E_j measurable);*

(b) *Step functions (functions of the form $\sum_{j=1}^n a_j \chi_{E_j}(x)$ with $n \in \mathbb{N}$ and the sets E_j intervals);*

(c) *Continuous functions that vanish at a and b;*

(d) *Periodic C^∞ functions of period $b - a$;*

(e) *C^∞ functions that are supported in (a,b).*

Above $\chi_E(x)$ denotes the characteristic function of the set E.

Proof. See Exercises A.1.13 and A.1.14. □

Exercise A.1.13. Let $\varepsilon > 0$ and $-\infty < a < b < \infty$. Let m be Lebesgue measure and $f : [a,b] \to \mathbb{R}$ be a Lebesgue-measurable function.

(a) Prove that there exists an $M \in [0, \infty)$ such that

$$m\{ x \in [a,b] \mid |f(x)| \geq M \} \leq \varepsilon.$$

(b) Assume that $f : [a,b] \to [c,C]$. Prove that there exists a simple function s such that $c \leq s(x) \leq C$ and $|f(x) - s(x)| \leq \varepsilon$ for all $x \in [a,b]$. A simple function is, by definition, of the form $\sum_{j=1}^{n} a_j \chi_{E_j}(x)$ with $n \in \mathbb{N}$ and the sets E_j measurable.

(c) Let $s : [a,b] \to [c,C]$ be a simple function. Prove that there exists a step function $g : [a,b] \to [c,C]$ such that the measure

$$m\{\, x \in [a,b] \mid s(x) \neq g(x) \,\} < \varepsilon.$$

A step function is, by definition, of the form $\sum_{i=1}^{n} a_i \chi_{E_i}(x)$ with $n \in \mathbb{N}$ and the sets E_i intervals.

(d) Let $g : [a,b] \to [c,C]$ be a step function. Prove that there exists a continuous function $h : [a,b] \to [c,C]$ such that $h(a) = h(b) = 0$ and

$$m\{\, x \in [a,b] \mid g(x) \neq h(x) \,\} < \varepsilon.$$

Exercise A.1.14. Let $-\infty < a < b < \infty$ and $1 \leq p < \infty$. Prove that the following sets of functions are dense in $L^p([a,b])$.

(a) simple functions

(b) step functions

(c) continuous functions that vanish at a and b

(d) periodic C^∞ functions of period $b - a$

(e) C^∞ functions that are supported in (a,b)

Definition A.1.15 (Basis)**.** Let $\mathcal{B}$ be a Banach space and $\mathcal{H}$ a Hilbert space.

(a) A subset S of $\mathcal{H}$ is an *orthonormal subset* if each vector in S is of length one and each pair of distinct vectors in S is orthogonal.

(b) An *orthonormal basis* (or complete orthonormal system) for $\mathcal{H}$ is an orthonormal subset of $\mathcal{H}$, which is maximal in the sense that it is not properly contained in any other orthonormal subset of $\mathcal{H}$.

(c) A *Schauder basis* for $\mathcal{B}$ is a sequence $\{\mathbf{e}_n\}_{n\in\mathbb{N}}$ of elements of $\mathcal{B}$ such that for each $v \in \mathcal{B}$ there is a unique sequence $\{\alpha_n\}_{n\in\mathbb{N}} \subset \mathbb{C}$ such that $v = \sum_{n=1}^{\infty} \alpha_n \mathbf{e}_n$.

(d) An *algebraic basis* (or Hamel basis) for $\mathcal{B}$ is a subset $\mathcal{S} \subset \mathcal{B}$ such that each $\mathbf{x} \in \mathcal{B}$ has a unique representation as a finite linear combination of elements of $\mathcal{S}$. This is the case if and only if every finite subset of $\mathcal{S}$ is linearly independent and each $\mathbf{x} \in \mathcal{B}$ has some representation as a finite linear combination of elements of $\mathcal{S}$.

Theorem A.1.16. *Every Hilbert space has an orthonormal basis.*

Proof. See [**RS80**, Theorem II.5] or [**Con90**, Proposition 4.2]. □

Theorem A.1.17. *Every vector space has an algebraic basis.*

Proof. This is Exercise A.1.19. □

Theorem A.1.18. *Let $\{\mathbf{e}_i\}_{i\in\mathcal{I}}$ be an orthonormal basis for the Hilbert space $\mathcal{H}$. Then, for each $\mathbf{x}\in\mathcal{H}$, $\{\, i\in\mathcal{I} \mid \langle \mathbf{e}_i,\mathbf{x}\rangle \neq 0 \,\}$ is countable[3] and*

$$\mathbf{x} = \sum_{i\in\mathcal{I}} \langle \mathbf{x},\mathbf{e}_i\rangle\, \mathbf{e}_i, \qquad \|\mathbf{x}\|^2 = \sum_{i\in\mathcal{I}} |\langle \mathbf{x},\mathbf{e}_i\rangle|^2.$$

(The right-hand sides converge independent of order.)

Conversely, if $\{c_i\}_{i\in\mathcal{I}} \subset \mathbb{C}$ and $\sum_{i\in\mathcal{I}} |c_i|^2 < \infty$, then $\sum_{i\in\mathcal{I}} c_i\,\mathbf{e}_i$ converges to an element of $\mathcal{H}$.

Proof. See [**RS80**, Theorem II.6] or [**Con90**, Theorem 4.13]. □

Exercise A.1.19. Prove that every vector space has an algebraic basis.

Hint: Use Zorn's lemma (which is equivalent to the axiom of choice). It says that if a nonempty set $\mathfrak{S}$
(1) is partially ordered and
(2) has the property that every linearly ordered subset has an upper bound
then $\mathfrak{S}$ has a maximal element.

Exercise A.1.20. Prove that

$$B = \left\{\mathbf{e}_n(x) = \tfrac{1}{\sqrt{2\pi}} e^{inx}\right\}_{n\in\mathbb{Z}}$$

is an orthonormal basis for $L^2([0,2\pi])$.

Definition A.1.21 (Separable). A metric space is said to be separable if it has a countable dense subset.

Lemma A.1.22. *A metric space $(\mathcal{M},d)$ fails to be separable if and only if there is an $\varepsilon > 0$ and an uncountable subset $\{m_i\}_{i\in\mathcal{I}} \subset \mathcal{M}$ with $d(m_i,m_j) \geq \varepsilon$ for all $i,j\in\mathcal{I}$ with $i\neq j$.*

Proof. This is Exercise A.1.23. □

Exercise A.1.23. Prove that a metric space $(\mathcal{M},d)$ fails to be separable if and only if there is an $\varepsilon > 0$ and an uncountable subset $\{m_i\}_{i\in\mathcal{I}} \subset \mathcal{M}$ with $d(m_i, m_j) \geq \varepsilon$ for all $i,j\in\mathcal{I}$ with $i\neq j$.

Theorem A.1.24. *Let $\mathcal{H}$ be a Hilbert space.*

(a) *$\mathcal{H}$ is separable if and only if it has a countable orthonormal basis.*
(b) *If $\dim\mathcal{H} = n \in \mathbb{N}$, then $\mathcal{H} \cong \mathbb{C}^n$.*
(c) *If $\mathcal{H}$ is separable but is not of finite dimension, then $\mathcal{H} \cong \ell^2$.*

[3] We include finite in countable.

Proof. See [**RS80**, Theorem II.7] or [**Con90**, Theorem 5.4 and Corollary 5.5]. □

Example A.1.25.

(a) As $L^2([0,2\pi])$ has a countable, orthonormal basis, it is separable and isomorphic to ℓ^2.

(b) ℓ^∞ is not separable. To see this define, for each subset $S \subset \mathbb{N}$,

$$x^{(S)} = \left(x_n^{(S)}\right)_{n\in\mathbb{N}} \in \ell^\infty$$

by

$$x_n^{(S)} = \begin{cases} 1 & \text{if } n \in S \\ 0 & \text{if } n \notin S \end{cases}$$

This is an uncountable family of elements of ℓ^∞ with $\|x^{(S)} - x^{(T)}\|_\infty = 1$ for all distinct subsets S, T of $\mathbb{N}$.

Definition A.1.26 (Orthogonal complement)**.** The orthogonal complement, $\mathcal{M}^\perp$, of any subset $\mathcal{M}$ of a Hilbert space $\mathcal{H}$, is defined to be

$$\mathcal{M}^\perp = \{\, \mathbf{y} \in \mathcal{H} \mid \langle \mathbf{x}, \mathbf{y}\rangle = 0 \text{ for all } \mathbf{x} \in \mathcal{M} \,\}.$$

Theorem A.1.27. *Let $\mathcal{M}$ be a linear subspace of a Hilbert space $\mathcal{H}$. Then*

(a) *$\mathcal{M}^\perp$ is a closed linear subspace of $\mathcal{H}$.*

(b) *$\mathcal{M} \cap \mathcal{M}^\perp = \{0\}$.*

(c) *$\left(\mathcal{M}^\perp\right)^\perp = \overline{\mathcal{M}}$ (the closure of $\mathcal{M}$).*

Proof. See [**RS80**, Problem 6 of Chapter 2] and [**Con90**, Corollary 2.9]. □

Theorem A.1.28 (Projection)**.** *Let $\mathcal{M}$ be a closed linear subspace of a Hilbert space $\mathcal{H}$. Then each $\mathbf{x} \in \mathcal{H}$ has a unique representation $\mathbf{x} = \mathbf{x}^\parallel + \mathbf{x}^\perp$ with $\mathbf{x}^\parallel \in \mathcal{M}$ and $\mathbf{x}^\perp \in \mathcal{M}^\perp$.*

Proof. See [**RS80**, Theorem II.3] or [**Con90**, Theorem 2.6]. □

Theorem A.1.29 (Complete orthonormal set)**.** *Let $\mathcal{H}$ be a separable Hilbert space and let $\{\mathbf{e}_j\}_{j=1}^\infty$ be an orthonormal set. If $\{\mathbf{e}_j\}$ is* complete *in the sense that the only $\mathbf{x} \in \mathcal{H}$ with $\langle \mathbf{x}, \mathbf{e}_j\rangle = 0$ for all j is $\mathbf{x} \equiv 0$, then $\{\mathbf{e}_j\}$ is an orthonormal basis of $\mathcal{H}$.*

Proof. Let $\mathcal{M}$ be the closure of the linear span of $\{\mathbf{e}_j\}$ (i.e., of the set of finite linear combinations of the $\mathbf{e}_j$). Then $\mathcal{M}$ is a closed linear subspace of $\mathcal{H}$, and the completeness condition ensures that $\mathcal{M}^\perp = \{0\}$. By Theorem A.1.28, $\mathcal{M} = \mathcal{H}$. □

A.2. Bounded linear operators

Definition A.2.1 (Linear operator). Let $\mathcal{B}$, $\tilde{\mathcal{B}}$ be Banach spaces and $\mathcal{H}$, $\tilde{\mathcal{H}}$ be Hilbert spaces.

(a) Let $\mathcal{D}$ be a linear subspace of $\mathcal{B}$. A map $A : \mathcal{D} \to \tilde{\mathcal{B}}$ is called a *linear operator* if it obeys

$$A(\alpha\mathbf{x} + \beta\mathbf{y}) = \alpha A(\mathbf{x}) + \beta A(\mathbf{y}) \qquad \text{for all } \alpha, \beta \in \mathbb{C} \text{ and } \mathbf{x}, \mathbf{y} \in \mathcal{D}$$

One usually denotes the image of $\mathbf{x}$ under A as $A\mathbf{x}$, rather than $A(\mathbf{x})$. The set $\mathcal{D}$ is called the domain of A and is generally denoted $\mathcal{D}(A)$. One often calls A a "linear operator on $\mathcal{B}$" even when its domain is a proper subset of $\mathcal{B}$.

(b) A linear operator $A : \mathcal{D} \to \tilde{\mathcal{B}}$ is said to be *bounded* if

$$\|A\| = \sup_{0 \neq \mathbf{x} \in \mathcal{D}} \frac{\|A\mathbf{x}\|_{\tilde{\mathcal{B}}}}{\|\mathbf{x}\|_{\mathcal{B}}} = \sup_{\substack{\mathbf{x} \in \mathcal{D} \\ \|\mathbf{x}\|_{\mathcal{B}} = 1}} \|A\mathbf{x}\|_{\tilde{\mathcal{B}}} \tag{A.1}$$

is finite. The set of all bounded, linear operators defined on $\mathcal{B}$ and taking values in $\tilde{\mathcal{B}}$ is denoted $\mathcal{L}(\mathcal{B}, \tilde{\mathcal{B}})$. With the norm (A.1), it is itself a Banach space. The set of all bounded, linear operators defined on $\mathcal{B}$ and taking values in $\mathcal{B}$ is denoted $\mathcal{L}(\mathcal{B})$.

(c) A *linear functional* on $\mathcal{B}$ is a linear operator $f : \mathcal{B} \to \mathbb{C}$. A *bounded linear functional* on $\mathcal{B}$ is a linear operator $f : \mathcal{B} \to \mathbb{C}$ for which

$$\sup_{0 \neq \mathbf{x} \in \mathcal{B}} \frac{|f(\mathbf{x})|}{\|\mathbf{x}\|_{\mathcal{B}}}$$

is finite.

(d) The *dual space* of a Banach space $\mathcal{B}$ is the space $\mathcal{B}'$ of all bounded linear functionals on $\mathcal{B}$. The dual space is itself a Banach space.

(e) Let $T : \mathcal{D}(T) \subset \mathcal{H} \to \tilde{\mathcal{H}}$ be a linear operator. Denote

$$\mathcal{D}(T^*) = \left\{ \varphi \in \tilde{\mathcal{H}} \mid \exists!\ \eta \in \mathcal{H} \quad \text{s.t.} \quad \langle \varphi, T\psi \rangle_{\tilde{\mathcal{H}}} = \langle \eta, \psi \rangle_{\mathcal{H}} \quad \forall \quad \psi \in \mathcal{D}(T) \right\}$$

If $\varphi \in \mathcal{D}(T^*)$ the corresponding η is denoted $T^*\varphi$. Thus $T^*\varphi$ is the unique vector in $\mathcal{H}$ such that

$$\langle \varphi, T\psi \rangle_{\tilde{\mathcal{H}}} = \langle T^*\varphi, \psi \rangle_{\mathcal{H}} \qquad \text{for all } \psi \in \mathcal{D}(T)$$

The operator T^* is called the adjoint of T.

Proposition A.2.2. *The normed vector space $\mathcal{L}(\mathcal{B}, \tilde{\mathcal{B}})$, with norm* (A.1)*, is a Banach space.*

Proof. See [**RS80**, Theorem III.2]. □

Lemma A.2.3. *Let $\mathcal{H}$ be an infinite-dimensional Hilbert space. Then there is a linear operator $W : \mathcal{H} \to \mathcal{H}$ which is defined on all of $\mathcal{H}$, but is **not** bounded.*

Proof. This is Exercise A.2.4. □

Excrcisc A.2.4. Lct $\mathcal{H}$ bc an infinitc-dimcnsional Hilbcrt spacc. Construct a linear operator $W : \mathcal{H} \to \mathcal{H}$ which is defined on all of $\mathcal{H}$, but is **not** bounded. (Hint: use an algebraic basis.)

Example A.2.5.

(a) *Matrices*: Let $n \in \mathbb{N}$. An $n \times n$ matrix $\left[M_{i,j}\right]_{1\le i,j\le n}$ is naturally associated to the operator $M : \mathbb{C}^n \to \mathbb{C}^n$ determined by

$$(M\mathbf{x})_i = \sum_{j=1}^{n} M_{i,j}x_j$$

The adjoint operator is associated to the matrix $\left[M^*_{i,j} = \overline{M_{j,i}}\,\right]_{1\le i,j\le n}$.

(b) *Multiplication operators*: Let $1 \le p < \infty$. Let $(X, \mathcal{M}, \mu)$ be a measure space and let $f : X \to \mathbb{C}$ be measurable. If the essential supremum of f is finite, then

$$\begin{aligned} M_f \, : \, L^p(X, \mathcal{M}, \mu) &\to L^p(X, \mathcal{M}, \mu) \\ \varphi(x) &\mapsto (f\varphi)(x) = f(x)\,\varphi(x) \end{aligned}$$

is a bounded linear operator with $\|M_f\| = \text{ess sup}_{x\in X} |f(x)|$. On the other hand, if the essential supremum of f is infinite, then M_f will not be defined on all of $L^p(X, \mathcal{M}, \mu)$ (as a map into $L^p(X, \mathcal{M}, \mu)$) and will not be bounded (as a map into $L^p(X, \mathcal{M}, \mu)$). In the case $p = 2$, $M^*_f = M_{\overline{f}}$.

(c) *Projection operators*: Let $\mathcal{H}$ be Hilbert space and let $\mathcal{M}$ be a nonempty, closed, linear subspace of $\mathcal{H}$. Define the map $P : \mathcal{H} \to \mathcal{H}$ by

$$P\mathbf{x} = \mathbf{x}^{\|},$$

where $\mathbf{x} = \mathbf{x}^{\perp} + \mathbf{x}^{\|}$ is the decomposition of Theorem A.1.28. It is a bounded linear operator with $\|P\| = 1$, called the orthogonal projection on $\mathcal{M}$. It obeys

$$P^2 = P \qquad P^* = P,$$

where P^* is the adjoint of P. We will see, in Lemma A.2.7, that conversely, if $P : \mathcal{H} \to \mathcal{H}$ is a bounded linear operator that obeys $P^2 = P$ and $P^* = P$, then P is orthogonal projection on $\mathcal{M} = \text{range}(P)$.

(d) *Integral operators*: Let $(X, \mathcal{M}, \mu)$ and $(Y, \mathcal{N}, \nu)$ be measure spaces and $T : X{\times}Y \to \mathbb{C}$ be a function that is measurable with respect to $\mathcal{M}{\otimes}\mathcal{N}$. Let $1 \le p \le \infty$ and $\varphi \in L^p(Y, \mathcal{N}, \nu)$. Define, for each $x \in X$ for which the function $y \mapsto T(x, y)\varphi(y)$ is in $L^1(Y, \mathcal{N}, \nu)$,

$$(T\varphi)(x) = \int_Y T(x, y)\varphi(y)\, d\nu(y) \tag{A.2}$$

(1) If

$$M_1 = \operatorname*{ess\,sup}_{x\in X} \int_Y |T(x,y)|\,d\nu(y) < \infty$$

$$M_2 = \operatorname*{ess\,sup}_{y\in Y} \int_X |T(x,y)|\,d\mu(x) < \infty$$

then (A.2) is a bounded operator $T : L^p(Y,\mathcal{N},\nu) \to L^p(X,\mathcal{M},\mu)$ with norm $\|T\| \le M_1^{1-\frac{1}{p}} M_2^{\frac{1}{p}}$.

(2) If the Hilbert–Schmidt norm

$$\|T\|_{\text{H.S.}} = \left[\int_{X\times Y} |T(x,y)|^2\, d\mu\times\nu(x,y)\right]^{1/2} < \infty$$

then (A.2) is a bounded operator $T : L^2(Y,\mathcal{N},\nu) \to L^2(X,\mathcal{M},\mu)$ with norm $\|T\| \le \|T\|_{\text{H.S.}}$.

In the case $p = 2$,

$$(T^*\psi)(y) = \int_X \overline{T(x,y)}\psi(x)\,d\mu(x).$$

(e) *Differential operators*: Let Ω be an open subset of $\mathbb{R}^n$ for some $n \in \mathbb{N}$. Recall that if $\alpha = (\alpha_1,\cdots,\alpha_n) \in \mathbb{N}_0^n$, where $\mathbb{N}_0 = \{0\}\cup\mathbb{N}$, we use $\partial^\alpha u(x)$ to denote the partial derivative $\frac{\partial^{\alpha_1}}{\partial x_1^{\alpha_1}} \cdots \frac{\partial^{\alpha_n}}{\partial x_n^{\alpha_n}} u(x)$ and $|\alpha| = \alpha_1+\cdots+\alpha_n$ to denote the order of this partial derivative. For any finite subset $\mathcal{I} \subset \mathbb{N}_0^n$ and any family $\{a_\alpha(x)\}_{\alpha\in\mathcal{I}}$ of bounded, measurable functions on Ω the map

$$\varphi(x) \mapsto \sum_{\alpha\in\mathcal{I}} a_\alpha(x)\,\partial^\alpha\varphi(x)$$

is a linear map on $C^\infty(\Omega) \subset L^2(\Omega)$. But it is not bounded as a map from $L^2(\Omega)$ to $L^2(\Omega)$.

Exercise A.2.6. Let $(X,\mathcal{M},\mu)$ and $(Y,\mathcal{N},\nu)$ be σ–finite measure spaces and $T : X\times Y \to \mathbb{C}$ be a function that is measurable with respect to $\mathcal{M}\otimes\mathcal{N}$.

(a) Assume that

$$M_1 = \operatorname*{ess\,sup}_{x\in X} \int_Y |T(x,y)|\,d\nu(y) < \infty$$

$$M_2 = \operatorname*{ess\,sup}_{y\in Y} \int_X |T(x,y)|\,d\mu(x) < \infty,$$

Let $1 \le p \le \infty$. Prove that

$$(T\varphi)(x) = \int_Y T(x,y)\varphi(y)\,d\nu(y)$$

defines a bounded operator $T : L^p(Y, \mathcal{N}, \nu) \to L^p(X, \mathcal{M}, \mu)$ with norm $\|T\| \le M_1^{1-\frac{1}{p}} M_2^{\frac{1}{p}}$.

(b) Assume that

$$\|T\|_{\text{H.S.}} = \Big[\int_{X\times Y} |T(x,y)|^2 \, d\mu \times \nu(x,y) \Big]^{1/2} < \infty.$$

Prove that $(T\varphi)(x) = \int_Y T(x,y)\varphi(y)\, d\nu(y)$ defines a bounded operator from $L^2(Y, \mathcal{N}, \nu)$ to $L^2(X, \mathcal{M}, \mu)$ with norm $\|T\| \le \|T\|_{\text{H.S.}}$.

Lemma A.2.7. *Let $\mathcal{H}$ be a Hilbert space. Let $P : \mathcal{H} \to \mathcal{H}$ be a bounded operator that obeys*

$$P^2 = P \qquad P^* = P.$$

Then P is orthogonal projection on the range of P.

Proof. This is Exercise A.2.8. □

Exercise A.2.8. Let $\mathcal{H}$ be a Hilbert space. Let $P : \mathcal{H} \to \mathcal{H}$ be a bounded linear operator that obeys

$$P^2 = P \qquad P^* = P$$

Prove that P is orthogonal projection on the range of P.

Lemma A.2.9. *Let E and F be orthogonal projections onto closed subspaces of a Hilbert space $\mathcal{H}$. Then $E + F$ is again an orthogonal projection if and only if $EF = FE = 0$.*

Proof. This is Exercise A.2.10. □

Exercise A.2.10. Let E and F be orthogonal projections onto closed subspaces of a Hilbert space $\mathcal{H}$. Prove that $E + F$ is again an orthogonal projection if and only if $EF = FE = 0$. What is the geometric significance of the condition $EF = FE = 0$?

Theorem A.2.11. *Let $\mathcal{V}$ and $\tilde{\mathcal{V}}$ be normed vector spaces and let $T : \mathcal{V} \to \tilde{\mathcal{V}}$ be a linear operator. The following are equivalent.*

(a) *T is continuous at every $\mathbf{x} \in \mathcal{V}$.*

(b) *T is continuous at one $\mathbf{x}_0 \in \mathcal{V}$.*

(c) *T is bounded.*

Proof. The proof is straightforward. □

Theorem A.2.12. *Let $\mathcal{B}$ be a Banach space.*

(a) *Let $\mathcal{S}$ be a subspace of $\mathcal{B}$ and $\lambda \in \mathcal{S}'$. Then there is a $\Lambda \in \mathcal{B}'$ such that $\|\Lambda\|_{\mathcal{B}'} = \|\lambda\|_{\mathcal{S}'}$ and $\Lambda(\mathbf{x}) = \lambda(\mathbf{x})$ for all $\mathbf{x} \in \mathcal{S}$.*

(b) *Let $\mathbf{x} \in \mathcal{B}$. There is a nonzero $\Lambda \in \mathcal{B}'$ such that $|\Lambda(\mathbf{x})| = \|\Lambda\|_{\mathcal{B}'} \|\mathbf{x}\|_{\mathcal{B}}$.*

(c) *Let $\mathcal{Y}$ be a subspace of $\mathcal{B}$ and $\mathbf{x} \in \mathcal{B}$ with the distance from $\mathbf{x}$ to $\mathcal{Y}$ being d. There is a $\Lambda \in \mathcal{B}'$ such that $\|\Lambda\|_{\mathcal{B}'} \leq 1$, $\Lambda(\mathbf{x}) = d$ and $\Lambda(\mathbf{y}) = 0$ for all $\mathbf{y} \in \mathcal{Y}$.*

(d) *Let $\mathbf{x} \in \mathcal{B}$. Then*

$$\|\mathbf{x}\|_{\mathcal{B}} = \sup_{\substack{\Lambda \in \mathcal{B}' \\ \|\Lambda\|_{\mathcal{B}'} = 1}} |\Lambda(\mathbf{x})|.$$

Proof. Part (a) is [**RS80**, Corollary 1 of Theorem III.6 (the Hahn–Banach theorem)]. The other parts follow easily from part (a) and, in the case of part (d), the definition of $\|\Lambda\|_{\mathcal{B}'}$ and part (b). □

Theorem A.2.13 (The B.L.T. theorem). *Let $\mathcal{V}$ be a dense linear subspace of a Banach space $\mathcal{B}$. Let $\tilde{\mathcal{B}}$ be a second Banach space and $T : \mathcal{V} \to \tilde{\mathcal{B}}$ be a bounded linear operator. Then there is a unique bounded linear operator $\tilde{T} : \mathcal{B} \to \tilde{\mathcal{B}}$ such that $T\mathbf{x} = \tilde{T}\mathbf{x}$ for all $\mathbf{x} \in \mathcal{V}$. Furthermore $\|T\| = \|\tilde{T}\|$.*

Proof. See [**RS80**, Theorem I.7]. □

Example A.2.14. We define the Fourier transform as a unitary operator $\mathcal{F} : L^2(\mathbb{R}) \to L^2(\mathbb{R})$. To start we define Schwartz space to be

$$\mathcal{S}(\mathbb{R}) = \left\{ \varphi : \mathbb{R} \to \mathbb{C} \mid \varphi \text{ is } C^\infty,\ \|\varphi\|_{n,m} < \infty \text{ for all integers } n, m \geq 0 \right\},$$

where $\|\varphi\|_{n,m} = \sup_{x \in \mathbb{R}} \left|x^n \frac{d^m \varphi}{dx^m}(x)\right|$. Next we define the Fourier transform and inverse Fourier transform on $\mathcal{S}(\mathbb{R})$ by

$$\hat{\varphi}(\xi) = \int_{-\infty}^{\infty} e^{-i\xi x} \varphi(x)\, dx$$

$$\check{\psi}(x) = \tfrac{1}{2\pi} \int_{-\infty}^{\infty} e^{i\xi x} \psi(\xi)\, d\xi$$

and verify that the linear operators $\varphi \mapsto \hat{\varphi}$ and $\psi \mapsto \check{\psi}$ each map $\mathcal{S}(\mathbb{R})$ into (in fact onto) $\mathcal{S}(\mathbb{R})$ and are inverses of each other and obey

$$\int_{-\infty}^{\infty} \varphi(x)\, \overline{\psi(x)}\, dx = \tfrac{1}{2\pi} \int_{-\infty}^{\infty} \hat{\varphi}(\xi)\, \overline{\hat{\psi}(\xi)}\, d\xi$$

for all $\varphi, \psi \in \mathcal{S}(\mathbb{R})$. Then the B.L.T. theorem provides us with the unique bounded extension of the map $\varphi \mapsto \hat{\varphi}$ to $L^2(\mathbb{R})$, which we call $\mathcal{F}$. For the details, see Appendix B.

Theorem A.2.15 (Riesz representation theorem). *Let $\mathcal{H}$ be a Hilbert space and $\lambda \in \mathcal{H}^*$ be a bounded linear functional on $\mathcal{H}$. Then there is a unique $\mathbf{y}_\lambda \in \mathcal{H}$ such that*

$$\lambda(\mathbf{x}) = \langle \mathbf{x}, \mathbf{y}_\lambda \rangle$$

for all $\mathbf{x} \in \mathcal{H}$. Furthermore $\|\lambda\|_{\mathcal{H}^*} = \|\mathbf{y}_\lambda\|_{\mathcal{H}}$.

Proof. See [**RS80**, Theorem II.4] or [**Con90**, Theorem 3.4]. □

Corollary A.2.16. *Let* $B : \mathcal{H} \times \mathcal{H} \to \mathbb{C}$ *and* $C \geq 0$ *obey*

(1) $B(\alpha\mathbf{x} + \beta\mathbf{y}, \mathbf{z}) = \alpha B(\mathbf{x}, \mathbf{z}) + \beta B(\mathbf{y}, \mathbf{z})$

(2) $B(\mathbf{x}, \alpha\mathbf{y} + \beta\mathbf{z}) = \bar{\alpha} B(\mathbf{x}, \mathbf{y}) + \bar{\beta} B(\mathbf{x}, \mathbf{z})$

(3) $|B(\mathbf{x}, \mathbf{y})| \leq C\|\mathbf{x}\|\,\|\mathbf{y}\|$

for all $\mathbf{x}, \mathbf{y}, \mathbf{z} \in \mathcal{H}$ *and* $\alpha, \beta \in \mathbb{C}$. *Then there is a unique* $A \in \mathcal{L}(\mathcal{H})$ *such that* $B(\mathbf{x}, \mathbf{y}) = \langle A\mathbf{x}, \mathbf{y}\rangle$ *for all* $\mathbf{x}, \mathbf{y} \in \mathcal{H}$. *Furthermore* $\|A\| \leq C$.

Corollary A.2.17. *Let* $\mathcal{H}$ *and* $\tilde{\mathcal{H}}$ *be Hilbert spaces and* $T : \mathcal{H} \to \tilde{\mathcal{H}}$ *be a bounded linear operator. Then the adjoint* T^* *of* T *is a bounded linear operator defined on all of* $\tilde{\mathcal{H}}$.

Proof. This is Exercise A.2.18. □

Exercise A.2.18. Let $\mathcal{H}$ and $\tilde{\mathcal{H}}$ be Hilbert spaces and $T : \mathcal{H} \to \tilde{\mathcal{H}}$ be a bounded linear operator. Prove that the adjoint, T^*, of T is a bounded linear operator defined on all of $\tilde{\mathcal{H}}$.

Definition A.2.19 (Operator topologies). Let $\mathcal{B}$ and $\tilde{\mathcal{B}}$ be Banach spaces. Let $T : \mathcal{B} \to \tilde{\mathcal{B}}$ and, for each $n \in \mathbb{N}$, $T_n : \mathcal{B} \to \tilde{\mathcal{B}}$ be bounded linear operators.

(a) The sequence of $\{T_n\}_{n\in\mathbb{N}}$ of operators is said to converge *uniformly* or *in norm* to T if

$$\lim_{n\to\infty} \|T - T_n\| = 0$$

(b) The sequence of $\{T_n\}_{n\in\mathbb{N}}$ of operators is said to converge *strongly* to T if

$$\lim_{n\to\infty} \|T\mathbf{x} - T_n\mathbf{x}\|_{\tilde{\mathcal{B}}} = 0 \qquad \text{for each } \mathbf{x} \in \mathcal{B}$$

(c) The sequence of $\{T_n\}_{n\in\mathbb{N}}$ of operators is said to converge *weakly* to T if

$$\lim_{n\to\infty} \ell\big(T_n\mathbf{x}\big) = \ell\big(T\mathbf{x}\big) \qquad \text{for each } \mathbf{x} \in \mathcal{B} \text{ and each } \ell \in \tilde{\mathcal{B}}'$$

In the event that $\tilde{\mathcal{B}}$ is a Hilbert space, this is equivalent to

$$\lim_{n\to\infty} \langle T_n\mathbf{x}, \mathbf{y}\rangle_{\tilde{\mathcal{B}}} = \langle T\mathbf{x}, \mathbf{y}\rangle_{\tilde{\mathcal{B}}} \qquad \text{for each } \mathbf{x} \in \mathcal{B} \text{ and each } \mathbf{y} \in \tilde{\mathcal{B}}.$$

Remark A.2.20 (Operator topologies). Since

$$|\ell((T_n - T)\mathbf{x})| \leq \|\ell\|_{\tilde{\mathcal{B}}'} \left\|(T_n - T)\mathbf{x}\right\|_{\tilde{\mathcal{B}}}$$

and

$$\left\|(T_n - T)\mathbf{x}\right\|_{\tilde{\mathcal{B}}} \leq \|T_n - T\|\,\|\mathbf{x}\|_{\mathcal{B}}$$

we have

norm convergence $\implies$ strong convergence $\implies$ weak convergence.

In general, the other implications are false unless $\mathcal{B}$ and $\tilde{\mathcal{B}}$ are finite-dimensional. This is illustrated by the following example.

Example A.2.21 (Operator topologies). Let $\mathcal{B} = \tilde{\mathcal{B}} = \ell^2$.

(a) Let

$$P_n(x_1, x_2, x_3 \cdots) = (\overbrace{0, \cdots, 0}^{n \text{ places}}, x_{n+1}, x_{n+2}, x_{n+3}, \cdots)$$

be projection on the orthogonal complement of the first n components. Then for each fixed $\mathbf{x} \in \ell^2$, $\lim_{n\to\infty} P_n\mathbf{x} = \mathbf{0}$ so that P_n converges strongly to 0 as $n \to \infty$. But, for any $n > m$,

$$(P_n - P_m)(x_1, x_2, x_3 \cdots) = (\overbrace{0, \cdots, 0}^{m \text{ places}}, x_{m+1}, x_{m+2}, \cdots, x_n, 0, \cdots)$$

so that there is a vector $\mathbf{x} \in \ell^2$ with $(P_n - P_m)\mathbf{x} = \mathbf{x}$. Consequently $\|P_n - P_m\| = 1$ and the sequence $\{P_n\}_{n\in\mathbb{N}}$ is not Cauchy and does not converge in norm.

(b) Let

$$R_n(x_1, x_2, x_3 \cdots) = (\overbrace{0, \cdots, 0}^{n \text{ places}}, x_1, x_2, x_3, \cdots)$$

be right shift by n places. For any $\mathbf{x}, \mathbf{y} \in \ell^2$

$$|\langle R_n\mathbf{x}, \mathbf{y}\rangle| = |\langle R_n\mathbf{x}, P_n\mathbf{y}\rangle| \le \|R_n\mathbf{x}\| \, \|P_n\mathbf{y}\| = \|\mathbf{x}\| \, \|P_n\mathbf{y}\| \xrightarrow{n\to\infty} 0$$

So R_n converges weakly to zero as $n \to \infty$. On the other hand, $\|R_n\mathbf{x}\| = \|\mathbf{x}\|$ for all $n \in \mathbb{N}$ and $\mathbf{x} \in \ell^2$. So the R_n does not converge strongly or in norm. (If R_n did converge either strongly or in norm to some R, the fact that

$$R_n \xrightarrow{\text{weakly}} 0$$

would force $R = 0$.)

Theorem A.2.22 (Neumann expansion). *Let T be a bounded linear operator on the Banach space $\mathcal{B}$ whose operator norm $\|T\| < 1$. Then* $\mathrm{Id} - T$ *is bijective and has a bounded inverse and furthermore*

$$(\mathrm{Id} - T)^{-1} = \sum_{n=0}^{\infty} T^n, \qquad \left\|(\mathrm{Id} - T)^{-1}\right\| \le \frac{1}{1 - \|T\|}, \qquad \left\|(\mathrm{Id} - T)^{-1} - \mathrm{Id}\right\| \le \frac{\|T\|}{1 - \|T\|}.$$

The series $\sum_{n=0}^{\infty} T^n$ converges in norm.

Theorem A.2.23 (Adjoints). *Let $\mathcal{H}$ be a Hilbert space and $S, T \in \mathcal{L}(\mathcal{H})$.*

(a) *The map* $A \mapsto A^*$ *is a conjugate linear isometric isomorphism of* $\mathcal{L}(\mathcal{H})$ *onto* $\mathcal{L}(\mathcal{H})$. *In particular*

$$(\alpha A + \beta B)^* = \overline{\alpha} A^* + \overline{\beta} B^* \qquad \|A^*\| = \|A\|$$

for all $A, B \in \mathcal{L}(\mathcal{H})$ *and all* $\alpha, \beta \in \mathbb{C}$.

(b) $(TS)^* = S^* T^*$

(c) $(T^*)^* = T$

(d) *If* T *has a bounded inverse, then* T^* *has a bounded inverse and* $(T^*)^{-1} = (T^{-1})^*$.

(e) *The map* $A \mapsto A^*$ *is continuous in the weak and uniform topologies. That is, if* $\{A_n\}_{n\in\mathbb{N}}$ *converges to* A *weakly (in norm), then* $\{A_n^*\}_{n\in\mathbb{N}}$ *converges to* A^* *weakly (in norm). The map* $A \mapsto A^*$*is continuous in the strong topology if and only if* $\mathcal{H}$ *is finite-dimensional.*

(f) $\|T^*T\| = \|T\|^2$

(g) *If* $T = T^*$, *then* $\|T\| = \sup\{\,|\langle T\mathbf{x}, \mathbf{x}\rangle|\,|\,\mathbf{x} \in \mathcal{H},\ \|\mathbf{x}\| = 1\,\}$.

Proof. See [**RS80**, Theorem VI.3] and Example A.2.26 and Proposition A.2.24, or [**Con90**, Propositions 2,6, 2.7, 2.13]. □

Proposition A.2.24. *Let* $\mathcal{H}$ *be a Hilbert space and* $T : \mathcal{H} \to \mathcal{H}$ *be a bounded linear operator.*

(a) *We have*

$$\|T\| = \sup_{\substack{\mathbf{x},\upsilon\mathbf{y}\in\mathcal{H}\\ \mathbf{x},\mathbf{y}\neq\mathbf{0}}} \frac{|\langle T\mathbf{x},\mathbf{y}\rangle|}{\|\mathbf{x}\|\,\|\mathbf{y}\|}$$

(b) *Assume in addition that* $T = T^*$. *Then*

$$\|T\| = \sup_{\substack{\mathbf{x}\in\mathcal{H}\\ \mathbf{x}\neq\mathbf{0}}} \frac{|\langle T\mathbf{x},\mathbf{x}\rangle|}{\|\mathbf{x}\|^2}$$

Proof. This is Exercise A.2.25. □

Exercise A.2.25. Let $\mathcal{H}$ be a Hilbert space and $T : \mathcal{H} \to \mathcal{H}$ be a bounded linear operator.

(a) Prove that

$$\|T\| = \sup_{\substack{\mathbf{x},\mathbf{y}\in\mathcal{H}\\ \mathbf{x},\mathbf{y}\neq\mathbf{0}}} \frac{|\langle T\mathbf{x},\mathbf{y}\rangle|}{\|\mathbf{x}\|\,\|\mathbf{y}\|}$$

(b) Assume in addition that $T = T^*$. Prove that

$$\|T\| = \sup_{\substack{\mathbf{x}\in\mathcal{H}\\ \mathbf{x}\neq\mathbf{0}}} \frac{|\langle T\mathbf{x},\mathbf{x}\rangle|}{\|\mathbf{x}\|^2}$$

(c) Find an example which shows that the equation of part (b) can fail if $T \neq T^*$.

Example A.2.26. Let $\mathcal{H} = \ell^2$ and define the right and left shift operators by

$$L(x_1, x_2, x_3, \cdots) = (x_2, x_3, \cdots),$$
$$R(x_1, x_2, x_3, \cdots) = (0, x_1, x_2, x_3, \cdots).$$

First observe that $\|L\| = \|R\| = 1$ and that

$$\langle L\mathbf{x}, \mathbf{y}\rangle = \sum_{j=1}^{\infty} (L\mathbf{x})_j \overline{y_j} = \sum_{j=1}^{\infty} x_{j+1} \overline{y_j} = \sum_{i=2}^{\infty} x_i \overline{y_{i-1}} = \sum_{i=1}^{\infty} x_i \overline{(R\mathbf{y})_i} = \langle \mathbf{x}, R\mathbf{y}\rangle$$

so that $L^* = R$ and $R^* = L$. Next observe that, for each $n \in \mathbb{N}$ and $\mathbf{x} \in \ell^2$,

$$\|L^n\mathbf{x}\|^2 = \sum_{m=n+1}^{\infty} |x_m|^2 \xrightarrow{n\to\infty} 0,$$
$$\|R^n\mathbf{x}\|^2 = \sum_{m=1}^{\infty} |x_m|^2 = \|\mathbf{x}\|^2.$$

Thus, as $n \to \infty$, L^n converges strongly to zero, but $(L^*)^n = R^n$ does not converge strongly to anything. On the other hand, $(L^*)^n$ does converge weakly to zero since, for all $\mathbf{x}, \mathbf{y} \in \ell^2$,

$$|\langle R^n\mathbf{x}, \mathbf{y}\rangle| = |\langle (L^*)^n\mathbf{x}, \mathbf{y}\rangle| = |\langle \mathbf{x}, L^n\mathbf{y}\rangle| \le \|\mathbf{x}\|\,\|L^n\mathbf{y}\| \xrightarrow{n\to\infty} 0.$$

Theorem A.2.27 (Principle of uniform boundedness etc.). *Unless otherwise stated, $\mathcal{X}$ and $\mathcal{Y}$ are Banach spaces and $T : \mathcal{X} \to \mathcal{Y}$ is linear and has domain $\mathcal{X}$.*

(a) *T is bounded if and only if*

$$T^{-1}\{\,\mathbf{y} \in \mathcal{Y} \mid \|\mathbf{y}\|_{\mathcal{Y}} \le 1\,\} = \{\,\mathbf{x} \in \mathcal{X} \mid \|T\mathbf{x}\|_{\mathcal{Y}} \le 1\,\}$$

has nonempty interior. ($\mathcal{X}, \mathcal{Y}$ need not be complete.)

(b) Principle of uniform boundedness: *Let $\mathcal{F} \subset \mathcal{L}(\mathcal{X}, \mathcal{Y})$.*

If, for each $\mathbf{x} \in X$,	*$\{\,\|T\mathbf{x}\| \mid T \in \mathcal{F}\,\}$ is bounded,*
then	*$\{\,\|T\| \mid T \in \mathcal{F}\,\}$ is bounded,*

($\mathcal{Y}$ need not be complete.)

(c) *If $B : \mathcal{X} \times \mathcal{Y} \to \mathbb{C}$ is bilinear and continuous in each variable separately (i.e., $B(\mathbf{x}, \mathbf{y})$ is continuous in $\mathbf{x}$ for each fixed $\mathbf{y}$ and vice versa), then $B(\mathbf{x}, \mathbf{y})$ is jointly continuous (i.e., if $\lim_{n\to\infty} \mathbf{x}_n = 0$ and $\lim_{n\to\infty} \mathbf{y}_n = 0$, then $\lim_{n\to\infty} B(\mathbf{x}_n, \mathbf{y}_n) = 0$).*

(d) Open mapping theorem: *If $T \in \mathcal{L}(\mathcal{X}, \mathcal{Y})$ is surjective (i.e., onto) and if $\mathcal{O}$ is an open subset of $\mathcal{X}$, then $T\mathcal{O} = \{\,T\mathbf{x} \mid \mathbf{x} \in \mathcal{O}\,\}$ is an open subset of $\mathcal{Y}$.*

(e) Inverse mapping theorem: *If $T \in \mathcal{L}(\mathcal{X}, \mathcal{Y})$ is bijective (i.e., 1–1 and onto), then T^{-1} is bounded.*

(f) Closed graph theorem: *The graph of T is defined to be*

$$\Gamma(T) = \{ (\mathbf{x}, \mathbf{y}) \subset \mathcal{X} \times \mathcal{Y} \mid \mathbf{y} = T\mathbf{x} \}.$$

Then

$$T \text{ is bounded } \iff \Gamma(T) \text{ is closed.}$$

In other words, T is bounded if and only if

$$\lim_{n\to\infty} \mathbf{x}_n = \mathbf{x}, \ \lim_{n\to\infty} T\mathbf{x}_n = \mathbf{y} \implies \mathbf{y} = T\mathbf{x}.$$

(g) Hellinger–Toeplitz theorem: *Let T be an everywhere defined linear operator on $\mathcal{H}$ that obeys $\langle \mathbf{x}, T\mathbf{y}\rangle = \langle T\mathbf{x}, \mathbf{y}\rangle$ for all $\mathbf{x}, \mathbf{y} \in \mathcal{H}$. Then T is bounded.*

Proof.

(a) See Proposition at the beginning of [**RS80**, §III.5].

(b) See [**RS80**, Theorem III.9] or [**Con90**, Theorem 14.1].

(c) See the Corollary to [**RS80**, Theorem III.9].

(d) See [**RS80**, Theorem III.10] or [**Con90**, Theorem 12.1].

(e) See [**RS80**, Theorem III.11] or [**Con90**, Theorem 12.5].

(f) See [**RS80**, Theorem III.12] or [**Con90**, Theorem 12.6].

(g) See the Corollary to [**RS80**, Theorem III.12]. □

We conclude this section with a discussion of Fréchet derivatives (see [**Hör03a**, §1.1] for more details).

Definition A.2.28 (Fréchet derivative). Let $\mathcal{B}, \tilde{\mathcal{B}}$ be Banach spaces and let $V \subset \mathcal{B}$ be an open set. A map $f : V \to \tilde{\mathcal{B}}$ is said to be *differentiable* at $\mathbf{x} \in \mathcal{B}$ if there is $f'(\mathbf{x}) = (Df)_{\mathbf{x}} \in \mathcal{L}(\mathcal{B}, \mathcal{B}')$ such that

$$\lim_{\|\mathbf{y}\|\to 0} \frac{f(\mathbf{x}+\mathbf{y}) - f(\mathbf{x}) - f'(\mathbf{x})\mathbf{y}}{\|\mathbf{y}\|} = 0.$$

The map $f'(\mathbf{x})$ is called the *Fréchet derivative* of f at $\mathbf{x}$.

The map f is said to be C^1 in V if the map $V \to \mathcal{L}(\mathcal{B}, \tilde{\mathcal{B}})$, $\mathbf{x} \mapsto f'(\mathbf{x})$ is continuous.

If f is C^1 and the map $V \to \mathcal{L}(\mathcal{B}, \tilde{\mathcal{B}})$, $\mathbf{x} \mapsto f'(\mathbf{x})$ is also C^1, we say that f is C^2. Then the second derivative of f is the map

$$V \to \mathcal{L}(\mathcal{B}, \mathcal{L}(\mathcal{B}, \tilde{\mathcal{B}})), \quad \mathbf{x} \mapsto f''(\mathbf{x}).$$

We may identify $f''(\mathbf{x})$ with the bilinear map

$$(D^2 f)_{\mathbf{x}} : \mathcal{B} \times \mathcal{B} \to \tilde{\mathcal{B}}, \quad (\mathbf{y}_1, \mathbf{y}_2) = (f''(\mathbf{x})(\mathbf{y}_1))(\mathbf{y}_2).$$

Similarly we can define C^k maps.

Definition A.2.29 (Higher Fréchet derivatives). Inductively, we say that a map $f : V \to \tilde{\mathcal{B}}$ is C^k if it is C^{k-1} and $f^{(k-1)}$ is C^1 on V. The kth Fréchet derivative $f^{(k)}$ can be identified with a k-linear map

$$(D^k f)_{\mathbf{x}} : \mathcal{B} \times \ldots \times \mathcal{B} \to \tilde{\mathcal{B}}.$$

We say that f is C^∞ if it is C^k for all k.

A.3. Compact operators

In this section we provide an introduction to compact linear operators on Banach and Hilbert spaces. These operators behave very much like familiar finite-dimensional matrices, without necessarily having finite rank. For more thorough treatments, see [**RS80**, §VI.5, VI.6] or [**Yos95**].

Definition A.3.1. Let $\mathcal{X}$ and $\mathcal{Y}$ be Banach spaces. A linear operator $C : \mathcal{X} \to \mathcal{Y}$ is said to be compact if for each bounded sequence $\{x_i\}_{i\in\mathbb{N}} \subset \mathcal{X}$, there is a subsequence of $\{Cx_i\}_{i\in\mathbb{N}}$ that is convergent.

Example A.3.2. Let $a < b$ and $c < d$. If $C : [c,d] \times [a,b] \to \mathbb{C}$ is continuous, then the integral operator

$$(Cf)(y) = \int_a^b C(y,x) f(x)\, dx$$

is compact as an operator from $\mathcal{X} = C[a,b]$, the space of continuous functions on $[a,b]$ with supremum norm, to $\mathcal{Y} = C[c,d]$.

Exercise A.3.3. Use the Arzelà–Ascoli theorem (Theorem 2.4.1) to prove that the operator C of Example A.3.2 is compact.

Example A.3.4 (Hilbert–Schmidt operators). Let $\langle X, \mu \rangle$ and $\langle Y, \nu \rangle$ be measure spaces and let $k(x,y)$ be a measurable function on $X \times Y$ with

$$\int_{X\times Y} |k(x,y)|^2 \, d\mu(x)\, d\nu(y) < \infty.$$

Then

$$(Kf)(x) = \int_Y k(x,y) f(y)\, d\nu(y)$$

is a compact map from $L^2(Y, d\nu)$ to $L^2(X, d\mu)$. Such an operator is called Hilbert–Schmidt.

Proof. Let $\{f_i\}_{i\in\mathbb{N}}$ be a bounded sequence in $L^2(Y, d\nu)$. By part (c) of Exercise A.3.5, $\{f_i\}_{i\in\mathbb{N}}$ has a weakly convergent subsequence. By throwing away all but this subsequence, we may assume that $\{f_i\}_{i\in\mathbb{N}}$ converges weakly to $f \in L^2(Y, d\nu)$.

We now show that $\{Kf_i\}_{i\in\mathbb{N}}$ converges strongly to $Kf \in L^2(X, d\mu)$. Since $\int_{X\times Y} |k(x,y)|^2\, d\mu(x)\, d\nu(y) < \infty$ we have that $\int_Y |k(x,y)|^2\, d\nu(y) < \infty$ for almost every $x \in X$. For any such $x \in X$,

$$\begin{aligned}\lim_{i\to\infty} \int_Y k(x,y) f_i(y)\, d\nu(y) &= \lim_{i\to\infty} \langle f_i\,,\, \overline{k(x,\,\cdot\,)}\rangle_{L^2(Y,d\nu)} \\ &= \langle f\,,\, \overline{k(x,\,\cdot\,)}\rangle_{L^2(Y,d\nu)} \\ &= \int_Y k(x,y) f(y)\, d\nu(y).\end{aligned}$$

Furthermore, by Cauchy–Schwarz,

$$\begin{aligned}|(Kf_i)(x)| &\le \int_Y |k(x,y) f_i(y)|\, d\nu(y) \\ &\le \|f_i\|_{L^2(Y,d\nu)} \sqrt{\int_Y |k(x,y)|^2\, d\nu(y)} \\ &\le \sup_i \|f_i\|_{L^2(Y,d\nu)} \sqrt{\int_Y |k(x,y)|^2\, d\nu(y)} =: H(x).\end{aligned}$$

Thus we have shown that $(Kf_i)(x)$ converges pointwise to $(Kf)(x)$ for almost every x and is bounded, for all i by the function $H(x)$ which is square integrable with respect to $d\mu(x)$. Thus, by the Lebesgue dominated convergence theorem,

$$\lim_{i\to\infty} \|Kf - Kf_i\|^2_{L^2(X,d\mu)} = \lim_{i\to\infty} \int_X |(Kf)(x) - (Kf_i)(x)|^2\, d\mu(x) = 0. \qquad \square$$

Exercise A.3.5. Let $\mathcal{H}$ be a Hilbert space. A sequence $\{f_i\}_{i\in\mathbb{N}} \subset \mathcal{H}$ is said to converge weakly to $f \in \mathcal{H}$ if

$$\lim_{i\to\infty} \langle f_i, g\rangle = \langle f, g\rangle$$

for all $g \in \mathcal{H}$.

(a) Give an example of a sequence that converges weakly but not strongly.

(b) Prove that if $\{f_i\}_{i\in\mathbb{N}}$ converges weakly to f, then $\|f\| \le \liminf_{i\to\infty} \|f_i\|$. Prove that if $\{f_i\}_{i\in\mathbb{N}}$ converges weakly to f and $\|f\| = \lim_{i\to\infty} \|f_i\|$, then $\{f_i\}_{i\in\mathbb{N}}$ converges strongly to f.

(c) Prove that $\mathcal{H}$ is weakly sequentially compact. That is, every bounded sequence in $\mathcal{H}$ has a weakly convergent subsequence.

Example A.3.6 (Nuclear operators). Let $\mathcal{X}$ and $\mathcal{Y}$ be Banach spaces and denote by $\mathcal{X}'$ the dual space of $\mathcal{X}$. That is, the space of bounded linear functionals on $\mathcal{X}$. If $\{x'_i\}_{i\in\mathbb{N}}$ is a bounded sequence in $\mathcal{X}'$, $\{y_i\}_{i\in\mathbb{N}}$ is a bounded sequence in $\mathcal{Y}$

and $\{c_i\}_{i\in\mathbb{N}}$ is a set of complex numbers obeying $\sum_i |c_i| < \infty$, then

$$Kx = \sum_{i=1}^{\infty} c_i\, x_i'(x)\, y_i$$

is called a nuclear operator from $\mathcal{X}$ to $\mathcal{Y}$. Since

$$\sum_{i=1}^{\infty} |c_i|\, |x_i'(x)|\, \|y_i\|_{\mathcal{Y}} \le \|x\|_{\mathcal{X}} \sup_i \|y_i\|_{\mathcal{Y}} \sup_i \|x_i'\|_{\mathcal{X}'} \sum_{i=1}^{\infty} |c_i|$$

the series defining Kx converges strongly and K is a bounded operator of norm at most $\sup_i \|y_i\|_{\mathcal{Y}} \sup_i \|x_i'\|_{\mathcal{X}'} \sum_{i=1}^{\infty} |c_i|$.

Exercise A.3.7. Prove that any nuclear operator is compact.

Proposition A.3.8. *Let $\mathcal{X}$, $\mathcal{Y}$, and $\mathcal{Z}$ be Banach spaces.*

(a) *If $C : \mathcal{X} \to \mathcal{Y}$ is a compact operator, then C is a bounded operator.*

(b) *If $C_1, C_2 : \mathcal{X} \to \mathcal{Y}$ are compact operators and $\alpha_1, \alpha_2 \in \mathbb{C}$, then the operator $\alpha_1 C_1 + \alpha_2 C_2$ is compact.*

(c) *If $C : \mathcal{X} \to \mathcal{Y}$ is a compact operator and $B_{\mathcal{X}} : \mathcal{Z} \to \mathcal{X}$ and $B_{\mathcal{Y}} : \mathcal{Y} \to \mathcal{Z}$ are bounded operators, then $CB_{\mathcal{X}}$ and $B_{\mathcal{Y}}C$ are compact.*

(d) *Let, for each $i \in \mathbb{N}$, $C_i : \mathcal{X} \to \mathcal{Y}$ be a compact operator. If all the C_i converge in operator norm to an operator $C : \mathcal{X} \to \mathcal{Y}$, then C is compact.*

Proof. Let $\{x_i\}_{i\in\mathbb{N}}$ be a bounded sequence in $\mathcal{X}$.

(a) This is Exercise A.3.9.

(b) Since C_1 is compact, there is a subsequence $\{x_{i_\ell}\}_{\ell\in\mathbb{N}}$ such that $C_1 x_{i_\ell}$ converges in $\mathcal{Y}$. Since C_2 is compact, there is a subsequence $\{x_{i_{\ell_m}}\}_{m\in\mathbb{N}}$ of the bounded sequence $\{x_{i_\ell}\}_{\ell\in\mathbb{N}}$ such that $C_2 x_{i_{\ell_m}}$ converges in $\mathcal{Y}$. Then $\alpha_1 C_1 x_{i_{\ell_m}} + \alpha_2 C_2 x_{i_{\ell_m}}$ also converges in $\mathcal{Y}$.

(c) Let $\{z_i\}_{i\in\mathbb{N}}$ be a bounded sequence in $\mathcal{Z}$. Since $B_{\mathcal{X}}$ is bounded, $\{B_{\mathcal{X}} z_i\}_{i\in\mathbb{N}}$ is a bounded sequence in $\mathcal{X}$. Since C is compact, there is a subsequence $\{B_{\mathcal{X}} z_{i_\ell}\}_{\ell\in\mathbb{N}}$ such that $CB_{\mathcal{X}} z_{i_\ell}$ converges in $\mathcal{Y}$.

Since C is compact, there is a subsequence $\{x_{i_\ell}\}_{\ell\in\mathbb{N}}$ such that Cx_{i_ℓ} converges in $\mathcal{Y}$. Since $C_{\mathcal{Y}}$ is bounded, $B_{\mathcal{Y}} C x_{i_\ell}$ converges in $\mathcal{Y}$.

(d) Let $\{x_j\}_{j\in\mathbb{N}}$ be a bounded sequence in $\mathcal{X}$ and set

$$X = \sup_j \|x_j\|_{\mathcal{X}}.$$

For each fixed $i \in \mathbb{N}$, $\{C_i x_j\}_{j\in\mathbb{N}}$ has a convergent subsequence, since C_i is compact by hypothesis. By taking subsequences of subsequences and using the diagonal trick, we can find a subsequence $\{x_{j_\ell}\}_{\ell\in\mathbb{N}}$ such that $\lim_{\ell\to\infty} C_i x_{j_\ell}$ exists for each $i \in \mathbb{N}$. It suffices for us to prove that $\{Cx_{j_\ell}\}_{\ell\in\mathbb{N}}$

is Cauchy. Let $\varepsilon > 0$. Since the C_i's converge in operator norm to C, there is an $I \in \mathbb{N}$ such that $\|C - C_i\| < \frac{\varepsilon}{3X}$ for all $i \geq I$. Since $\{C_I x_{j_\ell}\}_{\ell\in\mathbb{N}}$ is Cauchy, there is an $L \in \mathbb{N}$ such that

$$\left\|C_I x_{j_\ell} - C_I x_{j_m}\right\|_{\mathcal{Y}} < \frac{\varepsilon}{3}$$

for all $\ell, m > L$. Hence if $\ell, m > L$, then

$$\begin{aligned}
\left\|Cx_{j_\ell} - Cx_{j_m}\right\|_{\mathcal{Y}} &\leq \left\|Cx_{j_\ell} - C_I x_{j_\ell}\right\|_{\mathcal{Y}} + \left\|C_I x_{j_\ell} - C_I x_{j_m}\right\|_{\mathcal{Y}} + \left\|C_I x_{j_m} - Cx_{j_m}\right\|_{\mathcal{Y}} \\
&\leq X\|C - C_I\| + \left\|C_I x_{j_\ell} - C_I x_{j_m}\right\|_{\mathcal{Y}} + X\|C_I - C\| \\
&< X\frac{\varepsilon}{3X} + \frac{\varepsilon}{3} + X\frac{\varepsilon}{3X} \\
&= \varepsilon.
\end{aligned}$$

□

Exercise A.3.9. Prove that compact operators are necessarily bounded.

Proposition A.3.10. *Let $\mathcal{X}$ and $\mathcal{Y}$ be Banach spaces. Denote by $\mathcal{X}'$ and $\mathcal{Y}'$ their dual spaces. That is, $\mathcal{X}'$ (resp. $\mathcal{Y}'$) is the Banach space of bounded linear functionals on $\mathcal{X}$ (resp. $\mathcal{Y}$). The adjoint, $C^* : \mathcal{Y}' \to \mathcal{X}'$, of a bounded operator $C : \mathcal{X} \to \mathcal{Y}$ is determined by*

$$(C^*\eta)(x) = \eta(Cx) \qquad \text{for all } \eta \in \mathcal{Y}' \text{ and } x \in \mathcal{X}.$$

A bounded operator $C : \mathcal{X} \to \mathcal{Y}$ is compact if and only if C^* is compact.

Proof. First assume that C is compact. Let $\{\eta_i\}_{i\in\mathbb{N}}$ be a bounded subset of $\mathcal{Y}'$ and set

$$Y' = \sup_i \|\eta_i\|_{\mathcal{Y}'}.$$

Let $B = \left\{ x \in \mathcal{X} \mid \|x\|_{\mathcal{X}} \leq 1 \right\}$ be the unit ball in $\mathcal{X}$. Since C is compact, $\overline{CB}$, which is the closure of $\left\{ Cx \in \mathcal{X} \mid \|x\|_{\mathcal{X}} \leq 1 \right\}$, is a compact subset of $\mathcal{Y}$. We shall apply Arzelà–Ascoli (Theorem 2.4.1) to the sequence of functions

$$f_i : y \in \overline{CB} \mapsto \eta_i(y) \in \mathbb{C}.$$

Since

$$|f_i(y)| \leq Y'\|y\|_{\mathcal{Y}} \leq Y'\|C\|$$

the sequence is uniformly bounded. Since

$$|f_i(y) - f_i(\tilde{y})| \leq Y'\|y - \tilde{y}\|_{\mathcal{Y}}$$

it is equicontinuous. So, by Arzelà–Ascoli, there is a subsequence f_{i_ℓ} that converges uniformly on $\overline{CB}$. Since

$$\begin{aligned}
\|C^*\eta_i - C^*\eta_j\|_{\mathcal{X}'} &= \sup_{x\in B} |(C^*\eta_i)(x) - (C^*\eta_j)(x)| \\
&= \sup_{x\in B} |\eta_i(Cx) - \eta_j(Cx)| \\
&= \sup_{x\in B} |f_i(Cx) - f_j(Cx)| \\
&= \sup_{y\in CB} |f_i(y) - f_j(y)|,
\end{aligned}$$

the sequence $\{C^*\eta_{i_\ell}\}_{\ell\in\mathbb{N}}$ is Cauchy in $\mathcal{X}'$.

Conversely, assume that C^* is compact. Let $\{x_i\}_{i\in\mathbb{N}}$ be a bounded sequence in $\mathcal{X}$. By the implication that we have already proven, the adjoint, $C^{**} : \mathcal{X}'' \to \mathcal{Y}''$, of C^* is compact. We may naturally view $\mathcal{X}$ as a closed subspace of $\mathcal{X}''$ and $\mathcal{Y}$ as a closed subspace of $\mathcal{Y}''$. So we may view $\{x_i\}_{i\in\mathbb{N}}$ as a bounded sequence in $\mathcal{X}''$. Then $\{C^{**}x_i\}_{i\in\mathbb{N}}$ has a subsequence $\{C^{**}x_{i_\ell}\}_{\ell\in\mathbb{N}}$ that converges in $\mathcal{Y}''$. For any $\eta \in \mathcal{Y}'$ and $x \in \mathcal{X}$ (we'll write X for x, when we want to think of it as an element of $\mathcal{X}''$),

$$\begin{aligned}
(C^{**}X)(\eta) &= X(C^*\eta) && \text{by the definition of “adjoint”} \\
&= (C^*\eta)(x) && \text{by the identification of } \mathcal{X} \text{ with a subset of } \mathcal{X}'' \\
&= \eta(Cx) && \text{by the definition of “adjoint”.}
\end{aligned}$$

Thus $C^{**}x \in \mathcal{Y}''$ is $Cx \in \mathcal{Y}$, viewed as an element of $\mathcal{Y}''$ and $\{Cx_{i_\ell}\}_{\ell\in\mathbb{N}}$ converges in $\mathcal{Y}$. □

It is the spectral properties of compact operators that make them act very much like matrices. Perhaps it is more appropriate to say that the spectral properties of noncompact operators are often very different from those of matrices. A simple, yet typical, example of this is given in Exercise A.3.13. We start with careful definitions of “eigenvalue” like terms. For a thorough, but still readable, treatment of the spectral theory of self–adjoint operators on Hilbert spaces; see [**RS80**].

Definition A.3.11. Let $\mathcal{X}$ be a Banach space and $B : \mathcal{X} \to \mathcal{X}$ be a linear operator defined on $\mathcal{X}$.

(a) The number $\lambda \in \mathbb{C}$ is said to be in the resolvent set of B if the operator $B - \lambda\mathrm{Id}$ is bijective (one-to-one and onto) with bounded inverse. We shall use $\rho(B)$ to denote the resolvent set of B.

(b) The number $\lambda \in \mathbb{C}$ is said to be in the spectrum of B if it is not in the resolvent set of B. We write $\sigma(B) = \mathbb{C} \setminus \rho(B)$.

(c) The number $\lambda \in \mathbb{C}$ is said to be an eigenvalue of B if there is a nonzero vector $x \in \mathcal{X}$, called an eigenvector corresponding to λ, such that $Bx = \lambda x$. The set of all eigenvalues of B is called the point spectrum of B.

Proposition A.3.12. *Let $\mathcal{X}$ be a Banach space and $B : \mathcal{X} \to \mathcal{X}$ be a linear operator defined on $\mathcal{X}$.*

(a) *If $|\lambda| > \|B\|$, then $\lambda \in \rho(B)$.*

(b) *$\rho(B)$ is an open subset of $\mathbb{C}$.*

(c) *If λ is an eigenvalue of B, then $\lambda \in \sigma(B)$.*

Proof.

(a) Since $\frac{\|B\|}{|\lambda|} < 1$, the series $-\frac{1}{\lambda}\sum_{m=0}^{\infty}\left(\frac{B}{\lambda}\right)^m$ converges in operator norm to a bounded operator R on $\mathcal{X}$. As

$$(B - \lambda \mathrm{Id})R = R(B - \lambda \mathrm{Id}) = -\sum_{m=0}^{\infty}\left(\frac{B}{\lambda}\right)^{m+1} + \sum_{m=0}^{\infty}\left(\frac{B}{\lambda}\right)^m = \mathrm{Id},$$

$R = \left(B - \lambda \mathrm{Id}\right)^{-1}$ and $\lambda \in \rho(B)$.

(b) Let $\mu \in \rho(B)$ and denote by $\left(B - \mu \mathrm{Id}\right)^{-1}$ the inverse of $B - \mu \mathrm{Id}$. By hypothesis, this inverse is a bounded operator on $\mathcal{X}$. If

$$|\lambda - \mu| < \left\|\left(B - \mu \mathrm{Id}\right)^{-1}\right\|,$$

then the series $\left(B - \mu \mathrm{Id}\right)^{-1}\sum_{m=0}^{\infty}(\mu - \lambda)^m\left(B - \mu \mathrm{Id}\right)^{-m}$ converges in operator norm to a bounded operator $\tilde{R}$ on $\mathcal{X}$. As

$$\begin{aligned}(B - \lambda \mathrm{Id})\tilde{R} &= \tilde{R}(B - \lambda \mathrm{Id}) \\ &= \tilde{R}(B - \mu \mathrm{Id}) + (\mu - \lambda)\tilde{R} \\ &= \sum_{m=0}^{\infty}(\mu - \lambda)^m\left(B - \mu \mathrm{Id}\right)^{-m} + \sum_{m=0}^{\infty}(\mu - \lambda)^{m+1}\left(B - \mu \mathrm{Id}\right)^{-m-1} \\ &= \mathrm{Id},\end{aligned}$$

$\tilde{R}$ is the operator inverse of $\left(B - \lambda \mathrm{Id}\right)$ and $\lambda \in \rho(B)$. This shows that

$$\left\{\lambda \in \mathbb{C} \mid \lambda - \mu| < \left\|\left(B - \mu \mathrm{Id}\right)^{-1}\right\|\right\} \subset \rho(B)$$

and that $\rho(B)$ is open.

(c) If λ is an eigenvalue of B, then $B - \lambda \mathrm{Id}$ has a nontrivial kernel, namely all of the eigenvectors corresponding to λ. Thus $\lambda \notin \rho(B)$. □

The next example shows that, for operators acting on infinite-dimensional spaces, even nice operators, the bulk of the spectrum need not consist of eigenvalues.

Exercise A.3.13. Let $\mathcal{H} = L^2(X,\mu)$ for some measure space $\langle X,\mu\rangle$. Let $f : X \to \mathbb{C}$ be a bounded measurable function on X. Let A be the bounded linear operator on $\mathcal{H}$ given by multiplication by $f(x)$.

(a) Prove that $\lambda \in \sigma(A)$ if and only if

$$\forall \varepsilon > 0 \quad \mu\{\, x \in X \mid |f(x) - \lambda| < \varepsilon \,\} > 0.$$

(b) Prove that λ is an eigenvalue of A if and only if

$$\mu\{\, x \in X \mid f(x) = \lambda \,\} > 0.$$

(c) Let X be the open interval $(0,1)$, μ be Lebesgue measure on $(0,1)$ and $f(x) = x$. Find the spectrum of A, the operator on $\mathcal{H}$ given by multiplication by x. Also find all of the eigenvalues of A.

We next prove that if C is a compact operator, then $\sigma(C) \setminus \{0\}$ consists only eigenvalues of finite multiplicity. If there are infinitely many different eigenvalues, they must converge to zero. We first need the following technical lemma.

Lemma A.3.14. *Let $\mathcal{X}$ be a Banach space and $B : \mathcal{X} \to \mathcal{X}$ be a compact operator. If λ is a nonzero complex number, then the range of $C - \lambda\mathrm{Id}$ is a closed linear subspace of $\mathcal{X}$.*

Proof. Denote by $\mathcal{R}$ and $\mathcal{K}$ the range and kernel, respectively, of $C - \lambda\mathrm{Id}$. Let $y \in \overline{\mathcal{R}}$ and let $\{x_n\}_{n\in\mathbb{N}}$ be a sequence in $\mathcal{X}$ such that $(C - \lambda\mathrm{Id})x_n$ converges to y. Denote by ρ_n the distance from x_n to $\mathcal{K}$. For each $n \in \mathbb{N}$, there is a $z_n \in \mathcal{K}$ such that $\rho_n \le \|x_n - z_n\| < \rho_n + \frac{1}{n}$. Then $\tilde{x}_n = x_n - z_n$ obeys

$$\lim_{n\to\infty} (C - \lambda\mathrm{Id})\tilde{x}_n = \lim_{n\to\infty} (C - \lambda\mathrm{Id})x_n = y.$$

We first consider the case that $\{\rho_n\}_{n\in\mathbb{N}}$ is bounded. Then the sequence $\{\tilde{x}_n\}_{n\in\mathbb{N}}$ is bounded, and, since C is compact, there is a subsequence $\{\tilde{x}_{n_\ell}\}_{\ell\in\mathbb{N}}$ such that $C\tilde{x}_{n_\ell}$ converges in $\mathcal{X}$, say to $\tilde{y}$. Then

$$\tilde{x}_{n_\ell} = \tfrac{1}{\lambda}\big[C\tilde{x}_{n_\ell} - (C - \lambda\mathrm{Id})\tilde{x}_{n_\ell}\big]$$

converges in $\mathcal{X}$ to $x = \frac{1}{\lambda}(\tilde{y} - y)$. Since C is bounded, $\tilde{y} = Cx$ and $y = (C - \lambda\mathrm{Id})x \in \mathcal{R}$.

Finally, we consider the case that $\{\rho_n\}_{n\in\mathbb{N}}$ is not bounded. Then, possibly restricting to a subsequence, we may assume that $\lim_{n\to\infty} \rho_n = \infty$. As the sequence $\big\{\frac{\tilde{x}_n}{\|\tilde{x}_n\|}\big\}_{n\in\mathbb{N}}$ is bounded and C is still compact, there is a subsequence $\big\{\frac{\tilde{x}_{n_\ell}}{\|\tilde{x}_{n_\ell}\|}\big\}_{\ell\in\mathbb{N}}$ such that $C\frac{\tilde{x}_{n_\ell}}{\|\tilde{x}_{n_\ell}\|}$ converges in $\mathcal{X}$, say to $\tilde{z}$. As

$$\lim_{n\to\infty} (C - \lambda\mathrm{Id})\frac{\tilde{x}_{n_\ell}}{\|\tilde{x}_{n_\ell}\|} = \frac{y}{\lim_{n\to\infty} \|\tilde{x}_{n_\ell}\|} = 0$$

we have

$$\lim_{n\to\infty} \frac{\tilde{x}_{n_\ell}}{\|\tilde{x}_{n_\ell}\|} = \frac{1}{\lambda} \lim_{n\to\infty} \left[C\frac{\tilde{x}_{n_\ell}}{\|\tilde{x}_{n_\ell}\|} - (C - \lambda\mathrm{Id})\frac{\tilde{x}_{n_\ell}}{\|\tilde{x}_{n_\ell}\|}\right] = \frac{\tilde{z}}{\lambda}$$

and hence

$$(C - \lambda\mathrm{Id})\tilde{z} = \lambda \lim_{n\to\infty} (C - \lambda\mathrm{Id})\frac{\tilde{x}_{n_\ell}}{\|\tilde{x}_{n_\ell}\|} = 0.$$

In other words, $\tilde{z} \in \mathcal{K}$. This provides a contradiction, since $\tilde{x}_n$ is a distance ρ_n from $\mathcal{K}$ so that $\frac{\tilde{x}_n}{\|\tilde{x}_n\|}$ is a distance $\frac{\rho_n}{\|\tilde{x}_n\|} \geq \frac{\rho_n}{\rho_n+1/n}$ from $\mathcal{K}$. As $\lim_{n\to\infty} \frac{\rho_n}{\rho_n+1/n} = 1$, $\frac{\tilde{x}_{n_\ell}}{\|\tilde{x}_{n_\ell}\|}$ cannot converge to a point of $\mathcal{K}$. □

Proposition A.3.15 (The Fredholm alternative). *Let $C : \mathcal{X} \to \mathcal{X}$ be a compact operator on the Banach space $\mathcal{X}$. If λ is a nonzero complex number, then either λ is an eigenvalue of C or $\lambda \in \rho(C)$.*

Proof. Suppose that λ is not an eigenvalue of C. Then, by definition, $C - \lambda\mathrm{Id}$ is one-to-one. By Lemma A.3.14, the range of $C - \lambda\mathrm{Id}$ is closed. We now claim that the range of $C - \lambda\mathrm{Id}$ is all of $\mathcal{X}$. If not, $\mathcal{X}_1 = (C - \lambda\mathrm{Id})\mathcal{X}$ is a proper closed subspace of $\mathcal{X}$. Since the restriction of C to $\mathcal{X}_1$ is still compact, $\mathcal{X}_2 = (C-\lambda\mathrm{Id})\mathcal{X}_1$ is a closed subspace of $\mathcal{X}_1$. If $\mathcal{X}_2$ were not a proper subspace of $\mathcal{X}_1$, then for each $x \in \mathcal{X} \setminus \mathcal{X}_1$, there would be a vector $x' \in \mathcal{X}_1$ with $(C - \lambda\mathrm{Id})x' = (C - \lambda\mathrm{Id})x$ and this would contradict the assumption that $C - \lambda\mathrm{Id}$ is one-to-one. Thus $\mathcal{X}_2 = (C-\lambda\mathrm{Id})\mathcal{X}_1$ is a proper closed subspace of $\mathcal{X}_1$. Continuing in this way, we can generate a sequence $\{\mathcal{X}_n\}_{n\in\mathbb{N}}$ of subspaces of $\mathcal{X}$ with $\mathcal{X}_{n+1} = (C - \lambda\mathrm{Id})\mathcal{X}_n$ and $\mathcal{X}_{n+1}$ a proper closed subspace of $\mathcal{X}_n$. By Exercise A.3.16, there is, for each $n \in \mathbb{N}$, a unit vector $x_n \in \mathcal{X}_n \setminus \mathcal{X}_{n+1}$ whose distance from $\mathcal{X}_{n+1}$ is at least $\frac{1}{2}$. If $n > m$,

$$\frac{1}{\lambda}(Cx_m - Cx_n) = x_m - \tilde{x}_m$$

with

$$\tilde{x}_m = -\frac{1}{\lambda}(C-\lambda\mathrm{Id})x_m + \frac{1}{\lambda}Cx_n = -\frac{1}{\lambda}(C-\lambda\mathrm{Id})x_m + \frac{1}{\lambda}(C-\lambda\mathrm{Id})x_n + x_n \in \mathcal{X}_{m+1}.$$

Hence $\|Cx_m - Cx_n\| \geq \frac{|\lambda|}{2}$ for all $n > m$ and $\{Cx_n\}_{n\in\mathbb{N}}$ may not contain any convergent subsequence, contradicting the compactness of C.

So $C - \lambda\mathrm{Id}$ is both one-to-one and onto. The boundedness of the inverse map is an immediate consequence of the inverse mapping theorem (part (e) of Theorem A.2.27), But it is also easy to prove boundedness directly and we do that now. If $(C - \lambda\mathrm{Id})^{-1}$ is not bounded, there is a sequence of unit vectors $x_n \in \mathcal{X}$ such that

$$\lim_{n\to\infty} \left\|(C - \lambda\mathrm{Id})x_n\right\| = 0 \implies \lim_{n\to\infty} (C - \lambda\mathrm{Id})x_n = 0.$$

Since C is compact, there is a subsequence $\{x_{n_m}\}_{m\in\mathbb{N}}$ such that Cx_{n_m} converges, say to y. But then

$$\lim_{m\to\infty} x_{n_m} = \lim_{m\to\infty} \tfrac{1}{\lambda}Cx_{n_m} - \lim_{m\to\infty} \tfrac{1}{\lambda}(C-\lambda\mathrm{Id})x_{n_m} = \tfrac{y}{\lambda}$$

and

$$Cy = \lambda C \lim_{m\to\infty} x_{n_m} = \lambda y.$$

As $\|y\| = |\lambda| \neq 0$, this contradicts the assumption that λ is not an eigenvalue of C. □

Exercise A.3.16. Let $\mathcal{X}$ be a Banach space and $\mathcal{Y}$ a proper closed subspace of $\mathcal{X}$. Let $0 < \rho < 1$. Prove that there is a unit vector $x \in \mathcal{X} \setminus \mathcal{Y}$ whose distance from $\mathcal{Y}$ is at least ρ.

Exercise A.3.17. Let $\mathcal{X}$ be an infinite-dimensional Banach space. Prove that the identity operator on $\mathcal{X}$ is not compact.

Proposition A.3.18 (The spectrum of compact operators). *Let $C : \mathcal{X} \to \mathcal{X}$ be a compact operator on the Banach space $\mathcal{X}$. The spectrum of C consists of at most countably many points. For any $\varepsilon > 0$, $\{\, \lambda \in \sigma(C) \mid |\lambda| > \varepsilon \,\}$ is finite. If $0 \neq \lambda \in \sigma(C)$, then λ is an eigenvalue of C of finite multiplicity.*

Proof. We have already proven, in Proposition A.3.15, that any nonzero number in the spectrum of C is an eigenvalue and we have also already proven, in Proposition A.3.12, that $\sigma(C) \subset \{\, \lambda \in \mathbb{C} \mid |\lambda| \leq \|C\| \,\}$. Since eigenvectors corresponding to different eigenvalues are necessarily independent, it suffices to prove that there cannot exist a sequence $\{x_n\}_{n\in\mathbb{N}}$ of independent eigenvectors of C whose corresponding eigenvalues $\{\lambda_n\}_{n\in\mathbb{N}}$ converge to $\lambda \neq 0$.

Denote by $\mathcal{X}_n$ the span of $\{x_1, x_2, \cdots, x_n\}$. By Exercise A.3.16, there is, for each $n \geq 2$, a unit vector $y_n \in \mathcal{X}_n$ whose distance from $\mathcal{X}_{n-1}$ is at least $\frac{1}{2}$. If $n > m$,

$$\tfrac{1}{\lambda_n}Cy_n - \tfrac{1}{\lambda_m}Cy_m = y_n - \tilde{y}_n$$

with

$$\tilde{y}_n = -\tfrac{1}{\lambda_n}(C - \lambda_n\mathrm{Id})y_n + \tfrac{1}{\lambda_m}Cy_m \in \mathcal{X}_{n-1}$$

since $(C - \lambda_n\mathrm{Id})\mathcal{X}_n \subset \mathcal{X}_{n-1}$ and $C\mathcal{X}_m \subset \mathcal{X}_m \subset \mathcal{X}_{n-1}$. Hence

$$\left\|\tfrac{1}{\lambda_n}Cy_n - \tfrac{1}{\lambda_m}Cy_m\right\| \geq \tfrac{1}{2} \qquad \text{for all } n > m$$

By assumption $\lim_{n\to\infty} \lambda_n = \lambda \neq 0$, so that

$$\left\|Cy_n - Cy_m\right\| \geq \tfrac{|\lambda|}{4}$$

for all $n > m$ sufficiently large. Thus $\{Cy_n\}_{n\in\mathbb{N}}$ may not contain any convergent subsequence, contradicting the compactness of C. □

Exercise A.3.19. Let $\mathcal{X}$ be an infinite-dimensional Banach space and let $C : \mathcal{X} \to \mathcal{X}$ a compact operator. Prove that $0 \in \sigma(C)$.

Exercise A.3.20. Let $\mathcal{H}$ be a separable Hilbert space and let $\{e_n\}_{n\in\mathbb{N}}$ be an orthonormal basis for $\mathcal{H}$. Let $\{\mu_n\}_{n\in\mathbb{N}}$ be any sequence of complex numbers that converges to 0. Prove that the operator defined by

$$C\Big(\sum_{n=1}^{\infty} \alpha_n e_n\Big) = \sum_{n=1}^{\infty} \mu_n \alpha_n e_{n+1}$$

is compact and has $\sigma(C) = \{0\}$.

Appendix B

The Fourier Transform and Tempered Distributions

In this appendix, we provide a summary of the most basic definitions and results concerning the Fourier transform

$$\hat{f}(\xi) = \int_{\mathbb{R}^n} e^{-i\xi\cdot x} f(x)\, dx$$

and tempered distributions. For a more extensive treatment of Fourier transforms, see, for example, [**RS75**, §IX.1] or [**Hör03a**, §7]. For a more extensive treatment of tempered distributions see, for example, [**RS80**, §V.3]. We shall use the standard multi–index notation that if $\alpha = (\alpha_1, \cdots, \alpha_n) \in \mathbb{N}_0^n$, where $\mathbb{N}_0 = \{0\} \cup \mathbb{N}$, then x^α denotes $x_1^{\alpha_1} \cdots x_n^{\alpha_n}$ and $\partial^\alpha u(x)$ denotes the partial derivative $\frac{\partial^{\alpha_1}}{\partial x_1^{\alpha_1}} \cdots \frac{\partial^{\alpha_n}}{\partial x_n^{\alpha_n}} u(x)$. The order of this partial derivative is $|\alpha| = \alpha_1 + \cdots + \alpha_n$.

B.1. Schwartz space

Definition B.1.1.

(a) Schwartz space is the vector space

$$\mathcal{S}(\mathbb{R}^n) = \left\{ u \in C^\infty(\mathbb{R}^n) \mid \sup_{x\in\mathbb{R}^n} \left|(1 + |x|^m)\partial^\alpha u(x)\right| < \infty \ \forall\, m \in \mathbb{N}_0, \alpha \in \mathbb{N}_0^n \right\}$$

of all C^∞ functions on $\mathbb{R}^n$ all of whose derivatives (including the function itself) decay faster than any polynomial at infinity.

(b) Define, for each $\alpha, \beta \in \mathbb{N}_0^n$ and each $\varphi \in \mathcal{S}(\mathbb{R}^n)$

$$\|\varphi\|_{\alpha,\beta} = \sup_{x\in\mathbb{R}^n} |x^\alpha \partial^\beta \varphi(x)|.$$

Then

(1) $\|\varphi\|_{\alpha,\beta} \geq 0$

(2) $\|a\varphi\|_{\alpha,\beta} = |a|\,\|\varphi\|_{\alpha,\beta}$

(3) $\|\varphi + \psi\|_{\alpha,\beta} \leq \|\varphi\|_{\alpha,\beta} + \|\psi\|_{\alpha,\beta}$

for all $\varphi,\ \psi \in \mathcal{S}(\mathbb{R}^n)$ and $a \in \mathbb{C}$. These are precisely the defining conditions for $\|\cdot\|_{\alpha,\beta}$ to be a semi–norm. In order for $\|\cdot\|_{\alpha,\beta}$ to be a norm it must also obey $\|\varphi\|_{\alpha,\beta} = 0 \iff \varphi = 0$. This is the case if and only if $|\beta| = 0$. If $|\beta| \neq 0$ the constant function $\varphi(x) = 1$ has $\|\varphi\|_{\alpha,\beta} = 0$.

Example B.1.2.

(a) For any polynomial $P(x)$, the function $\varphi(x) = P(x)e^{-|x|^2}$ is in Schwartz space. This is because, firstly, for any $\alpha, \beta \in \mathbb{N}_0$, $x^\alpha\partial^\beta\varphi$ is again a polynomial times $e^{-|x|^2}$ and, secondly,

$$e^{-|x|^2} = \frac{1}{e^{|x|^2}} \leq \frac{1}{1 + |x|^2 + \frac{1}{2!}|x|^4 + \cdots + \frac{1}{p!}|x|^{2p}} \tag{B.1}$$

for every $p \in \mathbb{N}$. Consequently, $x^\alpha\partial^\beta\varphi$ is bounded.

(b) If φ is $C^\infty(\mathbb{R}^n)$ and of compact support then $\varphi \in \mathcal{S}(\mathbb{R}^n)$. One such function, with $n = 1$, is

$$\varphi(x) = \begin{cases} 0 & \text{if } |x| \geq 1 \\ e^{-\frac{1}{(x-1)^2}}e^{-\frac{1}{(x+1)^2}} & \text{if } -1 < x < 1. \end{cases}$$

The heart of the proof that this function really is C^∞ at $x = \pm 1$ is the observation that, for any $p \geq 0$,

$$\lim_{y\to 0} \frac{1}{|y|^p} e^{-\frac{1}{y^2}} = 0,$$

which follows immediately from (B.1) with $x = \frac{1}{y}$.

Next, we introduce a metric on $\mathcal{S}(\mathbb{R}^n)$ which is chosen so that φ and ψ are close together if and only if $\|\varphi - \psi\|_{\alpha,\beta}$ is small for every α, β. The details are given in the following.

Theorem B.1.3. *Define $d : \mathcal{S}(\mathbb{R}^n) \times \mathcal{S}(\mathbb{R}^n) \to \mathbb{R}$ by*

$$d(\varphi,\psi) = \sum_{\alpha,\beta\in\mathbb{N}_0^n} 2^{-|\alpha|-|\beta|} \frac{\|\varphi-\psi\|_{\alpha,\beta}}{1 + \|\varphi-\psi\|_{\alpha,\beta}}.$$

Then

(a) *$d(\varphi,\psi)$ is well–defined for all $\varphi, \psi \in \mathcal{S}(\mathbb{R}^n)$ and is a metric.*

(b) *With this metric, $\mathcal{S}(\mathbb{R}^n)$ is a complete metric space.*

(c) *In this metric $\varphi = \lim_{k\to\infty} \varphi_k$ if and only if $\lim_{k\to\infty} \|\varphi_k - \varphi\|_{\alpha,\beta} = 0$ for every $\alpha, \beta \in \mathbb{N}_0^n$.*

Proof.

(a) To prove that $\sum_{\alpha,\beta\in\mathbb{N}_0^n} 2^{-|\alpha|-|\beta|}\frac{\|\varphi-\psi\|_{\alpha,\beta}}{1+\|\varphi-\psi\|_{\alpha,\beta}}$ is well–defined it suffices to observe, firstly, that $\frac{A}{1+A} \leq 1$ for every $A \geq 0$ and, secondly, that

$$\sum_{\alpha,\beta\in\mathbb{N}_0^n}^{\infty} 2^{-|\alpha|-|\beta|} = \Big[\sum_{\ell=0}^{\infty} 2^{-\ell}\Big]^{2n}$$

converges because the geometric series converges.

- The metric axiom $d(\varphi, \psi) \geq 0$ is obvious.
- The metric axiom that $d(\varphi, \psi) = 0 \implies \varphi = \psi$ is obvious because $d(\varphi, \psi) = 0$ forces the $\alpha = \beta = 0$ term in its definition, namely $\frac{\|\varphi-\psi\|_{0,0}}{1+\|\varphi-\psi\|_{0,0}}$, to vanish. And that first term is zero if and only if its numerator $\|\varphi - \psi\|_{0,0} = \sup_{x\in\mathbb{R}^n} |\varphi(x) - \psi(x)|$ is zero.
- The metric axiom $d(\varphi, \psi) = d(\psi, \varphi)$ is obvious.
- The triangle inequality follows from

$$\frac{\|\varphi - \psi\|_{\alpha,\beta}}{1 + \|\varphi - \psi\|_{\alpha,\beta}} \leq \frac{\|\varphi - \zeta\|_{\alpha,\beta}}{1 + \|\varphi - \zeta\|_{\alpha,\beta}} + \frac{\|\zeta - \psi\|_{\alpha,\beta}}{1 + \|\zeta - \psi\|_{\alpha,\beta}},$$

which is proven as follows. We suppress the subscripts α, β. Because $\frac{x}{1+x} = 1 - \frac{1}{1+x}$ is an increasing function of x,

$$\begin{aligned}\frac{\|\varphi-\psi\|}{1+\|\varphi-\psi\|} &\leq \frac{\|\varphi-\zeta\|+\|\zeta-\psi\|}{1+\|\varphi-\zeta\|+\|\zeta-\psi\|} \\ &= \frac{\|\varphi-\zeta\|}{1+\|\varphi-\zeta\|+\|\zeta-\psi\|} + \frac{\|\zeta-\psi\|}{1+\|\varphi-\zeta\|+\|\zeta-\psi\|} \\ &\leq \frac{\|\varphi-\zeta\|}{1+\|\varphi-\zeta\|} + \frac{\|\zeta-\psi\|}{1+\|\zeta-\psi\|}.\end{aligned}$$

(b) For the "only if" part, assume that $\varphi = \lim_{k\to\infty} \varphi_k$ and let $\alpha, \beta \in \mathbb{N}_0$. Then

$$d(\varphi, \varphi_k) \geq 2^{-|\alpha|-|\beta|}\frac{\|\varphi - \varphi_k\|_{\alpha,\beta}}{1 + \|\varphi - \varphi_k\|_{\alpha,\beta}} \implies \lim_{k\to\infty} \frac{\|\varphi - \varphi_k\|_{\alpha,\beta}}{1 + \|\varphi - \varphi_k\|_{\alpha,\beta}} = 0.$$

For any $0 < \varepsilon < \frac{1}{2}$ and $x > 0$,

$$\frac{x}{1+x} < \varepsilon \implies x < \varepsilon(1 + x) \implies x - \varepsilon x < \varepsilon \implies x < \frac{\varepsilon}{1-\varepsilon} < 2\varepsilon.$$

Hence $\lim_{k\to 0} \|\varphi - \varphi_k\|_{\alpha,\beta} = 0$ too.

For the "if" part assume that $\lim_{k\to\infty} \|\varphi_k - \varphi\|_{\alpha,\beta} = 0$ for every $\alpha, \beta \in \mathbb{N}_0$. We must prove that, as a consequence, $\varphi = \lim_{k\to\infty} \varphi_k$. The idea is that, in the definition of $d(\varphi, \psi)$, the sum of all terms with $|\alpha|$ or $|\beta|$ large is small, regardless of what φ and ψ are. Precisely, write $\psi_k = \varphi - \varphi_k$ and note that, for every $M \in \mathbb{N}$,

$$\begin{aligned}
d(\varphi_k, \varphi) &= \sum_{\alpha,\beta\in\mathbb{N}_0^n} 2^{-|\alpha|-|\beta|} \frac{\|\psi_k\|_{\alpha,\beta}}{1+\|\psi_k\|_{\alpha,\beta}} \\
&= \sum_{\substack{\alpha,\beta\in\mathbb{N}_0^n \\ |\alpha|,|\beta|\le Mn}} 2^{-|\alpha|-|\beta|} \frac{\|\psi_k\|_{\alpha,\beta}}{1+\|\psi_k\|_{\alpha,\beta}} + \sum_{\substack{\alpha,\beta\in\mathbb{N}_0^n \\ |\alpha| \text{ or } |\beta|>Mn}}^{\infty} 2^{-|\alpha|-|\beta|} \frac{\|\psi_k\|_{\alpha,\beta}}{1+\|\psi_k\|_{\alpha,\beta}} \\
&\le \sum_{\substack{\alpha,\beta\in\mathbb{N}_0^n \\ |\alpha|,|\beta|\le Mn}} 2^{-|\alpha|-|\beta|} \frac{\|\psi_k\|_{\alpha,\beta}}{1+\|\psi_k\|_{\alpha,\beta}} + \sum_{\substack{\alpha,\beta\in\mathbb{N}_0^n \\ |\alpha| \text{ or } |\beta|>Mn}}^{\infty} 2^{-|\alpha|-|\beta|} \\
&\le \sum_{\substack{\alpha,\beta\in\mathbb{N}_0^n \\ |\alpha|,|\beta|\le Mn}} 2^{-|\alpha|-|\beta|} \frac{\|\psi_k\|_{\alpha,\beta}}{1+\|\psi_k\|_{\alpha,\beta}} + 2n\Big\{\sum_{m=M+1}^{\infty} 2^{-m}\Big\}\Big\{\sum_{m=0}^{\infty} 2^{-m}\Big\}^{2n-1} \\
&= \sum_{\substack{\alpha,\beta\in\mathbb{N}_0^n \\ |\alpha|,|\beta|\le Mn}} 2^{-|\alpha|-|\beta|} \frac{\|\psi_k\|_{\alpha,\beta}}{1+\|\psi_k\|_{\alpha,\beta}} + 2n\Big\{\frac{1}{2^M}\Big\}\{2\}^{2n-1}.
\end{aligned}$$

Let $\varepsilon > 0$ and choose M so that $\frac{1}{2^M} \le \frac{\varepsilon}{4n2^{2n-1}}$ and hence

$$2n\Big\{\frac{1}{2^M}\Big\}\{2\}^{2n-1} \le \frac{\varepsilon}{2}.$$

For each $\alpha, \beta \in \mathbb{N}_0$, $\lim_{k\to\infty} \|\psi_k\|_{\alpha,\beta} = 0$ so that there is a $K_{\alpha,\beta}$ for which $k \ge K_{\alpha,\beta}$ implies $\|\psi_k\|_{\alpha,\beta} < \frac{\varepsilon}{2^{2n+1}}$. Set

$$K = \max\big\{ K_{\alpha,\beta} \mid \alpha, \beta \in \mathbb{N}_0,\ |\alpha|, |\beta| \le M \big\}$$

If $k \ge K$, then

$$\begin{aligned}
d(\varphi_k, \varphi) &\le \sum_{\substack{\alpha,\beta\in\mathbb{N}_0 \\ |\alpha|,|\beta|\le Mn}} 2^{-|\alpha|-|\beta|} \frac{\|\psi_k\|_{\alpha,\beta}}{1+\|\psi_k\|_{\alpha,\beta}} + 2n\Big\{\frac{1}{2^M}\Big\}\{2\}^{2n-1} \\
&< \frac{\varepsilon}{2} + \sum_{\alpha,\beta\in\mathbb{N}_0} 2^{-|\alpha|-|\beta|} \frac{\varepsilon}{2^{2n+1}} \\
&= \varepsilon.
\end{aligned}$$

(c) Let $\{\varphi_k\}$ be a Cauchy sequence with respect to the metric d. Then, as in part (c), for each $\alpha, \beta \in \mathbb{N}_0$,

$$\lim_{k,k'\to\infty} \big\|\varphi_k - \varphi_{k'}\big\|_{\alpha,\beta} = 0.$$

In particular,

$$\lim_{k,k'\to\infty} \big\|\varphi_k - \varphi_{k'}\big\|_{0,0} = 0,$$

so that the sequence $\{\varphi_k\}$ is Cauchy in the set, $C(\mathbb{R}^n)$, of all bounded, continuous functions on $\mathbb{R}^n$ equipped with the uniform metric. Since $C(\mathbb{R}^n)$ is complete, there exists a continuous function φ such that $\{\varphi_k\}$ converges uniformly to φ. As well, for each $\beta \in \mathbb{N}_0$,

$$\lim_{k,k'\to\infty} \left\|\varphi_k - \varphi_{k'}\right\|_{0,\beta} = 0$$

so that the sequence $\{\partial^\beta\varphi_k\}$ of βth derivatives is Cauchy in $C(\mathbb{R}^n)$ and there exists a continuous function φ_β such that $\{\partial^\beta\varphi_k\}$ converges uniformly to φ_β. This ensures that φ is C^∞ with $\partial^\beta\varphi = \varphi_\beta$ for each $\beta \in \mathbb{N}_0$. Finally, we have that, for each $\alpha, \beta \in \mathbb{N}_0$, there is a $K_{\alpha,\beta}$ such that $|x^\alpha\partial^\beta\varphi_k(x) - x^\alpha\partial^\beta\varphi_{k'}(x)| < \varepsilon$ for all $k, k' \geq K_{\alpha,\beta}$ and all $x \in \mathbb{R}^n$. Consequently, if $k \geq K_{\alpha,\beta}$,

$$\begin{aligned}\left\|\varphi_k - \varphi\right\|_{\alpha,\beta} &= \sup_{x\in\mathbb{R}^n} |x^\alpha(\partial^\beta\varphi_k(x) - \partial^\beta\varphi(x))| \\ &= \sup_{x\in\mathbb{R}^n} \lim_{k'\to\infty} |x^\alpha||\partial^\beta\varphi_k(x) - \partial^\beta\varphi_{k'}(x)| \\ &\leq \sup_{x\in\mathbb{R}^n} \varepsilon = \varepsilon.\end{aligned}$$

So, by part (c), $\{\varphi_k\}$ converges to φ with respect to the metric d. □

Lemma B.1.4. *Let $\varphi, \psi \in \mathcal{S}(\mathbb{R}^n)$. Then*

(a) $a\varphi + b\psi \in \mathcal{S}(\mathbb{R}^n)$ *for all* $a, b \in \mathbb{C}$,

(b) $\partial^\gamma\varphi \in \mathcal{S}(\mathbb{R}^n)$ *for all* $\gamma \in \mathbb{N}_0^n$,

(c) $\varphi\zeta \in \mathcal{S}(\mathbb{R}^n)$ *for all C^∞ functions ζ that are polynomially bounded and have polynomially bounded derivatives, and*

(d) *the convolution* $(\varphi * \psi)(x) = \int_{\mathbb{R}^n} \varphi(y)\psi(x-y)\,dy$ *is in* $\mathcal{S}(\mathbb{R}^n)$.

Proof. These are all pretty obvious. Parts (a) and (b) are immediate consequences of the bounds

$$\begin{aligned}||a\varphi + b\psi||_{\alpha,\beta} &\leq |a|\,||\varphi||_{\alpha,\beta} + |\beta|\,||\psi||_{\alpha,\beta}, \\ ||\partial^\gamma\varphi||_{\alpha,\beta} &= ||\varphi||_{\alpha,\beta+\gamma},\end{aligned}$$

which are true for all $\alpha, \beta, \gamma \in \mathbb{N}_0$ and $a, b \in \mathbb{C}$.

For part (c), let $\alpha, \beta \in \mathbb{N}_0^n$. By hypothesis, there is an $L \in \mathbb{N}$ such that $\Big[\sum_{\substack{\alpha'\in\mathbb{N}_0^n\\|\alpha'|\leq L}} |x^{\alpha'}|\Big]^{-1}\partial^{\beta'}\zeta$ is uniformly bounded for all $\beta' \leq \beta$. (Here, $\beta' \leq \beta$ means that $\beta'_j \leq \beta_j$ for each $1 \leq j \leq n$. Also recall that, when α' is the zero vector, $x^{\alpha'} = 1$.) By the product rule

$$\partial^\beta(\varphi\zeta) = \sum_{\substack{\beta'\in\mathbb{N}_0^n\\ \beta'\leq\beta}} \binom{\beta}{\beta'}\,\partial^{\beta-\beta'}\varphi\,\partial^{\beta'}\zeta,$$

where $\binom{\beta}{\beta'} = \prod_{j=1}^{n} \frac{\beta_j!}{\beta'_j!(\beta-\beta')_j!}$, and part (c) follows from

$$\|\varphi\zeta\|_{\alpha,\beta} \le \sum_{\substack{\alpha',\beta'\in\mathbb{N}_0^n \\ |\alpha'|\le L,\ \beta'\le\beta}} \binom{\beta}{\beta'} \Big\| \Big[\sum_{\substack{\alpha''\in\mathbb{N}_0^n \\ |\alpha''|\le L}} |x^{\alpha''}| \Big]^{-1} \partial^{\beta'}\zeta \Big\|_{L^\infty(\mathbb{R}^n)} \|\varphi\|_{\alpha+\alpha',\beta-\beta'}.$$

The proof of part (d) is similar to that of part (c) but uses that

- the function $\Big[\sum_{\substack{\alpha'\in\mathbb{N}_0^n \\ |\alpha'|\le n+1}} |y^{\alpha'}| \Big]^{-1} \in L^1(\mathbb{R}^n)$
- $|x^\alpha| \le \sum_{\substack{\alpha'\in\mathbb{N}_0^n \\ \alpha'\le\alpha}} \binom{\alpha}{\alpha'} |y^{\alpha'}| |(x-y)^{\alpha-\alpha'}|$
- All derivatives of $\varphi(y)\psi(x-y)$ with respect to x are absolutely integrable with respect to y, so that we are allowed to move derivatives with respect to x inside the integral $\int_{\mathbb{R}^n} \varphi(y)\psi(x-y)\,dy$. □

B.2. The Fourier transform

Definition B.2.1. The Fourier transform $\hat f(\xi)$ of a function $f \in \mathcal{S}(\mathbb{R}^n)$ is defined by

$$\hat f(\xi) = \mathcal{F}f(\xi) = \int_{\mathbb{R}^n} e^{-i\xi\cdot x} f(x)\,dx. \tag{B.2a}$$

Since $f(x)$, and hence $e^{-i\xi\cdot x}f(x)$, is a continuous function of x which is bounded by a constant times $\frac{1}{1+|x|^{2n}}$, the integral exists and $\hat f(\xi)$ is a well–defined complex number for each $\xi \in \mathbb{R}^n$. We shall show in Theorem B.2.5 that the map $f \mapsto \hat f$ is a continuous, linear map from $\mathcal{S}(\mathbb{R}^n)$ to $\mathcal{S}(\mathbb{R}^n)$ and furthermore that this map is one-to-one and onto with the inverse map being the inverse Fourier transform given by

$$\check g(x) = \mathcal{F}^{-1}\{g\}(\xi) = \int_{\mathbb{R}^n} e^{i\xi\cdot x} g(\xi)\,\frac{d\xi}{(2\pi)^n}. \tag{B.2b}$$

The computational properties of the Fourier transform are given in the following theorem.

Theorem B.2.2. *Let $f, g \in \mathcal{S}(\mathbb{R}^n)$ and $a, b \in \mathbb{C}$. Then*

(a) *The Fourier transform of $af(x) + bg(x)$ is $a\hat f(\xi) + b\hat g(\xi)$.*

(b) *If $\beta \in \mathbb{N}_0$, then the Fourier transform of $\partial^\beta f(x)$ is $i^{|\beta|}\xi^\beta \hat f(\xi)$.*

(c) *The Fourier transform, $\hat f(\xi)$, of $f(x)$ is infinitely differentiable and, for each $\beta \in \mathbb{N}_0$, $\frac{\partial^\beta}{\partial\xi^\beta}\hat f(\xi)$ is the Fourier transform of $(-i)^{|\beta|}x^\beta f(x)$.*

(d) *Let $a \in \mathbb{R}^n$. The Fourier transform of the translated function $(T_a f)(x) = f(x-a)$ is $e^{-ia\cdot\xi}\hat f(\xi)$.*

(e) *The Fourier transform of* $f(x) = e^{-|x|^2/2}$ *is* $\hat{f}(\xi) = \left(2\pi\right)^{n/2} e^{-|\xi|^2/2}$.

(f) $\int_{\mathbb{R}^n} f(x)\,\overline{g(x)}\,dx = \int_{\mathbb{R}^n} \hat{f}(\xi)\,\overline{\hat{g}(\xi)}\,\frac{d\xi}{(2\pi)^n}$.

(g) *The Fourier transform of the convolution* $h = f * g$ *is* $\hat{h}(\xi) = \hat{f}(\xi)\hat{g}(\xi)$.

Proof.

(a) is obvious.

(b) By induction, it suffices to prove the case $|\beta| = 1$. We do so for $n = 1$. By integration by parts, the Fourier transform of the first derivative $f'(x)$ is

$$\int_{-\infty}^{\infty} e^{-i\xi x} f'(x)\,dx = -\int_{-\infty}^{\infty} f(x)\left(\tfrac{d}{dx} e^{-i\xi x}\right) dx = i\xi \int_{-\infty}^{\infty} e^{-i\xi x} f(x)\,dx$$
$$= i\xi \hat{f}(\xi).$$

The boundary terms vanished because

$$\lim_{x\to\infty} e^{-i\xi x} f(x) = \lim_{x\to-\infty} e^{-i\xi x} f(x) = 0.$$

(c) Again, by induction, it suffices to prove the case $|\beta| = 1$. Again, we do so for $n = 1$.

$$\tfrac{d}{d\xi}\hat{f}(\xi) = \tfrac{d}{d\xi}\int_{-\infty}^{\infty} e^{-i\xi x} f(x)\,dx = \int_{-\infty}^{\infty} \tfrac{\partial}{\partial \xi}\big(e^{-i\xi x} f(x)\big)\,dx$$
$$= \int_{-\infty}^{\infty} (-ix) e^{-i\xi x} f(x)\,dx$$

is indeed $-i$ times the Fourier transform of $xf(x)$. The second equality, in which the derivative with respect to ξ was moved past the integral sign is justified by Exercise B.2.3.

(d) is obvious—just make the change of variables $x' = x - a$ in the integral defining the Fourier transform of $T_a f$.

(e) Since the integral defining $\hat{f}(\xi)$ factorizes, it suffices to condsider $n = 1$. By part (c) of this Theorem, $\frac{d}{d\xi}\hat{f}(\xi)$ is the Fourier transform of $-ixf(x) = -ixe^{-x^2/2} = i\frac{d}{dx}e^{-x^2/2} = if'(x)$. Thus by parts (a) and (b) of this Theorem, $\frac{d}{d\xi}\hat{f}(\xi) = -\xi\hat{f}(\xi)$ and

$$\tfrac{d}{d\xi}\left\{\hat{f}(\xi)e^{\xi^2/2}\right\} = e^{\xi^2/2}\left\{\tfrac{d}{d\xi}\hat{f}(\xi) + \xi\hat{f}(\xi)\right\} = 0$$

for all $\xi \in \mathbb{R}$. Consequently $\hat{f}(\xi)e^{\xi^2/2}$ must be some constant, independent of ξ. Hence to determine $\hat{f}(\xi)$ we need only to determine the value of that constant, which we may do by computing $\hat{f}(\xi)e^{\xi^2/2}|_{\xi=0} = \hat{f}(0)$. Since $\hat{f}(0) = \int_{-\infty}^{\infty} e^{-x^2/2}\,dx > 0$, it is determined by

$$\begin{aligned}\hat{f}(0)^2 &= \Big[\int_{-\infty}^{\infty} e^{-x^2/2}\,dx\Big]^2\\ &= \Big[\int_{-\infty}^{\infty} e^{-x^2/2}\,dx\Big]\Big[\int_{-\infty}^{\infty} e^{-y^2/2}\,dy\Big]\\ &= \iint_{\mathbb{R}^2} e^{-(x^2+y^2)/2}\,dx\,dy.\end{aligned}$$

Changing to polar coordinates,

$$\hat{f}(0)^2 = \int_0^{\infty}\int_0^{2\pi} re^{-r^2/2}\,dr = 2\pi\int_0^{\infty} re^{-r^2/2}\,dr = 2\pi\Big[-e^{-r^2/2}\Big]_0^{\infty} = 2\pi.$$

Thus $\hat{f}(0) = \sqrt{2\pi}$, which tells us that $\hat{f}(\xi)e^{\xi^2/2} = \sqrt{2\pi}$ and hence that $\hat{f}(\xi) = \sqrt{2\pi}e^{-\xi^2/2}$ for all ξ.

(f) By the definition in (B.2a),

$$\begin{aligned}\int \hat{f}(\xi)\,\overline{\hat{g}(\xi)}\,\frac{d\xi}{(2\pi)^n} &= \int\Big[\int e^{-i\xi\cdot x} f(x)\,dx\Big]\,\overline{\hat{g}(\xi)}\,\frac{d\xi}{(2\pi)^n}\\ &= \iint e^{-i\xi\cdot x} f(x)\,\overline{\hat{g}(\xi)}\,\frac{d\xi}{(2\pi)^n}\,dx\\ &= \int f(x)\Big[\overline{\int e^{i\xi\cdot x}\hat{g}(\xi)\,\frac{d\xi}{(2\pi)^n}}\Big]dx\\ &= \int f(x)\,\overline{g(x)}\,dx.\end{aligned}$$

The last equality uses Theorem B.2.5.

(g) By the definition in (B.2a) and Lemma B.1.4,

$$\begin{aligned}\hat{h}(\xi) &= \int_{\mathbb{R}^n} e^{-i\xi\cdot x} h(x)\,dx\\ &= \iint e^{-i\xi\cdot y} f(y)\, e^{-i\xi\cdot(x-y)} g(x-y)\,dy\,dx\\ &= \int_{\mathbb{R}^n} e^{-i\xi\cdot y} f(y)\,dy \int_{\mathbb{R}^n} e^{-i\xi\cdot(x-y)} g(x-y)\,dx \qquad \text{by Fubini}\\ &= \int_{\mathbb{R}^n} e^{-i\xi\cdot y} f(y)\,dy \int_{\mathbb{R}^n} e^{-i\xi\cdot x'} g(x')\,dx' \qquad \text{with } x' = x-y\\ &= \hat{f}(\xi)\,\hat{g}(\xi).\end{aligned}$$

□

Exercise B.2.3. Let $f : (-\infty, \infty) \times [c, d] \to \mathbb{C}$ be continuous. Assume that $\frac{\partial f}{\partial y}$ exists and is continuous and that there is a constant C such that

$$|f(x, y)|, |\tfrac{\partial f}{\partial y}(x, y)| \le \tfrac{C}{1+x^2} \qquad \text{and} \qquad |\tfrac{\partial f}{\partial y}(x, y) - \tfrac{\partial f}{\partial y}(x, y')| \le C\tfrac{|y-y'|}{1+x^2}$$

for all $-\infty < x < \infty$ and $c \le y, y' \le d$. Prove that $g(y) = \int_{-\infty}^{\infty} f(x, y)\, dx$ is differentiable with $g'(y) = \int_{-\infty}^{\infty} \frac{\partial f}{\partial y}(x, y)\, dx$.

Exercise B.2.4. Let $f \in \mathcal{S}(\mathbb{R}^n)$. Set

$$r(x) = f(-x) \qquad c(x) = \overline{f(x)}.$$

Prove that

$$\hat{r}(\xi) = \overline{\hat{c}(\xi)} \qquad \hat{c}(\xi) = \overline{\hat{f}(-\xi)}.$$

Theorem B.2.5. *The maps*

$$f(x) \in \mathcal{S}(\mathbb{R}^n) \mapsto \hat{f}(\xi) = \int_{\mathbb{R}^n} e^{-i\xi\cdot x} f(x)\, dx,$$

$$g(\xi) \in \mathcal{S}(\mathbb{R}^n) \mapsto \check{g}(x) = \int_{\mathbb{R}^n} e^{i\xi\cdot x} g(\xi)\, \frac{d\xi}{(2\pi)^n}$$

are one-to-one, continuous, linear maps from $\mathcal{S}(\mathbb{R}^n)$ onto $\mathcal{S}(\mathbb{R}^n)$ and are inverses of each other.

Proof. That $\hat{f}$ is linear in f was Theorem B.2.2.

We now assume that $f \in \mathcal{S}(\mathbb{R}^n)$ and prove that $\hat{f}(\xi) \in \mathcal{S}(\mathbb{R}^n)$. For simplicity of notation, we assume that $n = 1$. Let α, β be nonnegative integers. By parts (b) and (c) of Theorem B.2.2 followed by the product rule, $\xi^\alpha \frac{d^\beta}{d\xi^\beta}\hat{f}(\xi)$ is the Fourier transform of

$$\begin{aligned}(-i)^\alpha \tfrac{d^\alpha}{dx^\alpha}\big[(-ix)^\beta f(x)\big] &= (-i)^{\alpha+\beta} \sum_{\ell=0}^{\min\{\alpha,\beta\}} \tbinom{\alpha}{\ell}\big(\tfrac{d^\ell}{dx^\ell} x^\beta\big)\big(\tfrac{d^{\alpha-\ell}}{dx^{\alpha-\ell}} f(x)\big) \\ &= (-i)^{\alpha+\beta} \sum_{\ell=0}^{\min\{\alpha,\beta\}} \tbinom{\alpha}{\ell} \tfrac{\beta!}{(\beta-\ell)!} x^{\beta-\ell} f^{(\alpha-\ell)}(x).\end{aligned}$$

Hence

(B.3)

$$
\begin{aligned}
\|\hat f(\xi)\|_{\alpha,\beta} &= \sup_{\xi\in\mathbb{R}} |\xi^\alpha \tfrac{d^\beta}{d\xi^\beta}\hat f(\xi)| \\
&= \sup_{\xi\in\mathbb{R}} \Big| \int_{-\infty}^{\infty} e^{-i\xi x}\Big[\sum_{\ell=0}^{\min\{\alpha,\beta\}} \tbinom{\alpha}{\ell}\tfrac{\beta!}{(\beta-\ell)!} x^{\beta-\ell} f^{(\alpha-\ell)}(x)\Big]dx\Big| \\
&\le \sum_{\ell=0}^{\min\{\alpha,\beta\}} \tbinom{\alpha}{\ell}\tfrac{\beta!}{(\beta-\ell)!} \int_{-\infty}^{\infty} |x^{\beta-\ell} f^{(\alpha-\ell)}(x)|dx \\
&= \sum_{\ell=0}^{\min\{\alpha,\beta\}} \tbinom{\alpha}{\ell}\tfrac{\beta!}{(\beta-\ell)!} \int_{-\infty}^{\infty} \tfrac{1}{1+x^2}\big\{|x|^{\beta-\ell} + |x|^{\beta-\ell+2}\big\}|f^{(\alpha-\ell)}(x)|dx \\
&\le \sum_{\ell=0}^{\min\{\alpha,\beta\}} \tbinom{\alpha}{\ell}\tfrac{\beta!}{(\beta-\ell)!} \big\{\|f\|_{\beta-\ell,\alpha-\ell} + \|f\|_{\beta-\ell+2,\alpha-\ell}\big\} \int_{-\infty}^{\infty} \tfrac{1}{1+x^2}\,dx \\
&= \sum_{\ell=0}^{\min\{\alpha,\beta\}} \pi\tbinom{\alpha}{\ell}\tfrac{\beta!}{(\beta-\ell)!} \big\{\|f\|_{\beta-\ell,\alpha-\ell} + \|f\|_{\beta-\ell+2,\alpha-\ell}\big\}.
\end{aligned}
$$

Since $f \in \mathcal{S}(\mathbb{R})$, the right-hand side is finite. The corresponding argument for general n proves that, when $f \in \mathcal{S}(\mathbb{R}^n)$, $\|\hat f\|_{\alpha,\beta}$ is finite for all $\alpha, \beta \in \mathbb{N}_0^n$, so that $\hat f \in \mathcal{S}(\mathbb{R}^n)$.

It also proves that the map $f \mapsto \hat f$ is continuous, since if the sequence $\{f_j\}_{j\in\mathbb{N}}$ converges to f in $\mathcal{S}(\mathbb{R}^n)$, then replacing f by $f - f_j$ in (B.3), or its analog for general n, shows that $\|\hat f - \hat f_j\|_{\alpha,\beta}$ converges to zero as $j \to \infty$, for all $\alpha, \beta \in \mathbb{N}_0^n$. So $\{\hat f_j\}_{j\in\mathbb{N}}$ converges to $\hat f$ in $\mathcal{S}(\mathbb{R})$ too.

The proof that the map $g(\xi) \mapsto \check g(x)$ is a continuous, linear map from $\mathcal{S}(\mathbb{R}^n)$ into $\mathcal{S}(\mathbb{R}^n)$ is similar.

We now assume that $f(x) \in \mathcal{S}(\mathbb{R}^n)$ and prove that the inverse Fourier transform of $\hat f(\xi)$ is $f(x)$. In symbols, we prove that

$$f(x) = \int_{\mathbb{R}^n} e^{i\xi\cdot x}\hat f(\xi)\,\frac{d\xi}{(2\pi)^n}. \tag{B.4}$$

We first prove the $(x = 0, n = 1)$ special case that

$$f(0) = \int_{-\infty}^{\infty} \hat f(\xi)\,\frac{d\xi}{2\pi}. \tag{B.5}$$

Write

$$f(x) = f(0)e^{-x^2/2} + xh(x) \quad \text{where } h(x) = \begin{cases} \frac{1}{x}\big(f(x) - f(0)e^{-x^2/2}\big) & \text{if } x \neq 0, \\ f'(0) & \text{if } x = 0. \end{cases}$$

By Exercise B.2.6, the function $h \in \mathcal{S}(\mathbb{R})$. So, by parts (e) and (c) of Theorem B.2.2,

$$\hat{f}(\xi) = \sqrt{2\pi} f(0) e^{-\xi^2/2} + i\tfrac{d}{d\xi}\hat{h}(\xi)$$

and

$$\tfrac{1}{2\pi}\int_{-\infty}^{\infty} \hat{f}(\xi)\,d\xi = \tfrac{f(0)}{\sqrt{2\pi}}\int_{-\infty}^{\infty} e^{-\xi^2/2}\,d\xi + \tfrac{i}{2\pi}\int_{-\infty}^{\infty} \tfrac{d}{d\xi}\hat{h}(\xi)\,d\xi.$$

The first term satisfies

$$\tfrac{f(0)}{\sqrt{2\pi}}\int_{-\infty}^{\infty} e^{-\xi^2/2}\,d\xi = f(0)$$

by the computation in the proof of Theorem B.2.2. The second term is $\frac{i}{2\pi}$ times

$$\int_{-\infty}^{\infty} \tfrac{d}{d\xi}\hat{h}(\xi)\,d\xi = \lim_{A,B\to\infty}\int_{-A}^{B} \tfrac{d}{d\xi}\hat{h}(\xi)\,d\xi = \lim_{A,B\to\infty}\big[\hat{h}(B) - \hat{h}(-A)\big] = 0.$$

Here we have used the fundamental theorem of calculus and the decay at $\pm\infty$ which follows from the fact that $\hat{h} \in \mathcal{S}(\mathbb{R})$, which, in turn, follows from $h \in \mathcal{S}(\mathbb{R})$. This completes the proof of (B.5). The proof of the analog of (B.5) for general n is similar. Replacing f by $T_{-x}f$ and using $f(x) = \big(T_{-x}f\big)(0)$ and $\widehat{T_{-x}f}(\xi) = e^{i\xi\cdot x}\hat{f}(\xi)$ gives (B.4).

The proof that

$$g(\xi) = \int_{\mathbb{R}^n} e^{-i\xi\cdot x}\check{g}(x)\,dx \tag{B.6}$$

is similar. The formulas (B.4) and (B.6) show that the maps $f(x) \mapsto \hat{f}(\xi)$ and $g(\xi) \mapsto \check{g}(x)$ are onto $\mathcal{S}(\mathbb{R})$ and are inverses of each other. □

Exercise B.2.6. Let $f \in \mathcal{S}(\mathbb{R})$ and define

$$h(x) = \begin{cases} \frac{1}{x}\big(f(x) - f(0)e^{-x^2/2}\big) & \text{if } x \neq 0 \\ f'(0) & \text{if } x = 0. \end{cases}$$

Prove that $h \in \mathcal{S}(\mathbb{R})$.

Theorem B.2.7 (Plancherel theorem). *The Fourier transform* (B.2a) *has a unique continuous extension to* $L^2(\mathbb{R}^n)$. *The inverse Fourier transform* (B.2b) *has unique continuous extension to* $L^2(\mathbb{R}^n)$. *The two extensions are inverses of each other.*

For $f, g \in L^2(\mathbb{R}^n)$ *one has the Plancherel (or Parseval) identities*

$$\int_{\mathbb{R}^n} f(x)\overline{g(x)}\,dx = \int_{\mathbb{R}^n} f(\xi)\overline{g(\xi)}\,\frac{d\xi}{(2\pi)^n},$$

$$\int_{\mathbb{R}^n} |f(x)|^2\,dx = \int_{\mathbb{R}^n} |\hat{f}(\xi)|^2\,\frac{d\xi}{(2\pi)^n}.$$

Proof. This is an immediate consequence of the B.L.T. theorem (Theorem A.2.13), Theorem B.2.5, Theorem B.2.2 (which implies that the Fourier and inverse Fourier transforms are bounded operators with respect to the $L^2(\mathbb{R}^n)$ norm) and a simple extension of Exercise A.1.14 (which implies that $\mathcal{S}(\mathbb{R}^n)$ is dense in $L^2(\mathbb{R}^n)$). □

Lemma B.2.8 (Riemann–Lebesgue lemma). *The Fourier transform* (B.2a) *extends uniquely to a bounded map from $L^1(\mathbb{R}^n)$ to $C_0(\mathbb{R}^n)$, the space of continuous functions on $\mathbb{R}^n$ that vanish at infinity.*

Proof. By Theorem B.2.5, the Fourier transform maps $\mathcal{S}(\mathbb{R}^n)$, which is dense in $L^1(\mathbb{R}^n)$ (by a simple extension of Exercise A.1.14), into $\mathcal{S}(\mathbb{R}^n) \subset C_0(\mathbb{R}^n)$. It now suffices to observe that

$$\|\hat{f}\|_{L^\infty(\mathbb{R}^n)} \leq \|f\|_{L^1(\mathbb{R}^n)}$$

and apply the B.L.T. theorem, Theorem A.2.13. □

Theorem B.2.9 (Paley–Wiener theorem). *Let f be a Schwartz function in $\mathbb{R}^n$. The following are equivalent.*

(a) *f is supported in the closed ball $\bar{B}_R = \{ x \in \mathbb{R}^n \mid |x| \leq R \}$.*

(b) *The Fourier transform $\hat{f}$ extends to a holomorphic function on $\mathbb{C}^n$, and for any N there is C_N such that*

$$|\hat{f}(\xi)| \leq \tfrac{C_N}{1+|\xi|^{2N}} e^{R|\operatorname{Im}\xi|}. \tag{B.7}$$

Exercise B.2.10. Prove Theorem B.2.9 by completing the following outline.

(a) Let $f \in C_c^\infty(\mathbb{R}^n)$ be supported in $\bar{B}_R$. Prove that $\hat{f}(\xi)$ extends to a holomorphic function on $\mathbb{C}^n$ and that, for each $N \in \mathbb{N}$, there is a constant C_N such that (B.7) holds.

(b) Assume that the Fourier transform $\hat{f}(\xi)$ of a function $f(x)$ extends to a holomorphic function on $\mathbb{C}^n$ and that, for each $N \in \mathbb{N}$, there is a constant C_N such that (B.7) holds. Let $\eta \in \mathbb{R}^n$. Prove that

$$f(x) = e^{-\eta\cdot x} \int e^{i\xi\cdot x} \hat{f}(\xi + i\eta)\, \tfrac{d\xi}{(2\pi)^n}.$$

(c) Prove that, under the hypotheses of part (b), $f(x)$ is supported in $\bar{B}_R$.

B.3. Tempered distributions

The theory of tempered distributions allows us to give a rigorous meaning to the Dirac delta function. It is "defined", on a handwaving level, by the properties that

(1) $\delta(x) = 0$ except when $x = 0$,

(2) $\delta(0)$ is "infinite", and

(3) the area under the graph of $\delta(x)$ is one.

Still on a handwaving level, if f is any continuous function, then the functions $f(x)\delta(x)$ and $f(0)\delta(x)$ are the same since they are both zero for every $x \neq 0$. Consequently

$$\int_{-\infty}^{\infty} f(x)\delta(x)\,dx = \int_{-\infty}^{\infty} f(0)\delta(x)\,dx = f(0)\int_{-\infty}^{\infty} \delta(x)\,dx = f(0). \tag{B.8}$$

That $\int_{-\infty}^{\infty} f(x)\delta(x)\,dx = f(0)$ is by far the most important property of the Dirac delta function. But there is no Riemann integrable function $\delta(x)$ that satisfies (B.8).

Exercise B.3.1. Prove that there is no Riemann integrable function $\delta(x)$ that satisfies (B.8).

The basic idea which allows us to make make rigorous sense of (B.8) is to generalize the meaning of "a function on $\mathbb{R}$". We shall call the generalization a "tempered distribution on $\mathbb{R}$". Of course a function on $\mathbb{R}$, in the conventional sense, is a rule which assigns a number to each $x \in \mathbb{R}$. A tempered distribution will be a rule which assigns a number to each nice function on $\mathbb{R}$. We will associate to the conventional function $f : \mathbb{R} \to \mathbb{C}$ the tempered distribution which assigns to the nice function $\varphi(x)$ the number $\int_{-\infty}^{\infty} f(x)\varphi(x)\,dx$. The tempered distribution which corresponds to the Dirac delta function will assign to $\varphi(x)$ the number $\varphi(0)$. The space of "nice functions", or *test functions*, used by tempered distributions is the Schwartz space of Definition B.1.1.

Definition B.3.2 (Tempered distributions). The space of all tempered distributions on $\mathbb{R}^n$, denoted $\mathcal{S}'(\mathbb{R}^n)$, is the dual space of $\mathcal{S}(\mathbb{R}^n)$. That is, it is the set of all functionals

$$f : \mathcal{S}(\mathbb{R}^n) \to \mathbb{C}$$

that are linear and continuous. One usually denotes by $\langle f, \varphi\rangle$ the value in $\mathbb{C}$ that the distribution $f \in \mathcal{S}'(\mathbb{R})$ assigns to $\varphi \in \mathcal{S}(\mathbb{R}^n)$.

Above,

- the linearity of f means that $\langle f, a\varphi + b\psi\rangle = a\langle f, \varphi\rangle + b\langle f, \psi\rangle$ for all $\varphi, \psi \in \mathcal{S}(\mathbb{R}^n)$ and all $a, b \in \mathbb{C}$.
- the continuity of f means that if $\varphi = \lim_{n\to\infty} \varphi_n$ in $\mathcal{S}(\mathbb{R}^n)$, then $\langle f, \varphi\rangle = \lim_{n\to\infty} \langle f, \varphi_n\rangle$.

Example B.3.3.

(a) Here is the motivating example for the whole subject. Let $f : \mathbb{R}^n \to \mathbb{C}$ be any function that is polynomially bounded (that is, there is a polynomial $P(x)$ such that $|f(x)| \leq P(x)$ for all $x \in \mathbb{R}^n$) and that is Riemann integrable on $[-M, M]^n$ for each $M > 0$. Then

$$f : \varphi \in \mathcal{S}(\mathbb{R}^n) \mapsto \langle f, \varphi \rangle = \int_{\mathbb{R}^n} f(x)\varphi(x)\, dx$$

is a tempered distribution. The integral converges because every $\varphi \in \mathcal{S}(\mathbb{R}^n)$ decays faster at infinity than one over any polynomial. See Exercise B.3.4. The linearity in φ of $\langle f, \varphi \rangle$ is obvious. The continuity in φ of $\langle f, \varphi \rangle$ follows easily from Exercise B.3.4 (generalized to $\mathbb{R}^n$) and Theorem B.3.5.

(b) The Dirac delta function, on $\mathbb{R}$, and more generally the Dirac delta function translated to $b \in \mathbb{R}$, are defined as tempered distributions by

$$\langle \delta, \varphi \rangle = \varphi(0), \qquad \langle \delta_b, \varphi \rangle = \varphi(b).$$

Once again, the linearity in φ is obvious and the continuity in φ is easily verified if one applies Theorem B.3.5.

(c) The derivative of the Dirac delta function δ_b is defined by

$$\left\langle \delta_b', \varphi \right\rangle = -\varphi'(b).$$

The reason for the name "derivative of the Dirac delta function" will be given in the section on differentiation, later. See Definition B.4.5.

(d) The principal value of $\frac{1}{x}$, with x running over $\mathbb{R}$, is defined by

$$\left\langle P\frac{1}{x}, \varphi \right\rangle = \lim_{\varepsilon \to 0+} \int_{|x|>\varepsilon} \frac{\varphi(x)}{x}\, dx.$$

The first thing that we have to do is verify that the limit above actually exists. This is not a trivial statement, because not only is $\frac{\varphi(x)}{x}$ not integrable on $[-1, 1]$ if $\varphi(0) \neq 0$ (because, for x near zero, $\frac{\varphi(x)}{x} \approx \frac{\varphi(0)}{x}$), but $\int_0^1 \frac{1}{x} dx$ and $\int_{-1}^0 \frac{1}{x}\, dx$ do not even exist as improper integrals:

$$\int_0^1 \frac{1}{x}\, dx = \lim_{\varepsilon \to 0+} \int_\varepsilon^1 \frac{1}{x}\, dx = \lim_{\varepsilon \to 0+} \ln \frac{1}{\varepsilon} = \infty,$$
$$\int_{-1}^0 \frac{1}{x}\, dx = \lim_{\varepsilon \to 0+} \int_{-1}^{-\varepsilon} \frac{1}{x}\, dx = \lim_{\varepsilon \to 0+} \ln \varepsilon = -\infty.$$

Here is the verification that the limit defining $\left\langle P\frac{1}{x}, \varphi\right\rangle$ exists:

$$\begin{aligned}
\lim_{\varepsilon\to 0+} \int_{|x|>\varepsilon} \frac{\varphi(x)}{x}\,dx
&= \lim_{\substack{\varepsilon\to 0+\\ M,M'\to\infty}} \left\{ \int_{\varepsilon}^{M} \frac{\varphi(x)}{x}\,dx + \int_{-M'}^{-\varepsilon} \frac{\varphi(x)}{x}\,dx \right\}\\
&= \lim_{\substack{\varepsilon\to 0+\\ M,M'\to\infty}} \left\{ \int_{\varepsilon}^{1} \frac{\varphi(x)}{x}\,dx + \int_{1}^{M} \frac{\varphi(x)}{x}\,dx + \int_{-M'}^{-1} \frac{\varphi(x)}{x}\,dx + \int_{-1}^{-\varepsilon} \frac{\varphi(x)}{x}\,dx \right\}\\
&= \lim_{\substack{\varepsilon\to 0+\\ M,M'\to\infty}} \left\{ \int_{\varepsilon}^{1} \frac{\varphi(x)-\varphi(-x)}{x}\,dx + \int_{1}^{M} \frac{\varphi(x)}{x}\,dx + \int_{-M'}^{-1} \frac{\varphi(x)}{x}\,dx \right\}.
\end{aligned}$$

The first integral converges because, by the mean value theorem, we have, for some c between x and $-x$,

$$\left|\frac{\varphi(x)-\varphi(-x)}{x}\right| = \left|\frac{\varphi'(c)\,2x}{x}\right| \le 2\|\varphi\|_{0,1}.$$

The second and third integrals converge because, for $|x| \ge 1$

$$\left|\frac{\varphi(x)}{x}\right| \le \frac{1}{x^2}|x\varphi(x)| \le \frac{1}{x^2}\|\varphi\|_{1,0}.$$

These bounds give both that $\left\langle P\frac{1}{x}, \varphi\right\rangle$ is well–defined and

$$\begin{aligned}
\left|\left\langle P\tfrac{1}{x}, \varphi\right\rangle\right| &\le 2\|\varphi\|_{0,1} \int_0^1 dx + \|\varphi\|_{1,0} \int_1^\infty \frac{1}{x^2}\,dx + \|\varphi\|_{1,0} \int_{-\infty}^{-1} \frac{1}{x^2}\,dx\\
&= 2\|\varphi\|_{0,1} + 2\|\varphi\|_{1,0}.
\end{aligned}$$

Linearity is again obvious. Continuity again follows by Theorem B.3.5.

Exercise B.3.4. Let $f : \mathbb{R} \to \mathbb{C}$ be Riemann integrable on $[-M, M]$ for all $M > 0$ and obey the bound $|f(x)| \le P(x)$ for all $x \in \mathbb{R}$, where $P(x)$ is the polynomial $P(x) = \sum_{m=N_-}^{N_+} a_m x^m$ and $N_\pm$ are nonnegative integers.

(a) Prove that there is a constant $C > 0$ such that

$$|f(x)|(1+x^2) \le C\big(|x|^{N_-} + |x|^{N_+ + 2}\big)$$

for all $x \in \mathbb{R}$.

(b) Prove that

$$\int_{-\infty}^{\infty} |f(x)\varphi(x)|\,dx \le \pi C\big(\|\varphi\|_{N_-,0} + \|\varphi\|_{N_+ + 2,0}\big)$$

for all $\varphi \in \mathcal{S}(\mathbb{R})$.

Theorem B.3.5 (Continuity test). *A linear map*

$$f : \varphi \in \mathcal{S}(\mathbb{R}^n) \mapsto \langle f, \varphi \rangle \in \mathbb{C}$$

is continuous if and only if there are constants $C > 0$ and $N \in \mathbb{N}$ such that

$$\big|\langle f, \varphi \rangle\big| \le C \sum_{\substack{\alpha,\beta \in \mathbb{N}_0^n \\ |\alpha|,|\beta| \le N}} \|\varphi\|_{\alpha,\beta}.$$

Proof.

$\Leftarrow$: Assume that

$$\big|\langle f, \varphi \rangle\big| \le C \sum_{\substack{\alpha,\beta \in \mathbb{N}_0^n \\ |\alpha|,|\beta| \le N}} \|\varphi\|_{\alpha,\beta}$$

and that the sequence $\{\varphi_k\}_{k \in \mathbb{N}}$ converges to φ in $\mathcal{S}(\mathbb{R}^n)$. Then

$$\big|\langle f, \varphi \rangle - \langle f, \varphi_k \rangle\big| = \big|\langle f, \varphi - \varphi_k \rangle\big| \le C \sum_{\substack{\alpha,\beta \in \mathbb{N}_0^n \\ |\alpha|,|\beta| \le N}} \|\varphi - \varphi_k\|_{\alpha,\beta}$$

converges to zero as $k \to \infty$. So f is continuous.

$\Rightarrow$: Assume that $f \in \mathcal{S}'(\mathbb{R}^n)$. In particular f is continuous at $\varphi = 0$. Then there is a $\delta > 0$ such that

$$d(\psi, 0) < \delta \implies \big|\langle f, \psi \rangle\big| < 1.$$

Choose N so that

$$\sum_{\substack{\alpha,\beta \in \mathbb{N}_0^n \\ |\alpha| \text{ or } |\beta| > N}} 2^{-|\alpha|-|\beta|} < \tfrac{\delta}{2}$$

and consider any $\psi \in \mathcal{S}(\mathbb{R}^n)$ that obeys

$$\sum_{\substack{\alpha,\beta \in \mathbb{N}_0^n \\ |\alpha|,|\beta| \le N}} \|\psi\|_{\alpha,\beta} \le \tfrac{\delta}{2}.$$

For any such ψ we have

$$\begin{aligned} d(\psi, 0) = \sum_{\alpha,\beta \in \mathbb{N}_0^n} 2^{-|\alpha|-|\beta|} \tfrac{\|\psi\|_{\alpha,\beta}}{1+\|\psi\|_{\alpha,\beta}} &\le \sum_{\substack{\alpha,\beta \in \mathbb{N}_0^n \\ |\alpha|,|\beta| \le N}} \|\psi\|_{\alpha,\beta} + \sum_{\substack{\alpha,\beta \in \mathbb{N}_0^n \\ |\alpha| \text{ or } |\beta| > N}} 2^{-|\alpha|-|\beta|} \\ \implies d(\psi, 0) < \delta \\ \implies \big|\langle f, \psi \rangle\big| < 1. \end{aligned}$$

Consequently, for any $0 \ne \varphi \in \mathcal{S}(\mathbb{R}^n)$, setting

$$\psi = \tfrac{\delta}{2} \Big[\sum_{\substack{\alpha',\beta' \in \mathbb{N}_0^n \\ |\alpha'|,|\beta'| \le N}} \|\varphi\|_{\alpha',\beta'} \Big]^{-1} \varphi$$

we have

$$\sum_{\substack{\alpha,\beta\in\mathbb{N}_0^n\\|\alpha|,|\beta|\le N}} \|\psi\|_{\alpha,\beta} = \sum_{\substack{\alpha,\beta\in\mathbb{N}_0^n\\|\alpha|,|\beta|\le N}} \frac{\delta}{2}\Bigg[\sum_{\substack{\alpha',\beta'\in\mathbb{N}_0^n\\|\alpha'|,|\beta'|\le N}} \|\varphi\|_{\alpha',\beta'}\Bigg]^{-1} \|\varphi\|_{\alpha,\beta} = \frac{\delta}{2}$$

and hence

$$|\langle f,\varphi\rangle| = \frac{2}{\delta}\Bigg[\sum_{\substack{\alpha',\beta'\in\mathbb{N}_0^n\\|\alpha'|,|\beta'|\le N}} \|\varphi\|_{\alpha',\beta'}\Bigg] |\langle f,\psi\rangle| < \frac{2}{\delta}\sum_{\substack{\alpha,\beta\in\mathbb{N}_0^n\\|\alpha|,|\beta|\le N}} \|\varphi\|_{\alpha,\beta}$$

as desired. □

B.4. Operations on tempered distributions

We now define a number of operations, for example, addition and differentiation, on tempered distributions. The motivation for all of these definitions comes from Example B.3.3 with $f \in \mathcal{S}(\mathbb{R}^n)$. Then we can view f both as a conventional function and as a tempered distribution. We will define each operation in such a way that when it is applied to $f \in \mathcal{S}(\mathbb{R}^n)$, viewed as a distribution, it yields the same answer as when the operation is applied to f viewed as an ordinary function, with the result viewed as a distribution. As a trivial example, suppose that we wish to define multiplication by 7. If $f \in \mathcal{S}(\mathbb{R}^n)$ is viewed as an ordinary function, applying the operation of multiplication by 7 to it gives the ordinary function $7f$. But $7f$ can again be viewed as the distribution $\langle 7f,\varphi\rangle = \int 7f(x)\,\varphi(x)\,dx = 7\,\langle f,\varphi\rangle$. So we would define the operation of multiplication by 7 applied to any distribution f as the distribution $7f$ defined by $\langle 7f,\varphi\rangle = 7\,\langle f,\varphi\rangle$.

Addition and scalar multiplication.

Motivation B.4.1. If $f, g \in \mathcal{S}(\mathbb{R}^n)$ and $a, b \in \mathbb{C}$, then

$$\begin{aligned}\int_{\mathbb{R}^n} \big[af(x) + bg(x)\big]\,\varphi(x)\,dx &= a\int_{\mathbb{R}^n} f(x)\,\varphi(x)\,dx + b\int_{\mathbb{R}^n} g(x)\,\varphi(x)\,dx\\ &= a\,\langle f,\varphi\rangle + b\,\langle g,\varphi\rangle\,.\end{aligned}$$

Definition B.4.2. If $f, g \in \mathcal{S}'(\mathbb{R}^n)$ and $a, b \in \mathbb{C}$, then define $af + bg \in \mathcal{S}'(\mathbb{R}^n)$ by

$$\langle af + bg, \varphi\rangle = a\,\langle f,\varphi\rangle + b\,\langle g,\varphi\rangle\,.$$

Theorem B.4.3. *If $f, g \in \mathcal{S}'(\mathbb{R}^n)$ and $a, b \in \mathbb{C}$, then $af + bg$, defined above, is a well-defined element of $\mathcal{S}'(\mathbb{R}^n)$. The operations of addition and scalar multiplication so defined obey the usual vector space axioms of Definition* A.1.1.

Proof. Trivial. □

Differentiation.

Motivation B.4.4. If $f \in \mathcal{S}(\mathbb{R})$, then, by integration by parts,

$$\int_{-\infty}^{\infty} f'(x)\,\varphi(x)\,dx = -\int_{-\infty}^{\infty} f(x)\,\varphi'(x)\,dx \quad \text{(the boundary terms vanish).}$$

More generally, if $f \in \mathcal{S}(\mathbb{R}^n)$, and $\gamma \in \mathbb{N}_0^n$,

$$\int_{\mathbb{R}^n} (\partial^\gamma f)(x)\,\varphi(x)\,dx = (-1)^{|\gamma|} \int_{\mathbb{R}^n} f(x)\,(\partial^\gamma \varphi)(x)\,dx.$$

Definition B.4.5. If $\gamma \in \mathbb{N}_0^n$, we define the γth derivative of $f \in \mathcal{S}'(\mathbb{R}^n)$ by

$$\langle \partial^\gamma f, \varphi \rangle = (-1)^{|\gamma|} \langle f, \partial^\gamma \varphi \rangle.$$

Since $\|\partial^\gamma \varphi\|_{\alpha,\beta} = \|\varphi\|_{\alpha,\beta+\gamma}$ the right-hand side gives a well-defined element of $\mathcal{S}'(\mathbb{R}^n)$.

Remark B.4.6. Note that *every* derivative of *every* distribution *always* exists.

Example B.4.7. The Heaviside function

$$H(x) = \begin{cases} 1 & \text{if } x \geq 0 \\ 0 & \text{if } x < 0 \end{cases}$$

(on $\mathbb{R}$) may also be viewed as the tempered distribution

$$\langle H, \varphi \rangle = \int_0^\infty \varphi(x)\,dx$$

(in $\mathcal{S}'(\mathbb{R})$) via Example B.3.3. The derivative of this distribution is

$$\langle H', \varphi \rangle = -\langle H, \varphi' \rangle = -\int_0^\infty \varphi'(x)\,dx = -\left[\varphi(x)\right]_0^\infty = \varphi(0) = \langle \delta, \varphi \rangle.$$

Thus H' is the Dirac delta function.

Fourier transform.

Motivation B.4.8. If f and φ are both in $\mathcal{S}(\mathbb{R}^n)$, then, writing $\varphi(\xi) = \overline{\psi(\xi)}$

$$\begin{aligned}
\langle \hat{f}, \varphi \rangle &= \int_{\mathbb{R}^n} \hat{f}(\xi)\varphi(\xi)\,d\xi \\
&= \int_{\mathbb{R}^n} \hat{f}(\xi)\overline{\psi(\xi)}\,d\xi \\
&= (2\pi)^n \int_{\mathbb{R}^n} f(x)\overline{\check{\psi}(x)}\,dx \qquad \text{(by Theorems B.2.2 and B.2.5)} \\
&= \int_{\mathbb{R}^n} f(x)\hat{\varphi}(x)\,dx
\end{aligned}$$

since

$$\begin{aligned}\overline{\check{\psi}(x)} &= \overline{\int_{\mathbb{R}^n} e^{i\xi\cdot x}\psi(\xi)\,\frac{d\xi}{(2\pi)^n}} \\ &= \int_{\mathbb{R}^n} e^{-i\xi\cdot x}\overline{\psi(\xi)}\,\frac{d\xi}{(2\pi)^n} \\ &= \int_{\mathbb{R}^n} e^{-i\xi\cdot x}\varphi(\xi)\,\frac{d\xi}{(2\pi)^n} \\ &= \frac{1}{(2\pi)^n}\hat{\varphi}(x).\end{aligned}$$

Definition B.4.9. The Fourier transform of $f \in \mathcal{S}'(\mathbb{R}^n)$ is the tempered distribution defined by

$$\langle \hat{f}, \varphi \rangle = \langle f, \hat{\varphi} \rangle.$$

It is well defined by Theorem B.2.5.

Example B.4.10. The Fourier transform of the Dirac delta function is given by

$$\langle \hat{\delta}, \varphi \rangle = \langle \delta, \hat{\varphi} \rangle = \hat{\varphi}(0) = \int_{-\infty}^{\infty} \varphi(x)\,dx = \langle 1, \varphi \rangle .$$

That is, $\hat{\delta}$ is the constant function 1.

Example B.4.11. The Fourier transform of the constant function 1, viewed as a tempered distribution, is

$$\langle \hat{1}, \varphi \rangle = \langle 1, \hat{\varphi} \rangle = \int_{-\infty}^{\infty} \hat{\varphi}(\xi)\,d\xi = 2\pi\varphi(0)$$

by (B.5). That is, the Fourier transform of the constant function 1 is $2\pi\delta(\xi)$.

Appendix C

Sobolev Spaces

This appendix provides some background concerning Sobolev spaces. More detailed presentations are in [**AF03, Bre11, Eva10, Leo17, Leo23, Tri83**]. The case of C^k domains is discussed in [**Gri85**].

In the first section, we define the spaces $H^s(\Omega)$, $W^{\ell,p}(\Omega)$, $H^s(\partial\Omega)$, and $H_0^s(\Omega)$. We start with Ω being an arbitrary open subset of $\mathbb{R}^n$, but later restrict Ω to being a bounded open subset of $\mathbb{R}^n$ with C^k boundary (see Definition C.1.33), which we denote $\partial\Omega$. Roughly speaking, the spaces $H^s(\Omega)$ and $H_0^s(\Omega)$ consist of all functions whose s^{th} order derivatives are L^2 and, in the case of $H_0^s(\Omega)$, vanish appropriately on $\partial\Omega$. If $\alpha = (\alpha_1, \cdots, \alpha_n) \in \mathbb{N}_0^n$, where $\mathbb{N}_0 = \{0\} \cup \mathbb{N}$, we shall use $\partial^\alpha u(x)$ to denote the partial derivative $\frac{\partial^{\alpha_1}}{\partial x_1^{\alpha_1}} \cdots \frac{\partial^{\alpha_n}}{\partial x_n^{\alpha_n}} u(x)$. The order of this partial derivative is $|\alpha| = \alpha_1 + \cdots + \alpha_n$.

There is more than one way to precisely implement this rough picture. We choose to define

- $H^\ell(\Omega)$, resp. $H_0^\ell(\Omega)$, with $\ell \in \mathbb{N}_0$, to be the completion of

  $$\left\{ u \in C^\ell(\Omega) \mid \|u\|_{H^\ell(\Omega)} < \infty \right\},$$

  resp. $C_c^\infty(\Omega)$, under the norm $\|\cdot\|_{H^\ell(\Omega)}$ satisfying

  $$\|u\|_{H^\ell(\Omega)}^2 = \sum_{|\alpha| \le \ell} \int_\Omega |\partial^\alpha u(x)|^2 \, dx$$

  (Definition C.1.1).
- $W^{\ell,p}(\Omega)$, with $\ell \in \mathbb{N}_0$, to be the completion of

  $$\left\{ u \in C^\ell(\Omega) \mid \|u\|_{W^{\ell,p}(\Omega)} < \infty \right\}$$

under the norm $\|\cdot\|_{W^{\ell,p}(\Omega)}$ satisfying

$$\|u\|^p_{W^{\ell,p}(\Omega)} = \sum_{|\alpha|\le\ell} \int_\Omega |\partial^\alpha u(x)|^p\, dx$$

(Definition C.1.1).

- and $H^\ell(\Omega)$, with ℓ a negative integer, to be the space $H_0^{-\ell}(\Omega)^*$ of bounded linear functionals on $H_0^{-\ell}(\Omega)$ (Definition C.1.25).

We shall make some natural identifications, like identifying $H^0(\Omega) = L^2(\Omega)$ with $L^2(\Omega)^*$, and show that

- if $\ell, \ell' \in \mathbb{Z}$ with $\ell \le \ell'$, then $H^{\ell'}(\Omega) \subset H^\ell(\Omega)$ and

$$\|v\|_{H^\ell(\Omega)} \le \|v\|_{H^{\ell'}(\Omega)}$$

for all $v \in H^{\ell'}(\Omega)$ (Remark C.1.28) and

- we can define ∂^α as a bounded linear map from $H^\ell(\Omega)$ to $H^{\ell-|\alpha|}(\Omega)$ for all $\ell \in \mathbb{Z}$ and all $\alpha \in \mathbb{N}_0^n$ (part (b) of Lemma C.1.29) and then
- if $\ell \in \mathbb{N}_0$,

$$H^\ell(\Omega) = \left\{\, u \in L^2(\mathbb{R}^n) \mid \partial^\alpha u \in L^2(\mathbb{R}^n) \text{ for all } \alpha \in \mathbb{N}_0^n \text{ with } |\alpha| \le \ell \,\right\}$$

(part (a) of Proposition C.2.24).

We shall equivalently characterize, in part (b) of Proposition C.2.24, $H^\ell(\Omega)$, $\ell \in \mathbb{N}_0$, as the set of all $u \in L^2(\Omega)$ for which there is a constant C such that

$$|\langle \partial^\alpha f, u\rangle_{L^2(\Omega)}| \le C\big\|f\big\|_{L^2(\Omega)} \tag{C.1}$$

for all $\alpha \in \mathbb{N}_0^n$ with $|\alpha| \le \ell$ and all $f \in C_c^\infty(\Omega)$. The map

$$D_u^\alpha : C_c^\infty(\Omega) \to \mathbb{C}, \quad f \mapsto (-1)^{|\alpha|}\langle \partial^\alpha f, u\rangle_{L^2(\Omega)}$$

is called the weak (or distributional) αth derivative of u. It is defined for all $u \in L^2(\Omega)$ and all $\alpha \in \mathbb{N}_0^n$, but is not in general a function on Ω. One says that the weak derivative D_u^α is in $L^2(\Omega)$ if there exists a $v_\alpha \in L^2(\Omega)$ such that $\langle f, v_\alpha\rangle_{L^2(\Omega)} = (-1)^{|\alpha|}\langle \partial^\alpha f, u\rangle_{L^2(\Omega)}$ for all $f \in C_c^\infty(\Omega)$. In the special case that $u \in C^{|\alpha|}(\Omega)$, we can take $v_\alpha = \partial^\alpha u$, by repeated integration by parts. By the Riesz representation theorem (Theorem A.2.15), the bound (C.1) provides a necessary and sufficient condition for the weak derivative D_u^α to be in $L^2(\Omega)$. So part (b) of Proposition C.2.24 characterizes $H^\ell(\Omega)$, as the set of all $u \in L^2(\Omega)$ all of whose weak derivatives of order up to ℓ are in $L^2(\Omega)$.

In the second section, we consider the problem of restricting functions on Ω, that a priori need only be defined almost everywhere, to $\partial\Omega$. We shall see that if the function u, on Ω, has $s > \frac{1}{2}$ derivatives in L^2, then one can define a natural restriction of u to $\partial\Omega$ and this restriction has $s - \frac{1}{2}$ derivatives in L^2. Of course, for this to make sense, we have define what it means for a function to have fractional derivatives.

Finally, in the third section, we prove a number of useful inequalities.

C.1. Definitions of Sobolev spaces

We wish to define spaces of functions on Ω that have s derivatives, in some reasonable sense. When s is a positive integer, this is easy.

Definition C.1.1 ($H^\ell(\Omega)$, $W^{\ell,p}(\Omega)$, $H_0^\ell(\Omega)$, $\ell \in \mathbb{N}_0$). Let Ω be an open subset of $\mathbb{R}^n$, $\ell \in \mathbb{N}_0$ and $1 \le p < \infty$. Define

$$||u||_{H^\ell(\Omega)} = \left\{ \sum_{|\alpha|\le\ell} \int_\Omega |\partial^\alpha u(x)|^2 \, dx \right\}^{1/2}$$

$$||u||_{W^{\ell,p}(\Omega)} = \left\{ \sum_{|\alpha|\le\ell} \int_\Omega |\partial^\alpha u(x)|^p \, dx \right\}^{1/p}$$

for each $u \in C^\ell(\Omega)$ for which the right-hand side makes sense. Then $H^\ell(\Omega)$ is the completion[1] of the pre–Hilbert space $\{ u \in C^\ell(\Omega) \mid ||u||_{H^\ell(\Omega)} < \infty \}$ equipped with the inner product

$$\langle u, v\rangle_{H^\ell(\Omega)} = \sum_{|\alpha|\le\ell} \int_\Omega \partial^\alpha u(x) \, \overline{\partial^\alpha v(x)} \, dx.$$

We sometimes drop the Ω from the notation. Similarly, $H_0^\ell(\Omega)$ is the completion of $C_c^\infty(\Omega)$ with respect to the inner product $\langle u, v\rangle_{H^\ell(\Omega)}$ and the Banach space $W^{\ell,p}(\Omega)$ is the completion of $\{ u \in C^\ell(\Omega) \mid ||u||_{W^{\ell,p}(\Omega)} < \infty \}$ with respect to the norm $||u||_{W^{\ell,p}(\Omega)}$.

For $p = \infty$, define

$$||u||_{W^{\ell,\infty}(\Omega)} = \sum_{|\alpha|\le\ell} ||\partial^\alpha u||_{L^\infty(\Omega)}.$$

We let $W^{\ell,\infty}(\Omega)$ be the completion of $\{ u \in C^\ell(\Omega) \mid ||u||_{W^{\ell,p}(\Omega)} < \infty \}$ with respect to the norm $||u||_{W^{\ell,\infty}(\Omega)}$.

Exercise C.1.2. Let $\Omega = (-1, 1)$ and $\ell = 1$. Think of $f(x) = |x|$ and $g(x) = \operatorname{sgn} x$ as functions in $L^2(\Omega)$. Find a sequence $\{f_n\}_{n\in\mathbb{N}} \subset C^1(\Omega)$ such that

- $\{f_n\}_{n\in\mathbb{N}}$ is Cauchy with respect to the norm $|| \; ||_{H^\ell(\Omega)}$,
- $\{f_n\}_{n\in\mathbb{N}}$ converges in $L^2(\Omega)$ to f,
- $\left\{\frac{df_n}{dx}\right\}_{n\in\mathbb{N}}$ converges in $L^2(\Omega)$ to g.

[1] Readers that are not familiar with completion should read [**RS80**, §I]. It is also common to define $H^\ell(\Omega)$ to be the space of distributions $u \in \mathcal{D}'(\Omega)$, all of whose derivatives $\partial^\alpha u$ of order $|\alpha| \le \ell$ are L^2 functions, not just distributions. We shall not assume that readers are familiar the distribution space $\mathcal{D}'(\Omega)$.

By definition, each element of $H^\ell(\Omega)$ is an equivalence class of Cauchy sequences. We can identify f with the Cauchy sequence $\{f_n\}_{n\in\mathbb{N}}$ and so think of f as an element of $H^\ell(\Omega)$.

Remark C.1.3. By definition, each element of $H^\ell(\Omega)$ is an equivalence class of Cauchy sequences. A sequence $\{u_i\}_{i\in\mathbb{N}} \subset \{\, u \in C^\ell(\Omega) \mid \|u\|_{H^\ell(\Omega)} < \infty \,\}$ is Cauchy with respect to the norm $\|\cdot\|_{H^\ell(\Omega)}$ if and only if $\{\partial^\alpha u_i\}_{i\in\mathbb{N}}$ is Cauchy with respect to the norm $\|\cdot\|_{L^2(\Omega)}$ for all $\alpha \in \mathbb{N}_0^n$ with $|\alpha| \le \ell$. Since $L^2(\Omega)$ is complete, this in turn is the case if and only if, for each $\alpha \in \mathbb{N}_0^n$ with $|\alpha| \le \ell$, the sequence $\partial^\alpha u_i$ converges in $L^2(\Omega)$ to some $v_\alpha \in L^2(\Omega)$. Thus we may equivalently view $H^\ell(\Omega)$ as the set of all $v \in L^2(\Omega)$ for which there is a sequence $\{u_i\}_{i\in\mathbb{N}} \subset \{\, u \in C^\ell(\Omega) \mid \|u\|_{H^\ell(\Omega)} < \infty \,\}$ obeying

- u_i converges to v in $L^2(\Omega)$ and
- for each $\alpha \in \mathbb{N}_0^n$ with $|\alpha| \le \ell$, the sequence $\partial^\alpha u_i$ converges in $L^2(\Omega)$.

In this way, we may view all of $H^\ell(\Omega)$, $H_0^\ell(\Omega)$, $\ell \in \mathbb{N}_0$ as subsets of $L^2(\Omega)$. Furthermore, if $\ell', \ell \in \mathbb{N}_0$ with $\ell' \ge \ell$, we have

$$H_0^{\ell'}(\Omega) \subset H^{\ell'}(\Omega) \subset H^\ell(\Omega) \subset L^2(\Omega)$$

and $\|u\|_{H^\ell(\Omega)} \le \|u\|_{H^{\ell'}(\Omega)}$ for all $u \in H^{\ell'}(\Omega)$. Similarly we may view $W^{\ell,p}(\Omega)$ as a subspace of $L^p(\Omega)$.

Exercise C.1.4. Prove that $C_c^\ell(\Omega) \subset H_0^\ell(\Omega)$.

Exercise C.1.5. Let $\ell \in \mathbb{N}_0$.

(a) Let $\varphi \in W^{\ell,\infty}(\Omega)$, so that φ and all of its derivatives of order at most ℓ are bounded. Prove that the map

$$u \in \{\, u \in C^\ell(\Omega) \mid \|u\|_{H^\ell(\Omega)} < \infty \,\} \mapsto \varphi u,$$

where, as you would expect, $(\varphi u)(x) = \varphi(x)u(x)$, has a unique extension to a bounded, linear map on $H^\ell(\Omega)$. Prove, in particular, that there is a constant $C_{\ell,n}$, depending only on ℓ and n, such that

$$\|\varphi u\|_{H^\ell(\Omega)} \le C_{\ell,n} \sum_{k=0}^{\ell} \|\varphi\|_{W^{k,\infty}(\Omega)} \|u\|_{H^{\ell-k}(\Omega)}$$

for all $u \in H^\ell(\Omega)$.

(b) Let $\alpha \in \mathbb{N}_0^n$ with $|\alpha| \le \ell$. Prove that the map

$$u \in \{\, u \in C^\ell(\Omega) \mid \|u\|_{H^\ell(\Omega)} < \infty \,\} \mapsto \partial^\alpha u$$

has a unique extension to a bounded linear map from $H^\ell(\Omega)$ to $H^{\ell-|\alpha|}(\Omega)$.

Exercise C.1.6. Let Ω be a bounded open subset of $\mathbb{R}^n$. Suppose that $F : \mathbb{R} \to \mathbb{R}$ is a C^1 function with bounded derivative. Let $H_{\mathbb{R}}^1(\Omega)$ denote the set of real valued functions in $H^1(\Omega)$.

(a) Prove that if $u \in C^1(\Omega) \cap H^1_{\mathbb{R}}(\Omega)$, then $F \circ u \in C^1(\Omega) \cap H^1_{\mathbb{R}}(\Omega)$.

(b) Prove that the map $u \mapsto F \circ u$, with domain $C^1(\Omega) \cap H^1_{\mathbb{R}}(\Omega)$, is continuous with respect to the topology of $H^1_{\mathbb{R}}(\Omega)$. That is, prove that if $\varepsilon > 0$ and $v \in C^1(\Omega) \cap H^1_{\mathbb{R}}(\Omega)$, then there is a $\delta > 0$ such that $\|F \circ u - F \circ v\|_{H^1(\Omega)} < \varepsilon$ for all $u \in C^1(\Omega) \cap H^1_{\mathbb{R}}(\Omega)$ obeying

$$\|u - v\|_{H^1(\Omega)} < \delta.$$

(c) Prove that the map $u \mapsto F \circ u$ has a continuous extension to $H^1_{\mathbb{R}}(\Omega)$.

Exercise C.1.7. Let Ω be a bounded open subset of $\mathbb{R}^n$. Prove that if $u \in H^1_{\mathbb{R}}(\Omega)$, then $|u|, \max\{0, u\} \in H^1_{\mathbb{R}}(\Omega)$.

Example C.1.8. Here is an example which shows that $H^\ell_0(\Omega)$ can be a strictly proper subspace of $H^\ell(\Omega)$. We take $n = 1$, $\Omega = (0, \infty)$. Start with $f \in C^\infty_c(\Omega)$. Since f is supported in Ω, we can think of it as being defined on all $\mathbb{R}$, taking value zero everywhere except in $(0, \infty)$. In particular $f^{(j)}(0) = 0$. So, for $x > 0$,

$$\begin{aligned}
|f^{(j)}(x)| &= |f^{(j)}(x) - f^{(j)}(0)| \\
&= \Big|\int_0^x f^{(j+1)}(t)\,dt\Big| \\
&= \Big|\int_0^\infty \chi_{[0,x]}(t) f^{(j+1)}(t)\,dt\Big| \\
&\le \Big[\int_0^\infty \chi_{[0,x]}(t)^2\,dt\Big]^{1/2} \Big[\int_0^\infty \big|f^{(j+1)}(t)\big|^2\,dt\Big]^{1/2} \\
&= \sqrt{x}\Big[\int_0^\infty \big|f^{(j+1)}(t)\big|^2\,dt\Big]^{1/2} \le \sqrt{x}\,\|f\|_{H^\ell(\Omega)}
\end{aligned}$$

provided $\ell \ge j + 1$. Now suppose that $f \in H^\ell_0(\Omega)$. It is a limit, in $H^\ell(\Omega)$, of functions $f_i \in C^\infty_c(\Omega)$. We can always choose the sequence so that $\|f_i\|_{H^\ell(\Omega)} \le 2\,\|f\|_{H^\ell(\Omega)}$ for every i. Thus, if $j \le \ell - 1$,

$$|f_i^{(j)}(x)| \le \sqrt{x}\,\|f_i\|_{H^\ell(\Omega)} \le 2\sqrt{x}\,\|f\|_{H^\ell(\Omega)}$$

for all i. We can think of this as saying that the first $\ell - 1$ derivatives of every $f \in H^\ell_0(\Omega)$ vanish on the boundary of Ω. Certainly, given any constant C, if $0 < x < \min\{1, \frac{1}{100C^2}\}$ and $|f(x)| \le C\sqrt{x}$ then $|e^{-x} - f(x)| \ge e^{-1} - \frac{1}{10} > 0$. So the function $e^{-x} \in H^\ell(\Omega)$ cannot be approximated in $L^2(\Omega)$, let alone $H^\ell(\Omega)$, by functions $f_i(x)$ that obey $|f_i(x)| \le C\sqrt{x}$.

Exercise C.1.9. Let $\Omega = (0, \infty)$.

(a) Prove that there is a constant C such that if $u \in C^1(\Omega)$ and $\|u\|_{H^1(\Omega)} < \infty$, then $\lim_{x\to 0} u(x)$ exists and obeys

$$|\lim_{x\to 0} u(x)| \le C\, \|u\|_{H^1(\Omega)}\,.$$

(b) Prove that there is a unique bounded linear map $B : H^1(\Omega) \to \mathbb{C}$ such that $Bu = \lim_{x\to 0} u(x)$ for all $u \in C^1(\Omega)$ with $\|u\|_{H^1(\Omega)} < \infty$.

(c) Prove that $H_0^1(\Omega)$ is of codimension one in $H^1(\Omega)$. This means that there exists a $u_0 \in H^1(\Omega)$ such that each $u \in H^1(\Omega)$ has a unique representation of the form $u = \alpha u_0 + u_1$ with $\alpha \in \mathbb{C}$ and $u_1 \in H_0^1(\Omega)$.

(d) Let $I = (a, b)$ be a finite open interval in $\mathbb{R}$. What is the codimension of $H_0^1(I)$ in $H^1(I)$?

An important special case of $H^\ell(\Omega)$ is $H^\ell(\mathbb{R}^n)$. Many properties of functions, $f(x)$, in this space become easy to understand when expressed in terms of their Fourier transforms

$$\hat{f}(\xi) = \int e^{-i\xi\cdot x} f(x)\, dx. \tag{C.2a}$$

For a summary of the most basic definitions and results concerning the Fourier transform, see Appendix B. The main properties of the Fourier transform that we shall use are:

(a) (Plancherel identity). The L^2 norm of a function is the same as the L^2 norm of its Fourier transform, up to some factors of 2π that depend on your Fourier transform conventions. For the definition (C.2a),

$$\int |f(x)|^2\, dx = \int |\hat{f}(\xi)|^2 \frac{d\xi}{(2\pi)^n}. \tag{C.2b}$$

(b) The inverse Fourier transform of $g(\xi)$ is

$$\check{g}(x) = \int e^{i\xi\cdot x} g(\xi)\, \frac{d\xi}{(2\pi)^n}. \tag{C.2c}$$

(c) The Fourier transform of $\frac{\partial}{\partial x_m} f(x)$ is $i\xi_m \hat{f}(\xi)$ so that the Fourier transform of $\partial^\alpha f(x)$ is $i^{|\alpha|}\xi^\alpha \hat{f}(\xi)$ where $\xi^\alpha = \xi_1^{\alpha_1} \cdots \xi_n^{\alpha_n}$.

(d) The Fourier transform (with respect to x) of the translate $f(x + z)$ is $e^{i\xi\cdot z}\hat{f}(\xi)$.

Using properties (a) and (c) of the Fourier transform, we have that, when $\Omega = \mathbb{R}^n$, the (square of the) norm of Definition C.1.1 is

$$\|u\|^2_{H^\ell(\mathbb{R}^n)} = \int_{\mathbb{R}^n} \Big(\sum_{|\alpha|\le\ell} \xi^{2\alpha}\Big) |\hat{u}(\xi)|^2 \frac{d\xi}{(2\pi)^n}.$$

It is common to replace the norm $\|u\|_{H^\ell(\mathbb{R}^n)}$ with

$$|||u|||_{H^\ell(\mathbb{R}^n)} = \left[\int_{\mathbb{R}^n} \left(1+|\xi|^2\right)^\ell |\hat{u}(\xi)|^2 \frac{d\xi}{(2\pi)^n} \right]^{1/2}.$$

There are constants c and C, depending only on n and ℓ, such that

$$\sum_{|\alpha|\le\ell} \xi^{2\alpha} \le c^2\left(1+|\xi|^2\right)^\ell, \qquad \left(1+|\xi|^2\right)^\ell \le C^2\Big(\sum_{|\alpha|\le\ell} \xi^{2\alpha}\Big),$$

for all $\xi \in \mathbb{R}^n$. Consequently the norms $\|\cdot\|_{H^\ell(\mathbb{R}^n)}$ and $|||\cdot|||_{H^\ell(\mathbb{R}^n)}$ are *equivalent* in the sense that

$$\|u\|_{H^\ell(\mathbb{R}^n)} \le c|||u|||_{H^\ell(\mathbb{R}^n)} \qquad |||u|||_{H^\ell(\mathbb{R}^n)} \le C\,\|u\|_{H^\ell(\mathbb{R}^n)}$$

for all $u \in H^\ell(\mathbb{R}^n)$. So a sequence of functions converges with respect to one of the norms if and only if it converges with respect to the other. For most purposes in this book, we can use either norm on $H^\ell(\mathbb{R}^n)$.

Since $\left(1+|\xi|^2\right)^s$ makes sense and is positive for all real s, the right-hand side of

$$|||u|||_{H^s(\mathbb{R}^n)} = \left[\int_{\mathbb{R}^n} \left(1+|\xi|^2\right)^s |\hat{u}(\xi)|^2 \frac{d\xi}{(2\pi)^n} \right]^{1/2}$$

makes sense (though may be $+\infty$) for all functions $u \in L^2(\mathbb{R}^n)$. So we may define the fractional Sobolev spaces as follows.

Definition C.1.10 ($H^s(\mathbb{R}^n)$, $s \in \mathbb{R}$)**.** The Sobolev space $H^s(\mathbb{R}^n)$ is the completion[2] of

$$\left\{\, u \in L^2(\mathbb{R}^n) \;\middle|\; |||u|||_{H^s(\mathbb{R}^n)} < \infty \,\right\}$$

equipped with the inner product

$$\langle u, v\rangle_{H^s(\mathbb{R}^n)} = \int_{\mathbb{R}^n} \left(1+|\xi|^2\right)^s \hat{u}(\xi)\overline{\hat{v}(\xi)} \frac{d\xi}{(2\pi)^n}.$$

Lemma C.1.11. *Let $s \in \mathbb{R}$.*

(a) *The map $u \in \mathcal{S}(\mathbb{R}^n) \mapsto \hat{u} \in \mathcal{S}(\mathbb{R}^n)$ has a unique extension to a bounded, linear map*

$$\mathcal{F}_s : H^s(\mathbb{R}^n) \to \Big\{ g : \mathbb{R}^n \to \mathbb{C},\ \textit{measurable} \;\Big|\; \int_{\mathbb{R}^n} \left(1+|\xi|^2\right)^s |g(\xi)|^2 \frac{d\xi}{(2\pi)^n} < \infty \Big\}.$$

[2] Readers already familiar with distributions can avoid taking a completion by defining $H^s(\mathbb{R}^n)$ to be the space of tempered distributions $u \in \mathcal{S}'(\mathbb{R}^n)$ whose Fourier transforms, $\hat{u}(\xi)$, obey $\int \left(1+|\xi|^2\right)^s |\hat{u}(\xi)|^2\, d\xi < \infty$.

This extension is unitary (that is, one-to-one, onto and inner product preserving) when the target space is equipped with the inner product

$$\langle f, g\rangle = \int_{\mathbb{R}^n} \left(1+|\xi|^2\right)^s f(\xi)\overline{g(\xi)}\,\frac{d\xi}{(2\pi)^n}.$$

We also denote $\mathcal{F}_s u = \hat{u}$.

(b) *Let* $\alpha \in \mathbb{N}_0^n$. *The map* $u \in \mathcal{S}(\mathbb{R}^n) \mapsto \partial^\alpha u \in \mathcal{S}(\mathbb{R}^n)$ *has a unique extension (that we persist in denoting* ∂^α*) to a bounded, linear map*

$$\partial^\alpha : H^s(\mathbb{R}^n) \to H^{s-|\alpha|}(\mathbb{R}^n).$$

Proof. The proofs of both parts are trivial applications of the B.L.T. theorem (Theorem A.2.13). □

Remark C.1.12. If $s > s'$, then there is a natural identification of $H^s(\mathbb{R}^n)$ with a subset of $H^{s'}(\mathbb{R}^n)$. So any element u of $H^s(\mathbb{R}^n)$ can also be viewed as an element of $H^{s'}(\mathbb{R}^n)$ for any $s' < s$. As $\mathcal{F}_{s'}u$ is independent of s', it is safe to use $\hat{u}$ to denote $\mathcal{F}_{s'}u$ for all $s' \le s$. similarly, the absence of "s" in the notation $\partial^\alpha u$ is harmless.

Exercise C.1.13. Let $s \in \mathbb{R}$. Prove that $C_c^\infty(\mathbb{R}^n)$ is dense in $H^s(\mathbb{R}^n)$.

Exercise C.1.14. Let $r < s < t$. Prove that, for each $\varepsilon > 0$, there is a $C > 0$, depending only on r, s, t and ε such that

$$|||f|||_{H^s(\mathbb{R}^n)} \le \varepsilon |||f|||_{H^t(\mathbb{R}^n)} + C|||f|||_{H^r(\mathbb{R}^n)}$$

for all $f \in H^t(\mathbb{R}^n)$.

Exercise C.1.15. Let $s \in \mathbb{R}$.

(a) Let $f \in \mathcal{S}(\mathbb{R}^n)$ and $\alpha \in \mathbb{N}_0^n$. By Exercise C.1.5, the map

$$u \in \mathcal{S}(\mathbb{R}^n) \mapsto fu \in \mathcal{S}(\mathbb{R}^n)$$

has a unique extension to a bounded, linear map on $H^s(\mathbb{R}^n)$. Prove that, for all $u \in H^s(\mathbb{R}^n)$,

$$\partial^\alpha(fu) = \sum_{\substack{\beta\in\mathbb{N}_0^n\\ \beta\le\alpha}} \binom{\alpha}{\beta}(\partial^\beta f)(\partial^{\alpha-\beta}u).$$

Here $\beta \le \alpha$ if and only if $\beta_i \le \alpha_i$ for all $1 \le i \le n$, $(\alpha-\beta)_i = \alpha_i - \beta_i$ for all $1 \le i \le n$ and $\binom{\alpha}{\beta} = \prod_{i=1}^n \frac{\alpha_i!}{\beta_i!(\alpha_i-\beta_i)!}$.

(b) Let $u, v \in H^s(\mathbb{R}^n)$ and let $\mathcal{O}$ be an open subset of $\mathbb{R}^n$. Make up a definition for "$u = v$ on $\mathcal{O}$".

(c) Prove that differentiation is local in the sense that if $u, v \in H^s(\mathbb{R}^n)$ and $\mathcal{O}$ is any open subset of $\mathbb{R}^n$ with $u = v$ on $\mathcal{O}$, then $\partial^\alpha u = \partial^\alpha v$ on $\mathcal{O}$ for all $\alpha \in \mathbb{N}_0^n$.

Exercise C.1.16 (Sobolev embedding theorem). Let $k \in \mathbb{N}_0$ and $s > k + \frac{n}{2}$. Prove that if $u \in H^s(\mathbb{R}^n)$, then $\partial^\alpha u$ is continuous whenever $\alpha \in \mathbb{N}_0^n$ and $|\alpha| \le k$, and there is a constant C, depending only on s, n and k, such that

$$\sum_{|\alpha|\le k} \|\partial^\alpha u\|_{L^\infty(\mathbb{R}^n)} \le C|||u|||_{H^s(\mathbb{R}^n)}.$$

Exercise C.1.17 (Logarithmic convexity of Sobolev norms). Let s and s' be real numbers and $0 \le \mu \le 1$. Prove that

$$|||u|||_{H^{\mu s+(1-\mu)s'}(\mathbb{R}^n)} \le |||u|||^{\mu}_{H^s(\mathbb{R}^n)}\ |||u|||^{1-\mu}_{H^{s'}(\mathbb{R}^n)}$$

for all $u \in H^{\max\{s,s'\}}(\mathbb{R}^n)$.

Exercise C.1.18. Let $\ell \in \mathbb{N}_0$, $\Omega \subset \mathbb{R}^n$ be open and $u \in H^\ell(\Omega)$.

(a) Let $K \subset \Omega$ be compact. Suppose that $u(x) = 0$ for all $x \in \Omega \setminus K$. Prove that $u \in H_0^\ell(\Omega)$.

(b) Let $K \subset \Omega$. Suppose that $u(x) = 0$ for all $x \in \Omega \setminus K$ and that, for each $R > 0$, $K \cap \{|x| \le R\}$ is compact. Prove that $u \in H_0^\ell(\Omega)$.

We proceed to define the Sobolev spaces $H^s(\Omega)$ for positive noninteger s via restriction from $\mathbb{R}^n$.

Definition C.1.19 ($H^s(\Omega)$, $s \ge 0$). If s happens to be an integer, we use Definition C.1.1. If $s \ge 0$ is not an integer, we define

$$H^s(\Omega) = \{\, u|_\Omega \mid u \in H^s(\mathbb{R}^n) \,\}$$

equipped with the quotient norm

$$\|u\|_{H^s(\Omega)} = \inf\{\, \|\tilde{u}\|_{H^s(\mathbb{R}^n)} \mid \tilde{u} \in H^s(\mathbb{R}^n) \text{ with } \tilde{u}|_\Omega = u \,\}.$$

If s is an integer, we could define a space $H^s_{\text{restriction}}(\Omega)$ via Definition C.1.19. However, if Ω has smooth boundary, the spaces $H^s(\Omega)$ and $H^s_{\text{restriction}}(\Omega)$ are equal and have equivalent norms. This follows from the existence of a bounded extension operator (Lemma C.2.19).

Definition C.1.20 ($H_0^s(\Omega)$, $s \ge 0$). If s happens to be an integer, we use Definition C.1.1. Otherwise, we define $H_0^s(\Omega)$ to be the completion of $C_c^\infty(\Omega)$ under the norm $|||\cdot|||_{H^s(\mathbb{R}^n)}$.

Exercise C.1.21 (Compact Sobolev embedding). Let $s < t$ and Ω be a bounded open subset of $\mathbb{R}^n$. Since $|||u|||_{H^s(\mathbb{R}^n)} \le |||u|||_{H^t(\mathbb{R}^n)}$, $H_0^t(\Omega) \subset H_0^s(\Omega)$. The goal of this problem is to prove Rellich's theorem, which states that any bounded subset of $H_0^t(\Omega)$ is precompact when viewed as a subset of $H_0^s(\Omega)$.

(a) Let ψ be a real valued function in $C_c^\infty(\mathbb{R}^n)$ that is identically one on Ω. Prove that if $u \in C_c^\infty(\Omega)$ then, for all $\alpha \in \mathbb{N}_0^n$,

$$|\partial_\xi^\alpha \hat{u}(\xi)| \le |||u|||_{H^t(\mathbb{R}^n)} \Big|\Big|\Big| \psi_\xi^\alpha \Big|\Big|\Big|_{H^{-t}(\mathbb{R}^n)},$$

where $\psi_\xi^\alpha(x) = x^\alpha e^{i\xi\cdot x}\psi(x)$.

(b) Prove that if $u \in H_0^t(\Omega)$ then, for all $\alpha \in \mathbb{N}_0^n$, $\partial_\xi^\alpha \hat{u}(\xi)$ exists, is continuous and obeys the bound of part (a).

(c) Let $r > 0$. Prove that if $\{u_i\}_{i\in\mathbb{N}} \subset \big\{ u \in H_0^t(\Omega) \mid |||u|||_{H^t(\mathbb{R}^n)} \le r \big\}$, then there is a subsequence u_{i_j} which converges in $H_0^s(\Omega)$.

Hints: (1) Let $R > 0$. Use the Arzelà–Ascoli theorem to prove the existence of a subsequence that converges uniformly on $|\xi| \le R$. (2) Bound the two terms in

$$\begin{aligned} &\big|\big|\big| u_{i_j} - u_{i_\ell} \big|\big|\big|_{H^s(\mathbb{R}^n)}^2 \\ &\quad = \int_{|\xi|\le R} (1+|\xi|^2)^s \big|\hat{u}_{i_j}(\xi) - \hat{u}_{i_\ell}(\xi)\big|^2 \frac{d\xi}{(2\pi)^n} \\ &\quad\quad + \int_{|\xi|>R} (1+|\xi|^2)^{-(t-s)} (1+|\xi|^2)^t \big|\hat{u}_{i_j}(\xi) - \hat{u}_{i_\ell}(\xi)\big|^2 \frac{d\xi}{(2\pi)^n} \end{aligned}$$

separately.

Our next job is to define $H^s(\Omega)$ for negative s. For motivation, we again find a characterization of $H^s(\mathbb{R}^n)$ that does not use the Fourier transform. Recall, from Definition A.2.1(c) that a bounded linear functional on a Banach space $\mathcal{B}$ is a linear map $\mathcal{L} : \mathcal{B} \to \mathbb{C}$ obeying

$$\|\mathcal{L}\| = \sup_{\substack{v\in\mathcal{B} \\ \|v\|\ne 0}} \frac{|\mathcal{L}v|}{\|v\|} < \infty$$

and, from Definition A.2.1(d), that the dual space, $\mathcal{B}^*$, of $\mathcal{B}$ is the Banach space of all bounded linear functionals on $\mathcal{B}$. The following Proposition proves that

$$H^s(\mathbb{R}^n)^* \cong H^{-s}(\mathbb{R}^n)$$

for all $s \in \mathbb{R}$.

Proposition C.1.22. *Let $s \in \mathbb{R}$.*

(a) *Let $u \in H^s(\mathbb{R}^n)$ and $v \in H^{-s}(\mathbb{R}^n)$. Then the map*

$$\begin{aligned} H^s(\mathbb{R}^n) \times H^{-s}(\mathbb{R}^n) &\to \mathbb{C}, \\ (u, v) &\mapsto \int \hat{u}(\xi)\overline{\hat{v}(\xi)} \frac{d\xi}{(2\pi)^n} \end{aligned}$$

is sesquilinear (i.e., linear in the first argument and conjugate linear in the second) and obeys

$$\Big| \int \hat{u}(\xi)\overline{\hat{v}(\xi)} \frac{d\xi}{(2\pi)^n} \Big| \le |||u|||_{H^s(\mathbb{R}^n)}\, |||v|||_{H^{-s}(\mathbb{R}^n)}.$$

(b) *If $\mathcal{L} \in H^s(\mathbb{R}^n)^*$, then there exists $v \in H^{-s}(\mathbb{R}^n)$ such that*

$$\mathcal{L}u = \int \hat{u}(\xi)\overline{\hat{v}(\xi)}\, \frac{d\xi}{(2\pi)^n}.$$

Furthermore $\|\mathcal{L}\| = |||v|||_{H^{-s}(\mathbb{R}^n)}$.

Proof.

(a) By Cauchy–Schwarz

$$\left|\int \hat{u}(\xi)\overline{\hat{v}(\xi)}\, \frac{d\xi}{(2\pi)^n}\right| \le \int \left|(1+|\xi|^2)^{s/2}\hat{u}(\xi)\right| \left|(1+|\xi|^2)^{-s/2}\hat{v}(\xi)\right| \frac{d\xi}{(2\pi)^n}$$
$$\le |||u|||_{H^s(\mathbb{R}^n)}\, |||v|||_{H^{-s}(\mathbb{R}^n)}.$$

The linearity in u and conjugate linearity in v are obvious.

(b) Let $\mathcal{L} \in H^s(\mathbb{R}^n)^*$. By the Riesz representation theorem (Theorem A.2.15) there is a $g \in H^s(\mathbb{R}^n)$ such that

$$\mathcal{L}u = \langle u, g\rangle_s = \int (1+|\xi|^2)^s\, \hat{u}(\xi)\, \overline{\hat{g}(\xi)}\, \frac{d\xi}{(2\pi)^n}.$$

Furthermore $\|\mathcal{L}\| = |||g|||_{H^s(\mathbb{R}^n)}$. Then the $v \in H^{-s}(\mathbb{R}^n)$ with

$$\hat{v}(\xi) = (1+|\xi|^2)^s\, \hat{g}(\xi),$$

using the Fourier transform of Lemma C.1.11 fulfills the requirements of the lemma. □

Exercise C.1.23.

(a) Let $s \in \mathbb{R}$ and $v \in H^{-s}(\mathbb{R}^n)$. Prove that

$$|||v|||_{H^{-s}(\mathbb{R}^n)} = \sup_{\substack{u \in H^s(\mathbb{R}^n) \\ |u|_{H^s(\mathbb{R}^n)} \le 1}} \left|\int \hat{u}(\xi)\overline{\hat{v}(\xi)}\, \frac{d\xi}{(2\pi)^n}\right|.$$

(b) Let $v \in L^2(\mathbb{R}^n)$ and $s \ge 0$. Prove that

$$|||v|||_{H^{-s}(\mathbb{R}^n)} = \sup_{\substack{u \in H^s(\mathbb{R}^n) \\ |u|_{H^s(\mathbb{R}^n)} \le 1}} \left|\langle u, v\rangle_{L^2(\mathbb{R}^n)}\right|.$$

(c) Let $t < s$ and $v \in H^t(\mathbb{R}^n)$. Prove that if

$$M = \sup\left\{ \left|\int \hat{u}(\xi)\overline{\hat{v}(\xi)}\, \frac{d\xi}{(2\pi)^n}\right| \;\middle|\; \hat{u} \in C_c^\infty(\mathbb{R}^n),\ |||u|||_{H^{-s}(\mathbb{R}^n)} \le 1 \right\} < \infty,$$

then $v \in H^s(\mathbb{R}^n)$ and $|||v|||_{H^s(\mathbb{R}^n)} = M$.

Corollary C.1.24 (Representation for $u \in H^{-\ell}$). *Let $\ell \in \mathbb{N}_0$. There are constants c, C, depending only on n and ℓ such that for each $u \in H^{-\ell}(\mathbb{R}^n)$ there are $f_\alpha \in L^2(\mathbb{R}^n)$, for all $\alpha \in \mathbb{N}_0^n$, with $|\alpha| \le \ell$, such that*

$$u = \sum_{\substack{\alpha\in\mathbb{N}_0^n \\ |\alpha|\le\ell}} \partial^\alpha f_\alpha, \qquad c|||u|||^2_{H^{-\ell}(\mathbb{R}^n)} \le \sum_{\substack{\alpha\in\mathbb{N}_0^n \\ |\alpha|\le\ell}} \|f_\alpha\|^2_{L^2(\mathbb{R}^n)} \le C|||u|||^2_{H^{-\ell}(\mathbb{R}^n)}.$$

Proof. By Lemma C.1.11, $(1+|\xi|^2)^{-\ell/2}\hat{u}(\xi) \in L^2(\mathbb{R}^n)$. Let L be any integer with $L \ge \frac{\ell}{2}$. Set $F(\xi) = (1+|\xi|^2)^{-L}\hat{u}(\xi)$ and write

$$(1+|\xi|^2)^L = \sum_{\substack{\alpha\in\mathbb{N}_0^n \\ |\alpha|\le\ell}} \xi^\alpha P_\alpha(\xi)$$

where $P_\alpha(\xi)$ is a polynomial in $\xi_1, \cdots, \xi_n$ of degree at most $2L-\ell$. There are many such representations. Pick any one you like. For each $\alpha \in \mathbb{N}_0^n$ with $|\alpha| \le \ell$ determine f_α by $\hat{f}_\alpha = (-i)^{|\alpha|}P_\alpha(\xi)F(\xi)$. Since $P_\alpha(\xi)^2$ is of degree at most $4L-2\ell$, there is a constant $\tilde{C}$, which depends on n, L and the P_α's we chose, but not on u, such that $P_\alpha(\xi)^2 \le \tilde{C}(1+|\xi|^2)^{2L-\ell}$. Hence

$$\begin{aligned}
\left\|f_\alpha\right\|^2_{L^2(\mathbb{R}^n)} &= \int P_\alpha^2(\xi)|F(\xi)|^2 \frac{d\xi}{(2\pi)^n} \\
&= \int \frac{P_\alpha^2(\xi)}{(1+|\xi|^2)^{2L}}|\hat{u}(\xi)|^2 \frac{d\xi}{(2\pi)^n} \\
&\le \tilde{C}\int \frac{|\hat{u}(\xi)|^2}{(1+|\xi|^2)^\ell}\frac{d\xi}{(2\pi)^n} \\
&= \tilde{C}|||u|||^2_{H^{-\ell}(\mathbb{R}^n)}
\end{aligned}$$

and

$$\sum_{\substack{\alpha\in\mathbb{N}_0^n \\ |\alpha|\le\ell}} (i\xi)^\alpha \hat{f}_\alpha(\xi) = \sum_{\substack{\alpha\in\mathbb{N}_0^n \\ |\alpha|\le\ell}} \xi^\alpha P_\alpha(\xi)F(\xi) = (1+|\xi|^2)^L F(\xi) = \hat{u}(\xi).$$

The final remaining bound is

$$
\begin{aligned}
|||u|||^2_{H^{-\ell}(\mathbb{R}^n)} &= \int \frac{|\hat{u}(\xi)|^2}{(1+|\xi|^2)^{\ell}} \frac{d\xi}{(2\pi)^n} \\
&= \int \frac{|\Sigma_\alpha (i\xi)^\alpha \hat{f}_\alpha(\xi)|^2}{(1+|\xi|^2)^{\ell}} \frac{d\xi}{(2\pi)^n} \\
&\le (\ell+1)^n \sum_{\substack{\alpha\in\mathbb{N}_0^n \\ |\alpha|\le\ell}} \int \frac{|\xi^{2\alpha}|}{(1+|\xi|^2)^{\ell}} |\hat{f}_\alpha(\xi)|^2 \frac{d\xi}{(2\pi)^n} \\
&\le (\ell+1)^n \sum_{\substack{\alpha\in\mathbb{N}_0^n \\ |\alpha|\le\ell}} \int |\hat{f}_\alpha(\xi)|^2 \frac{d\xi}{(2\pi)^n} \\
&= (\ell+1)^n \sum_{\substack{\alpha\in\mathbb{N}_0^n \\ |\alpha|\le\ell}} \|f_\alpha\|^2_{L^2(\mathbb{R}^n)}.
\end{aligned}
$$

We used Cauchy–Schwarz to get from the first line to the second line. □

Definition C.1.25 ($H^s(\Omega)$, $s < 0$). Let Ω be an open subset of $\mathbb{R}^n$. If $s < 0$, we define $H^s(\Omega) = H_0^{-s}(\Omega)^*$. The norm on $H_0^{-s}(\Omega)^*$ is denoted $\|\cdot\|_{H^s(\Omega)}$.

Let $s < 0$. By Exercise C.1.13, $H^{-s}(\mathbb{R}^n) = H_0^{-s}(\mathbb{R}^n)$. So Proposition C.1.22 states that $H^s(\mathbb{R}^n) \cong H^{-s}(\mathbb{R}^n)^* = H_0^{-s}(\mathbb{R}^n)^*$. Thus, Definition C.1.25, in the special case $\Omega = \mathbb{R}^n$, is consistent with our Definition C.1.10 of $H^s(\mathbb{R}^n)$. But in general, $H_0^{-s}(\Omega)$ can be a closed strict subspace of $H^{-s}(\Omega)$. By the Hahn–Banach theorem (Theorem A.2.12(a)), every bounded linear functional on $H_0^{-s}(\Omega)$ extends to a bounded linear functional on $H^{-s}(\Omega)$. But when $H_0^{-s}(\Omega)$ is a closed strict subspace of $H^{-s}(\Omega)$, every bounded linear functional on $H_0^{-s}(\Omega)$ can be extended to many different bounded linear functionals on $H^{-s}(\Omega)$ and $H^{-s}(\Omega)^*$ is much larger than $H_0^{-s}(\Omega)^*$. We shall prove, in Theorem C.1.26, that every bounded linear functional $\mathcal{L} \in H_0^{\ell}(\Omega)^*$, $\ell \in \mathbb{N}_0$, is of the form

$$\mathcal{L}u = \sum_{\substack{\alpha\in\mathbb{N}_0^n \\ |\alpha|\le\ell}} \langle \partial^\alpha u, f_\alpha\rangle_{L^2(\Omega)}$$

for some $f_\alpha \in L^2(\Omega)$, $\alpha \in \mathbb{N}_0^n$, $|\alpha| \le \ell$. The same representation, and even the same proof, also applies to $\mathcal{L} \in H^{\ell}(\Omega)^*$. The difference is that, when the f_α's are sufficiently regular and we integrate by parts to move the ∂^α's from the u to the f_α's, there are no boundary terms when $u \in H_0^{\ell}(\Omega)$ while there can be boundary terms when $u \in H^{\ell}(\Omega)$. This makes it more useful to define $H^s(\Omega) = H_0^{-s}(\Omega)^*$ rather than $H^s(\Omega) = H^{-s}(\Omega)^*$.

Theorem C.1.26 (Representation for $\mathcal{L} \in H^{-\ell}(\Omega)$). *Let $\ell \in \mathbb{N}_0$.*

(a) *Let $\mathcal{L} \in H_0^\ell(\Omega)^*$. There exist $f_\alpha \in L^2(\Omega)$, $\alpha \in \mathbb{N}_0^n$, $|\alpha| \leq \ell$ such that*

$$\mathcal{L}u = \sum_{\substack{\alpha\in\mathbb{N}_0^n \\ |\alpha|\leq\ell}} \langle \partial^\alpha u, f_\alpha \rangle_{L^2(\Omega)} \tag{C.3}$$

for all $u \in H_0^\ell(\Omega)$. Furthermore

$$\|\mathcal{L}\| = \min\Big\{\Big[\sum_{|\alpha|\leq\ell} \|f_\alpha\|^2_{L^2(\Omega)}\Big]^{1/2} \;\Big|\; (f_\alpha)_{|\alpha|\leq\ell} \text{ obeys (C.3)}\Big\}.$$

(b) *If $v \in L^2(\Omega)$ then*

$$\mathcal{L}_v u = \langle u, \bar{v} \rangle_{L^2(\Omega)}$$

defines a bounded linear functional on $H_0^\ell(\Omega)$ whose norm satisfies

$$\|\mathcal{L}_v\|_{H^{-\ell}(\Omega)} \leq \|v\|_{L^2(\Omega)}.$$

Furthermore $\{\, \mathcal{L}_v \mid v \in C_c^\infty(\Omega) \,\}$ is dense in $H_0^\ell(\Omega)^*$.

Proof.

(a) Let $N = \#\{\, \alpha \in \mathbb{N}_0^n \mid |\alpha| \leq \ell \,\}$ and L_N^2 be the Hilbert space

$$\bigoplus_{\substack{\alpha\in\mathbb{N}_0^n \\ |\alpha|\leq\ell}} L^2(\Omega).$$

The elements of L_N^2 are of the form $\big(f_\alpha\big)_{|\alpha|\leq\ell}$ with each component $f_\alpha \in L^2(\Omega)$. The inner product on L_N^2 is

$$\langle \big(f_\alpha\big)_{|\alpha|\leq\ell}, \big(g_\alpha\big)_{|\alpha|\leq\ell} \rangle = \sum_{|\alpha|\leq\ell} \langle f_\alpha, g_\alpha \rangle_{L^2(\Omega)}.$$

If $|\alpha| \leq \ell$, the map $v \in \big\{\, v \in C^\ell(\Omega) \mid \|v\|_{H^\ell(\Omega)} < \infty \,\big\} \mapsto \partial^\alpha v$ has a unique extension to a bounded, linear map from $H^\ell(\Omega)$ to $L^2(\Omega)$. Define the map

$$\begin{aligned} P : H_0^\ell(\Omega) &\to L_N^2(\Omega) \\ u &\mapsto \big(\partial^\alpha u\big)_{|\alpha|\leq\ell} \end{aligned}$$

This is a unitary map from $H_0^\ell(\Omega)$ to its range W, which is a closed linear subspace of $L_N^2(\Omega)$. Hence

$$\tilde{\mathcal{L}}(Pu) = \mathcal{L}(u)$$

defines a bounded linear functional on W with norm $\|\mathcal{L}\|$. As W is itself a Hilbert space, the Riesz representation theorem (Theorem A.2.15) implies that there is a vector $(f_\alpha)_{|\alpha|\leq\ell}$ in W such that

$$\tilde{\mathcal{L}}\big((v_\alpha)_{|\alpha|\leq\ell}\big) = \langle \big(v_\alpha\big)_{|\alpha|\leq\ell}, (f_\alpha)_{|\alpha|\leq\ell} \rangle = \sum_{|\alpha|\leq\ell} \langle v_\alpha, f_\alpha \rangle_{L^2(\Omega)}$$

and that $\|\tilde{\mathcal{L}}\| = \left\|(f_\alpha)_{|\alpha|\le\ell}\right\|_{L^2_N(\Omega)}$. Hence

$$\mathcal{L}(u) = \tilde{\mathcal{L}}(Pu) = \sum_{\substack{\alpha\in\mathbb{N}_0^n\\|\alpha|\le\ell}} \langle \partial^\alpha u, f_\alpha\rangle_{L^2(\Omega)}.$$

If $(\tilde{f}_\alpha)_{|\alpha|\le\ell}$ is any other vector in $L^2_N(\Omega)$ that obeys

$$\tilde{\mathcal{L}}\big((v_\alpha)_{|\alpha|\le\ell}\big) = \langle (v_\alpha)_{|\alpha|\le\ell}, (\tilde{f}_\alpha)_{|\alpha|\le\ell}\rangle = \sum_{|\alpha|\le\ell} \langle v_\alpha, \tilde{f}_\alpha\rangle_{L^2(\Omega)}$$

for all $(v_\alpha)_{|\alpha|\le\ell} \in W$, then

$$\langle (v_\alpha)_{|\alpha|\le\ell}, (\tilde{f}_\alpha - f_\alpha)_{|\alpha|\le\ell}\rangle = 0$$

for all $(v_\alpha)_{|\alpha|\le\ell} \in W$. This says that $(\tilde{f}_\alpha - f_\alpha)_{|\alpha|\le\ell} \in W^\perp$, the orthogonal complement of W in $L^2_N(\Omega)$. Consequently,

$$\left\|(\tilde{f}_\alpha)_{|\alpha|\le\ell}\right\|_{L^2_N(\Omega)} \ge \left\|(f_\alpha)_{|\alpha|\le\ell}\right\|_{L^2_N(\Omega)} = \|\tilde{\mathcal{L}}\| = \|\mathcal{L}\|.$$

(b) Let $v \in L^2(\Omega)$. Since $H_0^\ell(\Omega) \subset L^2(\Omega)$, $\mathcal{L}_v$ is a well–defined linear functional on $H_0^\ell(\Omega)$. The boundedness and the claimed bound follow from

$$|\mathcal{L}_v u| = |\langle u, \bar{v}\rangle_{L^2(\Omega)}| \le \|v\|_{L^2(\Omega)}\|u\|_{L^2(\Omega)} \le \|v\|_{L^2(\Omega)}\,\|u\|_{H^\ell(\Omega)}.$$

Now we prove the denseness. Since $H_0^\ell(\Omega)$ is a Hilbert space, the Riesz representation theorem (Theorem A.2.15) says that, for each $\mathcal{L} \in H_0^\ell(\Omega)^*$, there is a vector $u_{\mathcal{L}} \in H_0^\ell(\Omega)$ such that $\mathcal{L}w = \langle w, u_{\mathcal{L}}\rangle_{H_0^\ell(\Omega)}$ for all $w \in H_0^\ell(\Omega)$. The map $\mathcal{L} \mapsto u_{\mathcal{L}}$ is an isometry from $H_0^\ell(\Omega)^*$ to $H_0^\ell(\Omega)$. So if

$$\{\, \mathcal{L}_v \mid v \in C_c^\infty(\Omega) \,\}$$

is not dense in $H_0^\ell(\Omega)^*$, then

$$\{\, u_{\mathcal{L}_v} \mid v \in C_c^\infty(\Omega) \,\}$$

is not dense in $H_0^\ell(\Omega)$ and there is a nonzero vector $w \in H_0^\ell(\Omega)$ that is orthogonal to

$$\{\, u_{\mathcal{L}_v} \mid u \in C_c^\infty(\Omega) \,\}.$$

But

$$w \perp \{\, u_{\mathcal{L}_v} \mid v \in C_c^\infty(\Omega) \,\}$$
$$\implies 0 = \langle w, u_{\mathcal{L}_v}\rangle_{H_0^\ell(\Omega)} = \mathcal{L}_v(w) = \langle w, \bar{v}\rangle_{L^2(\Omega)} \text{ for all } v \in C_c^\infty(\Omega).$$

Since $C_c^\infty(\Omega)$ is dense in $L^2(\Omega)$, w must be zero as an element of $L^2(\Omega)$ and hence also as an element of $H_0^\ell(\Omega)$, which contradicts the assumption that $\{\, \mathcal{L}_v \mid v \in C_c^\infty(\Omega) \,\}$ is not dense in $H_0^\ell(\Omega)^*$. □

Remark C.1.27. For each $\ell \in \mathbb{N}$, we identify each $v \in L^2(\Omega)$ with the corresponding $\mathcal{L}_v \in H^{-\ell}(\Omega) = H_0^\ell(\Omega)^*$ of Theorem C.1.26. Using this identification, $H^{-\ell}(\Omega)$ is the completion of $L^2(\Omega)$ under the norm

$$\|v\|_{H^{-\ell}(\Omega)} = \sup_{0\neq u\in H_0^\ell(\Omega)} |\langle u, \bar{v}\rangle_{L^2(\Omega)}| / \|u\|_{H^\ell(\Omega)}\,.$$

We also identify each vector $v \in H^0(\Omega) = L^2(\Omega)$ with the corresponding $\mathcal{L}_v \in H_0^0(\Omega)^*$ of Theorem C.1.26, so that $L^2(\Omega) = H^0(\Omega) = H_0^0(\Omega)^*$.

Remark C.1.28. Let $\ell', \ell \in \mathbb{N}_0$ with $\ell' \le \ell$. We saw in Remark C.1.3 that $H_0^\ell(\Omega) \subset H_0^{\ell'}(\Omega)$ and $\|u\|_{H^{\ell'}(\Omega)} \le \|u\|_{H^\ell(\Omega)}$ for all $u \in H_0^\ell(\Omega)$. Now consider any $v \in H^{-\ell'}(\Omega) = H_0^{\ell'}(\Omega)^*$. It obeys

$$|v(u)| \le \|v\|_{H^{-\ell'}(\Omega)} \|u\|_{H^{\ell'}(\Omega)} \le \|v\|_{H^{-\ell'}(\Omega)} \|u\|_{H^\ell(\Omega)}$$

for all $u \in H_0^\ell(\Omega)$. Thus the restriction of v to $H_0^\ell(\Omega)$ is an element of $H^{-\ell}(\Omega) = H_0^\ell(\Omega)^*$, which we again call v, and $\|v\|_{H^{-\ell}(\Omega)} \le \|v\|_{H^{-\ell'}(\Omega)}$.

Combining this with the corresponding part of Remark C.1.3, we have that if $\ell, \ell' \in \mathbb{Z}$ with $\ell \le \ell'$, then $H^{\ell'}(\Omega) \subset H^\ell(\Omega)$ and $\|v\|_{H^\ell(\Omega)} \le \|v\|_{H^{\ell'}(\Omega)}$ for all $v \in H^{\ell'}(\Omega)$.

Lemma C.1.29. *Let $\ell \in \mathbb{Z}$.*

(a) *Let $\varphi \in W^{|\ell|,\infty}(\Omega)$, i.e., φ and all of its derivatives of order at most $|\ell|$ are in $L^\infty(\Omega)$. The map*

$$v \in \left\{ v \in C^{|\ell|}(\Omega) \;\middle|\; \|v\|_{H^{\max\{0,\ell\}}(\Omega)} < \infty \right\} \mapsto \varphi v$$

has a unique extension to a bounded, linear map on $H^\ell(\Omega)$ and there is a constant $C_{|\ell|,n}$, depending only on $|\ell|$ and n, such that

$$\|\varphi u\|_{H^\ell(\Omega)} \le C_{|\ell|,n} \|\varphi\|_{W^{|\ell|,\infty}(\Omega)} \|u\|_{H^\ell(\Omega)}$$

for all $u \in H^\ell(\Omega)$.

(b) *Let $\alpha \in \mathbb{N}_0^n$. In the case $0 < \ell < |\alpha|$, assume that*

$$\left\{ v \in C^{|\alpha|}(\Omega) \;\middle|\; \|v\|_{H^{|\alpha|}(\Omega)} < \infty \right\}$$

is dense in $H^\ell(\Omega)$. The map

$$v \in \left\{ v \in C^{\max\{\ell,|\alpha|\}}(\Omega) \;\middle|\; \|v\|_{H^{\max\{\ell,|\alpha|\}}(\Omega)} < \infty \right\} \mapsto \partial^\alpha v$$

has a unique extension to a bounded linear map from $H^\ell(\Omega)$ to $H^{\ell-|\alpha|}(\Omega)$.

Exercise C.1.30. Prove Lemma C.1.29.

Exercise C.1.31. This problem illustrates the need for the multiplier φ of part (a) of Lemma C.1.29 to be relatively smooth if the product φv is to be well-defined when v is quite unsmooth. In this problem, φ will be a characteristic function with a discontinuity at 0 and there will be two different v's. One will be a Dirac delta function and the other the derivative of a Dirac delta function. Both will be supported on the point 0. Let $\Omega = (-1,1)$. Let $w \in C_c^\infty(\mathbb{R})$ be supported in $\left[-\frac{1}{2},\frac{1}{2}\right]$. Assume that w is even and obeys $\int_{\mathbb{R}} w(x)\,dx = 1$ and $w'(x) \geq 0$ for $x < 0$. Define, for $\varepsilon > 0$ and $u \in C_c^\infty(\Omega)$,

$$\delta(u) = u(0),$$

$$\delta_\varepsilon(u) = \int_{-1}^{1} \frac{1}{\varepsilon}\, w\Big(\frac{x}{\varepsilon}\Big) u(x)\,dx, \qquad \delta_{\varepsilon,+}(u) = \int_{-1}^{1} \frac{1}{\varepsilon}\, w\Big(\frac{x-\varepsilon}{\varepsilon}\Big)\, u(x)\,dx,$$

$$\delta'(u) = -u'(0), \qquad \delta_\varepsilon'(u) = \int_{-1}^{1} \frac{1}{\varepsilon^2}\, w'\Big(\frac{x}{\varepsilon}\Big)\, u(x)\,dx.$$

(a) Prove that δ, δ_ε, and $\delta_{\varepsilon,+}$ all have unique continuous extensions to elements of $H^{-1}(\Omega)$. Prove that δ' and δ_ε' have unique continuous extensions to elements of $H^{-2}(\Omega)$.

(b) Prove that

$$\lim_{\varepsilon\to 0+} \delta_\varepsilon = \delta \qquad \text{and} \qquad \lim_{\varepsilon\to 0+} \delta_{\varepsilon,+} = \delta$$

in $H^{-2}(\Omega)$ and

$$\lim_{\varepsilon\to 0+} \delta_\varepsilon' = \delta'$$

in $H^{-3}(\Omega)$.

(c) Let φ be the characteristic function of $(0,1)$. Prove that

$$\lim_{\varepsilon\to 0+} \varphi\,\delta_\varepsilon = \tfrac{1}{2}\delta \qquad \text{and} \qquad \lim_{\varepsilon\to 0+} \varphi\,\delta_{\varepsilon,+} = \delta$$

in $H^{-2}(\Omega)$ and that $\lim_{\varepsilon\to 0+} \varphi\delta_\varepsilon'$ diverges in $H^{-\ell}(\Omega)$ for all $\ell \in \mathbb{N}$.

We will next give a multiplier result for fractional Sobolev spaces.

Lemma C.1.32 (Pointwise multipliers on $H^s(\Omega)$). *Let $\Omega \subset \mathbb{R}^n$ be an open set and let $s \in \mathbb{R}$. If $\varphi \in W^{\ell,\infty}(\Omega)$ for some integer ℓ with $\ell \geq |s|$, then the map*

$$u \mapsto \varphi u$$

extends to a bounded linear map $H^s(\Omega) \to H^s(\Omega)$ with

$$\|\varphi u\|_{H^s(\Omega)} \leq C_{\ell,n}\, \|\varphi\|_{W^{\ell,\infty}(\Omega)}\, \|u\|_{H^s(\Omega)}\,.$$

Proof. The result follows from Lemma C.1.29 applied in $H^\ell(\Omega)$ and $H^{-\ell}(\Omega)$, and the interpolation result in Lemma C.3.2. □

In the final part of this section, we want to define the space $H^\ell(\partial\Omega)$ in the same way as the space $H^\ell(\Omega)$, but for functions that are only defined on $\partial\Omega$. To do so, we will employ local coordinates and a partition of unity. The following notions have appeared already in §2.2.

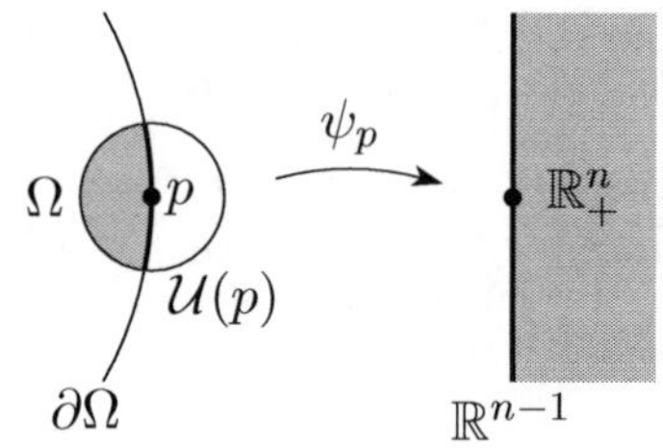

Figure C.1. Boundary flattening map ψ_p.

Definition C.1.33. Let $\Omega \subset \mathbb{R}^n$ be a bounded open set.

(a) We say that Ω has C^k boundary (or that $\partial\Omega$ is C^k or that $\partial\Omega \in C^k$) if, for each $p \in \partial\Omega$, there is an open neighborhood $\mathcal{U}(p)$ of p and a C^k diffeomorphism $\psi_p : \mathcal{U}(p) \to \mathbb{R}^n$ such that, as in Figure C.1,

$$\begin{aligned} \psi_p\big(\mathcal{U}(p) \cap \Omega\big) &= \big\{\, x \in \mathbb{R}^n \mid x_n > 0 \,\big\} = \mathbb{R}^n_+, \\ \psi_p\big(\mathcal{U}(p) \cap \partial\Omega\big) &= \big\{\, x \in \mathbb{R}^n \mid x_n = 0 \,\big\} = \mathbb{R}^{n-1}. \end{aligned} \tag{C.4}$$

(b) We say that Ω has smooth boundary (or that $\partial\Omega$ is smooth or that $\partial\Omega \in C^\infty$) if each ψ_p, $p \in \partial\Omega$, of (a) is a C^∞ diffeomorphism.

Exercise C.1.34. Let $\Omega \subset \mathbb{R}^n$ be a bounded open set and $k \in \mathbb{N}$. Prove that Ω has C^k boundary if and only if, for each $p \in \partial\Omega$, there is an open neighborhood $\mathcal{U}(p)$ of p and a C^k function $\phi_p : \mathcal{U}(p) \to \mathbb{R}$ such that $\nabla\phi_p(p) \neq 0$, $\mathcal{U}(p) \cap \partial\Omega = \phi_p^{-1}(0)$ and $\mathcal{U}(p) \cap \Omega = \phi_p^{-1}((0, \infty))$.

For the rest of this appendix, we assume that Ω is a bounded open subset of $\mathbb{R}^n$ with C^k boundary for some $k \geq 1$.

Definition C.1.35. We call a system $\big(\mathcal{U}(p), \psi_p\big)_{p \in \partial\Omega}$ as in Definition C.1.33, a coordinate system for $\partial\Omega$.

Definition C.1.36. We say that $f : \partial\Omega \to \mathbb{C}$ is C^k and write $f \in C^k(\partial\Omega)$ if there is a coordinate system $\big(\mathcal{U}(p), \psi_p\big)_{p \in \partial\Omega}$ such that $f \circ \psi_p^{-1} : \mathbb{R}^{n-1} \to \mathbb{C}$ is C^k.

Definition C.1.37 ($H^s(\partial\Omega)$, $-k + 1 \leq s \leq k$). Let $\big(\mathcal{U}(p), \psi_p\big)_{p \in \partial\Omega}$ be a coordinate system for $\partial\Omega$. Since $\partial\Omega$ is compact, there are p_i, $i = 1, \ldots, N$, so that $\partial\Omega \subset \cup_{i=1}^N \mathcal{U}(p_i)$. Choose for $1 \leq i \leq N$ a function $\chi_i \in C_c^\infty\big(\mathcal{U}(p_i)\big)$, taking values in $[0, 1]$, so that $\sum_{i=1}^N \chi_i = 1$ on some neighborhood of $\partial\Omega$.

(a) For $f \in C^k(\partial\Omega)$ we define

$$\|f\|^2_{H^s(\partial\Omega)} = \sum_{i=1}^N \|(\chi_i f) \circ \psi_{p_i}^{-1}\|^2_{H^s(\mathbb{R}^{n-1})}.$$

(b) We define $H^s(\partial\Omega)$ to be the completion of $C^k(\partial\Omega)$ under the norm $\|\cdot\|_{H^s(\partial\Omega)}$.

Exercise C.1.38. Prove that $\|\cdot\|_{H^s(\partial\Omega)}$, given in Definition C.1.37, is a norm. Prove further that

$$\langle f, g\rangle_{H^s(\partial\Omega)} = \sum_{i=1}^{N} \langle (\chi_i f)\circ \psi_{p_i}^{-1}, (\chi_i g)\circ\psi_{p_i}^{-1}\rangle_{H^s(\mathbb{R}^{n-1})}$$

is an inner product.

Remark C.1.39. The reason for omitting the case $-k \le s \le -k+1$ in Definition C.1.37 is due to the lack of C^k diffeomorphism invariance in Lemma C.1.40 for that case. One could define $H^s(\partial\Omega)$ for $-k \le s \le -k+1$ as the dual space of $H^{-s}(\partial\Omega)$, but this will not be needed.

We prove in Proposition C.1.42 that all of the norms, defined through different choices of partition of unity $\{\chi_i\}$ and coordinate systems $(\mathcal{U}(p), \psi_p)$ in Definition C.1.37, are equivalent. That is, if $\|\cdot\|$ and $\|\cdot\|'$ are two such norms, then there are constants c and c' such that

$$\|f\| \le c\|f\|', \qquad \|f\|' \le c'\|f\|.$$

So a sequence of functions converges with respect to one of the norms if and only if it converges with respect to the other.

Lemma C.1.40. *Let $\varphi \in C_c^k(\mathbb{R}^m)$ and let $\psi : \mathbb{R}^m \to \mathbb{R}^m$ be a C^k diffeomorphism. Then the map*

$$\begin{aligned} C_c^k(\mathbb{R}^m) &\to C_c^k(\mathbb{R}^m)\\ f &\mapsto (\varphi f)\circ\psi^{-1} \end{aligned}$$

extends to a bounded linear map on $H^s(\mathbb{R}^m)$ whenever $-k+1 \le s \le k$.

Proof. By the B.L.T. theorem (Theorem A.2.13) and Exercise C.1.13, it suffices to prove that there is a constant C such that

$$\|(\varphi f)\circ\psi^{-1}\|_{H^s(\mathbb{R}^m)} \le C|\!|\!|f|\!|\!|_{H^s(\mathbb{R}^m)}$$

for all $f \in C_c^k(\mathbb{R}^m)$. The case of integer s with $0 \le s \le k$ is Exercise C.1.41. The case $s \ge 0$ then follows by interpolation via Lemma C.3.2.

So let $-k+1 \le s < 0$. Recalling, from Exercise C.1.23, that

$$|\!|\!|v|\!|\!|_{H^s(\mathbb{R}^m)} = \sup_{\|u\|_{H^{-s}(\mathbb{R}^m)}\le 1} |\langle u, v\rangle_{L^2(\mathbb{R}^m)}|$$

for all $v \in L^2(\mathbb{R}^m)$, we have

$$\|(\varphi f)\circ\psi^{-1}\|_{H^s(\mathbb{R}^m)} = \sup_{\|u\|_{H^{-s}(\mathbb{R}^m)}\le 1} |\langle u, (\varphi f)\circ\psi^{-1}\rangle_{L^2(\mathbb{R}^m)}|$$
$$= \sup_{\|u\|_{H^{-s}(\mathbb{R}^m)}\le 1} \left| \int u(x)\overline{(\varphi f)(\psi^{-1}(x))}\, dx \right|.$$

Making the change of variables $x = \psi(y)$ and using $|\frac{\partial x}{\partial y}|$ to denote the associated Jacobian determinant, we obtain

$$\|(\varphi f)\circ\psi^{-1}\|_{H^s(\mathbb{R}^m)} = \sup_{\|u\|_{H^{-s}(\mathbb{R}^m)}\le 1} \left| \int |\frac{\partial x}{\partial y}(y)| u(\psi(y))\overline{(\varphi f)(y)}\, dy \right|.$$

Set $\tilde{\varphi}(x) = |\frac{\partial x}{\partial y}(\psi^{-1}(x))|\overline{\varphi(\psi^{-1}(x))}$ and observe that $\tilde{\varphi} \in C_c^{k-1}(\mathbb{R}^m)$, since $\varphi \in C_c^k(\mathbb{R}^m)$ and ψ is a C^k diffeomorphism of $\mathbb{R}^m$. By the $s > 0$ case of this Lemma, with ψ replaced by ψ^{-1},

$$\|(\varphi f)\circ\psi^{-1}\|_{H^s(\mathbb{R}^m)} = \sup_{\|u\|_{H^{-s}(\mathbb{R}^m)}\le 1} \left| \int (\tilde{\varphi} u)(\psi(y))\overline{f(y)}\, dy \right|$$
$$\le \sup_{\|u\|_{H^{-s}(\mathbb{R}^m)}\le 1} \|(\tilde{\varphi} u)\circ\psi\|_{H^{-s}(\mathbb{R}^m)}\, |||f|||_{H^s(\mathbb{R}^m)}$$
$$\le \sup_{\|u\|_{H^{-s}(\mathbb{R}^m)}\le 1} C|||u|||_{H^{-s}(\mathbb{R}^m)}\, |||f|||_{H^s(\mathbb{R}^m)}$$
$$\le C\, |||f|||_{H^s(\mathbb{R}^m)}.$$ □

Exercise C.1.41. Prove Lemma C.1.40 in the special case that s is an integer with $0 \le s \le k$.

Proposition C.1.42. *All of the norms, defined through different choices of partition of unity $\{\chi_i\}$ and coordinate systems $(\mathcal{U}(p), \psi_p)$ in Definition C.1.37, are equivalent.*

Proof. Let $(\mathcal{U}(p), \psi_p)$ and $(\tilde{\mathcal{U}}(p), \tilde{\psi}_p)$ be two coordinate systems as in Notation C.1.35 and let $\chi_i \in C_c^\infty(\mathcal{U}(p_i))$, $1 \le i \le N$ and $\tilde{\chi}_i \in C_c^\infty(\tilde{\mathcal{U}}(\tilde{p}_i))$, $1 \le i \le \tilde{N}$ be two partitions of unity as in Definition C.1.37. If $f \in C^k(\partial\Omega)$, then

$$(\chi_i f)\circ\psi_{p_i}^{-1} = \sum_{j=1}^{\tilde{N}} (\tilde{\chi}_j\chi_i f)\circ\psi_{p_i}^{-1} = \sum_{j=1}^{\tilde{N}} (\tilde{\chi}_j\chi_i f)\circ\tilde{\psi}_{\tilde{p}_j}^{-1}\circ\tilde{\psi}_{\tilde{p}_j}\circ\psi_{p_i}^{-1}$$

Note that $(\tilde{\chi}_j\chi_i)\circ\psi_{p_i}^{-1}(x)$ vanishes except when $x \in \psi_{p_i}(\tilde{\mathcal{U}}(\tilde{p}_j)\cap\mathcal{U}(p_i))$ and then $\psi_{p_i}^{-1}(x)$ is in the domain of $\tilde{\psi}_{\tilde{p}_j}$. Applying Lemma C.1.40, with

$$\varphi = \chi_i\circ\tilde{\psi}_{\tilde{p}_j}^{-1}, \psi^{-1} = \tilde{\psi}_{\tilde{p}_j}\circ\psi_{p_i}^{-1}$$

and f replaced by $(\tilde{\chi}_j f) \circ \tilde{\psi}_{\tilde{p}_j}^{-1}$,

$$\begin{aligned}
\|(\chi_i f) \circ \psi_{p_i}^{-1}\|_{H^s(\mathbb{R}^{n-1})} &\le \sum_{j=1}^{\tilde{N}} \|(\tilde{\chi}_j \chi_i f) \circ \tilde{\psi}_{\tilde{p}_j}^{-1} \circ \tilde{\psi}_{\tilde{p}_j} \circ \psi_{p_i}^{-1}\|_{H^s(\mathbb{R}^{n-1})} \\
&\le \sum_{j=1}^{\tilde{N}} C\|(\tilde{\chi}_j f) \circ \tilde{\psi}_{\tilde{p}_j}^{-1}\|_{H^s(\mathbb{R}^{n-1})} \\
&\le C\tilde{N} \max_{1\le j\le \tilde{N}} \|(\tilde{\chi}_j f) \circ \tilde{\psi}_{\tilde{p}_j}^{-1}\|_{H^s(\mathbb{R}^{n-1})}
\end{aligned}$$

and hence

$$\begin{aligned}
\sum_{i=1}^{N} \|(\chi_i f) \circ \psi_{p_i}^{-1}\|^2_{H^s(\mathbb{R}^{n-1})} &\le C^2\tilde{N}^2 N \max_{1\le j\le \tilde{N}} \|(\tilde{\chi}_j f) \circ \tilde{\psi}_{\tilde{p}_j}^{-1}\|^2_{H^s(\mathbb{R}^{n-1})} \\
&\le C^2\tilde{N}^2 N \sum_{j=1}^{\tilde{N}} \|(\tilde{\chi}_j f) \circ \tilde{\psi}_{\tilde{p}_j}^{-1}\|^2_{H^s(\mathbb{R}^{n-1})},
\end{aligned}$$

completing the proof. □

Exercise C.1.43. Prove that $\|\cdot\|_{H^0(\partial\Omega)}$ is equivalent to

$$\|f\|_{L^2(\partial\Omega)} = \sqrt{\int_{\partial\Omega} |f(x)|^2 \, dS(x)},$$

where $dS(x)$ is the surface measure on $\partial\Omega$.

Exercise C.1.44. Let $|s| \le k-1$. Prove that there are constants C_s and c_s such that

$$|\langle f, \bar{g}\rangle_{L^2(\partial\Omega)}| \le C_s \|f\|_{H^{-s}(\partial\Omega)} \|g\|_{H^s(\partial\Omega)}$$

for all $f, g \in C^k(\partial\Omega)$ and

$$\|f\|_{H^{-s}(\partial\Omega)} \le c_s \sup\{ |\langle f, \bar{g}\rangle_{L^2(\partial\Omega)}| \mid g \in C^k(\partial\Omega),\ \|g\|_{H^s(\partial\Omega)} \le 1 \}$$

for all $f \in C^k(\partial\Omega)$. Here, as you would expect,

$$\langle f, g\rangle_{L^2(\partial\Omega)} = \int_{\partial\Omega} f(x)\overline{g(x)} \, dS(x),$$

where $dS(x)$ is the surface measure on $\partial\Omega$.

Exercise C.1.45. Let $0 \le s \le k-1$. By Exercise C.1.44, if $f \in C^k(\partial\Omega)$ then

$$g \in C^k(\partial\Omega) \mapsto \mathcal{L}_f g = \langle f, \bar{g}\rangle_{L^2(\partial\Omega)}$$

extends to a bounded linear functional on $H^s(\partial\Omega)$ with norm bounded by $C_s \|f\|_{H^{-s}(\partial\Omega)}$. Prove that the map $f \mapsto \mathcal{L}_f$ has a unique continuous extension to an isomorphism

$$\mathcal{L} : H^{-s}(\partial\Omega) \to H^s(\partial\Omega)^*$$

and that $\{ \mathcal{L}_f \mid f \in C^\infty(\partial\Omega) \}$ is dense in $H^s(\partial\Omega)^*$.

Exercise C.1.46 (Sobolev embedding). Let $\ell \in \mathbb{N}_0$ and $s > \ell + \frac{n}{2}$, $s \leq k$. Prove that if $u \in H^s(\partial\Omega)$, then $u \in C^\ell(\partial\Omega)$ and there is a constant C, depending only on Ω and $|\ell|$, such that

$$\|u\|_{C^\ell(\partial\Omega)} \leq C \|u\|_{H^s(\partial\Omega)} .$$

Exercise C.1.47 (Logarithmic convexity). Let $|s|, |s'| \leq k-1$ and $0 \leq \mu \leq 1$. Let N be the number of neighborhoods in the cover of $\partial\Omega$ used in Definition C.1.37. Prove that

$$\|f\|_{H^{\mu s+(1-\mu)s'}(\partial\Omega)} \leq N \|f\|^{\mu}_{H^s(\partial\Omega)} \|u\|^{1-\mu}_{H^{s'}(\partial\Omega)}$$

for all $f \in H^{\max\{s,s'\}}(\partial\Omega)$.

We conclude this section with a technical lemma required in §C.2.

Lemma C.1.48. *Let $\ell \in \mathbb{N}_0$.*

(a) *Let $\mathcal{U}$ and $\mathcal{V}$ be bounded open sets in $\mathbb{R}^n$ and let $\psi : \mathcal{U} \to \mathcal{V}$ be a $C^{\max\{1,\ell\}}$ diffeomorphism with $\psi \in C^1(\overline{\mathcal{U}})$ and $\psi^{-1} \in C^{\max\{1,\ell\}}(\overline{\mathcal{V}})$. Then the map*

$$\{ u \in C^\ell(\mathcal{U}) \mid \|u\|_{H^\ell(\mathcal{U})} < \infty \} \to C^\ell(\mathcal{V}),$$
$$u \mapsto u \circ \psi^{-1},$$

extends to a bounded linear map from $H^\ell(\mathcal{U})$ to $H^\ell(\mathcal{V})$.

(b) *Let $\mathcal{U}$, $\mathcal{V}$, ψ be as in part (a), except that we do not require $\mathcal{U}$ or $\mathcal{V}$ to be bounded and we do not require $\psi \in C^1(\overline{\mathcal{U}})$ or $\psi^{-1} \in C^{\max\{1,\ell\}}(\overline{\mathcal{V}})$. Then if $\varphi \in C_c^\infty(\mathcal{U})$, the map*

$$C_c^\infty(\mathcal{U}) \to C_0^\ell(\mathcal{V})$$
$$u \mapsto (\varphi u) \circ \psi^{-1}$$

extends to a bounded linear map from $H_0^\ell(\mathcal{U})$ to $H_0^\ell(\mathcal{V})$.

(c) *Let $\mathcal{U}$, $\mathcal{V}$ and Ω be open subsets of $\mathbb{R}^n$ and let $\psi : \mathcal{U} \to \mathcal{V}$ be a $C^{\max\{1,\ell\}}$ diffeomorphism. Then if $\chi \in C_c^\infty(\mathcal{U})$, the map (see Figure C.2)*

$$\{ u \in C^\ell(\Omega) \mid \|u\|_{H^\ell(\Omega)} < \infty \} \to C^\ell\big(\psi(\Omega \cap \mathcal{U})\big)$$
$$u \mapsto (\chi u) \circ \psi^{-1},$$

extends to a bounded linear map from $H^\ell(\Omega)$ to $H^\ell\big(\psi(\Omega \cap \mathcal{U})\big)$.

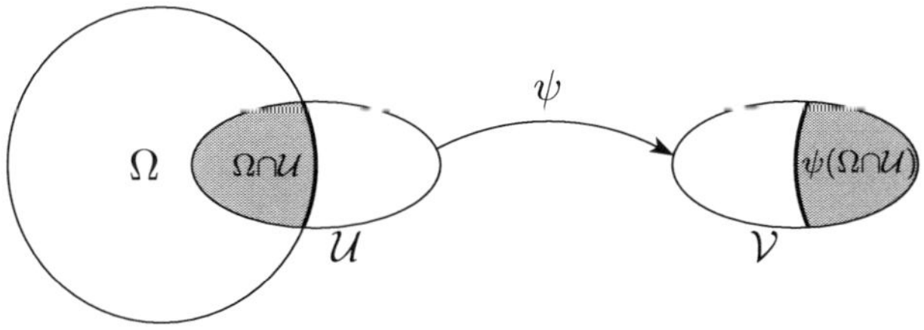

Figure C.2. The map ψ in Lemma C.1.48.

Proof.

(a) We start with $\ell = 0$. Making the change of variables $x = \psi(y)$ and using $|\frac{\partial x}{\partial y}|$ to denote the associated Jacobian determinant

$$\left\|u \circ \psi^{-1}\right\|^2_{H^0(\mathcal{V})} = \int_{\mathcal{V}} \left|u(\psi^{-1}(x))\right|^2 dx = \int_{\mathcal{U}} |u(y)|^2 \left|\frac{\partial x}{\partial y}(y)\right| dy.$$

Since $\psi \in C^1(\overline{\mathcal{U}})$, $\sup_{y\in\overline{\mathcal{U}}} \left|\frac{\partial x}{\partial y}(y)\right|$ is finite and

$$\left\|u \circ \psi^{-1}\right\|^2_{H^0(\mathcal{V})} \leq \sup_{y\in\overline{\mathcal{U}}} \left|\frac{\partial x}{\partial y}(y)\right| \|u\|^2_{H^0(\mathcal{U})}.$$

By definition, the set $\{\, u \in C^0(\mathcal{U}) \mid \|u\|_{H^0(\mathcal{U})} < \infty \,\}$ is dense in $H^0(\mathcal{U})$, so the B.L.T. theorem (Theorem A.2.13) yields the desired result.

Next we consider $\ell = 1$. By the chain rule

$$\frac{\partial}{\partial x_i} u \circ \psi^{-1}(x) = \sum_{j=1}^{n} \frac{\partial u}{\partial y_j}(\psi^{-1}(x)) \frac{\partial \psi_j^{-1}}{\partial x_i}(x).$$

Since $\psi^{-1} \in C^1(\overline{\mathcal{V}})$, $|\frac{\partial \psi_j^{-1}}{\partial x_i}(x)|$ is bounded on $\overline{\mathcal{V}}$ for all $1 \le j \le n$ so, by Cauchy–Schwarz,

$$\begin{aligned}
\int_{\mathcal{V}} \Big|\frac{\partial}{\partial x_i} u \circ \psi^{-1}(x)\Big|^2 dx &= \int_{\mathcal{V}} \Big|\sum_{j=1}^n \frac{\partial u}{\partial y_j}(\psi^{-1}(x)) \frac{\partial \psi_j^{-1}}{\partial x_i}(x)\Big|^2 dx \\
&\le \int_{\mathcal{V}} \Big[\sum_{j=1}^n \Big|\frac{\partial u}{\partial y_j}(\psi^{-1}(x))\Big|^2\Big]\Big[\sum_{j=1}^n \Big|\frac{\partial \psi_j^{-1}}{\partial x_i}(x)\Big|^2\Big] dx \\
&\le \sup_{x \in \overline{\mathcal{V}}} \Big[\sum_{j=1}^n \Big|\frac{\partial \psi_j^{-1}}{\partial x_i}(x)\Big|^2\Big] \int_{\mathcal{V}} \Big[\sum_{j=1}^n \Big|\frac{\partial u}{\partial y_j}(\psi^{-1}(x))\Big|^2\Big] dx \\
&= \sup_{x \in \overline{\mathcal{V}}} \Big[\sum_{j=1}^n \Big|\frac{\partial \psi_j^{-1}}{\partial x_i}(x)\Big|^2\Big] \int_{\mathcal{U}} \Big[\sum_{j=1}^n \Big|\frac{\partial u}{\partial y_j}(y)\Big|^2\Big] \Big|\frac{\partial x}{\partial y}(y)\Big| dy \\
&\le \sup_{x \in \overline{\mathcal{V}}} \Big[\sum_{j=1}^n \Big|\frac{\partial \psi_j^{-1}}{\partial x_i}(x)\Big|^2\Big] \sup_{y \in \overline{\mathcal{U}}} \Big|\frac{\partial x}{\partial y}(y)\Big| \, \|u\|^2_{H^1(\mathcal{U})} .
\end{aligned}$$

The proof for $\ell > 1$ is similar, since $\partial^\alpha[u \circ \psi^{-1}](x)$ is a finite sum of terms, each having one factor $(\partial^\beta u) \circ \psi^{-1}$, with $|\beta| \le |\alpha|$, and one or more factors $\partial^\gamma \psi_j^{-1}$ with $1 \le |\gamma| \le |\alpha|$.

(b) We already know, from part (a) of Exercise C.1.5, that multiplication by φ is a bounded operator on $H^\ell(\mathcal{U})$. Denote by Φ the support of φ. It is a compact subset of $\mathcal{U}$. Because ψ is continuous, $\psi(\Phi)$ is a compact subset of $\mathcal{V}$. For all $|\alpha| \le \ell$, $\partial^\alpha(\varphi u)(\psi^{-1}(x))$ vanishes unless $x \in \psi(\Phi)$. Thus when we replace u with φu in the arguments of part (a), all of the integrals $\int_{\mathcal{V}} \cdot\ dx$ are restricted to $\int_{\psi(\Phi)} \cdot\ dx$ and all of the integrals $\int_{\mathcal{U}} \cdot\ dy$ are restricted to $\int_\Phi \cdot\ dy$. Since $\psi(\Phi)$ is compact, $\psi^{-1}(x)$ and all of its derivatives to order ℓ are bounded on $\psi(\Phi)$, even if $\psi^{-1} \notin C^{\max\{1,\ell\}}(\overline{\mathcal{V}})$. Since Φ is compact, $\frac{\partial x}{\partial y}(y)$ is bounded on Φ even if $\psi \notin C^1(\overline{\mathcal{U}})$. So the arguments of part (a) still imply that

$$\Big\|(\varphi u) \circ \psi^{-1}\Big\|_{H^\ell(\mathcal{V})} \le \text{const}\, \|u\|_{\ell,\mathcal{U}}$$

and the B.L.T. theorem (Theorem A.2.13) implies that the map extends to the closure of $C_c^\infty(\mathcal{U})$, which is $H_0^\ell(\mathcal{U})$. By Exercise C.1.4, $C_0^\ell(\mathcal{V}) \subset H_0^\ell(\mathcal{V})$, so the image of $C_c^\infty(\mathcal{U})$ is contained in $H_0^\ell(\mathcal{V})$. Since $H_0^\ell(\mathcal{V})$ is a closed subset of $H^\ell(\mathcal{V})$, the image of the extended map is still contained in $H_0^\ell(\mathcal{V})$.

(c) The proof is similar to that of part (b). Let $\mathcal{W}$ denote $\psi(\Omega \cap \mathcal{U})$ and Φ denote the support of χ. As Φ is a compact subset of $\mathcal{U}$, $\overline{\Phi \cap \Omega}$ is also a compact subset of $\mathcal{U}$. Because ψ is continuous, $\psi(\overline{\Phi \cap \Omega})$ is a compact subset of $\mathcal{V}$. For all $x \in \mathcal{W}$ and $|\alpha| \le \ell$, $\partial^\alpha(\chi u)(\psi^{-1}(x))$ vanishes unless $x \in \psi(\Phi) \cap \psi(\Omega \cap \mathcal{U}) = \psi(\Phi \cap \Omega)$. Thus when we replace u with χu in the arguments of part (a)

- all of the integrals $\int_{\mathcal{V}} \cdot \, dx$ of part (a) become integrals $\int_{\mathcal{W}} \cdot \, dx$ and are restricted to $\int_{\psi(\Phi\cap\Omega)} \cdot \, dx$ and
- all of the integrals $\int_{\mathcal{U}} \cdot \, dy$ of part (a) become integrals $\int_{\Omega\cap\mathcal{U}} \cdot \, dy$ and are restricted to $\int_{\Phi\cap\Omega} \cdot \, dy$.

Since $\psi(\overline{\Phi \cap \Omega})$ is compact, $\psi^{-1}(x)$ and all of its derivatives to order ℓ are bounded on $\psi(\Phi \cap \Omega)$, even if $\psi^{-1} \notin C^{\max\{1,\ell\}}(\overline{\mathcal{V}})$. Since Φ is compact, $\frac{\partial x}{\partial y}(y)$ is bounded on $\Phi \cap \Omega$ even if $\psi \notin C^1(\overline{\mathcal{U}})$. So the arguments of part (a) still imply that

$$\left\|(\chi u) \circ \psi^{-1}\right\|_{H^\ell(\mathcal{W})} \leq \text{const} \|u\|_{H^\ell(\mathcal{U}\cap\Omega)} \leq \text{const} \|u\|_{H^\ell(\Omega)}$$

and the B.L.T. theorem (Theorem A.2.13) implies that the map extends to the closure of the domain of definition

$$\left\{ u \in C^\ell(\Omega) \mid \|u\|_{H^\ell(\Omega)} < \infty \right\},$$

which is $H^\ell(\Omega)$. □

C.2. Traces: Restriction to the boundary

The goal of this section is to prove the following theorem.

Theorem C.2.1. *Let $k \in \mathbb{N}$ and $1 \leq \ell \leq k$, and let Ω be a bounded open subset of $\mathbb{R}^n$ with C^k boundary. Define the restriction map*

$$r : C^k(\overline{\Omega}) \to C^k(\partial\Omega), \qquad u \mapsto u|_{\partial\Omega}.$$

There exists a unique map

$$R : H^\ell(\Omega) \to H^{\ell-1/2}(\partial\Omega)$$

and constants C, C' such that

(i) *R extends r. That is, $Ru = ru$ for all $u \in C^k(\overline{\Omega})$.*

(ii) *R is bounded. That is, $\|Ru\|_{H^{\ell-1/2}(\partial\Omega)} \leq C \|u\|_{H^\ell(\Omega)}$.*

(iii) *R is surjective (onto).*

(iv) *For each $f \in H^{\ell-1/2}(\partial\Omega)$, there is a $u \in H^\ell(\Omega)$ such that*

$$Ru = f \quad \text{and} \quad \|u\|_{H^\ell(\Omega)} \leq C' \|f\|_{H^{\ell-1/2}(\partial\Omega)}.$$

That is, R has a bounded right inverse.

(v) *R has kernel $H^1_0(\Omega) \cap H^\ell(\Omega)$.*

Remark C.2.2. Theorem C.2.1 is true for all real $\ell > \frac{1}{2}$ with $\ell \le k$, except that, when $\frac{1}{2} < \ell < 1$, R has kernel $H_0^\ell(\Omega)$. See, e.g., [**Gri85**, §1.5] or [**Tri83**, §3.3.3 and §3.4.3].

Remark C.2.3. There are three main parts to the proof of Theorem C.2.1, which is given following Lemma C.2.20.

- The first part is to show that r is bounded. Once this is done, and once we prove in Lemma C.2.20, that the domain of r is dense in $H^\ell(\Omega)$, we can extend r by continuity to all of $H^\ell(\Omega)$ and call the result R. Parts (i) and (ii) of the theorem and also the uniqueness of R are then automatic. For pedagogical purposes, we prove the boundedness of r first for $\Omega = \mathbb{R}^n$ with $\partial\Omega$ replaced by $\{x_n = 0\}$ (Lemma C.2.4), second for $\Omega = \mathbb{R}^n_+ = \{x_n > 0\}$, $\partial\Omega = \{x_n = 0\}$ (Proposition C.2.17) and finally for Ω as in the theorem.
- The second part is to prove that each $f \in H^{\ell-1/2}(\partial\Omega)$ can be extended to a $u \in H^\ell(\Omega)$ in a bounded way. This will prove part (iv), which implies part (iii). For pedagogical purposes, we prove extendibility first for $\Omega = \mathbb{R}^n$ with $\partial\Omega$ replaced by $\{x_n = 0\}$ (Lemma C.2.8), second for $\Omega = \mathbb{R}^n_+ = \{x_n > 0\}$, $\partial\Omega = \{x_n = 0\}$ (Proposition C.2.17) and finally for Ω as in the theorem.
- Since r vanishes on $C_c^\infty(\Omega)$, which is dense in $H_0^\ell(\Omega)$, the boundedness of r will also imply that the kernel of R contains $H_0^\ell(\Omega)$. Note that, when ℓ is large, many derivatives of any $f \in H_0^\ell(\Omega)$ must vanish on $\partial\Omega$. On the other hand, to be in the kernel of R only f itself—not its derivatives—must vanish on $\partial\Omega$. In fact, if $u \in H_0^1(\Omega) \cap H^\ell(\Omega)$, then there is a sequence of functions $u_j \in C_c^\infty(\Omega)$ that converge to u with respect to the norm $\|\cdot\|_{H^1(\Omega)}$. Since, by part (ii) of the theorem with $\ell = 1$,

$$\|Ru\|_{H^{1/2}(\partial\Omega)} = \left\|Ru - Ru_j\right\|_{H^{1/2}(\partial\Omega)} \le C\left\|u - u_j\right\|_{H^1(\Omega)}$$
$$\text{for all } j \implies Ru = 0,$$

the kernel of R (acting on $H^\ell(\Omega)$) contains all of $H_0^1(\Omega) \cap H^\ell(\Omega)$. To prove part (v), we have to show that any function in the kernel of R can be approximated, in the $H^1(\Omega)$ norm, by functions in $C_c^\infty(\Omega)$. For pedagogical purposes, we prove this first for

$$\Omega = \mathbb{R}^n_+ = \{x_n > 0\}, \quad \partial\Omega = \{x_n = 0\}$$

(Proposition C.2.18) and then for Ω as in the theorem.

We start with a particularly simple geometry, to illustrate what is going on. In the illustration, we replace Ω by $\mathbb{R}^n$ and $\partial\Omega$ by $\{x_n = 0\}$, which we identify with $\mathbb{R}^{n-1}$. To save writing, we denote

$$x' = (x_1, \cdots, x_{n-1}) \subset \mathbb{R}^{n-1} = \{x_n = 0\}.$$

Lemma C.2.4 (Restriction from $\mathbb{R}^n$ to $\mathbb{R}^{n-1}$). *Let $s > \frac{1}{2}$. The linear map r : $C_c^\infty(\mathbb{R}^n) \to C_c^\infty(\mathbb{R}^{n-1})$ defined by*

$$(ru)(x') = u(x', 0),$$

has a unique extension to a bounded linear map

$$R : H^s(\mathbb{R}^n) \to H^{s-\frac{1}{2}}(\mathbb{R}^{n-1}).$$

Proof. We shall prove that there is a constant C (depending only on s) such that

$$|||ru|||_{H^{s-1/2}(\mathbb{R}^{n-1})} \leq C|||u|||_{H^s(\mathbb{R}^n)}$$

for all $u \in C_c^\infty(\mathbb{R}^n)$. The Lemma will then follow by the B.L.T. theorem (Theorem A.2.13).

If $\xi \in \mathbb{R}^n$, write $\xi = (\xi', \xi_n)$ with $\xi' = (\xi_1, \cdots, \xi_{n-1})$. The definition of the Fourier transform gives

$$\int e^{i\xi'\cdot x'} (\widehat{ru})(\xi') \frac{d\xi'}{(2\pi)^{n-1}} = (ru)(x') = u(x', 0) = \int e^{i\xi'\cdot x'} \hat{u}(\xi', \xi_n) \frac{d\xi'}{(2\pi)^{n-1}} \frac{d\xi_n}{2\pi}$$

so that

$$(\widehat{ru})(\xi') = \int \hat{u}(\xi', \xi_n) \frac{d\xi_n}{2\pi}.$$

Thus, by Cauchy–Schwarz,

$$\begin{aligned}
&|(\widehat{ru})(\xi')|^2 \\
&= \left| \int \hat{u}(\xi', \xi_n) \left(1 + |\xi'|^2 + \xi_n^2\right)^{s/2} \left(1 + |\xi'|^2 + \xi_n^2\right)^{-s/2} \frac{d\xi_n}{2\pi} \right|^2 \\
&\leq \left[\int \frac{1}{\left(1 + |\xi'|^2 + \xi_n^2\right)^s} \frac{d\xi_n}{2\pi} \right] \left[\int |\hat{u}(\xi', \xi_n)|^2 \left(1 + |\xi'|^2 + \xi_n^2\right)^s \frac{d\xi_n}{2\pi} \right] \\
&= \left[\frac{1}{\left(1 + |\xi'|^2\right)^{s-1/2}} \int \frac{1}{(1 + p^2)^s} \frac{dp}{2\pi} \right] \left[\int |\hat{u}(\xi', \xi_n)|^2 \left(1 + |\xi'|^2 + \xi_n^2\right)^s \frac{d\xi_n}{2\pi} \right],
\end{aligned}$$

where

$$\xi_n = p\sqrt{1 + |\xi'|^2} = \frac{C^2}{\left(1 + |\xi'|^2\right)^{s-1/2}} \int |\hat{u}(\xi', \xi_n)|^2 \left(1 + |\xi'|^2 + \xi_n^2\right)^s \frac{d\xi_n}{2\pi}$$

where the constant $C^2 = \int \frac{1}{(1+p^2)^s} \frac{dp}{2\pi}$ depends only on s and is finite because $s > \frac{1}{2}$. Hence

$$\begin{aligned}
|||ru|||^2_{H^{s-1/2}(\mathbb{R}^{n-1})} &= \int (1 + |\xi'|^2)^{s-\frac{1}{2}} |\widehat{ru}(\xi')|^2 \frac{d\xi'}{(2\pi)^{n-1}} \\
&\le C^2 \int \left|\hat{u}(\xi', \xi_n)\right|^2 \left(1 + |\xi'|^2 + \xi_n^2\right)^s \frac{d\xi}{(2\pi)^n} \\
&= C|||u|||^2_{H^s(\mathbb{R}^n)}.
\end{aligned}$$

□

Exercise C.2.5. Let $k \in \mathbb{N}$ and $s \in \mathbb{R}$ obey $k < n$ and $s > \frac{k}{2}$. Identify $\mathbb{R}^{n-k}$ with $\{ x \in \mathbb{R}^n \mid x_{n-k+1} = \cdots = x_n = 0 \}$ and write $x \in \mathbb{R}^n$ as $x = (x', y)$ with $x' \in \mathbb{R}^{n-k}$ and $y \in \mathbb{R}^k$. Prove that the linear transformation $r : C_c^\infty(\mathbb{R}^n) \to C_c^\infty(\mathbb{R}^{n-k})$ defined by

$$(ru)(x') = u(x', 0)$$

has a unique extension to a bounded linear map

$$R : H^s(\mathbb{R}^n) \to H^{s-\frac{k}{2}}(\mathbb{R}^{n-k}).$$

Exercise C.2.6. Let $\frac{1}{2} < s \le \ell \in \mathbb{N}$ and let $u \in C^\ell(\mathbb{R}^n)$. Suppose that each derivative of u of order at most ℓ is bounded by a constant times $(1 + |x|)^{-\alpha}$ for some $\alpha > \frac{n}{2}$. Then $u \in H^\ell(\mathbb{R}^n) \subset H^s(\mathbb{R}^n)$ and $u(x', 0) \in H^\ell(\mathbb{R}^{n-1}) \subset H^{s-\frac{1}{2}}(\mathbb{R}^{n-1})$. Prove that $(Ru)(x') = u(x', 0)$, where $R : H^s(\mathbb{R}^n) \to H^{s-\frac{1}{2}}(\mathbb{R}^{n-1})$ is the map of Lemma C.2.4.

Exercise C.2.7. Let $s > \frac{1}{2}$. Define, for each $t \in \mathbb{R}$,

$$R_t : H^s(\mathbb{R}^n) \to H^{s-\frac{1}{2}}(\mathbb{R}^{n-1})$$

to be the unique bounded linear extension of the linear transformation $r_t : C_c^\infty(\mathbb{R}^n) \to C_c^\infty(\mathbb{R}^{n-1})$ defined by

$$(ru)(x') = u(x', t).$$

Prove that R_t is strongly continuous in t. That is, prove that

$$\lim_{t \to t_0} |||R_t f - R_{t_0} f|||^2_{H^{s-1/2}(\mathbb{R}^{n-1})} = 0$$

for each $t_0 \in \mathbb{R}$ and each $f \in H^s(\mathbb{R}^n)$.

Lemma C.2.8 (Extension from $\mathbb{R}^{n-1}$ to $\mathbb{R}^n$). *Let $s > \frac{1}{2}$. There is a constant C (depending only on n and s) such that for each $f \in H^{s-1/2}(\mathbb{R}^{n-1})$ there is a $u \in H^s(\mathbb{R}^n)$ obeying*

$$Ru = f \quad \text{and} \quad |||u|||_{H^s(\mathbb{R}^n)} \le C|||f|||_{H^{s-1/2}(\mathbb{R}^{n-1})},$$

where R is the map of Lemma C.2.4.

Proof. As in the last lemma, write $\xi = (\xi', \xi_n) \in \mathbb{R}^n$ with

$$\xi' - (\xi_1, \cdots, \xi_{n-1}).$$

We claim that if $f \in \mathcal{S}(\mathbb{R}^{n-1})$, then the u determined by

$$\hat{u}(\xi', \xi_n) = \hat{f}(\xi')\, 2\sqrt{\pi} e^{-\xi_n^2/(1+|\xi'|^2)}/\sqrt{1+|\xi'|^2}$$

does the job for that f. Once the claim is proven, we will be able to extend the map $f \mapsto u$ by continuity (i.e., by the B.L.T. theorem) to $H^{s-1/2}(\mathbb{R}^{n-1})$ and the lemma will follow, since R is itself continuous.

We first check that $\hat{u} \in \mathcal{S}(\mathbb{R}^n)$. Every derivative of

$$e^{-\xi_n^2/(1+|\xi'|^2)}/\sqrt{1+|\xi'|^2}$$

is a polynomial in ξ times $e^{-\xi_n^2/(1+|\xi'|^2)}(1+|\xi'|^2)^{-\ell/2}$ for some $\ell \in \mathbb{N}$. Since, for each $m \in \mathbb{N}_0$,

$$\begin{aligned} |\xi_n|^m e^{-\xi_n^2/(1+|\xi'|^2)} &= (1+|\xi'|^2)^{m/2}\Big[\,\xi_n^2/(1+|\xi'|^2)\Big]^{m/2} e^{-\xi_n^2/(1+|\xi'|^2)} \\ &\le \text{const}_m (1+|\xi'|^2)^{m/2}, \end{aligned}$$

each polynomial in ξ times $e^{-\xi_n^2/(1+|\xi'|^2)}(1+|\xi'|^2)^{-\ell/2}$ is bounded by a polynomial in ξ'. As $\hat{f} \in \mathcal{S}(\mathbb{R}^{n-1})$, this ensures that $\hat{u} \in \mathcal{S}(\mathbb{R}^n)$.

To verify that $Ru = f$, we need to verify that $u(x', 0) = f(x')$, or equivalently $\int \hat{u}(\xi', \xi_n)\, \frac{d\xi_n}{2\pi} = \hat{f}(\xi')$. But this is an immediate consequence of

$$2\sqrt{\pi} \int e^{-\xi_n^2/(1+|\xi'|^2)} \frac{d\xi_n}{2\pi\sqrt{1+|\xi'|^2}} = \frac{1}{\sqrt{\pi}} \int e^{-p^2}\, dp = 1.$$

We made the change of variables $\xi_n = p\sqrt{1+|\xi'|^2}$. The verification of the boundedness requirement is

$$\begin{aligned} |||u|||^2_{H^s(\mathbb{R}^n)} &= \int (1+|\xi|^2)^s |\hat{u}(\xi)|^2 \frac{d\xi}{(2\pi)^n} \\ &= 4\pi \int (1+|\xi|^2)^s (1+|\xi'|^2)^{-1} e^{-2\xi_n^2/(1+|\xi'|^2)} |\hat{f}(\xi')|^2 \frac{d\xi'}{(2\pi)^{n-1}} \frac{d\xi_n}{2\pi} \\ &= 2\Big[\int (1+p^2)^s e^{-2p^2}\, dp\Big] \int (1+|\xi'|^2)^{s-1+1/2} |\hat{f}(\xi')|^2 \frac{d\xi'}{(2\pi)^{n-1}} \\ &= \Big[\int 2(1+p^2)^s e^{-2p^2}\, dp\Big]\, |||f|||^2_{H^{s-1/2}(\mathbb{R}^{n-1})}. \end{aligned}$$

□

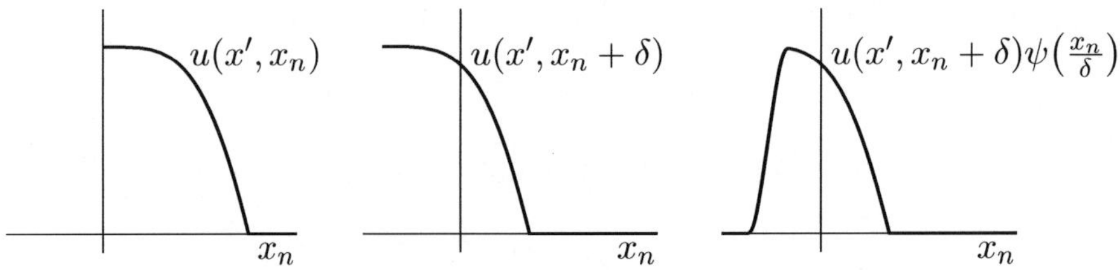

Figure C.3. The function $u_\delta(x)$.

We next consider another, slightly more realistic, but still simple, geometry. Namely,

$$\Omega = \mathbb{R}^n_+ = \{\, x \in \mathbb{R}^n \mid x_n > 0 \,\},$$
$$\partial\Omega = \mathbb{R}^{n-1} = \{\, x \in \mathbb{R}^n \mid x_n = 0 \,\}.$$

Lemma C.2.9. *Let $\ell \in \mathbb{N}_0$. The set of restrictions of functions in $C_c^\infty(\mathbb{R}^n)$ to $\mathbb{R}^n_+$ is dense in $H^\ell(\mathbb{R}^n_+)$. In particular, $C^\infty(\overline{\mathbb{R}^n_+}) \cap H^\ell(\mathbb{R}^n_+)$ is dense in $H^\ell(\mathbb{R}^n_+)$.*

Proof. Let $u \in H^\ell(\mathbb{R}^n_+)$ and $\varepsilon > 0$. Let, for any measurable function w on $\mathbb{R}^n$, $P_+ w$ be the restriction of w to $\mathbb{R}^n_+$. Observe that

$$\begin{aligned} \|P_+ w\|_{H^\ell(\mathbb{R}^n_+)} &= \Big\{ \sum_{|\alpha|\le\ell} \int_{\mathbb{R}^n_+} |\partial^\alpha w(x)|^2 \, dx \Big\}^{1/2} \\ &\le \Big\{ \sum_{|\alpha|\le\ell} \int_{\mathbb{R}^n} |\partial^\alpha w(x)|^2 \, dx \Big\}^{1/2} \\ &= \|w\|_{H^\ell(\mathbb{R}^n)}. \end{aligned}$$

Step 1. We first find a $v \in H^\ell(\mathbb{R}^n)$ such that $\|u - P_+ v\|_{H^\ell(\mathbb{R}^n_+)} < \frac{\varepsilon}{2}$. Let $\psi \in C^\infty(\mathbb{R})$ obey

$$\psi(t) = \begin{cases} 1 & \text{for } t \ge -\frac{1}{2}, \\ 0 & \text{for } t \le -\frac{3}{4}, \end{cases}$$

and define, for $\delta > 0$,

$$u_\delta(x) = \begin{cases} u(x', x_n + \delta)\psi(\frac{x_n}{\delta}) & \text{if } x_n > -\delta, \\ 0 & \text{if } x_n \le -\frac{3}{4}\delta. \end{cases}$$

The function u_δ is depicted in Figure C.3. Using the same procedure as in part (a) of Exercise C.1.5, we see that $u_\delta \in H^\ell(\mathbb{R}^n)$ (though $\|u_\delta\|_{H^\ell}$ may diverge as $\delta \to 0$). For $x_n \ge 0$, $u_\delta(x) = u(x', x_n + \delta)$ and hence

$$\|u - P_+ u_\delta\|^2_{H^\ell(\mathbb{R}^n_+)} = \sum_{|\alpha|\le\ell} \int_{\mathbb{R}^n_+} |\partial^\alpha u(x) - \partial^\alpha u(x', x_n + \delta)|^2 \, dx.$$

We will show in the next paragraph that translation is continuous in $L^2(\mathbb{R}^n_+)$ (this would also follow from approximation by C_c functions as in Theorem 2.1.4(d)). Hence we can pick $\delta_0 > 0$ so small that

$$\left\| u - P_+ u_{\delta_0} \right\|_{H^\ell(\mathbb{R}^n_+)} < \frac{\varepsilon}{2}$$

and set $v = u_{\delta_0}$.

To complete this step, we show that translation is continuous in $L^2(\Omega)$ for any open $\Omega \subset \mathbb{R}^n$. Let $f \in L^2(\Omega)$. Define f to be zero on $\mathbb{R}^n \setminus \Omega$. Then for all $t \in \mathbb{R}^n$

$$\begin{aligned}\int_\Omega |f(x) - f(x+t)|^2 \, dx &\le \int_{\mathbb{R}^n} |f(x) - f(x+t)|^2 \, dx \\ &= \int_{\mathbb{R}^n} |1 - e^{i\xi\cdot t}|^2 \, |\hat{f}(\xi)|^2 \, \frac{d\xi}{(2\pi)^n}\end{aligned}$$

by (C.2b) and the translation property of the Fourier transform. The integrand converges pointwise to zero as $t \to 0$ and is bounded uniformly by the integrable function $4|\hat{f}(\xi)|^2$. So, by the Lebesgue dominated convergence theorem,

$$\lim_{t\to 0} \int_\Omega |f(x) - f(x+t)|^2 \, dx = 0.$$

Step 2. By Exercise C.1.13 there exists a $w \in C_c^\infty(\mathbb{R}^n)$ such that

$$|||v - w|||_{H^\ell(\mathbb{R}^n)} < \tfrac{\varepsilon}{2}.$$

Hence

$$\begin{aligned}\|u - P_+ w\|_{H^\ell(\mathbb{R}^n_+)} &\le \|u - P_+ v\|_{H^\ell(\mathbb{R}^n_+)} + \|P_+ v - P_+ w\|_{H^\ell(\mathbb{R}^n_+)} \\ &\le \|u - P_+ v\|_{H^\ell(\mathbb{R}^n_+)} + |||v - w|||_{H^\ell(\mathbb{R}^n)} < \varepsilon\end{aligned}$$

and we are done. □

Exercise C.2.10. Let $\ell \in \mathbb{N}_0$, $\varepsilon, \varepsilon', R > 0$ and $u \in H^\ell(\mathbb{R}^n_+)$. Suppose that $u(x) = 0$ for all $|x| \ge R$. Show that there is a function $v \in C_c^\infty(\mathbb{R}^n)$ that obeys $v(x) = 0$ for all $|x| \ge R + \varepsilon'$ and $\|u - P_+ v\|_{H^\ell(\mathbb{R}^n_+)} < \varepsilon$, where P_+ is the restriction from $\mathbb{R}^n$ to $\mathbb{R}^n_+$.

Lemma C.2.11 (Extension from $\mathbb{R}^n_+$ to $\mathbb{R}^n$). *Let $\ell \in \mathbb{N}_0$. There exists a bounded linear operator*

$$E : H^\ell(\mathbb{R}^n_+) \to H^\ell(\mathbb{R}^n)$$

such that $Eu(x) = u(x)$ for all $u \in C^\infty(\overline{\mathbb{R}^n_+}) \cap H^\ell(\mathbb{R}^n_+)$, $x \in \overline{\mathbb{R}^n_+}$.

Proof. Let $u \in C^\infty(\overline{\mathbb{R}^n_+}) \cap H^\ell(\mathbb{R}^n_+)$. Write $x = (x', x_n) \in \mathbb{R}^n$ with

$$x' = (x_1, \cdots, x_{n-1}).$$

Define

$$\text{(C.5)} \qquad Eu(x) = \begin{cases} u(x) & \text{if } x_n \geq 0, \\ \sum_{j=1}^{\ell+1} \beta_j u(x', -\frac{x_n}{j}), & \text{if } x_n < 0, \end{cases}$$

where the β_j's are to be chosen, independent of u, so as to ensure that Eu is C^ℓ across $x_n = 0$. This will ensure that $Eu \in C^\ell(\mathbb{R}^n)$. Since

$$\text{(C.6)} \qquad \int_{x_n<0} \left| \frac{\partial^\alpha}{\partial x^\alpha} u\left(x', -\frac{x_n}{j}\right) \right|^2 dx = j\left(\frac{1}{j}\right)^{2\alpha_n} \int_{x_n>0} \left|\partial^\alpha u(x)\right|^2 dx$$

it will also ensure that E is bounded. It will then suffice to extend E by continuity to all of $H^\ell(\mathbb{R}^n_+)$.

To ensure that $\partial^\alpha Eu$ is continuous across $x_n = 0$, when $\alpha_n = 0$, it suffices to choose all the β_j so that $\sum_{j=1}^{\ell+1} \beta_j = 1$. To ensure that $\partial^\alpha Eu$ is continuous across $x_n = 0$, when $\alpha_n = 1$, it suffices to choose the β_j so that $\sum_{j=1}^{\ell+1} -\frac{\beta_j}{j} = 1$. In general, to ensure that $\partial^\alpha Eu$ is continuous across $x_n = 0$, when $\alpha_n = m$, it suffices to choose the β_j so that

$$\text{(C.7)} \qquad \sum_{j=1}^{\ell+1} \left(-\frac{1}{j}\right)^m \beta_j = 1.$$

The system of linear equations (C.7), for $0 \leq m \leq \ell$, has a solution provided

$$D := \det \begin{bmatrix} 1 & 1 & \cdots & 1 \\ -1 & -\frac{1}{2} & \cdots & -\frac{1}{\ell+1} \\ \vdots & \vdots & \ddots & \vdots \\ (-1)^\ell & \left(-\frac{1}{2}\right)^\ell & \cdots & \left(-\frac{1}{\ell+1}\right)^\ell \end{bmatrix} \neq 0.$$

But D is a Vandermonde determinant. By Exercise C.2.12, it has the value $\prod_{1\leq i<j\leq\ell+1} \left(-\frac{1}{j} + \frac{1}{i}\right)$, which is certainly not zero. □

Exercise C.2.12. Let $\{x_i\}_{i\in\mathbb{N}}$ be any sequence of complex numbers. Define the Vandermonde determinant

$$V_n = \det \begin{bmatrix} 1 & 1 & \cdots & 1 \\ x_1 & x_2 & \cdots & x_n \\ x_1^2 & x_2^2 & \cdots & x_n^2 \\ \vdots & \vdots & \ddots & \vdots \\ x_1^{n-1} & x_2^{n-1} & \cdots & x_n^{n-1} \end{bmatrix}.$$

(a) Use row and column operations to prove that

$$V_n = V_{n-1} \prod_{1\leq i<n} (x_n - x_i).$$

(b) Prove that

$$V_n = \prod_{1 \le i < j \le n} (x_j - x_i).$$

Exercise C.2.13. Let $\ell \in \mathbb{N}_0$ and $\ell' \in \mathbb{Z}$ with $\ell' \le \ell$. Denote by E_ℓ the operator of Lemma C.2.11. Prove that there is a constant C such that

$$\|E_\ell u\|_{H^{\ell'}(\mathbb{R}^n)} \le C \|u\|_{H^{\ell'}(\mathbb{R}^n_+)}$$

for all $u \in H^\ell(\mathbb{R}^n_+)$.

Exercise C.2.14. Let $\ell', \ell \in \mathbb{N}_0$ with $\ell' \le \ell$ and $\alpha \in \mathbb{N}_0^n$ with $\alpha_n = 0$. Denote by E_ℓ the operator of Lemma C.2.11. By Exercise C.2.13, E_ℓ has a unique bounded extension to $H^{\ell'}(\mathbb{R}^n_+)$ and it in turn has a unique bounded extension to $H^{\ell'-|\alpha|}(\mathbb{R}^n_+)$. We persist in denoting both of them E_ℓ. Recall, from Lemma C.1.29, that ∂^α is a bounded operator from $H^{\ell'}(\mathbb{R}^n_+)$ to $H^{\ell'-|\alpha|}(\mathbb{R}^n_+)$ and from $H^{\ell'}(\mathbb{R}^n)$ to $H^{\ell'-|\alpha|}(\mathbb{R}^n)$. Let $u \in H^{\ell'}(\mathbb{R}^n_+)$. Prove $\partial^\alpha E_\ell u = E_\ell \partial^\alpha u$ and that $\partial^\alpha u \in H^\ell(\mathbb{R}^n_+)$ if and only if $\partial^\alpha E_\ell u \in H^\ell(\mathbb{R}^n)$.

Exercise C.2.15.

(a) Let $\{a_\ell\}_{\ell \in \mathbb{N}_0}$ be any sequence of real numbers. Prove that there is a function $f \in C^\infty(\mathbb{R})$ such that $f^{(m)}(0) = a_m$ for all $m \in \mathbb{N}_0$.

(b) Let $\{a_\ell(x')\}_{\ell \in \mathbb{N}_0}$ be any sequence of C^∞ functions on $\mathbb{R}^{n-1}$. Prove that there is a function $f \in C^\infty(\mathbb{R}^n)$ such that $\frac{\partial^m f}{\partial x_n^m}(x', 0) = a_m(x')$ for all $m \in \mathbb{N}_0$ and $x' \in \mathbb{R}^{n-1}$.

(c) Prove that if $f \in C^\infty(\overline{\mathbb{R}^n_+})$, then there exists $F \in C^\infty(\mathbb{R}^n)$ such that $f(x) = F(x)$ for all $\mathbb{R}^n_+$.

We defined $H^\ell(\mathbb{R}^n_+)$, rather abstractly, as the completion of $C^\ell(\mathbb{R}^n_+)$ under a certain norm. Part (a) of the next proposition (with $\ell' = 0$) shows that $H^\ell(\mathbb{R}^n_+)$ is precisely the set of all $u \in L^2(\mathbb{R}^n_+)$ all of whose derivatives $\partial^\alpha u$ with order $|\alpha| \le \ell$ are again in $L^2(\mathbb{R}^n_+)$. We also defined (in part (b) of Lemma C.1.29) $\partial^\alpha : H^\ell(\mathbb{R}^n_+) \to H^{\ell-|\alpha|}(\mathbb{R}^n_+)$, rather abstractly, as the continuous extension of $\partial^\alpha : C^{\max\{\ell,|\alpha|\}}(\mathbb{R}^n_+) \to C^{\max\{\ell-|\alpha|,0\}}(\mathbb{R}^n_+)$. Part (b) of the next proposition replaces these derivatives with, more explicit, weak derivatives and shows that $H^\ell(\mathbb{R}^n_+)$ is precisely the set of all $u \in L^2(\mathbb{R}^n_+)$ all of whose weak derivatives of order at most ℓ are again in $L^2(\mathbb{R}^n_+)$. Recall that the weak (or distributional) α–derivative of u is the map

$$D_u^\alpha : C_c^\infty(\Omega) \to \mathbb{C},$$
$$f \mapsto (-1)^{|\alpha|} \langle \partial^\alpha f, u \rangle_{L^2(\Omega)},$$

and it is defined for all $u \in L^2(\Omega)$ and all $\alpha \in \mathbb{N}_0^n$. Recall also, from the discussion following (C.1), that one says that the weak derivative D_u^α is in $L^2(\Omega)$

if there exists a $v_\alpha \in L^2(\Omega)$ such that $\langle f, v_\alpha\rangle_{L^2(\Omega)} = (-1)^{|\alpha|}\langle \partial^\alpha f, u\rangle_{L^2(\Omega)}$ for all $f \in C_c^\infty(\Omega)$. These characterizations of $H^\ell(\mathbb{R}^n_+)$, as the set of all $u \in L^2(\mathbb{R}^n_+)$ all of whose derivatives of order at most ℓ are again in $L^2(\mathbb{R}^n_+)$, provide a good mental image for $H^\ell(\mathbb{R}^n_+)$.

Proposition C.2.16. *Let $\ell \in \mathbb{N}_0$.*

(a) *Let $u \in H^{\ell'}(\mathbb{R}^n_+)$ with $\ell' \in \mathbb{N}_0$ obeying $\ell' \le \ell$. Then $u \in H^\ell(\mathbb{R}^n_+)$ if and only if $\partial^\alpha u \in H^{\ell'}(\mathbb{R}^n_+)$ for all $\alpha \in \mathbb{N}_0^n$ with $|\alpha| \le \ell - \ell'$.*

(b) *Let $u \in L^2(\mathbb{R}^n_+)$. Then $u \in H^\ell(\mathbb{R}^n_+)$ if and only if there is a constant C such that*

$$|\langle \partial^\alpha f, u\rangle_{L^2(\mathbb{R}^n_+)}| \le C\big\|f\big\|_{L^2(\mathbb{R}^n_+)}$$

for all $\alpha \in \mathbb{N}_0^n$ with $|\alpha| \le \ell$ and all $f \in C_c^\infty(\mathbb{R}^n_+)$. Also, $u \in H^\ell(\mathbb{R}^n_+)$ if and only if, for each $\alpha \in \mathbb{N}_0^n$ with $|\alpha| \le \ell$, there is a $v_\alpha \in L^2(\mathbb{R}^n_+)$ such that

$$(-1)^{|\alpha|}\langle \partial^\alpha f, u\rangle_{L^2(\mathbb{R}^n_+)} = \langle f, v_\alpha\rangle_{L^2(\mathbb{R}^n_+)}$$

for all $f \in C_c^\infty(\mathbb{R}^n_+)$.

Proof.

(a) The implication that if $u \in H^\ell(\mathbb{R}^n_+)$, then $\partial^\alpha u \in H^{\ell'}(\mathbb{R}^n_+)$ for all $\alpha \in \mathbb{N}_0^n$ with $|\alpha| \le \ell - \ell'$ is an immediate consequence of part (b) of Lemma C.1.29. This lemma also implies that if $\partial^\alpha u \in H^{\ell'}(\mathbb{R}^n_+)$ for all $\alpha \in \mathbb{N}_0^n$ with $|\alpha| \le \ell - \ell'$ then $\partial^\alpha u \in L^2(\mathbb{R}^n_+)$ for all $\alpha \in \mathbb{N}_0^n$ with $|\alpha| \le \ell$, so it suffices to consider $\ell' = 0$.

So let $u \in L^2(\mathbb{R}^n_+)$ and assume that $\partial^\alpha u \in L^2(\mathbb{R}^n_+)$ for all $\alpha \in \mathbb{N}_0^n$ with $|\alpha| \le \ell$. Denote by E_ℓ the operator of Lemma C.2.11 and its extensions to $H^{\ell'}(\mathbb{R}^n_+)$ for $\ell' < \ell$, as in Exercise C.2.14. It suffices to prove that we have $E_\ell u \in H^\ell(\mathbb{R}^n)$, since u is the restriction of $E_\ell u$ to $\mathbb{R}^n_+$. To prove that $E_\ell u \in H^\ell(\mathbb{R}^n)$, it suffices to prove that $\partial^\alpha E_\ell u \in L^2(\mathbb{R}^n)$ for all $\alpha \in \mathbb{N}_0^n$ with $|\alpha| \le \ell$, since then $\xi^\alpha \widehat{E_\ell u}(\xi) \in L^2(\mathbb{R}^n)$ for all $\alpha \in \mathbb{N}_0^n$ with $|\alpha| \le \ell$ and this implies that $\big(1 + |\xi|^2\big)^{\ell/2}\widehat{E_\ell u}(\xi) \in L^2(\mathbb{R}^n)$. We have already proven, in Exercise C.2.14, that $\partial^\alpha E_\ell u \in L^2(\mathbb{R}^n)$ for all $\alpha \in \mathbb{N}_0^n$ with $|\alpha| \le \ell$ and $\alpha_n = 0$.

So assume that $\alpha_n > 0$ and set $\alpha' = \alpha - (0, \cdots, 0, \alpha_n)$. We need to prove that $\partial^\alpha E_\ell u = \frac{\partial^{\alpha_n}}{\partial x_n^{\alpha_n}} E_\ell \partial^{\alpha'} u \in L^2(\mathbb{R}^n)$. Recall, from (C.5), that, for $v \in C^\infty(\overline{\mathbb{R}^n_+}) \cap H^\ell(\mathbb{R}^n_+)$,

$$E_\ell v(x) = \begin{cases} v(x) & \text{if } x_n \ge 0, \\ \sum_{j=1}^{\ell+1} \beta_j v\big(x', -\frac{x_n}{j}\big) & \text{if } x_n < 0, \end{cases}$$

where $x = (x', x_n) \in \mathbb{R}^n$ with $x' = (x_1, \cdots, x_{n-1})$. Hence

$$\text{(C.8)} \qquad \frac{\partial^{\alpha_n}}{\partial x_n^{\alpha_n}} E_\ell v(x) = \begin{cases} \frac{\partial^{\alpha_n} v}{\partial x_n^{\alpha_n}}(x) & \text{if } x_n \geq 0, \\ \sum_{j=1}^{\ell+1} \beta_j \left(-\frac{1}{j}\right)^{\alpha_n} \frac{\partial^{\alpha_n} v}{\partial x_n^{\alpha_n}}\left(x', -\frac{x_n}{j}\right) & \text{if } x_n < 0. \end{cases}$$

Define, for $w \in C^\infty(\overline{\mathbb{R}^n_+}) \cap H^{\ell-\alpha_n}(\mathbb{R}^n_+)$,

$$\text{(C.9)} \qquad Fw(x) = \begin{cases} w(x) & \text{if } x_n \geq 0, \\ \sum_{j=1}^{\ell+1} \beta_j \left(-\frac{1}{j}\right)^{\alpha_n} w\left(x', -\frac{x_n}{j}\right) & \text{if } x_n < 0. \end{cases}$$

For the β_j chosen in Lemma C.2.11, Fw is $C^{\ell-\alpha_n}$ across $x_n = 0$ so that

$$Fw \in C^{\ell-\alpha_n}(\mathbb{R}^n)$$

and F is bounded as a map from $H^{\ell-\alpha_n}(\mathbb{R}^n_+)$ to $H^{\ell-\alpha_n}(\mathbb{R}^n_+)$. So we may extend F by continuity to all of $H^{\ell-\alpha_n}(\mathbb{R}^n_+)$. As in Exercise C.2.13, F has a unique bounded extension to an operator from $H^{\ell'}(\mathbb{R}^n_+)$ to $H^{\ell'}(\mathbb{R}^n)$ for each $\ell' \in \mathbb{Z}$ with $\ell' \leq \ell - \alpha_n$. Comparing (C.8) and (C.9), we see that $\frac{\partial^{\alpha_n}}{\partial x_n^{\alpha_n}} E_\ell v = F \frac{\partial^{\alpha_n} v}{\partial x_n^{\alpha_n}}$ for all $v \in C^\infty(\overline{\mathbb{R}^n_+}) \cap H^\ell(\mathbb{R}^n_+)$. By the continuity of all of the operators involved and the denseness of $C^\infty(\overline{\mathbb{R}^n_+}) \cap H^\ell(\mathbb{R}^n_+)$, $\frac{\partial^{\alpha_n}}{\partial x_n^{\alpha_n}} E_\ell v = F \frac{\partial^{\alpha_n} v}{\partial x_n^{\alpha_n}}$ for all $v \in H^{\ell'}(\mathbb{R}^n_+)$ with $\ell' \leq \ell$. Applying this with $v = \partial^{\alpha'} u$, we have that

$$\partial^\alpha E_\ell u = \frac{\partial^{\alpha_n}}{\partial x_n^{\alpha_n}} E_\ell \partial^{\alpha'} u = F \partial^\alpha u.$$

By assumption, $\partial^\alpha u \in L^2(\mathbb{R}^n_+)$. So $F\partial^\alpha u \in L^2(\mathbb{R}^n_+)$ and hence $\partial^\alpha E_\ell u \in L^2(\mathbb{R}^n_+)$.

(b) First, suppose that $u \in H^\ell(\mathbb{R}^n_+)$. By Lemma C.2.9, there is a sequence of functions $u_i \in C^\infty(\overline{\mathbb{R}^n_+})$ that converges in $H^\ell(\mathbb{R}^n_+)$ to u. In particular,

$$\lim_{i \to \infty} \|u_i\|_{H^\ell(\mathbb{R}^n_+)} = \|u\|_{H^\ell(\mathbb{R}^n_+)}$$

and, if $f \in C_c^\infty(\mathbb{R}^n_+)$,

$$\begin{aligned}
|\langle \partial^\alpha f, u\rangle_{L^2(\mathbb{R}^n_+)}| &= \lim_{i\to\infty} |\langle \partial^\alpha f, u_i\rangle_{L^2(\mathbb{R}^n_+)}| \\
&= \lim_{i\to\infty} \left| \int_{\mathbb{R}^n_+} \partial^\alpha f(x)\, \overline{u_i(x)}\, dx \right| \\
&= \lim_{i\to\infty} \left| \int_{\mathbb{R}^n_+} f(x)\, \overline{\partial^\alpha u_i(x)}\, dx \right| \\
&\le \limsup_{i\to\infty} \left\| f \right\|_{L^2(\mathbb{R}^n_+)} \left\| \partial^\alpha u_i \right\|_{L^2(\mathbb{R}^n_+)} \\
&\le \left\| f \right\|_{L^2(\mathbb{R}^n_+)} \limsup_{i\to\infty} \|u_i\|_{H^\ell(\mathbb{R}^n_+)} \\
&= \left\| f \right\|_{L^2(\mathbb{R}^n_+)} \|u\|_{H^\ell(\mathbb{R}^n_+)}
\end{aligned}$$

provided $|\alpha| \le \ell$. The Riesz representation theorem (Theorem A.2.15) provides the existence of the v_α required for the second statement.

For the converse, suppose that $u \in L^2(\mathbb{R}^n_+) = H^0(\mathbb{R}^n_+)$ obeys

$$|\langle \partial^\alpha f, u\rangle_{L^2(\mathbb{R}^n_+)}| \le C \left\| f \right\|_{L^2(\mathbb{R}^n_+)}$$

for all $\alpha \in \mathbb{N}_0^n$ with $|\alpha| \le \ell$ and all $f \in C_c^\infty(\mathbb{R}^n_+)$. (For the second statement, we can use

$$C = \max_{|\alpha|\le\ell} \left\| v_\alpha \right\|_{L^2(\mathbb{R}^n_+)}.)$$

Let $\alpha \in \mathbb{N}_0^n$ with $|\alpha| \le \ell$. By Lemma C.1.29(b), $\partial^\alpha u$ exists as an element of $H^{-|\alpha|}(\mathbb{R}^n_+) = H_0^{|\alpha|}(\mathbb{R}^n_+)^*$. For $v \in \{ v \in C^\ell(\mathbb{R}^n_+) \mid \|v\|_{H^\ell(\mathbb{R}^n_+)} < \infty \}$, $\partial^\alpha v$ is defined, as an element of $H^{-|\alpha|}(\mathbb{R}^n_+)$ by

$$(\partial^\alpha v)(f) = \langle f, \partial^\alpha \bar{v}\rangle_{L^2(\mathbb{R}^n_+)}$$

for all $f \in H_0^{|\alpha|}(\mathbb{R}^n_+)$. In particular, for $f \in C_c^\infty(\mathbb{R}^n_+)$,

$$\begin{aligned}
(\partial^\alpha v)(f) &= \langle f, \partial^\alpha \bar{v}\rangle_{L^2(\mathbb{R}^n_+)} \\
&= \int_{\mathbb{R}^n_+} f(x)\, \partial^\alpha v(x)\, dx \\
&= (-1)^{|\alpha|} \int_{\mathbb{R}^n_+} v(x)\, \partial^\alpha f(x)\, dx \\
&= (-1)^{|\alpha|} \langle \partial^\alpha f, \bar{v}\rangle_{L^2(\mathbb{R}^n_+)}.
\end{aligned}$$

By continuity,

$$(\partial^\alpha u)(f) = (-1)^{|\alpha|} \langle \partial^\alpha f, \bar{u}\rangle_{L^2(\mathbb{R}^n_+)}$$

for all $u \in L^2(\mathbb{R}^n_+)$ and $f \in H_0^{|\alpha|}(\mathbb{R}^n_+)$. Thus for our u and α,

$$|(\partial^\alpha u)(f)| \le C' \left\| f \right\|_{L^2(\mathbb{R}^n_+)}.$$

So, by the Riesz representation theorem (Theorem A.2.15), there is a function $u_\alpha \in L^2(\mathbb{R}^n_+)$ such that $(\partial^\alpha u)(f) = \langle f, \overline{u_\alpha} \rangle_{L^2(\mathbb{R}^n_+)}$ for all $f \in H_0^{|\alpha|}(\mathbb{R}^n_+)$. In other words $\partial^\alpha u \in L^2(\mathbb{R}^n_+)$. Since this is the case for all $\alpha \in \mathbb{N}_0^n$ with $|\alpha| \le \ell$, $u \in H^\ell(\mathbb{R}^n_+)$ by part (a). □

Proposition C.2.17. *Let $\ell \in \mathbb{N}$. Define the restriction map*

$$r : \left\{ u \in C^\infty(\overline{\mathbb{R}^n_+}) \mid u \text{ has bounded support} \right\} \to C_c^\infty(\mathbb{R}^{n-1}),$$
$$u \mapsto u|_{x_n=0}.$$

There exists a unique map

$$R : H^\ell(\mathbb{R}^n_+) \to H^{\ell-1/2}(\mathbb{R}^{n-1})$$

and constants C, C' such that

(i) *R extends r.*

(ii) *R is bounded. That is, $|||Ru|||_{H^{\ell-1/2}(\mathbb{R}^{n-1})} \le C\, \|u\|_{H^\ell(\mathbb{R}^n_+)}$.*

(iii) *R is surjective (onto).*

(iv) *For each $f \in H^{\ell-1/2}(\mathbb{R}^{n-1})$, there is a $u \in H^\ell(\mathbb{R}^n_+)$ such that*

$$Ru = f \quad \text{and} \quad \|u\|_{H^\ell(\mathbb{R}^n_+)} \le C' |||f|||_{H^{\ell-1/2}(\mathbb{R}^{n-1})}.$$

That is, R has a bounded right inverse.

Proof. By Lemma C.2.9, $\left\{ u \in C^\infty(\overline{\mathbb{R}^n_+}) \mid u \text{ has bounded support} \right\}$ is dense in $H^\ell(\mathbb{R}^n_+)$, so there is at most one bounded extension.

Let E_+ be the extension map from $\mathbb{R}^n_+$ to $\mathbb{R}^n$ of Lemma C.2.11 and let $\tilde{R}$ be the restriction map from $\mathbb{R}^n$ to $x_n = 0$ of Lemma C.2.4. Then

$$Ru = \tilde{R}E_+u$$

defines a bounded linear map from $H^\ell(\mathbb{R}^n_+)$ to $H^{\ell-1/2}(\mathbb{R}^{n-1})$. If $u \in C^\infty(\overline{\mathbb{R}^n_+})$ has bounded support, then $E_+u \in C_c^\ell(\mathbb{R}^n)$ (see (C.5)). By Exercise C.2.6, $(\tilde{R}E_+u)(x') = (E_+u)(x',0) = u(x',0)$, so R extends r. We have proven parts (i) and (ii).

Let E_0 be the extension map of Lemma C.2.8, with $s = \ell$. Let R_+u be the restriction of $u \in H^\ell(\mathbb{R}^n)$ to $\mathbb{R}^n_+$. It is a bounded map from $H^\ell(\mathbb{R}^n)$ to $H^\ell(\mathbb{R}^n_+)$. Then $\mathcal{E} = R_+E_0$ is a bounded map from $H^{\ell-1/2}(\mathbb{R}^{n-1})$ through $H^\ell(\mathbb{R}^n)$ to $H^\ell(\mathbb{R}^n_+)$. Furthermore, if $f \in \mathcal{S}(\mathbb{R}^{n-1})$, then $E_0 f \in \mathcal{S}(\mathbb{R}^n)$ (this was shown in the proof of Lemma C.2.8) and $E_+\mathcal{E}f = E_+R_+E_0f$ is in $C^\ell(\mathbb{R}^n)$ with derivatives that decay faster than the inverse of any polynomial (see (C.5)). Hence, by Exercise C.2.6,

$$(R\mathcal{E}f)(x') = (\tilde{R}E_+\mathcal{E}f)(x') = (E_+R_+E_0f)(x',0) = (E_0f)(x',0) = f(x').$$

Since $\mathcal{S}(\mathbb{R}^n)$ is dense in $H^{\ell-1/2}(\mathbb{R}^{n-1})$ and R and $\mathcal{E}$ are bounded, $R\mathcal{E}$ is the identity on $H^{\ell-1/2}(\mathbb{R}^{n-1})$. This proves parts (iii) and (iv). □

Proposition C.2.18. *Let $\ell \in \mathbb{N}$ and let R be the restriction operator of Proposition* C.2.17. *The kernel of R is $H^\ell(\mathbb{R}^n_+) \cap H^1_0(\mathbb{R}^n_+)$.*

Proof. If $u \in H^1_0(\mathbb{R}^n_+) \cap H^\ell(\mathbb{R}^n_+)$, then there is a sequence of functions $u_j \in C^\infty_c(\mathbb{R}^n_+)$ that converge to u with respect to the norm $\|\cdot\|_{H^1(\mathbb{R}^n_+)}$. Since, by part (ii) of Proposition C.2.17 with $\ell = 1$,

$$\begin{gathered}|||Ru|||_{H^{1/2}(\mathbb{R}^{n-1})} = |||Ru - Ru_j|||_{H^{1/2}(\mathbb{R}^{n-1})} \le C\left\|u - u_j\right\|_{H^1(\mathbb{R}^n_+)} \\ \text{for all } j \implies Ru = 0.\end{gathered}$$

Thus the kernel of R (acting on $H^\ell(\mathbb{R}^n_+)$) contains all of $H^1_0(\mathbb{R}^n_+) \cap H^\ell(\mathbb{R}^n_+)$.

To prove the converse, we have to show that any function in the kernel of R can be approximated, in the $H^1(\mathbb{R}^n_+)$ norm, by functions in $C^\infty_c(\mathbb{R}^n_+)$. Let $\psi \in C^\infty_c(\mathbb{R})$ take values in $[0,1]$ and obey $\psi(t) = 0$ for all $|t| \le \frac{1}{2}$ and $\psi(t) = 1$ for all $|t| \ge 1$.

Step 1. We first prove that if $u \in H^1(\mathbb{R}^n_+)$, $\varepsilon > 0$ and $u_\varepsilon(x) = u(x',x_n)\psi\left(\frac{x_n}{\varepsilon}\right)$, then

$$\begin{aligned}&\|u - u_\varepsilon\|^2_{H^1(\mathbb{R}^n_+)} \le \frac{2\Psi}{\varepsilon}\|Ru\|^2_{L^2(\mathbb{R}^{n-1})} + 2(1+\Psi)\,\|u\|^2_{H^1(\Omega_\varepsilon)}, \text{ where}\\ &\Omega_\varepsilon = \left\{\, x \in \mathbb{R}^n_+ \mid x_n < \varepsilon \,\right\},\\ &\Psi = \max\left\{\, \psi'(t)^2 \mid \tfrac{1}{2} \le t \le 1 \,\right\}.\end{aligned} \tag{C.10}$$

Multiplication by $\psi\left(\frac{x_n}{\varepsilon}\right)$ is a bounded operator on $H^1(\mathbb{R}^n_+)$ (possibly with a norm that diverges as $\varepsilon \to 0$). So it suffices to prove (C.10) for $u \in C^\infty(\overline{\mathbb{R}^n_+})$.

It is clear that

$$\|u - u_\varepsilon\|^2_{L^2(\mathbb{R}^n_+)} = \left\|\left[1 - \psi\left(\frac{x_n}{\varepsilon}\right)\right]u\right\|^2_{L^2(\mathbb{R}^n_+)} \le \int_{0<x_n<\varepsilon} |u(x)|^2\,dx \tag{C.11}$$

so we only need to bound the first derivative terms. If $1 \le j \le n-1$,

$$\frac{\partial}{\partial x_j}\left[\psi\left(\frac{x_n}{\varepsilon}\right)u(x)\right] = \psi\left(\frac{x_n}{\varepsilon}\right)\frac{\partial u}{\partial x_j}(x)$$

so that

$$\text{(C.12)}\qquad \begin{aligned} \left\|\frac{\partial}{\partial x_j}[u-u_\varepsilon]\right\|^2_{L^2(\mathbb{R}^n_+)} &= \left\|\left[1-\psi\left(\tfrac{x_n}{\varepsilon}\right)\right]\frac{\partial u}{\partial x_j}(x)\right\|^2_{L^2(\mathbb{R}^n_+)} \\ &\le \int_{0<x_n<\varepsilon}\left|\frac{\partial u}{\partial x_j}(x)\right|^2\,dx. \end{aligned}$$

Finally,

$$\frac{\partial}{\partial x_n}\left[u(x)-\psi\left(\tfrac{x_n}{\varepsilon}\right)u(x)\right] = \left[1-\psi\left(\tfrac{x_n}{\varepsilon}\right)\right]\frac{\partial u}{\partial x_n}(x) - \frac{1}{\varepsilon}\psi'\left(\tfrac{x_n}{\varepsilon}\right)u(x).$$

Since $(a+b)^2 \le 2a^2+2b^2$,

$$\begin{aligned} \left\|\frac{\partial}{\partial x_n}[u-u_\varepsilon]\right\|^2_{L^2(\mathbb{R}^n_+)} &\le 2\left\|\left[1-\psi\left(\tfrac{x_n}{\varepsilon}\right)\right]\frac{\partial u}{\partial x_n}(x)\right\|^2_{L^2(\mathbb{R}^n_+)} \\ &\quad + 2\left\|\frac{1}{\varepsilon}\psi'\left(\tfrac{x_n}{\varepsilon}\right)u(x)\right\|^2_{L^2(\mathbb{R}^n_+)}. \end{aligned}$$

The first term is bounded by

$$\text{(C.13)}\qquad 2\left\|\left[1-\psi\left(\tfrac{x_n}{\varepsilon}\right)\right]\frac{\partial u}{\partial x_n}(x)\right\|^2_{L^2(\mathbb{R}^n_+)} \le 2\int_{0<x_n<\varepsilon}\left|\frac{\partial u}{\partial x_n}(x)\right|^2\,dx.$$

Since $\psi'\left(\frac{x_n}{\varepsilon}\right)$ vanishes for $0 \le x_n \le \frac{\varepsilon}{2}$ and for $x_n \ge \varepsilon$,

$$\begin{aligned} 2\left\|\frac{1}{\varepsilon}\psi'\left(\tfrac{x_n}{\varepsilon}\right)u(x)\right\|^2_{L^2(\mathbb{R}^n_+)} &= \frac{2}{\varepsilon^2}\int_{\frac{\varepsilon}{2}<x_n<\varepsilon}\psi'\left(\tfrac{x_n}{\varepsilon}\right)^2|u(x)|^2\,dx \\ &= \frac{2}{\varepsilon}\int_{\frac{1}{2}<y<1}\psi'(y)^2|u(x',\varepsilon y)|^2\,dx'dy \\ &\le \frac{2\Psi}{\varepsilon}\int_{\frac{1}{2}<y<1}|u(x',\varepsilon y)|^2\,dx'dy, \end{aligned}$$

where $x_n = \varepsilon y$. By the fundamental theorem of calculus

$$\begin{aligned} u(x',\varepsilon y) &= u(x',0) + \int_0^{\varepsilon y}\frac{\partial u}{\partial x_n}(x',x_n)\,dx_n \\ &= (Ru)(x') + \int_0^{\varepsilon y}\frac{\partial u}{\partial x_n}(x',x_n)\,dx_n. \end{aligned}$$

So, by Cauchy–Schwarz and the fact that $(a+b)^2 \le 2a^2+2b^2$,

$$|u(x',\varepsilon y)|^2 \le 2|(Ru)(x')|^2 + 2\varepsilon\int_0^{\varepsilon}\left|\frac{\partial u}{\partial x_n}(x',x_n)\right|^2\,dx_n$$

for all $0 < y < 1$. Hence

$$2\left\|\frac{1}{\varepsilon}\psi'\left(\frac{x_n}{\varepsilon}\right)u(x)\right\|^2_{L^2(\mathbb{R}^n_+)} \le \frac{2\Psi}{\varepsilon}\int |(Ru)(x')|^2\,dx' + 2\Psi\int_{0<x_n<\varepsilon}\left|\frac{\partial u}{\partial x_n}(x)\right|^2\,dx. \tag{C.14}$$

Adding (C.11)–(C.14) gives (C.10).

Step 2. We now apply (C.10) to complete the proof. Let $u \in H^\ell(\mathbb{R}^n_+)$ be in the kernel of R. By (C.10),

$$\lim_{\varepsilon\to 0}\|u - u_\varepsilon\|^2_{H^1(\mathbb{R}^n_+)} \le 2(1+\Psi)\lim_{\varepsilon\to 0}\|u\|^2_{H^1(\Omega_\varepsilon)} = 0.$$

Since $u_\varepsilon(x)$ vanishes for $x_n < \frac{\varepsilon}{2}$, we have that $u_\varepsilon \in H^1_0(\mathbb{R}^n_+)$ by Exercise C.1.18(b). Since $H^1_0(\mathbb{R}^n_+)$ is complete, $u \in H^1_0(\mathbb{R}^n_+)$. □

We are now ready to deal with general bounded open sets Ω. We develop some generally useful results about $H^\ell(\Omega)$ and then give the proof of Theorem C.2.1.

Lemma C.2.19 (Extension operator). *Let $k \ge 1$ and $0 \le \ell \le k$, and let Ω be a bounded open subset of $\mathbb{R}^n$ with C^k boundary. Let $\mathcal{O}$ be an open subset of $\mathbb{R}^n$ with $\overline{\Omega} \cap \overline{\mathcal{O}} = \emptyset$. Then there exists a bounded linear operator $E : H^\ell(\Omega) \to H^\ell(\mathbb{R}^n)$ such that*

$$\begin{aligned} Eu(x) &= u(x) \quad \textit{a.e. for } x \in \Omega,\\ Eu(x) &= 0 \qquad\; \textit{a.e. for } x \in \mathcal{O}. \end{aligned}$$

Proof. Let N and, for $1 \le i \le N$, $\mathcal{U}(p_i)$, χ_i, ψ_{p_i} be as in Definition C.1.37. Let $\mathcal{V}$ be a neighborhood of $\partial\Omega$ on which $\sum_{i=1}^N \chi_i = 1$. Select a function $\varphi \in C^\infty_c(\Omega)$ that is identically one on $\Omega \setminus \mathcal{V}$. Since $\overline{\Omega}$ is compact, the distance from $\overline{\Omega}$ to $\overline{\mathcal{O}}$ is strictly positive and we may also choose a function $\chi \in C^\infty_c(\mathbb{R}^n \setminus \overline{\mathcal{O}})$ that is identically one on Ω.

Let $u \in H^\ell(\Omega)$. By construction, $u = \varphi u + \sum_{i=1}^N \chi_i(1-\varphi)u$. By Lemma C.1.48(c), for each $1 \le i \le N$, $v_i = \chi_i(1-\varphi)u \circ \psi_{p_i}^{-1} \in H^\ell(\mathbb{R}^n_+)$ and

$$\|v_i\|_{H^\ell(\mathbb{R}^n_+)} \le c_1\|u\|_{H^\ell(\Omega)}\,.$$

Let E_+ be the extension operator of Lemma C.2.11 and set

$$Eu = \chi\Big(\varphi u + \sum_{i=1}^N (E_+ v_i)\circ\psi_{p_i}\Big).$$

Since χ_i is supported in a compact subset of the domain, $\mathcal{U}(p_i)$, of the function ψ_{p_i} and ψ_{p_i} is continuous, there is an $r > 0$, independent of u, such that $v_i = \chi_i(1-\varphi)u \circ \psi_i^{-1}$ vanishes outside of $\{\, x \in \mathbb{R}^n_+ \mid |x| < r \,\}$. By Exercise C.2.10

and the construction (C.5) of Lemma C.2.11, there is an $R > 0$, such that $E_+ v_i$ vanishes outside of $\mathcal{U}'_R = \{ x \in \mathbb{R}^n \mid |x| < R \}$ for all u. Set $\mathcal{V}' = \psi_{p_i}^{-1}(\mathcal{U}'_{2R}) \subset \mathcal{U}(p_i)$. As both $\overline{\mathcal{U}'_{2R}}$ and $\overline{\mathcal{V}'}$ are compact, the restriction of $\psi_{p_i}^{-1}$ to $\mathcal{U}'_{2R}$ is in $C^\infty(\overline{\mathcal{U}'_{2R}})$ and the restriction of ψ_{p_i} to $\mathcal{V}'$ is in $C^k(\overline{\mathcal{V}'})$. By Lemma C.1.48(a), with $\psi = \psi_{p_i}^{-1}$, $\mathcal{U}$ replaced by $\mathcal{U}'_{2R}$ and $\mathcal{V}$ replaced by $\mathcal{V}'$, followed by Lemma C.2.11,

$$\begin{aligned} \|(E_+ v_i) \circ \psi_{p_i}\|_{H^\ell(\mathbb{R}^n)} &= \left\|(E_+ v_i) \circ \psi_{p_i}\right\|_{H^\ell(\mathcal{V}')} \\ &\le c_2 \|E_+ v_i\|_{H^\ell(\mathcal{U}'_{2R})} \\ &= c_2 |\!|\!| E_+ v_i |\!|\!|_{H^\ell(\mathbb{R}^n)} \\ &\le c_3 \|v_i\|_{H^\ell(\mathbb{R}^n_+)} \\ &\le c_1 c_3 \|u\|_{H^\ell(\Omega)} . \end{aligned}$$

It now suffices to recall that, by Exercise C.1.5, multiplication by φ is bounded on $H^\ell(\Omega)$ and multiplication by χ is bounded on $H^\ell(\mathbb{R}^n)$. □

Lemma C.2.20. *Let Ω be a bounded open subset of $\mathbb{R}^n$ with C^1 boundary. Then $C^\infty(\overline{\Omega})$ is dense in $H^\ell(\Omega)$ for any $\ell \in \mathbb{N}_0$.*

Proof. We first consider the model case where

$$\Omega = \{ (x', x_n) \in \mathbb{R}^n \mid x_n > h(x') \}$$

for some C^1 function $h : \mathbb{R}^{n-1} \to \mathbb{R}$ (in fact the result is valid even for continuous h). Let $u \in H^\ell(\Omega)$ have support contained in $\overline{B} \cap \overline{\Omega}$ for some ball B, and fix $\varepsilon > 0$. For $\delta > 0$ define the shifted domain

$$\Omega_\delta = \{ (x', x_n) \in \mathbb{R}^n \mid x_n > h(x') - \delta \},$$

and for $x = (x', x_n) \in \Omega_\delta$ let

$$u_\delta(x', x_n) = u(x', x_n + \delta).$$

Since translation is a continuous operation in $L^2(\Omega)$ (see the proof of Lemma C.2.9), for δ small enough we have

$$\|u - u_\delta\|_{L^2(\Omega)} < \frac{\varepsilon}{2}.$$

Now $\partial_j(u_\delta) = (\partial_j u)_\delta$, so repeating this argument for the derivatives of u shows that, for δ small,

$$\|u - u_\delta\|_{H^\ell(\Omega)} < \frac{\varepsilon}{2}.$$

If $\chi \in C_c^\infty(\mathbb{R}^n)$ satisfies $\chi = 1$ near $\overline{B} \cap \overline{\Omega}$ and $\text{supp}(\chi) \subset \Omega_{\delta/2}$, we have $\chi u_\delta \in H^\ell(\mathbb{R}^n)$. By Exercise C.1.13, there is $w \in C_c^\infty(\mathbb{R}^n)$ with

$$\|\chi u_\delta - w\|_{H^\ell(\mathbb{R}^n)} < \frac{\varepsilon}{2}.$$

It follows that

$$\|u - w|_\Omega\|_{H^\ell(\Omega)} \leq \|u - u_\delta\|_{H^\ell(\Omega)} + \|u_\delta - w\|_{H^\ell(\Omega)} < \varepsilon.$$

We have proved that any $u \in H^\ell(\Omega)$ where Ω has the form $\{x_n > h(x')\}$, with $\operatorname{supp}(u)$ bounded, can be approximated in $H^\ell(\Omega)$ by functions in $C^\infty(\overline{\Omega})$.

Now let Ω be a bounded open set with C^1 boundary, let $u \in H^\ell(\Omega)$ and fix $\varepsilon > 0$. Given any $x_0 \in \partial\Omega$ there is a ball B centered at x_0 such that, in new coordinates $x = (x', x_n)$ obtained by a rotation and translation,

$$\Omega \cap B = \{\, x \in B \mid x_n > h(x') \,\}$$

for some C^1 function $h : \mathbb{R}^{n-1} \to \mathbb{R}$. By compactness we can cover $\partial\Omega$ by finitely many such balls $B_1, \ldots, B_N$ with corresponding C^1 functions h_j. We use a partition of unity as in Theorem 2.1.9 and choose $\chi_j \in C_c^\infty(B_j)$ such that $\chi_1 + \ldots + \chi_N = 1$ near $\partial\Omega$. Now $\chi_j u$ is as in the first part of the proof, so there exist $w_j \in C^\infty(\overline{\Omega})$ with

$$\|\chi_j u - w_j\|_{H^\ell(\Omega)} < \frac{\varepsilon}{N+1}.$$

Define $\chi_0 = 1 - (\chi_1 + \ldots + \chi_N)$, so that $\chi_0 u|_\Omega \in H^\ell(\Omega)$ vanishes near $\partial\Omega$ and therefore its extension by zero is in $H^\ell(\mathbb{R}^n)$. By Exercise C.1.13 there is $w_0 \in C_c^\infty(\mathbb{R}^n)$ with

$$\|\chi_0 u - w_0\|_{H^\ell(\Omega)} < \frac{\varepsilon}{N+1}.$$

Writing $w = w_0 + \ldots + w_N$, we obtain

$$\|u - w\|_{H^\ell(\Omega)} \leq \|\sum(\chi_j u - w_j)\|_{H^\ell(\Omega)} < \varepsilon.$$

Since ε was arbitrary, this proves the result. □

Exercise C.2.21 (Sobolev embedding). Let Ω be a bounded open set with C^k boundary, let $\alpha \in \mathbb{N}_0^n$ and $\ell \in \mathbb{N}_0$ obey $\ell > |\alpha| + \frac{n}{2}$ and $\ell \leq k$. Prove that if $u \in H^\ell(\Omega)$, then $\partial^\alpha u$ is continuous on $\overline{\Omega}$ and there is a constant C, depending only on Ω, ℓ and $|\alpha|$, such that

$$\sup_{x \in \overline{\Omega}} |\partial^\alpha u(x)| \leq C \, \|u\|_{H^\ell(\Omega)} \, .$$

Lemma C.2.22 (Classical regularity). *Let Ω be a bounded open subset of $\mathbb{R}^n$ with smooth boundary. If $u \in H^\ell(\Omega)$ for all $\ell \in \mathbb{N}$, then $u \in C^\infty(\overline{\Omega})$.*

Proof. Just apply Exercise C.2.21 for all $\ell \in \mathbb{N}_0$. □

We can now prove the main result in this section.

Proof of Theorem C.2.1. Let N and, for $1 \leq i \leq N$, $\mathcal{U}(p_i)$, χ_i, ψ_{p_i} be as in Definition C.1.37.

(i), (ii) Let $u \in C^\infty(\overline{\Omega})$. Then, using R_+ to denote the restriction map from $H^\ell(\mathbb{R}^n_+)$ to $H^{\ell-\frac{1}{2}}(\mathbb{R}^{n-1})$,

$$\begin{aligned}
\|ru\|^2_{H^{\ell-1/2}(\partial\Omega)} &= \sum_{i=1}^N \|(\chi_i ru)\circ\psi_{p_i}^{-1}\|^2_{H^{\ell-1/2}(\mathbb{R}^{n-1})} && \text{by Definition C.1.37}\\
&= \sum_{i=1}^N \|R_+(\chi_i u\circ\psi_{p_i}^{-1})\|^2_{H^{\ell-1/2}(\mathbb{R}^{n-1})}\\
&\le \sum_{i=1}^N C^2 \left\|(\chi_i u)\circ\psi_{p_i}^{-1}\right\|^2_{H^\ell(\mathbb{R}^n_+)} && \text{by Proposition C.2.17}\\
&\le \sum_{i=1}^N C^2\tilde{C}^2\,\|u\|^2_{H^\ell(\Omega)} && \text{by Lemma C.1.48(c).}
\end{aligned}$$

The existence of a unique $R : H^\ell(\Omega) \to H^{\ell-1/2}(\partial\Omega)$ fulfilling the requirements of parts (i) and (ii) now follows by the B.L.T. theorem (Theorem A.2.13).

(iii), (iv) Let $f \in H^{\ell-1/2}(\partial\Omega)$. By definition, for each $1 \le i \le n$,

$$(\chi_i f)\circ\psi_{p_i}^{-1} \in H^{\ell-1/2}(\mathbb{R}^{n-1}).$$

By parts (iii) and (iv) of Proposition C.2.17, there exists $v_i \in H^\ell(\mathbb{R}^n_+)$ such that $R_+v_i = (\chi_i f)\circ\psi_{p_i}^{-1}$ and

$$\|v_i\|_{H^\ell(\mathbb{R}^n_+)} \le C'\|(\chi_i f)\circ\psi_{p_i}^{-1}\|_{H^{\ell-1/2}(\mathbb{R}^{n-1})}$$

For each $1 \le i \le N$, select a function $\varphi_i \in C^\infty(\overline{\mathbb{R}^n_+})$ that has bounded support and is identically one on the support of $(\chi_i f)\circ\psi_{p_i}^{-1}$ and set

$$u = \sum_{i=1}^N (\varphi_i v_i)\circ\psi_{p_i}.$$

By Lemma C.1.48(a) (with the $\mathcal{U}$ of Lemma C.1.48(a) being a bounded open subset of $\mathbb{R}^n_+$ that contains the support of φ_i and the ψ of Lemma C.1.48(a) being $\psi_{p_i}^{-1}$) and the boundedness of multiplication by C^∞ functions all of whose derivatives are bounded, the map

$$v \in H^\ell(\mathbb{R}^n_+) \mapsto (\varphi_i v)\circ\psi_{p_i} \in H^\ell(\Omega)$$

is bounded. Similarly, the map

$$g \in H^{\ell-1/2}(\mathbb{R}^{n-1}) \mapsto \big((R_+\varphi_i)g\big)\circ\psi_{p_i} \in H^{\ell-1/2}(\partial\Omega)$$

is bounded. Hence

$$R\big((\varphi_i v)\circ\psi_{p_i}\big) = (R_+\varphi_i)(R_+v)\circ\psi_{p_i}$$

is valid for all $v \in H^\ell(\mathbb{R}^n_+)$, not just $v \in C^\infty(\overline{\mathbb{R}^n_+})$. In particular,

$$R\big((\varphi_i v_i)\circ\psi_{p_i}\big) = (R_+\varphi_i)(R_+v_i)\circ\psi_{p_i} = (R_+\varphi_i)\left(\chi_i f\circ\psi_{p_i}^{-1}\right)\circ\psi_{p_i} = \chi_i f$$

and $Ru = \sum_{i=1}^N \chi_i f = f$.

For the bound of part (iv),

$$\begin{aligned}
\|u\|_{H^\ell(\Omega)} &\le \sum_{i=1}^N \Big\|(\varphi_i v_i)\circ\psi_{p_i}\Big\|_{H^\ell(\Omega)} \\
&\le \sum_{i=1}^N \|v_i\|_{H^\ell(\mathbb{R}^n_+)} \\
&\le \sum_{i=1}^N C'\,\|(\chi_i f)\circ\psi_{p_i}^{-1}\|_{H^{\ell-1/2}(\mathbb{R}^{n-1})} \\
&\le C'\sqrt{N}\Big[\sum_{i=1}^N \|(\chi_i f)\circ\psi_{p_i}^{-1}\|^2_{H^{\ell-1/2}(\mathbb{R}^{n-1})}\Big]^{1/2} \\
&= C'\sqrt{N}\,\|f\|_{H^{\ell-1/2}(\partial\Omega)}
\end{aligned}$$

by Cauchy–Schwarz and Definition C.1.37.

(v) Let $u \in H^\ell(\Omega)$ be in the kernel of R. Observe that

$$R_+\big((\chi_i u)\circ\psi_{p_i}^{-1}\big) = \big((R\chi_i)(Ru)\big)\circ\psi_{p_i}^{-1}\Big|_{\mathbb{R}^{n-1}} = 0.$$

Therefore, by Proposition C.2.18, $(\chi_i u)\circ\psi_{p_i}^{-1} \in H_0^1(\mathbb{R}^n_+)\cap H^\ell(\mathbb{R}^n_+)$ and there exist sequences $u_m^{(i)} \in C_c^\infty(\mathbb{R}^n_+)$ obeying

$$\lim_{m\to\infty} u_m^{(i)} = (\chi_i u)\circ\psi_{p_i}^{-1} \text{ in } H^1(\mathbb{R}^n_+)$$

Consequently $\lim_{m\to\infty} u_m^{(i)}\circ\psi_{p_1} = \chi_i u$ in $H^1(\Omega)$ and

$$\lim_{m\to\infty}\sum_{i=1}^N u_m^{(i)}\circ\psi_{p_1} = \sum_{i=1}^N \chi_i u \text{ in } H^1(\Omega).$$

As $u_m^{(i)}\circ\psi_{p_i} \in C_c^\infty(\Omega)$, $\sum_{i=1}^N \chi_i u \in H_0^1(\Omega)$. By Exercise C.1.18,

$$\Big(1-\sum_{i=1}^N \chi_i\Big)u \in H_0^\ell(\Omega) \subset H_0^1(\Omega).$$

Thus the kernel is contained in $H_0^1(\Omega)\cap H^\ell(\Omega)$. We have already observed, in Remark C.2.3, that the kernel contains all of $H_0^1(\Omega)\cap H^\ell(\Omega)$. So the proof is concluded. □

Exercise C.2.23. Let Ω be a bounded open subset of $\mathbb{R}^n$ with smooth boundary. Define the restriction map

$$r : C^\infty(\overline{\Omega}) \to C^\infty(\partial\Omega),$$
$$u \mapsto u|_{\partial\Omega}.$$

Prove that there does *not* exist a bounded map

$$R : L^2(\Omega) \to L^2(\partial\Omega)$$

such that $Ru = ru$ for all $u \in C^\infty(\overline{\Omega})$.

Here is the analog of Proposition C.2.16 for $H^\ell(\Omega)$.

Proposition C.2.24. *Let $\ell \in \mathbb{N}_0$ and let Ω be a bounded open subset of $\mathbb{R}^n$ with smooth boundary.*

(a) *Let $u \in H^{\ell'}(\Omega)$ with $\ell' \in \mathbb{N}_0$ obeying $\ell' \le \ell$. Then $u \in H^\ell(\Omega)$ if and only if $\partial^\alpha u \in H^{\ell'}(\Omega)$ for all $\alpha \in \mathbb{N}_0^n$ with $|\alpha| \le \ell - \ell'$.*

(b) *Let $u \in L^2(\Omega)$. Then $u \in H^\ell(\Omega)$ if and only if there is a constant C such that*

$$|\langle \partial^\alpha f, u\rangle_{L^2(\Omega)}| \le C\big\|f\big\|_{L^2(\Omega)}$$

for all $\alpha \in \mathbb{N}_0^n$ with $|\alpha| \le \ell$ and all $f \in C_c^\infty(\Omega)$. Also, $u \in H^\ell(\Omega)$ if and only if, for each $\alpha \in \mathbb{N}_0^n$ with $|\alpha| \le \ell$, there is a $v_\alpha \in L^2(\Omega)$ such that

$$(-1)^{|\alpha|}\langle \partial^\alpha f, u\rangle_{L^2(\Omega)} = \langle f, v_\alpha\rangle_{L^2(\Omega)}$$

for all $f \in C_c^\infty(\Omega)$.

Proof.

(a) The implication that if $u \in H^\ell(\Omega)$, then $\partial^\alpha u \in H^{\ell'}(\Omega)$ for all $\alpha \in \mathbb{N}_0^n$ with $|\alpha| \le \ell - \ell'$ is an immediate consequence of part (b) of Lemma C.1.29, so we just prove the converse.

Throughout the next two paragraphs, assume that α runs over the elements of $\mathbb{N}_0^n$ that obey $|\alpha| \le \ell - \ell'$. Let N and, for $1 \le i \le N$, $\mathcal{U}(p_i)$, χ_i, ψ_{p_i} be as in Definition C.1.37. Let $\mathcal{V}$ be a neighborhood of $\partial\Omega$ on which $\sum_{i=1}^N \chi_i = 1$. Select a function $\varphi \in C_c^\infty(\Omega)$ that is identically one on $\Omega \setminus V$. We prove that each of the $N+1$ terms on the right-hand side of

$$u = \varphi u + \sum_{i=1}^N \chi_i(1-\varphi)u$$

is in $H^\ell(\Omega)$. By the product rule and Lemma C.1.29(a), $\partial^\alpha(\varphi u) \in H^{\ell'}(\Omega)$. Since $\partial^\alpha(\varphi u)$ vanishes outside of a compact subset of Ω, $\partial^\alpha(\varphi u) \in H^{\ell'}(\mathbb{R}^n)$. Thus $\xi^\alpha\big(1+|\xi|^2\big)^{\ell'/2}\widehat{\varphi u}(\xi) \in L^2(\mathbb{R}^n)$ for all $\alpha \in \mathbb{N}_0^n$

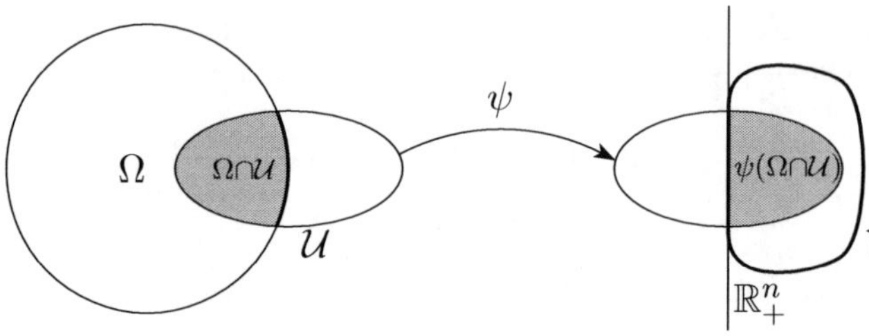

Figure C.4. The map ψ in the proof of Proposition C.2.24.

with $|\alpha| \le \ell - \ell'$. This implies that $\big(1+|\xi|^2\big)^{\ell/2}\widehat{\varphi u}(\xi) \in L^2(\mathbb{R}^n)$ and hence that $\varphi u \in H^\ell(\mathbb{R}^n)$ and that $\varphi u \in H^\ell(\Omega)$.

Fix any $1 \le i \le N$. Denote

$$\psi = \psi_{p_i} \qquad v = \chi_i(1-\varphi)u \qquad w = \chi_i(1-\varphi)u \circ \psi^{-1}$$

By the chain rule, product rule and part (c) of Lemma C.1.48, we have $\partial^\alpha w \in H^{\ell'}(\mathbb{R}^n_+)$. Hence, by part (a) of Lemma C.2.16, $w \in H^\ell(\mathbb{R}^n_+)$. Let $\mathcal{U}$ be an open neighborhood of the support of χ_i whose closure is contained in $\mathcal{U}(p_i)$. Then $\psi(\mathcal{U}) \cap \mathbb{R}^n_+ = \psi(\mathcal{U} \cap \Omega)$ is a bounded open subset of $\mathbb{R}^n$ and w vanishes outside of $\psi(\mathcal{U} \cap \Omega)$. Enlarge $\psi(\mathcal{U}) \cap \mathbb{R}^n_+$ to $\mathcal{V}$, a bounded open subset of $\mathbb{R}^n_+$ with smooth boundary. See Figure C.4.

By part (a) of Lemma C.1.48, with $\mathcal{U}$ replaced by $\mathcal{V}$, $\mathcal{V}$ replaced by $\psi^{-1}(\mathcal{V})$, and ψ replaced by ψ^{-1}, we have $v \in H^\ell(\Omega)$.

(b) The proof is virtually identical to that of part (b) of Proposition C.2.16. It suffices to replace $\mathbb{R}^n_+$ by Ω and to replace the reference to Lemma C.2.9 by a reference to Lemma C.2.20. □

C.3. Inequalities

In this section we prove a few useful inequalities, such as

- interpolation—Lemma C.3.2

$$|||A|||_{H^{(1-t)s_0+ts_1}(\mathbb{R}^n)} \le |||A|||^{1-t}_{H^{s_0}(\mathbb{R}^n)}|||A|||^{t}_{H^{s_1}(\mathbb{R}^n)}.$$

- Young's inequality—Lemma C.3.10

$$\|f * g\|_{L^p(\mathbb{R}^n)} \le \|f\|_{L^1(\mathbb{R}^n)}\|g\|_{L^p(\mathbb{R}^n)}.$$

We start with the three lines theorem of complex analysis and apply it prove an interpolation lemma that says that the operator norm on Sobolev spaces is log convex.

Lemma C.3.1 (Three lines theorem). *Let $F(z)$ be a function that is holomorphic in $0 < \operatorname{Re} z < 1$ and continuous and bounded on $0 \le \operatorname{Re} z \le 1$. Assume that*

$$\sup_{\operatorname{Re} z=0} |F(z)| \le M_0, \qquad \sup_{\operatorname{Re} z=1} |F(z)| \le M_1.$$

Then

$$\sup_{\operatorname{Re} z=t} |F(z)| \le M_0^{1-t} M_1^t$$

for all $0 \le t \le 1$.

Proof. Define, for $\varepsilon > 0$,

$$F_\varepsilon(z) = e^{\varepsilon z^2} F(z) M_0^{z-1} M_1^{-z}.$$

Since $\operatorname{Re}(z^2) = (\operatorname{Re} z)^2 - (\operatorname{Im} z)^2$,

$$|F_\varepsilon(z)| = e^{\varepsilon(\operatorname{Re} z)^2} e^{-\varepsilon(\operatorname{Im} z)^2} |F(z)| M_0^{\operatorname{Re} z-1} M_1^{-\operatorname{Re} z}.$$

Recalling that $|F(z)|$ is bounded, by hypothesis, we have

$$\sup_{\operatorname{Re} z=0} |F_\varepsilon(z)| \le 1, \quad \sup_{\operatorname{Re} z=1} |F_\varepsilon(z)| \le e^\varepsilon, \quad \lim_{|\operatorname{Im} z|\to\infty} \sup_{0\le \operatorname{Re} z\le 1} |F_\varepsilon(z)| = 0.$$

The maximum modulus of $F_\varepsilon(z)$ on the rectangle $\{0 \le \operatorname{Re} z \le 1, |\operatorname{Im} z| \le y\}$ is achieved on the boundary of the rectangle, since $F_\varepsilon(z)$ is holomorphic. Hence

$$\sup_{0\le \operatorname{Re} z\le 1} |F_\varepsilon(z)| = \lim_{y\to\infty} \sup_{\substack{|\operatorname{Im} z|\le y \\ 0\le \operatorname{Re} z\le 1}} |F_\varepsilon(z)| \le \lim_{y\to\infty} \max\Big\{ e^\varepsilon, \sup_{\substack{|\operatorname{Im} z|=y \\ 0\le \operatorname{Re} z\le 1}} |F_\varepsilon(z)| \Big\} = e^\varepsilon.$$

Taking the limit $\varepsilon \to 0+$ gives $|F(z) M_0^{\operatorname{Re} z-1} M_1^{-\operatorname{Re} z}| \le 1$ and hence

$$|F(z)| \le M_0^{1-\operatorname{Re} z} M_1^{\operatorname{Re} z},$$

which is what we want. □

Let $s_0 \in \mathbb{R}$ and suppose that A is a continuous linear operator on $H^{s_0}(\mathbb{R}^n)$. For any $s > s_0$, $H^s(\mathbb{R}^m) \subset H^{s_0}(\mathbb{R}^m)$ so that we may also view A as an operator defined on $H^s(\mathbb{R}^m)$, though possibly still taking values in $H^{s_0}(\mathbb{R}^m)$. The next lemma shows that if, in addition, we know that $AH^{s_1}(\mathbb{R}^m) \subset H^{s_1}(\mathbb{R}^m)$ for some $s_1 > s_0$, then we also have that $AH^s(\mathbb{R}^m) \subset H^s(\mathbb{R}^m)$ for all s between s_0 and s_1. The lemma also provides a bound on the operator norm

$$|||A|||_{H^s(\mathbb{R}^m)} = \sup_{\substack{f\in H^s(\mathbb{R}^m) \\ f\ne 0}} \frac{|||Af|||_{H^s(\mathbb{R}^m)}}{|||f|||_{H^s(\mathbb{R}^m)}}$$

of A, viewed as a map on $H^s(\mathbb{R}^m)$, in terms of its operator norms when viewed as maps on $H^{s_0}(\mathbb{R}^m)$ and on $H^{s_1}(\mathbb{R}^m)$.

Lemma C.3.2 (Interpolation). *Let* $s_0, s_1 \in \mathbb{R}$, $0 \le t \le 1$ *and set*

$$s = (1-t)s_0 + ts_1.$$

Let

$$A : H^{s_0}(\mathbb{R}^m) \to H^{s_0}(\mathbb{R}^m),$$
$$A : H^{s_1}(\mathbb{R}^m) \to H^{s_1}(\mathbb{R}^m)$$

be bounded. Then

$$A : H^{s}(\mathbb{R}^m) \to H^{s}(\mathbb{R}^m)$$

is bounded with

$$\|A\|_{H^s(\mathbb{R}^m)} \le \|A\|_{H^{s_0}(\mathbb{R}^n)}^{1-t} \|A\|_{H^{s_1}(\mathbb{R}^n)}^{t}.$$

Proof. By Exercise C.3.3, it suffices to prove that

$$\left| \int_{\mathbb{R}^m} \widehat{Af}(\xi)\overline{\hat{g}(\xi)} \frac{d\xi}{(2\pi)^m} \right| \le \|A\|_{H^{s_0}(\mathbb{R}^n)}^{1-t} \|A\|_{H^{s_1}(\mathbb{R}^n)}^{t} \|f\|_{H^s(\mathbb{R}^n)} \|g\|_{H^{-s}(\mathbb{R}^m)}$$

for all $f, g \in \mathcal{S}(\mathbb{R}^m)$. So let $f, g \in \mathcal{S}(\mathbb{R}^m)$ and set, for $z \in \mathbb{C}$,

$$\sigma(z) = (1-z)s_0 + zs_1$$

Observe that $\sigma(t) = s$. Define $f_z \in \mathcal{S}(\mathbb{R}^m)$ by

$$\hat{f}_z(\xi) = \left(1 + |\xi|^2\right)^{\frac{s-\sigma(z)}{2}} \hat{f}(\xi).$$

We shall apply the three lines theorem, Lemma C.3.1, to

$$F(z) = \int_{\mathbb{R}^m} \left(1 + |\xi|^2\right)^{\frac{\sigma(z)-s}{2}} \widehat{Af_z}(\xi)\overline{\hat{g}(\xi)} \frac{d\xi}{(2\pi)^m}$$

in order to derive a bound on

$$F(t) = \int_{\mathbb{R}^m} \widehat{Af}(\xi)\overline{\hat{g}(\xi)} \frac{d\xi}{(2\pi)^m}.$$

Inserting $1 = \left(1+|\xi|^2\right)^{\frac{\operatorname{Re}\sigma(z)}{2}} \left(1+|\xi|^2\right)^{-\frac{\operatorname{Re}\sigma(z)}{2}}$ into the integral defining $F(z)$ and applying Cauchy–Schwarz gives

$$\begin{aligned}
|F(z)| &\le \left[\int_{\mathbb{R}^m} \left(1+|\xi|^2\right)^{\operatorname{Re}\sigma(z)} \left|\widehat{Af_z}(\xi)\right|^2 \frac{d\xi}{(2\pi)^m}\right]^{1/2} \\
&\qquad \left[\int_{\mathbb{R}^m} \left(1+|\xi|^2\right)^{-\operatorname{Re}\sigma(z)} \left(1+|\xi|^2\right)^{\operatorname{Re}\sigma(z)-s} \left|\hat{g}(\xi)\right|^2 \frac{d\xi}{(2\pi)^m}\right]^{1/2} \\
&\le \left[\int_{\mathbb{R}^m} \left(1+|\xi|^2\right)^{\operatorname{Re}\sigma(z)} \left|\widehat{Af_z}(\xi)\right|^2 \frac{d\xi}{(2\pi)^m}\right]^{1/2} \\
&\qquad \left[\int_{\mathbb{R}^m} \left(1+|\xi|^2\right)^{-s} \left|\hat{g}(\xi)\right|^2 \frac{d\xi}{(2\pi)^m}\right]^{1/2} \\
&= |||Af_z|||_{H^{\operatorname{Re}\sigma(z)}(\mathbb{R}^m)}\, |||g|||_{H^{-s}(\mathbb{R}^m)} \\
&\le |||A|||_{H^{\operatorname{Re}\sigma(z)}(\mathbb{R}^m)}\, |||f_z|||_{H^{\operatorname{Re}\sigma(z)}(\mathbb{R}^m)}\, |||g|||_{H^{-s}(\mathbb{R}^m)} \\
&\le |||A|||_{H^{\operatorname{Re}\sigma(z)}(\mathbb{R}^m)}\, |||f|||_{H^{s}(\mathbb{R}^n)}\, |||g|||_{H^{-s}(\mathbb{R}^m)}
\end{aligned}$$

since $\left|\hat{f}_z(\xi)\right|^2 = \left(1+|\xi|^2\right)^{s-\operatorname{Re}\sigma(z)} |\hat{f}(\xi)|^2$. When $\operatorname{Re} z = 0$, $\operatorname{Re}\sigma(z) = s_0$ so that

$$|F(z)| \le |||A|||_{H^{s_0}(\mathbb{R}^n)}\, |||f|||_{H^{s}(\mathbb{R}^n)}\, |||g|||_{H^{-s}(\mathbb{R}^m)}.$$

When $\operatorname{Re} z = 1$, $\operatorname{Re}\sigma(z) = s_1$ so that

$$|F(z)| \le |||A|||_{H^{s_1}(\mathbb{R}^n)}\, |||f|||_{H^{s}(\mathbb{R}^n)}\, |||g|||_{H^{-s}(\mathbb{R}^m)}.$$

By Exercise C.3.4, $F(z)$ is an entire function of z so that the three lines theorem gives

$$\begin{aligned}
\left|\int_{\mathbb{R}^m} \widehat{Af}(\xi)\overline{\hat{g}(\xi)}\, \frac{d\xi}{(2\pi)^m}\right| &= |F(t)| \\
&\le |||A|||^{1-t}_{H^{s_0}(\mathbb{R}^n)}\, |||A|||^{t}_{H^{s_1}(\mathbb{R}^n)}\, |||f|||_{H^{s}(\mathbb{R}^n)}\, |||g|||_{H^{-s}(\mathbb{R}^m)}
\end{aligned}$$

and the lemma follows. □

Exercise C.3.3. Let $t < s$ and let A be a bounded linear map on $H^t(\mathbb{R}^m)$. Prove that if there is a constant M such that

$$\left|\int_{\mathbb{R}^m} \widehat{Af}(\xi)\overline{\hat{g}(\xi)}\, \frac{d\xi}{(2\pi)^m}\right| \le M\, |||f|||_{H^{s}(\mathbb{R}^n)}\, |||g|||_{H^{-s}(\mathbb{R}^m)}$$

for all $f, g \in \mathcal{S}(\mathbb{R}^m)$ then A is a bounded linear map $H^s(\mathbb{R}^m)$ with norm at most M.

Exercise C.3.4. Let $s \in \mathbb{R}$ and A be a bounded linear map on $H^s(\mathbb{R}^m)$. Let $f, g \in \mathcal{S}(\mathbb{R}^m)$ and set, for $z \in \mathbb{C}$, $\sigma(z) = (1-z)s_0 + zs_1$. Define

$$F(z) = \int_{\mathbb{R}^m} \left(1+|\xi|^2\right)^{\frac{\sigma(z)-s}{2}} \widehat{Af_z}(\xi)\overline{\hat{g}(\xi)}\, \frac{d\xi}{(2\pi)^m},$$

where $f_z \in \mathcal{S}(\mathbb{R}^m)$ is determined by

$$\hat{f}_z(\xi) = \left(1 + |\xi|^2\right)^{\frac{s-\sigma(z)}{2}} \hat{f}(\xi)$$

(a) Prove that $\frac{f_{z'} - f_z}{z' - z}$ converges in $H^s(\mathbb{R}^m)$ as $z' \to z$.

(b) Prove that $F(z)$ is an entire function of z.

Exercise C.3.5. Let Ω be an open subset of $\mathbb{R}^n$. Let $1 \leq p < q < r \leq \infty$. Prove that if $f \in L^p(\Omega) \cap L^r(\Omega)$, then $f \in L^q(\Omega)$ and

$$\|f\|_{L^q(\Omega)} \leq \|f\|_{L^p(\Omega)}^{1-t} \|f\|_{L^r(\Omega)}^t,$$

where $0 < t < 1$ is determined by $\frac{1}{q} = \frac{1-t}{p} + \frac{t}{r}$.

Exercise C.3.6. The goal of this problem is to prove the Riesz–Thorin interpolation theorem, which is as follows. Let Ω be an open subset of $\mathbb{R}^n$. Let $1 \leq p_0, p_1, q_0, q_1 \leq \infty$ and, for $0 \leq t \leq 1$, define p_t, q_t by

$$\frac{1}{p_t} = \frac{1-t}{p_0} + \frac{t}{p_1}, \qquad \frac{1}{q_t} = \frac{1-t}{q_0} + \frac{t}{q_1}.$$

Suppose that T is a linear transformation defined on $L^{p_0}(\Omega) \cap L^{p_1}(\Omega)$ and taking values in $L^{q_0}(\Omega) \cap L^{q_1}(\Omega)$ which satisfies

$$\|Tf\|_{L^{q_0}(\Omega)} \leq M_0 \|f\|_{L^{p_0}(\Omega)}, \qquad \|Tf\|_{L^{q_1}(\Omega)} \leq M_1 \|f\|_{L^{p_1}(\Omega)}$$

for all $f \in L^{p_0}(\Omega) \cap L^{p_1}(\Omega)$. Then, for each $f \in L^{p_0}(\Omega) \cap L^{p_1}(\Omega)$ and $0 < t < 1$, $Tf \in L^{q_t}(\Omega)$ and

$$\|Tf\|_{L^{q_t}(\Omega)} \leq M_0^{1-t} M_1^t \|f\|_{L^{p_t}(\Omega)}.$$

(a) Prove the Riesz–Thorin interpolation theorem in the special case that $p_0 = p_1$.

(b) Denote by

$$\Sigma(\Omega) = \Big\{ \sum_{j=1}^{m} a_j \chi_{E_j} \;\Big|\; m \in \mathbb{N},\ a_1, \cdots, a_m \in \mathbb{C}, \ E_1, \cdots, E_m \subset \Omega \text{ measurable and of finite measure} \Big\}$$

the set of simple functions on Ω. Prove that if $f \in \Sigma(\Omega)$, then

$$\|Tf\|_{L^{q_t}(\Omega)} = \sup\Big\{ \Big| \int_\Omega (Tf)(x) g(x)\, dx \Big| \;\Big|\; g \in \Sigma(\Omega),\ \|g\|_{L^{q_t'}(\Omega)} \leq 1 \Big\},$$

where q_t', the conjugate index to q_t, is determined by $\frac{1}{q_t} + \frac{1}{q_t'} = 1$.

(c)

$$\Big| \int_\Omega (Tf)(x) g(x)\, dx \Big| \leq M_0^{1-t} M_1^t$$

for all $f, g \in \Sigma(\Omega)$ with $\|f\|_{L^{p_t}(\Omega)}, \|g\|_{L^{q'_t}(\Omega)} \le 1$.

(d) Prove the Riesz–Thorin interpolation theorem.

Exercise C.3.7. Let $\ell, L \in \mathbb{Z}$ and $s \in \mathbb{R}$ with $\ell \le s \le L$. Let Ω be a bounded open subset of $\mathbb{R}^n$ with smooth boundary and assume that

$$\varphi \in C^{\max\{|\ell|,|L|\}}(\partial\Omega).$$

Prove that the map $v \in C^\infty(\partial\Omega) \mapsto \varphi v$ has a unique extension to a bounded, linear map on $H^s(\partial\Omega)$ and there is a constant C, depending only on Ω, $\max\{|\ell|, |L|\}$ and n, such that

$$\begin{aligned} \|\varphi u\|_{H^s(\partial\Omega)} &\le C \|\varphi\|_{C^{|\ell|}(\partial\Omega)}^{\frac{L-s}{L-\ell}} \|\varphi\|_{C^{|L|}(\partial\Omega)}^{\frac{s-\ell}{L-\ell}} \|u\|_{H^s(\partial\Omega)} \\ &\le C \|\varphi\|_{C^{\max\{|\ell|,|L|\}}(\partial\Omega)} \|u\|_{H^s(\partial\Omega)} \end{aligned}$$

for all $u \in H^s(\partial\Omega)$.

Exercise C.3.8 (Hausdorff–Young inequality). Let $1 \le p \le 2$ and $2 \le q \le \infty$ obey $\frac{1}{p} + \frac{1}{q} = 1$. Prove that if $f \in L^p(\mathbb{R}^n)$, then $\hat{f} \in L^q(\mathbb{R}^n)$ and

$$\|\hat{f}\|_{L^q(\mathbb{R}^n, \frac{d\xi}{(2\pi)^n})} \le \|f\|_{L^p(\mathbb{R}^n, dx)}.$$

Exercise C.3.9. Let $2 < p \le \infty$ and let $1 \le q < 2$ obey $\frac{1}{p} + \frac{1}{q} = 1$. Prove that there does *not* exist a constant C such that

$$\|\hat{f}\|_{L^q(\mathbb{R}^n, \frac{d\xi}{(2\pi)^n})} \le C\|f\|_{L^p(\mathbb{R}^n, dx)}$$

for all $f \in \mathcal{S}(\mathbb{R}^n)$. Hint: consider $f(x) = e^{-(a+ib)x^2}$ with $a > 0$.

Since $|\hat{f}(\xi)\hat{g}(\xi)| \le \|\hat{f}\|_{L^\infty(\mathbb{R}^n)} |\hat{g}(\xi)|$ for almost every $k \in \mathbb{R}^n$, we have

$$\|\hat{f}\hat{g}\|_{L^q(\mathbb{R}^n)} \le \|\hat{f}\|_{L^\infty(\mathbb{R}^n)} \|\hat{g}\|_{L^q(\mathbb{R}^n)}.$$

Young's inequality is the "Fourier transform" of this inequality.

Lemma C.3.10 (Young's inequality). *Let $1 \le p \le \infty$. If $f \in L^1(\mathbb{R}^n)$ and $g \in L^p(\mathbb{R}^n)$, then $f * g \in L^p(\mathbb{R}^n)$ and*

$$\|f * g\|_{L^p(\mathbb{R}^n)} \le \|f\|_{L^1(\mathbb{R}^n)} \|g\|_{L^p(\mathbb{R}^n)}.$$

Proof. First consider $1 < p < \infty$ and set $q = \left(1 - \frac{1}{p}\right)^{-1}$. By Hölder's inequality,

$$\begin{aligned}
|f * g(x)| &= \left| \int f(x-y)\, g(y)\, dy \right| \\
&\leq \int |f(x-y)|^{1/q}\, |f(x-y)|^{1/p} |g(y)|\, dy \\
&\leq \left[\int |f(x-y)|\, dy \right]^{1/q} \left[\int |f(x-y)||g(y)|^p\, dy \right]^{1/p} \\
&= \|f\|_{L^1(\mathbb{R}^n)}^{1/q} \left[\int |f(x-y)||g(y)|^p\, dy \right]^{1/p}.
\end{aligned}$$

Hence,

$$\begin{aligned}
\|f * g\|_{L^p(\mathbb{R}^n)}^p &\leq \|f\|_{L^1(\mathbb{R}^n)}^{p/q} \iint |f(x-y)||g(y)|^p\, dx\, dy \\
&= \|f\|_{L^1(\mathbb{R}^n)}^{p/q} \iint |f(x-y)||g(y)|^p\, dy\, dx \\
&= \|f\|_{L^1(\mathbb{R}^n)}^{1+p/q} \int |g(y)|^p\, dy \\
&= \|f\|_{L^1(\mathbb{R}^n)}^p \|g\|_{L^p(\mathbb{R}^n)}^p.
\end{aligned}$$

In the remaining cases, the desired inequality follows directly from

$$|f * g(x)| \leq \begin{cases} \int |f(x-y)||g(y)|\, dy & \text{if } p = 1, \\ \|f\|_{L^1(\mathbb{R}^n)} \|g\|_{L^\infty(\mathbb{R}^n)} & \text{if } p = \infty. \end{cases}$$

□

Exercise C.3.11.

(a) Let $1 \leq p, q, r \leq \infty$ obey $\frac{1}{p} + \frac{1}{q} + \frac{1}{r} = 2$. Prove that if $f \in L^p(\mathbb{R}^n)$, $g \in L^q(\mathbb{R}^n)$ and $h \in L^r(\mathbb{R}^n)$, then

$$\left| \iint f(x) g(x-y) h(y)\, dx\, dy \right| \leq \|f\|_{L^p(\mathbb{R}^n)} \|g\|_{L^q(\mathbb{R}^n)} \|h\|_{L^r(\mathbb{R}^n)}.$$

(b) Let $1 \leq p, q, r \leq \infty$ obey $\frac{1}{p} + \frac{1}{q} = 1 + \frac{1}{r}$. Prove that if $f \in L^p(\mathbb{R}^n)$ and $g \in L^q(\mathbb{R}^n)$, then $f * g \in L^r(\mathbb{R}^n)$ and

$$\|f * g\|_{L^r(\mathbb{R}^n)} \leq \|f\|_{L^p(\mathbb{R}^n)} \|g\|_{L^q(\mathbb{R}^n)}.$$

Appendix D

Elliptic Equations

In this appendix, we give the basic existence, uniqueness and regularity proofs for the second order elliptic boundary value problem

$$\text{(D.1)} \qquad \begin{aligned} -\sum_{i,j=1}^{n} \frac{\partial}{\partial x_i}\Big(\gamma^{i,j}(x)\frac{\partial u}{\partial x_j}(x)\Big) + \mu(x)u(x) = F(x) \qquad & \text{in } \Omega, \\ u(x) = f(x) \qquad & \text{on } \partial\Omega. \end{aligned}$$

Ellipticity is the requirement that, for a.e. $x \in \overline{\Omega}$, the matrix of coefficients, $\gamma(x)$, be a symmetric real, strictly positive definite matrix. That is, there is a constant $C > 0$ such that

$$\sum_{i,j=1}^{n} \gamma^{i,j}(x)\xi_i\xi_j \geq C|\xi|^2$$

for all $\xi \in \mathbb{R}^n$.

We have already discussed the basics of weak solutions in §2.5. This appendix is a more detailed presentation that also gives the proofs of the higher regularity results in §2.6. For pedagogical purposes we first discuss equations in $\mathbb{R}^n$, then in $\mathbb{R}^n_+$ and finally in bounded open sets with C^k boundary.

D.1. Equations in $\mathbb{R}^n$

To get warmed up, we go through the proof of existence, uniqueness and regularity of solutions to

$$\text{(D.2)} \qquad (L_{\gamma,\mu}u)(x) = -\sum_{i,j=1}^{n} \frac{\partial}{\partial x_i}\Big(\gamma^{i,j}(x)\frac{\partial u}{\partial x_j}(x)\Big) + \mu(x)u(x) = F(x)$$

on $\mathbb{R}^n$. The hypotheses on γ and μ are as follows.

Hypothesis D.1.1. *Assume that*

- $\gamma^{i,j}, \mu \in L^\infty(\mathbb{R}^n)$ are real valued,
- $\gamma^{i,j} = \gamma^{j,i}$ for all $i, j = 1, \dots, n$,
- $\sum_{i,j=1}^n \gamma^{i,j}(x)\xi_i\xi_j \geq c\,|\xi|^2$ *for a.e.* $x \in \mathbb{R}^n$ *and all* $\xi \in \mathbb{R}^n$,
- $\mu(x) \geq c$ *for a.e.* $x \in \mathbb{R}^n$,

for some constant $c > 0$.

Exercise D.1.2. Prove that, under Hypotheses D.1.1, there are constants $C_1 > 0$ and C_2 such that

$$C_1|\xi|^2 \leq \sum_{i,j=1}^n \gamma^{i,j}(x)\xi_i\overline{\xi_j} \leq C_2|\xi|^2 \tag{D.3}$$

for all $\xi \in \mathbb{C}^n$ and $x \in \mathbb{R}^n$.

In the special case that $\gamma(x)$ and $\mu(x)$ are independent of x, the Fourier transform of $L_{\gamma,\mu}u = F$ is

$$\Big(\mu + \sum_{i,j=1}^n \xi_i\gamma^{i,j}\xi_j\Big)\hat{u}(\xi) = \hat{F}(\xi).$$

As μ and the matrix γ are positive definite, the factor

$$(\mu + \sum_{i,j=1}^n \xi_i\gamma^{i,j}\xi_j) \geq \mu > 0$$

and we may solve for $\hat{u}(\xi)$ by simply dividing across.

When $\gamma(x)$ and $\mu(x)$ have a nontrivial dependence on x, Fourier transforming does not immediately help. By way of motivation for the strategy we shall use for handling the general case, first consider the case in which $\mu = 1$ and γ is the identity matrix. Observe that, if $u, v, F \in \mathcal{S}(\mathbb{R}^n)$ with $(-\Delta + 1)u = F$, then

$$\begin{aligned} \int v(x)F(x)\,dx &= \int v(x)(-\Delta + 1)u(x)\,dx \\ &= \int \big[\nabla v(x) \cdot \nabla u(x) + v(x)u(x)\big]\,dx \\ &= \langle v, \bar{u}\rangle_{H^1(\mathbb{R}^n)}. \end{aligned} \tag{D.4}$$

By the Riesz representation theorem, the linear map

$$u \in H^1(\mathbb{R}^n) \mapsto \mathcal{R}_u \in H^1(\mathbb{R}^n)^*$$

given by

$$\mathcal{R}_u(v) = \langle v, \bar{u} \rangle_{H^1(\mathbb{R}^n)}$$

is a well-defined, norm-preserving bijection. Recall from Proposition C.1.22 that the linear map $F \in H^{-1}(\mathbb{R}^n) \mapsto \mathcal{L}_F \in H^1(\mathbb{R}^n)^*$ given by

$$\mathcal{L}_F(v) = \int \hat{v}(\xi)\hat{F}(-\xi)\,\frac{d\xi}{(2\pi)^n}$$

is also a well-defined, norm-preserving bijection. We chose to express the right-hand side in terms of Fourier transforms because it is applicable to all $F \in H^{-1}(\mathbb{R}^n)$ and $v \in H^1(\mathbb{R}^n)$. When $v, F \in \mathcal{S}(\mathbb{R}^n)$,

$$\int \hat{v}(\xi)\hat{F}(-\xi)\,\frac{d\xi}{(2\pi)^n} = \int v(x)F(x)\,dx.$$

As $\mathcal{S}(\mathbb{R}^n)$ is dense in $H^1(\mathbb{R}^n)$, (D.4) says that

$$(-\Delta + 1)u = F \iff \mathcal{R}_u = \mathcal{L}_F$$

at least when $u, F \in \mathcal{S}(\mathbb{R}^n)$. By continuity, the same equivalence applies for all $u \in H^1(\mathbb{R}^n)$ and $F \in H^{-1}(\mathbb{R}^n)$. The existence and uniqueness of a solution $u \in H^1(\mathbb{R}^n)$ for any given $F \in H^{-1}(\mathbb{R}^n)$ is now assured by the Riesz representation theorem.

To implement this strategy for general γ and μ, we construct a variant, $\langle\ ,\ \rangle_{\gamma,\mu}$, of $\langle\ ,\ \rangle_{H^1(\mathbb{R}^n)}$, that is adapted to $L_{\gamma,\mu}$ and whose associated norm is equivalent to $\|\quad\|_{H^1(\mathbb{R}^n)}$. Then we prove a generalization of (D.4).

Lemma D.1.3. *There is a unique continuous sesquilinear form*

$$\langle\,\cdot\,,\,\cdot\,\rangle_{\gamma,\mu} : H^1(\mathbb{R}^n) \times H^1(\mathbb{R}^n) \to \mathbb{C}$$

such that

$$\langle u, v \rangle_{\gamma,\mu} = \sum_{i,j=1}^{n} \int \gamma^{i,j}(x)\frac{\partial u}{\partial x_i}(x)\overline{\frac{\partial v}{\partial x_j}(x)}\,dx + \int \mu(x)u(x)\overline{v(x)}\,dx$$

for all $u, v \in \mathcal{S}(\mathbb{R}^n)$. It is an inner product. The associated norm,

$$\|v\|_{\gamma,\mu} = \sqrt{\langle v, v \rangle_{\gamma,\mu}}$$

is equivalent to $\|\cdot\|_{H^1(\mathbb{R}^n)}$.

Proof. Setting

$$K_1 = \min\bigl\{C_1, \min_{x\in\mathbb{R}^n} \mu(x)\bigr\}, \qquad K_2 = \max\bigl\{C_2, \max_{x\in\mathbb{R}^n} \mu(x)\bigr\}$$

with the C_1, C_2 of Exercise D.1.2, we have

$$
\begin{aligned}
K_1 |||u|||^2_{H^1(\mathbb{R}^n)} &= K_1\big[\|u\|^2_{L^2(\mathbb{R}^n)} + \|\nabla u\|^2_{L^2(\mathbb{R}^n)}\big] \\
&\le \langle u, u\rangle_{\gamma,\mu} \\
&= \int_{\mathbb{R}^n} \mu(x)|u(x)|^2\,dx + \sum_{i,j=1}^n \int_{\mathbb{R}^n} \gamma^{i,j}(x)\frac{\partial u}{\partial x_i}(x)\overline{\frac{\partial u}{\partial x_j}(x)}\,dx \qquad \text{(D.5)}\\
&\le K_2\big[\|u\|^2_{L^2(\mathbb{R}^n)} + \|\nabla u\|^2_{L^2(\mathbb{R}^n)}\big] \\
&= K_2 |||u|||^2_{H^1(\mathbb{R}^n)}
\end{aligned}
$$

for all $u \in \mathcal{S}(\mathbb{R}^n)$. The map

$$(u, v) \in \mathcal{S}(\mathbb{R}^n) \times \mathcal{S}(\mathbb{R}^n) \mapsto \langle u, v\rangle_{\gamma,\mu}$$

is obviously linear in u, conjugate linear in v, symmetric and positive definite. Consequently the Cauchy–Schwarz inequality applies to it and, by (D.5), the map is also nondegenerate and continuous in the $H^1(\mathbb{R}^n)$ topology. The existence of a unique continuous extension to $H^1(\mathbb{R}^n)\times H^1(\mathbb{R}^n)$, that the extension is an inner product and the equivalence to the $\|\cdot\|_{H^1(\mathbb{R}^n)}$ norm are now all obvious. □

Exercise D.1.4. Prove that there is a unique continuous sesquilinear form

$$\langle\,\cdot\,,\,\cdot\,\rangle_{\gamma,0} : H^1(\mathbb{R}^n)\times H^1(\mathbb{R}^n) \to \mathbb{C}$$

such that

$$\langle u, v\rangle_{\gamma,0} = \sum_{i,j=1}^n \int \gamma^{i,j}(x)\frac{\partial u}{\partial x_i}(x)\overline{\frac{\partial v}{\partial x_j}(x)}\,dx$$

for all $u, v \in \mathcal{S}(\mathbb{R}^n)$. Is it an inner product? If so, is the associated norm equivalent to $\|\cdot\|_{H^1(\mathbb{R}^n)}$?

Lemma D.1.5. *There is a unique bounded linear map*

$$L_{\gamma,\mu} : H^1(\mathbb{R}^n) \to H^{-1}(\mathbb{R}^n)$$

such that

$$(L_{\gamma,\mu}u)(x) = -\sum_{i,j=1}^n \frac{\partial}{\partial x_i}\Big(\gamma^{i,j}(x)\frac{\partial u}{\partial x_j}(x)\Big) + \mu(x)u(x)$$

for all $u \in \mathcal{S}(\mathbb{R}^n)$. Furthermore

$$\mathcal{L}_{L_{\gamma,\mu}u}(v) = \langle v, \bar{u}\rangle_{\gamma,\mu}$$

for all $u, v \in H^1(\mathbb{R}^n)$.

Proof. We shall show that there is a constant C such that

$$\big|\langle v, L_{\gamma,\mu}\bar{u}\rangle_{L^2(\mathbb{R}^n)}\big| \le C|||u|||_{H^1(\mathbb{R}^n)}|||v|||_{H^1(\mathbb{R}^n)}$$

for all $u, v \in C_c^\infty(\mathbb{R}^n)$. Since, by Exercise C.1.13, $C_c^\infty(\mathbb{R}^n)$ is dense in $H^1(\mathbb{R}^n)$, this will imply that

$$|\langle v, L_{\gamma,\mu}\bar{u}\rangle_{L^2(\mathbb{R}^n)}| \le C|||u|||_{H^1(\mathbb{R}^n)}|||v|||_{H^1(\mathbb{R}^n)}$$

for all $u \in C_c^\infty(\mathbb{R}^n)$ and $v \in H^1(\mathbb{R}^n)$ and hence that

$$||L_{\gamma,\mu}u||_{H^{-1}(\mathbb{R}^n)} \le C|||u|||_{H^1(\mathbb{R}^n)}$$

by part (b) of Exercise C.1.23. Since $C_c^\infty(\mathbb{R}^n)$ is still dense in $H^1(\mathbb{R}^n)$, this will in turn imply that $L_{\gamma,\mu}$ has a unique continuous extension to all of $H^1(\mathbb{R}^n)$ by the B.L.T. theorem.

By the divergence theorem

$$\begin{aligned}
&\langle v, L_{\gamma,\mu}\bar{u}\rangle_{L^2(\mathbb{R}^n)} \\
&= -\int_{\mathbb{R}^n} v(x)\Big[\sum_{i,j=1}^n \frac{\partial}{\partial x_i}\Big(\gamma^{i,j}(x)\frac{\partial u}{\partial x_j}(x)\Big)\Big]\,dx + \int_{\mathbb{R}^n} v(x)\mu(x)u(x)\,dx \\
&= \sum_{i,j=1}^n \int_{\mathbb{R}^n} \gamma^{i,j}(x)\frac{\partial v}{\partial x_i}(x)\frac{\partial u}{\partial x_j}(x)\,dx + \int_{\mathbb{R}^n} \mu(x)v(x)u(x)\,dx \\
&= \langle v, \bar{u}\rangle_{\gamma,\mu}.
\end{aligned} \tag{D.6}$$

There are no boundary terms because u and v have compact support. Hence, by Cauchy–Schwarz and the equivalence of the norms $||\cdot||_{\gamma,\mu}$ and $||\cdot||_{H^1(\mathbb{R}^n)}$ that was proven in Lemma D.1.3,

$$|\langle v, L_{\gamma,\mu}\bar{u}\rangle_{L^2(\mathbb{R}^n)}| \le ||v||_{\gamma,\mu}||u||_{\gamma,\mu} \le C|||u|||_{H^1(\mathbb{R}^n)}|||v|||_{H^1(\mathbb{R}^n)}.$$

By (D.6),

$$\mathcal{L}_{L_{\gamma,\mu}u}(v) = \langle v, L_{\gamma,\mu}\bar{u}\rangle_{L^2(\mathbb{R}^n)} = \langle v, \bar{u}\rangle_{\gamma,\mu}$$

for all $u, v \in C_c^\infty(\mathbb{R}^n)$. Since both $\mathcal{L}_{L_{\gamma,\mu}u}(v)$ and $\langle v, \bar{u}\rangle_{\gamma,\mu}$ are continuous in both u and v with respect to the $H^1(\mathbb{R}^n)$ topology and $C_c^\infty(\mathbb{R}^n)$ is dense in $H^1(\mathbb{R}^n)$, we have that $\mathcal{L}_{L_{\gamma,\mu}u}(v) = \langle v, \bar{u}\rangle_{\gamma,\mu}$ for all $u, v \in H^1(\mathbb{R}^n)$. □

Exercise D.1.6. Prove that there is a unique bounded linear map

$$L_{\gamma,0} : H^1(\mathbb{R}^n) \to H^{-1}(\mathbb{R}^n)$$

such that

$$(L_{\gamma,0}u)(x) = -\sum_{i,j=1}^n \frac{\partial}{\partial x_i}\Big(\gamma^{i,j}(x)\frac{\partial u}{\partial x_j}(x)\Big)$$

for all $u \in \mathcal{S}(\mathbb{R}^n)$.

Theorem D.1.7. *For each $F \in H^{-1}(\mathbb{R}^n)$ there is a unique $u \in H^1(\mathbb{R}^n)$ such that $L_{\gamma,\mu}u = F$. Furthermore, there is a constant C such that*

$$F \in H^{-1}(\mathbb{R}^n),\ u \in H^1(\mathbb{R}^n),\ L_{\gamma,\mu}u = F \implies |||u|||_{H^1(\mathbb{R}^n)} \le C|||F|||_{H^{-1}(\mathbb{R}^n)}.$$

Thus $L_{\gamma,\mu}$ is an isomorphism (1–1, onto, bounded with bounded inverse).

Proof.

Boundedness: was proven in Lemma D.1.5.

Injectiveness: If $u \in H^1(\mathbb{R}^n)$ and $L_{\gamma,\mu}u = 0$, then, by Lemma D.1.5,

$$0 = \mathcal{L}_{L_{\gamma,\mu}u}(\bar{u}) = \langle \bar{u}, \bar{u} \rangle_{\gamma,\mu}$$

so that $\bar{u} = 0$ and hence $u = 0$.

Surjectiveness: Let $F \in H^{-1}(\mathbb{R}^n)$. Since the norms $\| \quad \|_{H^1(\mathbb{R}^n)}$ and $\| \quad \|_{\gamma,\mu}$ are equivalent, $H^1(\mathbb{R}^n)$ equipped with the inner product $\langle\ ,\ \rangle_{\gamma,\mu}$ is a Hilbert space and, by the Riesz representation theorem, there exists $u \in H^1(\mathbb{R}^n)$ such that

$$\mathcal{L}_F(v) = \langle v, \bar{u} \rangle_{\gamma,\mu}$$

for all $v \in H^1(\mathbb{R}^n)$. By Lemma D.1.5, $\mathcal{L}_F = \mathcal{L}_{L_{\gamma,\mu}u}$. As $\mathcal{L}$ is injective, $F = L_{\gamma,\mu}u$.

Boundedness of the inverse: We must show there is a constant C such that, if $u \in H^1(\mathbb{R}^n)$ and $F = L_{\gamma,\mu}u$, then

$$|||u|||_{H^1(\mathbb{R}^n)} \leq C|||F|||_{H^{-1}(\mathbb{R}^n)}.$$

Since $\mathcal{L}$ is norm preserving,

$$\begin{aligned} |||F|||_{H^{-1}(\mathbb{R}^n)} &= \sup_{\substack{v \in H^1(\mathbb{R}^n) \\ v \neq 0}} \frac{|\mathcal{L}_F(v)|}{|||v|||_{H^1(\mathbb{R}^n)}} \\ &= \sup_{\substack{v \in H^1(\mathbb{R}^n) \\ v \neq 0}} \frac{|\langle v, \bar{u} \rangle_{\gamma,\mu}|}{|||v|||_{H^1(\mathbb{R}^n)}} \\ &= \sup_{\substack{v \in H^1(\mathbb{R}^n) \\ v \neq 0}} \frac{|\langle \bar{v}, u \rangle_{\gamma,\mu}|}{\|\bar{v}\|_{H^1(\mathbb{R}^n)}} \\ &= \sup_{\substack{v \in H^1(\mathbb{R}^n) \\ v \neq 0}} \frac{|\langle v, u \rangle_{\gamma,\mu}|}{|||v|||_{H^1(\mathbb{R}^n)}}. \end{aligned}$$

By the equivalence of the norms $\| \quad \|_{H^1(\mathbb{R}^n)}$ and $\| \quad \|_{\gamma,\mu}$,

$$|||F|||_{H^{-1}(\mathbb{R}^n)} \geq C^{-1} \sup_{\substack{v \in H^1(\mathbb{R}^n) \\ v \neq 0}} \frac{|\langle v, u \rangle_{\gamma,\mu}|}{\|v\|_{\gamma,\mu}} = C^{-1}\|u\|_{\gamma,\mu}. \qquad \square$$

We next prove (Proposition D.1.11) that if the coefficients $\gamma^{i,j}$ and μ have sufficient regularity, then the solution, $u \in H^1(\mathbb{R}^n)$, of

$$L_{\gamma,\mu}u = F$$

is more regular than F by two derivatives. That is, if $F \in H^{\ell-2}(\mathbb{R}^n)$ then $u \in H^{\ell}(\mathbb{R}^n)$. This is called elliptic regularity. We shall control $\partial^{\alpha}u$ through difference quotients $\Delta_H^{\alpha}u$ that approximate $\partial^{\alpha}u$.

If $u(x)$ is a function on $\mathbb{R}^n$, $0 \neq h \in \mathbb{R}$ and $\alpha \in \mathbb{N}_0^n$ with $|\alpha| = 1$, we define

$$\Delta_h^\alpha u(x) = \frac{1}{h}[u(x+h\alpha) - u(x)] \tag{D.7}$$

More generally, if $\alpha \in \mathbb{N}_0^n$ with $|\alpha| > 1$ and all components of

$$H = \left(h_{m,j}\right)_{\substack{1\le m\le n \\ 1\le j\le \alpha_m}} \in \mathbb{R}^{|\alpha|}$$

are nonzero, we define

$$\Delta_H^\alpha u(x) = \prod_{m=1}^{n} \prod_{j=1}^{\alpha_m} \Delta_{h_{m,j}}^{e_m} u(x), \tag{D.8}$$

where e_m is the multi-index in $\mathbb{N}_0^n$ all of whose components are zero except for the mth, which is one.

Exercise D.1.8. Let $h, h' \in \mathbb{R}$ be nonzero and $\alpha, \alpha' \in \mathbb{N}_0^n$ with $|\alpha| = |\alpha'| = 1$. Prove that the operators Δ_h^α and $\Delta_{h'}^{\alpha'}$ commute.

Lemma D.1.9. *Let $s \in \mathbb{R}$, $\alpha \in \mathbb{N}_0^n$ with $|\alpha| \geq 1$ and $u \in H^s(\mathbb{R}^n)$. Then $\partial^\alpha u \in H^s(\mathbb{R}^n)$ if and only if $\limsup_{H\to 0} \|\Delta_H^\alpha u\|_{H^s(\mathbb{R}^n)}$ is finite. In this case $\|\Delta_H^\alpha u\|_{H^s(\mathbb{R}^n)} \leq \|\partial^\alpha u\|_{H^s(\mathbb{R}^n)}$ for all H and*

$$\|\partial^\alpha u\|_{H^s(\mathbb{R}^n)} = \lim_{H\to 0} \|\Delta_H^\alpha u\|_{H^s(\mathbb{R}^n)}$$

Proof. If $\alpha = e_m$, then by the translation property (e) of the Fourier transform (C.2),

$$\widehat{\Delta_h^\alpha u}(\xi) = \frac{1}{h}[e^{i\xi\cdot h e_m} - 1]\hat{u}(\xi) = i e^{ih\xi_m/2}\, \frac{\sin(h\xi_m/2)}{h/2}\, \hat{u}(\xi).$$

Hence for general α

$$\widehat{\Delta_h^\alpha u}(\xi) = \hat{u}(\xi) \prod_{m=1}^{n} \prod_{j=1}^{\alpha_m} i \frac{\sin(h_{m,j}\xi_m/2)}{h_{m,j}/2}\, e^{ih_{m,j}\xi_m/2}$$

and

$$|\widehat{\Delta_h^\alpha u}(\xi)| = |\hat{u}(\xi)| \prod_{m=1}^{n} \prod_{j=1}^{\alpha_m} \left| \frac{\sin(h_{m,j}\xi_m/2)}{h_{m,j}/2} \right|.$$

Since $|\sin x| \leq |x|$ and $\lim_{x\to 0} \frac{\sin x}{x} = 1$,

$$|\widehat{\Delta_h^\alpha u}(\xi)| \leq |\xi^\alpha \hat{u}(\xi)| \qquad \text{and} \qquad \lim_{H\to 0} |\widehat{\Delta_h^\alpha u}(\xi)| = |\xi^\alpha \hat{u}(\xi)|$$

for all $\xi \in \mathbb{R}^n$.

If $\partial^\alpha u \in H^s(\mathbb{R}^n)$, then $\|\partial^\alpha u\|^2_{H^s(\mathbb{R}^n)} = \int_{\mathbb{R}^n} (1+|\xi|^2)^s |\xi^\alpha \hat{u}(\xi)|^2 \frac{d\xi}{(2\pi)^n} < \infty$ and, by the Lebesgue dominated convergence theorem,

$$\|\Delta_H^\alpha u\|_{H^s(\mathbb{R}^n)} \leq \|\partial^\alpha u\|_{H^s(\mathbb{R}^n)}$$

for all H and

$$\lim_{H\to 0} \|\Delta_H^\alpha u\|_{H^s(\mathbb{R}^n)} = \|\partial^\alpha u\|_{H^s(\mathbb{R}^n)}.$$

On the other hand, even if $\int_{\mathbb{R}^n} (1+|\xi|^2)^s |\xi^\alpha \hat{u}(\xi)|^2 \frac{d\xi}{(2\pi)^n} = \infty$, the Lebesgue dominated convergence theorem still implies that, for each $R > 0$,

$$\lim_{H\to 0} \int_{|\xi|<R} (1+|\xi|^2)^s \left|\widehat{\Delta_h^\alpha u}(\xi)\right|^2 \frac{d\xi}{(2\pi)^n} = \int_{|\xi|<R} (1+|\xi|^2)^s |\xi^\alpha \hat{u}(\xi)|^2 \frac{d\xi}{(2\pi)^n}.$$

Consequently,

$$\begin{aligned}\limsup_{H\to 0} \|\Delta_H^\alpha u\|^2_{H^s(\mathbb{R}^n)} &\geq \limsup_{H\to 0} \int_{|\xi|<R} (1+|\xi|^2)^s \left|\widehat{\Delta_h^\alpha u}(\xi)\right|^2 \frac{d\xi}{(2\pi)^n} \\ &= \int_{|\xi|<R} (1+|\xi|^2)^s |\xi^\alpha \hat{u}(\xi)|^2 \frac{d\xi}{(2\pi)^n}.\end{aligned}$$

As this is true for all $R > 0$,

$$\begin{aligned}\limsup_{H\to 0} \|\Delta_H^\alpha u\|^2_{H^s(\mathbb{R}^n)} &\geq \lim_{R\to\infty} \int_{|\xi|<R} (1+|\xi|^2)^s |\xi^\alpha \hat{u}(\xi)|^2 \frac{d\xi}{(2\pi)^n} \\ &= \|\partial^\alpha u\|^2_{H^s(\mathbb{R}^n)}.\end{aligned}$$

So $\partial^\alpha u \in H^s(\mathbb{R}^n)$ whenever $\limsup_{H\to 0} \|\Delta_H^\alpha u\|_{H^s(\mathbb{R}^n)}$ is finite. □

Lemma D.1.10. *Let $s \geq 1$, let m be an integer with $m \geq s$, and let $\alpha \in \mathbb{N}_0^n$ with $|\alpha| = 1$.*

(a) *Let $\varphi \in W^{m,\infty}(\mathbb{R}^n)$. Also use φ to denote the operator of multiplication by φ. Then there is a $C > 0$ (depending on φ, s, m, n and α, but independent of h) such that $[\Delta_h^\alpha, \varphi] = \Delta_h^\alpha \varphi - \varphi \Delta_h^\alpha$ is a bounded operator from $H^{s-1}(\mathbb{R}^n)$ to $H^{s-1}(\mathbb{R}^n)$ with norm at most C.*

(b) *Let $Lu = \sum_{i,j} \partial_i(a^{i,j}(x)\partial_j u)$ be a differential operator of order two, where $a^{i,j} \in W^{m,\infty}(\mathbb{R}^n)$. Then there is a $C > 0$, independent of h, such that $[\Delta_h^\alpha, L]$ is a bounded operator from $H^{s-2}(\mathbb{R}^n)$ to $H^s(\mathbb{R}^n)$ with norm at most C.*

Proof.

(a) Since $|\alpha| = 1$, we have

$$\begin{aligned}([\Delta_h^\alpha, \varphi]u)(x) &= (\Delta_h^\alpha(\varphi u))(x) - (\varphi \Delta_h^\alpha u)(x) \\ &= \frac{1}{h}\big[\varphi(x+h\alpha)u(x+h\alpha) - \varphi(x)u(x) - \varphi(x)u(x+h\alpha) + \varphi(x)u(x)\big] \\ &= \frac{1}{h}\big[\varphi(x+h\alpha)u(x+h\alpha) - \varphi(x)u(x+h\alpha)\big] \\ &= \Phi_h(x)u_h(x).\end{aligned}$$

where $\Phi_h = \Delta_h^\alpha \varphi$ and $u_h(x) = u(x+h\alpha)$ is a translate of u. By Lemma C.1.32, since $m-1 \ge s-1 \ge 0$,

$$\|[\Delta_h^\alpha, \varphi]u\|_{H^{s-1}} \le C_{n,m}\,\|\Phi_h\|_{W^{m-1,\infty}}\,\|u_h\|_{H^{s-1}} \le C_{n,m}\|\varphi\|_{W^{m,\infty}}\|u_h\|_{H^{s-1}}.$$

In the last step we used the mean value theorem in the form

$$\partial^\alpha \Phi_h = \nabla \partial^\alpha \varphi(x + th\alpha) \cdot \alpha$$

for some $t = t(x,h) \in [0,1]$. The proof is completed by observing that $\|u_h\|_{H^{s-1}} = \|u\|_{H^{s-1}}$, since $\hat{u}_h(\xi) = e^{ih\xi\cdot\alpha}\hat{u}(\xi)$.

(b) As the differential operator ∂_i commutes with translation, it also commutes with Δ_h^α. Consequently,

$$[\Delta_h^\alpha, L]u = \sum_{i,j} \partial_i([\Delta_h^\alpha, a^{i,j}]\partial_j u).$$

So, by part (a) of this lemma

$$\begin{aligned}\|[\Delta_h^\alpha, L]u\|_{H^{s-2}} &\le \sum_{i,j} \|[\Delta_h^\alpha, a^{i,j}]\partial_j u\|_{H^{s-1}}\\ &\le \sum_{i,j} C\,\|\partial_j u\|_{H^{s-1}}\\ &\le C\|u\|_{H^s}.\end{aligned}$$

□

Proposition D.1.11 (Elliptic regularity in $\mathbb{R}^n$). *Let $\ell \ge 1$, let γ and μ obey Hypothesis* D.1.1 *and additionally*

$$\gamma^{i,j} \in W^{\ell-1,\infty}(\mathbb{R}^n), \qquad \mu \in W^{\max\{\ell-2,0\},\infty}(\mathbb{R}^n).$$

There is a constant C, depending only on ℓ, γ, and μ such that, for all $u \in H^1(\mathbb{R}^n)$ and $F \in H^{\ell-2}(\mathbb{R}^n)$ obeying $L_{\gamma,\mu}u = F$ we have $u \in H^\ell(\mathbb{R}^n)$ and

$$\|u\|_{H^\ell(\mathbb{R}^n)} \le C\,\|F\|_{H^{\ell-2}(\mathbb{R}^n)}.$$

Proof. We use induction. We proved the result for $\ell = 1$ in Theorem D.1.7. Assume that we have proven the desired result for some $\ell \ge 1$. We now prove it for $\ell + 1$. That is, we assume that

$$\gamma^{i,j} \in W^{\ell,\infty}(\mathbb{R}^n), \qquad \mu \in W^{\ell-1,\infty}(\mathbb{R}^n)$$

and that for all $u \in H^1(\mathbb{R}^n)$ and $F \in H^{\ell-2}(\mathbb{R}^n)$ obeying $L_{\gamma,\mu}u = F$ we have $u \in H^\ell(\mathbb{R}^n)$ and $\|u\|_{H^\ell(\mathbb{R}^n)} \le C\,\|F\|_{H^{\ell-2}(\mathbb{R}^n)}$. We must prove that if $u \in H^1(\mathbb{R}^n)$, $F \in H^{\ell-1}(\mathbb{R}^n)$ and $L_{\gamma,\mu}u = F$, then $u \in H^{\ell+1}(\mathbb{R}^n)$ and $\|u\|_{H^{\ell+1}(\mathbb{R}^n)} \le C'\,\|F\|_{H^{\ell-1}(\mathbb{R}^n)}$.

So let $u \in H^1(\mathbb{R}^n)$ and $F \in H^{\ell-1}(\mathbb{R}^n)$ with $L_{\gamma,\mu}u = F$. By the inductive hypothesis, $u \in H^\ell(\mathbb{R}^n)$ with $\|u\|_{H^\ell(\mathbb{R}^n)} \le C\,\|F\|_{H^{\ell-2}(\mathbb{R}^n)}$. For each $\alpha \in \mathbb{N}_0^n$ with $|\alpha| = 1$ and each $h \in \mathbb{R}\setminus\{0\}$, set $u_{\alpha,h} = \Delta_h^\alpha u$. Then

$$L_{\gamma,\mu}u_{\alpha,h} = F_{\alpha,h},$$

where

$$F_{\alpha,h} = \Delta_h^\alpha F + [L_{\gamma,0}, \Delta_h^\alpha]u + [\mu, \Delta_h^\alpha]u.$$

We wish to estimate the different terms in $F_{\alpha,h}$. First by Lemma D.1.9, with $s = \ell - 2$,

$$\left\|\Delta_h^\alpha F\right\|_{H^{\ell-2}} \leq \|\partial^\alpha F\|_{H^{\ell-2}} \leq \|F\|_{H^{\ell-1}}.$$

By Lemma D.1.10(b) with $s = m = \ell$, we have

$$\begin{aligned}\left\|[L_{\gamma,0}, \Delta_h^\alpha]u\right\|_{H^{\ell-2}} &\leq C_1 \|u\|_{H^\ell} \leq C_1 C \, \|F\|_{H^{\ell-2}} \\ &\leq C_1 C \, \|F\|_{H^{\ell-1}}.\end{aligned}$$

If $\ell \geq 2$, Lemma D.1.10(a) with $s = m = \ell - 1$ yields

$$\left\|[\mu, \Delta_h^\alpha]u\right\|_{H^{\ell-2}} \leq C\|u\|_{H^{\ell-2}} \leq C\|F\|_{H^{\ell-1}},$$

whereas for $\ell = 1$ we use the duality statement $H^{-1} \cong (H^1)^*$ and the computation $[\Delta_h^\alpha, \mu]u = (\Delta_h^\alpha \mu)u_h$ from the proof of Lemma D.1.10 to obtain

$$\begin{aligned}\left\|[\mu, \Delta_h^\alpha]u\right\|_{H^{-1}} &= \sup_{\|v\|_{H^1} \leq 1} |\langle(\Delta_h^\alpha \mu)u_h, v\rangle| \\ &= \sup_{\|v\|_{H^1} \leq 1} |\langle \Delta_h^\alpha \mu, u_h v\rangle| \\ &= \sup_{\|v\|_{H^1} \leq 1} |\langle \mu, u\Delta_{-h}^\alpha v\rangle - \langle \mu, (\Delta_h^\alpha u)v\rangle| \\ &\leq \sup_{\|v\|_{H^1} \leq 1} \|\mu\|_{L^\infty} \left(\|u\|_{L^2} \left\|\Delta_{-h}^\alpha v\right\|_{L^2} + \left\|\Delta_h^\alpha u\right\|_{L^2} \|v\|_{L^2}\right) \\ &\leq C \, \|\mu\|_{L^\infty} \, \|u\|_{H^1} \\ &\leq C\|F\|_{L^2}.\end{aligned}$$

Combining these estimates gives $F_{\alpha,h} \in H^{\ell-2}(\mathbb{R}^n)$ with

$$\left\|F_{\alpha,h}\right\|_{H^{\ell-2}(\mathbb{R}^n)} \leq C\|F\|_{H^{\ell-1}(\mathbb{R}^n)}$$

and, by the inductive hypothesis, $u_{\alpha,h} = \Delta_h^\alpha u \in H^\ell(\mathbb{R}^n)$ with

$$\left\|\Delta_h^\alpha u\right\|_{H^\ell(\mathbb{R}^n)} \leq C \, \left\|F_{\alpha,h}\right\|_{H^{\ell-2}(\mathbb{R}^n)} \leq C \, \|F\|_{H^{\ell-1}(\mathbb{R}^n)}.$$

By Lemma D.1.9, with $s = \ell$, $\partial^\alpha u \in H^\ell(\mathbb{R}^n)$ with

$$\|\partial^\alpha u\|_{H^\ell(\mathbb{R}^n)} \leq C\|F\|_{H^{\ell-1}(\mathbb{R}^n)}.$$

Hence $u \in H^{\ell+1}(\mathbb{R}^n)$ and

$$\begin{aligned}\|u\|_{H^{\ell+1}(\mathbb{R}^n)}^2 &= \|u\|_{H^\ell(\mathbb{R}^n)}^2 + \sum_{\substack{\alpha \in \mathbb{N}_0^n \\ |\alpha|=1}} \|\partial^\alpha u\|_{H^\ell(\mathbb{R}^n)}^2 \\ &\leq C^2(\|F\|_{H^{\ell-2}(\mathbb{R}^n)}^2 + \|F\|_{H^{\ell-1}(\mathbb{R}^n)}^2) \\ &\leq C^2\|F\|_{H^{\ell-1}(\mathbb{R}^n)}^2.\end{aligned}$$

□

Exercise D.1.12. Let $\gamma(x)$ obey Hypotheses D.1.1 with $\partial^\alpha\gamma \in L^\infty$ for all α, and let $\ell \in \mathbb{N}$. Prove that there is a constant C, depending only on ℓ and γ such that, for all $u \in H^1(\mathbb{R}^n)$ and $F \in H^{\ell-2}(\mathbb{R}^n)$ obeying $L_{\gamma,0}u = F$ (setting $\mu = 0$ eliminates the $\mu(x)u(x)$ term from the differential equation) we have $u \in H^\ell(\mathbb{R}^n)$ and

$$\begin{aligned}\|\!|u|\!\|_{H^\ell(\mathbb{R}^n)} &\le C\Big(\|\!|F|\!\|_{H^{\ell-2}(\mathbb{R}^n)} + \|\!|u|\!\|_{H^{-1}(\mathbb{R}^n)}\Big)\\ &\le C\Big(\|\!|F|\!\|_{H^{\ell-2}(\mathbb{R}^n)} + \|\!|u|\!\|_{H^{1}(\mathbb{R}^n)}\Big).\end{aligned}$$

Exercise D.1.13. In this problem we explore the need for the $\mu(x)u(x)$ term in $L_{\gamma,\mu}u = F$.

(a) Find a $\gamma(x)$ obeying Hypotheses D.1.1 and an $F \in H^{-1}(\mathbb{R}^n)$ such that **no** $u \in H^1(\mathbb{R}^n)$ obeys $L_{\gamma,0}u = F$.

(b) Find a $\gamma(x)$ obeying Hypotheses D.1.1 such that there does **not** exist a constant C that obeys $\|\!|u|\!\|_{H^1(\mathbb{R}^n)} \le C\|\!|F|\!\|_{H^{-1}(\mathbb{R}^n)}$ for all $F \in H^{-1}(\mathbb{R}^n)$ and $u \in H^1(\mathbb{R}^n)$ with $L_{\gamma,0}u = F$.

(c) Let $\gamma(x)$ obey Hypotheses D.1.1. Prove that the map

$$L_{\gamma,0} : H^1(\mathbb{R}^n) \to H^{-1}(\mathbb{R}^n)$$

is bounded and injective, but that it is NOT surjective. Prove that the inverse map $L_{\gamma,0}^{-1}$ (defined on the range of $L_{\gamma,0}$) is NOT bounded.

(d) Let $\gamma(x)$ obey Hypotheses D.1.1 with $\partial^\alpha\gamma \in L^\infty$ for all α, let $\ell \in \mathbb{N}$ and let Ω be a bounded open subset of $\mathbb{R}^n$. Prove that there is a constant C, depending only on ℓ, γ and Ω such that, for all $F \in H^{\ell-2}(\mathbb{R}^n)$ and all $u \in H^1(\mathbb{R}^n)$ that vanish outside of Ω and obey $L_{\gamma,0}u = F$ we have $u \in H^\ell(\mathbb{R}^n)$ and

$$\|\!|u|\!\|_{H^\ell(\mathbb{R}^n)} \le C\|\!|F|\!\|_{H^{\ell-2}(\mathbb{R}^n)}.$$

D.2. Equations in $\mathbb{R}^n_+$

We now introduce a simple boundary and study the boundary value problem

$$\begin{aligned}-\sum_{i,j=1}^{n} \frac{\partial}{\partial x_i}\Big(\gamma^{i,j}(x)\frac{\partial u}{\partial x_j}(x)\Big) + \mu(x)u(x) = F(x) \quad &\text{in} \quad \mathbb{R}^n_+,\\ u(x) = f(x) \quad \text{on} \quad \partial\mathbb{R}^n_+ = \mathbb{R}^{n-1}&.\end{aligned} \tag{D.9}$$

Throughout this section we assume that $\gamma(x)$ and $\mu(x)$ are restrictions to $\mathbb{R}^n_+$ of functions satisfying Hypotheses D.1.1.

Lemma D.2.1. *There is a unique continuous sesquilinear form*

$$\langle\,\cdot\,,\,\cdot\,\rangle_{\gamma,\mu,+} : H^1(\mathbb{R}^n_+) \times H^1(\mathbb{R}^n_+) \to \mathbb{C}$$

such that

$$\langle u, v\rangle_{\gamma,\mu,+} = \sum_{i,j=1}^{n} \int_{\mathbb{R}^n_+} \gamma^{i,j}(x)\frac{\partial u}{\partial x_i}(x)\overline{\frac{\partial v}{\partial x_j}(x)}\, dx + \int_{\mathbb{R}^n_+} \mu(x)u(x)\overline{v(x)}\, dx$$

for all u, v that are restrictions to $\mathbb{R}^n_+$ of functions in $C_c^\infty(\mathbb{R}^n)$. It is an inner product. The associated norm,

$$\|v\|_{\gamma,\mu,+} = \sqrt{\langle v, v\rangle_{\gamma,\mu,+}}$$

is equivalent to $\|\cdot\|_{H^1(\mathbb{R}^n_+)}$.

Proof. The proof is virtually identical to that of Lemma D.1.3. □

Lemma D.2.2. *There is a unique bounded linear map*

$$L_{\gamma,\mu,+} : H^1(\mathbb{R}^n_+) \to H^{-1}(\mathbb{R}^n_+) = H^1_0(\mathbb{R}^n_+)^*$$

such that

$$(L_{\gamma,\mu,+}u)(v) = -\sum_{i,j=1}^{n} \int_{\mathbb{R}^n_+} v(x)\frac{\partial}{\partial x_i}\Big(\gamma^{i,j}(x)\frac{\partial u}{\partial x_j}(x)\Big)\, dx + \int_{\mathbb{R}^n_+} \mu(x)u(x)v(x)\, dx$$

for all u that are restrictions to $\mathbb{R}^n_+$ of functions in $C_c^\infty(\mathbb{R}^n)$ and all functions $v \in C_c^\infty(\mathbb{R}^n_+)$. Furthermore

$$(L_{\gamma,\mu,+}u)(v) = \langle v, \bar{u}\rangle_{\gamma,\mu,+} \tag{D.10}$$

for all $u \in H^1(\mathbb{R}^n_+)$ and all $v \in H^1_0(\mathbb{R}^n_+)$.

Proof. The proof is virtually identical to that of Lemma D.1.5. □

Theorem D.2.3. *Let $L_{\gamma,\mu,+} : H^1(\mathbb{R}^n_+) \to H^{-1}(\mathbb{R}^n_+)$ be as in Lemma* D.2.2 *and let $R : H^1(\mathbb{R}^n_+) \to H^{1/2}(\mathbb{R}^{n-1})$ be the restriction map of Proposition* C.2.17. *Then the map*

$$\begin{aligned} H^1(\mathbb{R}^n_+) &\to H^{-1}(\mathbb{R}^n_+) \oplus H^{\frac{1}{2}}(\mathbb{R}^{n-1}) \\ u &\mapsto \big(L_{\gamma,\mu,+}u\,,\ Ru\big) \end{aligned}$$

is an isomorphism (1–1, onto, bounded with bounded inverse). That is, for each $F \in H^{-1}(\mathbb{R}^n_+)$ and $f \in H^{1/2}(\mathbb{R}^{n-1})$ there is a unique $u \in H^1(\mathbb{R}^n_+)$ such that

$$-\sum_{i,j=1}^{n} \frac{\partial}{\partial x_i}\Big(\gamma^{i,j}(x)\frac{\partial u}{\partial x_j}(x)\Big) + \mu(x)u(x) = F(x) \qquad \text{in} \quad \mathbb{R}^n_+,$$

$$u(x) = f(x) \qquad \text{in} \quad \partial\mathbb{R}^n_+ = \mathbb{R}^{n-1}.$$

Furthermore, there is a constant C such that

$$\|u\|^2_{H^1(\mathbb{R}^n_+)} \le C\big(\,\|F\|^2_{H^{-1}(\mathbb{R}^n_+)} + |||f|||^2_{H^{1/2}(\mathbb{R}^{n-1})}\big)$$

for all $F \in H^{-1}(\mathbb{R}^n_+)$ and $f \in H^{1/2}(\mathbb{R}^{n-1})$.

Proof. *Boundedness:*

$$\left\| L_{\gamma,\mu,+} u \right\|_{H^{-1}(\mathbb{R}^n_+)} \le c_1 \, \|u\|_{H^1(\mathbb{R}^n_+)}$$

by Lemma D.2.2 and $|||Ru|||_{H^{1/2}(\mathbb{R}^{n-1})} \le c_2 \, \|u\|_{H^1(\mathbb{R}^n_+)}$ by part (ii) of Proposition C.2.17.

Injectiveness: If $Ru = 0$, then $u \in H^1_0(\mathbb{R}^n_+)$ by Proposition C.2.18. If $u \in H^1_0(\mathbb{R}^n_+)$ and $L_{\gamma,\mu,+}u = 0$, then, by Lemma D.2.2,

$$0 = (L_{\gamma,\mu,+}u)(\bar{u}) = \langle \bar{u}, \bar{u} \rangle_{\gamma,\mu,+}$$

so that $\bar{u} = 0$ and hence $u = 0$.

Surjectiveness: Let $F \in H^{-1}(\mathbb{R}^n_+) = H^1_0(\mathbb{R}^n_+)^*$ and $f \in H^{1/2}(\mathbb{R}^{n-1})$. Since the norms $\| \ \|_{H^1(\mathbb{R}^n_+)}$ and $\| \ \|_{\gamma,\mu,+}$ are equivalent, $H^1_0(\mathbb{R}^n_+)$, equipped with the inner product $\langle \ , \ \rangle_{\gamma,\mu,+}$, is still a Hilbert space and F is still a bounded linear functional on this Hilbert space. By the Riesz representation theorem, there exists $w_1 \in H^1_0(\mathbb{R}^n_+)$ such that

$$Fv = \langle v, w_1 \rangle_{\gamma,\mu,+}$$

for all $v \in H^1_0(\mathbb{R}^n_+)$. By Lemma D.2.2, $F = L_{\gamma,\mu,+}u_1$ for $u_1 = \bar{w}_1$.

By part (iii) of Proposition C.2.17, R is onto. So there is a $w_2 \in H^1(\mathbb{R}^n_+)$ such that $Rw_2 = f$. Since $H^1(\mathbb{R}^n_+)$ is still a Hilbert space when equipped with the inner product $\langle \ , \ \rangle_{\gamma,\mu,+}$ and $H^1_0(\mathbb{R}^n_+)$ is still a closed linear subspace of this Hilbert space, there exist unique $w_0 \in H^1_0(\mathbb{R}^n_+)$ and $u_2 \in H^1_0(\mathbb{R}^n_+)^{\gamma,\mu\perp}$ such that $w_2 = u_2 + w_0$. Here $H^1_0(\mathbb{R}^n_+)^{\gamma,\mu,\perp}$ is the orthogonal complement of $H^1_0(\mathbb{R}^n_+)$ in $H^1(\mathbb{R}^n_+)$ with respect to the inner product $\langle \ , \ \rangle_{\gamma,\mu,+}$.

Set $u = u_1 + u_2$. By (D.10), $L_{\gamma,\mu,+}$ annihilates $H^1_0(\mathbb{R}^n_+)^{\gamma,\mu,\perp}$ so that $L_{\gamma,\mu,+}u = L_{\gamma,\mu,+}u_1 = F$. As R annihilates $H^1_0(\mathbb{R}^n_+)$,

$$Ru = Ru_1 + Ru_2 = Ru_1 + R(w_2 - w_0) = Rw_2 = f.$$

Boundedness of the inverse: We must prove the existence of a constant C such that if $u \in H^1(\mathbb{R}^n_+)$, $F = L_{\gamma,\mu,+}u$ and $f = Ru$ then

$$\|u\|^2_{H^1(\mathbb{R}^n_+)} \le C\Big(\|F\|^2_{H^{-1}(\mathbb{R}^n_+)} + |||f|||^2_{H^{1/2}(\mathbb{R}^{n-1})} \Big).$$

Write $u = u_1 + u_2$ with $u_1 \in H^1_0(\mathbb{R}^n_+)$ and $u_2 \in H^1_0(\mathbb{R}^n_+)^{\gamma,\mu,\perp}$. Again, $H^1_0(\mathbb{R}^n_+)^{\gamma,\mu,\perp}$ is the orthogonal complement of $H^1_0(\mathbb{R}^n_+)$ in $H^1(\mathbb{R}^n_+)$ with respect to the inner product $\langle\ ,\ \rangle_{\gamma,\mu,+}$. By (D.10),

$$\begin{aligned}
|||F|||_{H^{-1}(\mathbb{R}^n_+)} &= \sup_{\substack{v\in H^1_0(\mathbb{R}^n_+)\\ v\neq 0}} \frac{|Fv|}{||v||_{H^1(\mathbb{R}^n_+)}} \\
&= \sup_{\substack{v\in H^1_0(\mathbb{R}^n_+)\\ v\neq 0}} \frac{|\langle v,\bar{u}\rangle_{\gamma,\mu,+}|}{||v||_{H^1(\mathbb{R}^n_+)}} \\
&= \sup_{\substack{v\in H^1_0(\mathbb{R}^n_+)\\ v\neq 0}} \frac{|\langle \bar{v},u\rangle_{\gamma,\mu,+}|}{||\bar{v}||_{H^1(\mathbb{R}^n_+)}} \\
&= \sup_{\substack{v\in H^1_0(\mathbb{R}^n_+)\\ v\neq 0}} \frac{|\langle v,u\rangle_{\gamma,\mu,+}|}{||v||_{H^1(\mathbb{R}^n_+)}}.
\end{aligned}$$

By the equivalence of the norms $|||\ |||_{H^1(\mathbb{R}^n_+)}$ and $||\ ||_{\gamma,\mu,+}$

$$\begin{aligned}
||F||_{H^{-1}(\mathbb{R}^n_+)} &\geq c_3 \sup_{\substack{v\in H^1_0(\mathbb{R}^n_+)\\ v\neq 0}} \frac{|\langle v,u\rangle_{\gamma,\mu,+}|}{||v||_{\gamma,\mu,+}} \\
&= c_3 \sup_{\substack{v\in H^1_0(\mathbb{R}^n_+)\\ v\neq 0}} \frac{|\langle v,u_1\rangle_{\gamma,\mu,+}|}{||v||_{\gamma,\mu,+}} \\
&= c_3 ||u_1||_{\gamma,\mu,+}.
\end{aligned}$$

By part (iv) of Proposition C.2.17, which gives the existence of a bounded right inverse for R, there is a $w \in H^1(\mathbb{R}^n_+)$ such that $Rw = f$ and

$$||w||_{\gamma,\mu,+} \leq c_4\, ||w||_{H^1(\mathbb{R}^n_+)} \leq c_5 |||f|||_{H^{1/2}(\mathbb{R}^{n-1})}.$$

As $Ru_2 = Ru = f$ and $Rw = f$, $w - u_2$ is in the kernel of R, which is $H^1_0(\mathbb{R}^n_+)$. As $w - u_2$ and u_2 are perpendicular to each other with respect to the inner product $\langle\ ,\ \rangle_{\gamma,\mu,+}$,

$$||w||^2_{\gamma,mu,+} = ||u_2||^2_{\gamma,\mu,+} + ||w - u_2||^2_{\gamma,\mu,+} \geq ||u_2||^2_{\gamma,\mu,+}$$

so that $||u_2||^2_{\gamma,\mu,+} \leq c_5^2 |||f|||_{H^{1/2}(\mathbb{R}^{n-1})}$. All together

$$\begin{aligned}
||F||^2_{H^{-1}(\mathbb{R}^n_+)} + |||f|||^2_{H^{1/2}(\mathbb{R}^{n-1})} &\geq c_3^2 ||u_1||^2_{\gamma,\mu,+} + \frac{1}{c_5^2} ||u_2||^2_{\gamma,\mu,+} \\
&\geq \min\left\{c_3^2, \frac{1}{c_5^2}\right\} \left[||u_1||^2_{\gamma,\mu,+} + ||u_2||^2_{\gamma,\mu,+}\right] \\
&= \min\{c_3^2, \frac{1}{c_5^2}\} ||u||^2_{\gamma,\mu,+} \\
&\geq c_6\, ||u||^2_{H^1(\mathbb{R}^n_+)}\,. \qquad \square
\end{aligned}$$

Remark D.2.4. By (D.10), the kernel of $L_{\gamma,\mu,+}$ is $H^1_0(\mathbb{R}^n_+)^{\gamma,\mu,\perp}$, the orthogonal complement of $H^1_0(\mathbb{R}^n_+)$ in $H^1(\mathbb{R}^n_+)$ with respect to the inner product $\langle\ ,\ \rangle_{\gamma,\mu,+}$. So, if $f \in H^{1/2}(\mathbb{R}^{n-1})$, then there exists a unique $u \in H^1_0(\mathbb{R}^n_+)^\perp$ such that

$$L_{\gamma,\mu,+}u = 0, \qquad Ru = f.$$

The kernel of R is $H^1_0(\mathbb{R}^n_+)$. So if $F \in H^{-1}(\mathbb{R}^n_+)$, then there exists a unique $u \in H^1_0(\mathbb{R}^n_+)$ such that

$$L_{\gamma,\mu,+}u = F, \qquad Ru = 0.$$

In preparation for proving elliptic regularity in $\mathbb{R}^n_+$, we prove analogs of Lemmas D.1.9 and D.1.10.

Lemma D.2.5. *Let $\ell \in \mathbb{N}_0$, $u \in H^\ell(\mathbb{R}^n_+)$ and $\alpha \in \mathbb{N}_0^n$ with $|\alpha| \geq 1$ and $\alpha_n = 0$. Then $\partial^\alpha u \in H^\ell(\mathbb{R}^n_+)$ if and only if $\limsup_{H\to 0} \|\Delta^\alpha_H u\|_{H^\ell(\mathbb{R}^n_+)}$ is finite. There is a constant C, depending only on ℓ, α and n such that, in this case, $\|\Delta^\alpha_H u\|_{H^\ell(\mathbb{R}^n_+)} \leq C\,\|\partial^\alpha u\|_{H^\ell(\mathbb{R}^n_+)}$ for all H and*

$$\|\partial^\alpha u\|_{H^\ell(\mathbb{R}^n_+)} \leq C \limsup_{H\to 0} \|\Delta^\alpha_H u\|_{H^\ell(\mathbb{R}^n_+)}\,.$$

Proof. Let $E : H^\ell(\mathbb{R}^n_+) \to H^\ell(\mathbb{R}^n)$ be the extension operator of Lemma C.2.11. First suppose that $\limsup_{H\to 0} \|\Delta^\alpha_H u\|_{H^\ell(\mathbb{R}^n_+)}$ is finite. Then, by Lemma C.2.11, $\limsup_{H\to 0} \|E\Delta^\alpha_H u\|_{H^\ell(\mathbb{R}^n)}$ is also finite. Since $\alpha_n = 0$, the construction of (C.5) gives $E\Delta^\alpha_H u = \Delta^\alpha_H Eu$. Thus $\limsup_{H\to 0} \|\Delta^\alpha_H Eu\|_{H^\ell(\mathbb{R}^n)}$ is finite too. By Lemma D.1.9, $\partial^\alpha Eu \in H^\ell(\mathbb{R}^n)$ and

$$\begin{aligned}\|\partial^\alpha Eu\|_{H^\ell(\mathbb{R}^n)} &= \lim_{H\to 0} \|\Delta^\alpha_H Eu\|_{H^\ell(\mathbb{R}^n)}\\ &= \lim_{H\to 0} \|E\Delta^\alpha_H u\|_{H^\ell(\mathbb{R}^n)}\\ &\leq C \limsup_{H\to 0} \|\Delta^\alpha_H u\|_{H^\ell(\mathbb{R}^n_+)}\end{aligned}$$

by Lemma C.2.11. By Exercise C.2.14, $\partial^\alpha u \in H^\ell(\mathbb{R}^n_+)$ and

$$\|\partial^\alpha u\|_{H^\ell(\mathbb{R}^n_+)} \leq \|E\partial^\alpha u\|_{H^\ell(\mathbb{R}^n)} = \|\partial^\alpha Eu\|_{H^\ell(\mathbb{R}^n)} \leq C \limsup_{H\to 0} \|\Delta^\alpha_H u\|_{H^\ell(\mathbb{R}^n_+)}\,.$$

Conversely, suppose that $\partial^\alpha u \in H^\ell(\mathbb{R}^n_+)$. By Exercise C.2.14, we have $\partial^\alpha Eu \in H^\ell(\mathbb{R}^n)$ and $E\partial^\alpha u = \partial^\alpha Eu$. Then, since $\alpha_n = 0$,

$$\begin{aligned}\|\Delta^\alpha_H u\|_{H^\ell(\mathbb{R}^n_+)} &\leq \|\!|E\Delta^\alpha_H u|\!\|_{H^\ell(\mathbb{R}^n)}\\ &= \|\!|\Delta^\alpha_H Eu|\!\|_{H^\ell(\mathbb{R}^n)}\\ &\leq \|\!|\partial^\alpha Eu|\!\|_{H^\ell(\mathbb{R}^n)} \quad \text{(by Lemma D.1.9)}\\ &= \|\!|E\partial^\alpha u|\!\|_{H^\ell(\mathbb{R}^n)}\\ &\leq C\,\|\partial^\alpha u\|_{H^\ell(\mathbb{R}^n_+)}\,.\end{aligned}$$

□

Lemma D.2.6. *Let $s \geq 1$, let m be an integer with $m \geq s$, and let $\alpha \in \mathbb{N}_0^n$ with $|\alpha| = 1$ and $\alpha_n = 0$.*

(a) *Let $\varphi \in W^{m,\infty}(\mathbb{R}_+^n)$. Also use φ to denote the operator of multiplication by φ. Then there is a $C > 0$ (depending on φ, s, m, n and α, but independent of h) such that $[\Delta_h^\alpha, \varphi] = \Delta_h^\alpha \varphi - \varphi \Delta_h^\alpha$ is a bounded operator from $H^{s-1}(\mathbb{R}_+^n)$ to $H^{s-1}(\mathbb{R}_+^n)$ with norm at most C.*

(b) *Let $Lu = \sum_{i,j} \partial_i(a^{i,j}(x)\partial_j u)$ be a differential operator of order two where $a^{i,j} \in W^{m,\infty}(\mathbb{R}_+^n)$. Then there is a $C > 0$, independent of h, such that $[\Delta_h^\alpha, L]$ is a bounded operator from $H^{s-2}(\mathbb{R}_+^n)$ to $H^s(\mathbb{R}_+^n)$ with norm at most C.*

Proof.

(a) Note that if v is a function on $\mathbb{R}_+^n$, then $\Delta_h^\alpha v$ is well defined on $\mathbb{R}_+^n$ since $\alpha_n = 0$. We have already seen, in the proof of Lemma D.1.10, that

(D.11) $$([\Delta_h^\alpha, \varphi]u)(x) = \Phi_h^\alpha(x) u_h(x),$$

where $\Phi_h^\alpha = \Delta_h^\alpha \varphi$ and $u_h(x) = u(x + h\alpha)$ is a translate of u. By Lemma C.1.32, since $m - 1 \geq s - 1 \geq 0$,

$$\begin{aligned} \|[\Delta_h^\alpha, \varphi]u\|_{H^{s-1}(\mathbb{R}_+^n)} &\leq C_{n,m} \|\Phi_h\|_{W^{m-1,\infty}(\mathbb{R}_+^n)} \|u_h\|_{H^{s-1}(\mathbb{R}_+^n)} \\ &\leq C_{n,m} \|\varphi\|_{W^{m,\infty}(\mathbb{R}_+^n)} \|u_h\|_{H^{s-1}(\mathbb{R}_+^n)}. \end{aligned}$$

In the last step we used the mean value theorem in the form

$$\partial^\alpha \Phi_h = \nabla \partial^\alpha \varphi(x + th\alpha) \cdot \alpha$$

for some $t = t(x, h) \in [0, 1]$. So the proof is completed by observing that

(D.12) $$\|u_h\|_{H^\ell(\mathbb{R}_+^n)} = \|u\|_{H^\ell(\mathbb{R}_+^n)}$$

when $\alpha_n = 0$.

(b) The proof is identical to the proof of Lemma D.1.10(b). □

Proposition D.2.7 (Elliptic regularity in $\mathbb{R}_+^n$). *Let $\ell \geq 1$ and suppose that*

$$\gamma^{i,j} \in W^{\ell-1,\infty}(\mathbb{R}_+^n), \qquad \mu \in W^{\max\{\ell-2,0\},\infty}(\mathbb{R}_+^n).$$

There is a constant C, depending only on ℓ, γ and μ such that, for all $u \in H^1(\mathbb{R}_+^n)$, $F \in H^{\ell-2}(\mathbb{R}_+^n)$ and $f \in H^{\ell-\frac{1}{2}}(\mathbb{R}^{n-1})$ obeying

$$L_{\gamma,\mu,+}u = F, \qquad Ru = f,$$

we have $u \in H^\ell(\mathbb{R}_+^n)$ and

$$\|u\|_{H^\ell(\mathbb{R}_+^n)} \leq C\Big(\|F\|_{H^{\ell-2}(\mathbb{R}_+^n)} + |||f|||_{H^{\ell-1/2}(\mathbb{R}^{n-1})}\Big).$$

Proof. We proceed by induction. We proved the result for $\ell = 1$ in Theorem D.2.3. Assume that we have proven the desired result for some $\ell \geq 1$. We now prove it for $\ell + 1$. That is, we assume that

$$\gamma \in W^{\ell,\infty}(\mathbb{R}^n_+), \qquad \mu \in W^{\ell-1,\infty}(\mathbb{R}^n_+)$$

and that for all $u \in H^1(\mathbb{R}^n_+)$, $F \in H^{\ell-2}(\mathbb{R}^n_+)$ and $f \in H^{\ell-\frac{1}{2}}(\mathbb{R}^{n-1})$ obeying

$$L_{\gamma,\mu,+}u = F \qquad Ru = f$$

we have $u \in H^\ell(\mathbb{R}^n_+)$ and

$$\|u\|_{H^\ell(\mathbb{R}^n_+)} \leq C\big(\|F\|_{H^{\ell-2}(\mathbb{R}^n_+)} + |||f|||_{H^{\ell-1/2}(\mathbb{R}^{n-1})}\big).$$

We must prove that if $u \in H^1(\mathbb{R}^n_+)$, $F \in H^{\ell-1}(\mathbb{R}^n_+)$, $f \in H^{\ell+\frac{1}{2}}(\mathbb{R}^{n-1})$, $L_{\gamma,\mu,+}u = F$ and $Ru = f$, then $u \in H^{\ell+1}(\mathbb{R}^n_+)$ and

$$\|u\|_{H^{\ell+1}(\mathbb{R}^n_+)} \leq C'\big(\|F\|_{H^{\ell-1}(\mathbb{R}^n_+)} + |||f|||_{H^{\ell+1/2}(\mathbb{R}^{n-1})}\big).$$

So let $u \in H^1(\mathbb{R}^n_+)$, $F \in H^{\ell-1}(\mathbb{R}^n_+)$, $f \in H^{\ell+\frac{1}{2}}(\mathbb{R}^{n-1})$ with $L_{\gamma,\mu,+}u = F$ and $Ru = f$. By the inductive hypothesis, $u \in H^\ell(\mathbb{R}^n_+)$ with

$$\|u\|_{H^\ell(\mathbb{R}^n_+)} \leq C\big(\|F\|_{H^{\ell-2}(\mathbb{R}^n_+)} + |||f|||_{H^{\ell-1/2}(\mathbb{R}^{n-1})}\big).$$

For each $\alpha \in \mathbb{N}^n_0$ with $|\alpha| = 1$, $\alpha_n = 0$ and each $h \in \mathbb{R} \setminus \{0\}$, set $u_{\alpha,h} = \Delta^\alpha_h u$ and $f_{\alpha,h} = \Delta^\alpha_h f$. Then

$$L_{\gamma,\mu,+}u_{\alpha,h} = F_{\alpha,h}, \qquad Ru_{\alpha,h} = f_{\alpha,h},$$

where

$$F_{\alpha,h} = \Delta^\alpha_h F + [L_{\gamma,0}, \Delta^\alpha_h]u + [\mu, \Delta^\alpha_h]u.$$

We estimate the different terms in $F_{\alpha,h}$. By Lemma D.2.5,

$$\left\|\Delta^\alpha_h F\right\|_{H^{\ell-2}(\mathbb{R}^n_+)} \leq C_2\, \|\partial^\alpha F\|_{H^{\ell-2}(\mathbb{R}^n_+)} \leq C_2\, \|F\|_{H^{\ell-1}(\mathbb{R}^n_+)}\,.$$

By part (b) of Lemma D.2.6 with $m = s = \ell$, we have that

$$\begin{aligned}\left\|[L_{\gamma,0}, \Delta^\alpha_h]u\right\|_{H^{\ell-2}(\mathbb{R}^n_+)} &\leq C\, \|u\|_{H^\ell(\mathbb{R}^n_+)} \\ &\leq C\big(\|F\|_{H^{\ell-2}(\mathbb{R}^n_+)} + |||f|||_{H^{\ell-1/2}(\mathbb{R}^{n-1})}\big) \\ &\leq C\big(\|F\|_{H^{\ell-1}(\mathbb{R}^n_+)} + |||f|||_{H^{\ell+1/2}(\mathbb{R}^{n-1})}\big).\end{aligned}$$

If $\ell \geq 2$, Lemma D.2.6(a) with $s = m = \ell - 1$ yields

$$\begin{aligned}|||[\mu, \Delta^\alpha_h]u|||_{H^{\ell-2}(\mathbb{R}^n_+)} &\leq C|||u|||_{H^{\ell-2}(\mathbb{R}^n_+)} \\ &\leq C\big(\|F\|_{H^{\ell-2}(\mathbb{R}^n_+)} + |||f|||_{H^{\ell-1/2}(\mathbb{R}^{n-1})}\big),\end{aligned}$$

whereas for $\ell = 1$ we use the duality statement $H^{-1}(\mathbb{R}^n_+) \cong (H^1_0(\mathbb{R}^n_+))^*$, the computation $[\Delta^\alpha_h, \mu]u = (\Delta^\alpha_h \mu)u_h$ as in the proof of Lemma D.2.6, and the fact that for $\alpha_n = 0$ we have

$$\langle \Delta^\alpha_h w, v\rangle_{L^2(\mathbb{R}^n_+)} = \langle w, \Delta^\alpha_{-h} v\rangle_{L^2(\mathbb{R}^n_+)},$$

to obtain

$$\begin{aligned}
\left\|[\mu, \Delta^\alpha_h]u\right\|_{H^{-1}(\mathbb{R}^n_+)} &= \sup_{\|v\|_{H^1_0(\mathbb{R}^n_+)}\le 1} |\langle(\Delta^\alpha_h\mu)u_h, v\rangle_{L^2(\mathbb{R}^n_+)}| \\
&= \sup_{\|v\|_{H^1_0(\mathbb{R}^n_+)}\le 1} |\langle\Delta^\alpha_h\mu, u_h v\rangle_{L^2(\mathbb{R}^n_+)}| \\
&= \sup_{\|v\|_{H^1_0(\mathbb{R}^n_+)}\le 1} |\langle\mu, u\Delta^\alpha_{-h} v\rangle_{L^2(\mathbb{R}^n_+)} - \langle\mu, (\Delta^\alpha_h u)v\rangle_{L^2(\mathbb{R}^n_+)}| \\
&\le \sup_{\|v\|_{H^1_0(\mathbb{R}^n_+)}\le 1} \|\mu\|_{L^\infty} \left(\|u\|_{L^2}\left\|\Delta^\alpha_{-h}v\right\|_{L^2} + \left\|\Delta^\alpha_h u\right\|_{L^2}\|v\|_{L^2}\right) \\
&\le C\,\|\mu\|_{L^\infty(\mathbb{R}^n_+)}\,\|u\|_{H^1(\mathbb{R}^n_+)} \\
&\le C\big(\|F\|_{H^{-1}(\mathbb{R}^n_+)} + \|\!|f|\!\|_{H^{1/2}(\mathbb{R}^{n-1})}\big).
\end{aligned}$$

Combining these estimates gives $F_{\alpha,h} \in H^{\ell-2}(\mathbb{R}^n_+)$ with

$$\left\|F_{\alpha,h}\right\|_{H^{\ell-2}(\mathbb{R}^n_+)} \le C\big(\|F\|_{H^{\ell-1}(\mathbb{R}^n_+)} + \|\!|f|\!\|_{H^{\ell+1/2}(\mathbb{R}^{n-1})}\big).$$

On the other hand, by Lemma D.1.9, with $s = \ell - \frac{1}{2}$,

$$\begin{aligned}
\|\!|f_{\alpha,h}|\!\|_{H^{\ell-1/2}(\mathbb{R}^{n-1})} &= \|\!|\Delta^\alpha_h f|\!\|_{H^{\ell-1/2}(\mathbb{R}^{n-1})} \\
&\le \|\!|\partial^\alpha f|\!\|_{H^{\ell-1/2}(\mathbb{R}^{n-1})} \\
&\le \|\!|f|\!\|_{H^{\ell+1/2}(\mathbb{R}^{n-1})}.
\end{aligned}$$

Hence by the inductive hypothesis, $u_{\alpha,h} = \Delta^\alpha_h u \in H^\ell(\mathbb{R}^n_+)$ with

$$\begin{aligned}
\left\|\Delta^\alpha_h u\right\|_{H^\ell(\mathbb{R}^n_+)} &\le C\Big(\left\|F_{\alpha,h}\right\|_{H^{\ell-2}(\mathbb{R}^n_+)} + \|\!|f_{\alpha,h}|\!\|_{H^{\ell-1/2}(\mathbb{R}^{n-1})}\Big) \\
&\le C\big(\|F\|_{H^{\ell-1}(\mathbb{R}^n_+)} + \|\!|f|\!\|_{H^{\ell+1/2}(\mathbb{R}^{n-1})}\big).
\end{aligned}$$

By Lemma D.2.5, $\partial^\alpha u \in H^\ell(\mathbb{R}^n_+)$ with

$$\|\partial^\alpha u\|_{H^\ell(\mathbb{R}^n_+)} \le C\big(\|F\|_{H^{\ell-1}(\mathbb{R}^n_+)} + \|\!|f|\!\|_{H^{\ell+1/2}(\mathbb{R}^{n-1})}\big). \tag{D.13}$$

So far, we have assumed that $\alpha_n = 0$. So we now know that $\partial^\beta u \in L^2(\mathbb{R}^n_+)$ for all $\beta \in \mathbb{N}^n_0$ with $|\beta| \le \ell + 1$ and $\beta_n \le \ell$, and that $\left\|\partial^\beta u\right\|_{L^2(\mathbb{R}^n_+)}$ is bounded by the right-hand side of (D.13). By part (a) of Proposition C.2.16, it remains only to prove the same bound for $\partial^\gamma u$ with $\gamma = (0, \cdots, 0, \ell+1)$. From the differential equation

$$L_{\gamma,\mu,+}u = F$$

we have

$$\frac{\partial}{\partial x_n}\Big(\gamma^{n,n}\frac{\partial u}{\partial x_n}\Big) = -\sum_{\substack{1\le i,j\le n\\ (i,j)\ne(n,n)}} \frac{\partial}{\partial x_i}\Big(\gamma^{i,j}\frac{\partial u}{\partial x_j}\Big) + \mu u - F$$

and hence

$$\frac{\partial^2 u}{\partial x_n^2} = \frac{1}{\gamma^{n,n}}\Big\{-\sum_{\substack{1\le i,j\le n\\ (i,j)\ne(n,n)}} \frac{\partial}{\partial x_i}\Big(\gamma^{i,j}\frac{\partial u}{\partial x_j}\Big) + \mu u - F - \frac{\partial\gamma^{n,n}}{\partial x_n}\frac{\partial u}{\partial x_n}\Big\}.$$

By (D.13) and Lemma C.1.32, the $\|\cdot\|_{H^{\ell-1}(\mathbb{R}^n_+)}$ norm of each term on the right-hand side is bounded by a constant times the right-hand side of (D.13). This is all we need. □

Exercise D.2.8. Let $\gamma(x)$ obey Hypotheses D.1.1, let $\partial^\alpha\gamma \in L^\infty(\mathbb{R}^n_+)$ for all α, and let $\ell \in \mathbb{N}$.

(a) Prove that there is a constant C, depending only on ℓ and γ, such that for all $u \in H^1(\mathbb{R}^n_+)$, $F \in H^{\ell-2}(\mathbb{R}^n_+)$ and $f \in H^{\ell-\frac{1}{2}}(\mathbb{R}^{n-1})$ obeying

$$L_{\gamma,0,+}u = F, \qquad Ru = f,$$

we have $u \in H^\ell(\mathbb{R}^n_+)$ and

$$\|u\|_{H^\ell(\mathbb{R}^n_+)} \le C\Big(\|F\|_{H^{\ell-2}(\mathbb{R}^n_+)} + |||f|||_{H^{\ell-1/2}(\mathbb{R}^{n-1})} + \|u\|_{H^{-1}(\mathbb{R}^n_+)}\Big).$$

(b) Let Ω be a bounded open subset of $\mathbb{R}^n$. Prove that there is a constant C, depending only on ℓ, γ and Ω such that, for all functions $F \in H^{\ell-2}(\mathbb{R}^n_+)$, all $f \in H^{\ell-\frac{1}{2}}(\mathbb{R}^{n-1}_+)$ and all $u \in H^1(\mathbb{R}^n_+)$ that vanish outside of Ω and obey

$$L_{\gamma,0,+}u = F, \qquad Ru = f,$$

we have $u \in H^\ell(\mathbb{R}^n_+)$ and

$$\|u\|_{H^\ell(\mathbb{R}^n_+)} \le C\Big(\|F\|_{H^{\ell-2}(\mathbb{R}^n_+)} + |||f|||_{H^{\ell-1/2}(\mathbb{R}^{n-1})}\Big).$$

D.3. Equations on bounded domains

Let Ω be a bounded open subset of $\mathbb{R}^n$. Consider a second order differential operator L, acting on functions u on Ω, given by

$$Lu = -\sum_{i,j=1}^{\infty} \frac{\partial}{\partial x_i}\left(\gamma^{i,j}\frac{\partial u}{\partial x_j}\right) + \mu u. \tag{D.14}$$

The coefficients are assumed to satisfy the following conditions:

$$\begin{cases} \gamma^{i,j}, \mu \in L^\infty(\Omega) \text{ are real valued,} \\ \gamma^{i,j} = \gamma^{j,i} \text{ for all } i,j = 1,\dots,n, \\ \sum_{j,k=1}^n \gamma^{i,j}(x)\xi_i\xi_j \ge c\,|\xi|^2 \text{ for a.e. } x\in\Omega \text{ and all } \xi\in\mathbb{R}^n, \end{cases} \tag{D.15}$$

where $c > 0$. The last condition is an ellipticity condition for L.

We already proved the basic facts on weak solutions in the Dirichlet problem for L in §2.5. This was first done in the case $\mu \geq 0$ by precisely the same method as in §D.1–§D.2, via introducing an equivalent inner product on $H_0^1(\Omega)$ associated with L and by using the Riesz representation theorem.

In the present setting we can formulate the result in the case of C^1 domains as follows. As in Lemma D.2.2, L may be considered as an operator $H^1(\Omega) \to H^{-1}(\Omega) = (H_0^1(\Omega))^*$.

Theorem D.3.1 (Weak solutions for $\mu \geq 0$). *Let $\Omega \subset \mathbb{R}^n$ be a bounded open set with C^1 boundary, let L be as in* (D.14)–(D.15), *and let $R : H^1(\Omega) \to H^{1/2}(\partial\Omega)$ be the restriction map of Theorem* C.2.1. *Assume that*

$$\mu \geq 0 \text{ a.e. in } \Omega.$$

Then the map

$$\begin{aligned} H^1(\Omega) &\to H^{-1}(\Omega) \oplus H^{\frac{1}{2}}(\partial\Omega), \\ u &\mapsto \qquad (Lu\,,\, Ru), \end{aligned}$$

is an isomorphism (1–1, onto, bounded with bounded inverse).

In other words, for each $F \in H^{-1}(\Omega)$ and $f \in H^{1/2}(\partial\Omega)$ there is a unique $u \in H^1(\Omega)$ such that $Lu = F$ in Ω and $u|_{\partial\Omega} = f$. Furthermore, there is a constant C such that

$$\|u\|_{H^1(\Omega)} \leq C\big(\|F\|_{H^{-1}(\Omega)} + \|f\|_{H^{1/2}(\partial\Omega)} \big)$$

for all $F \in H^{-1}(\Omega)$ and $f \in H^{1/2}(\partial\Omega)$.

Proof. A version of this result was proved in Theorem 2.5.6, for any bounded open set $\Omega \subset \mathbb{R}^n$ and with the trace operator R replaced by the abstract trace map

$$\tilde{R} : H^1(\Omega) \to H^1(\Omega)/H_0^1(\Omega), \quad u \mapsto [u] = \{\, u + \varphi \mid \varphi \in H_0^1(\Omega) \,\}.$$

The kernel of $\tilde{R}$ is $H_0^1(\Omega)$.

When Ω has C^1 boundary, we have defined the space $H^{1/2}(\partial\Omega)$ and the trace operator $R : H^1(\Omega) \to H^{1/2}(\partial\Omega)$ in Theorem C.2.1. Since the kernel of R is also $H_0^1(\Omega)$, it follows from Theorem 2.5.6 that for any $F \in H^{-1}(\Omega)$ the equation

$$Lu = F, \qquad Ru = 0,$$

has a unique solution $u \in H^1(\Omega)$ (and then $u \in H_0^1(\Omega)$). Moreover, there is $C > 0$ independent of F such that

$$\|u\|_{H^1(\Omega)} \leq C \, \|F\|_{H^{-1}(\Omega)} \, .$$

For any $f \in H^{1/2}(\partial\Omega)$, we may use the right inverse of the trace map in Theorem C.2.1 to find $e_f \in H^1(\Omega)$ with

$$\left\| e_f \right\|_{H^1(\Omega)} \leq C \, \|f\|_{H^1(\Omega)} \, .$$

Given $F \in H^{-1}(\Omega)$, we see that u solves

$$Lu = F, \qquad Ru = f,$$

if and only if $v = u - e_f$ solves

$$Lv = F - Le_f, \qquad Rv = 0.$$

By the first part of this proof there is a unique solution $v \in H_0^1(\Omega)$ with

$$\begin{aligned} \|v\|_{H^1} &\leq C(\|F\|_{H^{-1}} + \left\|Le_f\right\|_{H^{-1}}) \\ &\leq C(\|F\|_{H^{-1}} + \left\|e_f\right\|_{H^1}) \\ &\leq C(\|F\|_{H^{-1}} + \|f\|_{H^{1/2}(\partial\Omega)}). \end{aligned}$$

This shows that there is a unique solution $u \in H^1(\Omega)$ that satisfies

$$\|u\|_{H^1} \leq C(\|F\|_{H^{-1}} + \|f\|_{H^{1/2}(\partial\Omega)}).$$

The result follows. □

Next we consider the case of a general lower order coefficient $\mu \in L^\infty(\Omega)$. As we saw in §2.5, in this case one has to account for the existence of Dirichlet eigenvalues. The proof proceeds by considering $L+\kappa$ where κ is a constant such that $\mu + \kappa \geq c > 0$, applying Theorem D.3.1 to $L + \kappa$, and using the fact that the inverse of $L + \kappa$ is a compact operator by compact Sobolev embedding so that one can apply the spectral theorem for compact operators and the Fredholm alternative.

Theorem D.3.2 (Weak solutions). *Let $\Omega \subset \mathbb{R}^n$ be a bounded open set with C^1 boundary, and let L be as in* (2.5), (2.6).

(a) *There is a set of real numbers*

$$\mathrm{Spec}(L) = \{\lambda_j\}_{j=1}^\infty$$

with $\lambda_1 \leq \lambda_2 \leq \ldots \to \infty$, such that whenever $\lambda \notin \mathrm{Spec}(L)$ the boundary value problem

$$\begin{cases} Lu = \lambda u + F & \text{in } \Omega, \\ u = f & \text{on } \partial\Omega \end{cases}$$

has a unique solution $u \in H^1(\Omega)$ for any $F \in H^{-1}(\Omega)$ and $f \in H^{1/2}(\partial\Omega)$.

(b) *If $\lambda \notin \mathrm{Spec}(L)$, then the map*

$$\begin{aligned} H^1(\Omega) &\to H^{-1}(\Omega) \oplus H^{\frac{1}{2}}(\partial\Omega), \\ u &\mapsto \quad (Lu - \lambda u\,,\, Ru) \end{aligned}$$

is an isomorphism (1–1, onto, bounded with bounded inverse). There is a constant C independent of F and f (but depending on λ) such that

$$\|u\|_{H^1(\Omega)} \leq C(\,\|F\|_{H^{-1}(\Omega)} + \|f\|_{H^{1/2}(\partial\Omega)}\,).$$

(c) *If* $\lambda \in \operatorname{Spec}(L)$, *there is a nontrivial solution* $u \in H_0^1(\Omega)$ *to the Dirichlet problem*

$$\begin{cases} Lu = \lambda u & \text{in } \Omega, \\ u = 0 & \text{on } \partial\Omega. \end{cases}$$

The space of such solutions is finite-dimensional.

(d) *If* $a \in \mathbb{R}$ *is a constant such that*

$$\mu(x) \geq a \quad \text{a.e. in } \Omega,$$

then $\operatorname{Spec}(L) \subset (a, \infty)$.

Proof. This is precisely Theorem 2.5.9, except that the abstract trace operator has been replaced by the trace map R of Theorem C.2.1 as in the proof of Theorem D.3.1. □

We proceed to the main new result in this section, concerning higher regularity of solutions.

Proposition D.3.3 (Elliptic regularity). *Let* Ω *be a bounded open subset of* $\mathbb{R}^n$ *and let* L *be as in* (D.14)–(D.15). *Let* $\ell \geq 1$ *be an integer and assume additionally that* Ω *has* C^ℓ *boundary and*

$$\gamma \in W^{\ell-1,\infty}(\Omega), \qquad \mu \in W^{\max\{\ell-2,0\},\infty}(\Omega).$$

There is a constant $C > 0$ *such that for all* $u \in H^1(\Omega)$, $F \in H^{\ell-2}(\Omega)$ *and* $f \in H^{\ell-\frac{1}{2}}(\partial\Omega)$ *obeying*

$$Lu = F, \qquad Ru = f,$$

we have $u \in H^\ell(\Omega)$ *and*

$$\|u\|_{H^\ell(\Omega)} \leq C\Big(\|F\|_{H^{\ell-2}(\Omega)} + \|f\|_{H^{\ell-1/2}(\partial\Omega)} + \|u\|_{H^1(\Omega)} \Big).$$

Moreover, if 0 *is not a Dirichlet eigenvalue of* L *in* Ω, *there is* $C > 0$ *such that in the above setting*

$$\|u\|_{H^\ell(\Omega)} \leq C\Big(\|F\|_{H^{\ell-2}(\Omega)} + \|f\|_{H^{\ell-1/2}(\partial\Omega)} \Big).$$

Proof. We may assume that μ satisfies the positivity condition

$$\mu \geq c > 0 \text{ a.e. in } \Omega. \tag{D.16}$$

In fact, if this is not the case, we may replace μ by $\mu + \lambda$ for a sufficiently large constant $\lambda > 0$ by considering the equivalent Dirichlet problem

$$Lu + \lambda u = F + \lambda u, \qquad Ru = f.$$

A regularity result for this new problem, showing that $u \in H^1$, $F \in L^2$ and $f \in H^{3/2}(\partial\Omega)$ implies $u \in H^2$, yields an analogous regularity result for the original problem. Proceeding inductively, we obtain the result for all $\ell \geq 1$. Thus we will assume (D.16). Then in particular 0 is not a Dirichlet eigenvalue.

We proceed by induction. The case $\ell = 1$ follows immediately from Theorem D.3.2. Assume that we have proven the desired bound for some $\ell \geq 1$. We now prove it for $\ell + 1$. That is, we assume that

$$\gamma \in W^{\ell,\infty}(\Omega), \qquad \mu \in W^{\ell-1,\infty}(\Omega),$$

that $F \in H^{\ell-1}(\Omega)$ and $f \in H^{\ell+\frac{1}{2}}(\Omega)$, and that we already know $u \in H^{\ell}(\Omega)$ and

$$\|u\|_{H^{\ell}(\Omega)} \leq C\Big(\|F\|_{H^{\ell-2}(\Omega)} + \|f\|_{H^{\ell-1/2}(\partial\Omega)} \Big). \tag{D.17}$$

We are to prove that $u \in H^{\ell+1}(\Omega)$ and

$$\|u\|_{H^{\ell+1}(\Omega)} \leq C'\Big(\|F\|_{H^{\ell-1}(\Omega)} + \|f\|_{H^{\ell+1/2}(\partial\Omega)} \Big).$$

Let N and, for $1 \leq i \leq N$, $\mathcal{U}(p_i)$, χ_i, ψ_{p_i} be as in Definition C.1.37. In particular, each $\chi_i \in C_c^{\infty}\big(\mathcal{U}(p_i)\big)$. Let $\mathcal{V}$ be a neighborhood of $\partial\Omega$ on which $\sum_{i=1}^{N} \chi_i = 1$. Select a function $\varphi \in C_c^{\infty}(\Omega)$ that is identically one on $\Omega \setminus \mathcal{V}$. We prove that each of the $N + 1$ terms on the right-hand side of

$$u = \varphi u + \sum_{i=1}^{N} \chi_i(1 - \varphi)u$$

is in $H^{\ell+1}(\Omega)$ and obeys the desired bound.

The term φu: Let $v = \varphi u$. We find a $G \in H^{\ell-1}(\mathbb{R}^n)$ obeying

$$Lv = G$$

and

$$|||G|||_{H^{\ell-1}(\mathbb{R}^n)} \leq c\Big(\|F\|_{H^{\ell-1}(\Omega)} + \|f\|_{H^{\ell+1/2}(\partial\Omega)} \Big).$$

We also find, in Exercise D.3.4, a $\gamma' \in W^{\ell,\infty}(\mathbb{R}^n)$ and $\mu' \in W^{\ell-1,\infty}(\mathbb{R}^n)$ that satisfy Hypothesis D.1.1 and coincide with γ and μ on the support of φ. It will then suffice to apply Proposition D.1.11 with $F = G$, u replaced by v, γ replaced by γ', μ replace by μ' and ℓ replaced by $\ell+1$. To find G, we just have to compute

$$
\begin{aligned}
Lv &= -\sum_{i,j=1}^{n} \frac{\partial}{\partial x_i}\big[\gamma^{i,j}\frac{\partial}{\partial x_j}(\varphi u)\big] + \mu(\varphi u) \\
&= -\sum_{i,j=1}^{n} \frac{\partial}{\partial x_i}\big[\gamma^{i,j}\varphi\frac{\partial u}{\partial x_j}\big] - \sum_{i,j=1}^{n} \frac{\partial}{\partial x_i}\big[\gamma^{i,j}u\frac{\partial \varphi}{\partial x_j}\big] + \mu\varphi u \\
&= \varphi L u - \sum_{i,j=1}^{n} \gamma^{i,j}\frac{\partial \varphi}{\partial x_i}\frac{\partial u}{\partial x_j} \\
&\qquad - \sum_{i,j=1}^{n} \gamma^{i,j}\frac{\partial u}{\partial x_i}\frac{\partial \varphi}{\partial x_j} - \sum_{i,j=1}^{n} u\frac{\partial}{\partial x_i}\big[\gamma^{i,j}\frac{\partial \varphi}{\partial x_j}\big] \\
&= \varphi F - 2\sum_{i,j=1}^{n} \gamma^{i,j}\frac{\partial \varphi}{\partial x_i}\frac{\partial u}{\partial x_j} - \sum_{i,j=1}^{n} u\frac{\partial}{\partial x_i}\big[\gamma^{i,j}\frac{\partial \varphi}{\partial x_j}\big] \\
&:= G.
\end{aligned}
\tag{D.18}
$$

Recalling that $\varphi \in C_c^\infty(\Omega)$ and that, by Lemmas C.1.32 and C.1.29,

- multiplication by any function in $W^{|k|,\infty}(\Omega)$ is a bounded operator on $H^k(\Omega)$ for any $k \in \mathbb{Z}$,
- $\frac{\partial}{\partial x_j}$ is a bounded operator from $H^k(\Omega)$ to $H^{k-1}(\Omega)$ for all $k \in \mathbb{Z}$ and $1 \le j \le n$,

we have $G \in H^{\ell-1}(\mathbb{R}^n)$ with

$$
\begin{aligned}
|||G|||_{H^{\ell-1}(\mathbb{R}^n)} &= \|G\|_{H^{\ell-1}(\Omega)} \\
&\le c_1\Big(\|F\|_{H^{\ell-1}(\Omega)} + \|u\|_{H^{\ell}(\Omega)} + \|u\|_{H^{\ell-1}(\Omega)} \Big) \\
&\le 2c_1 C\Big(\|F\|_{H^{\ell-1}(\Omega)} + \|F\|_{H^{\ell-2}(\Omega)} + \|f\|_{H^{\ell-1/2}(\partial\Omega)} \Big) \text{ by (D.17)} \\
&\le C\Big(\|F\|_{H^{\ell-1}(\Omega)} + \|f\|_{H^{\ell-1/2}(\partial\Omega)} \Big).
\end{aligned}
$$

The term $\chi_{i_0}(1-\varphi)u$ – algebra: Fix $1 \le i_0 \le N$ and write $\chi_{i_0} = \chi$ and $\psi_{p_{i_0}} = \psi$. The strategy for treating this term is analogous to that we applied to φu, but this time we construct a boundary value problem on $\mathbb{R}^n_+$ and apply Proposition D.2.7. Recall that ψ is a diffeomorphism from $\mathcal{U}(p_{i_0})$ to $\mathbb{R}^n$ that maps $\mathcal{U}(p_{i_0}) \cap \Omega$ to $\mathbb{R}^n_+$. We use x to denote coordinates in $\mathcal{U}(p_{i_0})$ and y to denote coordinates in $\mathbb{R}^n$. Set $v = \chi(1-\varphi)u$, $w = v \circ \psi^{-1}$ and

$$
\tilde{\gamma}^{i,j}(y) = \sum_{k,r=1}^{n} \gamma^{k,r}(\psi^{-1}(y))\frac{\partial \psi_i}{\partial x_k}(\psi^{-1}(y))\frac{\partial \psi_j}{\partial x_r}(\psi^{-1}(y)).
$$

By the chain rule

$$
\psi_m^{-1}(\psi(x)) = x_m \implies \sum_{j=1}^{n} \frac{\partial \psi_j}{\partial x_r}(x)\frac{\partial \psi_m^{-1}}{\partial y_j}(\psi(x)) = \begin{cases} 1 & \text{if } r = m \\ 0 & \text{if } r \ne m \end{cases}
$$

and

$$\text{(D.19)}\quad \sum_{j=1}^{n} \frac{\partial \psi_j}{\partial x_r}(\psi^{-1}(y))\frac{\partial}{\partial y_j}v(\psi^{-1}(y))$$
$$= \sum_{j,m=1}^{n} \frac{\partial \psi_j}{\partial x_r}(\psi^{-1}(y))\frac{\partial \psi_m^{-1}}{\partial y_j}(y)\frac{\partial v}{\partial x_m}(\psi^{-1}(y)) = \frac{\partial v}{\partial x_r}(\psi^{-1}(y)).$$

Hence

$$\begin{aligned}
-L_{\tilde{\gamma},0,+}w &= \sum_{i,j=1}^{n} \frac{\partial}{\partial y_i}\Big(\tilde{\gamma}^{i,j}(y)\frac{\partial}{\partial y_j}v\circ\psi^{-1}(y)\Big)\\
&= \sum_{i,k,r=1}^{n} \frac{\partial}{\partial y_i}\Big(\gamma^{k,r}(\psi^{-1}(y))\frac{\partial \psi_i}{\partial x_k}(\psi^{-1}(y))\frac{\partial v}{\partial x_r}(\psi^{-1}(y))\Big)\\
&= \sum_{k,r=1}^{n} \big[\frac{\partial}{\partial x_k}\gamma^{k,r}(x)\frac{\partial v}{\partial x_r}(x)\big]_{x=\psi^{-1}(y)}\\
&\quad + \sum_{i,k,r=1}^{n} \gamma^{k,r}(\psi^{-1}(y))\frac{\partial v}{\partial x_r}(\psi^{-1}(y))\frac{\partial}{\partial y_i}\frac{\partial \psi_i}{\partial x_k}(\psi^{-1}(y)).
\end{aligned}$$

For the final equality, we used the product rule (using $\frac{\partial \psi_i}{\partial x_k}(\psi^{-1}(y))$ and

$$\gamma^{k,r}(\psi^{-1}(y))\frac{\partial v}{\partial x_r}(\psi^{-1}(y))$$

as the two factors) followed by a second application of (D.19) with $v(x)$ replaced by $\gamma^{k,r}(x)\frac{\partial v}{\partial x_r}(x)$, j replaced by i and r replaced by k. Substitute in

$$\begin{aligned}
&-\sum_{k,r=1}^{n} \frac{\partial}{\partial x_k}\gamma^{k,r}(x)\frac{\partial v}{\partial x_r}(x) + \mu(x)v(x)\\
&= -\sum_{k,r=1}^{n} \frac{\partial}{\partial x_k}\gamma^{k,r}\chi(1-\varphi)\frac{\partial u}{\partial x_r} - \sum_{k,r=1}^{n} \frac{\partial}{\partial x_k}\gamma^{k,r}u\frac{\partial}{\partial x_r}\chi(1-\varphi) + \mu v\\
&= \chi(1-\varphi)Lu - \sum_{k,r=1}^{n} \gamma^{k,r}\frac{\partial u}{\partial x_r}\frac{\partial}{\partial x_k}\chi(1-\varphi)\\
&\quad - \sum_{k,r=1}^{n} \gamma^{k,r}\frac{\partial u}{\partial x_k}\frac{\partial}{\partial x_r}\chi(1-\varphi) - \sum_{k,r=1}^{n} u\frac{\partial}{\partial x_k}\gamma^{k,r}\frac{\partial}{\partial x_r}\chi(1-\varphi)\\
&= \chi(1-\varphi)F - 2\sum_{i,j=1}^{n} \gamma^{i,j}\frac{\partial u}{\partial x_i}\frac{\partial}{\partial x_j}\chi(1-\varphi) - \sum_{k,r=1}^{n} u\frac{\partial}{\partial x_k}\gamma^{k,r}\frac{\partial}{\partial x_r}\chi(1-\varphi)
\end{aligned}$$

to give, with $\tilde{\mu}(y) = \mu(\psi^{-1}(y))$,

$$L_{\tilde{\gamma},\tilde{\mu},+}w = G, \qquad w|_{\mathbb{R}^{n-1}} = \big(\chi(1-\varphi)f\big)\circ\psi^{-1} = (\chi f)\circ\psi^{-1}$$

(recall that φ is supported in the interior of Ω) where

$$\begin{aligned} G(y) &= \chi(1-\varphi)F\circ\psi^{-1} \\ &\quad - 2\sum_{i,j=1}^{n}\big[\gamma^{i,j}(x)\frac{\partial u}{\partial x_i}(x)\frac{\partial}{\partial x_j}\chi(x)(1-\varphi(x))\big]_{x=\psi^{-1}(y)} \\ &\quad - \sum_{k,r=1}^{n}\big[u\frac{\partial}{\partial x_k}\gamma^{k,r}\frac{\partial}{\partial x_r}\chi(1-\varphi)\big]_{x=\psi^{-1}(y)} \\ &\quad - \sum_{i,j,k=1}^{n}\gamma^{k,j}(\psi^{-1}(y))\frac{\partial v}{\partial x_j}(\psi^{-1}(y))\frac{\partial}{\partial y_i}\frac{\partial \psi_i}{\partial x_k}(\psi^{-1}(y)). \end{aligned}$$

The term $\chi_{i_0}(1-\varphi)u$ – bounds: By part (c) of Lemma C.1.48, Lemma C.1.32 and (D.17), we get the following bounds on the various parts of G:

$$\begin{aligned} \left\|\chi(1-\varphi)F\circ\psi^{-1}\right\|_{H^{\ell-1}(\mathbb{R}^n_+)} &\le c_2\,\|F\|_{H^{\ell-1}(\Omega)}, \\ \left\|[u\frac{\partial}{\partial x_k}\gamma^{k,r}\frac{\partial}{\partial x_r}\chi(1-\varphi)]\circ\psi^{-1}\right\|_{H^{\ell-1}(\mathbb{R}^n_+)} &\le c_3\,\|u\|_{H^{\ell-1}(\Omega)} \le c_3\,\|u\|_{H^{\ell}(\Omega)} \\ &\le c_3C\Big(\,\|F\|_{H^{\ell-2}(\Omega)} + \|f\|_{H^{\ell-1/2}(\partial\Omega)}\Big), \\ \left\|\left[\frac{\partial u}{\partial x_i}(x)\frac{\partial}{\partial x_j}\chi(x)(1-\varphi(x))\right]\circ\psi^{-1}\right\|_{H^{\ell-1}(\mathbb{R}^n_+)} & \\ &\le c_3\left\|\frac{\partial u}{\partial x_i}\right\|_{H^{\ell-1}(\Omega)} \le c_3\,\|u\|_{H^{\ell}(\Omega)} \\ &\le c_3C\Big(\,\|F\|_{H^{\ell-2}(\Omega)} + \|f\|_{H^{\ell-1/2}(\partial\Omega)}\Big), \\ \left\|\frac{\partial v}{\partial x_j}\circ\psi^{-1}\right\|_{H^{\ell-1}(\mathbb{R}^n_+)} &\le c_3\left[\left\|\frac{\partial u}{\partial x_j}\right\|_{H^{\ell-1}(\Omega)} + \|u\|_{H^{\ell-1}(\Omega)}\right] \\ &\le 2c_3\,\|u\|_{H^{\ell}(\Omega)} \\ &\le 2c_3C\Big(\,\|F\|_{H^{\ell-2}(\Omega)} + \|f\|_{H^{\ell-1/2}(\partial\Omega)}\Big). \end{aligned}$$

Consequently, by part (a) of Lemma C.1.29, $G \in H^{\ell-1}(\mathbb{R}^n_+)$ with

$$\begin{aligned} \|G\|_{H^{\ell-1}(\mathbb{R}^n_+)} &\le c_4\Big(\,\|F\|_{H^{\ell-1}(\Omega)} + \|F\|_{H^{\ell-2}(\Omega)} + \|f\|_{H^{\ell-1/2}(\partial\Omega)}\Big) \\ &\le 2c_4\Big(\,\|F\|_{H^{\ell-1}(\Omega)} + \|f\|_{H^{\ell-1/2}(\partial\Omega)}\Big). \end{aligned}$$

Let $\mathcal{U}$ be an open neighborhood of the support of χ whose closure is contained in $\mathcal{U}(p_{i_0})$. Then $\psi(\mathcal{U})\cap\mathbb{R}^n_+ = \psi(\mathcal{U}\cap\Omega)$ is a bounded open subset of $\mathbb{R}^n$

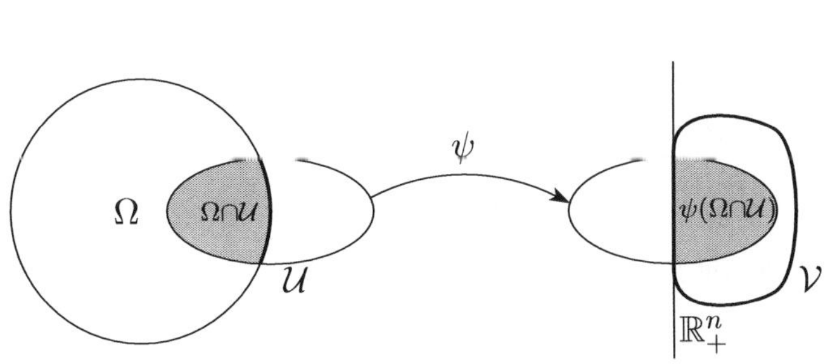

Figure D.1. The map ψ.

and w vanishes outside of $\psi(\mathcal{U} \cap \Omega)$. Enlarge $\psi(\mathcal{U}) \cap \mathbb{R}^n_+$ to $\mathcal{V}$, a bounded open subset of $\mathbb{R}^n_+$ with smooth boundary. See Figure D.1.

In Exercise D.3.4, we construct suitable extensions $\tilde{\gamma}'$ and $\tilde{\mu}'$ to $\mathbb{R}^n$ that coincide with $\tilde{\gamma}$ and $\tilde{\mu}$ on $\mathcal{V}$. Applying Proposition D.2.7 with $F = G$, $u = w$, γ replaced by $\tilde{\gamma}'$, μ replaced by $\tilde{\mu}'$, f replaced by $(\chi f) \circ \psi^{-1}$ and ℓ replaced by $\ell + 1$ gives

$$\begin{aligned}\|w\|_{H^{\ell+1}(\mathbb{R}^n_+)} &\le c_5\big(\|G\|_{H^{\ell-1}(\mathbb{R}^n_+)} + \|\chi f \circ \psi^{-1}\|_{H^{\ell+1/2}(\mathbb{R}^{n-1})}\big)\\ &\le c_6\big(\|F\|_{H^{\ell-1}(\Omega)} + \|f\|_{H^{\ell-1/2}(\partial\Omega)} + \|\chi f \circ \psi^{-1}\|_{H^{\ell+1/2}(\mathbb{R}^{n-1})}\big).\end{aligned}$$

By Definition C.1.37,

$$\|w\|_{H^{\ell+1}(\mathbb{R}^n_+)} \le 2c_6\big(\|F\|_{H^{\ell-1}(\Omega)} + \|f\|_{H^{\ell+1/2}(\partial\Omega)} \big).$$

Finally, by part (a) of Lemma C.1.48, with $\mathcal{U}$ replaced by $\mathcal{V}$, $\mathcal{V}$ replaced by $\psi^{-1}(\mathcal{V})$, ψ replaced by ψ^{-1} and ℓ replaced by $\ell + 1$,

$$\begin{aligned}\Big\|\chi_{i_0}(1-\varphi)u\Big\|_{H^{\ell+1}(\Omega)} &= \|w \circ \psi\|_{H^{\ell+1}(\psi^{-1}(\mathcal{V}))}\\ &\le \|w\|_{H^{\ell+1}(\mathcal{V})}\\ &= \|w\|_{H^{\ell+1}(\mathbb{R}^n_+)}\\ &\le 2c_6\big(\|F\|_{H^{\ell-1}(\Omega)} + \|f\|_{H^{\ell+1/2}(\partial\Omega)} \big). \qquad \square\end{aligned}$$

Exercise D.3.4. Let $\ell \ge 1$, let Ω be a bounded open subset of $\mathbb{R}^n$ with C^ℓ boundary, and let $\gamma \in W^{\ell,\infty}(\Omega)$, $\mu \in W^{\ell-1,\infty}(\Omega)$ obey (D.15) together with $\mu \ge c > 0$ in Ω. Prove that there exist $\gamma' \in W^{\ell,\infty}(\mathbb{R}^n)$ and $\mu' \in W^{\ell-1,\infty}(\mathbb{R}^n)$ that satisfy Hypothesis D.1.1 and coincide with γ and μ on $\overline{\Omega}$.

The next proposition shows that the regularity of u in some region, in the interior of its domain of definition, is really just determined by the regularity of F in essentially the same region.

Proposition D.3.5 (Interior regularity). *Let Ω and Ω' be bounded open subsets of $\mathbb{R}^n$ with $\overline{\Omega} \subset \Omega'$. Let $\ell \ge 2$ and assume that γ, μ satisfy* (D.15) *in Ω'*

with $\gamma \in W^{\ell-1,\infty}(\Omega')$ and $\mu \in W^{\ell-2,\infty}(\Omega')$. There is a constant C, depending only on ℓ, Ω, Ω', γ and μ, such that for all $u \in H^1(\Omega')$ and $F \in H^{\ell-2}(\Omega')$ obeying

$$Lu = F \text{ in } \Omega',$$

we have $u \in H^\ell(\Omega)$ and

$$\|u\|_{H^\ell(\Omega)} \leq C\Big(\|F\|_{H^{\ell-2}(\Omega')} + \|u\|_{L^2(\Omega')} \Big).$$

Proof. As in the proof of Proposition D.3.3, after possibly replacing L by $L+\lambda$ for a suitable constant $\lambda > 0$, we may assume the positivity condition $\mu \geq c > 0$ in Ω'. Similarly, applying induction in $\ell \geq 1$, we may assume that $u \in H^{\ell-1}(\Omega'')$ whenever $\overline{\Omega}'' \subset \Omega'$.

Let $\chi \in C_c^\infty(\mathbb{R}^n)$, $0 \leq \chi \leq 1$ be such that $\chi \equiv 1$ on Ω and $\operatorname{supp}(\chi) \subset \Omega'$. Set $v = \chi u$. Now we may replace Ω' by a slightly smaller domain, still containing $\operatorname{supp}(\chi)$, that has smooth boundary. We have already found, in Exercise D.3.4, a $\gamma' \in W^{\ell-1,\infty}(\mathbb{R}^n)$ and $\mu' \in W^{\ell-2,\infty}(\mathbb{R}^n)$ that satisfy Hypotheses D.1.1 and coincide with γ and μ on the support of χ. We denote by L the operator in $\mathbb{R}^n$ with coefficients γ' and μ'. As in the proof of Proposition D.3.3, we shall shortly find a $G \in H^{\ell-2}(\mathbb{R}^n)$ obeying

$$Lv = G \text{ in } \mathbb{R}^n$$

and

$$|||G|||_{H^{\ell-2}(\mathbb{R}^n)} \leq c\big(\|F\|_{H^{\ell-2}(\Omega')} + \|u\|_{H^{\ell-1}(\Omega')} \big).$$

Then by Proposition D.1.11 with $F = G$ and u replaced by v, we have

$$\begin{aligned} \|u\|_{H^\ell(\Omega)} &\leq |||v|||_{H^\ell(\mathbb{R}^n)} \\ &\leq C|||G|||_{H^{\ell-2}(\mathbb{R}^n)} \\ &\leq C\Big(\|F\|_{H^{\ell-2}(\Omega')} + \|u\|_{H^{\ell-1}(\Omega')} \Big). \end{aligned}$$

This proves the result.

To find G, we just have to replace φ by χ in (D.18). This gives

$$Lv = \chi F - 2\sum_{i,j=1}^n \gamma^{i,j} \frac{\partial \chi}{\partial x_i}\frac{\partial u}{\partial x_j} - \sum_{i,j=1}^n u \frac{\partial}{\partial x_i}\big[\gamma^{i,j}\frac{\partial \chi}{\partial x_j}\big] =: G.$$

Recalling that $\chi \in C_c^\infty(\Omega')$ and that, by Lemmas C.1.32 and C.1.29,

- multiplication by any function in $W^{|k|,\infty}(\Omega)$ is a bounded operator on $H^k(\Omega)$ for any $k \in \mathbb{Z}$,
- $\frac{\partial}{\partial x_j}$ is a bounded operator from $H^k(\Omega)$ to $H^{k-1}(\Omega)$ for all $k \in \mathbb{Z}$ and $1 \leq j \leq n$,

we have $G \in H^{\ell-2}(\mathbb{R}^n)$ such that whenever $\overline{\Omega}'' \subset \Omega'$,

$$|||G|||_{H^{\ell-2}(\mathbb{R}^n)} = \|G\|_{H^{\ell-2}(\Omega'')} \lesssim C\Big(\|F\|_{H^{\ell-2}(\Omega'')} + \|u\|_{H^{\ell-1}(\Omega'')} \Big). \qquad \square$$

Finally we prove, as a corollary to Proposition D.3.3, regularity up to the boundary, in the classical sense, of solutions of the Dirichlet problem.

Theorem D.3.6 (Regularity up to the boundary). *Let $\Omega \subset \mathbb{R}^n$ be a bounded open set with smooth boundary and let $\gamma, \mu \in C^\infty(\overline{\Omega})$ satisfy* (D.15). *Let $F \in C^\infty(\overline{\Omega})$ and $f \in C^\infty(\partial\Omega)$. Then the solution u to*

$$L_\gamma u = F, \qquad Ru = f,$$

is in $C^\infty(\overline{\Omega})$.

Proof. The conclusion follows immediately from Lemma C.2.22, since we already know, from Proposition D.3.3, that $u \in H^\ell(\Omega)$ for all $\ell \in \mathbb{N}$. □

Bibliography

[AC91] M. J. Ablowitz and P. A. Clarkson, *Solitons, nonlinear evolution equations and inverse scattering*, London Mathematical Society Lecture Note Series, vol. 149, Cambridge University Press, Cambridge, 1991, DOI 10.1017/CBO9780511623998. MR1149378

[AS64] M. Abramowitz and I. A. Stegun, *Handbook of mathematical functions with formulas, graphs, and mathematical tables*, National Bureau of Standards Applied Mathematics Series, No. 55, U. S. Government Printing Office, Washington, DC, 1964. MR167642

[AF03] R. A. Adams and J. J. F. Fournier, *Sobolev spaces*, 2nd ed., Pure and Applied Mathematics (Amsterdam), vol. 140, Elsevier/Academic Press, Amsterdam, 2003. MR2424078

[AH21] A. Adler and D. Holder (eds.), *Electrical Impedance Tomography: methods, history and applications*, 2nd ed., CRC Press, Boca Raton, 2021, DOI 10.1201/9780429399886.

[Agm75] S. Agmon, *Spectral properties of Schrödinger operators and scattering theory*, Ann. Scuola Norm. Sup. Pisa Cl. Sci. (4) **2** (1975), no. 2, 151–218. MR397194

[AÇL+20] J. P. Agnelli, A. Çöl, M. Lassas, R. Murthy, M. Santacesaria, and S. Siltanen, *Classification of stroke using neural networks in electrical impedance tomography*, Inverse Problems **36** (2020), no. 11, 115008, 26, DOI 10.1088/1361-6420/abbdcd. MR4173583

[Ale88] G. Alessandrini, *Stable determination of conductivity by boundary measurements*, Appl. Anal. **27** (1988), no. 1-3, 153–172, DOI 10.1080/00036818808839730. MR922775

[Ale90] G. Alessandrini, *Singular solutions of elliptic equations and the determination of conductivity by boundary measurements*, J. Differential Equations **84** (1990), no. 2, 252–272, DOI 10.1016/0022-0396(90)90078-4. MR1047569

[Ale07] G. Alessandrini, *Open issues of stability for the inverse conductivity problem*, J. Inverse Ill-Posed Probl. **15** (2007), no. 5, 451–460, DOI 10.1515/jiip.2007.025. MR2367859

[AdHG17] G. Alessandrini, M. V. de Hoop, and R. Gaburro, *Uniqueness for the electrostatic inverse boundary value problem with piecewise constant anisotropic conductivities*, Inverse Problems **33** (2017), no. 12, 125013, 24, DOI 10.1088/1361-6420/aa982d. MR3738499

[AK12] G. Alessandrini and K. Kim, *Single-logarithmic stability for the Calderón problem with local data*, J. Inverse Ill-Posed Probl. **20** (2012), no. 4, 389–400, DOI 10.1515/jip-2012-0014. MR2984474

[ARRV09] G. Alessandrini, L. Rondi, E. Rosset, and S. Vessella, *The stability for the Cauchy problem for elliptic equations*, Inverse Problems **25** (2009), no. 12, 123004, 47, DOI 10.1088/0266-5611/25/12/123004. MR2565570

[AV05] G. Alessandrini and S. Vessella, *Lipschitz stability for the inverse conductivity problem*, Adv. in Appl. Math. **35** (2005), no. 2, 207–241, DOI 10.1016/j.aam.2004.12.002. MR2152888

[AU04] H. Ammari and G. Uhlmann, *Reconstruction of the potential from partial Cauchy data for the Schrödinger equation*, Indiana Univ. Math. J. **53** (2004), no. 1, 169–183, DOI 10.1512/iumj.2004.53.2299. MR2048188

[AT18] J. Andersson and L. Tzou, *Stability for a magnetic Schrödinger operator on a Riemann surface with boundary*, Inverse Probl. Imaging **12** (2018), no. 1, 1–28, DOI 10.3934/ipi.2018001. MR3810147

[AFGS20] P. Angulo, D. Faraco, L. Guijarro, and M. Salo, *Limiting Carleman weights and conformally transversally anisotropic manifolds*, Trans. Amer. Math. Soc. **373** (2020), no. 7, 5171–5197, DOI 10.1090/tran/8072. MR4127874

[Arr99] S. R. Arridge, *Optical tomography in medical imaging*, Inverse Problems **15** (1999), no. 2, R41–R93, DOI 10.1088/0266-5611/15/2/022. MR1684463

[AS09] S. R. Arridge and J. C. Schotland, *Optical tomography: forward and inverse problems*, Inverse Problems **25** (2009), no. 12, 123010, 59, DOI 10.1088/0266-5611/25/12/123010. MR3460046

[AAS23] G. S. Alberti, Á. Arroyo, and M. Santacesaria, *Inverse problems on low-dimensional manifolds*, Nonlinearity **36** (2023), no. 1, 734–808, DOI 10.1088/1361-6544/aca73d. MR4521957

[AS19] G. S. Alberti and M. Santacesaria, *Calderón's inverse problem with a finite number of measurements*, Forum Math. Sigma **7** (2019), Paper No. e35, 20, DOI 10.1017/fms.2019.31. MR4016502

[AS22a] G. S. Alberti and M. Santacesaria, *Calderón's inverse problem with a finite number of measurements II: independent data*, Appl. Anal. **101** (2022), no. 10, 3636–3654, DOI 10.1080/00036811.2020.1745192. MR4446595

[AS22b] G. S. Alberti and M. Santacesaria, *Infinite-dimensional inverse problems with finite measurements*, Arch. Ration. Mech. Anal. **243** (2022), no. 1, 1–31, DOI 10.1007/s00205-021-01718-4. MR4359448

[AGTU13] P. Albin, C. Guillarmou, L. Tzou, and G. Uhlmann, *Inverse boundary problems for systems in two dimensions*, Ann. Henri Poincaré **14** (2013), no. 6, 1551–1571, DOI 10.1007/s00023-012-0229-1. MR3085925

[AFR16] K. Astala, D. Faraco, and K. M. Rogers, *Unbounded potential recovery in the plane* (English, with English and French summaries), Ann. Sci. Éc. Norm. Supér. (4) **49** (2016), no. 5, 1027–1051, DOI 10.24033/asens.2302. MR3581809

[AIM09] K. Astala, T. Iwaniec, and G. Martin, *Elliptic partial differential equations and quasiconformal mappings in the plane*, Princeton Mathematical Series, vol. 48, Princeton University Press, Princeton, NJ, 2009. MR2472875

[ALP13] K. Astala, M. Lassas, and L. Päivärinta, *Calderón's inverse problem: imaging and invisibility*, Inverse problems and applications: inside out. II, Math. Sci. Res. Inst. Publ., vol. 60, Cambridge Univ. Press, Cambridge, 2013, pp. 1–54. MR3098655

[ALP16] K. Astala, M. Lassas, and L. Päivärinta, *The borderlines of invisibility and visibility in Calderón's inverse problem*, Anal. PDE **9** (2016), no. 1, 43–98, DOI 10.2140/apde.2016.9.43. MR3461301

[AP06a] K. Astala and L. Päivärinta, *Calderón's inverse conductivity problem in the plane*, Ann. of Math. (2) **163** (2006), no. 1, 265–299, DOI 10.4007/annals.2006.163.265. MR2195135

[AP06b] K. Astala and L. Päivärinta, *A boundary integral equation for Calderón's inverse conductivity problem*, Collect. Math. **Vol. Extra** (2006), 127–139. MR2264207

[APL05] K. Astala, L. Päivärinta, and M. Lassas, *Calderón's inverse problem for anisotropic conductivity in the plane*, Comm. Partial Differential Equations **30** (2005), no. 1-3, 207–224, DOI 10.1081/PDE-200044485. MR2131051

[BBR01] J. A. Barceló, T. Barceló, and A. Ruiz, *Stability of the inverse conductivity problem in the plane for less regular conductivities*, J. Differential Equations **173** (2001), no. 2, 231–270, DOI 10.1006/jdeq.2000.3920. MR1834116

[BFR07] T. Barceló, D. Faraco, and A. Ruiz, *Stability of Calderón inverse conductivity problem in the plane* (English, with English and French summaries), J. Math. Pures Appl. (9) **88** (2007), no. 6, 522–556, DOI 10.1016/j.matpur.2007.07.006. MR2373740

[BC85] R. Beals and R. R. Coifman, *Multidimensional inverse scatterings and nonlinear partial differential equations*, Pseudodifferential operators and applications (Notre Dame, Ind., 1984), Proc. Sympos. Pure Math., vol. 43, Amer. Math. Soc., Providence, RI, 1985, pp. 45–70, DOI 10.1090/pspum/043/812283. MR812283

[BC89] R. Beals and R. R. Coifman, *Linear spectral problems, nonlinear equations and the $\bar{\partial}$-method*, Inverse Problems **5** (1989), no. 2, 87–130, DOI 10.1088/0266-5611/5/2/002. MR991913

[Bel03] M. I. Belishev, *The Calderon problem for two-dimensional manifolds by the BC-method*, SIAM J. Math. Anal. **35** (2003), no. 1, 172–182, DOI 10.1137/S0036141002413919. MR2001471

[Ber58] Yu. M. Berezanski, *The uniqueness theorem in the inverse problem of spectral analysis for the Schrödinger equation* (Russian), Trudy Moskov. Mat. Obšč. **7** (1958), 1–62. MR101377

[BKSU25] S. Bhattacharyya, K. Krupchyk, S. K. Sahoo, and G. Uhlmann, *Inverse problems for third-order nonlinear perturbations of biharmonic operators*, Comm. Partial Differential Equations **50** (2025), no. 3, 407–440, DOI 10.1080/03605302.2024.2444972. MR4870992

[BM08] J. Bikowski and J. L. Mueller, *2D EIT reconstructions using Calderón's method*, Inverse Probl. Imaging **2** (2008), no. 1, 43–61, DOI 10.3934/ipi.2008.2.43. MR2375322

[BIY15] E. Blåsten, O. Yu. Imanuvilov, and M. Yamamoto, *Stability and uniqueness for a two-dimensional inverse boundary value problem for less regular potentials*, Inverse Probl. Imaging **9** (2015), no. 3, 709–723, DOI 10.3934/ipi.2015.9.709. MR3406415

[BPS14] E. Blåsten, L. Päivärinta, and J. Sylvester, *Corners always scatter*, Comm. Math. Phys. **331** (2014), no. 2, 725–753, DOI 10.1007/s00220-014-2030-0. MR3238529

[BTW20] E. Blåsten, L. Tzou, and J.-N. Wang, *Uniqueness for the inverse boundary value problem with singular potentials in 2D*, Math. Z. **295** (2020), no. 3-4, 1521–1535, DOI 10.1007/s00209-019-02436-0. MR4125698

[Bor02] L. Borcea, *Electrical impedance tomography*, Inverse Problems **18** (2002), no. 6, R99–R136, DOI 10.1088/0266-5611/18/6/201. MR1955896

[Bra16] T. Brander, *Calderón's problem for p-laplace type equations*, University of Jyväskylä, 2016.

[Bre11] H. Brezis, *Functional analysis, Sobolev spaces and partial differential equations*, Universitext, Springer, New York, 2011. MR2759829

[Bro01] R. M. Brown, *Recovering the conductivity at the boundary from the Dirichlet to Neumann map: a pointwise result*, J. Inverse Ill-Posed Probl. **9** (2001), no. 6, 567–574, DOI 10.1515/jiip.2001.9.6.567. MR1881563

[BS06] R. M. Brown and M. Salo, *Identifiability at the boundary for first-order terms*, Appl. Anal. **85** (2006), no. 6-7, 735–749, DOI 10.1080/00036810600603377. MR2232419

[BT03] R. M. Brown and R. H. Torres, *Uniqueness in the inverse conductivity problem for conductivities with 3/2 derivatives in L^p, $p > 2n$*, J. Fourier Anal. Appl. **9** (2003), no. 6, 563–574, DOI 10.1007/s00041-003-0902-3. MR2026763

[BU97] R. M. Brown and G. A. Uhlmann, *Uniqueness in the inverse conductivity problem for nonsmooth conductivities in two dimensions*, Comm. Partial Differential Equations **22** (1997), no. 5-6, 1009–1027, DOI 10.1080/03605309708821292. MR1452176

[Buk08] A. L. Bukhgeim, *Recovering a potential from Cauchy data in the two-dimensional case*, J. Inverse Ill-Posed Probl. **16** (2008), no. 1, 19–33, DOI 10.1515/jiip.2008.002. MR2387648

[BU02] A. L. Bukhgeim and G. Uhlmann, *Recovering a potential from partial Cauchy data*, Comm. Partial Differential Equations **27** (2002), no. 3-4, 653–668, DOI 10.1081/PDE-120002868. MR1900557

[Cal80] A.-P. Calderón, *On an inverse boundary value problem*, Seminar on Numerical Analysis and its Applications to Continuum Physics (Rio de Janeiro, 1980), Soc. Brasil. Mat., Rio de Janeiro, 1980, pp. 65–73. MR590275

[Cal08] A. P. Calderón, *Selected papers of Alberto P. Calderón*, American Mathematical Society, Providence, RI, 2008. With commentary; Edited by Alexandra Bellow, Carlos E. Kenig and Paul Malliavin. MR2435328

[CDJ63] J. R. Cannon, J. Douglas Jr., and B. F. Jones Jr., *Determination of the diffusivity of an isotropic medium* (English, with French, German, Italian, and Russian summaries), Internat. J. Engrg. Sci. **1** (1963), 453–455, DOI 10.1016/0020-7225(63)90002-8. MR160045

[Car39] T. Carleman, *Sur un problème d'unicité pur les systèmes d'équations aux dérivées partielles à deux variables indépendantes* (French), Ark. Mat. Astr. Fys. **26** (1939), no. 17, 9. MR334

[Car11] P. Caro, *On an inverse problem in electromagnetism with local data: stability and uniqueness*, Inverse Probl. Imaging **5** (2011), no. 2, 297–322, DOI 10.3934/ipi.2011.5.297. MR2805355

[CDSFR14] P. Caro, D. Dos Santos Ferreira, and A. Ruiz, *Stability estimates for the Radon transform with restricted data and applications*, Adv. Math. **267** (2014), 523–564, DOI 10.1016/j.aim.2014.08.009. MR3269187

[CDSFR16] P. Caro, D. Dos Santos Ferreira, and A. Ruiz, *Stability estimates for the Calderón problem with partial data*, J. Differential Equations **260** (2016), no. 3, 2457–2489, DOI 10.1016/j.jde.2015.10.007. MR3427672

[CLLZ21] P. Caro, R.-Y. Lai, Y.-H. Lin, and T. Zhou, *Boundary determination of electromagnetic and Lamé parameters with corrupted data*, Inverse Probl. Imaging **15** (2021), no. 5, 1171–1198, DOI 10.3934/ipi.2021033. MR4301287

[CM16] P. Caro and K. Marinov, *Stability of inverse problems in an infinite slab with partial data*, Comm. Partial Differential Equations **41** (2016), no. 4, 683–704, DOI 10.1080/03605302.2015.1127967. MR3488251

[COS09] P. Caro, P. Ola, and M. Salo, *Inverse boundary value problem for Maxwell equations with local data*, Comm. Partial Differential Equations **34** (2009), no. 10-12, 1425–1464, DOI 10.1080/03605300903296272. MR2581979

[CR16] P. Caro and K. M. Rogers, *Global uniqueness for the Calderón problem with Lipschitz conductivities*, Forum Math. Pi **4** (2016), e2, 28, DOI 10.1017/fmp.2015.9. MR3456182

[CF24] C. I. Cârstea and A. Feizmohammadi, *Two uniqueness results in the inverse boundary value problem for the weighted p-Laplace equation*, Preprint, `arXiv:2405.04123`, 2024.

[CFK+21] C. I. Cârstea, A. Feizmohammadi, Y. Kian, K. Krupchyk, and G. Uhlmann, *The Calderón inverse problem for isotropic quasilinear conductivities*, Adv. Math. **391** (2021), Paper No. 107956, 31, DOI 10.1016/j.aim.2021.107956. MR4300916

[CFO23] C. I. Cârstea, A. Feizmohammadi, and L. Oksanen, *Remarks on the anisotropic Calderón problem*, Proc. Amer. Math. Soc. **151** (2023), no. 10, 4461–4473, DOI 10.1090/proc/16455. MR4643331

[CGU23] C. I. Cârstea, T. Ghosh, and G. Uhlmann, *An inverse problem for the porous medium equation with partial data and a possibly singular absorption term*, SIAM J. Math. Anal. **55** (2023), no. 1, 162–185, DOI 10.1137/21M1465573. MR4538897

[CLT24] C. I. Cârstea, T. Liimatainen, and L. Tzou, *The Calderón problem on Riemannian surfaces and of minimal surfaces*, Preprint, `arXiv:2406.16944`, 2024.

[CW18] C. I. Cârstea and J.-N. Wang, *Uniqueness for the two dimensional Calderón's problem with unbounded conductivities*, Ann. Sc. Norm. Super. Pisa Cl. Sci. (5) **18** (2018), no. 4, 1459–1482. MR3829752

[CJSZ23] S. Cen, B. Jin, K. Shin, and Z. Zhou, *Electrical impedance tomography with deep Calderón method*, J. Comput. Phys. **493** (2023), Paper No. 112427, 14, DOI 10.1016/j.jcp.2023.112427. MR4643423

[CCPR97] K. Chadan, D. Colton, L. Päivärinta, and W. Rundell, *An introduction to inverse scattering and inverse spectral problems*, SIAM Monographs on Mathematical Modeling and Computation, Society for Industrial and Applied Mathematics (SIAM), Philadelphia, PA, 1997, DOI 10.1137/1.9780898719710. MR1445771

[Cha90] S. Chanillo, *A problem in electrical prospection and an n-dimensional Borg-Levinson theorem*, Proc. Amer. Math. Soc. **108** (1990), no. 3, 761–767, DOI 10.2307/2047798. MR998731

[Cho09] M. Choulli, *Une introduction aux problèmes inverses elliptiques et paraboliques* (French), Mathématiques & Applications (Berlin) [Mathematics & Applications], vol. 65, Springer-Verlag, Berlin, 2009, DOI 10.1007/978-3-642-02460-3. MR2554831

[Chu14a] F. J. Chung, *A partial data result for the magnetic Schrödinger inverse problem*, Anal. PDE **7** (2014), no. 1, 117–157, DOI 10.2140/apde.2014.7.117. MR3219502

[Chu14b] F. J. Chung, *Partial data for the Neumann-Dirichlet magnetic Schrödinger inverse problem*, Inverse Probl. Imaging **8** (2014), no. 4, 959–989, DOI 10.3934/ipi.2014.8.959. MR3295954

[COST18] F. J. Chung, P. Ola, M. Salo, and L. Tzou, *Partial data inverse problems for Maxwell equations via Carleman estimates* (English, with English and French summaries), Ann. Inst. H. Poincaré C Anal. Non Linéaire **35** (2018), no. 3, 605–624, DOI 10.1016/j.anihpc.2017.06.005. MR3778644

[CST17] F. J. Chung, M. Salo, and L. Tzou, *Partial data inverse problems for the Hodge Laplacian*, Anal. PDE **10** (2017), no. 1, 43–93, DOI 10.2140/apde.2017.10.43. MR3611013

[CT20] F. J. Chung and L. Tzou, *Partial data inverse problem with $L^{n/2}$ potentials*, Trans. Amer. Math. Soc. Ser. B **7** (2020), 97–132, DOI 10.1090/btran/39. MR4147582

[CFR10] A. Clop, D. Faraco, and A. Ruiz, *Stability of Calderón's inverse conductivity problem in the plane for discontinuous conductivities*, Inverse Probl. Imaging **4** (2010), no. 1, 49–91, DOI 10.3934/ipi.2010.4.49. MR2592783

[CK19] D. Colton and R. Kress, *Inverse acoustic and electromagnetic scattering theory*, Applied Mathematical Sciences, vol. 93, Springer, Cham, 2019. Fourth edition of [MR1183732], DOI 10.1007/978-3-030-30351-8. MR3971246

[Con90] J. B. Conway, *A course in functional analysis*, 2nd ed., Graduate Texts in Mathematics, vol. 96, Springer-Verlag, New York, 1990. MR1070713

[CGLA09] E. L. V. Costa, R. Gonzalez Lima, and M. B. P. Amato, *Electrical impedance tomography*, Curr. Opin. Crit. Care **15** (2009), no. 1, 18–24.

[Cov25] G. Covi, *The inverse problem for the fractional conductivity equation: A survey*, Trends Math. **11** (2025), 159–165, DOI 10.1007/978-3-031-87213-6_20.

[DHKN24] T. Daudé, B. Helffer, N. Kamran, and F. Nicoleau, *Global counterexamples to uniqueness for a Calderón problem with C^k conductivities*, Preprint, `arXiv:2406.14063`, 2024.

[DKN19a] T. Daudé, N. Kamran, and F. Nicoleau, *Non-uniqueness results for the anisotropic Calderón problem with data measured on disjoint sets* (English, with English and French summaries), Ann. Inst. Fourier (Grenoble) **69** (2019), no. 1, 119–170, DOI 10.5802/aif.3240. MR3973447

[DKN19b] T. Daudé, N. Kamran, and F. Nicoleau, *On the hidden mechanism behind non-uniqueness for the anisotropic Calderón problem with data on disjoint sets*, Ann. Henri Poincaré **20** (2019), no. 3, 859–887, DOI 10.1007/s00023-018-00755-2. MR3916964

[DKN20] T. Daudé, N. Kamran, and F. Nicoleau, *On nonuniqueness for the anisotropic Calderón problem with partial data*, Forum Math. Sigma **8** (2020), Paper No. e7, 17, DOI 10.1017/fms.2020.1. MR4061977

[DSFKS13] D. Dos Santos Ferreira, C. E. Kenig, and M. Salo, *Determining an unbounded potential from Cauchy data in admissible geometries*, Comm. Partial Differential Equations **38** (2013), no. 1, 50–68, DOI 10.1080/03605302.2012.736911. MR3005546

[DSFKSU09] D. Dos Santos Ferreira, C. E. Kenig, M. Salo, and G. Uhlmann, *Limiting Carleman weights and anisotropic inverse problems*, Invent. Math. **178** (2009), no. 1, 119–171, DOI 10.1007/s00222-009-0196-4. MR2534094

[DSFKSU07] D. Dos Santos Ferreira, C. E. Kenig, J. Sjöstrand, and G. Uhlmann, *Determining a magnetic Schrödinger operator from partial Cauchy data*, Comm. Math. Phys. **271** (2007), no. 2, 467–488, DOI 10.1007/s00220-006-0151-9. MR2287913

[DSFKSU09] D. Dos Santos Ferreira, C. E. Kenig, J. Sjöstrand, and G. Uhlmann, *On the linearized local Calderón problem*, Math. Res. Lett. **16** (2009), no. 6, 955–970, DOI 10.4310/MRL.2009.v16.n6.a4. MR2576684

[DSFKLS16] D. Dos Santos Ferreira, Y. Kurylev, M. Lassas, and M. Salo, *The Calderón problem in transversally anisotropic geometries*, J. Eur. Math. Soc. (JEMS) **18** (2016), no. 11, 2579–2626, DOI 10.4171/JEMS/649. MR3562352

[Dru82] V. Druskin, *The unique solution of the inverse problem of electrical surveying and electrical well-logging for piecewise-continuous conductivity*, Earth Physics **18** (1982), no. 1.

[Dru85] V. L. Druskin, *Uniqueness of the determination of three-dimensional underground structures from surface measurements for a stationary or monochromatic field source* (Russian), Izv. Akad. Nauk SSSR Ser. Fiz. Zemli **3** (1985), 63–69. MR788076

[DZ19] S. Dyatlov and M. Zworski, *Mathematical theory of scattering resonances*, Graduate Studies in Mathematics, vol. 200, American Mathematical Society, Providence, RI, 2019, DOI 10.1090/gsm/200. MR3969938

[ER92] G. Eskin and J. Ralston, *Inverse backscattering: Festschrift on the occasion of the 70th birthday of Shmuel Agmon*, J. Anal. Math. **58** (1992), 177–190, DOI 10.1007/BF02790363. MR1226942

[ER95] G. Eskin and J. Ralston, *Inverse scattering problem for the Schrödinger equation with magnetic potential at a fixed energy*, Comm. Math. Phys. **173** (1995), no. 1, 199–224. MR1355624

[ER97] G. Eskin and J. Ralston, *Inverse scattering problems for Schrödinger operators with magnetic and electric potentials*, Inverse problems in wave propagation (Minneapolis, MN, 1995), IMA Vol. Math. Appl., vol. 90, Springer, New York, 1997, pp. 147–166, DOI 10.1007/978-1-4612-1878-4_7. MR1491671

[ER02] G. Eskin and J. Ralston, *On the inverse boundary value problem for linear isotropic elasticity*, Inverse Problems **18** (2002), no. 3, 907–921, DOI 10.1088/0266-5611/18/3/324. MR1910209

[ER03] G. Eskin and J. Ralston, *Inverse boundary value problems for systems of partial differential equations*, Recent development in theories & numerics, World Sci. Publ., River Edge, NJ, 2003, pp. 105–113, DOI 10.1142/9789812704924_0009. MR2088194

[Eva10] L. C. Evans, *Partial differential equations*, 2nd ed., Graduate Studies in Mathematics, vol. 19, American Mathematical Society, Providence, RI, 2010, DOI 10.1090/gsm/019. MR2597943

[Fad56] L. D. Faddeev, *Uniqueness of solution of the inverse scattering problem* (Russian), Vestnik Leningrad. Univ. **11** (1956), no. 7, 126–130. MR80533

[Fad16] L. D. Faddeev, *Increasing solutions of the Schroedinger equation*, Fifty years of mathematical physics, World Sci. Ser. 21st Century Math., vol. 2, World Sci. Publ., Hackensack, NJ, 2016, pp. 34–36. Translated from Dokl. Akad. Nauk SSSR **165** (1965), no. 3, 514–517. MR3524286

[Fad74] L. D. Faddeev, *The inverse problem in the quantum theory of scattering. II* (Russian), Current problems in mathematics, Vol. 3 (Russian), Itogi Nauki i Tekhniki, Akad. Nauk SSSR Vsesojuz. Inst. Naučn. i Tehn. Informacii, Moscow, 1974, pp. 93–180, 259. (loose errata). MR523015

[FP18] D. Faraco and M. Prats, *Characterization for stability in planar conductivities*, J. Differential Equations **264** (2018), no. 9, 5659–5712, DOI 10.1016/j.jde.2018.01.013. MR3765762

[FR13] D. Faraco and K. M. Rogers, *The Sobolev norm of characteristic functions with applications to the Calderón inverse problem*, Q. J. Math. **64** (2013), no. 1, 133–147, DOI 10.1093/qmath/har039. MR3032092

[Fat07] I. K. Fathallah, *Stability for the inverse potential problem by the local Dirichlet-to-Neumann map for the Schrödinger equation*, Appl. Anal. **86** (2007), no. 7, 899–914, DOI 10.1080/00036810701497067. MR2355545

[FGKU21] A. Feizmohammadi, T. Ghosh, K. Krupchyk, and G. Uhlmann, *Fractional anisotropic Calderón problem on closed Riemannian manifolds*, To appear in J. Differential Geom., arXiv:2112.03480, (2021).

[FO20] A. Feizmohammadi and L. Oksanen, *An inverse problem for a semi-linear elliptic equation in Riemannian geometries*, J. Differential Equations **269** (2020), no. 6, 4683–4719, DOI 10.1016/j.jde.2020.03.037. MR4104456

[FA84] A. S. Fokas and M. J. Ablowitz, *On the inverse scattering transform of multidimensional nonlinear equations related to first-order systems in the plane*, J. Math. Phys. **25** (1984), no. 8, 2494–2505, DOI 10.1063/1.526471. MR751539

[Fol99] G. B. Folland, *Real analysis*, 2nd ed., Pure and Applied Mathematics (New York), John Wiley & Sons, Inc., New York, 1999. Modern techniques and their applications; A Wiley-Interscience Publication. MR1681462

[Fre00] I. Frerichs, *Electrical impedance tomography (EIT) in applications related to lung and ventilation: a review of experimental and clinical activities*, Physiol. Meas. **21** (2000), no. 2, R1–21.

[GZ16] A. García and G. Zhang, *Reconstruction from boundary measurements for less regular conductivities*, Inverse Problems **32** (2016), no. 11, 115015, 22, DOI 10.1088/0266-5611/32/11/115015. MR3627054

[GGKM67] C. S. Gardner, J. M. Greene, M. D. Kruskal, and R. M. Miura, *Method for Solving the Korteweg-deVries Equation*, Phys. Rev. Lett. **19** (1967), 1095–1097, DOI 10.1103/PhysRevLett.19.1095.

[Geb08] B. Gebauer, *Localized potentials in electrical impedance tomography*, Inverse Probl. Imaging **2** (2008), no. 2, 251–269, DOI 10.3934/ipi.2008.2.251. MR2395143

[GP17] A. Girouard and I. Polterovich, *Spectral geometry of the Steklov problem [reprint of MR3662010]*, Shape optimization and spectral theory, De Gruyter Open, Warsaw, 2017, pp. 120–148, DOI 10.1515/9783110550887-005. MR3681149

[GRSU20] T. Ghosh, A. Rüland, M. Salo, and G. Uhlmann, *Uniqueness and reconstruction for the fractional Calderón problem with a single measurement*, J. Funct. Anal. **279** (2020), no. 1, 108505, 42, DOI 10.1016/j.jfa.2020.108505. MR4083776

[GSU20] T. Ghosh, M. Salo, and G. Uhlmann, *The Calderón problem for the fractional Schrödinger equation*, Anal. PDE **13** (2020), no. 2, 455–475, DOI 10.2140/apde.2020.13.455. MR4078233

[GT01] D. Gilbarg and N. S. Trudinger, *Elliptic partial differential equations of second order*, Classics in Mathematics, Springer-Verlag, Berlin, 2001. Reprint of the 1998 edition. MR1814364

[GKLU09] A. Greenleaf, Y. Kurylev, M. Lassas, and G. Uhlmann, *Invisibility and inverse problems*, Bull. Amer. Math. Soc. (N.S.) **46** (2009), no. 1, 55–97, DOI 10.1090/S0273-0979-08-01232-9. MR2457072

[GLS+18] A. Greenleaf, M. Lassas, M. Santacesaria, S. Siltanen, and G. Uhlmann, *Propagation and recovery of singularities in the inverse conductivity problem*, Anal. PDE **11** (2018), no. 8, 1901–1943, DOI 10.2140/apde.2018.11.1901. MR3812861

[GLU03a] A. Greenleaf, M. Lassas, and G. Uhlmann, *Anisotropic conductivities that cannot be detected by EIT*, Physiological Measurement **24** (2003), no. 2, 413, DOI 10.1088/0967-3334/24/2/353.

[GLU03b] A. Greenleaf, M. Lassas, and G. Uhlmann, *On nonuniqueness for Calderón's inverse problem*, Math. Res. Lett. **10** (2003), no. 5-6, 685–693, DOI 10.4310/MRL.2003.v10.n5.a11. MR2024725

[GLU03c] A. Greenleaf, M. Lassas, and G. Uhlmann, *The Calderón problem for conormal potentials. I. Global uniqueness and reconstruction*, Comm. Pure Appl. Math. **56** (2003), no. 3, 328–352, DOI 10.1002/cpa.10061. MR1941812

[GU93] A. Greenleaf and G. Uhlmann, *Recovering singularities of a potential from singularities of scattering data*, Comm. Math. Phys. **157** (1993), no. 3, 549–572. MR1243710

[GU01] A. Greenleaf and G. Uhlmann, *Local uniqueness for the Dirichlet-to-Neumann map via the two-plane transform*, Duke Math. J. **108** (2001), no. 3, 599–617, DOI 10.1215/S0012-7094-01-10837-5. MR1838663

[GS94] A. Grigis and J. Sjöstrand, *Microlocal analysis for differential operators: An introduction*, London Mathematical Society Lecture Note Series, vol. 196, Cambridge University Press, Cambridge, 1994, DOI 10.1017/CBO9780511721441. MR1269107

[GN95] P. G. Grinevich and R. G. Novikov, *Transparent potentials at fixed energy in dimension two. Fixed-energy dispersion relations for the fast decaying potentials*, Comm. Math. Phys. **174** (1995), no. 2, 409–446. MR1362172

[Gri85] P. Grisvard, *Elliptic problems in nonsmooth domains*, Monographs and Studies in Mathematics, vol. 24, Pitman (Advanced Publishing Program), Boston, MA, 1985. MR775683

[GSB08] C. Guillarmou and A. Sá Barreto, *Scattering and inverse scattering on ACH manifolds*, J. Reine Angew. Math. **622** (2008), 1–55, DOI 10.1515/CRELLE.2008.064. MR2433611

[GST11] C. Guillarmou, M. Salo, and L. Tzou, *Inverse scattering at fixed energy on surfaces with Euclidean ends*, Comm. Math. Phys. **303** (2011), no. 3, 761–784, DOI 10.1007/s00220-011-1224-y. MR2786216

[GT10] C. Guillarmou and L. Tzou, *Calderón inverse problem for the Schrödinger operator on Riemann surfaces*, The AMSI-ANU Workshop on Spectral Theory and Harmonic Analysis, Proc. Centre Math. Appl. Austral. Nat. Univ., vol. 44, Austral. Nat. Univ., Canberra, 2010, pp. 129–141. MR2655386

[GT11a] C. Guillarmou and L. Tzou, *Calderón inverse problem with partial data on Riemann surfaces*, Duke Math. J. **158** (2011), no. 1, 83–120, DOI 10.1215/00127094-1276310. MR2794369

[GT11b] C. Guillarmou and L. Tzou, *Identification of a connection from Cauchy data on a Riemann surface with boundary*, Geom. Funct. Anal. **21** (2011), no. 2, 393–418, DOI 10.1007/s00039-011-0110-2. MR2795512

[GT13] C. Guillarmou and L. Tzou, *The Calderón inverse problem in two dimensions*, Inverse problems and applications: inside out. II, Math. Sci. Res. Inst. Publ., vol. 60, Cambridge Univ. Press, Cambridge, 2013, pp. 119–166. MR3098657

[Hab15a] B. E. Haberman, *Inverse problems with rough data*, ProQuest LLC, Ann Arbor, MI, 2015. Thesis (Ph.D.)–University of California, Berkeley. MR3427278

[Hab15b] B. Haberman, *Uniqueness in Calderón's problem for conductivities with unbounded gradient*, Comm. Math. Phys. **340** (2015), no. 2, 639–659, DOI 10.1007/s00220-015-2460-3. MR3397029

[Hab18] B. Haberman, *Unique determination of a magnetic Schrödinger operator with unbounded magnetic potential from boundary data*, Int. Math. Res. Not. IMRN **4** (2018), 1080–1128, DOI 10.1093/imrn/rnw263. MR3801456

[HT13] B. Haberman and D. Tataru, *Uniqueness in Calderón's problem with Lipschitz conductivities*, Duke Math. J. **162** (2013), no. 3, 496–516, DOI 10.1215/00127094-2019591. MR3024091

[Had53] J. Hadamard, *Lectures on Cauchy's problem in linear partial differential equations*, Dover Publications, New York, 1953. MR51411

[Häh96] P. Hähner, *A periodic Faddeev-type solution operator*, J. Differential Equations **128** (1996), no. 1, 300–308, DOI 10.1006/jdeq.1996.0096. MR1392403

[HSPG14] M. Hallaji, A. Seppänen, and M. Pour-Ghaz, *Electrical impedance tomography-based sensing skin for quantitative imaging of damage in concrete*, Smart Materials and Structures **23** (2014), no. 8, 085001.

[Har12] B. Harrach, *Simultaneous determination of the diffusion and absorption coefficient from boundary data*, Inverse Probl. Imaging **6** (2012), no. 4, 663–679, DOI 10.3934/ipi.2012.6.663. MR2998710

[Har19] B. Harrach, *Uniqueness and Lipschitz stability in electrical impedance tomography with finitely many electrodes*, Inverse Problems **35** (2019), no. 2, 024005, 19, DOI 10.1088/1361-6420/aaf6fc. MR3900209

[Har23] B. Harrach, *The Calderón problem with finitely many unknowns is equivalent to convex semidefinite optimization*, SIAM J. Math. Anal. **55** (2023), no. 5, 5666–5684, DOI 10.1137/23M1544854. MR4650908

[HLW07] H. Heck, X. Li, and J.-N. Wang, *Identification of viscosity in an incompressible fluid*, Indiana Univ. Math. J. **56** (2007), no. 5, 2489–2510, DOI 10.1512/iumj.2007.56.3037. MR2360617

[HW06] H. Heck and J.-N. Wang, *Stability estimates for the inverse boundary value problem by partial Cauchy data*, Inverse Problems **22** (2006), no. 5, 1787–1796, DOI 10.1088/0266-5611/22/5/015. MR2261266

[HW16] H. Heck and J.-N. Wang, *Optimal stability estimate of the inverse boundary value problem by partial measurements*, Rend. Istit. Mat. Univ. Trieste **48** (2016), 369–383, DOI 10.13137/2464-8728/13164. MR3592451

[HM07] G. Henkin and V. Michel, *On the explicit reconstruction of a Riemann surface from its Dirichlet-Neumann operator*, Geom. Funct. Anal. **17** (2007), no. 1, 116–155, DOI 10.1007/s00039-006-0590-7. MR2306654

[HM08] G. Henkin and V. Michel, *Inverse conductivity problem on Riemann surfaces*, J. Geom. Anal. **18** (2008), no. 4, 1033–1052, DOI 10.1007/s12220-008-9035-x. MR2438910

[HN11] G. M. Henkin and R. G. Novikov, *On the reconstruction of conductivity of a bordered two-dimensional surface in* $\mathbb{R}^3$ *from electrical current measurements on its boundary*, J. Geom. Anal. **21** (2011), no. 3, 543–587, DOI 10.1007/s12220-010-9158-8. MR2810843

[HS10] G. Henkin and M. Santacesaria, *On an inverse problem for anisotropic conductivity in the plane*, Inverse Problems **26** (2010), no. 9, 095011, 18, DOI 10.1088/0266-5611/26/9/095011. MR2671796

[HS12] G. Henkin and M. Santacesaria, *Gel′fand-Calderón's inverse problem for anisotropic conductivities on bordered surfaces in* $\mathbb{R}^3$, Int. Math. Res. Not. IMRN **4** (2012), 781–809, DOI 10.1093/imrn/rnr046. MR2889157

[Hör03a] L. Hörmander, *The analysis of linear partial differential operators. I*, Classics in Mathematics, Springer-Verlag, Berlin, 2003. Distribution theory and Fourier analysis; Reprint of the second (1990) edition [Springer, Berlin; MR1065993 (91m:35001a)], DOI 10.1007/978-3-642-61497-2. MR1996773

[Hör03b] L. Hörmander, *The analysis of linear partial differential operators. I*, Classics in Mathematics, Springer-Verlag, Berlin, 2003. Distribution theory and Fourier analysis; Reprint of the second (1990) edition [Springer, Berlin; MR1065993 (91m:35001a)], DOI 10.1007/978-3-642-61497-2. MR1996773

[Hör85a] L. Hörmander, *The analysis of linear partial differential operators. III: Pseudodifferential operators*, Grundlehren der mathematischen Wissenschaften [Fundamental Principles of Mathematical Sciences], vol. 274, Springer-Verlag, Berlin, 1985. MR781536

[Hör85b] L. Hörmander, *The analysis of linear partial differential operators. IV: Fourier integral operators*, Grundlehren der mathematischen Wissenschaften [Fundamental Principles of Mathematical Sciences], vol. 275, Springer-Verlag, Berlin, 1985. MR781537

[Ike00] M. Ikehata, *Reconstruction of the support function for inclusion from boundary measurements*, J. Inverse Ill-Posed Probl. **8** (2000), no. 4, 367–378, DOI 10.1515/jiip.2000.8.4.367. MR1816720

[Ike01] M. Ikehata, *The enclosure method and its applications*, Analytic extension formulas and their applications (Fukuoka, 1999/Kyoto, 2000), Int. Soc. Anal. Appl. Comput., vol. 9, Kluwer Acad. Publ., Dordrecht, 2001, pp. 87–103, DOI 10.1007/978-1-4757-3298-6_7. MR1830379

[Ike10] M. Ikehata, *A note on the enclosure method for an inverse obstacle scattering problem with a single point source*, Inverse Problems **26** (2010), no. 10, 105006, 17, DOI 10.1088/0266-5611/26/10/105006. MR2679470

[IUY10] O. Yu. Imanuvilov, G. Uhlmann, and M. Yamamoto, *The Calderón problem with partial data in two dimensions*, J. Amer. Math. Soc. **23** (2010), no. 3, 655–691, DOI 10.1090/S0894-0347-10-00656-9. MR2629983

[IUY11] O. Yu. Imanuvilov, G. Uhlmann, and M. Yamamoto, *Inverse boundary value problem by measuring Dirichlet data and Neumann data on disjoint sets*, Inverse Problems **27** (2011), no. 8, 085007, 26, DOI 10.1088/0266-5611/27/8/085007. MR2819949

[IUY12] O. Imanuvilov, G. Uhlmann, and M. Yamamoto, *Partial Cauchy data for general second order elliptic operators in two dimensions*, Publ. Res. Inst. Math. Sci. **48** (2012), no. 4, 971–1055, DOI 10.2977/PRIMS/94. MR2999548

[IUY15] O. Yu. Imanuvilov, G. Uhlmann, and M. Yamamoto, *The Neumann-to-Dirichlet map in two dimensions*, Adv. Math. **281** (2015), 578–593, DOI 10.1016/j.aim.2015.03.026. MR3366847

[IY12a] O. Yu. Imanuvilov and M. Yamamoto, *Inverse boundary value problem for Schrödinger equation in two dimensions*, SIAM J. Math. Anal. **44** (2012), no. 3, 1333–1339, DOI 10.1137/11083736X. MR2982714

[IY12b] O. Yu. Imanuvilov and M. Yamamoto, *Inverse problem by Cauchy data on an arbitrary subboundary for systems of elliptic equations*, Inverse Problems **28** (2012), no. 9, 095015, 30, DOI 10.1088/0266-5611/28/9/095015. MR2972464

[IY13a] O. Yu. Imanuvilov and M. Yamamoto, *Inverse boundary value problem for the Schrödinger equation in a cylindrical domain by partial boundary data*, Inverse Problems **29** (2013), no. 4, 045002, 8, DOI 10.1088/0266-5611/29/4/045002. MR3041540

[IY13b] O. Y. Imanuvilov and M. Yamamoto, *Uniqueness for inverse boundary value problems by Dirichlet-to-Neumann map on subboundaries*, Milan J. Math. **81** (2013), no. 2, 187–258, DOI 10.1007/s00032-013-0205-3. MR3129784

[IY14] O. Yu. Imanuvilov and M. Yamamoto, *Calderón problem for Maxwell's equations in cylindrical domain*, Inverse Probl. Imaging **8** (2014), no. 4, 1117–1137, DOI 10.3934/ipi.2014.8.1117. MR3295960

[IY15] O. Yu. Imanuvilov and M. Yamamoto, *Global uniqueness in inverse boundary value problems for the Navier-Stokes equations and Lamé system in two dimensions*, Inverse Problems **31** (2015), no. 3, 035004, 46, DOI 10.1088/0266-5611/31/3/035004. MR3319370

[IY16] O. Y. Imanuvilov and M. Yamamoto, *Calderón problem for Maxwell's equations in two dimensions*, J. Inverse Ill-Posed Probl. **24** (2016), no. 3, 351–355, DOI 10.1515/jiip-2015-0042. MR3505723

[IY17] O. Imanuvilov and M. Yamamoto, *On Calderón's problem for a system of elliptic equations*, Publ. Res. Inst. Math. Sci. **53** (2017), no. 1, 141–186, DOI 10.4171/PRIMS/53-1-5. MR3615044

[II89] D. Isaacson and E. L. Isaacson, *Comment on A.-P. Calderón's paper: "On an inverse boundary value problem" [in Seminar on Numerical Analysis and its Applications to Continuum Physics (Rio de Janeiro, 1980), 65–73, Soc. Brasil. Mat., Rio de Janeiro, 1980; MR0590275 (81k:35160)]*, Math. Comp. **52** (1989), no. 186, 553–559, DOI 10.2307/2008482. MR962208

[Isa88] V. Isakov, *On uniqueness of recovery of a discontinuous conductivity coefficient*, Comm. Pure Appl. Math. **41** (1988), no. 7, 865–877, DOI 10.1002/cpa.3160410702. MR951742

[Isa91] V. Isakov, *Completeness of products of solutions and some inverse problems for PDE*, J. Differential Equations **92** (1991), no. 2, 305–316, DOI 10.1016/0022-0396(91)90051-A. MR1120907

[Isa93] V. Isakov, *On uniqueness in inverse problems for semilinear parabolic equations*, Arch. Rational Mech. Anal. **124** (1993), no. 1, 1–12, DOI 10.1007/BF00392201. MR1233645

[Isa07] V. Isakov, *On uniqueness in the inverse conductivity problem with local data*, Inverse Probl. Imaging **1** (2007), no. 1, 95–105, DOI 10.3934/ipi.2007.1.95. MR2262748

[Isa17] V. Isakov, *Inverse problems for partial differential equations*, 3rd ed., Applied Mathematical Sciences, vol. 127, Springer, Cham, 2017, DOI 10.1007/978-3-319-51658-5. MR3616276

[IS94] V. Isakov and J. Sylvester, *Global uniqueness for a semilinear elliptic inverse problem*, Comm. Pure Appl. Math. **47** (1994), no. 10, 1403–1410, DOI 10.1002/cpa.3160471005. MR1295934

[Iso97] H. Isozaki, *Inverse scattering theory for Dirac operators* (English, with English and French summaries), Ann. Inst. H. Poincaré Phys. Théor. **66** (1997), no. 2, 237–270. MR1449193

[Iso20] H. Isozaki, *Inverse spectral and scattering theory—an introduction*, SpringerBriefs in Mathematical Physics, vol. 38, Springer, Singapore, 2020, DOI 10.1007/978-981-15-8199-1. MR4177841

[IM01] T. Iwaniec and G. Martin, *Geometric function theory and non-linear analysis*, Oxford Mathematical Monographs, The Clarendon Press, Oxford University Press, New York, 2001. MR1859913

[JNS23] D. Johansson, J. Nurminen, and M. Salo, *Inverse problems for semilinear elliptic PDE with a general nonlinearity $a(x,u)$*, Preprint, arXiv:2312.12196, 2023.

[Jos00] M. S. Joshi, *Explicitly recovering asymptotics of short range potentials*, Comm. Partial Differential Equations **25** (2000), no. 9-10, 1907–1923, DOI 10.1080/03605300008821572. MR1778785

[JM00] M. S. Joshi and S. R. McDowall, *Total determination of material parameters from electromagnetic boundary information*, Pacific J. Math. **193** (2000), no. 1, 107–129, DOI 10.2140/pjm.2000.193.107. MR1748184

[JSB98] M. S. Joshi and A. Sá Barreto, *Recovering asymptotics of short range potentials*, Comm. Math. Phys. **193** (1998), no. 1, 197–208, DOI 10.1007/s002200050324. MR1620321

[JSB99] M. S. Joshi and A. Sá Barreto, *Determining asymptotics of magnetic fields from fixed energy scattering data*, Asymptot. Anal. **21** (1999), no. 1, 61–70, DOI 10.3233/asy-1999-364. MR1718632

[KR23] B. Kaltenbacher and W. Rundell, *Inverse problems for fractional partial differential equations*, Graduate Studies in Mathematics, vol. 230, American Mathematical Society, Providence, RI, 2023, DOI 10.1090/gsm/230. MR4604107

[KY02] H. Kang and K. Yun, *Boundary determination of conductivities and Riemannian metrics via local Dirichlet-to-Neumann operator*, SIAM J. Math. Anal. **34** (2002), no. 3, 719–735, DOI 10.1137/S0036141001395042. MR1970890

[KBK$^+$08] T.-J. Kao, G. Boverman, B. S. Kim, D. Isaacson, G. J Saulnier, J. C Newell, M. H Choi, R. H Moore, and D. B Kopans, *Regional admittivity spectra with tomosynthesis images for breast cancer detection: preliminary patient study*, IEEE Transactions on Medical Imaging **27** (2008), no. 12, 1762–1768.

[KSL$^+$10] K. Karhunen, A. Seppänen, A. Lehikoinen, P. J. M. Monteiro, and J. P. Kaipio, *Electrical resistance tomography imaging of concrete*, Cement and Concrete Research **40** (2010), 137–145.

[KKL01] A. Katchalov, Y. Kurylev, and M. Lassas, *Inverse boundary spectral problems*, Chapman & Hall/CRC Monographs and Surveys in Pure and Applied Mathematics, vol. 123, Chapman & Hall/CRC, Boca Raton, FL, 2001, DOI 10.1201/9781420036220. MR1889089

[KS13] C. Kenig and M. Salo, *The Calderón problem with partial data on manifolds and applications*, Anal. PDE **6** (2013), no. 8, 2003–2048, DOI 10.2140/apde.2013.6.2003. MR3198591

[KS14] C. Kenig and M. Salo, *Recent progress in the Calderón problem with partial data*, Inverse problems and applications, Contemp. Math., vol. 615, Amer. Math. Soc., Providence, RI, 2014, pp. 193–222, DOI 10.1090/conm/615/12245. MR3221605

[KSU11a] C. E. Kenig, M. Salo, and G. Uhlmann, *Inverse problems for the anisotropic Maxwell equations*, Duke Math. J. **157** (2011), no. 2, 369–419, DOI 10.1215/00127094-1272903. MR2783934

[KSU11b] C. E. Kenig, M. Salo, and G. Uhlmann, *Reconstructions from boundary measurements on admissible manifolds*, Inverse Probl. Imaging **5** (2011), no. 4, 859–877, DOI 10.3934/ipi.2011.5.859. MR2852376

[KSU07] C. E. Kenig, J. Sjöstrand, and G. Uhlmann, *The Calderón problem with partial data*, Ann. of Math. (2) **165** (2007), no. 2, 567–591, DOI 10.4007/annals.2007.165.567. MR2299741

[KKU23] Y. Kian, K. Krupchyk, and G. Uhlmann, *Partial data inverse problems for quasilinear conductivity equations*, Math. Ann. **385** (2023), no. 3-4, 1611–1638, DOI 10.1007/s00208-022-02367-y. MR4566701

[Kir11] A. Kirsch, *An introduction to the mathematical theory of inverse problems*, 2nd ed., Applied Mathematical Sciences, vol. 120, Springer, New York, 2011, DOI 10.1007/978-1-4419-8474-6. MR3025302

[Knu06] K. Knudsen, *The Calderón problem with partial data for less smooth conductivities*, Comm. Partial Differential Equations **31** (2006), no. 1-3, 57–71, DOI 10.1080/03605300500361610. MR2209749

[KS07] K. Knudsen and M. Salo, *Determining nonsmooth first order terms from partial boundary measurements*, Inverse Probl. Imaging **1** (2007), no. 2, 349–369, DOI 10.3934/ipi.2007.1.349. MR2282273

[KT04] K. Knudsen and A. Tamasan, *Reconstruction of less regular conductivities in the plane*, Comm. Partial Differential Equations **29** (2004), no. 3-4, 361–381, DOI 10.1081/PDE-120030401. MR2041600

[KRS21] H. Koch, A. Rüland, and M. Salo, *On instability mechanisms for inverse problems*, Ars Inven. Anal. (2021), Paper No. 7, 93. MR4462475

[KLQ$^+$17] I. Kocyigit, R.-Y. Lai, L. Qiu, Y. Yang, and T. Zhou, *Applications of CGO solutions to coupled-physics inverse problems*, Inverse Probl. Imaging **11** (2017), no. 2, 277–304, DOI 10.3934/ipi.2017014. MR3625583

[KV84a] R. Kohn and M. Vogelius, *Determining conductivity by boundary measurements*, Comm. Pure Appl. Math. **37** (1984), no. 3, 289–298, DOI 10.1002/cpa.3160370302. MR739921

[KV84b] R. Kohn and M. Vogelius, *Identification of an unknown conductivity by means of measurements at the boundary*, Inverse problems (New York, 1983), SIAM-AMS Proc., vol. 14, Amer. Math. Soc., Providence, RI, 1984, pp. 113–123, DOI 10.1002/cpa.3160370302. MR0773707

[KV85] R. V. Kohn and M. Vogelius, *Determining conductivity by boundary measurements. II. Interior results*, Comm. Pure Appl. Math. **38** (1985), no. 5, 643–667, DOI 10.1002/cpa.3160380513. MR803253

[KLU12a] K. Krupchyk, M. Lassas, and G. Uhlmann, *Determining a first order perturbation of the biharmonic operator by partial boundary measurements*, J. Funct. Anal. **262** (2012), no. 4, 1781–1801, DOI 10.1016/j.jfa.2011.11.021. MR2873860

[KLU12b] K. Krupchyk, M. Lassas, and G. Uhlmann, *Inverse problems with partial data for a magnetic Schrödinger operator in an infinite slab and on a bounded domain*, Comm. Math. Phys. **312** (2012), no. 1, 87–126, DOI 10.1007/s00220-012-1431-1. MR2914058

[KU14] K. Krupchyk and G. Uhlmann, *Uniqueness in an inverse boundary problem for a magnetic Schrödinger operator with a bounded magnetic potential*, Comm. Math. Phys. **327** (2014), no. 3, 993–1009, DOI 10.1007/s00220-014-1942-z. MR3192056

[KU16a] K. Krupchyk and G. Uhlmann, *Inverse boundary problems for polyharmonic operators with unbounded potentials*, J. Spectr. Theory **6** (2016), no. 1, 145–183, DOI 10.4171/JST/122. MR3484382

[KU16b] K. Krupchyk and G. Uhlmann, *The Calderón problem with partial data for conductivities with 3/2 derivatives*, Comm. Math. Phys. **348** (2016), no. 1, 185–219, DOI 10.1007/s00220-016-2666-z. MR3551265

[KU17] K. Krupchyk and G. Uhlmann, *Absolute continuity of the periodic Schrödinger operator in transversal geometry*, J. Eur. Math. Soc. (JEMS) **19** (2017), no. 2, 531–550, DOI 10.4171/JEMS/673. MR3605024

[KU20] K. Krupchyk and G. Uhlmann, *Partial data inverse problems for semilinear elliptic equations with gradient nonlinearities*, Math. Res. Lett. **27** (2020), no. 6, 1801–1824, DOI 10.4310/MRL.2020.v27.n6.a10. MR4216606

[KLU18] Y. Kurylev, M. Lassas, and G. Uhlmann, *Inverse problems for Lorentzian manifolds and nonlinear hyperbolic equations*, Invent. Math. **212** (2018), no. 3, 781–857, DOI 10.1007/s00222-017-0780-y. MR3802298

[Lan33] R. E. Langer, *An inverse problem in differential equations*, Bull. Amer. Math. Soc. **39** (1933), no. 10, 814–820, DOI 10.1090/S0002-9904-1933-05752-X. MR1562734

[LLLS21] M. Lassas, T. Liimatainen, Y.-H. Lin, and M. Salo, *Inverse problems for elliptic equations with power type nonlinearities* (English, with English and French summaries), J. Math. Pures Appl. (9) **145** (2021), 44–82, DOI 10.1016/j.matpur.2020.11.006. MR4188325

[LLS20] M. Lassas, T. Liimatainen, and M. Salo, *The Poisson embedding approach to the Calderón problem*, Math. Ann. **377** (2020), no. 1-2, 19–67, DOI 10.1007/s00208-019-01818-3. MR4099622

[LLS22] M. Lassas, T. Liimatainen, and M. Salo, *The Calderón problem for the conformal Laplacian*, Comm. Anal. Geom. **30** (2022), no. 5, 1121–1184, DOI 10.4310/cag.2022.v30.n5.a6. MR4564032

[LTU03] M. Lassas, M. Taylor, and G. Uhlmann, *The Dirichlet-to-Neumann map for complete Riemannian manifolds with boundary*, Comm. Anal. Geom. **11** (2003), no. 2, 207–221, DOI 10.4310/CAG.2003.v11.n2.a2. MR2014876

[LU01] M. Lassas and G. Uhlmann, *On determining a Riemannian manifold from the Dirichlet-to-Neumann map* (English, with English and French summaries), Ann. Sci. École Norm. Sup. (4) **34** (2001), no. 5, 771–787, DOI 10.1016/S0012-9593(01)01076-X. MR1862026

[Lee18] J. M. Lee, *Introduction to Riemannian manifolds*, Graduate Texts in Mathematics, vol. 176, Springer, Cham, 2018. Second edition of [MR1468735]. MR3887684

[LU89] J. M. Lee and G. Uhlmann, *Determining anisotropic real-analytic conductivities by boundary measurements*, Comm. Pure Appl. Math. **42** (1989), no. 8, 1097–1112, DOI 10.1002/cpa.3160420804. MR1029119

[Leo06] U. Leonhardt, *Optical conformal mapping*, Science **312** (2006), no. 5781, 1777–1780, DOI 10.1126/science.1126493. MR2237569

[Leo17] G. Leoni, *A first course in Sobolev spaces*, 2nd ed., Graduate Studies in Mathematics, vol. 181, American Mathematical Society, Providence, RI, 2017, DOI 10.1090/gsm/181. MR3726909

[Leo23] G. Leoni, *A first course in fractional Sobolev spaces*, Graduate Studies in Mathematics, vol. 229, American Mathematical Society, Providence, RI, 2023, DOI 10.1090/gsm/229. MR4567945

[Ler19] N. Lerner, *Carleman inequalities*, Grundlehren der mathematischen Wissenschaften [Fundamental Principles of Mathematical Sciences], vol. 353, Springer, Cham, 2019. An introduction and more, DOI 10.1007/978-3-030-15993-1. MR3932103

[LRLR22a] J. Le Rousseau, G. Lebeau, and L. Robbiano, *Elliptic Carleman estimates and applications to stabilization and controllability. Vol. I. Dirichlet boundary conditions on Euclidean space*, Progress in Nonlinear Differential Equations and their Applications, vol. 97, Birkhäuser/Springer, Cham, 2022. PNLDE Subseries in Control, DOI 10.1007/978-3-030-88674-5. MR4436025

[LRLR22b] J. Le Rousseau, G. Lebeau, and L. Robbiano, *Elliptic Carleman estimates and applications to stabilization and controllability. Vol. II. General boundary conditions on Riemannian manifolds*, Progress in Nonlinear Differential Equations and their Applications, vol. 98, Birkhäuser/Springer, Cham, 2022. PNLDE Subseries in Control, DOI 10.1007/978-3-030-88670-7. MR4424908

[Li07] X. Li, *On the inverse problem for the Dirac operator*, Inverse Problems **23** (2007), no. 3, 919–932, DOI 10.1088/0266-5611/23/3/005. MR2329924

[LU10] X. Li and G. Uhlmann, *Inverse problems with partial data in a slab*, Inverse Probl. Imaging **4** (2010), no. 3, 449–462, DOI 10.3934/ipi.2010.4.449. MR2671106

[LW07] X. Li and J.-N. Wang, *Determination of viscosity in the stationary Navier-Stokes equations*, J. Differential Equations **242** (2007), no. 1, 24–39, DOI 10.1016/j.jde.2007.07.008. MR2361101

[LL24] Y.-H. Lin and H. Liu, *Inverse problems for integro-differential operators*, Springer Cham, 2024.

[Liu97] L. Liu, *Stability estimates for the two-dimensional inverse conductivity problem*, ProQuest LLC, Ann Arbor, MI, 1997. Thesis (Ph.D.)–University of Rochester. MR2696545

[MOS66] W. Magnus, F. Oberhettinger, and R. P. Soni, *Formulas and theorems for the special functions of mathematical physics*, Third enlarged edition, Die Grundlehren der mathematischen Wissenschaften, Band 52, Springer-Verlag New York, Inc., New York, 1966. MR232968

[Man01] N. Mandache, *Exponential instability in an inverse problem for the Schrödinger equation*, Inverse Problems **17** (2001), no. 5, 1435–1444, DOI 10.1088/0266-5611/17/5/313. MR1862200

[MH03] J. C. Mason and D. C. Handscomb, *Chebyshev polynomials*, Chapman & Hall/CRC, Boca Raton, FL, 2003. MR1937591

[McD97] S. R. McDowall, *Boundary determination of material parameters from electromagnetic boundary information*, Inverse Problems **13** (1997), no. 1, 153–163, DOI 10.1088/0266-5611/13/1/012. MR1435874

[McL00] W. McLean, *Strongly elliptic systems and boundary integral equations*, Cambridge University Press, Cambridge, 2000. MR1742312

[Mel95] R. B. Melrose, *Geometric scattering theory*, Stanford Lectures, Cambridge University Press, Cambridge, 1995. MR1350074

[MU08] R. Melrose and G. Uhlmann, *Generalized backscattering and the Lax-Phillips transform*, Serdica Math. J. **34** (2008), no. 1, 355–372. MR2414425

[Mer18] C. J. Meroño, *Recovery of singularities in inverse scattering*, Universidad Autónoma de Madrid, 2018. PhD thesis.

[MS12] J. L. Mueller and S. Siltanen, *Linear and nonlinear inverse problems with practical applications*, Computational Science & Engineering, vol. 10, Society for Industrial and Applied Mathematics (SIAM), Philadelphia, PA, 2012, DOI 10.1137/1.9781611972344. MR2986262

[Nac88] A. I. Nachman, *Reconstructions from boundary measurements*, Ann. of Math. (2) **128** (1988), no. 3, 531–576, DOI 10.2307/1971435. MR970610

[Nac92] A. I. Nachman, *Inverse scattering at fixed energy*, Mathematical physics, X (Leipzig, 1991), Springer, Berlin, 1992, pp. 434–441, DOI 10.1007/978-3-642-77303-7_48. MR1386440

[Nac96] A. I. Nachman, *Global uniqueness for a two-dimensional inverse boundary value problem*, Ann. of Math. (2) **143** (1996), no. 1, 71–96, DOI 10.2307/2118653. MR1370758

[NA84] A. I. Nachman and M. J. Ablowitz, *A multidimensional inverse-scattering method*, Stud. Appl. Math. **71** (1984), no. 3, 243–250, DOI 10.1002/sapm1984713243. MR769078

[NRT20] A. Nachman, I. Regev, and D. Tataru, *A nonlinear Plancherel theorem with applications to global well-posedness for the defocusing Davey-Stewartson equation and to the inverse boundary value problem of Calderón*, Invent. Math. **220** (2020), no. 2, 395–451, DOI 10.1007/s00222-019-00930-0. MR4081134

[NS10] A. Nachman and B. Street, *Reconstruction in the Calderón problem with partial data*, Comm. Partial Differential Equations **35** (2010), no. 2, 375–390, DOI 10.1080/03605300903296322. MR2748629

[NSU95] G. Nakamura, Z. Q. Sun, and G. Uhlmann, *Global identifiability for an inverse problem for the Schrödinger equation in a magnetic field*, Math. Ann. **303** (1995), no. 3, 377–388, DOI 10.1007/BF01460996. MR1354996

[NT01] G. Nakamura and K. Tanuma, *Local determination of conductivity at the boundary from the Dirichlet-to-Neumann map*, Inverse Problems **17** (2001), no. 3, 405–419, DOI 10.1088/0266-5611/17/3/303. MR1843272

[NT00] G. Nakamura and T. Tsuchida, *Uniqueness for an inverse boundary value problem for Dirac operators*, Comm. Partial Differential Equations **25** (2000), no. 7-8, 1327–1369, DOI 10.1080/03605300008821551. MR1765140

[NU94] G. Nakamura and G. Uhlmann, *Global uniqueness for an inverse boundary problem arising in elasticity*, Invent. Math. **118** (1994), no. 3, 457–474, DOI 10.1007/BF01231541. MR1296354

[NU95] G. Nakamura and G. Uhlmann, *Inverse problems at the boundary for an elastic medium*, SIAM J. Math. Anal. **26** (1995), no. 2, 263–279, DOI 10.1137/S0036141093247494. MR1320220

[New62] R. G. Newton, *Construction of potentials from the phase shifts at fixed energy*, J. Mathematical Phys. **3** (1962), 75–82, DOI 10.1063/1.1703790. MR138364

[New85] R. G. Newton, *A Faddeev-Marchenko method for inverse scattering in three dimensions*, Inverse Problems **1** (1985), no. 2, 127–132, DOI 10.1088/0266-5611/1/2/004. MR792587

[NW73] L. Nirenberg and H. F. Walker, *The null spaces of elliptic partial differential operators in* $\mathbf{R}^n$*: Collection of articles dedicated to Salomon Bochner*, J. Math. Anal. Appl. **42** (1973), 271–301, DOI 10.1016/0022-247X(73)90138-8. MR320821

[Nov88] R. G. Novikov, *A multidimensional inverse spectral problem for the equation* $-\Delta\psi + (v(x) - Eu(x))\psi = 0$ (Russian), Funktsional. Anal. i Prilozhen. **22** (1988), no. 4, 11–22, 96, DOI 10.1007/BF01077418; English transl., Funct. Anal. Appl. **22** (1988), no. 4, 263–272. MR0976992

[Nov92] R. G. Novikov, *The inverse scattering problem on a fixed energy level for the two-dimensional Schrödinger operator*, J. Funct. Anal. **103** (1992), no. 2, 409–463, DOI 10.1016/0022-1236(92)90127-5. MR1151554

[Nov94] R. G. Novikov, *The inverse scattering problem at fixed energy for the three-dimensional Schrödinger equation with an exponentially decreasing potential*, Comm. Math. Phys. **161** (1994), no. 3, 569–595. MR1269391

[Nov08] R. G. Novikov, *The* $\bar{\partial}$*-approach to monochromatic inverse scattering in three dimensions*, J. Geom. Anal. **18** (2008), no. 2, 612–631, DOI 10.1007/s12220-008-9015-1. MR2393272

[Nov22] R. G. Novikov, *Multidimensional inverse scattering for the Schrödinger equation*, Mathematical analysis, its applications and computation, Springer Proc. Math. Stat., vol. 385, Springer, Cham, 2022, pp. 75–98, DOI 10.1007/978-3-030-97127-4_3. MR4461039

[NK87] R. G. Novikov and G. M. Khenkin, *The* $\bar{\partial}$*-equation in the multidimensional inverse scattering problem* (Russian), Uspekhi Mat. Nauk **42** (1987), no. 3(255), 93–152, 255. MR896879

[OSSU24] L. Oksanen, M. Salo, P. Stefanov, and G. Uhlmann, *Inverse problems for real principal type operators*, Amer. J. Math. **146** (2024), no. 1, 161–240, DOI 10.1353/ajm.2024.a917541. MR4691487

[Ola94] P. Ola, *Boundary integral equations for the scattering of electromagnetic waves by a homogeneous chiral obstacle*, J. Math. Phys. **35** (1994), no. 8, 3969–3980, DOI 10.1063/1.530836. MR1284623

[OPS01] P. Ola, L. Päivärinta, and V. Serov, *Recovering singularities from backscattering in two dimensions*, Comm. Partial Differential Equations **26** (2001), no. 3-4, 697–715, DOI 10.1081/PDE-100001768. MR1842046

[OPS93] P. Ola, L. Päivärinta, and E. Somersalo, *An inverse boundary value problem in electrodynamics*, Duke Math. J. **70** (1993), no. 3, 617–653, DOI 10.1215/S0012-7094-93-07014-7. MR1224101

[OS96] P. Ola and E. Somersalo, *Electromagnetic inverse problems and generalized Sommerfeld potentials*, SIAM J. Appl. Math. **56** (1996), no. 4, 1129–1145, DOI 10.1137/S0036139995283948. MR1398411

[OD09] M. Oristaglio and A. Dorozynski, *A sixth sense*, 1st ed., Harry N. Abrams, 2009.

[PSU10] L. Päivärinta, M. Salo, and G. Uhlmann, *Inverse scattering for the magnetic Schrödinger operator*, J. Funct. Anal. **259** (2010), no. 7, 1771–1798, DOI 10.1016/j.jfa.2010.06.002. MR2665410

[PSU23] G. P. Paternain, M. Salo, and G. Uhlmann, *Geometric inverse problems—with emphasis on two dimensions*, Cambridge Studies in Advanced Mathematics, vol. 204, Cambridge University Press, Cambridge, 2023. MR4520155

[PSS06] J. B. Pendry, D. Schurig, and D. R. Smith, *Controlling electromagnetic fields*, Science **312** (2006), no. 5781, 1780–1782, DOI 10.1126/science.1125907. MR2237570

[Per14] P. A. Perry, *Miura maps and inverse scattering for the Novikov-Veselov equation*, Anal. PDE **7** (2014), no. 2, 311–343, DOI 10.2140/apde.2014.7.311. MR3218811

[PT17] V. Pohjola and L. Tzou, *Inverse scattering for the magnetic Schrödinger operator on surfaces with Euclidean ends*, Comm. Math. Phys. **356** (2017), no. 1, 107–142, DOI 10.1007/s00220-017-2982-y. MR3694024

[PMRT22] L. Potenciano-Machado, A. Ruiz, and L. Tzou, *Stability estimates for the magnetic Schrödinger operator with partial measurements*, J. Differential Equations **321** (2022), 475–521, DOI 10.1016/j.jde.2022.02.051. MR4399221

[RSSN$^+$24] O. Rajabi Shishvan, G. Saulnier, J. Newell, D. Isaacson, A. Keck, E. DeBoer, J. Hoppe, and J. Mueller, *EIT images of ventilation and perfusion correlate with spirometry in patients with cystic fibrosis and healthy controls*, Am. J. Respir. Crit. Care Med. **209** (2024), A6464.

[RS20a] Rakesh and M. Salo, *The fixed angle scattering problem and wave equation inverse problems with two measurements*, Inverse Problems **36** (2020), no. 3, 035005, 42, DOI 10.1088/1361-6420/ab23a2. MR4068234

[RS20b] Rakesh and M. Salo, *Fixed angle inverse scattering for almost symmetric or controlled perturbations*, SIAM J. Math. Anal. **52** (2020), no. 6, 5467–5499, DOI 10.1137/20M1319309. MR4170189

[RU14] Rakesh and G. Uhlmann, *Uniqueness for the inverse backscattering problem for angularly controlled potentials*, Inverse Problems **30** (2014), no. 6, 065005, 24, DOI 10.1088/0266-5611/30/6/065005. MR3224125

[Ram88] A. G. Ramm, *Recovery of the potential from fixed-energy scattering data*, Inverse Problems **4** (1988), no. 3, 877–886, DOI 10.1088/0266-5611/4/3/020. MR965652

[RS80] M. Reed and B. Simon, *Methods of modern mathematical physics. I: Functional analysis*, 2nd ed., Academic Press, Inc. [Harcourt Brace Jovanovich, Publishers], New York, 1980. MR751959

[RS75] M. Reed and B. Simon, *Methods of modern mathematical physics. II. Fourier analysis, self-adjointness*, Academic Press [Harcourt Brace Jovanovich, Publishers], New York-London, 1975. MR493420

[RS79] M. Reed and B. Simon, *Methods of modern mathematical physics. III: Scattering theory*, Academic Press [Harcourt Brace Jovanovich, Publishers], New York-London, 1979. MR529429

[RS78] M. Reed and B. Simon, *Methods of modern mathematical physics. IV. Analysis of operators*, Academic Press [Harcourt Brace Jovanovich, Publishers], New York-London, 1978. MR493421

[Rud87] W. Rudin, *Real and complex analysis*, 3rd ed., McGraw-Hill Book Co., New York, 1987. MR924157

[RS19] A. Rüland and M. Salo, *Quantitative Runge approximation and inverse problems*, Int. Math. Res. Not. IMRN **20** (2019), 6216–6234, DOI 10.1093/imrn/rnx301. MR4031236

[RS20] A. Rüland and M. Salo, *The fractional Calderón problem: low regularity and stability*, Nonlinear Anal. **193** (2020), 111529, 56, DOI 10.1016/j.na.2019.05.010. MR4062981

[Sab66] P. C. Sabatier, *Asymptotic properties of the potentials in the inverse-scattering problem at fixed energy*, J. Mathematical Phys. **7** (1966), 1515–1531, DOI 10.1063/1.1705062. MR207328

[Sal04] M. Salo, *Inverse problems for nonsmooth first order perturbations of the Laplacian*, Ann. Acad. Sci. Fenn. Math. Diss. **139** (2004), 67. Dissertation, University of Helsinki, Helsinki, 2004. MR2105191

[Sal06] M. Salo, *Semiclassical pseudodifferential calculus and the reconstruction of a magnetic field*, Comm. Partial Differential Equations **31** (2006), no. 10-12, 1639–1666, DOI 10.1080/03605300500530420. MR2273968

[Sal08] M. Salo, *Calderón problem*, 2008. Lecture notes, University of Helsinki, `http://users.jyu.fi/~salomi/lecturenotes/calderon_lectures.pdf`.

[Sal13] M. Salo, *The Calderón problem on Riemannian manifolds*, Inverse problems and applications: inside out. II, Math. Sci. Res. Inst. Publ., vol. 60, Cambridge Univ. Press, Cambridge, 2013, pp. 167–247. MR3098658

[Sal17a] M. Salo, *The fractional Calderón problem*, Journées équations aux dérivées partielles (2017), 1–8.

[Sal17b] M. Salo, *The Calderón problem and normal forms*, Preprint, `arXiv:1702.02136`, 2017.

[Sal20] M. Salo, *Applications of microlocal analysis in inverse problems*, Mathematics **8** (2020), no. 7, 1184.

[Sal24] M. Salo, *On geometric inverse problems and microlocal analysis*, Microlocal analysis and inverse problems in tomography and geometry, Radon Ser. Comput. Appl. Math., vol. 30, De Gruyter, Berlin, 2024, pp. 77–138. MR4815453

[ST09] M. Salo and L. Tzou, *Carleman estimates and inverse problems for Dirac operators*, Math. Ann. **344** (2009), no. 1, 161–184, DOI 10.1007/s00208-008-0301-9. MR2481057

[ST10] M. Salo and L. Tzou, *Inverse problems with partial data for a Dirac system: a Carleman estimate approach*, Adv. Math. **225** (2010), no. 1, 487–513, DOI 10.1016/j.aim.2010.03.003. MR2669360

[SZ12] M. Salo and X. Zhong, *An inverse problem for the p-Laplacian: boundary determination*, SIAM J. Math. Anal. **44** (2012), no. 4, 2474–2495, DOI 10.1137/110838224. MR3023384

[Sch82] A. G. Schlumberger, *Schlumberger adventure*, 1st ed., Arco Publishing House New York, 1982.

[Sha11] V. A. Sharafutdinov, *The geometric problem of electrical impedance tomography in the disk* (Russian, with Russian summary), Sibirsk. Mat. Zh. **52** (2011), no. 1, 223–238, DOI 10.1134/S0037446606010198; English transl., Sib. Math. J. **52** (2011), no. 1, 178–190. MR2810264

[Sli33] L. B. Slichter, *The interpretation of the resistivity prospecting method for horizontal structures*, Physics **4** (1933), no. 9, 307-322, `https://doi.org/10.1063/1.1745198`, DOI 10.1063/1.1745198.

[Sob99] A. V. Sobolev, *Absolute continuity of the periodic magnetic Schrödinger operator*, Invent. Math. **137** (1999), no. 1, 85–112, DOI 10.1007/s002220050324. MR1703339

[Ste92] P. Stefanov, *Generic uniqueness for two inverse problems in potential scattering*, Comm. Partial Differential Equations **17** (1992), no. 1-2, 55–68, DOI 10.1080/03605309208820834. MR1151256

[Sun93] Z. Q. Sun, *An inverse boundary value problem for Schrödinger operators with vector potentials*, Trans. Amer. Math. Soc. **338** (1993), no. 2, 953–969, DOI 10.2307/2154438. MR1179400

[Sun96] Z. Sun, *On a quasilinear inverse boundary value problem*, Math. Z. **221** (1996), no. 2, 293–305, DOI 10.1007/BF02622117. MR1376299

[SU97] Z. Sun and G. Uhlmann, *Inverse problems in quasilinear anisotropic media*, Amer. J. Math. **119** (1997), no. 4, 771–797. MR1465069

[SU03] Z. Sun and G. Uhlmann, *Anisotropic inverse problems in two dimensions*, Inverse Problems **19** (2003), no. 5, 1001–1010, DOI 10.1088/0266-5611/19/5/301. MR2024685

[Syl90] J. Sylvester, *An anisotropic inverse boundary value problem*, Comm. Pure Appl. Math. **43** (1990), no. 2, 201–232, DOI 10.1002/cpa.3160430203. MR1038142

[Syl96] J. Sylvester, *Inverse boundary value problems. An overview*, Algebra i Analiz **8** (1996), no. 2, 195–204; English transl., St. Petersburg Math. J. **8** (1997), no. 2, 333–340. MR1392032

[SU86] J. Sylvester and G. Uhlmann, *A uniqueness theorem for an inverse boundary value problem in electrical prospection*, Comm. Pure Appl. Math. **39** (1986), no. 1, 91–112, DOI 10.1002/cpa.3160390106. MR820341

[SU87] J. Sylvester and G. Uhlmann, *A global uniqueness theorem for an inverse boundary value problem*, Ann. of Math. (2) **125** (1987), no. 1, 153–169, DOI 10.2307/1971291. MR873380

[SU88] J. Sylvester and G. Uhlmann, *Inverse boundary value problems at the boundary—continuous dependence*, Comm. Pure Appl. Math. **41** (1988), no. 2, 197–219, DOI 10.1002/cpa.3160410205. MR924684

[TNH$^+$25] D. Tanyu, J. Ning, A. Hauptmann, B. Jin, and P. Maass, *Electrical Impedance Tomography: A fair comparative study on deep learning and analytic-based approaches*, Data-driven Models in Inverse Problems, Radon Ser. Comput. Appl. Math., vol. 31, De Gruyter, Berlin, 2025, pp. 437–470, DOI 10.1515/9783111251233-013.

[Tay23] M. E. Taylor, *Partial differential equations II. Qualitative studies of linear equations*, Applied Mathematical Sciences, vol. 116, Springer, Cham, 2023. Third edition [of 1395149], DOI 10.1007/978-3-031-33700-0. MR4703939

[Tej17] J. Tejero, *Reconstruction and stability for piecewise smooth potentials in the plane*, SIAM J. Math. Anal. **49** (2017), no. 1, 398–420, DOI 10.1137/16M1085048. MR3606422

[Tho73] L. E. Thomas, *Time dependent approach to scattering from impurities in a crystal*, Comm. Math. Phys. **33** (1973), 335–343. MR334766

[Tih49] A. N. Tihonov, *On the uniqueness of the solution of the problem of electric prospecting* (Russian), Doklady Akad. Nauk SSSR (N.S.) **69** (1949), 797–800. MR33409

[TPD$^+$24] J. Toivanen, A. Paldanius, B. Dekdouk, V. Candiani, A. Hänninen, T. Savolainen, D. Strbian, N. Forss, N. Hyvönen, J. Hyttinen, and V. Kolehmainen, *Simulation-based feasibility study of monitoring of intracerebral hemorrhages and detection of secondary hemorrhages using electrical impedance tomography*, J. Medical Imaging **11** (2024), no. 1, 014502–014502.

[Tri83] H. Triebel, *Theory of function spaces*, Monographs in Mathematics, vol. 78, Birkhäuser Verlag, Basel, 1983, DOI 10.1007/978-3-0346-0416-1. MR781540

[Tzo17] L. Tzou, *The reflection principle and Calderón problems with partial data*, Math. Ann. **369** (2017), no. 1-2, 913–956, DOI 10.1007/s00208-017-1525-3. MR3694665

[Tzo20] L. Tzou, *Inverse scattering for Dirac operators on surfaces with Euclidean ends*, J. Math. Anal. Appl. **491** (2020), no. 1, 124232, 24, DOI 10.1016/j.jmaa.2020.124232. MR4102769

[Uhl92] G. Uhlmann, *Inverse boundary value problems and applications*, Astérisque **207** (1992), 6, 153–211. Méthodes semi-classiques, Vol. 1 (Nantes, 1991). MR1205179

[Uhl99] G. Uhlmann, *Developments in inverse problems since Calderón's foundational paper*, Harmonic analysis and partial differential equations (Chicago, IL, 1996), Chicago Lectures in Math., Univ. Chicago Press, Chicago, IL, 1999, pp. 295–345. MR1743870

[Uhl03] G. Uhlmann, *Inverse boundary problems in two dimensions*, Function spaces, differential operators and nonlinear analysis (Teistungen, 2001), Birkhäuser, Basel, 2003, pp. 183–203. MR1984168

[Uhl09] G. Uhlmann, *Electrical impedance tomography and Calderón's problem*, Inverse Problems **25** (2009), no. 12, 123011, 39, DOI 10.1088/0266-5611/25/12/123011. MR3460047

[Uhl14a] G. Uhlmann, *30 years of Calderón's problem*, Séminaire Laurent Schwartz—Équations aux dérivées partielles et applications. Année 2012–2013, Sémin. Équ. Dériv. Partielles, École Polytech., Palaiseau, 2014, pp. Exp. No. XIII, 25. MR3381003

[Uhl14b] G. Uhlmann, *Inverse problems: seeing the unseen*, Bull. Math. Sci. **4** (2014), no. 2, 209–279, DOI 10.1007/s13373-014-0051-9. MR3228575

[UV02] G. Uhlmann and A. Vasy, *Fixed energy inverse problem for exponentially decreasing potentials*, Methods Appl. Anal. **9** (2002), no. 2, 239–247, DOI 10.4310/MAA.2002.v9.n2.a2. MR1957487

[Wed91] R. Weder, *Global uniqueness at fixed energy in multidimensional inverse scattering theory*, Inverse Problems **7** (1991), no. 6, 927–938, DOI 10.1088/0266-5611/7/6/012. MR1140323

[Wed07] R. Weder, *Completeness of averaged scattering solutions and inverse scattering at a fixed energy*, Comm. Partial Differential Equations **32** (2007), no. 4-6, 675–691, DOI 10.1080/03605300500530370. MR2334828

[WY05] R. Weder and D. Yafaev, *On inverse scattering at a fixed energy for potentials with a regular behaviour at infinity*, Inverse Problems **21** (2005), no. 6, 1937–1952, DOI 10.1088/0266-5611/21/6/009. MR2183660

[Yaf10] D. R. Yafaev, *Mathematical scattering theory: Analytic theory*, Mathematical Surveys and Monographs, vol. 158, American Mathematical Society, Providence, RI, 2010, DOI 10.1090/surv/158. MR2598115

[Yor01] T. York, *Status of electrical tomography in industrial application*, J. Electronic Imaging **10** (2001), 608–619.

[Yos95] K. Yosida, *Functional analysis*, Classics in Mathematics, Springer-Verlag, Berlin, 1995. Reprint of the sixth (1980) edition, DOI 10.1007/978-3-642-61859-8. MR1336382

[Zwo12] M. Zworski, *Semiclassical analysis*, Graduate Studies in Mathematics, vol. 138, American Mathematical Society, Providence, RI, 2012, DOI 10.1090/gsm/138. MR2952218

Index

Selected Published Titles in This Series

253 **Joel Feldman, Mikko Salo, and Gunther Uhlmann,** The Calderón Problem, 2025
251 **Eric Carlen,** Inequalities in Matrix Algebras, 2025
250 **David Eisenbud and Joe Harris,** The Practice of Algebraic Curves, 2024
249 **David Damanik and Jake Fillman,** One-Dimensional Ergodic Schrödinger Operators, 2024
248 **J. I. Hall,** Introduction to Lie Algebras, 2024
247 **Michel L. Lapidus and Goran Radunović,** An Invitation to Fractal Geometry, 2024
246 **Ale Jan Homburg and Jürgen Knobloch,** Bifurcation Theory, 2024
245 **Bennett Chow and Yutze Chow,** Lectures on Differential Geometry, 2024
244 **John M. Lee,** Introduction to Complex Manifolds, 2024
243 **J. M. Landsberg,** Quantum Computation and Quantum Information, 2024
242 **Jayadev S. Athreya and Howard Masur,** Translation Surfaces, 2024
241 **Thorsten Theobald,** Real Algebraic Geometry and Optimization, 2024
240 **Nam Q. Le,** Analysis of Monge–Ampère Equations, 2024
239 **K. Cieliebak, Y. Eliashberg, and N. Mishachev,** Introduction to the h-Principle, Second Edition, 2024
238 **Julio González-Díaz, Ignacio García-Jurado, and M. Gloria Fiestras-Janeiro,** An Introductory Course on Mathematical Game Theory and Applications, Second Edition, 2023
237 **Michael Levitin, Dan Mangoubi, and Iosif Polterovich,** Topics in Spectral Geometry, 2023
236 **Stephanie Alexander, Vitali Kapovitch, and Anton Petrunin,** Alexandrov Geometry, 2024
235 **Bennett Chow,** Ricci Solitons in Low Dimensions, 2023
234 **Andrea Ferretti,** Homological Methods in Commutative Algebra, 2023
233 **Andrea Ferretti,** Commutative Algebra, 2023
232 **Harry Dym,** Linear Algebra in Action, Third Edition, 2023
231 **Luís Barreira and Yakov Pesin,** Introduction to Smooth Ergodic Theory, Second Edition, 2023
230 **Barbara Kaltenbacher and William Rundell,** Inverse Problems for Fractional Partial Differential Equations, 2023
229 **Giovanni Leoni,** A First Course in Fractional Sobolev Spaces, 2023
228 **Henk Bruin,** Topological and Ergodic Theory of Symbolic Dynamics, 2022
227 **William M. Goldman,** Geometric Structures on Manifolds, 2022
226 **Milivoje Lukić,** A First Course in Spectral Theory, 2022
225 **Jacob Bedrossian and Vlad Vicol,** The Mathematical Analysis of the Incompressible Euler and Navier-Stokes Equations, 2022
224 **Ben Krause,** Discrete Analogues in Harmonic Analysis, 2022
223 **Volodymyr Nekrashevych,** Groups and Topological Dynamics, 2022
222 **Michael Artin,** Algebraic Geometry, 2022
221 **David Damanik and Jake Fillman,** One-Dimensional Ergodic Schrödinger Operators, 2022
220 **Isaac Goldbring,** Ultrafilters Throughout Mathematics, 2022
219 **Michael Joswig,** Essentials of Tropical Combinatorics, 2021
218 **Riccardo Benedetti,** Lectures on Differential Topology, 2021
217 **Marius Crainic, Rui Loja Fernandes, and Ioan Mărcuţ,** Lectures on Poisson Geometry, 2021
216 **Brian Osserman,** A Concise Introduction to Algebraic Varieties, 2021
215 **Tai-Ping Liu,** Shock Waves, 2021
214 **Ioannis Karatzas and Constantinos Kardaras,** Portfolio Theory and Arbitrage, 2021
213 **Hung Vinh Tran,** Hamilton–Jacobi Equations, 2021